D1519837

INTRODUCTION TO MULTIMEDIA COMMUNICATIONS

INTRODUCTION TO MULTIMEDIA COMMUNICATIONS
Applications, Middleware, Networking

K. R. Rao
University of Texas at Arlington

Zoran S. Bojkovic
Dragorad A. Milovanovic
University of Belgrade, Serbia and Montenegro

A JOHN WILEY & SONS, INC., PUBLICATION

Copyright © 2006 by John Wiley & Sons, Inc. All rights reserved.

Published by John Wiley & Sons, Inc., Hoboken, New Jersey.
Published simultaneously in Canada.

No part of this publication may be reproduced, stored in a retrieval system, or transmitted in any form or by any means, electronic, mechanical, photocopying, recording, scanning, or otherwise, except as permitted under Section 107 or 108 of the 1976 United States Copyright Act, without either the prior written permission of the Publisher, or authorization through payment of the appropriate per-copy fee to the Copyright Clearance Center, Inc., 222 Rosewood Drive, Danvers, MA 01923, 978-750-8400, fax 978-646-8600, or on the web at www.copyright.com. Requests to the Publisher for permission should be addressed to the Permissions Department, John Wiley & Sons, Inc., 111 River Street, Hoboken, NJ 07030, (201) 748-6011, fax (201) 748-6008, or online at http://www.wiley.com/go/permission.

Limit of Liability/Disclaimer of Warranty: While the publisher and author have used their best efforts in preparing this book, they make no representations or warranties with respect to the accuracy or completeness of the contents of this book and specifically disclaim any implied warranties of merchantability or fitness for a particular purpose. No warranty may be created or extended by sales representatives or written sales materials. The advice and strategies contained herein may not be suitable for your situation. You should consult with a professional where appropriate. Neither the publisher nor author shall be liable for any loss of profit or any other commercial damages, including but not limited to special, incidental, consequential, or other damages.

For general information on our other products and services please contact our Customer Care Department within the U.S. at 877-762-2974, outside the U.S. at 317-572-3993 or fax 317-572-4002.

Wiley also publishes its books in a variety of electronic formats. Some content that appears in print, may not be available in electronic format. For more information about Wiley products, visit our website at www.wiley.com

Library of Congress Cataloging-in-Publication Data:

Rao, K. R.
  Introduction to multimedia communications : applications, middleware, networking / K.R. Rao, Z.S. Bojkovic, and D.A. Milovanovic.
    p. cm.
  ISBN 13 978-0-471-46742-7 (cloth)
  ISBN 10 0-471-46742-1 (cloth)
  1. Multimedia communications. I. Bojkovic, Z. S. II. Milovanovic, Dragorad A. III. Title.
  TK5105.15.R36 2006
  621.382--dc22
                                                                    2005001259

Printed in the United States of America

10 9 8 7 6 5 4 3 2 1

Copyrights release for ISO/IEC

All the figures and tables obtained from ISO/IEC used in this book are subject to the following: "The terms and definitions taken from the Figures and Tables ref. ISO/IEC IS11172, ISO/IEC11172-2, ISO/IEC 11172-73, ISO/IEC IS13818-2, ISO/IEC IS13818-6, ISO/IEC JTC1/SC29WG11 Doc. N2196, ISO/IEC JTC1/SC29/WG11 N3536, ISO/MPEG N2502 ISO/IEC JTC1/SC29/WG1 Doc. N1595, ISO/IEC JTC SC29/WG11 Doc. 2460, ISO/IEC JTC SC29/WG11 Doc. 3751, ISO/MPEG N2501, ISO/IEC JTC-1 SC29/WG11 M5804, ISO/IEC JTC SC29/WG11, Recomm.H.262 ISO/IEC 13818-2, ISO/IEC IS 13818-1, ISO/MPEG N3746, ISO/IEC Doc. N2502, ISO/IEC JTC1/SC29/WG11 Doc. N2424 are reproduced with permission of International Organization for Standardization, ISO. These standards can be obtained from any ISO member and from the Web site of the ISO Central Secretariat at the following address: http//www.iso.org. Non-exclusive copyright remains with ISO."

"The terms and definitions taken from the Figure ref. ISO/IEC 14496-1:1999, ISO/IEC 14496-2:1999, ISO/IEC 14496-3:1999, ISO/IEC 14496-4:2000, ISO/IEC 14496-5:2000, ISO/IEC 14496-10 AVC: 2003 are reproduced with the permission of the International for Standardization, ISO. The standards can be obtained from any ISO member and from the Web site of ISO Central Secretariat at the following address: www.iso.org. Non-exclusive copyright with ISO." The editors and authors are grateful to ISO/IEC for giving the permission.

CONTENTS

Preface xix

Acknowledgments xxiii

Acronyms xxv

1 Introduction 1

2 Multimedia Communications 5

    2.1 Introduction, 5
    2.2 Human Communication Model, 9
        2.2.1 Physical System, 12
        2.2.2 Symbol Encoding, 13
        2.2.3 Feeling, 14
        2.2.4 Memory, 15
        2.2.5 Cognitive System, 16
        2.2.6 Mission System, 16
    2.3 Evolution and Convergence, 17
        2.3.1 Convergence of Telecommunications and Computing, 18
        2.3.2 Architectures for Networked Applications, 20
        2.3.3 Networked Computers, 21
        2.3.4 Integration, 22
        2.3.5 Transportable Computation, 24
        2.3.6 Intelligent Agents, 25
        2.3.7 Convergence, 26
    2.4 Technology Framework, 26
        2.4.1 Multimedia Technologies, 27
            SONET (Synchronous Optical Network) and BISDN (Broadband Integrated Service Digital Network), 27
            Computer Networks, 28

       2.4.2 Multimedia Networking, 32
       2.4.3 Multimedia Conferencing, 34
           Centralized Conferencing Architectures, 35
           Decentralized Conferencing Architectures, 35
           Hybrid Class, 36
       2.4.4 Multicasting, 36
       2.4.5 Technologies for e-Content, 38
           Digital Signal Processing, 38
           Digital Networks, 38
           The Value Chain, 39
           Information Presentation, 40
           Content Description, Identification, and Protection, 40
  2.5 Standardization Framework, 41
       2.5.1 Research and Regulation, 43
       2.5.2 Technology and Education, 45
       2.5.3 Convergence and Regulatory Issues, 45
       2.5.4 Manufacturing and Marketing, 47
       2.5.5 Digital Video/Audio Coding Standards and
            Multimedia Industry, 49
           ITU-T Standardization of Audiovisual
             Communication Systems, 50
           Video Coding Standards, 53
           Speech Coding Standards, 55
           Multimedia Multiplex and Synchronization Standards, 56
           MPEG Standards: Role in Multimedia Communications, 57
           Still Image Coding Standards, 61
References, 63

3 Frameworks for Multimedia Standardization   69

  3.1 Introduction, 70
  3.2 Standardization Activities, 71
       3.2.1 Reasons for Seeking a Standard, 72
       3.2.2 Interfaces to be Standardized, 72
       3.2.3 Time to Standardize, 73
       3.2.4 How Will Standardization Take Place?, 75
       3.2.5 Who Will be Conducting the Standardization Activity? 76
       3.2.6 Where Will the Standardization Be Used? 76
       3.2.7 Technology Cycle and the Product or Service
            to be Standardized, 77
  3.3 Standards to Build a New Global Information
      Infrastructure (GII), 78
       3.3.1 Concept, 79
       3.3.2 Performance Purposes, 80
           GII Performance Parameters, 81

3.3.3 Functional Model, 83
3.3.4 Implementation Model, 84
3.3.5 Scenarios Approach, 85
3.3.6 Standardization Efforts, 87
3.3.7 ITU-T Standardization Sector Program for GII, 89
3.3.8 Future Directions, 92
3.4 Standardization Processes on Multimedia Communications, 92
3.4.1 ITU-T Strategies, 92
3.4.2 ISO/IEC JTC1 Directions in Multimedia Standards, 94
The ISO/IEC Approach, 95
JTC1 Standardization Project, 96
3.4.3 IETF Standards, 98
3.4.4 ETSI Standardization Projects for Applications,
Middleware and Networks, 99
3.4.5 GII-Related Activities, 101
3.5 ITU-T Mediacom2004 Framework for Multimedia
Communications, 102
3.5.1 Scope, 103
3.5.2 Multimedia Framework Study Areas, 103
3.5.3 Study Group 16 Work Program, 106
3.6 ISO/IEC MPEG-21 Multimedia Framework, 106
3.6.1 User Model, 108
3.6.2 Digital Items, 109
3.6.3 Work Plan, 109
Vision, Technology, and Strategy, 110
Digital Item Declaration, 110
Digital Item Identification, 110
Intellectual Property Management and Protection (IPMP), 111
Rights Expression Language (REL), 111
Rights Data Dictionary (RDD), 112
Digital Item Adaptation, 113
Reference Software, 114
File Format, 114
3.6.4 MPEG-21 Use Case Scenario, 114
Universal Multimedia Access, 114
Text News Distribution to Web Sites, 115
Video News Distribution, 116
E-Commerce and Channel Partner Distribution, 116
Real-Time Financial Data Distribution, 116
Picture Archive Access, 117
3.6.5 User Requirements, 117
3.7 IETF Multimedia Internet Standards, 119
3.8 Industrial Fora and Consortia, 121
ATM Forum Industry Cooperation, 121

Telecommunication Information Network
    Architecture (TINA), 124
Quality of Service in TINA, 129
International Multimedia Telecommunications
    Consortium (IMTC), 133
MPEG-4 Industry Forum (M4IF), 134
Internet Streaming Media Alliance (ISMA), 135
Third Generation Partnership Project (3GPP/3GPP2), 136
Digital Video Broadcasting (DVB), 137
Digital Audio Visual Council (DAVIC), 141
Advanced Television Systems Committee (ATSC), 143
European Broadcasting Union (EBU), 143
References, 144

4 Applications Layer   147

4.1 Introduction, 147
4.2 ITU Applications, 157
   4.2.1 Multimedia Services and Systems, 160
       Multimedia Applications and Services, 161
       Multimedia Systems, Terminals, and Data Conferencing, 162
       Multimedia Over Packet Networks Using H.323
          Systems, 163
       Video and Data Conferencing Using Internet Supported
          Services, 163
   4.2.2 Integrated Broadband Cable Networks and Television and
          Sound Transmission, 164
   4.2.3 Interactivity in Broadcasting, 166
       Interactive Multimedia, 171
4.3 MPEG Applications, 171
   4.3.1 Multimedia PC, 172
       MPEG-1 Systems, 172
       MPEG-1 Video, 173
       MPEG-1 Audio, 174
       MPEG-1 Constraints, 174
   4.3.2 Digital TV and Storage Media, 175
       MPEG-2 Systems, 176
       MPEG-2 Video, 178
       MPEG-2 Audio, 180
       MPEG-2 DSM-CC, 181
       MPEG-2 Profiling, 183
   4.3.3 Multimedia Conferencing, Streaming Media, and Interactive
          Broadcasting, 184
       MPEG-4 Systems, 186
       System Decoder Model, 187

Scene Description, 189
Multiplexing, 190
MPEG-4 Visual, 191
Natural Video, 193
Synthetic Video, 196
MPEG-4 Audio, 198
Natural Audio, 198
Synthetic Audio, 199
MPEG-4 Profiling, 200
   4.3.4 Media Description, Searching, and Retrieval, 200
MPEG-7 Tools, 202
MPEG-7 Descriptors, 202
MPEG-7 Description Schemes, 203
MPEG-7 Description Definition Language (DDL), 204
MPEG-7 Systems, 205
   4.3.5 Media Distribution and Consumption, 205
MPEG-21 Parts, 206
MPEG-21 in a Universal Multimedia Access (UMA) Environment, 208
MPEG-21 Digital Item Adaptation, 209
MPEG-21 and MEDIACOM2004, 210
4.4 Digital Broadcasting, 211
   4.4.1 Convergence of Telecommunications and Broadcast Infrastructure, 213
Mobile Telecommunications, 213
Broadcasting, 214
Application Scenario, 215
   4.4.2 Digital Radio Broadcasting, 216
Audio Coding, 218
Perceptual Audio Coding, 220
Variable Bit Rate Coding and Constant Bit Rate Transmission, 222
Joint Audio Coding, 224
Embedded and Multistream Audio Coding, 226
Terrestrial Digital Audio Broadcasting, 228
Evolution of IBOC, 230
Other Terrestrial Systems, 230
In-Band On-Channel (IBOC) FM Systems, 232
In-Band On-Channel (IBOC) AM Systems, 232
Satellite Digital Audio Radio Services, 234
Status and Systems Evolution, 237
   4.4.3 Digital Video Broadcasting, 237
DVB Interoperabilities, 240
MPEG-2 and DVB, 240
DVB Project, 241

DVB 2.0, 243
Commercial Strategy for the Future of DVB, 244
Architectural Framework for the Future DVB, 246
4.4.4 DVB/ATSC System, 248
DVB System, 249
DVB Baseband Processing, 253
ATSC Digital Television, 260
DVB Services over IP-Based Networks, 269
Home Reference Model, 272
Modules for Home Network Elements, 274
Services, 274
Requirements for Service Authentication, Authorization and Accounting, 278
4.4.5 DVB-MHP Application and Interactive Television, 278
MHP Layers, 281
Elements of the DVB-MHP, 281
Applications, 285
Multimedia Car Platform, 288
Migration Process and Future Operational Issues, 288
DVB and Interactivity, 290
Interactive Television, 291
4.4.6 Interactive Services in Digital Television Infrastructure, 296
Interactive Broadcast Data Services, 298
Data Carousel Concept, 300
DVB with Return Channel via Satellite (DVB-RCS), 301
DVB Terrestrial Return Channel System (DVB-RCT), 304
Video-On-Demand (VoD) Systems in Broadcast Environment, 305
4.4.7 DVB and Internet, 306
IP Multicast, 308
Audio/Video Streaming, 309
Data Streaming, 312
Linking Broadcasting Services to the Internet, 313
Multimedia Content over Broadcast Network, 314
IP Datacast Forum (IPDC), 314
Datacasting, 317
Internet Broadcast Based on IP Simulcast, 320
Integrated Multimedia Mailing System, 324
4.5 Mobile Services and Applications, 329
4.5.1 Mobile Services, 330
4.5.2 Types of Mobile Services, 332
4.5.3 Business Models, 334
4.5.4 Value Chain Structure and Coordination, 338
4.5.5 Challenges and Obstacles in Adopting Mobile Applications, 340

    4.5.6 Research Activities, 342
  4.6 Universal Multimedia Access, 344
    4.6.1 Existing and Emerging Technologies, 346
    4.6.2 Content Representation, 347
    4.6.3 User Environment Description, 348
    4.6.4 Content Adaptation, 349
    4.6.5 Intellectual Property and Protection, 350
    4.6.6 Presentation Conditions, 466
    4.6.7 Mobile and Wearable Devices, 351
    4.6.8 Content Delivery, 352
References, 352
Bibliography, 360

5 Middleware Layer   363

  5.1 Introduction, 363
  5.2 Middleware for Multimedia, 364
    5.2.1 GII Standardization Projects, 365
    5.2.2 EII Standardization Projects, 370
    5.2.3 ITU-T SG16 Work Program, 371
    5.2.4 MEDIACOM2004 Middleware Design, 372
  5.3 Media Coding, 374
    5.3.1 Multimedia Content Representation, 374
    5.3.2 Core Compression Technologies, 380
       H.264/AVC (MPEG-4), 382
          Robust (Error Resilient) Video Transmission, 382
          Network Friendliness, 383
          Support for Different Bit Rates, Buffer Sizes,
             and Start-Up Delays of the Buffer, 384
          Improved Prediction, 386
          Improved Fractional Accuracy, 386
          Significant Data Compression, 387
          Better Coding Efficiency, 387
          Overlay Coding Technique, 388
          Better Video Quality, 388
          Group Capabilities, 388
          Basic Architecture of the Standard, 389
          Intraprediction, 393
          Intraprediction for 4×4 Luma Block, 393
          Intraprediction for Chroma, 394
          Interprediction, 395
          Deblocking Filter, 397
          Reference Frames, 398
          Motion Vector/Estimation/Compensation, 398

Entropy Coding, 399
Decoding Process, 400
Profiles and Levels, 400
Main Profile, 401
Extended Profile, 402
Context-Based Adaptive Binary Arithmetic Coding (CABAC), 402
Fractional Pel Accuracy, 403
Chroma Sample Interpolation, 403
4×4 Integer Transform, 404
Comparative Studies, 406
Fine Granularity Scalability (FGS) Video Coding Techniques, 407
H.264/AVC Over IP, 411
H.264/AVC in Wireless Environment, 412
  5.3.3 Transcoding Architectures and Technologies, 418
    Bit Rate Reduction, 420
    Transcoding Architectures, 421
    Multipoint Video Bridging, 426
  5.3.4 Multimedia Implementation, 428
5.4 Media Streaming, 431
  5.4.1 MPEG-4 Delivery Framework, 432
  5.4.2 Streaming Video Over the Internet, 436
    Video Compression, 437
    Various Requirements Imposed by Streaming Applications, 439
    Application-Layer QoS Control, 441
    Continuous Media Distribution Services, 447
    Streaming Servers, 450
    Media Synchronization, 450
    Protocols for Streaming Media, 453
  5.4.3 Challenges for Transporting Real-Time Video Over the Internet, 456
  5.4.4 End-to-End Architecture for Transporting MPEG-4 Video Over the Internet, 459
  5.4.5 Broadband Access, 461
    Passive Optical Networks, 464
    Wireless Access, 465
    Digital Subscriber Line, 465
    Cellular Radio Networks, 467
  5.4.6 Quality of Service Framework, 469
    Quality of Service Signaling for IP-Based Mobile Networks, 470
    End-to-End QoS Issues in the IP Wired and Wireless Environment, 472

Individual QoS in Cellular Networks, 473
  5.4.7 Security of Multimedia Systems, 475
    Important Aspects of the Copyright, 476
    Conditional Access Systems, 477
    Copy Protection in Home Networks, 480
    Streaming Content Security, 481
5.5 Infrastructure for Multimedia Content Distribution, 482
  5.5.1 Content Description, 484
    Structural Aspects of Content, 484
  5.5.2 Multimedia Content Management, 489
  5.5.3 Multimedia Authentication Technologies, 489
    Hard Authentication, 492
    Soft Authentication, 494
    Speech Authentication, 496
    Future Directions, 496
  5.5.4 Watermarking Frameworks and Technologies, 497
    Basic Watermarking Principles, 499
    Text Documents Watermarking, 501
    Image Watermarking, 502
    Video Watermarking, 503
    Audio Watermarking, 504
    Watermarking of Other Multimedia Data, 505
  5.5.5 Digital Rights Management Systems and Security Information for Content Distribution, 506
    Data Hiding System, 506
    Copyright Protection, 508
    Item Identification, 510
  5.5.6 Multimedia Advances in Media Commerce, 512
    Business Models, 515
5.6 Middleware Technologies for Multimedia Networks, 517
  5.6.1 Middleware to Support Sensor Network Applications, 518
  5.6.2 Middleware for Wireless Sensor Networks, 519
  5.6.3 Peer-to-Peer Middleware, 521
References, 523

6 Network Layer — 535

6.1 Introduction, 536
6.2 Network Aspects of Standardization Projects, 536
  6.2.1 Technological Building Blocks of the GII, 537
    Transport, 537
    Network Technologies, 538
    Radio and Satellite Systems, 538
  6.2.2 ETSI EII Network-Related Projects, 539
  6.2.3 Network Design in the MEDIACOM2004 Project, 540

6.3 Network Functions, 542
    6.3.1 ISO Reference Model, 543
    6.3.2 ISO Transport Layer, 545
    6.3.3 ISO Network Management Framework, 546
    6.3.4 Advanced Control, 547
        Rate Control Function, 549
        Rate Control in MPEG-4 Visual and H.264 Standards, 551
    6.3.5 Signaling in Communications Networks, 554
    6.3.6 Network Management, 555
        Evolution of Network Management, 557
        Digital Subscriber Lines (DSL) Management, 564
        Management of Location Information, 566
        Resource Management for Quality of Presentation (QoP)-Based Multimedia Applications in Mobile Networks, 570
        Mobility Management in a 4G System, 574
    6.3.7 Transport Network Layered Architecture, 575
        Transport Mechanisms, 577
    6.3.8 Multicast Protocols Classification, 578
        Examples of Multicast Protocols, 581
    6.3.9 Routing Procedure for Multimedia Communications, 585
        Routing and Quality of Service (QoS) Requirements, 587
        Internet Routing Protocols, 588
    6.3.10 Security Issues, 589
        Network Security Services, 591
        General Security Analysis, 594
        Security Criteria, 595
        Security Aspects of ATM Networks, 596
        Security Aspects of IP Networks, 597
        Internet Infrastructure Security, 598
        General Packet Radio Service (GPRS) Security, 601
        Security Mechanisms Provided with Session Initiation Protocol (SIP), 603
6.4 Network Traffic Analysis, 604
    6.4.1 Traffic Engineering, 606
    6.4.2 Multimedia Traffic Management, 608
    6.4.3 Connection Admission Control, 609
        CAC Based on Peak Rate Allocation, 610
        CAC Based on Rate Envelope Multiplexing (REM), 610
    6.4.4 Resource Allocation, 612
    6.4.5 Bandwidth Allocation, 613
    6.4.6 Congestion Control for Multicast Communications, 615
        Congestion Optimization Problem, 617
    6.4.7 Traffic Modeling, 617
        On/Off Model, 618

Markov Modulated Model, 619
Gaussian Autoregressive Model, 619
Self-Similar Models, 620
Pseudo-Self-Similar Models, 621
Modeling the Aggregated Traffic, 622
6.5 Quality of Service (Qos) in Network Multimedia Systems, 622
    6.5.1 Selection and Configuration of QoS Mechanisms, 625
        QoS Provision, 625
        QoS Control, 626
        QoS Management, 627
    6.5.2 QoS Architecture, 628
    6.5.3 OSI QoS Framework, 629
    6.5.4 QoS from Providers' and Customers' Viewpoints, 631
    6.5.5 Quality of Service Parameters, 633
    6.5.6 Quality of Service Classes, 635
    6.5.7 QoS Maintenance and Monitoring, 638
    6.5.8 Framework for QoS-Based Routing, 638
    6.5.9 IP Oriented Quality of Service, 642
        QoS in the Internet Backbone, 643
        QoS in `DiffServ` IP Networks, 648
        QoS in Best-Effort Networks, 650
        QoS in Ethernet Networks, 652
        QoS in IP-Over-WDM Networks, 654
        QoS for Voice Over IP Technology, 658
        QoS in Next Generation Networks, 659
        QoS in 3G Multimedia Mobile Applications, 661
        Trends in QoS for 4G Wireless Systems, 663
6.6 Generic Networks, 664
    6.6.1 Layered Media Streams, 667
    6.6.2 Error Resilience Approach, 670
        Error-Resilient Encoding, 671
        Decoder Error Concealment, 674
        Error-Resilient Entropy Code, 675
6.7 Access Broadband Networks, 675
    6.7.1 DSL Access Networks, 676
        High Bit Rate DSL (HDSL), 677
        Symmetric DSL (SDSL), 677
        ISDN DSL (IDSL), 678
        Asymmetric DSL (ADSL), 678
        Rate-Adaptive DSL (RADSL), 679
        Very High Data Rate DSL (VDSL), 680
    6.7.2 Cable Access Networks, 681
        Hybrid Fiber-Coax, 682
    6.7.3 Wireless Access Networks, 686
        LMDS (Local Multipoint Distribution System), 687

MMDS (Multichannel Multipoint Distribution System), 688
U-NII (Unlicensed National Information Infrastructure), 688
Third Generation (3G), 689
Direct Broadcast Satellite (DBS), 689
6.8 Core Broadband Networks, 690
6.8.1 SONET/SDH, 690
6.8.2 Future of Asynchronous Transfer Mode (ATM), 693
6.8.3 Trends in Wireless Broadband Networking, 695
6.8.4 Toward the Fourth Generation (4G) System, 697
Service Framework Conception, 699
Hybrid QoS Management, 701
6.9 Content Delivery Networks, 703
6.9.1 Content Delivery Evolution, 705
6.9.2 Content Delivery Network (CDN) Functions, 707
Content Generation Tier, 707
Integration Tier, 709
Content Assembly and Delivery Tier, 709
6.10 Concluding Remarks, 709
References, 717

Index   **727**

PREFACE

Multimedia communications have emerged as a major research and development area. In particular, computers in multimedia open a wide range of possibilities by combining different types of digital media such as text, graphics, audio, and video. The emergence of the World Wide Web (WWW), two decades ago, has fuelled the growth of multimedia computing. Because the number of multimedia users is increasingly daily, there is a strong need for books on multimedia systems and communications. Generally speaking, the books can be divided into two major categories. In the first category, the books are purely technical, providing detailed theories of multimedia engineering with an emphasis on signal processing. In the second category, the books on multimedia are primarily about content creation and management. Today, there is a strong need for books somewhere between these two extremes.

This unique book intends to fill this gap by explaining multimedia communications in three areas: applications, middleware, and networking. In this way, the volume will be useful for readers who are carrying out research and development in systems area such as television engineering and storage media. It will also provide readers with the protocol information needed to support a wide variety of multimedia services.

The book reflects the latest work in the field of multimedia communications, providing both underlying theory and today's best design techniques, by

- systematically addressing aspects of recent trends and standardization activities in multimedia communications, and
- covering the layered structure of multimedia communication systems.

BOOK OBJECTIVES AND INTENDED AUDIENCE

Anyone who seeks to learn the core multimedia communication technologies will need this book. The practicing engineering or scientist working in the area of multimedia communication is forced to own a number of different texts and journals to ensure a satisfactory coverage of the essential ideas and techniques of the field. The pressing need for a comprehensive book on important topics in multimedia communications is apparent. Our first objective for the handbook is to be the source of information on important topics in multimedia communications, including the standardization process. The organization of standard bodies is currently vertical, and this should be changed to a horizontal one. There should be one body addressing the delivery layer issues, possibly structured along different delivery media, one body for the application layer, and one body for middleware. The only thing that digital technologies leave as specific to the individual industries is the delivery layer.

One of the proposed book's objectives is a distillation from the extensive literature of the central ideas and primary methods of analysis, design, and implementation of multimedia communication systems. The book also points the reader to the primary reference sources that give details of design and analysis methods.

To conclude, the objective of the book is not only to formalize the reader with this field, but also to provide the underlying theory, concepts, and principles related to the power and practical utility of the topics.

ORGANIZATION OF THE BOOK

Following an Introduction, Chapter 2 of the handbook begins by introducing the reader to the convergence of communications and computing for integrated multimedia applications and services. Following the introduction, we provide technologies for multimedia communications. Next, we invoke the multimedia database. We then discuss trends in multimedia standardization.

Chapter 3 covers frameworks for multimedia standardization. With the increasing demand for multimedia services, multimedia standardization has drawn tremendous levels of attention. First, the ITU MEDIACOM2004 framework for multimedia standardization will be presented. A framework includes an application layer, a middleware layer, and a network layer. After that, we will present an overview of ETSI standardization and describe various standardization projects with regard to applications, middleware, and networks. Next, we will discuss the key elements defined in the MPEG-21 multimedia framework: digital item declaration, identification and description, content handling and usage, intellectual property management and protection, content representation, and event reporting.

Chapter 4 is devoted to the application layer. We cover ITU applications, ISO MPEG applications (multimedia videoconferencing, interactive video and broadcast, streaming content over the Internet, browsing and searching), IETF multimedia applications over IP. Finally, ETSI Digital video broadcasting (interactive TV, Multimedia Home Platform) and ATSC Digital TV will be analyzed.

Chapter 5 provides an overview of the middleware layer. We will describe media coding and conversion, and clarify multimedia protocol architecture. After that, we deal with a model for distributed system multimedia services, together with quality of service and end-to-end performance in multimedia systems. We continue with accessibility to multimedia systems and services. The problem of security of multimedia systems and services will be demonstrated, too. This chapter will conclude with media-streaming middleware.

Chapter 6 concentrates on the network (distribution/delivery) layer. It includes many issues relating to the quality of service and network performance, traffic analysis, and management. We will also present modeling video source and network traffic. Traffic management will be analyzed, together with routing procedures, in multimedia communication. Signalling in communication networks will be discussed. Security issues in networks with Internet access will be provided as well. We will conclude this chapter with a description of multimedia transport and distribution, together with generic networks, broadband access networks, wireless communication networks, and streaming media networks. The goal of this chapter is to pave the way for future developments in the field and for a better understanding of its potential in today's world.

Each chapter has been organized so that it can be covered in one to two weeks when this handbook is used as a principal reference or text in a senior or graduate course at a university. It is generally assumed that the reader has prior exposure to the fundamentals of multimedia communication systems.

The bibliographic references will be grouped according to the various chapters. Special efforts will be taken to make this list as up to date and exhaustive as possible.

A major challenge during the preparation of this book was the rapid pace of development. Many specific applications have been realized in the past few years. We have tried to keep pace by including many of these latest developments. In this way it is hoped that the book is timely and will appeal to a wide audience in the engineering, scientific, and technical communication fields. In addition we have included more that 250 figures and over 570 references. Although this book is primarily for graduate students, it can also be very useful and suitable for advanced-level courses in multimedia communications (for academia, researchers, scientists, and engineers dealing with multimedia communications). Also, this is a well-needed addition to professional reference research. We feel that this book will serve as an essential and indispensable resource for many years.

K. R. Rao
Zoran S. Bojkovic
Dragorad A. Milovanovic

ACKNOWLEDGEMENT

This book is the product of many years of work that resulted in multimedia communication pages, but also in many lifetime friendships among people all around the world. Thus, it is a pleasure to acknowledge the help received from colleagues associated with various universities, research labs, and industry. This help was in the form of technical papers and reports, valuable discussions, information, brochures, the review of various sections of the manuscript, computer programs, invited lectures, and more. We give sincere and special thanks to the following people:

- Yuan Baozong (Northern Jiaotong University, Institute of Information Science, Beijing, China)
- Monica Borda (Technical University of Cluj, Faculty of Electronics and Telecommunications, Cluj-Napoca, Romania)
- Ling-Gee Chen (National Taiwan University, Institute of Electronics Engineering, Taipei, Taiwan)
- Walter Geisselhardt (Gerhard-Mercator University, Duisburg, Germany)
- Muhamed Hamza (IASTED, Calgary, Alberta, Canada)
- Anand Injeti (EM Solutions, LLC, Dallas, TX, USA)
- Nikos Mastorakis (WSEAS, Athens, Greece)
- Valeri Mladenov (Technical University–Sofia, Faculty of Automatics, Sofia, Bulgaria)
- Soontorn Oraintara (University of Texas at Arlington, Electrical Engineering Department, Arlington, TX, USA)
- Jonel Stancu (Sprint PCS, Houston, TX, USA)
- Cristos Stremmenos (University of Bologna, Bologna, Italy)
- Jan Turan (University of Kosice, Department of Electronics and Multimedia Telecommunications, Slovak Republic)

Branka Vucetic (University of Sydney, Department of Electrical Engineering, Sydney, Australia)

K. R. R.
Z. S. B.
D. A. M.

ACRONYMS

3GPP	Third Generation Partnership Project
A/V	Audio/Video
AAC	Advanced Audio Coding
AAL	ATM Adaptation Layer
AAP	ATM Access Point Process
ABME	All Binary Motion Estimation
ABR	Available Bit Rate
ACELP	Algebraic Code-Excited Linear Prediction
ACF	Autocovariance Function
ACK	Positive Acknowledgment
ADM	Add/Drop Multiplexer
ADSL	Asymmetrical Digital Subscriber Loop
AES	Advanced Encryption Standard
AGI	Active Group Integrity
AH	Authentication Header
AIR	Adaptive Intra Refresh
AL	Adaptation Layer
ALF	Application Level Framing
AM	Amplitude Modulation/Accounting Management
AMAC	Americas Market Awareness Committee
AMR	Adaptive Multirate
ANS	Announcement Server
ANSI	American National Standards Institute
API	Application Programming Interface
APMAC	Asia-Pacific Market Awareness Committee
AQM	Active Queue Management
ARMA	Autoregressive Moving Average
ARS	Actual Route Selection
AS	Autonomous System
ASIC	Application Specific Integrated Circuits
ASN	Abstract Syntax Notation

ASO	Arbitrary Slice Order
ASP	Application Service Provider/Active Service Pages
ATM	Asynchronous Transfer Mode
ATSC	Advanced Television System Committee
AuC	Authentication Center
AVC	Advanced Video Coding
AVO	Audio Visual Object
AVT	Audio Video Transport
B2B	Business to Business
B2C	Business to Consumer
B2E	Business to Employee
BB	Bandwidth Broker
BER	Bit Error Rate
BGP	Border Gateway Protocol
B-ICI	Broadband Inter Carrier Interface
BIFS	Binary Format for Scenes
BISDN	Broadband ISDN
BIU	Broadband Interface Unit
BLES	Broadband Loop Emulation Services
BMAP	Batch Markovian Arrival Process
BRAN	Broadband Radio Access Network
BS	Bearer Service
BSC	Base Station Subsystem
BTS	Base Transciever Station
BUS	Broadcast and Unknown Server
CA	Conditional Access
CAB	Charging Accounting and Billing
CABAC	Context-based Adaptive Binary Arithmetic Coding
CAC	Connection Admission Control
CAT	Conditional Access Table
CATS	Consortium for Audiographics Teleconferencing Standards
CATV	Cable Television
C-BAR	Capacity-balanced Alternate Routing
CBR	Constant Bit Rate
CBT	Core-Based Tree
CC	Copyright Compliant
CCS	Computation/Communication Sensing
CD	Compact Disc
CDMA	Code Division Multiple Access
CDN	Content Delivery Network
CD-ROM	CD – Read Only Memory
CDV	Cell Delay Variation
CDVT	Call Delay Variation Tolerance
CEA	Consumer Electronic Association
CELP	Code Excited Linear Prediction

CEMA	Consumer Electronics Manufacturers Association
CGI	Common Gateway Interface
cHTML	Compact HTML
CIF	Common Intermediate Format
CL	Connectionless
CLIP	Calling Line Identification Presentation
CLP	Cell Loss Priority
CLR	Cell Loss Rate (Ratio)
CM	Cable Modem/Configuration Management
CMIP	Common Management Information Protocol
CMOS	Complementary Metal Oxide Semiconductor
CMS	Call Management Server
CMTS	Cable Modem Terminations System
CN	Core Network
CO	Connection – Oriented
CODFM	Coded Orthogonal Frequency Division Multiplex
COPS	Common Open Policy Service
COPS-PP	COPS for Policy Provisioning
CORBA	Common Object Requests Broker Architecture
COTS	Commercial Off-the-Shelf
CPB	Coded Picture Buffer
CPE	Customer Premises Equipment
CPU	Central Processing Unit
CR	Code Rate
CRC	Cyclic Redundancy Code
CR-LDP	Constraint-based Routing Label Distribution Protocol
CS	Coding Scheme
CSA	Carrier Serving Area
CSMA/CD	Carrier Sense Multiple Access with Collision Detection
CSP	Content Service Provider
CSS	Content Scrambling System
CTD	Cell Transfer Delay
CUG	Closed User Group
CVoDSL	Channelized Voice over DSL
D+R	Drop and Repeat
DAB	Digital Audio Broadcast
DAI	Delivery Application Interface
DAVIC	Digital Audio Visual Council
DBS	Digital Broadcast Satellite
DCP	Dynamic Configuration Protocol
DCT	Discrete Cosine Transform
DDL	Description Definition Language
DDoS	Distributed DoS
DES	Data Encryption Standard
DI	Digital Item

DIA	Digital Item Adaptation	
DID	Digital Item Declaration	
DiffServ	Differentiated Service	
DII	Digital Item Identification	
DLC	Digital Loop Carrier	
DMIF	Delivery Multimedia Integration Framework	
DNG	Delivery Network Gateway	
DNH	Digital Home Network	
DNI	DMIF Network Interface	
DNS	Direct Name System	
DoS	Denial of Service	
DPB	Decoded Picture Buffer	
DPE	Distributed Processing Environment	
DRM	Digital Rights Management	
DS	Description Scheme	
DSL	Digital Subscriber Line	
DSLAM	DSL Access Multiplexer	
DSM	Digital Storage Media/Dynamic Spectrum Management	
DSM-CC	Digital Storage Media – Command and Control	
DSNG	Digital Satellite News Gathering	
DSR	Digital Satellite Radio	
DSSS	Direct Sequence Spread Spectrum	
DTH	Direct-to-Home	
DTS	Decoding Time Stamp	
DTTB	Digital Terrestrial Television Broadcasting	
DVB	Digital Video Broadcasting	
DVB-RCS	DVB – Return Channel Satellite	
DVB-RCT	DVB – Terrestrial Return Channel	
DVD	Digital Video Disk	
DVI	Digital Video Interface	
DVT	Digital Television	
DWT	Discrete Wavelet Transform	
EACEM	European Association of Consumer Electronics Manufacturers	
EBU	European Broadcast Union	
EC	Error Concealment	
ECM	Encryption Controlled Message	
ECN	Explicit Congestion Notification	
EDGE	Enhanced Data rates for GSM Evolution	
EIA	Electronics Industries Association	
EII	European Information Infrastructure	
EIR	Equipment Identity Register	
EMEA	Europe Middle East Africa Market Awareness	
EMM	Encryption Management Message	
EOB	End of Blocks	
EPG	Electronic Program Guide	

EPII	European Project on Information Infrastructure
EREC	Error Resilient Entropy Code
ES	Elementary Stream
ESI	Elementary Stream Interface
ESP	Encapsulating Security Payload
ETSI	European Telecommunications Standards Institute
FAP	Facial Animation Parameters
FAR	Fixed Alternate Routing
FBM	Fractional Brownian Motion
FCAPS	Fault Configuration Accounting Performance Security
FCC	Federal Communications Commission
FCS	Feadback Control Signaling
FDD	Frequency Division Duplex
FDDI	Fiber Distributed Data Interface
FDDS	Fiber Distributed Data Service
FDL	Fiber Delay Line
FDP	Facial Definition Parameters
FEC	Forward Error Correction
FFT	Fast Fourier Transform
FIFO	First-In First-Out
FM	Frequency Modulation
FMO	Flexible Macroblock Ordering
FPLS	Fair Packet Loss Sharing
FR	Fixed Routing
FRF	Frame Relay Forum
FSA	Framework Study Area
FSAN	Full Service Access Network
FTP	File Transfer Protocol
FTTC	Fiber To The Curb
FTTH	Fiber To The Home
FWA	Fixed Wireless Access
G2C	Government-to-Citizen
GDMO	Guidelines for the Definition of Managed Objects
GGSN	Gateway GPRS Support Node
GI	Guard Interval
GIF	Graphic Interchange Format
GII	Global Information Infrastructure
GIOP	General Inter-Operability Protocol
GMM	Global Multimedia Mobility
GOP	Group of Pictures
GPRS	General Packet Radio System
GPS	Global Positions Systems
GRM	Generic Relationship Model
GSC	Global Standards Collaboration
GSM	Global System for Mobile

GSTN	General Switched Telephone Network
GTP	GPRS Tunneling Protocol
GUI	Graphical User Interface
HAN	Home Access Network
HAS	Human Audible System
HDLC	High-bit-rate Data Link Control
HDSL	High-data-rate DSL
HDTV	High Definition Television
HFC	Hybrid Fiber Coaxial
HI	Hearing Impaired
HLN	Home Local Network
HLR	Home Location Register
HRD	Hypothetical Reference Decoder
HSS	Home Subscriber System
HTML	Hypertext Mark-up Language
HVXC	Harmonic Vector Excitation Coding
IAB	Internet Architecture Board
IANA	Internet Assigned Number Authority
IBAC	In-Band Adjacent-Channel
IBOC	In-Band On-Channel
IBT	Intrinsic Burst Tolerance
ICANN	Internet Corporation for Assigned Names and Numbers
ICTSB	Information and Communication Technology Standardization Board
IDB	Interactive Data Broadcast
IDL	Interface Definition Language
IDR	Instantaneous Decoder Refresh
IEC	International Electrotechnical Committee
IEEE	Institute of Electrical and Electronics Engineers
IESG	Internet Engineering Steering Group
IETF	Internet Engineering Task Force
IGMP	Internet Group Management Protocol
IGP	Interior Gateway Protocol
IHDN	In-Home Digital Network
IIF	Integrated Intermedia Format
IIOP	Internet Interoperability Protocol
IISP	Information Infrastructure Standards Panel
IMT	International Mobile Telecommunication
IMTC	Internet Multimedia Telecommunications Consortium
IN	Intelligent Network
INAP	Intelligent Network Application Protocol
INF	Interworking Function
IntServ	Integrated Services
IOR	Interoperable Reference
IP	Internet Protocol
IPCDN	IP over Cable Data Network

IPDC	Internet Protocol Datacast
IPI	Internet Protocol Infrastructure
IPMP	Intellectual Property Management and Protection
IPR	Intellectual Property Rights
IPSEC	Internet Protocol Security
IPv6	Internet Protocol version 6
iQoS	Individual Quality of Service
IRD	Integrated Receiver Decoder
ISD	Independent Segment Decoding
ISDN	Integrated Services Digital Network
IS-IS	Intermediate System to Intermediate System
ISM	Industrial, Scientific, Medical
ISMA	Internet Streaming Media Aliance
ISO	International Organization for Standardization
ISOC	Internet Society
ISP	Internet Service Provider
ITSC	Interregional Telecommunications Standards Conference
ITU	International Telecommunication Union
ITU-R	ITU — Radio-communication Standardization Sector
ITU-T	ITU — Telecommunications Standardization Sector
IVB	Interactive Video Broadcast
JIDM	Joint Inter Domain Management
JPEG	Joint Photographic Experts Group
JSP	Java Service Pages
JTC	Joint Technical Committee
JVT	Joint Video Team
Kbps	Kilobits per second
KDC	Key Distribution Center
LAN	Local Area Network
CAVLC	Context-based Adaptive Variable Length Coding
LCS	Location Service Server
LEC	Local Exchange Carriers
LGMP	Local Group Multicast Protocol
LIFO	Last-In First-Out
LMDS	Local Multipoint Distribution Service
LOS	Line-of-Site
LSB	Last Significant Bit
LSI	Large Scale Integration
LSP	Label Switch Path
M4IF	MPEG-4 Industry Forum
MA	Multimedia Authentication
MAC	Medium Access Control/Message Authentication Code
MAN	Metropolitan Area Network
MAP	Manufacturing Automation Protocol
MB	Macroblock

MBN	Multiservice Broadband Network
MBONE	Multicast Backbone
Mbps	Megabits per second
MC	Motion Compensation
MCC	Mobile Competence Centre
MCCOI	Multimedia Communications Community of Interest
MCN	Multimedia Communication Network
MCP	Multimedia Car Platform
MCR	Maximum Cell Rate
MCU	Multipoint Central Unit
MD	Metering Device
MDC	Multiple Description Coding
MDS	Multipoint Distribution System/Multimedia Description Scheme
MG	Media Gateway
MGCP	Media Gateway Control Protocol
MHEG	Multimedia and Hypermedia Experts Group
MHP	Multimedia Home Platform
MIB	Management Information Base
MIDI	Musical Instrument Digital Interface
MIME	Multipurpose Internet Mail Extension
MIT	Manager Information Tree
M-JPEG	Motion JPEG
ML	Main Level
MMDS	Microwave Multipoint Distribution Systems
MMM	Multimedia Mailing
MMPP	Markov Modulated Poisson Process
MMS	Multimedia Messaging Services
MMUSIC	Multiparty Multimedia Session Control
MN	Mobile Node
MOM	Message Oriented Middleware
MOS	Mean Opinion Score
MoU	Memorandum of Understanding
MP	Main Profile
MPEG	Motion Picture Expert Group
MPLS	Multiprotocol Label Switching
MRO	Maintenance Repair and Operation
MS	Mobile Station
MSB	Most Significant Bit
MSC	Mobile Switching Center
MSE	Mean Square Error
MSPN	Multimedia Services over Packet Networks
MTA	Multimedia Terminal Adapter
MTU	Maximum Transit Unit
MVDS	Multipoint Video Distribution System
MW	Middleware

NAB	National Association of Broadcasters
NACK	Negative Acknowledgment
NAL	Network Abstraction Layer
NAT	Network Address Translation
NBC	Nonbackward Compatible
NCCE	Network Computing and Communication Environment
NCCI	Network Control Center Interface
NCR	Network Clock Reference
NCS	Network-based Cell Signaling
NCTA	National Cable Television Association
NGBAN	Next Generation Broadband Access Network
NGM	Network Management Forum
NGN	Next Generation Network
NHPR	Next Hop Resolution Protocol
NIC	Network Interface Card
NISDN	Narrow-band ISDN
NIU	Network Interface Unit
NMAP	Network Management Application Process
NOC	Network Operating Center
NRIM	Network Resource Information Model
NRSC	National Radio System Committee
NSAP	Network Service Access Point
NSP	Network Service Provider
NTSC	National Television System Committee
OBS	Optical Burst Switching
OCI	Object Content Information
OD	Object Descriptor
ODL	Object Definition Language
ODP	Open Distributed Processing
OE	Optical-to-Electrical
OFDM	Orthogonal Frequency Division Multiplexing
OLT	Optical Line Terminator
OMG	Object Management Group
ONU	Optical Network Unit
OPS	Optical Packet Switching
OS	Operating System
OSA	Open Service Access
OSI	Open System Interconnection
OSIE	Open System Interconnection Environment
OSI-SM	OSI System Management
OSPF	Open Shortest Path First
OSS	Operational Support System
OTB	Object Time Base
P2P	Peer-to-Peer
PAC	Perceptual Audio Coder

PAL	Phase Alternating Line
PAS	Publicity Available Specification
PAT	Program Association Table
PCR	Program Clock Reference
PCT	Peak Cell Rate
PCWG	Personal Conferencing Work Group
PDA	Personal digital Assistant
PDH	Plesiochronous Digital Hierarchy
PDP	Packet Data Protocol
PES	Packetized Elementary Stream
PGM	Pragmatic General Multicast
PHB	Per Hop Behavior
PID	Packet Identifier
PIM	Protocol Independent Multicast
PIM-DM	PIM – Dense Mode
PIM-SM	PIM – Sparse Mode
PLMN	Public Land Mobile Network
PM	Performance Management
PMT	Program Map Table
PN	Pseudo Noise
PNG	Portable Network Graphics
POCS	Projection Onto Convex Set
PON	Passive Optical Network
POTS	Plain Old Telephone Service
PPC	Preliminary Path Caching
PPP	Point-to-Point Protocol
PRBS	Pseudo-Random Binary Sequence
PSI	Program Specific Information
PSK	Phase-Shift-Keying
PSN	Packed Switched Networks
PSO	Protocol Supporting Organization
PSTN	Public Switch Telephone Network
PTR	Priority Token Ring
PTS	Presentation Time Stamps
PVC	Permanent Virtual Connection
PVR	Personal Video Recording
QAM	Quadrature Amplitude Modulation
QCIF	Quarter Common Intermediate Format
QoS	Quality of Service
QoS-A	Quality of Service – Architecture
QP	Quantization Parameters
QPSK	Quadrature Phase-Shift Keying
RA	Radio Assembly
RAM	Resource Availability Matrix
RC	Rate Control

RCSPP	Reconfiguration Control and Service Provision Platform
RD	Rate Distortion
RDD	Rights Data Dictionary
RDS	Requests for Detailed Specifications
RED	Random Early Detection
REL	Rights Expression Language
REM	Rate Envelope Multiplexing
RFC	Request for Comments
RfP	Request for Proposal
RfQ	Request for Quotes
RIP	Routing Information Protocol
RKS	Record Keeping Server
RMI	Remote Method Invocation
RMODP	Reference Model of Open Distributed Processing
RMP	Reliable Multicast Protocol
RMTP	Reliable Multicast Transport Protocol
RPC	Remote Procedure Call/Resource Provisioning Cycle
RS	Reed-Solomon
RSVP	Resource Reservation Protocol
RTE	Run Time Engine
RTP	Real-Time Protocol
RTSP	Real-Time Streaming Protocol
RTT	Round-Trip Time
RUI	Routing Update Interval
RVS	Remote Video Surveillance
RWA	Routing and Wavelength Assignment
SA	Secure Agent
SCN	Switched Circuit Network
SCP	Service Control Point
SCR	Sustainable Cell Rate/System Clock Reference
SD	Source Destination
SDARS	Satellite Digital Audio Radio Services
SDES	Source Description
SDH	Synchronous Digital Hierarchy
SDL	Specification and Description Language
SDM	System Decoder Model
SDMI	Secure Digital Music Initiative
SDO	Standards Development Organization
SDP	Session Description Protocol
SDR	Software Defined Radio
SDSL	Single Line DSL
SDU	Service Data Unit
SFC	Stream Flow Connection
SFN	Single Frequency Network
SG	Study Group

SG-MMS	Study Group for Multimedia Services
SGSN	Serving GPRS Support Node
SHD	Super High Definition
SI	Service Information
SIG	Special Interest Group
SIP	Session Initiation Protocol
SL	Synchronization Layer
SLA	Service Level Agreement
SM	Security Management
SMFA	Specific Management Functional Area
SMG	Statistical Multiplexing Gain
SMI	Structure of Management Information
SMPTE	Society of Motion Picture and Television Engineers
SMS	Short Message Service
SMTP	Simple Mail Transfer Protocol
SMUX	Synchronous Multiplexing
SNMP	Simple Network Management Protocol
SNR	Signal-to-Noise Ratio
SOAP	Simple Object Access Protocol
SOHO	Small Office/Home Office
SONET	Synchronous Optical Network
SP	Simple Profile
SQF	Service Quality Function
SRC	Strategic Review Committee
SRLG	Shared Risk Link Group
SRM	Scalable Reliable Multicast
SrvMgt	Service Management
SS7	Signaling System No. 7
SSG	Special SG
SSL	Secure Sockets Layer
SSM	Serial Storage Media
SSP	Service Switching Point/Stream Synchronization Protocol
SSRC	Synchronization SouRCe Identifier
STB	Set-Top-Box
STD	Systems Target Decoder
STM	Synchronous Transport Module
STS	Synchronous Transport Signal
STU	Subscriber Terminal Unit
SUI	Synchronization Interval Unit
SVC	Switched Virtual Connection
TCP	Transmission Control Protocol
TD	Technical Direction
TDD	Time Division Duplex
TDM	Time Division Multiplex
TFB	Tandem-Free Bridge

TFC	Tandem-Free Conferencing
TGCP	Trunking Gateway Control Protocol
TGS	Ticket Granting Server
TINA	Telecommunication Information Network Architecture
TLD	Top Level Domain
TLS	Transport Level Security
TM	Traffic Management
TMF	Tele-Management Forum
TMN	Telecommunication Management Network
TN	Transport Network
ToS	Type of Service
TPDU	Transport Protocol Data Unit
TS	Transport Stream
TSAG	Telecommunication Standardization Advisory Group
TTC	Telecommunication Technology Committee
TTS	Text-to-Speech
TVoD	True VoD
UBR	Unspecified Bit Rate
UDDI	Universal Description Discovery and Integration
UDP	User Datagram Protocol
UEP	Unequal Error Protection
UMA	Universal Multimedia Access
UME	Universal Multimedia Experience
UML	Unified Modeling Language
UMTS	Universal Mobile Telecommunication System
U-N	User–Network
UNI	User–Network Interface
UPC	Usage Parameter Control
UPnP	Universal Plug and Play
UPO	Updated Path Ordering
UPS	United Parcel Service
URI	Uniform Resource Identifier
URL	Uniform Resource Locator
USO	Universal Service Obligation
UTRA	Universal Terrestrial Radio Access
U-U	User–User
UVLC	Universal Variable Length Coding
VAR	Variance
VASP	Value-Added Service Provider
VBI	Vertical Blanking Interval
VBR	Variable Bit Rate
VC	Virtual Connection/Virtual Circuit
VCC	Virtual Circuit Connection
VCEG	Video Coding Experts Group
VCI	Virtual Connection Identifier

VCL	Video Coding layer
VCR	Video Cassette Recorder
VDSL	Very-high-rate DSL
VESA	Video Electronics Standards Association
VFW	Video for Windows
VHS	Video Home System
VLC	Variable Length Coding
VLD	Variable Length Decoder
VM	Virtual Machine
VO	Video Object
VoD	Video-on-Demand
VoIP	Voice over IP
VOL	Video Object Layer
VOP	Video Object Plane
VP	Virtual Path
VPI	Virtual Path Identifier
VPN	Virtual Private Network
VQ	Vector Quantization
VRML	Virtual Reality Markup Language
VSB	Vestigal Side Band
VTC	Visual Texture Coding
W3C	World Wide Web Consortium
WA	Wavelength Allocation
WAN	Wide Area Network
WAP	Wireless Applications Protocol
WARC	World Administrative Radio Conference
WATM	Wireless ATM
WCDMA	Wideband Code-Division Multiple Access
WCS	Wireless Communications Services
WDM	Wavelength Division Multiplex
WG	Working Group
W-HDN	Wireless Home Distribution Network
WLAN	Wireless LAN
WP	Working Party
WR	Wavelength Routing
WSDL	Web Service Description Language
WSN	Wireless Sensor Network
WSS	Wide Sense Stationary
WTSA	World Telecommunication Standardization Assembly
WWW	World Wide Web
XML	Extensible Markup Language
XTP	Xpress Transport Protocol

1

INTRODUCTION

Multimedia – an interactive presentation of speech, audio, video, graphics, and text, has become a major theme in today's information technology that merges the practices of communications, computing, and information processing into an interdisciplinary field. In recent years, there has been a tremendous amount of activity in the area of multimedia communications: applications, middleware, and networking. A variety of techniques from various disciplines such as image and video processing, computer vision, audio and speech processing, statistical pattern recognition, learning theory, and data-based research have been employed.

This volume has the intention of providing a resource that covers introductory, intermediate, and advanced topics in multimedia communications, which can respond to user requirements in terms of mobility, easy of use, flexibility of systems, as well as end-to-end interoperability with specific quality requirements. We have initiated this handbook considering:

- the continuing trend in digitization
- the rapid growth of digital networks and, in particular, the Internet
- the increasing computational power in personal computers and lap tops, PDAs, and mobile phones
- the convergence of various technologies, including communications, broadcasting, information technology, and home electronics
- emerging new communication services and applications that are a result of the growth of the Internet and wireless technologies

Introduction to Multimedia Communications, By K. R. Rao, Zoran S. Bojkovic, and Dragorad A. Milovanovic
Copyright © 2006 John Wiley & Sons, Inc.

2 INTRODUCTION

- that with the emergence of high-speed, high-quality networks, society will request real-time multimedia communications as an extension of existing monomedia systems
- that the end-to-end performance of multimedia systems and services defined in the International Telecommunication Union (ITU) and elsewhere should be categorized
- that it will be necessary to study the interfaces in the Information Appliance environment as consumer devices increasingly perform multimedia functions.

Networked multimedia communication originated in a confluence of two technological trends. The first involved the development of multimedia computing, and the second involved advances in networking that allowed reliable widespread delivery of digital data at relatively high bandwidths. Advances continue in both of these areas. Multimedia is now a driving force rather than just a way of using already existing bandwidth.

In multimedia communications where human speech involved, technology and standardization frameworks are particularly significant. Multicast communication services are expended to become one of the most popular applications, with daily publications, software distribution, data base replication in future multimedia communication networks.

Global standardization is becoming more and more important in achieving a global information infrastructure. Telecommunication and information standardization organizations such as ITU and the International Organization for Standardization/International Electrotechnical Commission (ISO/IEC) have each set up special groups to focus on determining standards for global information infrastructure.

The challenge of multimedia communications is to provide applications that integrate text, sound, image and video information and to do it in a way that preserves the ease of use and interactivity. First of all ITU and MPEG applications from the integration, interactivity, storage, streaming, retrieval, and media distribution must be taken into account. Also, digital broadcasting, together with broadcasting infrastructures, are of the most interest. Mobile services and applications as well as research activities, such as usability, user interfaces, mobile access to databases and agent technologies represent today a modern area in multimedia communications from the applications layer point of view.

Much of the complexity and cost of building networked applications can be alleviated by the use of highly flexible, efficient, dependable and secure middleware, which is systems software that resides between the applications and the underlying operating systems and networks. It provides reusable services that can be composed, configured, and deployed to create multimedia applications rapidly and robustly. In particular, due to the explosive growth and great success of the Internet, as well as increasing demand for multimedia services, streaming media over the Internet has drawn tremendous attention from both academia and industry.

There are many issues relating to networking. The development of a network layer on a large scale presents numerous technical challenges in a variety of multimedia areas. As progress continue to be made in multimedia communications, it is necessary to ascertain consumer interest in this new for m of communications. In the network layer the emphasis is on connection admission control, resource and bandwidth allocation, congestion control for multicast communications, as well as traffic modeling. The issues concerning quality of service (QoS) in networked multimedia systems, specially Internet protocol (IP) oriented is of interest, too. Quality of service parameters and requirements must be in accordance with QoS guarantees. At the heart of the broadband networks are technologies that functions as high-speed, high-performance bit pumps, pumping data through high-capacity pipes between different points in the network.

2

MULTIMEDIA COMMUNICATIONS

In multimedia communications where human speech is involved, the human communication model, technology, and standardization frameworks are particularly significant. This chapter explains why multimedia communication technology is more than putting together text, audio, images, and video. After a short presentation of the human communication model, we deal with evolution and convergence of multimedia communications. This chapter also reviews a recent trend concerning the technology framework for multimedia communications. The emphasis is on multimedia networking, conferencing, multicasting, as well as e-content. The chapter concludes with the standardization framework, dealing with research, regulation, education, manufacturing, and marketing. Digital video/audio coding standards and their impact on multimedia industry are outlined, too.

2.1 INTRODUCTION

Networked multimedia communication originated in a confluence of two technological trends. The first involved the development of multimedia computing, and the second involved advances in networking that allowed reliable widespread delivery of digital data at relatively high bandwidths. Advances continue in both of these areas. Multimedia is now a driving force rather than just a way of using already existing bandwidth. Recognizing that various messages are best conveyed in different media, multimedia computing developers desired to add the richness of

Introduction to Multimedia Communications, By K. R. Rao, Zoran S. Bojkovic, and Dragorad A. Milovanovic
Copyright © 2006 John Wiley & Sons, Inc.

real-world sounds and pictures, thereby expanding the scope of applications, and also bring these applications to users for whom traditional text-based interfaces are not acceptable. The personal computer becomes a central communication node replacing a large number of other devices or systems, including television receivers, typewriters, fax machines, telephones, and mail delivery. Nevertheless, the motivation to connect user devices to multimedia networks remains. Networking enables communication with other people and indeed generally allows the user to experience live, remote events. Applications for live, remote access include conferencing, education, entertainment, and monitoring of remote locations.

Multimedia communication is being encouraged by the following trends in communication networks:

- seamless transition from centralized network control to distributed control
- seamless interconnection of different networks at affordable cost
- seamless integrated transmission of multimedia with multiple QoS levels (including multicast).

Multicast communication services are expected to become one of the most popular applications, with daily publications, software distribution, database replication, and so on in future multimedia communication networks. To provide multicast communication services in heterogeneous communication environments, it is necessary to develop a QoS guaranteed mechanism that can provide an appropriate QoS to each host according to its available bandwidth and resources. To utilize network resources effectively, nonreal-time multicasting is also important in addition to real-time multicasting. In this case, multiple QoS class support and the capability for dynamic bandwidth reservation will also be required.

As a result of the integration of broadband integrated services digital networks (BISDN), multimedia communication has become possible in addition to high-speed computer communications and conventional voice and video communication. These media are handled flexibly and interactively in the multimedia communication environment.

Communication services are generally specified as guaranteed type services or best effort type services from the viewpoints of communication quality assurance and routing mechanisms [1]. The service classification in terms of QoS for circuit switching, packet switching, frame relay and asynchronous transfer mode (ATM) switching is shown in Table 2.1.

Connection-oriented (CO) type packet switching is used to provide reliable communication services when the transmission line's quality cannot be guaranteed. With the introduction of optical fibers, packet switching is being replaced by frame relay switching, which can eliminate retransmission control within the network and hence realize more efficient packet data transmission. On the other hand, the connectionless (CL) type packet-switching service is deemed a best effort service since it does not handle retransmission in the network layer when error occurs, and error recovery is conducted in the upper transport layer such as in the transmission control

Table 2.1 Service classification on QoS

Switching Type	Connection Type	QoS Guarantee
Circuit switching	Connection-oriented	Guaranteed type
Packet switching	Connection-oriented	Guaranteed type (virtual circuit)
	Connectionless	Best effort type (datagram, IP packet)
Frame relay	Connection-oriented	Best effort type
ATM switching	Connection-oriented	Guaranteed type Best effort type

protocol (TCP). The connectionless function is considered to be useful for accommodating the multimedia database servers expected in the access network. The conventional telephone network service is a circuit-switching type and supports bandwidth guaranteed services. An asynchronous transfer mode (ATM) network can support both the guaranteed type and the best effort type. More specifically, it can support constant bit rate (CBR), variable bit rate (VBR), and unspecified bit rate (UBR) for the best effort type. They can be selected to match the user's requirements, for example, VBR for video high-definition image applications and CBR for voice applications. It is expected that most ATM users will use UBR and available bit rate (ABR), because it is difficult to specify strictly QoS for VBR, taking applications and higher layer protocols into account. Related subjects will be discussed in further chapters.

To understand the ATM technological background, we should also consider the higher layer standard communication protocol for a multimedia communication network. The network is expected to provide a reliable delivery mechanism (such as transmission control protocol, TCP) as well as an unreliable delivery mechanism (such as user datagram protocol, UDP) according to the user requirements of cost and reliability. The reliable delivery mechanisms typically use retransmission to guarantee delivery of packets, whereas the unreliable delivery mechanisms make a best effort to deliver packets. Without the overhead and latency of retransmission, the communication fee can be reduced. As for connectionless best effort service with QoS requests, for example, Internet Engineering Task Force (IETF) is investigating the concept of the resource reservation protocol (RSVP). It determines whether the router has sufficient resources to support the requested QoS and whether the user has administrative permission to make the reservation. If both checks succeed, RSVP sets parameters in the router packet classifier and scheduler to obtain the desired QoS. The RSVP is expected to be applied to small intranets including Internet telephony for the advantage of dynamic resource negotiations.

As for transmission technology, the options include asymmetric digital subscriber line (ADSL), which transmits digital signals over existing twisted-pair wiring, fiber-to-the home, fiber-to-the-curb with coaxial cable for the drop and wireless systems. Existing transmission systems utilize the basic pulse transmission technique over optical fiber or twisted-wire pair cable and the plesiochronous digital hierarchy (PDH) technique. The synchronous digital hierarchy (SDH) fiber optic systems or North American defined Synchronous Optical Networks (SONET) employing network synchronization techniques are currently being widely introduced. There are two advantages that SDH and SONET have over PDH. The former are superior because they can offer higher speeds and direct multiplexing without intermediate multiplexing stages. This technology is achieved through the use of pointers in the multiplexing overhead that directly identify the position of the payload. In particular, the virtual path (VP) concept exploits the benefits of ATM capabilities and provides the network with a powerful transport mechanism. In an ATM network, VP bandwidth and route can be controlled separately. The independence of the VP route and capacity establishment simplifies path bandwidth control. In addition, the direct multiplexing of VPs into transmission links eliminates hierarchical post multiplexing and eliminates hierarchical digital cross-connect systems. As for the large-scale integration (LSI) technology required for SDH systems, line interface LSI chips for 54/150/600 Mbps SDH transmission have been developed using 0.8 μm Bi-CMOS and 1 μm bipolar technology [2].

By making use of the multimedia communication network, multimedia applications such as high-speed inter-LAN communications, large-capacity file transfer, super-high-definition image transfer, remote medical diagnostics, and remote education have been tested. An example of the high-speed computer communication system is shown in Figure 2.1. Supercomputer-generated visualization data are sent over the ATM network and displayed in real time as sequential moving images. Other popular applications in the experimental network are electronic mail, inter local area network (LAN) communication, multimedia information retrieval using WWW, on-line transaction processing such as banking, and wide band video conferencing. Substantial multimedia systems such as an Internet-based digital network library system, a multimedia teleconferencing system, a video-on-demand (VoD) system, a super-high-definition (SHD) image system and a collaborative conferencing system are already in the stage of commercial introduction [3].

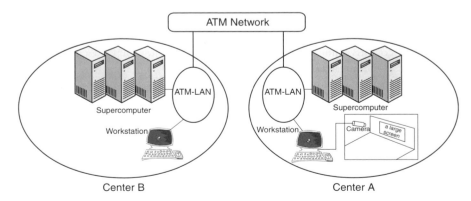

Figure 2.1 An example of the high-speed computer communication system.

The evolution of multimedia computing and its integration into business and personal life are dependent on two primary factors: access to desktop computers or other computer devices, and access to an appropriate communications infrastructure [4].

This chapter starts with a model of human communication. Then, it seeks to provide a convergence of telecommunications, computing, and consumer electronics. A brief description of technologies for multimedia communications is also presented. Research, standardization development, and regulation are emphasized. An outline of digital video/audio coding standards and their role in multimedia communications conclude the chapter.

2.2 HUMAN COMMUNICATION MODEL

To motivate the human communication model, let us recognize some obvious and apparent components of human communication. In any conversation between people, two channels of meaningful communication are usually open at any time: a *cognitive* one and an *emotional* one. Cognitive communication follows a logical structure, perhaps constructed along the patterns of rational language. Cognitive communication depends on the definitions of words, phrases, and sentences. It is public and understandable by almost anyone. Emotional communication may lie outside the language, and may include facial gestures, voice tone, word choice, and other signs of feeling. Emotional communication carries feeling messages. Cognitive communication carries more logical messages. These two message types are different in form and content. When we understand another person, we understand both of these messages and their relation to each other. Each of these channels of communication carries different, sometimes conflicting, information. The rules by which we communicate on these two channels differ. So these two channels parallel each other, and in many ways are independent of each other.

Another factor in communication systems is memory or its shared manifestation *culture*. Each of us has memory. It provides the environment for thinking and culture is a shared environment. Memory embodies our expectations by encoding our history: words, symbols, feelings, experience. No communication can occur without memory. Communication depends on expectation. The protocols of expectation follow neither cognitive nor feeling protocols. Both of these depend on memory's expectation. Cognition requires memory.

The *mission* is another area of human occupation and communication behavior. Missions refer to what we have organized for the purpose of our organization. We form a set of concepts, a set of rules about organization; each of us has a place in the organization, and a function within it. This function is bound by the communication rules within the mission that are worked out for that mission only. Mission-based communication contains memory (the past), cognitive and emotional information, and something else too – the structure of the mission itself.

Finally, there are separable areas of human communication. One of these *levels* is the level of symbolic encoding and data transmission – language and speech – which includes writing and other symbolic communication. This level follows the protocols (rules) of language: its usage and practice.

The physical level underlies all of the previous functions. By the manipulation of our physical environment, we transfer information between ourselves. The rules by which we manipulate the physical world to communicate differ from other levels of communication, and so we separate it from other levels.

Human information processing takes place in a hierarchical environment, where later-appearing (in an evolutionary sense) neural functions make use of preexisting functions. So, the functions are layered in the sense that communication functions are layered in concept and in design. Human information processing functions that we called out above connect with each other as peer processes. This means that a cognitive process in one person connects or communicates with another cognitive process in another. Cross-coupling also occurs, but usually does not result in efficient communication. Emotional messages make little cognitive sense. Protocols or communication rules are illustrated in Figure 2.2. The protocols govern the encoding and decoding of information, the types as well as amount of information that is possessed over these peer channels. Concepts in the person on the left being encoded, feeling being encoded, and both of these sets of information being encoded into words are shown. The words pass between the people, and are decoded, words first then feeling, then conceptual information. The physical path of information is top left to bottom left, to bottom right, and upward to top right; the virtual channels of information follow the arrows.

Protocols, which are sets of rules, serve as information pathways or virtual circuits between peer functions. If a set of common encoding rules exists for language between two people, they can exchange words. The common rules or protocols connect the word encoding and decoding process. The layers fall into six levels. Each connection between the peer communication processes, illustrated by a double-ended arrow, consists of a set of rules by which information is encoded and decoded—understood by each of the processes. An emotional process in a speaker encodes feeling, which follows shared feeling rules, and by which the listener understands the feeling message. The same process links the cognitive, speech, and memory processes. For example, processes of a type cognitive or thinking processes, are linked by the protocols that we share. When we do not share these rules, we cannot share missions, thinking, feeling, memory, speech, or even the physical world of the significant levels of human information processing, namely, the physical level (hearing, seeing, touch, smell, and taste for inputs), behavior in general as outputs, the emotional level (care, power, attachment, empathy), the level of memory (history

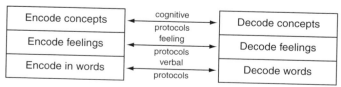

Figure 2.2 Protocol messages between people [5]. (©1998 IEEE.)

and culture) and the level of cognition (thinking together), and the mission level (what we are doing here). Then, communication refers to the dialogs between these levels of our information processing hierarchy in each of us. This point of view shows the way to structure human communication systems. Communication channels are represented in Figure 2.3. This figure shows communication protocols between two people modeled as the set of functions each uses to process information. The numbers refer to the levels. This figure represents the structure of the virtual circuits connecting the human information processing functions, and therefore illustrates the architecture of human communication.

The information transmission system in a two-person communication architecture is shown in Figure 2.4. The symbol encoding and decoding functions, represented as boxes, have bold lines and appear with an oval. The oval encompasses the peer symbol encoding processes and the communication protocols linking them. The oval represents the information encoding (for transmission) system. The boxes represent the ability to speak, hear, and otherwise physically transmit information between and among people. The two-way arrows connecting the boxes represent protocols, or rules that link the two people's physical information transmission functions into a synchronized system that actually feeds input information upward to higher levels of communication (feelings, memory, and cognition), and takes information from these levels and sends it to the listener. The rules for doing this exist at the physical and acoustical levels (behavior and language), and must be the same, or nearly so, for speaker and listener.

In the case of the *feeling channel*, people derive meaning from their feelings about each other, creating a relationship system. The relationship system, or the emotional system, gives significance to information. The relationship system of a group constitutes a system in itself, with each person *belonging* to the system to some extent, and partly remaining private. The reason it exists as a system is due to the separate neural mechanisms that give rise to feelings and emotions in individuals, generalizing in groups to constitute a body of options, feelings, and beliefs that arise in a group, but are located in any individual. These *feelings* circulate among group members, seeming to hang in the air among them. This system seeks to take on a life on its own, frequently controlling what happens at other levels of a group and individual function. If we encapsulate the similar functions and protocols

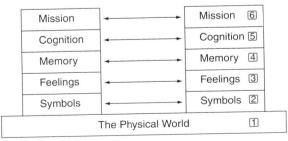

Figure 2.3 Human communication channels [5]. (©1998 IEEE.)

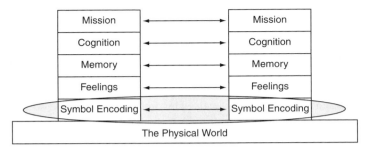

Figure 2.4 Information transmission system in a two-person communication architecture [5]. (©1998 IEEE.)

of each person, and encapsulate them as systems in themselves, the group can be depicted as shown in Figure 2.5.

The separate *symbol encoding* boxes, along with the protocols that connect them, have been incorporated into one long box labeled *Information Transmission System*. The same holds for the higher levels. In this way, the group appears as a higher level abstraction – a collection of people, unified by communication processes. The group itself appears to be a hierarchically organized information processing system. In this model the process that collects one person's thoughts, memories, feelings, and behavior with another is communication – the observation of one person's behavior by another. The implementation of each level of the information processing system will determine the effectiveness of fusing individual abilities and dependencies into the group's abilities. If a group suffers from a weak relationship system, the other functions will suffer. People rely on all of these abilities, and the systems that support them to work together. The lack of one of them severely disables a person or a group.

2.2.1 Physical System

Objects may be neutral in themselves, but they are used to encode information relevant to other layers of the communication system, and may be changed with

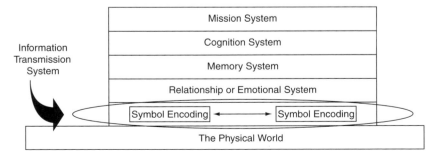

Figure 2.5 The group information processing architecture [5]. (©1998 IEEE.)

meaning. The transmission system depends on the physical system. The shorter the distance, the more likely we are to send and receive messages. Distance precludes energy or object exchange, and must be overcome with some expense and effort.

Group productivity requires managing physical isolation and connection. Since physical protocols consist of the roles and practices that govern the exchange or display of energy and objects, managing the physical system allows us to regulate the timing of exchanges of behavior and objects. Objects are also used as emotional symbols, and so serve to establish and maintain relationships. Many protocols of higher levels of human communication derive their meaning from the physical notions of distance and isolation. The memory system of a group frequently uses books, a memo system, and other implementations of recording and retrieval. Depending on the effectiveness of placement and access of the *memory objects*, a memory system might be a useless island outside a group's activities. The memory system's success depends on physically linking the information recorded there to users.

The cognitive system depends less overtly on the physical system for its power. However, the suggestive power of physical objects, their relationships and images of the physical worlds form an interesting set of suggested models for thinking. In engineering processes, specifications are frequently passed between the engineering group and the manufacturing group, which must understand the specification. Understanding improves if part of the specification includes a physical model of the item as well as a written description. The physical level of communication serves as the basis for other levels. Communication at the level of cognitive protocols and mission protocols becomes abstract. However, the physical levels of communication can make the meaning of higher level communication concrete, injecting a needed element of *reality* into sometimes airy tasks.

2.2.2 Symbol Encoding

A group's information transmission system encodes communication messages, provides links to route and deliver them, transmits them over the links, and establishes processes that synchronize them. This system injects noise (meaningless or irrelevant information) into the relevant information and distorts it to a greater or lesser degree. Four areas of the encodings we use are the rules by which we exchange these encodings, the organizations we build to facilitate information exchange, and the inevitable noise and distortion that inhibits transmission.

Effective human-to-human information conveys a maximum of information (meaningful data) with a minimum of critical human resource dedication. This usually translates to "short time, easy recognition, high information content symbols" used for encoding information. For example, between people, graphical encodings usually exceed the efficiency of text encodings because graphics contain more usable information at higher cognitive bandwidths than text. Between people, information may be easily encoded in multifont text, images (representational pictures), graphics (geometrical abstractions), voice, and material demonstration.

The importance of text dominates at the beginning and end of an engineering development process. For example, most projects begin with plans and specifications, and documentations dominate at the end. In the middle of the process, other forms of encodings rise and fall in importance, due to the specialized design and implementation methods of engineering. Graphics maintains its importance throughout the process.

We adopt encoding rules that differ with each person or group. Often we can detect the identity of the hidden party on a telephone conversation talking to a close friend. We do this by picking up clues in voice tone, word choice, and conversational style. There are protocols of information transmission for voice encodings. Parallels exist for written, graphics (drawn), object-based encodings, or other behavioral encodings. Rules exist for meetings, conversation, letters, telephone calls, and so on. These protocols determine what is transmitted (the message), as well as how and when it is transmitted. Information transmission protocols treat only encoding, formatting, routing, and delivery of messages. Emotional or cognitive content relates to their respective functions, which reside in overlaying layers. Underneath the information transmission function, the physical layer provides the objects and energy used to encode the messages of the information transmission layer.

Organization structure constitutes one kind of routing and delivery structure. Organization structure is a communication system designed to make certain types of communication simple, and other kinds more difficult. The organization structure or network describes the links between people. A chief task of an organization establishes, maintains, and destroys communication links at appropriate times. A *process* organizes the timing and synchronization of information flow. Communication processes establish when certain communications will take place, and between whom they will occur. Communication linkage between people and the synchronization of messages of various types occur at the level of information transmission. How a group implements these links, and processes determines how effective this level can function.

Noise and distortion affect information transmission. Both distort the truth, and to some extent, both are always present. However, isolation from irrelevant information (noise) and feedback processes among people suppress these effects somewhat. Noise in system is introduced by processes other than the relevant ones. The primary noise suppression method is isolation – getting rid of irrelevant information. Isolation means eliminating irrelevant links to sources of irrelevancy. Distortion occurs when information becomes transformed in content. The basic technique to suppress distortion is feedback. Feedback processes allow the comparison of information at an origination point to judge the fidelity with which a process occurred. This comparison produces information that can be used to reduce the amount of distortion within the process itself. Rarely does distortion disappear, but it can be reduced.

2.2.3 Feeling

We communicate differently with each person because we have a different relation to each of them. That is, we have different feelings about each of them. In human organizations, the relationship system generalizes the feeling experience of individuals, and is sometimes called the emotional system of group. The relationship system serves as a security system that authenticates messages, adding or taking away information. Relationships band people together into meaningful configurations that provide one of the basic needs we have in life. The relationship system develops communication protocols that condition the type of information passing through it.

Every meaningful message carries cognitive and feeling components. The feeling components of messages follow the relationships much more closely than the cognitive components. Feeling messages establish and stabilize the group's opinions and beliefs. This system of group-owned information traps individuals in it so they lack independence of judgment – they cannot think independently.

Several effects of the emotional system directly affect productivity. The analyzed effect causes conforming behavior of individuals, even in the face of facts so obvious that experimental results occasionally seem incredible. The relationship system coheres a collection of individuals into a group. Groups can do things individuals cannot.

2.2.4 Memory

Knowledge consists of authenticated facts that relate to each other in a network. A knowledge system refers to the collection of facts, relations, and associated information useful in this kind of work. Knowledge may lie in personal memory, the library, magazine subscriptions, drawings, and specifications.

Knowledge resides in memory – in individuals and groups. Groups store memory in archival systems and historical records, as well as in the personal memories of group members. Memory is the basis of perception, not communication and not thinking. Memory does not guarantee successful thinking. Effective memory systems must fill two requirements: contents must be storable and retrievable in simple ways that relate to the uses of memory. In computer systems, we have constructed a set of memory systems that store increasing amounts of information at the cost of longer access times. So knowledge acquisition, organization, storage, and retrieval mechanisms of organized, relevant information are important, to put all relevant information into *local memory* to provide the richest noise-free environment for thinking. Memory contains the *knowledge* of knowledge work. The protocols, or rules of connecting and communicating memories, together center around sets of rules we might call "access methods" – ways of storing and retrieving memory contents according to strategies formulated to make access usable in a special case.

Group memory is a knowledge that may be accessed by other persons in the group. In Figure 2.6, a person's knowledge lies within an oval of a simple *Venn diagram*. To the right is shown the knowledge of three people, which overlaps to some extent.

Memory benefits if we can associate requests with some "hook" on the memory itself. The ability to encode requests into voice, text, and pictures, and the respondent's ability to encode and answer might elicit more useful information than telephone calls or electronic mail requests.

16 MULTIMEDIA COMMUNICATIONS

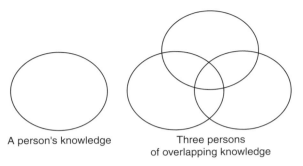

Figure 2.6 Pearson's knowledge represented using a *Venn diagram*.

The memory system provides the ambience of the cognitive systems. A memory system design for knowledge work includes a system of storage and retrieval for internal communications, a library, with extensions to outside libraries, and databases of relevance. In addition, coupling people who know with those who do not may be extremely useful.

2.2.5 Cognitive System

A cognitive system provides structure for people thinking together. In a sense, all the levels below the cognitive level support the cognitive level and make it productive. The cognitive system gives rise to ideas and develops them, authenticates new ideas, recognizes existing, useful knowledge, discards useless knowledge and ideas, transforms information, and architects networks of knowledge that support the group's mission. The communications that support cognitive group work consists of two parts: a collection of words and concepts (the content), and the rules of control (protocols) that govern the interchange, synchronization, and formulation of messages. What separates the successful from the unsuccessful groups in the control of their internal communications is their use of protocols. By specifying how messages are exchanged, when they are exchanged, and what they may and may not contain, cognitive communication may be designed to perform specific tasks productively (and enjoyably).

Some designs include brainstorming to elicit new ideas. Organizing research groups somewhat isolated from development groups allows gestation of a new idea. A group problem-solving meeting might carefully traverse several steps covering problem definition through selection of a solution, and an implementation plan. Here, the goal of the communication rules is to perform the thinking steps one at a time as a group.

2.2.6 Mission System

A *mission* describes the goals and objectives of a group: what it has to do. Most groups have some goals and these goals constitute the reason for a group's existence.

They also provide a measure of how efficiently the group is organized – that is, how effectively they communicate. The mission system involves the rules and information of a craft or skill that a particular mission involves. For testing project management – all the activities of particular missions usually contain a specialized language, technology, history, and other information specific to that mission.

Each person within particular mission performs a piece of the mission, which must be accepted by each person and communicated to the others. Most missions involve an interplay of two interleaved systems – the human system (described above) and the mission system. Typically, the mission system consists of values and rules that arise from the restraints and demands of the technology or the world outside the human concerns of the participants. Engineering projects require handling high complexity, leading to *levels* of designers who carry out complicated handoffs of design to one another.

These specific rules lead to protocols specific to a mission type as well as the people, time, and circumstances under which the mission takes place. This evidence indicates that the mission itself suggests an efficient communication structure.

2.3 EVOLUTION AND CONVERGENCE

Multimedia itself denotes the integrated manipulation of at least some information represented as continuous media data, as well as some information encoded as discrete media data (such as text and graphics). The manipulation refers to the act of capturing, processing, communicating, presenting, and/or storing. We understand continuous media data as time-dependent data in multimedia systems (such as audio and video data), which is manipulated in well-defined parts per time interval according to a contract. Hence, multimedia communications deal with the transfer, protocols, services, and mechanisms of/for discrete and continuous media in/over digital networks. The transmission of digital video data over a dedicated television (TV) distribution network is not multimedia as long as it does not allow the transfer of some type of discrete media data as well. A protocol designed to reserve capacity for continuous media type over an asynchronous transfer mode local area network (ATM-LAN) is certainly a communication issue [6].

For many years, the commitment of telecommunications companies to the bare transport of information, by means of digital transmission and switching facilities, far expanded its scope by enhancing network control and management capabilities. The emphasis of upcoming telecommunication networks is on multimedia distributed services and operators are exploiting technologies and network architecture capable of offering value-added network-centric services.

The computer industry's view of the network is driven mainly by the impressive success of the Internet. The Internet protocol based network is being used to transmit hypermedia information worldwide. The capability of the Internet to support real-time voice and video information exchange over the current best-effort IP connectionless protocol, namely, IP version 4 (IPv4), is subject to the availability of large spare network resources (bandwidth and buffers) [7]. Hence, a new version of the IP, namely, IP version 6 (IPv6) is being finalized by the IETF to support integrated services, together with other service extensions (security and mobility) [8]. IPv6 is complemented by soft reservation protocols (RSVP) capable of supporting the allocation of resources at the logical network level provided by the IP [9, 10].

In the telecommunication model, functional complexity is located in the network switching nodes and the introduction of a new service often requires updates in the network nodes' hardware and software. In contrast, the Internet approach is based on concentration of intelligent functions in the hosts, at the network edges, and their coordination with the interworking routers software. In this case, the network is seen as a dumb data-transport medium (the telecommunication pipe). This approach assures the separation of the service from the underlying network architecture. New services can be introduced almost independently of the evolution of the underlying network. It is recognized that the support of some services, however, such as those implying mobility, is easily achieved when the intelligent functionalities are offered by the network nodes [11].

In networked multimedia applications, various entities typically cooperate in order to provide the mentioned real-time guarantees to allow data to be presented at the user interface. These requirements are most often defined in terms of quality of service (QoS). We distinguish four layers of QoS: user QoS, application QoS, system QoS and network QoS [6]. The user QoS parameters describe requirements for the perception of multimedia data at the user interface. The application QoS parameters describe requirements for the application services possibly specified in terms of media quality (like end-to-end delay) and media relations (like inter/intra stream synchronization). The system QoS parameters describe requirements on the communications services resulting from the application QoS. These may be specified in terms of both quantitative (like bits per second, or task processing time) and quantitative (like multicast, interstream synchronization, error recovery, or ordered delivery of data) criteria. The network QoS parameters describe requirements on network services (like network load or network performance). Multimedia applications negotiate a desired QoS during the connection setup phase either with the system layer or possibly directly with the network layer, if the system is not able to provide QoS for the application.

2.3.1 Convergence of Telecommunications and Computing

The convergence of telecommunications and computing has been noted and commented on for some time. At present, there is a much richer interrelationship than at any other time in the past. These fields will become virtually distinguishable in the very near future. The convergence has, and will continue to have, a profound impact on technology, industry, and their larger society. Telecommunications and computing have already been irreparably changed by the other, and, as we argue below, will be even more substantially recast in the future. Much more profound changes are forthcoming, changes no less weighty than the rapid disintegration of the vertically integrated industrial model. While computing in the absence of communications has led to new applications and made substantive changes to leisure and work life, computing in conjunction with communications will have a profoundly greater impact on society. This is because communications are at the heart of what makes a society and civilization, and the convergence with computing will revolutionize the nature of those communications [12]. The classical terminology of telecommunications and computing is no longer useful. In light of this, it is appropriate to define a more transparent classification of networked applications that is media-blind and focuses on the functions provided by the user.

As an aid in understanding, we will adopt the three-level model for an information network and its services and applications, as shown in Figure 2.7. We define an application as a collection of functionality that provides value to a user (a person). We will be concerned here with networked applications, implying that they are distributed across a distributed telecommunications and computing environment. A service is defined as functionality of a generic or supportive nature, provided as a part of computing and telecommunications infrastructure, which is available for use in building all applications. Bit ways are network mechanisms for transporting bits from one location to another. Examples of networked applications are electronic mail, telephony, database access, file transfer, World Wide Web browsing, and video conferencing. Examples of services would be audio or video transport, file–system management, printing, electronic payment mechanisms, encryption and key distribution, and reliable data delivery. Examples of bit ways with sufficient flexibility for integrated multimedia applications are ATM [13–15] or Internets interfaced with the Internet Protocol (IP) [16].

Two categories of networked applications relate to the functionality [14]:

- User-to-user applications in which two (or more) users each participate in some shared environment
- User-to-information-server applications, in which a user (or sometimes two or more users) interacts with a remote system to access, receive, or interact with information stored on that system

Each user in a networked application interacts with a local terminal, which communicates in turn with remote computers or terminals across the network. We can also separate networked applications into two classes with respect to

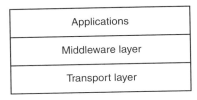

Figure 2.7 A three-level model for an information network and its services and applications.

the temporal relationship in the interaction of the user with a server or with another user:

- *Immediate*, meaning a user is interacting with a server or another user in real time, typically with requirements on the maximum latency or delay.
- *Deferred*, meaning a user is interacting with another user or a server in a manner that implies no fixed temporal relationship and for which the delay is typically not critical.

2.3.2 Architectures for Networked Applications

Networked applications are physically realized by terminal nodes or just terminals interconnected by bit ways. Functionally, there are two basic architectures available for networked services: peer-to-peer and client–server. This is shown in Figure 2.8. In peer-to-peer architecture, two or more peer terminals, each associated with a local user, communicate over a bit way to provide a user-to-user networked application. The networked communications component between peers is often symmetrical in terms of both functionality and bit way resources. In client–server architecture, a client terminal associated with a user communicates over the bit way with a server computer, which is not associated directly with a user, but rather realizes an information-server function. The functionality is often asymmetric, with the server embodying the primary functionality or database access and the client terminal focusing on the user interface. The communication components are also often asymmetric, with the server-to-client direction typically requiring much higher bandwidth [18, 19].

The three-level architecture of Figure 2.7 is a logical separation of functionality, where application functionality will physically reside in the terminals, and services functionality may reside in the terminals or somewhere within the bit way.

A user-to-information server application is always realized with the client–server architecture. As shown, many clients will typically access a single server, which provides functionally separated but time-shared services to the clients. As for a user-to-user application, it can be realized in either the peer-to-peer or client–server architectures, as shown in Figure 2.9 for two users. In the client–server architecture, the two clients are communicating through the server, which may be realizing

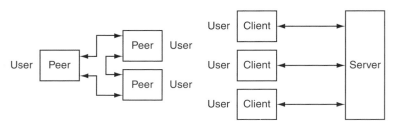

Figure 2.8 Peer-to-peer and client–server architecture.

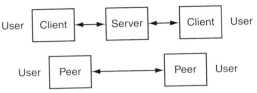

Figure 2.9 A user-to-user application realized by either the client–server or the peer-to-peer architecture.

additional applications or control functionality. The client–server architecture is particularly appropriate for deferred user-to-user applications, since the server provides a convenient point for the necessary buffering, with guaranteed availability regardless of the state of another peer.

Although clients and peers serve a similar user–interface functionality, there are some basic differences. For example, many clients will connect to a single server, whereas a peer must be prepared to connect to any other peer. To establish a new instance of an application, a server must always be prepared to respond to an establishment request from a client, while a client may originate an establishment request. A peer must be able to either originate or respond to establishment requests, and in this sense is a hybrid between a client and a server. A client can rely on the server for some functionality, whereas a peer must be self-contained. The biggest differences are in scalability to large numbers of users, interactive delay and interoperability.

2.3.3 Networked Computers

The networked computer provides a ready, large-scale market for new applications, thereby reducing the barriers to entry for new applications developers. Computer networking, like computer control before it, was widely adopted in the telecommunications industry as the basis of signaling and control. This signaling function was originally realized in-band on the same voice channel but was replaced by a signaling computer network called common-channel interface signaling (CCIS) [19]. The CCIS enabled the advance from simple circuit-connection functions to much more advanced features, and ultimately will provide terminal-to-terminal signaling capabilities.

Up to this point, there remained an infrastructure for computing that emphasized data-oriented media (graphics, animation) and a relatively separate telecommunications infrastructure that focused on continuous-media signals (voice and video). These converged in a relatively superficial way, at the physical and link layers, where telephone and videoconferencing and computer networks shared a common technology base for the physical layer transport of bits across geographical distances. The telecommunications industry made extensive use of computer and software technologies in the implementation of the configuration and control of the network. The computer industry made use of the telecommunications infrastructure to network computers, which enable networked applications.

The expanding importance of programmability flows from extraordinary advances in the cost/performance of the underlying electronics and communications technologies. In the context of any single application (like control, voice, audio, video, and so on), the performance requirements in relationship to the capabilities of the underlying technology pass through three stages:

- Initially, the application is very expensive to implement, and cost-effectiveness dictates customized hardware design. In this stage, efficiency (in metrics like processing power, bandwidth, etc.) is critical to cost-effectiveness, and hence commercial exploitation.
- Next, programmable software – defined implementations become feasible, and eventually cost-effective. At this point, efficiency remains a dominant consideration, but the lower design cost and lower time to market afforded by a software definition can often overcome the manufacturing-cost penalties of general-purpose hardware.
- Finally, technology advances far enough that software-defined implementations become the norm. At this point, the greater efficiency of a custom hardware implementation is definitely overcome by its lower volume of manufacture, greater design costs, and greater time to market.

A software-defined solution, as the final stage, has an important implication. Namely, the basic functionality need not be included or defined at the time of manufacture, but rather can be modified and extended later.

The modern trend is toward adaptability, a capability that usually builds on programmability and adds the capability to adjust the environment. For example, in a heterogeneous environment, it is helpful for each element to adapt to the capabilities of other system elements (bandwidth, processing, resolution, etc.).

2.3.4 Integration

There are two architectural models for provisioning networked applications, as shown in Figure 2.10. In the most extreme form of vertical integration, a dedicated infrastructure is used to realize each application. In contrast, the horizontal integration model is characterized by

- One or more integrated bit ways that transport integrated data and stream media like audio and video with configurable QoS parameters.
- A set of services, such as middleware services (directory, electronic funds transfer, key management, etc.) and media services (audio, video, etc.) that are made available to all applications.
- A diverse set of applications made available to the user.

A key advantage of the horizontal model is that it allows the integration of different media within each application, as well as different applications within the bit way.

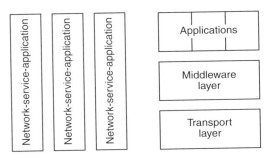

Figure 2.10 Two models for networked applications: vertical and horizontal.

For this reason, in the telecommunication industry, this is often called an integrated-services network.

Another useful distinction among networks is whether or not they are content-aware, and whether or not they are application-aware. Vertically integrated networks are frequently application-aware, meaning they are cognizant of the applications they are carrying (e.g., videoconferencing vs. file transfer), whereas horizontal bit ways are often application-blind (e.g., the current Internet). Vertically integrated networks are frequently even content-aware (e.g., a video-on-demand network that is cognizant of what movie is requested).

An important feature of horizontal integration is the open interface, which has several properties. It has a freely available specification, during acceptance, and allows a diversity of implementations that are separated from the specification. Another desirable property is the ability to add new or closed functionality. We define closed functionality as not published or extensible by other parties. Proprietary functionality may be published and extensible, but is subject to intellectual property protection. Open interfaces enforce modularity and thus allow a diversity of implementations and approaches to coexist and evolve on both sides of the interface. Some of the most important open horizontal interfaces in the computer industry are represented in Figure 2.11. The IP is an open standard for interconnecting bit ways below it, where those bit ways may incorporate a diverse set of technologies including ATM. The IP also allows for a diverse set of media types and applications

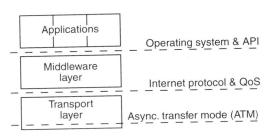

Figure 2.11 An example of open horizontal interfaces.

to reside above it. Another critical interface is the operating system (OS) application program interface, which allows a diverse set of applications to coexist on the same bit ways and services infrastructure, while hiding as much as feasible the details from that infrastructure. Horizontal interfaces also exist for control and signaling, which supports computer–telephony integration [20, 21].

One purpose of open horizontal interfaces is to contribute to the modularity by the separation or independence of the definition of the application, which we call platform independence. Increasingly, applications can be simultaneously developed for multiple target platforms, by generating distinct platform representations from a common functional description, based on appropriate software toolsets.

An even more powerful concept is *middleware*, which is a horizontal layer residing on top of a set of networked computers, providing a set of distributed services with standard programming interfaces and communication protocols, even though the modeling hosts and OS may be heterogeneous [22].

The computer industry is well along in the evolution to horizontal integration. The networked desktop computer resulted in the division of the industry into distinct horizontal segment (software, network, OS, and application). The telecommunications industry was vertically integrated with a focus on provisioning a single application with a dedicated network, such as voice telephony, or video conferencing, or cable television. Today this industry is also moving toward architectural horizontal integration at the bit way level, with ATM bit ways that flexibly mix different media. However, it remains largely vertically integrated at the services and applications layers, as bit way providers aspire to value-added applications such as video-on-demand and differentiated terminals such as set-top-box.

The separation of the applications from bit ways and services best serves the user by encouraging a diversity of applications, including many defined for specialized as well as widely popular purposes. Vertical integration discourages this diversity because a dedicated infrastructure demands a large market, and because users do not want to deal with multiple providers. Horizontal integration lowers the barriers to entry for application developers since most of the infrastructure (bit ways and services and even programmable terminals) is available. Applications can be defined in software and coexist in the same programmable terminals with other applications, reducing the distribution cost and the incremental cost of a new application.

2.3.5 Transportable Computation

Transportable computation offers four important advantages [12]:

- *Scalability.* It allows both memory (transient and persistent) and computation to be located on whatever terminal or computer is most advantageous. For example, in a client–server application, it allows computation to be shifted from the server to its clients, thus avoiding overload of the server.
- *Latency.* Executing the program in the local peer or client as opposed to a remote host eliminates interactive latency due to network transport delay.

- *Interoperability.* If the programs associated with a distributed application originate from a common source, it can be assured that they are interoperable, meaning that they properly coordinate their operations. The conventional approach to interoperability and standardization is, by comparison, cumbersome and time-consuming. This is doubtless the most important advantage.
- *Locality of data access.* A transportable program can access and modify data stored on any computer to which it can be transported.

Transportable computation facilitates the dynamic network deployment of applications. That is, a distributed application can be copied over the network during establishment, transparently and invisibly to the user, with guaranteed interoperability, as illustrated in Figure 2.12, for a client–server architecture. A repository of client applets can reside in the server, waiting to be dynamically deployed over the network to the client on demand. The client application code is stored in a repository in the server, to be loaded dynamically across the network into the client when the user invokes that application.

2.3.6 Intelligent Agents

Dynamic deployment benefits from broadband networking, since application executables will sometimes be large. This will be an important driver for broadband access to the bit way, just as low-latency downloading of executables is a primary driver for broadband local-area bit ways. Dynamic deployment does not exploit the full power of transportable computation, which is embodied in the more general concept of an intelligent agent. An intelligent agent is a transportable program that includes four attributes and capabilities:

- *Autonomy.* It contains all information necessary for its execution.
- *Social ability.* It can interact with other agents or its environment.
- *Reactivity.* It can exert actions based on attributes of its environment.
- *Proactive.* It can initiate actions by itself.

The capabilities of the intelligent agent open up a number of possibilities. Intelligent agent technology originated in artificial intelligence, where one can imagine sophisticated human-like qualities, such as adaptation to the environment and

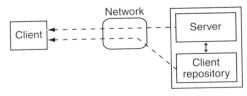

Figure 2.12 Guaranteed interoperability for a client–server architecture.

higher-level cognitive functions. Here, we can conceptualize more applications that provide useful generalization of user–driver information retrieval or even, on a basic level, as e-mail [23, 24]. In this application domain agents can act as *Internet-assistants*, which are not restricted to particular servers, but cruise the network gathering or disseminating information.

2.3.7 Convergence

At the beginnings of convergence, telecommunications and computing shared many common technologies, but were distinguished in two primary ways. First, telecommunications focused on immediate user-to-user applications (telephony, videoconferencing), while computing focused on initially stand-alone, subsequently deferred user-to-user and immediate user-to-information server applications, both on their dedicated and separate infrastructures. Secondly, telecommunications focused on continuous-media like audio and video, while computing focused on information storage, retrieval, and manipulation. These distinctions are no longer useful for two reasons. First, all applications and media will share a common horizontally integrated infrastructure. Secondly, largely as a result of this common infrastructure, network applications no longer can be neatly segmented.

The dynamic deployment of user-to-user and user-to-information server multimedia applications in a horizontally integrated terminal and network environment represents the pinnacle of convergence. Networked applications that freely mix the constituent elements traditional to telecommunications and computing will become commonplace. At this point, there no longer exist any technological or intellectual differences that distinguish telecommunication from computing. At this point, the dynamism and rate of progress in user-to-user applications becomes as great as has been recently experienced in user-to-information server applications.

2.4 TECHNOLOGY FRAMEWORK

Multimedia and multimedia communication can be globally viewed as a hierarchical system. The multimedia software and applications provide a direct interactive environment for users. When a computer requires information from remote computers or servers, multimedia information must travel through computer networks. Since the amount of information involved in the transmission of video and audio can be substantial, the multimedia information must be compressed before it can be sent through the network, in order to reduce the communication delay. Constraints, such as limited delay and jitter, are used to ensure reasonable video and audio effects at the receiving end. Therefore, communication networks are undergoing constant improvements in order to provide for multimedia communication capabilities. Local area networks are used to connect local computers and other equipment, and wide area networks and the Internet connect the local area networks together. Better standards are constantly being developed, in order to provide a global information superhighway over which multimedia information will travel.

2.4.1 Multimedia Technologies

There are a number of widely accepted multimedia technologies or products, such as QuickTime and Video for Windows. QuickTime technology was developed by Apple Computer for Apple computers, Microsoft Windows, and UNIX platforms. It is a proprietary technology and is the most widely accepted format for multimedia storage and communication. It supports MPEG and ISO compression standards, Indeo (IntelVideo), Kodak's PhotoCD format, and MIDI. Video for Windows (VFW) was developed by Microsoft for a Windows environment. The AVI (Audio Video Interleave) file format interleaves the audio and video. It is designed to playback videos in a window using software. It uses a run-length coding method to compress information and it supports other video compression algorithms, such as QuickTime and Indeo. Indeo technology was originally developed by the David Sarnoff Research Center for converting an NTSC analog signal into digital video in order to store and play back video in a personal computer (PC). Software-only Indeo is the choice of compression scheme in QuickTime and Video for Windows. One major difference between the above two technologies and the hardware Indeo is that Indeo uses hardware for compression and play back to achieve higher frame rate and resolution, whereas the other two technologies only support software-based compression. Indeo requires a PC-size card using digital video interface (DVI) technology. This proprietary compression method, using vector quantization and run-length coding, can achieve 160:1 compression ratio and store 70 minutes of video on a CD-ROM.

There is an ISO standard entitled MHEG (Multimedia Hypermedia Information Coding Expert Group). This is a standard ISO/IEC JTC1/SC29 WG12 set by the International Standards Organization for a set of object classes used to control the presentation of multimedia and hypermedia information. MPEG-5 is the standard developed to support the distribution of interactive multimedia applications in a multivendor client/server environment [25]. Application developers who use this standard developed applications only once, since information representation is interchangeable in a multimedia computer and network environment. In comparison to the document description capabilities of HTML (Hyper Text Markup Language), the MHEG provides the additional multimedia handling capabilities. The ISO's Abstract Syntax Notation (ASN.1) is used for representation between different computer platforms, operating systems and languages. It uses space (three spatial coordinates) and time to synchronize the presentation.

SONET (Synchronous Optical Network) and BISDN (Broadband Integrated Service Digital Network)

The SONET was developed by Bellcore for wide area networking ANSI (American National Standards Institute) and used in the United States and Canada. The European counterpart, SDH (Synchronous Digital Hierarchy), is an ITU-T standard. The SONET and SDH are physically compatible at synchronous transport signal (STS-3C 155.52 Mb/s) and synchronous transport module (STM-1). The SONET uses a frame to carry lower-speed information (tributaries: video, audio, and data). The bandwidth for each tributary is guaranteed. The SONET uses synchronous multiplexing (SMUX) and add/drop multiplexers (ADM) to insert a signal from a source and to extract a signal at the destination.

The BISDN uses ATM over the SONET to multiplex signals. The cells are packed in the SONET frame's payload. The BISDN provides LAN and metropolitan area network (MAN) connections, including 802.X.

Computer Networks

The computer network is the means by which the multimedia are transported. We will describe the major characteristics of computer networks.

802.X Local Area Networks (LANs). The 802.X standard defined by IEEE contains several LAN standards:

- 802.1 – including the relationship between 802.X and the open system (OSI) interconnect reference model, interconnecting 802.X by bridges and network management
- 802.2 – logical link control for addressing multiple service points to multiple simultaneous applications and for connection-oriented, connectionless, and acknowledged operations
- 802.3 – Ethernet standard
- 802.4 – token bus standard manufacturing automations protocol (MAP)
- 802.5 – token ring standard
- 802.12 – 100VG-AnyLAN standard.

The Ethernet is the most widely deployed LAN and includes the newly developed fast Ethernet standard. The token bus has very limited deployment for manufacturing. Token ring contents stations as a ring and uses a token to resolve contention.

Fast Ethernet and 100VG-AnyLAN. The Ethernet was originally designed to connect stations by coaxial cable, later, it evolved to twisted pairs to connect stations to a hub. The transport rate is 10 Mb/s. To improve the rate, the fast Ethernet was developed to accommodate 100 Mb/s and must use a hub to connect stations.

The 100VG-AnyLAN (802.12) was designed to handle multimedia communication using a priority demand scheme. It assigns high priority to multimedia, in contrast to the Ethernet (802.3), which uses the carrier sense multiple access with collision detection (SMA/CD) and cannot assign high priority to time-critical traffic, such as isochronous (video and audio) data. The hubs of a 100VG-AnyLAN can be connected in a hierarchical star geometry with three layers maximum, and token ring LAN's can be connected as part of the network. The same kind of cables used in the fast Ethernet can be used in the 100VG-AnyLAN.

Switched LANs. The shared medium in the LANs presented above has scalability problems. As the number of stations increases, the effective bandwidth for each station decreases. This problem can be alleviated by using a switched LAN that interconnects stations through a switching hub. The communication cannot be blocked if there is no contention for the same destination. Hence, the performance of the network is scalable. The switch can learn the location of a LAN station using the medium access control (MAC) and address in a LAN station in a manner similar to that used by a bridge. The layer 2 switch performs at Layer 2 of the OSI reference model and it behaves like a bridge. It can connect multiple LANs and stations. The layer 2 switch is much faster than the bridge because it uses application-specific integrated circuits (ASIC) to switch traffic. The layer 3 switch is designed to improve the performance and lower the cost of the router that connects the network at Layer 3 of the OSI reference model. A router must perform two functions: (1) generating the routing table for the next hop by collecting the route information from neighboring routers, as well as (2) switching the frame to the appropriate output port based on the routing table. The layer 3 switch is designed to perform the second function using ASICs. The routing table can be obtained from a route server that is designed to perform the first function. Several switches can share one route to reduce the cost and the ASICs can switch faster than a RISC (Reduced Instruction Set Computer) based router.

Asynchronous Transfer Mode (ATM). The ATM performs the following functions:

- Asynchronously multiplexes small packets called *cells*, each consisting of five octets of headers and 48 octets of payload going from a member of information sources to various destinations into seats in a constantly flowing train, if there is a seat for the specified destination with the required QoS.
- Switches cells during transport, if necessary, like changing trains in train station.
- Lets cells jump off the train and destination.

The ATM uses hardware switching to achieve much faster switching speeds in comparison to other communication switches. The QoS that can be negotiated between applications and the ATM provides the required bandwidth for video and audio communications. The ATM can also transfer bursty or asynchronous data and therefore is designed to accommodate all kinds of communications.

The ATM uses connection-oriented operations. It establishes a sequence of switches, so that a connection is made from the source to the destination. Such a connection is called *virtual circuit connection* (VCC). The switches can be established to perform simplex, duplex, multicast, and broadcast communications. A virtual connection (VC) is a connection between a switching node and the next node. Hence, a VCC consists of a series of VCs. There are two kinds of VCs: a permanent VC (PVC) for a leased line and a switched VC (SVC) for a dynamically established connection. To simplify the management of VCs, a number of VCs with the same starting and encoding node is grouped together as a virtual path (VP). To identify a VP or VC, a number is used as the identifier and is labeled VPI/VCI (VP identifier/VC identifier).

To adapt to different characteristics of the traffic, ATM provides five types of adaptation:

- Type 1 is for circuit emulation at constant bit rate (CBR) service for isochronous data
- Type 2 is for variable bit rate (VBR), connection-oriented service for isochronous data
- Type 3 for connection-oriented data service
- Type 4 is for connectionless data service
- Type 5 is for LAN emulation and all other possible traffic.

The ATM Forum also defines available bit rate (ABR) and unspecified bit rate (UBR) similar to ABR, but does not guarantee a minimum rate, and cells may be lost due to congestion. As for ABR, it guarantees a minimum rate, but delay may vary.

LAN emulation (LANE) was developed by the ATM Forum for interconnecting LANs and ATM networks. LANE works at Layer 2 of the OSI reference model. LANE uses LAN emulation server (LES), LANE configuration server (LECS), and broadcast and unknown serve (BUS) for configuration, address resolution, broadcast, and the resolution of unknown addresses.

The ATM is suitable for multimedia communication because it provides a guaranteed QoS. The cost of ATM is decreasing and there are expanding demands for multimedia communication. The ATM can also increase bandwidth efficiency by buffering and statistically multiplexing burst traffic at the expense of cell delay and loss [26].

High Definition TV (HDTV). The Federal Communications Commission (FCC) solicited and received proposals in 1988 for HDTV. The Grand Alliance, consisting of General Instruments, Philips, David Sarnoff Research Center, Thomson Consumer Products, AT&T, and Zenith, was formed in 1993 [27]. The Grand Alliance selected the MPEG-2 Main Profile and High Level for video encoding and transport, and the AC-3 developed by Dolby Labs for audio encoding, while MPEG-2 TS packetization and multiplexing are used. The bandwidth of the transmission channel is 6 MHz, the payload rate is 19.4 Mb/s, while the MPEG packets are scrambled and error correction information is added.

Cable TV (CATV). The new services from CATV include telephony, fax, video on demand, and computer access to the Internet. A cable modem, which must be used in order to connect to the Internet, is connected between the cable feed and the Ethernet network interface cards (NIC) inside a computer. The cable's bandwidth is shared by users connected to the same cable from CATV, as is done with the Ethernet. Depending upon the cable modem vendor, the downstream from headend to home bandwidth is from 4 to 27 Mb/s and the upstream from home to headend bandwidth is from 96 kb/s to 27 Mb/s. CATV implements telephone services by using SONET to connect headend offices to the public switched telephone network (PSTN).

Digital Subscriber Loop (DSL). To compete with CATV for VoD and Internet access services, the telephone companies are deploying DSL developed by Bellcore [28]. The DSL includes ISDN, asymmetrical digital subscriber loop (ADSL), high data rate DSL (HDSL), single-line DSL (SDSL) and very high data rate DSL (VDSL). The DSL supports analog telephone as well as ISDN basic rate interface (BRI), that is, two 64 kb/s B channels and one D channel for control and, also, 384 kb/s for high-quality video conferencing. DSL can be deployed to many users. Special electronic circuits are used for modulation and demodulation of DSL. The DSL technology will make multimedia communication affordable. The DSL bandwidth is dedicated to a single user, whereas the bandwidth of cable modems is shared.

Internet. The Internet has evolved from a classical, closed-community data network into the infrastructure for the global information society. Users are coming to trust the Internet for business, including not only the burgeoning use of the World Wide Web and traditional applications such as e-mail and file transfer, but increasingly also *real-time* multimedia such as audio, video, and shared applications (such as whiteboard or meeting planner or shared document editor). The original design goal of the Internet was to serve as a highly fault-tolerant data network for the defense community. To this end, the amount of data shared between the network and end systems is minimized, being merely sufficient information to calculate a set of routes, and for each packet sent by an end system to determine the new current route. In its original implementation, a great deal of attention was paid to this distributed, dynamic route calculation, and not too much to the performance aspects of packet forwarding. Early router implementations used *first in–first out* (FIFO) queries for all traffic, whose behavior under overhead conditions was simply to drop the latest arriving packets. The only reason for this was simplicity. Over time, it became apparent that traffic management for different applications, for different users, and for overloaded networks would be essential. A number of stages in the evolution of the protocols and implementations of the Internet have followed.

The first service definition was by application, and the idea of type of service (ToS) forwarding was proposed [29]. Conceptually, routers distinguish traffic either by examining special bits in the Internet Protocol (IP) packet header or by application protocol.

The file transfer protocol (FTP) and the World Wide Web's application protocol, hypertext transfer protocol, all operate over the transmission control protocol (TCP), which enhances the inherently unreliable IP, to provide ordered, reliable delivery. The effect of ToS routing and TCP congestion control is to offer a network that provides a fair share to the set of concurrent users. Because all TCP users achieve only a fair share of the capacity available at any moment, it has been common not to collect time-based usage changes, since the time to transfer any item of data depends on the number of active users. An access leasing system is operated by most commercial Internet service providers. This leads to very efficient cost recovery [29].

Group communication is very useful for applications that involve multiple simultaneous senders and receivers. The multicast backbone (MBone) is a virtual overlay on the current Internet that provides multicast, which results in a massive reduction of load on the network for such applications [30]. Initially, it was expected that this might be used for replicated transactions, but in the absence of transport (end-to-end) protocols to support these, what was deployed more quickly was a family of applications based on the UDP, namely, the audio and video conferencing programs.

Typical applications based on UDP offer an approximately fixed rate of packets to the network. To support arbitrary numbers of users of these applications and not suffer overload, the Internet had to be enhanced in some way. A number of possibilities exist, such as resource reservation, and usage-based charging. The first approach may work for a large class of applications, but many researchers and practitioners believe that there will always be some applications that can dominate the network capacity. The second of these approaches has received the most attention in the protocol standards development arena. The third approach is perceived as complex in the Internet community and possibly very hard to deploy [31].

The multicast backbone (MBone) was adopted by the IETF for time-critical communications. The MBone requires facilities such as an IP multicast router and a T1 (1.544 Mb/s) line. It is being used for video conferencing. The resource reservation protocol (RSVP) set by IETF provides QoS for the IP. Hence, isochronous communications over the Internet are enabled. The IP version 6 (IPv6) providing 128 bits of IP address (compared with the current 32-b address of IPv4) will have a flow label to identify the application. This label can be used to provide the QoS, as is done with VPI/VCI in ATM. The real-time protocol (RTP) is for real-time applications in the Internet and can be run on RSVP or ATM [32, 33].

2.4.2 Multimedia Networking

Many applications, such as video mail, video conferencing, and collaborative work systems, require networked multimedia. In these applications, the multimedia objects are stored at a server and played back at the client's sites. Such applications might require broadcasting multimedia data to various remote locations or accessing large depositories of multimedia sources. Multimedia networks require a very high transfer rate or bandwidth, even when the data is compressed. Traditional networks are used to provide error-free transmission. However, most multimedia applications can tolerate errors in transmission due to corruption or packet loss without retransmission or correction. In some cases, to meet real-time delivery requirements or to achieve synchronization, some packets are even discarded. As a result, we can apply lightweight transmission protocols to multimedia networks. These protocols cannot accept retransmission, since that might introduce unacceptable delays [34].

Multimedia networks must provide the low latency required for interactive operation. Since multimedia data must be synchronized when it arrives at the destination site, networks should provide synchronized transmission with low jitter.

In multimedia networks, most communications are multipoint as opposed to traditional point-to-point communication. For example, conferences involving more than two participants need to distribute information in different media to each participant. Conference networks use multicasting and bridging distribution methods. Multicasting replicates a single input signal and delivers it to multiple destinations. Bridging combines multiple input signals into one or more output signals, which then deliver to the participants [35].

Traditional networks do not suit multimedia Ethernet, which provides only 10 Mbps, its access time is not bounded, and its latency and jitter are unpredictable. Token-ring networks provide 16 Mbps and are deterministic. From this point of view, they can handle multimedia. However, the predictable worst case access latency can be very high.

A fiber distributed data interface (FDDI) network provides 100 Mb/s bandwidth, sufficient for multimedia. In the synchronized mode, FDDI has a low access latency and low jitter. It also guarantees a bounded access delay and a predictable average bandwidth for synchronous traffic. However, due to the high cost, FDDI networks are used primarily for backbone networks, rather than networks of workstations.

Less expensive alternatives include enhanced traditional networks. Fast Ethernet, for example, provides up to 100 Mb/s bandwidth. Priority token ring is another system. In priority token ring networks, the multimedia traffic is separated from regular traffic by priority. Figure 2.13 shows a priority token ring. The bandwidth manager plays a crucial role by tracking sessions, determining ratio priority, and registering multimedia sessions. Priority token ring (PTR) works on existing networks and does not require configuration control. The admission control in PTR guarantees bandwidth to multimedia sessions. However, regular traffic experiences delays.

Example 2.1 Assume a priority token ring network at 16 Mb/s that connects 32 nodes. When no priority scheme is set, each node gets an average of 0.5 Mb/s of bandwidth. When half the bandwidth (8 Mb/s) is dedicated to multimedia, the network can handle about 5 MPEG

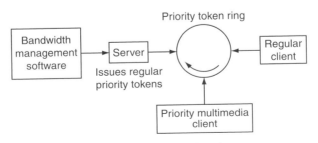

Figure 2.13 Priority token ring.

sessions (at 1.5 Mb/s). In that case, the remaining 27 nodes can expect about 8 Mb/s divided by 27 or 296 Mb/s, about half of what they would get without priority enabled.

They are three priority ring schemes for their applicability to video conferencing applications [36]

1. Equal priority for video and asynchronous packets.
2. Permanent high priority for video packets and permanent low priority for asynchronous packets.
3. Time-adjusted high priority for packets and permanent low priority for asynchronous packets.

The first scheme, which entails direct competition between videoconference and synchronous stations, achieves the lowest network delay for asynchronous frame. However, it reduces the videoconference quality. The second scheme, in which videoconference stations have permanent high priority, produces no degradation in conference quality, but increases the asynchronous network delay. Finally, the time-adjusted priority system provides a trade-off between the first two schemes. The quality of video conferencing is better than in the first scheme, while the asynchronous network delays are shorter than in the second scheme [26].

Present optical network technology can support the broadband integrated services digital networks (BISDN) standard and has become the key network for multimedia applications. The two B channels of the ISDN basic access provide 2×64 kb/s of composite bandwidth. Conferences can use part of this capacity for wideband speech, saving the remainder for purposes such as control, meeting data, and compressed video [37].

BISDN networks are in either synchronous transfer mode (STM) or asynchronous transfer mode (ATM) to handle both constant and variable bit rate traffic applications. STM provides fixed bandwidth channels and therefore is not flexible enough to handle the different types of traffic typical in multimedia applications.

2.4.3 Multimedia Conferencing

Multimedia conferencing enables a number of participants to exchange various multimedia information via voice and data networks. Each participant has a multimedia workstation, linked to the other workstations over high-speed networks. Each participant can send and receive video, audio, and data, and can perform certain collaborative activities. The multimedia conference uses the concept of the shared virtual workspace, which describes the part of the display replicated at every workstation.

The biggest performance challenge in multimedia conferencing occurs when conference participants continuously transmit video and voice streams. Research focuses on mixing these streams together to form a composite stream consisting of video and audio streams. Technologies and protocols for media indexing and an optimal communication architecture for multimedia conferencing are proposed [38].

Multimedia conferencing systems must provide a number of functions, such as multiple-call setup, conference status transmission, real-time control of audio and video, dynamic allocation of network resources, multipoint data transfer, synchronization of shared workspace, and graceful degradation under fault conditions.

Conferencing system architectures can be distinguished by the location of their audio mixing functions and their connection topologies. This leads to two general classes of architectures, namely centralized and decentralized, as well as a third hybrid class. Systems from these families can be evaluated in terms of perceived quality, scalability, controllability, and compatibility with existing standards and practices [39].

Centralized Conferencing Architectures

Traditional *meet-me* teleconferencing has been provided by centralized conference bridges, to which conferees dial in at a prearranged time. The endpoints establish one-to-one media and signaling connections with the bridge. The bridge establishes voice paths between endpoints by summing the input signals together and returning the summed signals to the conferences. To prevent echo, the conferees receive a tailored audio signal comprising the sum of all conferees' voices except their own. The bridge reduces background noise and the probability of hybrid echo by including only M out of N active talkers in the conference sums. This use of speaker selection implies that $M + 1$ sums are formed: one sum for each of the M talkers plus one sum for the $N - M$ unselected conferees (the listeners).

Decentralized Conferencing Architectures

In a decentralized conference, media are exchanged between endpoints without using a centralized bridge. Improved speech quality is inherent since the absence of the bridge eliminates tandeming. However, the endpoints must have the ability to receive and mix multiple streams. Distributing the speech processing functions across the endpoints implies that no single quality requires as much computing power as a conventional VoIP bridge. Decentralized conferencing is represented by the full mesh and multicast conferencing models.

In full mesh conferencing, a full duplex media connection is set up between every part of participants, resulting in *mesh* connections. Each endpoint transmits a copy of its stream to the $N - 1$ other endpoints, and receives $N - 1$ streams in return, each on its own port. Each pair of endpoints can communicate with any mutually supported codec type. Signaling control is centralized at a server, so that a consistent view of the conference state is maintained, wherein the conference state could comprise the conference membership or requests for supplementary audio services [40].

Multicast conferencing is synonymous with wide-area conferencing over the MBone. In a multicast conference, each endpoint transmits a single copy of its stream to the conference multicast address, and receives $N - 1$ streams in return. From a receiver point of view, nothing changes from the full mesh scenario except that the streams arrive on one port. Multicast conferencing is another form of a *meet-me* conference. Instead of connecting to a conference bridge, endpoints join the conference by subscribing to the conference multicast address. This address could be advertised by one of the endpoints or by a central server, or distributed to the conferees prior to the conference.

Hybrid Class

The tandem-free conferencing (TFC) architecture, proposed by Burns [41] and Rabipour and Coverdale [42], is a hybrid between traditional centralized and decentralized approaches. The model uses a tandem-free bridge (TFB), which is a multitalker select-and-forward conference bridge. The TFB selects M current speakers and forwards their compressed signals back to the $N-M$ endpoints, where they are decoded and mixed. If $M = 2$, the primary speaker receives the signal of the secondary speaker, and vice versa.

The TFC architecture has some interesting properties. The system eliminates tandeming, operates independently of the speech codec, and reduces the computational demands of the bridge. The disadvantages are that protocol extensions are necessary for carrying the TFC data, while endpoints must support multiple stream terminal and mixing.

2.4.4 Multicasting

The Internet has witnessed a phenomenal growth, which has, in turn, unleashed the development of newer and more sophisticated applications, which require richer network functionality. Owing to the advent of broadband, wireless, and Web technologies, it is becoming increasingly viable to design and implement large-scale heterogeneous networks that can support content distribution, teleconferencing, media streaming, distance learning, collaborative workspace, and *push* applications. As part of this trend, multicasting has become an enabling technology that plays an important role in the design, development, and operation of many current and next-generation applications and services that rely on the efficient delivery of packets to multiple destinations. Although the concept of multicasting seems very attractive from the surface, it is significantly more difficult to make it really work for the Internet service providers (ISPs). This is evident from the fairly moderate adoption of the technology since the inception of IP multicast over the MBone in 1989. The ISPs are waiting to see sophisticated applications that demand multicast from the business standpoint, whereas users or application developers are waiting for wide deployment of multicast support for them to exploit the technology. This trend is slowly changing as multicast emerges as an enabling technology not only for group communication support, but also for a wide variety of newer applications (content distribution, distributed databases, distributed games, and military applications) that can greatly benefit from exploiting multicasting. Moreover, in recent years, there has been a burst of active research on investigating alternatives to IP paradigms of multicasting, primarily focusing on deployment and scalability issues [43].

With the advances in digital video technology, video-on-demand (VoD) service has come into practice in recent years. A large-scale VoD system is required to store several hundreds of videos and support several thousands of concurrent customers. In order to develop a cost-effective solution, many researchers have been working on various resources, sharing policies by exploiting the multicast capability of modern communication networks. In recent years, different broadcasting schemes such as pyramid [44], harmonic [45], and skyscraper [46] have been proposed to reduce the start-up delay and the bandwidth requirements by the use of a receiver buffer. In addition to broadcasting, the multicast scheme is also used to serve a batch of customers to minimize the system requirement. Unlike the broadcast scheme, in which the bandwidth is bounded, the bandwidth requirement for the multicast scheme is dependent on the arrival rate of the videos. Among many different batching approaches in this multicast system, batching-by-timeout and batching-by-size [47] are two of the most commonly used policies, for their simplicity. Some novel batching schemes such as adaptive piggybacking [48] and patching [49] have therefore been proposed to provide time on-demand services. Obviously, broadcast transmission will be the most efficient if the video is very popular. However, if the arrival rate for a video is not high enough, some resources will be wasted, and in this case, the batching scheme will be more efficient.

Large-scale deployment of VoD service is still uncommon. One of the reasons is the high cost in provisioning large-scale interactive VoD service. The true VoD (TVoD) model calls for a dedicated channel, both at the server and at the network, for each active user during the entire duration of millions of subscribers, and the required infrastructure investment is immense. The central theme is the use of network-level multicast to enable sharing of transmitted data among a large number of users, thereby drastically reducing resource requirements when scaling up the system. The challenges to applying multicast to VoD applications are threefold. First, one needs to design a multicast transmission schedule to maximize resource sharing, while at the same time minimize startup latency. Secondly, as users arrive at random time instants, one would need ways to group them together so that they can share just a few multicast transmissions. Thirdly, to provide service comparable to traditional TVoD service, one would also need to find ways to support interactive controls such as pause–resume, slow motion, seeking, and so on, during video playback [49–51].

To conclude, with the vast improvements in multimedia technologies, VoD will become the key residential service in the emerging high-speed networks. In fact, the growth in the number of VoD providers and the operators they work with is not showing any sign of slowing and is a clear testament to the strength of the market [52]. As the number of customer requests increases, the quality of the service can be maintained only by increasing server resources and network bandwidth, which ultimately leads to an expensive to operate, and nonscalable system [53].

Finally, an efficient implementation of multicast technology permits the simultaneous servicing of many users without overloading either the network or the server resources, and thus providing an effective-cost and large-scalable video-on-demand system, known as near video-on-demand (VoD) [54, 55]. Near VoD makes use of multicast delivery to service more than one customer with a single set of resources to substantially reduce the system cost and achieve scalability. Unified video-on-demand (VoD) system unifies the existing true VoD and near VoD systems by integrating unicast with multicast transmissions [49]. In this system, requests arriving after the beginning of a multicast channel will be immediately served by a unicast stream instead of being scheduled for the upcoming multicast channel. This system can reduce the startup delay in a multicast environment.

2.4.5 Technologies for e-Content

A new technology coming to the fore has often meant a shake-up of the affected industry. The shake-up of our age is the result of the combined effect of three main technologies: signal processing, which transforms analog signals into digital ones; information technology, which allows the processing of data in digital form; and telecommunications, which allows the instantaneous transfer of digital data from anywhere to anywhere in the world. Content that is digital and reaches the end user is called e-content.

Digital Signal Processing

In the early 1980s Philips and Sony introduced a new type of musical disc – the Compact Disc or CD or CD-Audio – which used sophisticated opto-electronic technologies to yield *studio quality* sound with no fluctuations of quality with time. The CD has been an extremely successful device, and the number of CD players existing in the world is several hundred million units. Unlike the compact cassette, the average consumer could not copy the CD bit-by-bit, but could only copy the analog output.

In the first half of the 1990s, the CD-ROM, a computer peripheral utilizing the same digital basis of the CD, became a popular means to store computer data and programs. By the mid-1990s the CD-ROM had became a became a standard peripheral for the PC and, soon after, with the diffusion of writable CDs, it became possible for millions of users to make copies of their CDs, preserving the crisp clear sound of their originals or bit-by-bit copies of their CD-ROMs. At about the same time, audiovisual compression technologies achieved maturity and have been standardized. Millions of video CDs capable of playing an hour of movie on the same physical carrier as the CD, digital television set top boxes, and DVD, a new generation of CD capable of storing two hours of high-quality movies, went into use, much to the satisfaction of users, who could have smaller devices with more functionalities and much better pictures and sound.

Digital Networks

In the mid-1990s, the Internet achieved a level of usage that extended much beyond the original academic/research/defense environment. Hypertext markup language (HTML), a page-formatting standard with the possibility to define hyperlinks, often to parts of the page or other pages located anywhere *on the net*, created the basis of the so-called World Wide Web, where millions of *pages* of hyperlinked text with graphics and still pictures could be accessed by PCs, and displayed using a *browser*. The ease of use and immediacy of obtaining the results prompted several companies to add other media such as video and audio. However, the low bit rate available to the majority of users (a few tens of kb/s) and the high bit rate required by standard-quality audio and video, could only allow very reduced-quality audio and video.

The Value Chain

Business models are in constant evolution, sometimes enabled and sometimes rendered impractical by the evolution of technology. The typical use of the Web today is for advertisement of products or services to offer. The Web page itself is usually *copyrighted* in the sense that the author does not allow the page, as a literary and artistic work, to be used by others. On the other hand, the person who commissioned the page is more than happy if the page is copied as widely as possible so as to maximize the reach of his message.

The Web has created interesting opportunities to diversify the way different businesses are implemented. One of the latest attempts has used the narrow-band interactivity of the Web with the wide-band broadcasting of satellite. The Web is used as a means to post textual descriptions of videos on sale to professional buyers of video programs subscribed to a service. When a video is selected, the request is transmitted to the Web server. This triggers the transmission of the selected video via satellite in encrypted form at lower-than-original quality so that the perspective buyer can browse through the video. If the video meets the needs of the buyer, a video tape is sent to him by regular mail. The basic nature of the Web eliminates the *where* for the source of content. It also eliminates the *where* of the target. The receiver might be a mobile device, or an enterprise network that spans a mobile device, or an enterprise network that spans nations.

Figure 2.14 depicts graphically the actors in the value chain. The center of the figure highlights the three principle actors: the provider of content, the retailers, and the consumer device. Added to these are four major actors:

- *Financial services*, sitting between any two boxes to settle transactions between them.
- *Directory services*, providing additional information of content and assisting users to search for contents of interest.

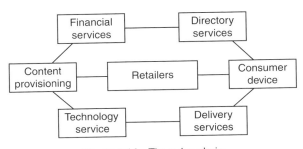

Figure 2.14 The value chain.

- *Technology services*, which are required to represent, fixate, distribute, consume content or data about content.
- *Delivery services*, which move contents between users.

In general, it can be said that between every two boxes there is a flow of content and data associated with the content, a flow of usage rights associated with the content, a flow of technology rights that enable the carriage or consumption of the content, and a financial flow that settles the transfer of content and associated rights. Further, regulation often plays a major role in enabling or disabling some of the functionalities of some of the boxes.

Information Presentation

Standards such as MPEG-1 and MPEG-2 give the possibility to represent high-quality audio and video streams with a number of bits/s that is a good match for many delivery systems in practical use. Further, MPEG-4 supports the encoding of the individual objects in a scene and the compositing of the resulting objects in a virtual space for presentation on physical audio and video presentation devices. By extending virtual reality modeling language (VRML) to support efficient file compression and composition of both synthetic and natural objects with real-time characteristics, MPEG-4 can be used to create content that represents traditional audio and video scenes, as well as virtual spaces populated with synthetic humans and other general synthetic audio and video objects.

Content Description, Identification, and Protection

Describing a piece of content is a very general and modified problem, because it depends on the intended use of the description. Authors and performers have an interest in making sure that their role is recognized. Producers have an interest in keeping track of the roles played by the different contributors in the work and the corresponding rights. Application developers need descriptions to support their work of making new content out of scattered pieces of content. Retailers want to be able to offer content with all the descriptive information that can entice perspective users to access and consume the content. End users want to have access to all kinds of information that enable an informed choice.

Most of the information can be textual (title of a movie, names of actors, place where the work was recorded, and so on). But there are other types of description that are of great interest to the user. An example is the description of a person that would be of interest to his friend.

There is another *description* associated with content that deserves to be treated separately. In today's physical world a book can be read, resold, and rented, a CD can be played, usually not be rented and not resold, a video cassette is for private consumption and may not be copied. Usage of broadcast material can take many forms: public broadcast because a yearly license fee has been paid; commercial broadcast because the user undertakes to watch commercials; pay-TV broadcast because a monthly subscription has been paid; pay per view broadcast because the event has been paid for.

Content identification methods exist for various types of physical content: ISBN (books), ISSN (periodicals), ISRC and ISMN (music), ISAN (audio-visual), and so on. MPEG has developed identification methods for such types of content in digital form as MPEG-2 (audio and video) and MPEG-4 (multimedia objects). These standards already have provisions for adding information about the authority with which the work has been registered, followed by a number giving the specific registration number of work assigned by that authority. This is a first step towards a solution for content identification, but is not fully satisfactory, as the data kept by the different authorities is not homogeneous. Moreover, the system has the major weakness that people with malicious intentions can easily erase all these pieces of information.

Content protection has been adopted by some industries as the way to make sure only those who are entitled to access content in digital form can indeed do so. On the physical distribution side, the DVD Forum has issued specifications that make use of encrypted audio and video content. Digital pay-TV services have done the same, but so far technology limitations have been such that service providers had to hard-wire their protection systems on the set-top-box (STB) and give it away to their subscribers. The consequence is that a user wishing to access multiple service providers must get an STB from each provider, unless two of them decide to use a so-called *simulcrypt* system where subscribers to one provider are recognized by the other. Technology evolution has made open access to protected content possible. This is achieved by injecting security at two levels: the first level is at the basic device hardware.

2.5 STANDARDIZATION FRAMEWORK

As new technologies offer ever greater functionality and performance, the need for standards to reduce the enormous number of possible permutations and combinations becomes increasingly important. Without standards, real-time services suffer, because encoders and decoders may not be able to communicate with each other. Nonreal-time services using stored bit streams may also be disadvantaged because of either service providers' unwillingness to encode their content in a variety of formats to match customer capabilities, or the reluctance of customers themselves to install a large number of decoder types to be able to handle a plethora of data formats [56].

Standardization for communication systems once as produced by specific industries in a vertically integrated fashion. Multimedia, which merges different forms of communication, requires a new approach – horizontal or layered – to develop standards. This implies a radical change in the way standards are produced. Increasing the bandwidth to end users also forces a change in the role of public authorities in regulating the use of technology to provide services [57].

There are many definitions of standardization. We can define it as the process by which individuals recognize the advantage of doing certain things in an agreed way and codify that agreement in a contract. In contrast, we compromise between what we want to do and what others want to do. Although it constraints our freedom, we usually enter into such an agreement because the perceived advantages exceed the perceived disadvantages. Like contracts, standards are – or should be – regulated by the civil code. Behind the word *standard* there should no mystery – these basic principles should apply to individuals, companies, countries, even the organizations that participate in international organizations.

All industries developed vertically integrated systems because various media represented information differently. Recent years have seen considerable market evolution, mostly driven by users' needs, towards unification of information representation.

Television broadcasting has changed with satellite television, which has a regional and often a worldwide scope. Multiple-standard receivers are becoming common in countries exposed to programs broadcast in multimedia standards. Similarly, the telephone system, originally built to carry 3.1 kHz bandwidth speech signals transparently, now carries graphics (facsimile) and even real-time video (video telephone) information transparently.

Computer applications, developed for specific user needs and matched to specific hardware capabilities when they were used on mainframes, now address general user needs in the personal computer environment (word processing, spreadsheets, databases, Internet browsing, and games). Programs can produce and consume hardware-independent data formats. Once autonomous and able to interact via keyboard and screen only, computers needed the ability to talk to other computers, and not just of the same make and model. Whereas standards have been created by agreements between manufacturers, service providers, and regulations, the market has indicated the party neglected in these agreements, the users.

The new digital arrangement requires new standards – and new ways of creating them. When the need for standards arose in the old analog world, representatives from the interested industries met and found a way to incorporate new technology without too much yelling. Technology changes quickly, but people and, in particular, organizations remain anchored to old paradigms.

What we must do is reform the standards bodies themselves. The five relevant international standards entities (ITU-T, ITU-R, ISO, IEC, and JTC1) have different traditions, are often at odds with each other, and have inherent areas of work. The reluctance of formal standards bodies to reform their ways led to establishing the Digital Audio-Visual Council (DAVIC), based on the principle that audiovisual standards should aim for end-to-end interoperability across countries, services, and applications. We should create *a priori* in the digital domain what had (partially) resulted from a long *a posteriori* process in the analog domain. Implementing this principle requires changing the nature of standards from *system* standards to *component* or *toolkit* standards. System integrators must currently assemble the tools to build systems that suit their needs. If *tools* are the object of standardization in the new environment, we must devise a new process that will produce meaningful tools. The following practical steps should yield the desired results:

- select target applications
- list the functionalities needed by each

- break down the functionalities into components of sufficiently reduced complexity that can be identified in the different applications
- identify the components common across the systems of interest
- specify tools that support these components
- verify that these tools can be used to assemble the target systems and provide the desired functionalities.

A work based on these steps cannot be implemented using the work organization of existing standards bodies. Technical work should not be organized by industry (that is, vertically) but rather should be based on horizontal layers. For instance, this means, that there should be one group for applications (all of them), and another for the physical layer (all of them preferably, but not mandatory). This work in each layer must be supplemented by an internal function that keeps track of the consistency of tools and verifies their usability for building the systems desired. Standards committees should strictly adhere to a *one functionality, one tool* principle. More options require more investments, making retail systems more expensive and less interoperable.

2.5.1 Research and Regulation

In audiovisual communications today, the current drive seems to be towards integrating the technology in high-functionality terminals and to be able to provide the information to the end-user at any location at any time. This is what we here will term multimedia and thus be in line with the presentation of the ETSI Program Advisor Committee Expert Group 5 (PAC5) in their Global Multimedia Mobility (GMM) concept.

There are several issues that have to be addressed to be able to succeed in these developments, where the technical innovations only represent one of them. The driving forces can be summarized as [58]

- *technical developments*
- *politics* (licensing, standardization, regulations)
- *marketing/economics* (operators, forecast).

Audiovisual communications involve a broad range of problems. At the heart of the set is the information representation and coding. The dimension of coder performance can be applied to source coding, channel coding, and modulation by indication of the appropriate units on each axis. This coder performance is shown in Figure 2.15. In fact, the four-dimensional space, indicating regions of theoretically available areas and design trade-offs can serve as a model for the research community as well as standardization bodies and regulatory authorities. The difficulties in use are of course defining and agreeing on the quality evolution criteria to run by and the measurable units of complexity, efficiency, and delay. The technical approaches to addressing the challenges will be based on integrated system design methodology,

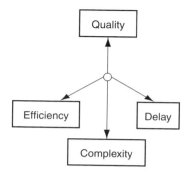

Figure 2.15 The dimension of coder performance.

interaction with the networking community, increased attention to the man–machine interface, and inclusion of advanced system functionalities. The technical challenges for the future will be the interaction of all the players and to find a common approach to system design. Typical initiatives in such a way are the design of the MPEG family of standards, the ITU-T H.xxx series and the current multimedia issues in Europe (European Project on Information Infrastructure – EPII, Information and Communication Technology Standardization Board – ICTSB) and globally by ITU-T (Study Group for Multimedia Services and Applications – SGMMS).

The nucleus of business for a regulator can be described as being into the nature of

- effective supervision and market control to ensure fair competition
- licensing
- frequency management
- standardization/type – approval.

To be able to perform these tasks the persons will typically have backgrounds in

- law (regulations and directives)
- economics (market applications)
- telecommunications engineering (frequency management, technical directives, standards).

The split will vary from country to country. The implications on research and education are on the system engineering level rather than on a detailed design level, particularly in considering the two important issues of

- system efficiency
- system interference (radio in particular).

2.5.2 Technology and Education

Multimedia communication systems are a well-balanced interplay between the users–markets–regulations and technology. This is becoming more and more clear and is setting demands on the various groupings. A good example of this is development of the third-generation mobile communication system.

For academia, an obvious demand is in the future development of education through timelines and a modern approach to setting the curriculum. Communications education has to be broadened to include general knowledge of merging the signal processing/transmission part with network aspects. The danger is losing out on detail and ability to bring forward new knowledge necessary to initiate evolutions, or possibly revolutions, in the communication industry. Academia also has clear responsibilities for developing knowledge as well as overview, for instance, to be able to provide insight in trends and foresight for industry, regulations, and standardization. Research strategies should be set with this in mind. A good approach and interworking with the other bodies will give the necessary strength to convince skeptics at all levels of introducing new technology and ultimately designing better systems.

2.5.3 Convergence and Regulatory Issues

There is no universal definition of convergence. For example, convergence can mean

- provision of various communication services like text, data, image, and video over the existing infrastructure
- development of new infrastructure for handling multimedia transmission
- management of technologically and commercially distinct markets such as broadcasting, publishing, cable TV, fixed voice, cellular mobile services, and Internet services.

There exists convergence of technologies, convergence of services, and regulatory convergence. By convergence of technologies, it is meant a common platform to deliver voice, data, and video services. Convergence of services represents delivery of multimedia services to the end users over the same medium/network. Finally, regulatory convergence understands establishing a simple regulatory authority by blurring the regulatory boundaries for telecommunications, information technology, and broadcasting.

As for genesis and manifestation of convergence, they are driven by technological innovations, progressive integration of distinct markets such as IT, telecommunications, broadcasting, cable TV into a single value chain regulation as an enabler to convergence.

There are four drivers of convergence, that is, technology, market, customers, and regulation. Technology includes

- digitization of transmission and switching networks
- global networks based on packet switching and open standard

- increase in processing power of computers
- emergence of new applications leveraging enhanced software capabilities
- evolution of broadband technologies (xDSL, hybrid fiber coax HFC, broadband wireless-fixed/mobile/satellite).

Market comprises the following:

- new markets and services such as multimedia services, video on demand, interactive TV, pay-TV, cable telephony, unified messaging services, Internet telephony, and so on
- integration of content service providers with access providers
- emergence of new market players.

As for customers, they have

- captions for new value-added services in a modular fashion
- cheaper access to communications
- single information socket for phone, data, video.

Finally, regulations permit

- reduction in cost of regulations by optimum utilization of regulatory resources
- ease of regulation and interconnection.

Challenges to convergence consist of migration to converged license requirements, technical standards, fair and nondiscriminatory access, competition issues, the frequency spectrum, as well as universal service obligation and bridging the digital divide.

Migration to converged licensing requirements consists of

- maintenance of a level playing field between existing and new players
- service-based licenses with heterogeneous terms and conditions of licenses
- license mapping for all current licenses from old to new license categories
- license transitions process
- license structure design and drafting.

From the point of view of technical standards some problems often arise, such as

- standards supporting convergence are still evolving
- availability of multiple standards by different agencies (ITU, IETF, ETSI, ANSI, IEEE)
- interworking among various standards

- standards for QoS have not yet matured
- need for technology central regulations.

Challenges to convergence concerning fair and nondiscriminatory access are

- complex regulations governing access to bottleneck facilities
- transparent, nondiscriminatory and cost-based access
- access to contents with program access regime.

By competition issues, we mean

- removal of industry-specific regulations and their replacement with general competitive law applied equally to all industries
- legal framework aimed at promoting competition and prohibiting a range of anticompetitive practices, including anticompetitive agreements, anticompetitive mergers, and misuse of market power
- cross-border mergers and cross-media ownership restrictions in certain parts of the media
- foreclosure of markets by anticompetitive conduct.

As for the frequency spectrum, the following must be taken into account:

- possible reallocation of frequency spectrum
- optimum allocation of spectrum to the new licenses
- no precedence for allocation of the spectrum in the new regime
- delicensing of frequency bands for certain applications.

The influence of universal service obligation (USO) can be summarized as

- redefining the scope of *universal service*
- identifying contributors for USO
- formulation of regulatory policy to bridge the digital divide.

To conclude, while technology changes quickly, regulation changes slowly. On the other hand, legacy regulation is not relevant in the convergence era. In addition, prioritization of regulatory challenges is important for optimum utilization of regulatory resources. Also, there is a need for an efficient and timely dispute resolution mechanism for the converged licensing regime.

2.5.4 Manufacturing and Marketing

There are similar demands on the manufacturers and regulators in deploying new applications and services in the communication market sector. The need for technically skilled workers is beyond doubt. It is also the responsibility of the manufacturers, together with research and standardization bodies to contribute to the making of strategies and polices for the future. Issues cover

- the competitive converging market
- whether regulators fit the future
- competition and convergence worldwide.

There are a lot of benefits of bringing together manufacturers, researchers, and academia to focus on a common goal. A major benefit is the partners getting to know each other. Also, it is of importance to manage to reach a common set of targets of interest to all partners from different environments, setting up the ideal situation for reaching an open, harmonized state-of-the-art standard for the next generation still image coder.

The communications market is experiencing rapid growth after complete liberalization of the sector in January 1998. In addition, telecommunications is seeking a complete change of both technology and services offered. This development and massive growth creates a strong demand for highly skilled personnel at all levels and in all disciplines of communications and computing. Such personnel is a scarce commodity, making an expansion of the capacity at educational institutes instrumental for continuing success and growth. Crucial to these developments is a close link between education and the research institutions to keep pace in the market sector as well as in the definition of new technology and services. The development of fiber technology is enabling affordable high-capacity links over great distances in telecommunications. New services (Internet, audiovisual communications, multimedia, and so on) are able to utilize this capacity. The access to the fixed network's transport layer will therefore be a key factor for the flexible use of the services.

Communications in the future will integrate different technologies and services in open and intelligent systems that give access to information from any point at any time. Multimedia communications over hierarchical and seamless networks will become a reality where the boundaries between fixed and mobile networks will disappear. Mobile communications have capacity limitation and are by nature highly variable with regards to quality and reliability compared to the fixed networks. Technologies for improved capacity, tailored quality, and reliability through the use of adaptive and flexible radio access solutions are therefore key factors in any mobile system.

Industry and research communities have a strong position within radio communications, which make it natural to focus on radio communications as a prioritized center of excellence in the area of communications for future research and education. This is a field with strong potential for new products and services for mobile and fixed users. In addition, it includes a number of generic elements spanning from component to system level. Crucial to these developments is a close link between education, industry, and the research institutions to keep pace in the market sector as well as in the definition of new technology and services.

The environment for communications is changing drastically and quickly, as well as both politically and technically. Deregulation of the communications market is being scheduled or planned for in many countries. This deregulation will affect the process of standardization among international organizations. The emerging multimedia technologies and global information infrastructure (GII) require broader collaboration among not only *de jure* standardization organizations, but also with the *de facto* industry standardization bodies.

2.5.5 Digital Video/Audio Coding Standards and Multimedia Industry

Audiovisual services provide real-time communication of speech, together with visual information, between two or more end users. The visual information is typically moving pictures, but may be still pictures, graphics, or any other form [59]. The ITU-T Study Group 15 has been standardizing the audiovisual communication systems in various network environments. The first set of such standards, called Recommendations in the ITU-T, was formally established in December 1990 for narrow-band ISDN (NISDN), which provides digital channels of 64 kb/s (B channel), 384 kb/s (H_0 channel) and 1536/1920 kb/s (H_{11}/H_{12} channel). Recomm. H.320 [60] describes a total system stipulating several other Recomm. to which respective constituent elements, such as audio coding, video coding, multimedia multiplexing, and system control, should conform. Its major target applications are videoconferencing and video telephony, although other applications are not excluded. Since then, multipoint and security enhancement of NISDN systems have been developed [61]. In parallel with this, a new standardization activity was initiated in July 1990 towards broadband and high-quality audiovisual systems by forming an expert group, as had previously been done successfully for the development of Recomm. H.261 [62–64]. This new group was initially charged with asynchronous transfer mode (ATM) video coding, and collaborated with ISO/IEC JTC1/SC29/WG1 (MPEG) [65].

The success of MPEG is based on a number of concurrent elements. The most important is probably the timing of its establishment. MPEG appeared at a time when coding algorithms of audio and video were reaching asymptotic performance, and the capability of digital signal processing was matching algorithm complexity. MPEG also succeeded in enlisting the participations of all industries that are now claimed to be *converging*. By rallying the support in terms of technical expertise, of all industries interested in digital audio and video applications, MPEG contributed to the practical acceptance of the audiovisual representation layer, independent of the delivery system. A last element of success has been the focus on the decoder instead of the traditional encoder–decoder approach. Therefore, MPEG could provide the standard solution to the major players who were considering the use of digital coding of audio and video for innovative mass-market products and allow a faster achievement of a critical mass that would have been possible without it [66]. For multimedia industry there are two major standards organizations: ITU-T and the ISO. Recent video coding standards defined by these two organizations are summarized in Table 2.2 [67]. These standards differ mainly in the operating bit rates due to the applications that they were originally designed for. All standards can essentially be used for all applications at a wide range of bit rates. In terms of coding algorithms, all standards follow a similar framework.

ITU-T Standardization of Audiovisual Communication Systems

The General Switched Telephone Network (GSTN) system has been studied by a separate experts group for low-bit-rate coding. It has produced not only the total system recommendation H.324 [68], but also the improved video coding Recomm. H.263 [69], the improved audio coding Recomm. G.723.1 [70], and the multiplexing scheme defined in Recomm. H.223 [71]. A general protocol stack model of an H-series audiovisual communication terminal is shown in Figure 2.16.

ITU-T H.320 Standard. For all of the ITU standards, interoperability with the H.320 standard is mandatory. This interoperability is achieved through a gateway that, in some cases, must perform translations between different signaling protocols, different compression standards, and different multiplexing schemes. The variations of signaling, compression, and multiplexing for the various standards are due to the different characteristics of the underlying networks to which each standard applies. The H.320 standard defines a central conference server called Multipoint Control Unit (MCU) to enable multiple calls [72].

ITU-T H.310 and H.321 Standard. The high bandwidth of the ATM network provides a capability of low delay for conversational services. NISDN audiovisual systems use H.261 video coding, which incurs a buffering delay of at least four times the frame period (133 ms) plus any display delay due to picture skipping. The ATM audiovisual systems should significantly improve the end-to-end delay, so a target of less than about 150 ms has been set. This value corresponds to the acceptable-for-most-user-applications level of specification in ITU-T Recomm. G.114 for one-way transmission time [73]. To meet the previous requirements, ITU-T SG15 has developed the following two recommendations for audiovisual communication systems in ATM environments:

- H.321 – adaptation of H.320 visual telephone terminals to broadband ISDN environments
- H.310 – broadband audiovisual communication systems and terminals [74].

Recomm. H.321 specifies the adaptation of H.320 visual telephone terminals to BISDN environments, thus satisfying the requirement that ATM terminals should interwork with those connected to NISDN. Recomm. H.310 includes the H.320/H.321 interoperation mode, which takes advantages of the opportunities provided by ATM, to provide higher quality audiovisual communication systems.

ITU-T H.322 Standard. The proposal that ITU should have a recommendation covering the provision for local area networks (LANs) and for the video telephony and video conferencing facilities, equivalent to those specified by Recomm. H.320 for NISDN, was made at the September 1993 meeting of the SG15 Experts Group for ATM Video Coding. At a subsequent meeting the Working Party mandated the Experts Group to begin studies and to produce a draft recommendation under

Table 2.2 Audiovisual communication systems

			Network		
	GSTN	NISDN	Guaranteed QoS LANs	Nonguaranteed QoS LANs	ATM (B-ISDN, ATM LANs)
Channel capacity	Up to 28.8 kbit/s	Up to 1536 or 1920 kbit/s	Up to 6/16 Mbit/s	Up to 10/100 Mbit/s	Up to 600 Mbit/s
Characteristics	Ubiquitous	Circuit-based (existing)	Similar to NISDN	Packet loss prone	Future basic network
Total system (date of first approval)	H.324 (96/03)	H.320 (90/12)	H.322 (96/03)	H.323 (96/11)	H.310 (96/11) H.321 (96/03)
Audio coding	G.723.1	G.711 G.722 G.728	G.711 G.722 G.728	G.711 G.722 G.723.1 G.728	G.711 G.722 G.728
Video coding	H.261 H.263	H.261	H.261	H.261 H.263	ISO/IEC 11172-3 H.261 H.262
Data	T.120, etc.	T.120, etc.	T.120, etc.	T.120, etc.	T.120, etc.
System control	H.245	H.242	H.242	H.245	H.242 (for H.321) H.245 (for native H.310)
Multimedia multiplex and synchronization	H.223	H.221	H.221	H.225.0 TCP/IP, etc.	H.222.0 H.222.1
Call setup signaling	National standards	Q.931	Q.931	Q.931 H.225.0	Q.2931

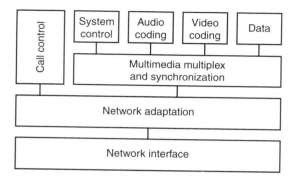

Figure 2.16 General protocol stack of H-series audiovisual communication terminal.

the working number H.322. The H.322 gateway unit is not restricted to serve only H.322 terminals on its LAN side, but can equally handle generic ISDN terminals. The first H.322 gateways will have NISDN interfaces, but as BISDN becomes widespread, interfaces to that will become more prevalent. Such a configuration will then permit good to high-quality video at 2 Mb/s to be sent to and received from the remote locations across the public BISDN. However, H.322 does not mandate any minimum number of simultaneous calls that it can support or any maximum number of simultaneous calls to or from the ISDN.

ITU-T H.323 Standard. This recommendation specifies equipment and systems for visual telephony on nonguaranteed QoS LANs. It covers those situations where the transmission path includes one or more LANs that may not provide a guaranteed QoS equivalent to that of NISDN. Examples of this type of LAN include Ethernet, Fast Ethernet, Token ring [75] and Fiber Distributed Data Service [76].

The primary design considerations in the development of H.323 were the following:

- interoperability, especially with NSIDN and H.320
- control of the access to the LAN to avoid congestion
- multipoint call models
- scalability from small to medium-sized networks.

The scope of H.323 does not include the LAN itself or the transport layer that may be used to connect various LANs. Only elements needed for interaction with the Switched Circuit Network (SCN) are within the scope of H.323. The combination of the H.323 gateway, the H.323 terminal and the out-of-scope LAN appears on the SCN as an H.320, H.310, and H.324 terminal. Recommendation H.323 describes the total system and components, including terminals, gateways, gatekeepers, multipoint controllers, multipoint processors, and multipoint control units (MCUs).

Video Coding Standards

There are two approaches to understanding video coding standards (H.261, H.263, and H.26L). One approach is to focus on the bit-stream syntax and to try to understand what each layer of the syntax represents and what each bit in the bit stream indicates. This approach is very important for manufacturers who need to build equipment that is compliant to the standard. The other approach is to focus on coding algorithms that can be used to generate standard-compliant bit streams and to try to understand that each component does not specify any coding algorithms. The latter approach provides a better understanding of video-coding techniques as a whole.

ITU-T H.261 Standard. This standard is defined by ITU-T SG15 for video telephony and videoconferencing applications [77]. H.261 emphasizes low bit rates and the low coding delay. The coding algorithm used in H.261 is basically a hybrid of motion compensation to remove temporal redundancy and transform coding to reduce spatial redundancy. Such a framework focuses the basis of all video coding standards that were developed later. Therefore, H.261 has a very significant influence on many other existing and evolving video-encoding standards. H.261 is designed for video telephony and videoconferencing, in which typical source material is composed of scenes of talking persons, so-called head and shoulder sequences, rather than general TV programs that contain a lot of motion and scene changes. As for compression of video data, it is typically based on two principles: the reduction of spatial redundancy and the reduction of temporal redundancy. H.261 uses the DCT to remove spatial redundancy [78] and motion compensation to remove temporal redundancy [79]. As in all video coding standards, H.261 specifies only the bit stream syntax and how a decoder should interpret the bit stream to decode the video. Therefore, it specifies only the design of the decoder, not how the encoding should be done. Also, H.261 has been included in several ITU-T H-series terminal standards for various network environments. One example is H.320, which is mainly designed for narrow-band ISDN terminals [80]. H.320 defines the systems and terminal equipment that use H.261 for video coding.

ITU-T H.263 Standard. During the development of H.263, it was identified that the near-term goal would be to enhance H.261 using the same general framework, and the long-term goal would be to design a video coding standard that may be fundamentally different from H.261 in order to achieve further improvement in coding efficiency. As the standardization activities moved along, the near-term effort became H.263 and H.263 Version 2, while the long-term effort is referred to as H.26L/H.264 [81]. In essence, H.263 combines the features of H.261 together with MPEG, and is optimized for very low bit rates. In terms of signal-to-noise ratio (SNR), H.263 can provide a 3 to 4 dB gain over H.261 at bit rates below 64 kb/s. In fact, H.263 provides superior coding efficiency to that of H.261 at all bit rates. Because H.263 was built on top of H.261, the main structure of the two standards is essentially the same. The major differences include the following:

- H.263 supports more picture formats
- H.263 uses half-pel motion compensation, but does not use filtering as in H.261

- H.263 uses 3D VLC for coding DCT coefficients
- in addition to the basic coding algorithm, four options in H.263 that are negotiable between the encoder and the decoder provide improved performance
- H.263 allows the quantization step size to change at each macroblock (MB) with less overhead.

A major difference between H.261 and H.263 is the half-pel prediction in the motion compensation. This concept is also used in MPEG.

As in H.261, H.263 can be used in several terminal standards for different network environments. One example is H.324 [82] which defines audiovisual terminals for the traditional PSTN. In H.324, a telephone terminal uses H.263 as the video codec. The ITU-T H.261/H.263 video compression standards were designed for real-time coding and decoding for videoconferencing across constant bit rate connections.

ITU-T H.263+ Standard. H.263+ (H.263 Version 2) is a revision of the original 1999 version of the H.263 standard [83]. H.263+ contains approximately 12 new features that do not exist in H.263. These include new coding modes that improve compression efficiency, support for scalable bit streams, several new features to support packet networks and error-prone environments, added functionality, and support for a variety of video formats. Among the new features of H.263+, one of several that correct design inefficiencies of the original H.263 recommendation is modified quantization mode. Also, one of the modifications of the original H.263 is the motion vector range. When the H.263+ mode is invoked, the range is generally larger and depends on the frame size. Another modification to the original H.263 recommendation is the addition of a rounding term to the equation for half-pel interpolation. Finally, H.263+ supports a wider variety of input video formats than H.263. In addition to five standard sizes, arbitrary frame sizes, in multiples of four from (32×32) to (2048×1152), can be supported. Another round of H.263 extensions has created a third generation of the H.263 syntax, informally called H.263++.

H.263++ Standard Development. H.263++ development effort is intended for near-term standardization of enhancements to produce a third version of the H.263 video codec for real-time communications and related nonconversational services [84]. Key technical areas showing potential for performance gain of H.263++ are the following [85, 86]:

- error-resilient data partitioning
- 4×4 block-size motion compensation.

By error-resilient data partitioning, we mean creation of a data partitioned and layered protection structure for the coded data and a larger resynchronization codeword to improve the detectability and to reduce the probability of false detection.

For 4×4 block-size motion compensation, we can include longterm picture memories, rate-distortion optimization alternations, motion optimization algorithms, a new type of deblocking filter, a new type of intraspatial prediction, as well as some VLC alternations for transform coefficients, motion vectors, and coded block pattern.

H.264/AVC Standard. Although H.264 [87] is based on the conventional block-based motion compensated predictive-transform framework used in H.261, H.262, and H.263, significant changes such as loop filtering, (4 × 4) integer transform and its exact matched inverse transform, directional spatial prediction for intracoding, multiple (up to 5) reference picture motion compensated prediction, enhanced variable length coding, network friendliness, parameter set concept, switched P and I pictures, and so on, have contributed to nearly 50% bit rate savings at the same quality levels compared to previous standards. H.264 is aimed at wide range of bit rates (64 kbps to 240 Mbps), quality levels, and spatial/temporal resolutions with real-time, low-end-to-end delay coding for a variety of source materials. By introducing profiles and levels, this standard can be applied to video streaming, broadcast video, conversational services, and so on. All these features and coding gains are at the cost of increased complexity. While the main encoding (actually coded bit stream) has been standardized, other parts such as conformance bit streams, reference software, 4 : 4 : 4 format, increased bit depths (up to 12 bpp), verification testing, intellectual property licensing, example encoding, industry interpretative testing of implementation, and so on, are under various stages of development.

Speech Coding Standards

The ITU has standardized three speech coders, which are applicable to low-bit-rate multimedia communications. ITU recommendation G.729 8 kb/s CD-ACELP has a 15 ms algorithmic codec delay and provides network-quality speech [88]. It was originally designed for wireless applications, but is applicable to multimedia communication as well. Annex A of recommendation G.729 is a reduced-complexity version of the CD-ACELP coder. It was designed explicitly for simultaneous voice and data applications. These two coders use the same bit stream format and can be interoperable. The ITU recommendation G.723.1 6.3 and 5.3 kb/s speech coder for multimedia communications was designed originally for low-bit-rate videophones. Its frame size of 30 ms and one-way algorithmic codec delay of 37.5 ms allow for a further reduction in bit rate compared to the G.729 coder.

An enormous number of new speech coders have been standardized. For example, in the 1995–1996 time period, three new international standards (ITU G.729, G.729A, and G.723.1) and three new regional standards have emerged. There are enhanced full-rate coders for European and North American mobile systems.

Speech quality as produced by a speech coder is a function of bit rate, complexity, delay, and bandwidth. Hence, when considering speech coders, it is important to review all these attributes. Also, we have to realize that there is strong interaction between all these attributes and thus they can be traded off against one another.

Multimedia Multiplex and Synchronization Standards

ITU-T Recommendation H.222. Elementary streams such as audio, video, data, video frame synchronous control and indication signals, each of which may be internationally standardized or private, are multiplexed into a serial packet stream according to H.222.0, H.222.1/H.222.0 functions. These include multiplexing time base recovery, media synchronization, jitter removal, buffer management, security and access control, subchannel signaling, and trick modes, which are mechanisms to support video recorder-like control functionality. Recommendation H.222.1 specifies elements and procedures from the generic H.222.0 for their use in ATM environments and also specifies code points and procedures for ITU-T defined elementary streams [89]. H.222.1 allows the use of both the H.222.0 program streams and the H.222.0 transport stream.

ITU-T Recommendation H.221. This recommendation is entitled *Frame structure for a 64–1920 kb/s channel in audiovisual teleservices* [90]. H.221 is the multiplex and bending protocol for H.320 terminals. Up to 30 ISDN B channels can be bundled together to form a superchannel with a bit rate of n∗64 kb/s. The media channels for audio and video information (H.221) do not perform any error control, but rely completely on the error resilience of the media coding.

ITU-T Recommendation H.223. This recommendation is entitled *Multiplexing protocol for low bit rate multimedia communication* [91]. Three different types of adaptation layers (ALs) are available, which have different characteristics in terms of error probability and delay. Low-delay channels allow higher error rates, and reliable channels might have indefinitely long delays. AL1 and AL2 serve different duties. AL3 is designed for use with coded video.

ITU-T Recommendation H.225. Recommendation H.225.0 describes the means by which audio, video, data, and control are associated, coded, and packetized for transport between H.323 terminals on a nonguaranteed QoS LAN, or between H.323 terminals and H.323 gateways, which in turn may be connected to H.320, H.324, or H.310/H.321 terminals on NISDN, GSTN, or BISDN, respectively. This gateway terminal configuration and procedures are described in H.323, and H.225.0 covers protocols and message formats. The scope of H.225.0 communication is between H.323 terminals and H.323 gateways on the same LAN, using the same transport protocol.

Common Control Protocol H.245. This recommendation defines messages and procedures for the exchange of control information between multimedia terminals. H.245 specifies terminal-to-terminal signaling to determine the coding and decoding capabilities of the remote terminal, following the establishment of the network connection, and to coordinate the assignment and release of terminal resources throughout the call. H.245 has been defined as a generic recommendation that is suitable for use in a range of multimedia terminal applications. H.245 was structured by defining three main sections: syntax, semantics, and procedures. Interaction between the different protocol entities is only through communication with the H.245 user. H.245 provides a number of different services to the H.245 user. Services may by applicable to a specific terminal recommendation. Some of these services are capability exchange, logical channels signaling procedures, control and indication signals.

MPEG Standards: Role in Multimedia Communications

The success of MPEG is based on a number of concurrent elements. Probably the most important is the timing at which it was established. Namely, MPEG appeared at a time when often coding algorithms of audio and video were reaching asymptotic performance. MPEG also succeeded in enlisting the participation of all industries that are now claimed to be *converging* by rallying the support in terms of technical expertise of all industries interested in digital audio and video applications. MPEG contributed to the practical acceptance of the audiovisual representation layer, independent of the delivery system. A last element of success has been the focus on the decoder, instead of the traditional encoder–decoder approach. Therefore, MPEG could provide the standard solution to the major players who were considering the use of digital coding of audio and video for innovative mass-market products and allow a faster achievement of a critical mass than would have been possible without it [66].

Each MPEG standard starts by identifying its scope and issuing a call for proposals. It then enters into two major stages. The first stage is a competitive stage that involves testing and evaluation of the candidate proposals to select a few top-performing proposals, components of which are then used as the starting basis for the second stage. The second stage involves collaborative development of these components via interactive refinement of the experimentation model (coding description). The format for representing data is referred to as the syntax and can be based to construct various kinds of valid data streams referred to as the bit streams. The rules for interpreting the data (bit streams) are called the decoding semantics. An ordered set of decoding semantics is referred to as the decoding process. Given audio and/or video data to be compressed, an encoder must follow an ordered set of steps called the encoding process. However, this encoding process is not standardized and varies because encoders of different complexities may be used in different applications. The only constraint is that the output of the encoding process results in a syntactically correct bit stream that can be interpreted according to the decoding semantics by a standards-compliant decoder.

MPEG-1 Standard. MPEG-1 is the standard for storage and retrieval of moving pictures and audio on a digital storage medium. The original target for the MPEG-1 standard was good quality video and audio at about 1.4 Mb/s for compact disc application. A number of primary requirements are listed as follows [92]:

- coding of video with good quality at 1 to 1.5 Mb/s and audio with good quality at 128 to 256 kb/s
- random access to a frame in limited time
- capability for fast forward and fast reverse, enabling seek and play forward or backward at several times the normal speed
- a system for synchronized playback and access to audiovisual data.

Besides the preceding requirements, a number of other requirements also arose, such as support for a number of picture resolutions, robustness to errors, coding quality trade-off with coding delay (150 ms to 1 s), and the possibility of real-time encoders at reasonable costs. The MPEG-1 standard consists of the following parts: Systems, Video, Audio, Conformance, and Software.

The audio part of the MPEG-1 standard has become the key component for *radio broadcasting at CD quality* MPEG-1 Audio Layer II and more recently, Layer III have become the standard form for music distribution on the Web. The full MPEG-1 standard (audio-video-systems) is the standard format for distribution of moving video material over the Web.

MPEG-1 provided the first concrete opportunity for the microelectronics industry to invest in digital video technology. Today, MPEG-1 decoder chips are produced by multiple sources, some of which incorporate the electronics needed to read bits from a CD. Devices and the growing number of personal computers are creating the conditions for the popularization of multimedia contents production.

MPEG-2 Standard. The MPEG-2 standard was designed to provide the capability for compressing, coding, and transmitting high-quality, multichannel multimedia signals over terrestrial broadcast, satellite distribution, and broadband networks, for example, using ATM protocols. The MPEG-2 standard specifies the requirements for video coding, audio coding, and systems coding for combining coded audio and video with user-defined private data streams, conformance testing to verify that bit streams and decoders meet the requirements, and software simulation for encoding and decoding of both programs and transport streams. Designed as a transmission standard, MPEG-2 supports a variety of packet formats, including long and variable-length packets of from 1 up to 64 kb [93]. Also, it provides error correction capability that is suitable for transmission over cable TV and satellite links [56]. The MPEG-2 standard consists of the following parts: Systems, Video, Audio, Conformance, Software, Digital storage media and command and control (DSM-CC), Advanced audio coding (AAC), 10-bit Video, Real-time interface, Conformance of DSM-CC.

MPEG-4 Standard. The MPEG-4 is standard for multimedia applications [94]. It is aimed at Internet and intranet video, wireless video, interactive home shopping, video e-mail and home movies, virtual reality games, simulation, and training of media object databases. Because these application areas have several key requirements beyond those supported by the previous standards, the MPEG-4 addresses the following functionalities [95, 96]:

- content-based interactivity
- universal accessibility
- improved compression.

The MPEG-4 standard consists of the following parts: Systems, Video, Audio, Conformance, Software, and Delivery multimedia framework.

The conceptual architecture of MPEG-4 comprises three layers: the compression layer, the sync layer, and the delivery layer. The compression layer is media aware and delivery unaware. The sync layer is media unaware and delivery unaware. As for the delivery layer, it is media unaware and delivery aware [56].

The design of MPEG-4 is centered around a basic unit of content called the audio-visual object (AVO). Each AVO is represented separately, and becomes the basis for an independent stream.

MPEG-7 Standard. MPEG-7 is the content representation standard for multimedia information search, filtering, management, and processing [97]. A goal of MPEG-7 is to enable search for multimedia on the Internet and improve the current situation caused by proprietary solutions by standardizing an interface for descriptors and description schemes that may be associated with the content itself to facilitate fast and efficient search. Thus, audiovisual content with associated MPEG-7 metadata may easily be indexed and searched. MPEG-7 aims to address not only finding content of interest in *pull* applications, such as that of database retrieval, but also in *push* applications such as selection and filtering to extract content of interest within broadcast channels. It is expected that MPEG-7 will work not only with MPEG but also with non-MPEG coded content.

A number of traditional as well as upcoming application areas that employ search and retrieval, in which MPEG-7 is applicable are [98, 99]:

- *significant events* – historical, political
- *educational* – scientific, medical, geographic
- *business* – real estate, financial, architectural
- *entertainment and information* – movie archives, news archives
- *social and games* – dating service, interactive games
- *leisure* – sport, shopping, travel
- *legal* – investigate criminal and missing persons.

The MPEG-7 descriptors describe various types of multimedia information. This description will be associated with the content itself, to allow fast and efficient searching for material of a user's interest. Audiovisual material that has MPEG-7 data associated with it can be indexed and searched. This material may include still pictures, graphics, 3D models, audio, speech, video, and information about how these elements are combined in a multimedia presentation. Special cases of these general data types may include facial expressions and personal characteristics. Although MPEG-7 does not standardize the feature extraction, the MPEG-7 description is based on the output of feature extraction, and although it does not standardize the search engine, the resulting description is consumed by the search engine.

MPEG-21 Multimedia Framework. The aims of MPEG-21 are the following:

- To understand if and how various components fit together
- To discuss which new standards may be required, if gaps in the infrastructure exist and when the above two points have been reached
- To accomplish the integration of different standards.

The MPEG-21 project was started with the goal to enable transparent and augmented use of multimedia resources across a wide range of networks and devices.

The work carried out so far has identified some technologies that are needed to achieve the MPEG-21 goals. They include the following [100]:

- *Digital item declaration* – uniform and flexible abstraction and interoperable schema for declaring digital items.
- *Content representation* – how the data is represented as different media.
- *Digital item identification and description* – a framework for identification and description of any entity regardless of its nature, type, or granularity.
- *Content management and usage* – the provision of interfaces and protocols that enable creation, manipulation, search, access, storage, delivery, and (re)use of content across the content distribution and consumption value chain.
- *Intellectual property management and protection* – the means to enable content to be persistently and reliably managed and protected across a wide range of networks and devices.
- *Terminals and networks* – the ability to provide interoperable and transparent access to content across networks and terminal installations.
- *Event reporting* – the metrics and interfaces that enable users to understand precisely the performance of all reportable events within the framework.

The meaning of a user in MPEG-21 is very broad and is by no means restricted to the end-user. Therefore, an MPEG-21 user can be anybody who creates content, provides content, archives content, rates content, enhances and delivers content, aggregates content, syndicates content, sells content to end-users, consumes content, subscribes to content, regulates content, or facilitates or regulates transactions that occur from any of the previous examples.

Still Image Coding Standards
With increasing use of multimedia communication systems, image compression requires higher performance and new features. Image compression must not only

reduce the necessary storage and bandwidth requirements, but also allow extraction for editing, processing, and targeting of particular devices and applications. In order to present an analytical study of the corresponding functionalities, the following standards are mentioned: MPEG-4 VTC [101], JPEG [102], JPEG-LS [103] and PNG [104]. JPEG2000 supports coding of bilevel and paletted color images. Other image coding standards are JBIG [105] and JBIG2 [106]. These are known for providing good performance for bilevel images, but they do not support an efficient coding of continuous tone images with a large enough number of levels [107].

MPEG-4 VTC. MPEG-4 VTC is the algorithm used in the MPEG-4 standard in order to compress the texture information in photo-realistic 3D models. Because the texture in a 3D model is similar to a still picture, this algorithm can also be used for compression of still images [101]. MPEG-4 VTC supports coding of arbitrary shaped objects by means of a shape-adaptive discrete wavelet transform (DWT), but does not support lossless coding. Several objects can be encoded separately and then composited at the decoder to obtain the final decoded image.

JPEG. There are several modes defined for JPEG, including baseline, lossless, progressive, and hierarchical. Baseline mode is the most popular and supports lossy coding only. Progressive mode encodes the quantized coefficients by a mixture of spectral selection and successive approximation. The lossless mode is based on a predictive scheme and Huffman coding [108].

JPEG-LS. This is the ISO/ITU-T standard for lossless coding of still images [103]. It also provides for near-lossless compression. It is based on adaptive prediction and context modeling. Near-lossless compression is achieved by allowing a fixed maximum sample error. This algorithm was designed for low complexity, while providing lossless compression ratios. However, it does not provide support for scalability, error resilience, or any such functionality.

PNG. Portable network graphics (PNG) is a World Wide Web consortium recommendation for coding of still images. It is based on a predictive scheme and entropy coding [104]. PNG is capable of lossless compression only and supports grayscale, paletted color and true color, an optimum alpha plane, interlacing, and other features.

JPEG2000. The JPEG2000 is a standard for still image compression. It is not only intended to provide rate distortion and subject image quality performance superior to existing standards but also to provide functionalities that current standards can either not address efficiency or not address at all [109]. The compression advantages of JPEG2000 are a direct result of the inclusion into the standard of a number of advanced and attractive features, including progressive recovery, lossy/lossless compression, and region of interest capabilities. These features lay the foundation for JPEG2000 to provide tremendous benefits to a range of industries. Some of the applications that will benefit directly from JPEG2000 are image archiving,

Internet, Web browsing, document imaging, digital photography, medical imaging, and remote sensing. Fundamentally, JPEG2000 includes many advanced features, such as

- compression precision of 1 to 16 bits/sample (signed or unsigned)
- components that may each have a different precision and subsampling factor
- use of image data that may be stored compressed or uncompressed
- lossy and lossless compression
- progressive recovery by fidelity or resolution
- tiling
- error resilience
- region-of-interest coding
- random access to an image in a spatial domain
- security.

Image compression must not only reduce the necessary storage and bandwidth requirements, but also have functionalities for editing, processing, and targeting particular devices and applications.

REFERENCES

1. W. C. Lee, M. G. Hluchy, and P. A. Humblet, Routing subject to quality of service constraints in integrated communication networks, *IEEE Network Mag.*, 9, 46–55 (1995).
2. T. Koinuma and N. Miyaho, ATM in B-ISDN communication systems and VLSI realization, *IEEE J. Solid State Circuits*, 30, 341–347 (1995).
3. I. Namiaki and T. Kuneda, Broadband network applications, *NTT Review, Special Features (2)*, 8, 62–67 (1996).
4. M. Tatipamula and B. Khasnabish (Eds.), *Multimedia Communication Networks: Technologies and Services*, Artech House, Norwood, MA, 1998.
5. M. Smith, A model of human communication, *IEEE Comm. Magazine*, 26, 5–14 (1988).
6. R. Steinmetz and K. Nahrstedt, *Multimedia: Computing, Communications and Applications*, Prentice Hall, Englewood Cliffs, NJ, 1995.
7. I. Postel, *Internet protocol*, IETF RFC791, September 1981.
8. S. Deering and B. Hinden, *Internet protocol, version 6 specification*, IETF RFC1883, January 1996.
9. L. Zhang et al., RSVP a new resource reservation protocol, *IEEE Network Mag.*, 7, 8–18 (1993).
10. R. Braden et al., *Resource reservation protocol (RSVP) Version 1, Functional specification*, IETF RFC2205, September 1997.
11. M. Decina and V. Trecordi, Convergence of telecommunications and computing to networking models for integrated services and applications, *Proc. of the IEEE*, 85, 1887–1914 (1997).

12. D. G. Messerschmitt, The convergence of telecommunications and computing: What are the implications today?, *Proc. IEEE*, 84, 1167–1186 (1996).
13. A. Iwata et al., ATM connection and traffic management schemes for multimedia internet working, *Comm. ACM*, 38, 72–89, (1995).
14. F. Fluckiger, *Understanding Networked Multimedia: Applications and Technology*, Prentice Hall, Englewood Cliffs, NJ, 1995.
15. K. R. Rao and Z. S. Bojkovic, *Packet Video Communications Over ATM Networks*, Prentice Hall PTR, Upper Saddle River, NJ, 2000.
16. S. Carl-Mitchell, The new Internet protocol, *UNIX Review*, 13, 31–34 (1995).
17. A. Sinha, Client–server computing: current technology review, *Comm. ACM*, 35, 77–98 (1992).
18. S. Broadhead, Client–server: the past, present and future, *Network Computing*, 4, 38–43 (1995).
19. C. A. Dahlbon and J. S. Ryan, Common channel interoffice signaling: history and description of a new signaling system, *Bell Sys. Tech. J.*, 57, 225–250 (1978).
20. R. Walters, *Computer Telephony Integration*, Artech, London, UK, 1993.
21. P. Strauss, Welcome to client–server PBX computing, *Datamation*, 40, 49–52, (1994).
22. P. A. Bernstein, Middleware: a model for distributed system services, *Comm. ACM*, 39, 86–98 (1996).
23. O. Etziom and D. S. Weld, Intelligent agents on the Internet: fact, function and forecast, *IEEE Expert*, 10, 44–49 (1995).
24. V. Vittore, Intelligent agents may jump start PDAs, *American's Network*, 98, 28–29 (1994).
25. ISO/IEC Standard DIS 13522-5, *MHEG-5*, December 1995.
26. C. M. Woodruft and R. Kositpaiboon, Multimedia traffic management principles for guaranteed ATM network performance, *IEEE J. Selected Areas in Comm.*, 8, 437–446 (1990).
27. K. Challapalli et al., The grand alliance system for US HDTV, *Proc. IEEE*, 83, 158–174 (1995).
28. D. W. Lin, C. T. Chen, and T. R. Hang, Video on phone lines: technology and applications, *Proc. IEEE*, 83, 175–193 (1995).
29. J. Wakeman et al., Implementing real time packet forwarding policies using streams, *Proc. Usenix Conf*, pp 71–82, New Orleans, LA, January 1995.
30. P. P. White and J. Crowcroft, The integrated services in the Internet: state of the art, *Proc. IEEE*, 85, 1934–1946 (1997).
31. S. Shanker et al., "Pricing in computer networks: reshaping and research agenda," in L. W. McKnight and J. P. Bailey, Eds., *International Economics*, MIT Press, Cambridge, MA, 1996.
32. D. E. Comer, *Interworking with TCP/IP*, vols. 1–3, Prentice Hall, Englewood Cliffs, NJ, 1995.
33. S. A. Thomas, *IPNG and the TCP/IP Protocols*, Wiley, New York, NY, 1996.
34. B. Furth, Multimedia systems: an overview, *IEEE Multimedia Magazine*, 1, 47–59 (1994).
35. S. R. Ahuja and J. R. Eksor, Coordination and control for multimedia conferencing, *IEEE Comm. Magazine*, 30, 58–43 (1992).

36. S. M. Grimmius, Analysis for video conferencing on a token ring local area network, *Proc. ACM Multimedia 93*, pp. 301–310, ACM Press, New York, 1993.
37. W. J. Clark, Multipoint multimedia conferencing, *IEEE Comm. Magazine*, 30, 44–50 (1992).
38. S. Ramanathan et al., *Optimal communication architecture for multimedia conferencing in distributed systems*, Tech. Report No. CS91-213, University of California, San Diego, Computer Science and Engineering Department, October 1991.
39. P. J. Smith et al., Tandem-free VoIP conferencing: a bridge to next-generation networks, *IEEE Comm. Magazine*, 41, 136–145 (2003).
40. ITU-T Recommendation H.323, Packet-based multimedia communication systems, November 2000.
41. N. K. Burns, P. K. Edholm, and F. F. Siniard, *Apparatus and method for packet-based media communications*, Canadian patent application no. 2,319,655, June 2001; U.S. patent application no. 09/475,047 December 1999.
42. R. Rabipour and P. Coverdale, Tandem-free VoX conferencing, internal memo, Nortel Networks, Montreal, Canada, August 1999.
43. B. Li and J. Lin, Multimedia video multicast over the Internet: an overview, *IEEE Network*, 17, 24–29 (2003).
44. S. Viswanathan and T. Imielinski, Metropolitan area video-on-demand service using pyramid broadcasting, *Multimedia Systems*, 4, 197–208 (1996).
45. L. S. Juhn and L. M. Tseng, Harmonic broadcasting for video-on-demand service, *IEEE Trans. Broadcasting*, 43, 268–271 (1997).
46. K. A. Hua and S. Shen, Skyscraper broadcasting: a new broadcasting scheme for metropolitan video-on-demand systems, *Proc. SIG-COM*, pp. 89–100, Cannes, France, September 1997.
47. A. Dan, K. Sitaran, and P. Shalabuddin, Dynamic batching polices for an on-demand video server, *Multimedia Systems*, 4, 112–121 (1996).
48. L. Golubchik, C. S. Lui, and R. R. Muntz, Adaptive piggy-backing: a novel technique for data sharing in video-on-demand storage servers, *Multimedia Systems*, 4, 140–155 (1996).
49. J. Y. B. Lee, UVoD – a unified architecture for video-on-demand services, *IEEE Comm. Lett.*, 3, 277–279 (1999).
50. S. Ramesh, J. Rhae, and K. Greo, Multicast with cache: an adaptive zero-delay video-on-demand services, *IEEE Trans. CSVT*, 11, 440–456 (2001).
51. J. B. B. Lee, On a unified archtecture for video-on-demand services, *IEEE Trans. Multimedia*, 4, 38–47 (2002).
52. T. Taleb, N. Kato, and Y. Nemoto, Neighbors-buffering-based video-on-demand architecture, *Signal Processing: Image Comm.*, 18, 515–526 (2003).
53. V. O. K. Li, Performance model of interactive video-on-demand systems, *IEEE J. Selected Areas Comm.*, 14, 1099–1109 (1996).
54. K. C. Almeroh, The use of multicast delivery to provide a scalable and interactive video-on-demand services, *IEEE J. Selected Areas Comm.*, 14, 1102–1122 (1996).
55. L. Golubuchik and Y. C. S. Lui, Adaptive piggybacking: a novel technique for data sharing in video-on-demand storage servers, *Multimedia Systems*, 4, 140–155 (1996).

56. B. G. Haskell et al., Image and video coding – emerging standards and beyond, *IEEE Trans. CSVT*, 8, 814–837 (1998).
57. L. Chiariglione, The challenge of multimedia standardization, *IEEE Multimedia*, 4, 79–83 (1997).
58. N. Jayant, Signal compression: technology targets and research directions, *IEEE J. Selected Areas Comm.*, 10, 796–818 (1992).
59. M. Yamashita, N. D. Kenyon and S. Okubo, Standardization of audiovisual systems in CCITT, *Proc. IMAGECOM*, 42–47, Bordeaux, France, November 1990.
60. ITU-T Rec. H.320, Narrow-band ISDN Visual telephone systems and terminal equipment, 1996.
61. M. Yamashita, N. D. Kenyon, and S. Okubo, Standardization of multipoint audiovisual systems in CCITT, *Proc. IMAGECOM*, 154–159, Bordeaux, France, March 1993.
62. S. Okubo, Reference model methodology – a tool for the collaborative creation of video coding standards, *Proc. of the IEEE*, 83, 139–150 (1995).
63. ITU-T Rec. H.261, Video codec for audiovisual services at px64 kb/s, 1993.
64. S. Okubo et al., Hardware trials for verifying Recomm. H.261 on px64 kb/s video codec, *Signal Processing: Image Communication*, 3, 71–78 (1991).
65. S. Okubo et al., ITU-T standardization of audiovisual communication systems in ATM and LAN environments, *IEEE J. Selected Areas Comm.*, 15, 965–982 (1997).
66. L. Chiariglione, Impact of MPEG standards on multimedia industry, *Proc. IEEE*, 86, 1222–1227 (1998).
67. K. R. Rao, Z. S. Bojkovic, and D. A. Milovanovic, *Multimedia Communication Systems: Techniques, Standards and Networks*, Prentice Hall PTR, Upper Saddle River, NJ, 2002.
68. ITU-T Rec. H.324, Terminal for low bit rate multimedia communication, 1996.
69. ITU-T Rec. H.263, Video coding for low bit rate multimedia communications, 1996.
70. ITU-T Rec. G.723.1, Dual rate speech coder for multimedia communication transmitting at 5.3 and 6.3 kb/s, 1996.
71. ITU-T Rec. H.223, Multiplexing protocol for low bit rate multimedia communication, 1996.
72. M. Y. Willebeek-LeMair and Z. Y. Shae, Videoconferencing over packet-based networks, *IEEE J. Selected Areas Comm.*, 15, 1101–1114 (1997).
73. ITU-T Rec. G.114, *One-way transmission time*, 1993.
74. ITU-T Rec. H.310, Broadband audio-visual communications systems and terminal equipment, 1995.
75. ISO/IEC 8802-5 (ANSI/IEEE Std. 802.5 – 1992), Information technology – local and metropolitan area networks – Part 5: Token ring access method and physical layer specifications, 1992.
76. IEEE Standard 802.1i, Local area network MAC bridges-fiber distributed data interface (FDDI), 1992.
77. ITU-T Rec. H.261, Video codec for audiovisual services at px64 kb/s, Geneva 1990, revised in Helsinki, March 1993.
78. K. R. Rao and P. Yip, *Discrete Cosine Transform*, Academic Press, New York, NY, 1990.

79. A. N. Netravali and J. D. Robinson, Motion-compensated television coding: Part 1, *Bell Systems Technical J*, 58, 631–670 (1979).
80. ITU-T Rec. H.320, Narrow-band visual telephone systems and terminal equipment, March 1996.
81. ITU-T Rec. H.263, Video coding for low bit rate communication, March 1996.
82. ITU-T Rec. H.324, Terminal for low bit rate multimedia communications, 1995.
83. ITU-T Draft. Rec. H.263 Version 2, H.263+ Video coding for low bit rate communication, September 1997.
84. ITU-T, Study Group 16, Video coding experts group (Question 15), Doc.Q15F09, Report of the ad hoc committee H.263++ development, Seoul, Korea, November 1998.
85. ITU-T, Study Group 16, Video coding experts group (Question 15), Doc.Q15D62, Recommended simulation conditions for H.263v3, Tampere, Finland, April 1998.
86. ITU-T, Study Group 16, Video coding experts group (Question 15), Doc.Q15D65, Video codec test model, Near-term, version 10 (TMN10), Draft 1, Tampere, Finland, April 1998.
87. I. E. G. Richardson, *H.264 and MPEG-4 Video Compression*, Wiley, Hoboken, NJ, 2003.
88. R. V. Cox and P. Kroon, Low bit-rate coders for multimedia communications, *IEEE Comm. Magazine*, 34, 34–41 (1996).
89. ITU-T Rec. H.222.1, Multimedia multiplex and synchronization for audiovisual communication in ATM environments, 1996.
90. ITU-T Rec. H.221, Frame structure for a 64–1920 kb/s channel in audiovisual teleservices, 1997.
91. ITU-T Rec. H.223, Multiplexing protocol for low bit rate multimedia communication, 1997.
92. A. Puri, Video coding using the MPEG-1 compression standard, *Proc. International Symp. of Society for Information Display*, pp. 123–126, Boston, MA, May 1992.
93. A. Puri, Video coding using the MPEG-2 compression standard, *Proc. SPIE Visual Communications and Image Processing*, pp. 1701–1703, 1993.
94. ISO/IEC JTC1/SC29/WG11 N2459, R. Koenen et al. (Eds.), *Overview of the MPEG-4 Standard: Requirements, Audio, Delivery, SNHC, Systems, Video and Tests*, Atlantic City, October 1998.
95. B. G. Haskell, A. Puri, and A. N. Netravali, *Digital Video: An Introduction to MPEG-2*, Chapman and Hall, New York, NY, 1997.
96. ISO/IEC JTC1/SC29/WG11 N2456, MPEG-4 Requirements Group, MPEG-4 Requirements document version 10, December 1998.
97. ISO/IEC JTC1/SC29/WG11 N4582, MPEG-7 Requirements Group, Report of ad hoc group on MPEG-7 evaluation logistics, Seoul, March 1999.
98. ISO/IEC JTC1/SC29/WG11 N2728, MPEG-7 Requirements Group, MPEG-7 applications document, Seoul, March 1999.
99. ISO/IEC JTC1/SC29/WG11 N2724, MPEG-7 Requirements Group, MPEG-7 context, objectives and technical roadmap, Seoul, March 1999.
100. ISO/IEC N4041S, MPEG-21 Overview, Singapore, March 2001.
101. ISO/IEC 14496–2, Information technology – Coding of audio visual object – Part 2: Visual, December 1999.

102. ISO/IEC N4824, New options in radix-255 arithmetic coder, March 1997.
103. ISO/IEC 14495-1, Information technology – Lossless compression of continuous-tone still images: Baseline, December 1999.
104. W3C, PNG (Portable Network Graphics) Specification, October 1996.
105. ISO/IEC 11544-1, Information technology, Coded representation of picture and audio information – Progressive bi-level image compression, March 1993.
106. ISO/IEC FCD14492, Information technology, Coded representation of picture and audio information-Lossy/lossless coding of bi-level images, July 1999.
107. D. Santa-Cruz and T. Ebrahimi, An analytical study of JPEG2000 functionalities, *Proc. IEEE ICIP* 2, 49–52, Vancouver, Canada, September 2000.
108. W. B. Pennebaker and J. L. Mitchell, *JPEG Still Image Data Compression Standard*, Van Nostrand Reinhold, New York NY, 1993.
109. A. N. Skodras, C. Christopoulos, and T. Ebrahimi, JPEG2000: the upcoming still image compression standard, *Proc. 11th Portuguese Conference on Pattern Recognition*, pp. 359–366, Porto, Portugal, May 2000.
110. J. Y. B. Lee and C. H. Lee, Design, performance analysis, and implementation of a super-scalar video-on-demand system, *IEEE Trans. CSVT*, 12, 983–997 (2002).
111. W. F. Poon, K. T. Lo, and J. Feng, Determination of efficient transmission scheme for video-on-demand (VoD) services, *IEEE CSVT*, 13, 188–192 (2003).
112. G. Makimaran and P. Mahapatra, Multicasting: an enabling technology, *IEEE Network*, 17, 6–7 (2002).
113. M. Tatipamula and B. Khanabish (Eds.), *Multimedia Communication Networks, Technologies and Services*, Artech House, Boston, London, 1998.
114. ITU-T, Study Group 16, Video coding experts group (Question 15), Doc.Q15F10, Report of the ad hoc committee H.26L development, Seoul, Korea, November 1998.
115. K. Andre, C. Chakrabarti, and T. Acharya, A high performance JPEG2000 architecture, *IEEE Trans. CSVT*, 13, 209–218 (2003).

3

FRAMEWORKS FOR MULTIMEDIA STANDARDIZATION

Global standardization is becoming more and more important in achieving a global information infrastructure (GII). Telecommunication and information standardization organizations such as ITU and ISO/IEC have each set up special groups to focus on determining standards for GII. This chapter seeks to provide the activities of these organizations as they relate to GII. It also reviews the current status of standards for multimedia communications. Key achievements toward deployment of the GII are described. After that, the Internet Engineering Task Force (IETF) is presented as a standardization body focused on the development of protocols used on IP-based networks. It is shown that IETF consists of many working groups and is managed by the Internet Engineering Steering Group, Internet Architecture Board, and Internet Society. We continue with ETSI standardization process for application, middleware, and networks. We also introduce the MEDIACOM2004 to establish a framework for multimedia standardization for use both inside and external to the ITU. This framework will support the harmonized and coordinated development of the global multimedia communication standards across ITU-T and ITU-R study groups. Next, we describe the MPEG-21 multimedia framework standard with strong emphasis on electronic commerce and protection of media content. Finally, we give an overview of industrial fora and consortia, including the ATM Forum, Telecommunication Information Network Architecture (TINA), International Multimedia Telecommunications Consortium (IMTC), MPEG-4 Industry Forum (M4IF), Internet Streaming Media Alliance (ISMA), Third Generation Partnership Project (3GPP/3GPP2), Digital Video Broadcasting (DVB), Digital Audio

Introduction to Multimedia Communications, By K. R. Rao, Zoran S. Bojkovic, and Dragorad A. Milovanovic
Copyright © 2006 John Wiley & Sons, Inc.

Visual Council (DAVIC), Advanced Television Systems Committee and (ATSC), and European Broadcasting Union (EBU).

3.1 INTRODUCTION

A long time ago, telecommunications standardization was the province of international organizations such as the International Telecommunications Union (ITU), International Organization for Standardization (ISO), International Electrotechnical Committee (IEC), and Joint Technical Committee 1 (JTC1). Now, these activities are also being addressed by regional standardization bodies such as the European Telecommunications Standards Institute (ETSI), T1 Committee, and Telecommunication Technology Committee (TTC). Coordination among these bodies is important to ensure that resources are not wasted in getting to the final goal, that is, global standards. Such coordination activities were informally started with the formation of the International Telecommunications Standards Conference (ITSC) in 1989. The ITSC was reorganized as the Global Standards Collaboration (GSC) Group in 1994 and is continuing its activities. In areas of high interest, such as asynchronous transfer mode (ATM) and multimedia, fora and consortia have been established and have achieved most of their objectives very quickly. Thus, the international organizations are reengineering themselves so that they can act more quickly [1].

Telecommunications liberalization, the opening of local and long-distance market competition in the mid-1980s in the United States, Europe, and Asia, together with technical advances in digital communications, mobile communications, and global positioning systems since then, has resulted in significant changes in our socioeconomic and technical environments. The technologies and industries of telecommunication, information, entertainment, and consumer electronics are converging through digitization and bringing about of multimedia applications in these areas. In such new environments, standards that are being developed by industry groups and fora/consortia are becoming more and more important. In general, industry standards are developed very smoothly. Standardization organization such as ITU-T, ISO/IEC, and JTC1 are now required to move more quickly. The ITU-T has established collaborative relationships with relevance for consortia in the appropriate areas [2].

Global Information Infrastructure (GII) is a focal point of the converging technologies and industries, and will provide various multimedia applications. Global interconnectivity and interoperability is one of the key issues of GII and requires global standards. The ETSI, T1 Committee, and TTC of Japan established special committees for studying standards aspects of multimedia communications and GII in 1994. ITU-T formed the Joint Rapporteurs Group on Global Information Infrastructure (JRG-GII) in 1995. The ISO/IEC JTC1 established a special working group on GII (SWG-GII) in 1996. Also, there have been many joint activities related to GII standards for identifying the issues to be covered and establishing harmonious relationships among organizations for international, regional, and domestic standards.

Multimedia communication straddles telecommunications, information processing, and broadcasting technologies. As a consequence, information technology (IT) deals with the *preparation, collection, transport, retrieval, storage, access, presentation*, and *transformation* of signals in many forms. These forms include speech, audio, and video signals, graphics, texts, still images, video, data, and their combinations. Digital information is now accessible and tradable in customized formats, while the end users of information systems can be people, machines or a combination of both. The standardization of telecommunications and IT needs to extend beyond technical details to include business perspectives and management interests. Producers attempt to balance the desire for standards with the need to protect their markets. Service providers seek to reduce their costs through standardized products. Public authorities try to encourage competitiveness and protect the common good. In what follows, we will try to cover many angles of technical and socioeconomic aspects of standards, and management of standardization. Also, we will represent a wide range of viewpoints and highlight the challenges in making fair, transparent, and open standards. The starting point will be the fact that standards are technical products that demand a sophisticated level of strategic planning, extensive technical preparation, and superb interpersonal skills, particularly at the international level.

3.2 STANDARDIZATION ACTIVITIES

Standards are the only realistic means of maintaining compatibility in an increasingly complex multimedia environment. Success in standardization requires an understanding of the general environment in which it takes place. The development of communication systems (mobile telephony, Internet, multimedia) requires new specific agreements among the parties involved: equipment manufacturers, network operators, service providers, and end users. Many new systems tend to merge elements from telecommunications, information processing, and broadcasting technologies. Today, the majority of standards are adopted during, and sometimes before, product design. As a consequence, not only are many more specifications needed, but also the time intervals for production are becoming shorter. This had led first to an increase in the number of national and regional standard bodies, then to a dramatic increase in the number of consortia. Standardization, however, demands time and resources to understand what really needs to be standardized and how to achieve harmonization among the conflicting interests. Successful transfer of knowledge from research to practical applications usually requires experience gained through trial implementations. If the standard is about new concepts, urgency can distract from aspects that may not be apparent at the onset [3].

A framework for the standardization processes in telecommunication and information technology can help address issues of strategy and tactics. The strategic factors are reasons for seeking a standard, interfaces to be standardized, and time to standardize. On the other hand, tactical considerations relate to the following questions: How will standardization take place? Who will be conducting the standardization activity? And where will the standard be used? [4–7].

3.2.1 Reasons for Seeking a Standard

Standardization is a business decision. In the case of mature technologies, a commercial organization will accept and use standards if they help expand the market. However, standardization may be a way to serve the public interest, or for entrants to oppose the dominant firms. Standardization may also help legitimize a new technology and allow the organization to enable a new technology to gain a central position. From the user side, standards ensure the availability of a component or equipment from several sources, particularly if they allow mixing and matching of products from several suppliers.

The decision not to standardize has two implications. The producer may be seeking a unique and possibly controllable market. Pressure to standardize suggests that products and consumers seek some stability for the long term. For example, the global market for smart cards in electronic commerce requires a series of standards for the operating systems, commands, and interfaces, and so forth to encourage the development of applications.

3.2.2 Interfaces to be Standardized

Figure 3.1 illustrates the interfaces to be organized in a layered format. They include standards for reference, similarity, compatibility and flexibility. In addition, standards for performance and quality can be at all layers.

Reference standards provide measures to describe general entities in terms of reference units. The classical example is the unit standards for measurable physical qualities (ohm, volt, watt, dBm). Other examples in the information and telecommunications fields include the ASCII character set, various standardized high-level computer languages, open system interconnect (OSI) model, and addressing plan for international telephone service and the Internet.

Similarity standards define aspects that have to be identical on both sides of the communicating link as well as the allowed variations or tolerances, if any. Similarity is essential to establish the linkage successfully. Examples include the normal values of signal levels, the shapes (or masks) of current pulses, source coding algorithms for interactive speech and video, algorithms for live coding transmission links, encryption algorithms, and computer operating systems.

Compatibility standards coordinate the production of flows at the transmitting end and their reception at the other end. These standards are referred to as profiles, functional standards, interface templates, user agreements, or implementation agreements. Profiles, functional standards, or interface templates designate a fixed set of options to perform a given service. For example, architectural standards show the many possible arrangements of components and building blocks to achieve a specific function at both sides. Another example consists of the interface protocols

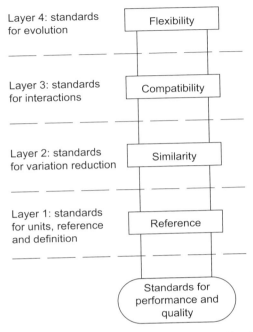

Figure 3.1 Technical standards with layered architecture.

that define common blocks the transmitter and receiver pairs must have to ensure successful communication, even though both sides do not have to be identical.

Flexible standards focus on compatible heterogeneity, the capability of a single platform to interoperate with different systems, and its upward/downward compatibility. For example, the negotiation procedures used in the modem recommendations of the ITU-T enable the exchange of user data with legacy modems through common ground rules.

3.2.3 Time to Standardize

There are two models for the timing of standards: a descriptive model and a predictive model.

The descriptive model describes the timing relationship between the standard and the product cycle. It helps us understand the *a posteriori* relationship of the standard to market development, but does not guide in making decisions regarding the timing or details of standardization. The position of the standard in the product cycle is shown in Figure 3.2. With respect to the product of service life cycle, standards can be anticipatory, participatory, or responsive. Anticipatory standards are those standards that are essential for widespread acceptance of a device or service. They are crucial to interoperable communication systems, which explain why

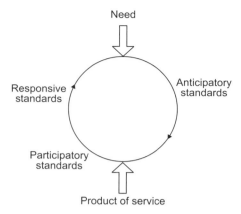

Figure 3.2 The position of the standard in the product cycle [3]. (©2001 IEEE.)

there are many anticipatory standards (e.g., V.32 mode, X.25 packet interface, ISDN, TCP/IP, Recomm. H.323 for multimedia terminals, the Secure Socket Layer (SSL) for electronic commerce). The best anticipatory standards must have well-defined scope and objectives and offer a minimum set of features to stimulate the market. Participatory standards proceed in lock step with implementations to test the specifications before adopting them. Some participatory standards are the speech and voice coding algorithms of ITU Recomm. G.726, G.724, G.728, and G.729. The various Internet applications above the TCP layer may also be viewed as participatory standards (e.g., MBONE, MMUSIC). This incidental benefit can be an important factor in spurring incremental innovation (for example the development of the G.728 ITU-T 16 kb/s speech coding algorithm led to a major breakthrough in voice coding [4]). Responsive standards codify a product or service that has been sold with the same success, or define the expected quality of a service and performance level of the technology. In such a case, services have already provided sufficient evidence that the technology or market interest justifies the work on such standards. Responsive standards offer a systematic way to distill scientific information and available data into useful technical products. They expedite the consolidation of knowledge and provide avenues for sharing technical know-how. Some examples of responsive standards are the V.90 modem, the various methods for the evaluation of voice quality through objective and subjective means (e.g., E.510, G.120, P.84), and the measures for the overall quality of services (e.g., E.800, X.140).

The predictive model matches the timing of standards to the intrinsic capabilities of the technology using the technology S-curve. Timing of standards in relation to the technology S-curve is shown in Figure 3.3. The ordinate axis represents the technology performance as perceived by the end user of the technology [3]. The timing of anticipatory standards is at the introduction of the technology. They should specify the production system of the new technology. Participatory standards are

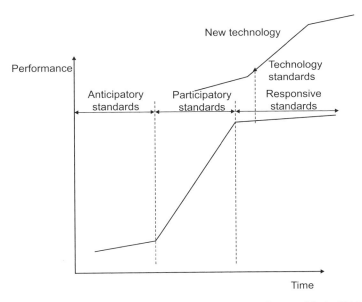

Figure 3.3 Timing of standards in relation to the technology S-curve [3]. (©2001 IEEE.)

generated while the performance of the innovation is improving exponentially, the knowledge of the technology is improving experimentally, the knowledge of the technology is diffused, and initial products are commercialized. Participatory standards may also specify the behavior of application systems to ensure a flexible evolution of the whole ensemble to a new state of the operation. A responsive standard relates to the manifestation of the technology in a service system: a completed and connected set of transformational technology systems used in communicating and transacting operations. Responsive standards mean that the initial innovators will contribute to the standards development while continuing their product development. Finally, it can be seen that standards could facilitate the transition process during a technology discontinuity.

3.2.4 How Will Standardization Take Place?

Firms approach standardization in several ways that can be listed from least to most responsive as follows:

- Do nothing and wait until a trend appears.
- Passive participation in standards bodies to collect the necessary documents and understand the background of the standard.
- Active participation in an existing standard group to collect the necessary information and understand the significance of the specifications, and influence the direction of the standardization process.

- Building or joining alliances, for example to have indirect influence through other participation and to obtain information on the specifications.
- Starting a new consortium and attracting other interested parties to approve or promote a given technology.

Voluntary standards depend on consensus, that is, the possibility of accommodating all technical viewpoints, provided that all parties negotiate in good faith [5].

3.2.5 Who Will be Conducting the Standardization Activity?

Active contribution and participation in the standards debate, which is a significant determination of the standards process, relies on thorough technical knowledge. This leads to some interesting problems, such as

- Who decides the standards strategy; who translates the business objectives into a technical goal?
- Who runs the tactical plan; who determines allies and adversaries and selects and evaluates the performance of the firm's representatives?
- Who determines the amount of information participants are allowed to reveal?

3.2.6 Where Will the Standardization Be Used?

The applicability of a standard can be assumed in terms of market, industry, or geography. Governmental authorities adopted standards and then enforced them over a defined geographic area. The explosion in applications of information technologies has stimulated the proliferation of standards, regional organizations, consortia, and fora.

There are many reasons for the development consortia. Working in closed consortia has the advantage of focusing the work on well-defined projects. Furthermore, it provides companies with more control over the release of proprietary information. As multimedia applications become incorporated in many business processes in various sectors such as banking and health applications, users and manufacturers may join industry groups to harmonize their views and requirements. One possible way to attract attention to a new technology or application is to start a consortium focused on a narrow area. There are many types of consortia: proof of concept consortia, implementation consortia, application consortia.

Proof of concept consortia expedite the transfer of new technologies and concepts to trial implementations. They provide a way to share innovations among researchers associated with competing firms as they develop anticipatory standards. Their task is to distill the results of research investigations and amounts of data into architectural standards and technical reports.

Implementation consortia provide an avenue for a wider dissemination of the technology to an industry sector. Among their activities are development of formal specifications, development of tools, testing implementations for conformance to specifications, and/or promotion of the technology.

Application consortia provide a neutral framework for the negotiations among the various parties by defining user requirements and deployment architectures. Their main focus is on systems, promoting innovations, and producing implementation agreements.

The choice of a particular standard development organization to develop a standard depends on many additional factors, such as:

- The working methods and procedures.
- The structure, composition, and decision-making process of the organization usually indicate the timing of standardization within the product cycle.
- The procedures to avoid blockages that may arise from attempts to assert intellectual property rights (IPR) [8].
- The work load of the organization, its expertize on the subject, resources, and the explicit and implicit power structure.

3.2.7 Technology Cycle and the Product or Service to be Standardized

In an attempt to provide a better understanding of the roles of various organizations, the relation among standard development organizations, product type, and technology is shown in Figure 3.4. It illustrates the correspondence between the standardization and the particular phase of the innovation cycle. When the technology is new and the product is new, small group discussions are more appropriate. In this case, proof of concept consortia seem to be most suitable for discussion. This was demonstrated by the success of TCP/IP standardization in the early phases of the Internet (late 1970s and early 1980s) in an environment akin to that of proof of concept consortia. As the Internet grew to worldwide acceptance, the Internet Engineering Task Force (IETF) added some aspects of implementations and applications consortia.

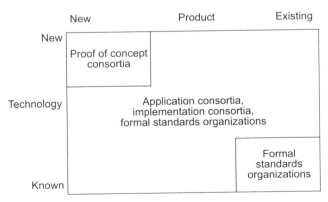

Figure 3.4 A map for the relation between the technology cycle and the product or service to be standardized [3]. (©2001 IEEE.)

Another output of a proof of concept consortium is the development of Common Object Request Broker Architecture (CORBA) under the auspices of the Object Management Group (OMG). On the other hand, format standardization codifies created and diffused knowledge for future reference. In the extreme case, it takes place when the technology is known, the product is already commercialized, and the outcome of market competition has been settled.

3.3 STANDARDS TO BUILD A NEW GLOBAL INFORMATION INFRASTRUCTURE (GII)

Many activities have been directed towards achieving Global Information Infrastructure (GII) since it was first advocated in March 1994. Standardization organizations committed themselves to GII by establishing special committees or groups to identify tasks for achieving GII and to coordinate these tasks with their organization and with the other organizations. In regional and domestic standardization areas, there has also been effort to support National Information Infrastructure (NII) in some regions and countries. In November 1994, the T1 Committee established an *ad hoc* Project Group in its Technical Subcommittee T1P1. This group issued a report that identified key items for functions, and function models are needed to ensure interoperability, interworking, integration, and operation of various media [9].

In Europe, Strategic Review Committee 6 (SRC6) was established in ETSI in September 1994 to define the coordinated work program on standards for the European Information Infrastructure (EII). The report of SRC6, which was approved at the ETSI General Assembly in June 1995, identifies key issues for EII, such as the characterization of EII, conceptual models, fields to be standardized, and management of standardization by ETSI [10]. The High Level Strategy Group (HLSG) was organized by four European information and communication technology (ICT) industry organizations: European Association of Consumer Electronics Manufacturers (EACEM), European Telecommunications and Professional Electronics Industry (ECTEL), European Public Telecommunications Network Operations Association (ETNO), and European Association of Business Machines Manufacturers and Information Technology Industry (EURO-BIT). The idea was to provide total coordination in covering ICT standards.

GII standards are categorizated as follow:

1. *Application areas*: Home entertainment, electronic commerce, interbusiness processs, personal communications, publice service communications, publications (text + multimedia) (ISO TCs, DAVIC, ETSI, InfoTEST, OpenGroup, etc.).
2. *GII services*: Data interchange, graphic, audio, data processing, data management, cost billing, network control (ITU-T, IETF, JTC1 SC6, SC18, SC21, SC24, SC29).
   Building blocks: access methods, addressing, compression, cost, quotation, data navigation, data portability, identification, internationalization, inter-

operable testing, latency control, nomadicity/mobility, priority management, privacy/ownership, quality of service, route selection, search, security.
3. *Service implementation tools*: Communications (broadcast, multicast and unicast); data structures for transport of information; user interaction (ITU-T, IETF, JTC1 SC6, SC18, SC21, SC24, SC29).

The HLSG plays a similar role in the United States. One objective of the HLSG is to identify obstacles blocking the implementation of GII and to give advice to the relevant organizations as to how they can be eliminated. Other objectives include identifying areas that require immediate standardization and to do whatever is necessary to help provide this standardization. The following key projects to facilitate these objectives are identified as

- interoperability of broadband networks
- city information services
- electronic commerce for small and medium-sized enterprises (SMEs).

In Japan, the Telecommunication Technology Committee (TTC) organized the Information Infrastructure Task Group (IITG) in March 1995, under the Strategic Research and Planning Committee. A report issued by IITG clarifies the role of TTC in promising standardization for GII and NII in Japan. The report identifies the basic principles to be incorporated into TTC standards for GII and NII in Japan. A two-layered approach for standards is proposed [11]. Low-layer standards are those for network infrastructure. They should be based on *de jure* standards. High-layered standards are related to application and services. They should primarily adopt the specifications produced by fora and consortia. In order to follow the global GII-related activities and to relate with other standardization organizations, a Special Working Group on Information Infrastructure was established in April 1996, replacing IITG.

In order to coordinate the international, regional, and domestic standardization organizations, the Global Standards Collaboration Groups (GSCG) was established in 1994, originally as the Interregional Telecommunication Standardization Conference (ISC). Members include ITU, T1, ETSI, TTC, the Telecommunication Technology Association (TTA) of Korea, and other standardization organizations. In the second GSC meeting in June 1995, aspects of the GII were discussed. Also, it was agreed that the GII standardization should be developed by ITU-Telecommunications Standardization Sector (ITU-T).

3.3.1 Concept

While visions of the GII vary enormously from country to country, there are several key elements on which they seem to agree. For a start, the networks and services that make up the fabric of the GII are digital. These networks incorporate native intelligence. It means that they depend largely on software for their functionality. Also, they are able to be reprogrammed to dynamically meet changing demands.

The types of services that run over the GII are different. Based on interactive applications, the networks need to carry information in many different formats – no longer just plain voice and data, but images, voice and picture, e-mail, real-time video and audio clips. Finally, through the third generation capabilities offered by wireless access systems, they need to support time, anywhere, anytime mobility.

Telecommunications networks and user equipment are developed and implemented in many different ways around the world, for perfectly legitimate reasons. Different traffic and usage patterns or geographical features mean that particular countries may have very special requirements for their telecommunications networks. However, national networks rapidly lose their usefulness if they cannot incorporate a range of equipment from different manufacturers, and connect with networks and systems in use in other countries around the world.

For the GII to be truly revolutionary in its capabilities, it needs to be much more than simply a bigger or faster telecommunications network. It will need to cope with changing patterns of telecommunications case, support new types of signals and protocols, and, most importantly, address issues of basic access for the world's population. Tomorrow's *Information Superhighway* will be formed as new technologies and applications are integrated into the existing network infrastructure, creating a hybrid network whose old and new elements need to be able to *talk* to one another and exchange information rapidly and efficiently.

3.3.2 Performance Purposes

Evolution of the GII and its performance description are strongly impacted by privatization, deregulation, and internationalization of the telecommunications industry. Around the world, governments and telecommunications industry organizations are opening up markets. Monopoly providers are being subjected to competition. Countering these trends, multinational companies are having success in offering end-to-end services.

The simultaneous fragmentation and consolidation of ownership in the telecommunications industry fundamentally change performance specification and allocation. Even for homogeneous parts of the GII, changes in ownership will create new jurisdictional boundaries within countries where as a practical matter, performance will be allocated by bilateral or multilateral agreements.

The GII will affect the number, placement, and nature of performance description boundaries in many recommendations. The end user interfaces will be extended to incorporate the performance of middleware into the GII. The performance of operating systems, communications protocol implementations, Web browsers, database management tools, e-mail, and file transfer agents will be included in the GII performance. GII performance will also span audio, image, and data bridges, and playback/display tools. The GII will not include application-specific software or the communicated content itself. From the user's point of view, presentation hardware (phones, speakers, microphones, video sources, televisions, sound and video cards) are part of the GII. From the information provider's point of view, hardware and software supporting directory assistance, browsing, and keyword searching

3.3 STANDARDS TO BUILD A NEW GLOBAL INFORMATION INFRASTRUCTURE (GII)

functions are GII components. GII reference connections should include the new hardware elements identified above and their middleware. Bounding the GII for performance description purposes, that is, the performance boundary for the GII, is illustrated in Figure 3.5 [12].

The end user boundaries are extended to include middleware with complicated performance specification and measurement by referencing end-system-internal events. The detection and interpretation of such events can be quite difficult when user information is coded and embedded in several nested layers of protocols. Concatenated networks may interwork through either protocol mapping or protocol encapsulation. The interworking points will be natural performance description boundaries if protocol mapping is involved. Performance boundaries will not be used to define encapsulation interfaces unless a jurisdictional transfer occurs in the same point. When protocol conversions take place, it will be necessary to define parameters for end-to-end communications on the basis of different sets of reference events. This could be done by either translating the input reference events to their output counterparts or defining the performance-significant reference events at both interfaces in protocol-independent terms. Protocol and media conversions may complicate the GII performance specification by creating uncertainties in the timing of reference events. When the input information is segmented differently at the output, the precise instant of the output event could be ambiguous.

GII Performance Parameters

The new parameters defined for the GII are largely determined by the new GII functions. Definition of speed, accuracy, and dependability parameters for

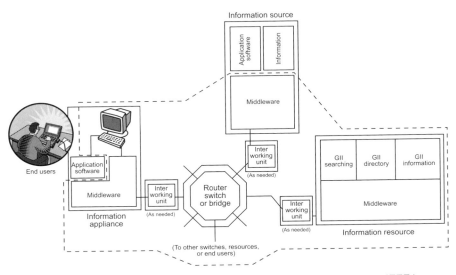

Figure 3.5 A proposed performance boundary for the GII [12]. (©1998 IEEE.)

the new GII functions will follow existing practices. Many proposed GII functions can be represented in the generic 3×3 matrix framework by adding rows.

The elaboration of GII services could make it desirable to define new generic performance criteria, in addition to the outcome-based speed, accuracy, and dependability criteria, to provide a more comprehensive performance specification framework. New generic criteria that have been suggested include security and ease of use. Security will be addressed primarily as a design issue. Where possible, specific security functions will be evaluated in terms of speed, accuracy, and dependability within the 3×3 matrix framework. GII ease-of-use issues should also be described using the objective criteria of speed, accuracy, and dependability, where possible. In some cases, this could be done by applying performance measures to user response functions and rating the GII on the basis of user's performance. Human factors studies could assist in, and possibly suggest ways to automate this process.

The new functions envisioned for the GII may have either discrete or nondiscrete outcomes. New GII functions with discrete outcomes will include new call processing functions, information storage and retrieval, format and media conversions, user authentication, and address translation. Nondiscrete outcomes are graduated or qualitative in some respect. Their results include an array of partial success outcomes, with varying degrees of associated user benefit. GII functions with nondiscrete outcomes include searching and browsing costs, requests for capacity, and importantly, quality of service (QoS) negotiation.

Defining the GII performance parameters, we can invoke primary parameters and availability parameters. The proliferation of technology-specific performance impairments in GII will be addressed by developing objective parameters measured at the customer interface rather than by attempting to quantify each particular technology effect. This will require the protocol independent reference events discussed previously. When the outcomes are nondiscrete, definition of primary parameters will be complex. In general, such parameters will characterize the outcomes in terms of distribution thresholds or statistics.

The GII will affect availability specification by motivating reconsideration of the outage criteria. In conventional definitions, the unavailable state is recognized by the almost complete inability to perform basic functions for a period of time. Traditionally, blocking probability is very low and a high call failure rate is a suitable criterion for declaring an outage. Implementation of handoff protocols and other rapid restorable technologies in the GII will affect availability specification by replacing hard outages with transient events that have a substantially less severe effect on the shortest events and the creation of new parameters to separately count the short interruption events. The development of GII outage criteria should continue to distinguish transient events from events. On the other side, as the duration of automatic restoration moves from minutes to millisecounds, the defined minimum period of unavailability may need to be reduced as well.

3.3.3 Functional Model

The GII functional model has been defined as a basis for format standardization. This model is an abstract description of a system and is developed in such a way that it is independent of any implementation of the system. Its objectives are

- To allow freedom in methods of implementation without affecting the operation of the overall system.
- To allow large-scale functional integration inside one equipment or software module while retaining a manageable and scalable description of the equipment or software module.
- To allow the dynamic creation of services that can be tailored to the needs of users.

The use of the functional model is finding more widespread use in both the telecommunications and IT industries. There are many examples of functional models and functional modeling methodologies, including

- The information and computational viewpoints of the open distributed processing model.
- The use of service-independent building blocks in intelligent networks.
- The layered description of the synchronous digital hierarchy interface description and the use of functional blocks in the equipment specifications.
- The extensive use of layers of application programming interfaces in operating the software support system architecture.

Figure 3.6 illustrates the basic types of functions in the GII and the basic types of logical interface in the GII. These functions are: applications functions, middleware functions, and baseware functions. By logical entities of applications we mean application functions. The entities are usually implemented in software and normally called application objects.

Middleware functions are logical entities of middleware, including service control functions and management functions:

- Service control functions are the middleware functions, which are the logical entities that allow building services from service components and the associated resources, and control the interaction of the user with the service. Sometimes, these functions are associated with session control.
- The middleware functions are the logical entities that perform the management of all other functions.

Baseware functions are the logical entities that allow application and middleware functions to run, communicate with other functions by interfacing with network functions and interface to users. These include network functions, processing and storage functions, and human–computer interfacing functions.

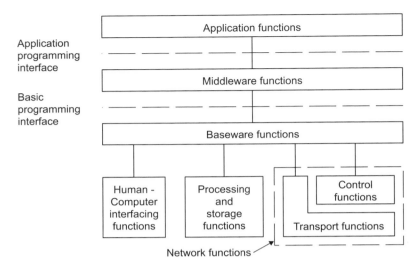

Figure 3.6 Types of GII functions with the basic types of interfaces.

Network functions are the baseware functions that are the logical entities that support communication between separate locations in the GII and include transport functions and control functions.

Processing and storage functions are the baseware functions that are the logical entities that execute middleware and application components and store information.

Human–computer interfacing functions are the baseware functions that are the logical entities that allow application components to present information to and gain input from a human user.

The relationship between functions and the role in which they perform their function are determined by the needs of the role. For example, a communications and networking of information role will require many middleware functions in order to have the resources to be able to offer the range of service components associated with this role.

3.3.4 Implementation Model

An implementation model describes how the functions of the functional model are implemented in equipment and owned by operators. Reasons for developing an implementation model include

- The differentiation and characterization of interfaces that are important for standardization (e.g., those that form interfaces between operators and between equipment from different vendors), thereby enabling priority to be given to the standardization of those interfaces.
- The preparation of a set of examples to illustrate how the system performance can be affected by implementation.

3.3 STANDARDS TO BUILD A NEW GLOBAL INFORMATION INFRASTRUCTURE (GII)

An implementation model shows which functions are implemented in which segment or in which equipment. It also identifies all the protocols passing across an interface between equipment. The GII implementation model involves connecting together a number of implementation components, including

- information appliances
- middleware software modules
- application software modules
- segments of telecommunication networks.

Each of these are segments in the implementation module and are interconnected by interfaces. The interfaces between information appliances and the access segments of the telecommunications network are physical telecommunications interfaces, as are those between access segments and core segments. Other interfaces are either programming interfaces, which are internal to information appliances, or protocols, which are transparent across telecommunications networks. Each segment consists of two functions: control functions and transport functions.

By information appliances, we mean elements that allow users to gain access to the GII and/or can install, invoke, and handle software modules, including those that are databases and video libraries.

By middleware software modules, we mean software modules that contain middleware functions. Middleware software modules run on information appliances.

Application software modules are software modules that contain application functions. Application software modules run on information appliances.

Segments of the telecommunications network connect together information appliances and allow middleware and applications functions installed in different information appliances to communicate with one another. These segments include access segments, core segments, enhanced service provisioning segments, and management segments.

Figure 3.7 presents an example of information appliances connected by a telecommunications network made up of different segments. The telecommunications network segments can each be operated by a different organization and will generally be interconnected as illustrated. Each segment is technology- and implementation-dependent. The delivery of a particular service will require a certain set of GII functions to be implemented in each of these segments.

3.3.5 Scenarios Approach

In some cases the GII will comprise components from different industry sectors (telecommunications, information technology, entertainment). As the convergence between the industry sectors increases, more diverse service delivery technologies are emerging. In order that scenarios can be developed and analyzed in a consistent fashion, a methodology that can be understood by different sectors is required. The provision of various services by a variety of service providers, over a variety of

86 FRAMEWORKS FOR MULTIMEDIA STANDARDIZATION

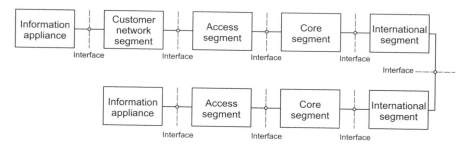

Figure 3.7 Information appliances connected by a telecommunication network.

network technologies, from different industry sectors, makes provision of end-to-end service a significant system integration issue. In order to understand the related standards issues, it is necessary to understand the interdependencies among all components in the system as well as those between immediately adjacent components at a single interface. This need can only be met by examininig the interrelationships among all components within a given scenario. The purpose of a scenario is to enable specific arrangements of certain GII components to be illustrated. Interesting cases occur when technologies from different industries come together and/or where services offered by particular providers over particular technologies are offered by nontraditional providers over nontraditional technologies. Examples of this are telephone service over a cable TV network, or video service over local loop networks. Other interesting cases involve the interconnection of technologies not previously interconnected and/or the nontraditional use of certain appliances.

The primary purposes of a scenario are

- Identification of points that form key interconnection interfaces, access interfaces, or appliance interfaces in a configuration involving a set of providers of services, networks, and appliances.
- Identification of the set of standards that could be applied at each key interface point.
- Identification of the key standards development organizations and/or industry consortia that might desire to be involved in the standards-related system integration issues.

Also, a scenario provides a means to

- facilitate classification of interfaces by type
- facilitate identification of services that can be carried across interfaces
- facilitate classification of services by type
- facilitate identification of endpoints for service delivery

- accommodate a profile of all protocols involved, either directly or indirectly, at a given interface
- document other related issues.

The scenario approach will

- provide a means of checking for completeness of a solution
- facilitate the development of common solutions
- facilitate comparison among solutions
- provide a catalog of standardized solutions to avoid unnecessary reinvention
- help to identify gaps in the standards repertoire
- identify joint interests among standards development organizations and areas where collaboration is required
- facilitate the investigation of interrelationships among all elements depicted in a given scenario.

Example 3.1 In the scenario for the provision of voice/data/video services over existing infrastructures, the downstream channel for delivery of video is achieved from the video server to the customer premises either directly via the satellite or terrestrial broadcast facilities, via the *Bvideo* interface between the video server and the headend of the cable distribution network. Here, *Bvideo* represents interface point notation between video service function and access network.

Example 3.2 A scenario for the use of asymmetrical digital subscriber loop (ADSL) or very-high-rate digital subscriber loop (VDSL) to provide video bandwidth over copper pairs demonstrates the use of these two technologies for delivering video bandwidths. Techniques have been developed for transmitting relatively high bandwidths (1.5–50 MHz) over the existing copper local network. This works only for relatively short distances. Standardized ADSL systems have downstream (to the subscriber) bit rates from 1.536 to 6.144 Mbit/s and upstream rates of 16 to 64 kbit/s. For 2 Mbit/s downstream rate, the range may be as far as 5 km depending on cable gauge, with a reduction of range with increase in bit rate. VDLS is being developed for the range 25–50 Mb/s (downstream), but for much shorter distances (50–500 m). In this case, fiber is used as transport to a convenient cross-connect in the local network, before conversion to copper for the remainder of the connection.

3.3.6 Standardization Efforts

A society can be considered to consist of a socioeconomic part and an industrial part, each with a number of specific roles. The Information Society operates on the roles of and between these two parts. The role of the socioeconomic part is

- establishing lifestyles courtesy, and customs (teleshopping practices, etiquette for a virtual society)
- creating culture and arts (multimedia books, teleorchestration)
- building regulations, laws, and provisions (medical laws in telemedicine).

On the other hand, the role of the industry part can be described as

- creating applications and services (teleshopping, video services)
- building information networks (telecommunications networks, cable television, satellites)
- manufacturing facilities (terminals, transmission systems).

The GII is seen as the means for every citizen to access and be part of the Information Society. It will also encourage economic growth, competitiveness, and sociocultural development. This is followed by advances in support infrastructures such as transportation facilities, communications, and higher level of general and professional education. In the same way, the development of the GII leads to increased and improved person-to-person communications. This facilitates an Information Society in which individuals have secure global access to all kinds of information and services. In fact, real and very significant standardization challenges are posed by each of the three component terms encompassed by the GII: global standards, information, and infrastructure.

Global standards are required for the information and infrastructure components of the GII. The globalization of business, the ease of information access, and the ease of personal mobility ensure that there is no longer any particular national or regional way of doing business.

The purpose of the global infrastructure is to enable users to globally manage the creation, storage, delivery, and use of information. Adequate global standards for the representation of and secure context-specific access to or exchange of information independent of the location of the information provider, and information user are needed to realize the benefits of the GII.

The technological convergence and interconnection of telecommunications equipment, computers, and much of consumer electronics has led to new demands on the communication infrastructure by information providers and users.

A key objective of the GII standardization work is to avoid overlap of work and for the various bodies to collaborate on preparation of the necessary standards. GII is one of the areas identified by the Global Standards Collaboration between the national, regional, and international standards bodies in support of the work of the ITU. These meetings provide a framework for the exchange of information, establishing objectives to accelerate the process of global telecommunications standards development, and promotion of interconnectivity and interoperability.

ITU standardization activities are carried out through its three sectors: Telecommunication Standardization (ITU-T), Radiocommunication (ITU-R), and Development (ITU-D). Recognizing the importance of impartial, globally accepted standards as a building block of the GII, the ITU has initiated a project-based approach to GII standardization and has agreed on the principles and framework architecture of the GII. A fundamental objective of the GII is that it will be a seamlessly interconnected *federation of networks*, which together will provide for end-to-end interoperability of applications and services. Standardization efforts related to the GII must permit new network capabilities and technologies to be introduced.

3.3 STANDARDS TO BUILD A NEW GLOBAL INFORMATION INFRASTRUCTURE (GII)

3.3.7 ITU-T Standardization Sector Program for GII

ITU-T is organized into a number of Study Groups, each responsible for a general area of work. Those involved in GII work in different areas are

- Study Group 2 – network and service operation
- Study Group 3 – tariff and accounting principles, including related telecommunications economic and policy issues
- Study Group 4 – telecommunications management network and network maintenance
- Study Group 7 – data networks and open system communications
- Study Group 8 – characteristics of telematics systems
- Study Group 10 – languages and general software aspects for telecommunication systems
- Study Group 11 – signaling requirements and protocols
- Study Group 12 – end-to-end transmission performance of networks and terminals
- Study Group 13 – general networks aspects
- Study Group 15 – transport networks, systems and equipment
- Study Group 16 – multimedia services and systems.

Study Group 13 has prepared the standards for ISDN and BISDN, and is also responsible for the overall GII broad concepts and principles. In addition, ITU-T Study Group 13 has been appointed as lead Study Group for overall project management, responsible not only for developing its own sets of Recommendations, but for coordinating inputs from other study groups and a wide range of external organizations.

In developing the ITU-T program for GII, the ITU-T is concentrating on four main areas:

- the development of standards to faster interoperability
- the location of sufficient radio frequency spectrum to the new mobile services, which will form integral parts of the infrastructure
- the provision of policy and technical assistance to developing nations
- the establishment of an ongoing, impartial form where government and industry representatives can formulate the strategies needed to turn the dream of a GII into a reality as quickly as possible.

The five technological building blocks of the GII are

- transport technologies
- network technologies
- user interfaces
- multimedia

- radio and satellite systems
- the transport network.

The transport technologies that will underpin information exchange across the GII include X.25, frame relay, ISDN and BISDN, as well as ATM. One of the oldest transport systems still in use is X.25. It will remain an important element of systems where accuracy is vital, such as financial transactions. Frame relay, as defined by ITU-T Recomm. I.233, makes use of existing network technologies to send data at higher speeds by reducing the protective overhead imposed by X.25 needed to guarantee information accuracy. Frame relay finds growing popularity as a bridge between older, slower systems and very fast transmission technologies like BISDN and ATM. ITU study groups continue to finetune the technology and to work on ways of integrating it into the vision of future global networks. ISDN is an ITU developed standard that is finding wider appeal because of its support for higher bandwidth data transfer and improved call quality. Because it excels in transferring large amounts of data very quickly from point to point, it is highly suitable for use in data-intensive applications, videoconferencing, and other multimedia applications. Broadband ISDN is a faster version of the original flavor, supporting standardized data rates of up to 155 Mbit/s over fiber optical networks. Both standards continue to be upgraded within the study groups of the ITU-T. Furthermore, transport capacity has dramatically increased, reaching several gigabits per second. A cell-based fast packet-switching system called asynchronous transfer mode (ATM) is certain to form one of the key elements of the GII. Its ability to move a range of different traffic types along multiple virtual traffic paths at up to several gigabits per second transmission speeds makes it the ideal medium for delivering true multimedia applications – real-time video, very large data files, and new audio and voice-based applications.

The network transmission system known as Synchronous Digital Hierarchy (SDH) is one of the most important elements of the emerging GII. This technology increases the usable bandwidth in telecommunications networks and allows them to be flexibly reconfigured to adapt to changing traffic loads and user demands. Under SDH, network intelligence will be *decentralized* across the networks, which will also be better able to handle different types of the traffic traveling at different speeds. SDH can also greatly improve *network bill* by more efficiently packing traffic into the available bandwidth. The ITU has already developed a wide range of standards for SDH, covering system operation and architectures, the performance and management capability of SDH networks and their interfaces, SDH multiplexing element and international interconnection between SDH and the older networks. Improved network management is also an important consideration because of the increasing complexity of equipment and the need to integrate an ever-widening range of maintenance routines as new types of equipment are added to the network. To address this issue, the ITU standardized Telecommunication Management Network model continues to upgrade to meet the needs of the world's evolving network technologies. The Union is

3.3 STANDARDS TO BUILD A NEW GLOBAL INFORMATION INFRASTRUCTURE (GII)

also working on Human Computer Interfaces to deal with the complex task of managing the large global networks, which will be intrinsic to proper functioning of the GII. Development of these new interfaces will address the need for large amounts of real-time user interaction with the network.

When we speak about the user interfaces to the GII, most people immediately think of the Internet and World Wide Web. Indeed, the growth of Internet use provides an excellent indication of future demand for GII-based services. The ITU is currently involved in a number of projects for reflection and recognition of the importance of Internet Protocol (IP) in today's networks, including telecom networks. Several ITU study groups are now examining aspects of IP-based networks such as multimedia over IP, interaction between Web-based services and Public Switched Telephone Network (PSTN) and ISDN services, end-to-end interoperability and security, billing systems, IP addressing, naming systems and the use of IP in mobile and satellite networks.

Some of the ITU's greatest milestones involve standards developed to facilitate delivery of multimedia traffic – one of the principal components of GII [3]. Under the leading ITU-T Study Group 16, the ITU is currently working in the areas of specification of service requirements, infrastructure components, application programming interfaces (API), and tools and operational procedures needed to bring fast multimedia to the average computer desktop. Standardized sound and image formats such as JPEG and ITU H.262/H.222.0 (corresponding to MPEG) will help application developers to ensure consistent performance and interoperability with products from other manufacturers. In addition, the first version of H.323 of the ITU Recommendation covering voice, data and video traffic over IP networks has met with enormous enthusiasm from the industry, including leading players such as Microsoft and Intel. This crucial standard, which has been reviewed and updated, seems certain to become the foundation of Internet telephony as well as of a wide range of other interactive applications that combine audio, video, and text. The ITU has also developed important standards in all these areas to keep pace with the best new developments from within the industry.

The ITU plays a vital role in management of the world's radio frequency spectrum, an increasingly scarce resource due to the rapid development and popularity of a wide range of mobile communication systems. The development of new kinds of satellite systems will certainly represent an important component of the future global information infrastructure. Global Mobile Personal Communications by Satellite (GMPCS) systems offer not only mobile voice and data services from almost anywhere on the surface of the planet, but also multimedia and so-called *Internet in the Sky* capabilities. GMPCS systems may also be used to deliver broadcasting applications such as interactive television and point-casting. The Union has an ongoing role in developing the GMPCS Memorandum of Understanding, which governs the worldwide implementation of such systems, as well as the future allocation of spectrum for the growing number of GMPCS services.

3.3.8 Future Directions

The ITU-T has agreements with a number of fora and consortia as well as with other standards bodies for the exchange of information, recognition of results, and referencing of their standards in ITU-T recommendations. An important area for collaboration is that of the evaluation of the Internet and its relation to GII. Experts in the various ITU-T Study Groups involved in GII work are increasingly becoming involved in the work of the Internet Engineering Task Force (IETF), which will help ensure that the necessary convergence is achieved.

Telecommunications networks are providing voice and data services worldwide with a high level of reliability and defined QoS and are based on different network technologies with interworking among them. Extension of the networks to include broadband capabilities is based on the ATM technology. Networks based on IP provide a platform that allows users connected to different network infrastructures to have a common set of applications and to exchange data with an undefined QoS. The IP protocol suite is evolving to include voice, data, and video applications with defined QoS. Additionally, terrestrial radio, cable, and satellite networks are providing local broadcast entertainment services and are also evolving to provide interactive voice, data, and video services [3].

3.4 STANDARDIZATION PROCESSES ON MULTIMEDIA COMMUNICATIONS

Audiovisual multimedia will play a major role in many applications of communication services. Multimedia owes much of its growth to high-performance computer and high-speed digital transmission technologies. Consequently, multimedia services are being improved daily with these technical advancements. In order to implement media services, various standardization organizations are collaborating. The international organizations send and receive various liaisons, exchange documents, and accept existing standards and specifications from related fora and consortia in order to speed up and improve the efficiency of standardization activities.

3.4.1 ITU-T Strategies

ITU-T has been continuously adapting its working methods and Study Group (SG) structure in order to maintain the prime role of ITU-T in the development of global standards. The World Telecommunication Standardization Assembly 2000 (WTSA2000) approved the following working method to cope with the new environment.

First, the time for standards approval was decreased. The recommendations are approved by unanimous consensus, with the member states having the right to vote following the Traditional Approved Procedure (TAP) at Study Group meetings or Assembly. A recommendation on the Alternative Approval Procedure (AAP) was approved, which enables four-week approvals in the minimum case, after approval at the level of the responsible Study Group. The AAP will apply to technical deliverables. The private sector plays a key role in producing and agreeing on these technical deliverables. A challenge is to maintain the quality of deliverables. The average interval between Study Group meetings is 10 months, and may be reduced if special projects need approval. Questions are approved if at least five ITU-T members declare their willingness to contribute. The current structure of ITU-T Groups corresponds to GII work, that is, to the program for GII.

A new ITU-T Study Group, Special Study Group (SSG) was established at WTSA2000 to deal with standards on International Mobile Telecommunications in 2000 (IMT-2000) and beyond. SSG not only deals with existing IMT-2000 issues, but also initiates and investigates the service and network capability requirements beyond IMT-2000 in close collaboration with the Third-Generation Partnership Project (3GPP) [13].

ITU-T has traditionally close relations with other standards development organizations (SDOs) such as ISO and IEC. Common texts have been developed and procedures have been agreed upon in order to cooperate, while respecting their different roles. The Telecommunication Standardization Advisory Group (TSAG) reviews priorities, programs, operations, financial matters, and strategies for the ITU-T, follows up on the accomplishment of the work programs, restructures and establishes ITU-T Study Groups, provides guidelines to the Study Groups, advises the ITU-T director, and elaborates A-series Recommendations on organization and working procedures. It established ITU-T Recommendation A.23, *Collaboration with the International Organization for Standardization (ISO) and the International Electrotechnical Commission (IEC) on Information Technology* (1996) about collaboration with other SDOs by exchanging liaisons and referencing relevant text in ITU-T Recommendations.

The Internet Society (ISOC) has been accepted as an ITU-T sector member for close cooperation. ITU Council-99 endorsed ITU-T participation in the Protocol Supporting Organization (PSO) of the Internet Corporation for Assigned Names and Numbers (ICANN). It also endorsed ITU management of the international top-level domain. Following a meeting between Internet Engineering Task Force (IETF) area directors and chairs of ITU-T Study Groups, an analysis was made by the TSAG of the scope of ITU-T Study Groups and related IETF Working Groups (WGs). The results of this analysis are shown in Table 3.1. The numbers represent the numbers of IETF WGs and ITU-T SGs concerned with the same areas. Some ITU-T SGs are counted repeatedly in different areas because ITU-T SGs have a much wider coverage than IETF WGs, while some SGs are related to multiple areas.

Collaboration has also been established with different regional/domestic organizations. For example, Telecommunications and Internet Protocol Harmonization Over Networks (TIPHON) is one of ETSI's projects dedicated to the development of IP telephony. ITU-T has a close relationship with this project, since SG2 gave the global code allocation for IP telephony.

The organizational structure of ITU-T is shown in Figure 3.8.

Table 3.1 The numbers of IETF WGs and ITU-T SGs concerned with the same areas [13]

	IETF Working Groups	ITU-T Study Groups
Applications	17	6
General area	2	2
Internet area	13	7
Operations and management area	17	6
Routing area	12	5
Security area	12	3
Transport area	14	5

3.4.2 ISO/IEC JTC1 Directions in Multimedia Standards

The International Organization for Standardization/International Electrotechnical Commission (ISO/IEC) is a primary source and arbiter of the key standards required for GII and multimedia communications. The range of ISO and IEC deliverables, the quality of the open development ballot processes, and the cooperative concerned work in information technology standardization offer the variety, quality, and assurances essential in a global environment. The challenge facing all standards development organizations is to ensure that business and market needs for standards within their scope are understood and met in an effective and timely manner. Coordination between different players and interests is essential.

In 1987, ISO/IEC Joint Technical Committee 1 (JTC1) was formed to bring together in one committee the information technology standards activities of ISO

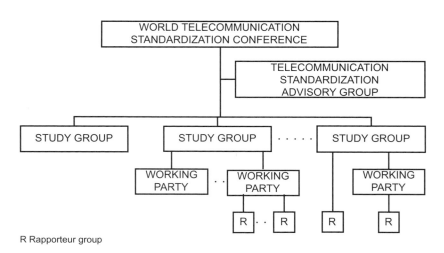

Figure 3.8 Organizational structure of ITU-T [14]. (©2002 ISO/IEC.)

and IEC to ensure greater coordination of this extremely complex and rapidly advancing area of technology standardization. JTC1 immediately established a Technical Study Group (TSG-1) to study the *standards necessary to define interfaces for application probability*. More than 10 years later, many of the issues that led to the TSG-1 Report have been resolved. However, the view points are very different and other issues have surfaced and intensified with the growth in multimedia communications [15].

ISO/IEC technology standards have a major recognized role in world trade. Important benefits are improvement of the suitability of products, processes, and services for their intended purposes, prevention of barriers to trade, and facilitation of technological cooperation. Use of agreed standards increases market size for products and services by providing clear criteria to identify and select from multiple compatible hardware, software, and service providers. However, standards are like other human artifacts – subject to change, growth, and decline. Given the rapid rate and extent of technology developments in digital multimedia and communications technology, standardization of these technologies is complex and has some special characteristics.

The ISO/IEC Approach

An open session on *Multimedia Challenge for International Standardization* was organized and held jointly by ISO/IEC/ITU in 1994. This was the first time that the three apex formal international standards organizations held such a joint session. Throughout the next sessions, the following key principles consistently emerged, and have been used to redirect subsequent activities in the standards development process and community: business and market, cooperation, participation, and program management.

The ISO is a nongovernmental organization that performs standardization in industrial fields other than electricity and electronics. The ISO has technical committees (TCs) and bodies from different countries. The IEC performs standardization in the industrial fields of electricity and electronics. It is also a nongovernmental organization, consisting of national committees, each with a representative from each member country. The IEC has TCs including the International Special Committee on Radio Interference. In 1996, the IEC undertook a large-scale restructuring to speed up its decision-making process to keep pace with the increasing speed of technical innovation, to simplify its management of technology, and to create a decision-making organization that can execute high-level judgments related to assessment conformity. To speed up the process of international standardization, the ISO and IEC adopted a *fast-track procedure* in which the examination is eliminated for standards that have been accepted by other organization. They also created the *technical report* to provide early disclosure of important technical information when the draft for an international standard is not accepted during the approval stage when the standard is still technically in the development stage so it is too early for international standardization, or when data is collected that differ from that in the standard.

ISO/IEC Technical Committees produce international standards using a process with some stages. Each stage is distinct, and progression between the stages is tracked and is conditional on the successful accomplishment of specific criteria. If a document with a certain degree of maturity is available at the start of a standardization project, for example, a standard or specification developed by another organization, it is possible to omit certain stages. For a *fast-track* submission where the document has been developed by an internationally recognized standardizing body, the document is submitted directly for approval as a Draft International Standard. The operation of ISO/IEC standards committees is carefully regulated by directives used by ISO/IEC [16].

JTC1 Standardization Project

In 1987, Joint-Technical Committee (JTC1) of the ISO/IEC was set up by integrating the technical committees of the ISO and IEC to eliminate the overlap of activities between these two organizations in the field of information standardization. JTC1 performs standardization in the industrial fields of information technology. Each member of JTC1 is a *national body* representing a country. Its work is divided among subcommittees. To meet the needs of the multimedia era, JTC1 is streamlining its operations. It has implemented a fast-track procedure, by which *existing standards* can be directly voted on for acceptance as *draft international standards*. The International Standardization Profile (ISP), a technical standards document, was introduced for easier implementation. The Publicly Available Specification (PAS) process was introduced for adopting de facto standards that have the same open characteristics as standards actually distributed in the information area. The aim of this process is to speed up development of international standards.

JTC1 has established several Technical Directions (TDs) as outlined in Figure 3.9. Technical Directions (TDs) are synergistic groupings of JTC1 Subcommittees (Sci) and Working Groups in areas considered to be strategic and market-relevant. JTC1 TDs are not organizational entities. They are a conceptual tool used to identify specific areas of importance for both standardizers and users. JTC1 defines TDs and may create, redefine, or eliminate them as external circumstances demand. JTC1 National Bodies and SCs may propose changes to the title or description of existing TDs and propose new directions for consideration and approval by JTC1.

Dialog is encouraged between multiple JTC1 subgroups doing work that falls under a common technical direction. The Technical Direction Multimedia and Representation, formed by pairing the existing JTC1 SCs 29 and 24 is already beginning to result in joint business plan update and greater collaboration on related standards development, such as work on the next version of Virtual Reality Markup Language (VRML) and its relationship with MPEG-4 and MPEG-7.

JTC1 has a strict market approach to standards. Work on standards development can commerce only after JTC1 national body approval of the project, and the active participation of at least five national bodies. A Business Team is a strategic planning activity of JTC1 for new and emerging markets, designed to reach outside the community of technologists who develop international standards within JTC1.

3.4 STANDARDIZATION PROCESSES ON MULTIMEDIA COMMUNICATIONS

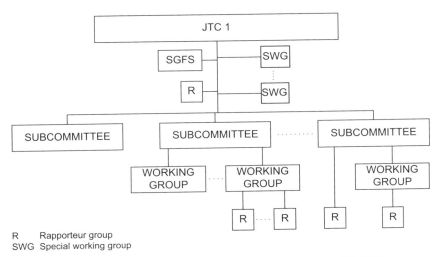

R   Rapporteur group
SWG Special working group

Figure 3.9 Organizational structure of JTC1 [14]. (©2002 ISO/IEC.)

Participation in a JTC1 Business Team is open to everyone. Its purpose is to actively involve users of standards, leaders in business and government, as well as trade associations, consortia, and other standards developing organizations to identify opportunities for creating new international standards. Let us mention here The Business Team on Electronic Commerce, The Imaging and Graphics Business Team (IGBT), which is charted to complete its work for presentation to the JTC1 Plenary meeting 1999. The IGBT scope includes all aspects of standard information formats and interfaces used for interaction and presentation, including but not limited to computer graphics, image processing, and single media – including animation, audio, video, and multimedia.

Collaboration with other organizations active or interested in aspects of IT standardization has long been a hallmark of ISO/IEC. A large number and range of liaisons are established at the TC/SC/WG levels with other bodies. ISO/IEC Technical Committees cover basic standards (ISO/IEC JTC1) and application-oriented standards. These ISO/IEC Committees coordinate work within their own constituencies and with the other standards development organizations and related major consortia and user groups from industry and consumer fora.

ISO/IEC JTC1 works well with other ISO and IEC TCs, ITU-T and ITU Radiocommunication Standardization sector (ITU-R), and industry fora. The agreement [17, 18] between ITU-T and JTC1 has resulted in many common text and identical standards [19]. The business and practical criteria developed by ISO/IEC for effective liaisons are clear. Liaisons must operate in both directions with suitable reciprocal arrangements. ISO/IEC liaison organizations include manufacturer associations, commercial associations, industrial consortia, user groups, and professional and scientific societies. They must agree to ISO/IEC procedures and have a sufficient degree of representation within their defined area of competence within the scope of the relevant ISO/IEC Committees.

JTC1 has experimented with electronic balloting and the ISO/IEC Central Secretariat has done a feasibility study on a full Web-based electronic document environment. The JTC1 Rapporteur Group on Implementing Information technology is responsible for assisting JTC1 and its technical subgroups to implement information technology in their working processes. The Group was formed in 1998 and is also responsible for representing JTC1 at the various fora designed to coordinate a consistent method for sharing the development of standards and specifications using electronic workflow, online document management, and group electronic conferencing.

ISO/IEC standards development is an entirely voluntary activity, depending on resources provided by standards-aware corporations, users, and small businesses. The program manages the deliverables within each SC and their overall coordination within and across JTC1. This internal task is compounded by the need for adequate coordination with many different standards development organizations in order to help eliminate duplication and improve synergy for standards development.

With the growth in use of the Internet due to increases in networking capabilities, higher desktop computing power, and improved display technologies, IT user expectations for high-quality multimedia exchanges have greatly increased. The next result is that digital multimedia technology is almost becoming synonymous with information technology. Thus, many key JTC1 standards are relevant to multimedia communications. JTC1 is actively developing standards for multimedia communications such as JPEG-2000, MPEG-4, MPEG-7, and Virtual Reality Modeling Language (VRML) [20].

3.4.3 IETF Standards

The Internet Engineering Task Force (IETF) verifies and examines the standardization of Internet protocols. The process of deciding on standards is managed by the Internet Engineering Task Group (IETG), which consists of members elected from the IETF whose activities are conducted in working groups (WGs) in the following areas: applications, IP next generation, Internet, network management, operational requirements, routing, security, transport, user services, and miscellaneous. The Request for Comments (RFC) document describes the protocol specification of an IETF standard and has some versions like the following:

- the *Standard Track* version is used for three kinds of standards – proposed, draft, and final
- the *Experimental* version is for protocols created for experiments or ones that are not mature
- the *Informational* version is for protocols for open systems interconnection (OSI), restricted vendors, or implementation
- the *Historic* version is for standards that have been replaced by newer versions.

Working group discussions are open. The IETF is addressing several important topics, such as [21]:

- the IP version 6 (IPv6)
- the next hop resolution protocol (NHRP)
- interdomain multicasting routing
- IP security
- resource reservation setup protocol (RSVP).

The aim of IPv6 is to increase address capacity and to reduce the cost of calculating the routing control table of the core router.

The next hop resolution protocol determines the address (Internet working layer address and nonbroadcast multiaccess, subnetwork address) for the next hop along the route from a sender (host or router) to the destination.

The mission of interdomain multicast routing is to develop a scalable routing protocol for multicasting and to propose standards. Promising draft methods include protocol-independent multicast sparse mode (PIM-SM), protocol-independent multicast, dense mode (PIM-DM), and core-based tree (CBT).

The IP security WG is specifying all aspects of IP-level security, including authentication, integrity, access control, and confidentiality. The protocols include secure IP header and encapsulation, and key exchange. They allow different encryption algorithms. The encryption can take place between hosts or between routers.

A resource reservation setup protocol (RSVP) is developed for use when the host requires the network quality of service (QoS) in order to specify the data stream, which is done by reserving resources on the network.

3.4.4 ETSI Standardization Projects for Applications, Middleware, and Networks

Many components go together to form the information infrastructure: telecommunications networks based on a variety of different technologies and operated by competing organizations, computing platforms using different hardware and software, and distributed information processing facilities. One goal of EII standardization is to allow all of these components to work together as a federation. The Enterprise Model is a means of identifying these federable components. The success of the Information Structure will require, in addition to the more technical aspects, the emergence of a framework for Rights and Obligations relating to the participations in the information industry.

The functions involved in interactions in the information infrastructure can be regarded as the kernel of all relationships between any combination of structural and infrastructural roles. These functions are as follows:

- *Application invoking and handling functions* – enable the invocation, use, and control of applications (by human users or other means).

- *Application supporting platforms* – provide support for the above functions.
- *Basic and enhanced telecommunications functions* – correspond to existing and future telecommunications networks, including intelligent networks and their network management capabilities.
- *Distributed information processing and storage functions* – correspond to distributed processing platforms, independently of the underlying data communication facilities.

The first two functions, that is, application invoking and handling functions as well as application supporting platforms, in combination support the structural roles of the Enterprise Model (for example, a client or server in a particular usage of the information infrastructure). An important element of the model is the existence of application independent elements. These are *building blocks* that provide well-identified, general-purpose functions that can be used by many applications. Examples include

- *Remote file access*, providing access to data held on another system.
- *Remote procedure call*, providing access to processing capabilities at a remote system.
- *Key registration and management*, providing the necessary information in a trustworthy fashion for authentication of information transferred between the various roles.

The internal structure of the functions and the interfaces between them consists of *domains*: distinct organizations, often in competition with each other but needing also to cooperate as part of the federation components that provide the function.

Domains are divided into *segments*: individual elements of the domain, cooperating to provide the services of the domain. Examples are the individual processing elements in the distributed processing platform, or the individual access and core networks forming up a complete telecommunications network.

The following standardization and specification highways have been identified, using the Enterprise Model:

- network-oriented standardization/specification highway
- middleware-oriented standardization/specification highway
- applications-oriented standardization/specification highway
- architecture-oriented standardization/specification highway

Each highway has been refined in terms of a number of specific *Standardization Projects*. For each project, detailed Terms of Reference have been defined, which include descriptions of deliverables and target dates. For the definition of these Standardization Projects, a pragmatic approach has been chosen, which has been successful in the recent past by other standardization and specification projects such as DAVIC and the MPEG projects in ISO/IEC. The approach consists of the following steps:

- collaborative definition of standardization requirements
- competitive proposals for projects
- collaborative endorsement of the terms of reference, choice of leading body and required deliverables
- building effective task forces
- timely collaborative development of standards.

Network-related projects concern the development of standards for provision of basic and enhanced telecommunications services required by EII. This area is covered by the existing expertise of ETSI members and experts. It covers aspects of network technology and interconnection, including naming and addressing, and telecommunication management network (TMN). Given the worldwide importance of the Internet, a key collaborative body is the IETF. This is a technical working body, which develops the specifications used in the Internet itself and for applications that run over it.

The middleware-related projects relate to the general-purpose computing support for EII/GII applications. The standards covered by the projects are of general utility and are not directed to any particular application. There are many bodies in Europe and worldwide that are involved in producing standards for this area.

The applications-related projects are in support of a particular applications area pertinent to the EII/GII. The number of applications of the EII is potentially limitless.

The architecture-related projects establish a framework into which the other projects can be fitted, so that their consistency can be ensured.

3.4.5 GII-Related Activities

Many activities have been directed towards achieving a global information infrastructure (GII). Standardization organizations have committed themselves to GII by establishing special committees or groups to identify tasks for achieving GII and by coordinating these tasks. In some regional and domestic standardization areas, there has also been some movement to develop a national information infrastructure (NII).

The information infrastructure standards panel (IISP) was established in 1994 with the primary objective to coordinate the activities of ANSI, the T1 Committee, and the IEEE for standardizing NII in the United States. The T1 Committee established an *ad hoc* project group in its Technical Subcommittee T1P1. This group issued a report identifying key items for NII/GII, such as the required information and telecommunication functions and functional models for ensuring interoperability, interworking, integration, and operation of various media [22].

In Europe, Strategic Review Committee 6 (SRC6) was established in the ETSI to define a coordinated work program on standards for the European Information Infrastructure (EII). Their report, which was approved at the ETSI General Assembly, identifies key issues for EII, such as the characterization of EII, conceptual models, fields to be standardized, and management of standardization by the ETSI [23].

In Japan, the Telecommunication Technology Committee (TTC) organized the Information Infrastructure task Force (IITF) under the umbrella of the Strategic Research and Planning Committee. A report issued by the IITF clarified the role of the TTC in promoting standardization for GII and NII in Japan in 1995. A two-layered approach is proposed. Low-layer standards (i.e., those for the network infrastructure) should be based on *de jure* standards, and higher-layer standards (i.e., those related to applications and services) should primarily adopt the specifications produced by fora and consortia as pre-TTC standards. To focus global GII-related activities and to coordinate with other standardization organizations, a Special Working Group on Information Infrastructure was established in 1996, replacing the IITF.

3.5 ITU-T MEDIACOM2004 FRAMEWORK FOR MULTIMEDIA COMMUNICATIONS

The rapidly developing world of multimedia technologies and standards requires a framework to develop standards for applications, services, and systems, which can respond to users' requirements in terms of mobility, ease of use, flexibility of systems, and end-to-end interoperability with specific quality requirements.

ITU-T SG16, the lead SG for multimedia, is working on project MEDIA-COM2004 (Multimedia Communications 2004), taking into account:

- the continuing trend in digitization
- the rapid growth of digital networks and in particular the Internet
- the increasing computational power in personal computers
- the convergence of various technologies including communications, broadcasting, information technology, and home electronics
- that multimedia topics are discussed in ITU-TSGs, ITU-T SGs, international and regional SDOs, and also in external organizations
- the merging new communication services and applications that are a result of the growth of the Internet and wireless technologies
- that with the emergence of high-speed high-quality networks, society will request real-time multimedia communications as an extension of existing monomedia systems
- that the end-to-end performance of multimedia systems and services defined in the ITU and elsewhere should be dimensioned
- that it will be necessary to study the interfaces in the information appliance environment as consumer devices increasingly perform multimedia functions.

3.5 ITU-T MEDIACOM2004 FRAMEWORK FOR MULTIMEDIA COMMUNICATIONS

The objective of the MEDIACOM2004 Project is to establish a framework for multimedia standardization for use both inside and external to the ITU. This framework will support the harmonized and coordinated development of global multimedia communication standards across all ITU-T and ITU-R Study Groups, and in close cooperation with other regional and international standards development organizations [24].

3.5.1 Scope

The scope of the MEDIACOM2004 includes the following applications:

- End-to-end multimedia systems and services over all network types including the Internet, that is, videophone/videoconference, distance learning, telemedicine, interactive TV services, Web-casting, MBone, including their distribution within home entertainment, and so on.
- End-to-end multimedia systems and services over wireless access systems – in this environment, computer or consumer information appliance devices will be used.
- Security system for/using multimedia systems (watermark in the video contents, individual authentication, etc.).
- Multimedia broadcasting systems that interactively handle audio and video.
- The extension of e-mail and WWW for the transmission of multimedia documents.

3.5.2 Multimedia Framework Study Areas

Multimedia applications/services consists of many aspects, for example, media coding, network interface, and optional tools for security and privacy, and so on. The development environment of the individual application is shown in Figure 3.10.

The main idea in the development of an application is the integration or assembling of the tools considering the customer's requirements, market trends, applicable network features, and so on. Figure 3.11 demonstrates the development process for

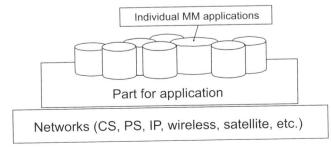

Figure 3.10 Layered environment in multimedia development process.

multimedia (MM) applications and services. The components consist of three blocks [25]:

(A) system design and integration
(B) common media and tools
(C) network interface.

Each multimedia application can be developed by A, considering the design document and using the technologies of B and C. For the development of MM systems, the role of block A (system design and integration/assembling group) is large. Arrows, *a*, *b*, *c*, and *d* show the flow of the MM application developing process.

Block A encompasses the transmission and/or usage protocols/procedure necessary for the media contents, that is, how to convey the media contents to the partner terminals with a reliable process. The technology includes the connection establishment, device confirmation, error free/flow control, and transmission confirmation procedures.

Interoperability of the Recommendations will be tested through the terminals/ devices implemented by the appropriate organizations.

One of the scopes of the Framework Study Areas (FSAs) is to define the architectural framework for future multimedia services and applications. MM applications and services need to be independent of the network they operate across. One impetus for this has been the growth in the use of the Internet with its concept of available anywhere and anytime at minimal cost. The shortfalls of the public Internet in terms of speed, bandwidth, and real-time needs are now being addressed. With the concept of an open network architecture, the means of ensuring that new

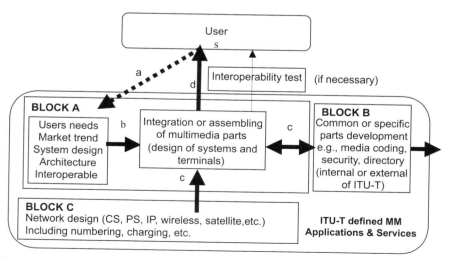

Figure 3.11 Development process for multimedia applications and services [25]. (©2002 ITU-T.)

multimedia applications can be readily configured, using available network resources, to meet user needs in a network-independent way, are investigated.

Convergence is an essential issue in the MEDIACOM2004 concerning MM framework study areas. We observe convergence in the fields of information technology, telecommunications, radio communications, broadcasting, and the interactive computing environment, which is coupled with the development of new networks, especially networks based on IP technology and mobile networks. The concept is that multimedia mobility will become a reality in the very near future. The aim of the FSAs is to generate a *convergence roadmap* to show how diverse existing and planned MM services and applications can converge so as to provide seamless interworking from the user viewpoint. Generally speaking, the aim is to ensure that the architecture is applicable to various services, independently from the networks that support them.

Interoperability can be considered in terms of reliable end-to-end multimedia operation across a number of different networks. There is an alternative view in terms of different applications and services (either network or end system-based) interoperating efficiently and reliably in a given multimedia environment. Support of such interoperability requires agreement on a framework within which common tasks can locate and establish communications with their peers, while dissimilar tasks can exchange media streams of mutual interest.

The work on the media coding area starts from the identified emerging services and applications and the corresponding information elements (media) that need to be encoded/decoded/represented. For example, in the video telephony service only the audio and video signals need to be coded, but in more sophisticated services and applications (multimedia database search, telemedicine, etc.), other kinds of information elements should be considered. It is relevant to note the applicability of speech recognition and speaker verification standardization. If new media are identified, beyond the traditional ones (video, audio, and data) included in existing services, studies should be started on the appropriate media coding techniques. Moreover, security aspects related to media coding will require additional media coding techniques.

From the point of view of QoS and performance, the aim of the FSA will be to

- Ensure that the required QoS levels for various media types are established and defined.
- Ensure that the necessary mechanisms and protocols for providing multimedia QoS levels are provided.

The FSA will identify suitable end-to-end performance guidelines to assist the implementation of new multimedia systems and services.

There are a number of security considerations that need to be addressed when developing an architecture for the multimedia information infrastructure. Such considerations include end-to-end privacy of data, authentication (user identification), anonymous access, access control, detection, electronic signature, and encryption.

Within a telecommunications context, security issues can be grouped in terms of user, network, and operator. The aim of this multimedia FSA is to try to ensure maximum consistency and interoperability across the range of MM services and applications.

The capabilities to handle different information media and control actions vary within wide boundaries among telecommunications and multimedia services. This variation may have a cause in age-related functional limitations. With the ageing populations in large parts of the world, many telecom users will have sensory and motor limitations. The work influences many aspects of multimedia systems, such as enabling multiple media streams, such as video, text, and voice. Adequate user control over services and devices should be possible without full use of all human senses and motor capabilities. This user control should be possible in alternative ways, assuring that information is provided in alternative media. The aim is to ensure that standardized solutions improving human accessibility will be identified, designed, and promoted.

3.5.3 Study Group 16 Work Program

Study Group (SG) 16 addresses the harmonization of the end-to-end media interfaces, services, and applications. In particular, SG16 is responsible for the protocol architecture higher than the Network Layer, MM architecture, audio/video components and interoperability. SG 16 is also responsible for the Middleware Design (MM Services, Media Coding, MM Protocol Architecture).

The current topics considered by SG16 are the following:

- modem and facsimile terminals
- multimedia platforms and interworking
- media coding
- multimedia framework
- MM applications over IP, cable television, home area wireless network
- wireless interfaces
- security aspects related to video coding.

Table 3.2 shows the SG16 work program for 2001–2004.

3.6 ISO/IEC MPEG-21 MULTIMEDIA FRAMEWORK

MPEG-21 is the project to incorporate a multimedia framework standard with a strong emphasis on electronic commerce and protection of media content. It addresses the entire chain of content value from generation to consumption. ISO/IEC JTC1/SC29 has initiated the MPEG-21 project.

The scope of MPEG-21 is the integration of the critical technologies enabling transparent and augmented use of multimedia resources across a wide range of networks and devices. The aim is to support functions such as content creation, content

Table 3.2 The Study Group SG16 work program for 2001–2004 [25]

Q. No.	Title
A	MEDIACOM2004
B	MM architecture
C	MM applications and services
D	Interoperability of MM systems and services
E	Media coding
F	Quality of service in MM systems
G	Security of MM systems and services
H	Accessibility to MM systems and services
1	Multimedia systems, terminals and data conferencing
2	Multimedia over packet networks using H.323 systems
3	Infrastructure and interoperability for multimedia over packet networks
4	Video and data conferencing using Internet-supported services
5	Mobility for MM systems
6	Advanced video coding
7	Wideband coding of speech at around 16 kbit/s
8	Encoding of speech signals at bit rates around 4 kbit/s
9	Variable bit rate coding of speech signals
10	Software tools for signal processing standardization activities and maintenance of existing voice coding standards
11	Voice band modems: specification and performance evaluation
12	DTE-DCE protocols for PSTN and ISDN
13	DTE-DCE interfaces and protocols
14	Facsimile terminals
15	Distributed speech recognition and speaker verification

production, content distribution, content consumption and usage, content packing, intellectual property management and protection, content identification and description, financial management, user privacy, terminals and network resource abstraction, content representation, and event reporting.

The MPEG-21 multimedia framework has identified seven key architectural elements in its *Vision, Technologies and Strategy* Technical Report (ISO/IEC 21000-1) that are needed to support the multimedia delivery chain, and is in the process of defining the relationships between the operations supported by them. The key architectural elements defined in MPEG-21 are [25]:

- digital item declaration
- digital item identification and description
- content handling and usage
- intellectual property management and protection
- terminals and networks
- content representation
- event reporting.

MPEG-21 recommendations will be determined by interoperability requirements and their levels of detail may vary for each architectural element.

Solutions with advanced multimedia functionality are becoming increasingly important as individuals are producing more and more digital media, not only for professional use but also for their personal use. All these *content providers* have many of the same concerns: management of content, re-purposing content based on consumer and device capabilities, protection of rights, protection from unauthorized access/modification, protection of privacy of providers and consumers, and so on. The need for technological solutions to these challenges is motivating the MPEG-21 Multimedia Framework initiatives that aim to enable the transparent and augmented use of multimedia resources across a wide range of networks and devices.

Based on the above observations, MPEG-21 aims at defining a normative open framework for multimedia delivery and consumption for use by all the players in the delivery and consumption chain. This open framework will provide content creators, producers, distributors, and service providers with equal opportunities in the MPEG-21 enabled open market. MPEG-21 is based on two essential concepts: the definition of fundamental unit of distribution and transaction and the concept of users interacting with digital items. The goal of MPEG-21 can thus be rephrased to defining the technology needed to support users to exchange, access, consume, trade, and otherwise manipulate digital items in an efficient, transparent, and interoperable way. MPEG-21 identifies and defines the mechanisms and elements needed to support the multimedia delivery chain as described as well as the relationships between the operations supported by them. Within the parts of MPEG-21, these elements are elaborated by defining the syntax and semantics of their characteristics, such as interfaces to the elements.

3.6.1 User Model

A user is any entity that interacts in the MPEG-21 environments or makes use of a digital item. Such users include individuals, consumers, communities, organizations, corporations, consortia, governments, and other standards bodies and initiatives around the world. Users are identified specifically by their relationship to another user for a certain interaction. At its most basic level, MPEG-21 provides a framework in which one user, interacts with another user, and the object of that interaction is a digital item commonly called content. Some such interactions are creating content, providing content, archiving content, rating content, enhancing and delivering content, aggregating content, syndicating content, retail selling of content, consuming content, subscribing to content, regulating content, facilitating transactions that occur from any of the above, and regulating transactions that occur from any of the above.

3.6.2 Digital Items

Within any system (such as MPEG-21) that proposes to facilitate a wide range of actions involving *digital items*, there is a need for a very precise description for defining exactly what constitutes such an *item*. There are many kinds of content, and probably just as many possible ways of describing it to reflect its context of use. This presents a strong challenge to develop a powerful and flexible model for digital items that can accommodate the forms that content can take and the new forms it will assume in the future. Such a model is only useful if it yields a format that can be used to represent any digital items defined within the model unambiguously and communicate them, and information about them, successfully.

Example 3.3 Consider a simple *Web page* as a digital item. A Web page typically consists of a hypertext mark-up language (HTML) document with embedded *links* to various image files and possibly some layout information. In this simple case, it is a straightforward exercise to inspect the HTML document and deduce that this digital item consists of the HTML document itself, thus all of the other resources upon which it depends.

Example 3.4 Let the *Web page* contain some custom scripted logic to determine the preferred language of the viewer (among some predefined set of choices) and to either build/display the page in that language, or to revert to a default choice of the preferred translation. The key point in this example is that the presence of the language logic clouds the question of exactly what constitutes the digital item now and how this can be unambiguously determined.

3.6.3 Work Plan

MPEG-21 has established a work plan for future standardization. Nine parts of the standardization within the multimedia framework have already been started:

- vision, technologies, and strategy
- digital item declaration
- digital item identification
- intellectual property management and protection
- rights expression language
- rights data dictionary
- digital item adaptation
- reference software
- file format.

In addition to these specifications, MPEG maintains a document containing the consolidated requirements for MPEG-21. This document will continue to evolve during the development of the various parts of MPEG-21 to reflect new requirements and change in existing requirements [26, 27].

Vision, Technology, and Strategy

A *technical report* has been written to describe the multimedia framework and its architectural elements together with the functional requirements for their specifications, which was finally approved in September 2001. The fundamental purposes of the technical report *Vision, Technology and Strategy* are to

- Define a *vision* for a multimedia framework to enable transparent and augmented use of multimedia resources across a wide range of networks and devices to meet the needs of all users.
- Achieve the *integration* of components and standards to facilitate harmonization of technologies for creation, management, transport, manipulation, distribution, and consumption of digital items.
- Define a *strategy* for achieving a multimedia framework by the development of specifications and standards based on well-defined functional requirement through a collaboration with the bodies.

Digital Item Declaration

The purpose of the digital item declaration (DID) specification is to describe a set of abstract terms and concepts to form a useful model for defining digital items. Within this model, a digital item is the digital representation of *a work* and, as such, it is the thing that is acted upon within the model. The goal of this model is to be as flexible and general as possible, while providing for the *hooks* that enable higher-level functionality. This, in turn, will allow the model to serve as a key foundation in the building of higher-level models in other MPEG-21 elements (such as Identification and Description or Intellectual Property Management and Protection IPMP). This model helps to provide a common set of abstract concepts and terms that can be used to define such a scheme, or to perform mappings between existing schemes capable of digital item declaration for comparison purposes. The DID technology is described in three normative sections: Model, Representation, and Schema.

The DID Model describes a set of abstract terms and concepts to form a useful model for defining digital items. Within this model, a digital item is the digital representation of *a work*, and as such, it is the thing that is acted upon within the Representation means of normative description of the syntax and semantics of each of the DID elements, as represented in extensible markup language (XML).

By schema, we mean normative XML schema comprising the entire grammar of the DID representation in XML.

Digital Item Identification

The scope of the digital item identification (DII) specification includes

- how to uniquely identify digital items and parts thereof (including resources)
- how to uniquely identify IP related to the digital items (and parts thereof), for example abstractions
- how to uniquely identify description schemes

- how to use identifiers to link digital items with related information such as descriptive metadata
- how to identify different types of digital items.

The DII specification does not specify new identification systems for the elements for which identification and description schemes already exist and are in use. Identifiers covered by this specification can be associated with digital items by including them in a specific place in the DID.

Digital items and their parts within the MPEG-21 framework are identified by encapsulating uniform resource identifiers (URI) into the identification description scheme (DS). A URI is a compact string of characters for identifying an abstract or physical resource, where a resource is defined as *anything that has identity*. The requirement that an MPEG-21 DII be a URI is also consistent with the statement that the MPEG-21 identifier may be a uniform resource locator (URL). The term URL refers to a specific subset of URI that is in use today as pointers to information on the Internet. It allows for long-term to short-term transition depending on the business case.

Intellectual Property Management and Protection (IPMP)
MPEG-21 defines an interoperable framework for intellectual property management and protection (IPMP). A new project on more interoperable IPMP systems and tools includes standardized ways of retrieving IPMP tools from remote locations, exchanging messages between IPMP tools and between these tools and terminal. It also addresses authentication of IPMP tools and has provisions for integration rights expressions according to the rights data dictionary and the rights expression language. Efforts are currently ongoing to define the requirements for the management and protection of intellectual property in the various parts of the MPEG-21 standard.

Rights Expression Language (REL)
A rights expression language is seen as a machine-readable language that can declare rights and permissions using the terms as defined in the rights data dictionary. The REL is intended to provide flexible, interoperable mechanisms to support transparent and augmented use of digital resources in publishing, distribution, and consuming of digital movies, digital music, electronic books, broadcasting, interactive games, computer software, and other creations in digital form, in a way that protects digital content and honors the rights, conditions, and fees specified for digital contents. It is also intended to support specification of access and user control for digital content in cases where financial exchange is not part of the terms of use, and to support exchange of sensitive or private digital control.

The REL is also intended to provide a flexible interoperable mechanism to ensure that personal data are processed in accordance with individual rights and to meet the requirement for users to be able to express their rights and interests in a way that addresses issues of privacy and use of personal data. A standard REL should be able to support guaranteed end-to-end interoperability, consistency, and reliability between different systems and services. To do so, it must offer richness and extensibility in declaring rights, conditions, and obligations, ease and persistence in identifying and associating these with digital contents, and flexibility in supporting multiple usage/business models. MPEG REL adopts a simple and extensive data model for many of its key concepts and elements. This data model for a rights expression consists of four basic entities and the relationship among those entities. This basic relationship is defined by the MPEG REL assertion *grant*, consisting structurally the following:

- the principal to whom the grant is issued
- the right that the grant specifies
- the resource to which the right in the grant applies
- the condition that must be met before the right can be exercised.

The REL data model is shown in Figure 3.12. The entities in the model – principal, right, resource, and condition – can correspond to user, including terminal, right, digital item, and condition in the MPEG-21 terminology.

Rights Data Dictionary (RDD)

The specification of a rights data dictionary (RDD) began in December 2001. A Committee Draft was published in July 2002. The RDD composes a set of clear, consistent, structured, integrated, and minimally identified Terms to support the MPEG-21 REL.

The structure of the dictionary is specified, along with a methodology for creating the dictionary. The means by which further Terms may be defined is also explained. The RDD also supports the circumstance that the same name may have different meanings under different authorities.

RDD recognizes legal definitions as and only as Terms from other authorities that can be mapped into the RDD. Therefore, Terms that are directly authorized by RDD neither define nor prescribe intellectual property rights or other legal entities. The RDD specifications are designed to support the mapping and transformation of

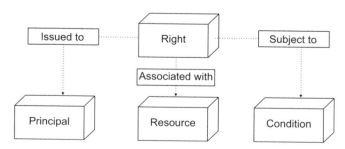

Figure 3.12 The rights expression language (REL) data model [26]. (©2002 ISO/IEC.)

metadata from the terminology of one namespace into that of another namespace in an automated or partially automated way, with the minimum ambiguity or loss of semantic integrity. The dictionary is based on a logical model, the Context Model, which is the basis of the dictionary ontology.

Digital Item Adaptation

The concept of digital item adaptation is illustrated in Figure 3.13. As is shown, a digital item is subject to a resource adaptation engine, as well as a descriptor adaptation engine, which produce together the adapted digital item.

The adaptation engines themselves are non-normative tools of digital item adaptation. Descriptions and format-independent mechanisms that provide support for digital item adaptation in terms of resource adaptation, descriptor adaptation, and/or quality of service management are within the scope of the requirements. The specific items targeted for standardization are user characteristics, terminal capabilities, network characteristics, natural environment characteristics, resource adaptability, and session mobility.

By user characteristics we mean description tools that specify the characteristics of a user, including preferences to particular media resources, preferences regarding the presentation of media resources, and the mobility characteristics of a user. Additionally, we have description tools to support the accessibility of digital items to various users, including those with audiovisual impairments to be considered.

By terminal capabilities we mean description tools that specify the capability of terminals, including media resource encoding and decoding capability, hardware, software, and system-related specifications, as well as communication protocols that are supported by the terminal.

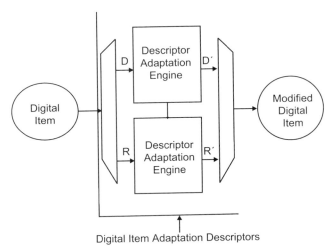

Figure 3.13 Conceptual architecture of digital item adaptation [26]. (©2002 ISO/IEC.)

Network characteristics are description tools that specify the capabilities and conditions of a network, including bandwidth utilization, delay, and error characteristics.

Natural environment characteristics are description tools that specify the location and time of user in a given environment, as well as audiovisual characteristics of the natural environment, which may include auditory noise levels and illumination properties.

Resource adaptability includes tools to assist with the adaptation of resources including the adaptation of binary resources in a generic way and metadata adaptation, and also tools that assist in making resource–complexity trade-offs and making associations between descriptions and resource characteristics for targeted quality of service.

By session mobility, we mean tools that specify how to transfer the state of digital items from one user to another. The capture, transfer, and reconstruction of state information are specified, also.

Reference Software

The reference software will form the first of what is envisaged to be a number of systems-related specifications in MPEG-21. Other candidates for specification are likely to include a binary representation of the Digital Declaration and an MPEG-21 file format. Reference software will be based on the requirements that have been defined for an architecture for processing digital items.

File Format

An MPEG-21 digital item can be a complex collection of information. Both still and dynamic media (e.g., images and movies) can be included, as well as digital item information, metadata layout information, and so on. It can include both textual data (e.g., XML) and binary data (e.g., an MPEG-4 presentation or a still picture). For this reason, the MPEG-21 file format will inherit several concepts from MPEG-4, in order to make *multipurpose* files possible. For example, a dual-purpose MPEG-4 and MPEG-21 file would play just the MPEG-4 data on an MP4 player, and would play the MEPG-21 data on an MP21 player. Requirements have been established with respect to the file format.

3.6.4 MPEG-21 Use Case Scenario

This clause presents the most relevant MPEG-21 use case scenario that has been used in MPEG-21. The next subclauses present use case scenarios grouped by topic. This grouping can be revised and is open for improvement.

Universal Multimedia Access

Digital resources (including audiovisual content) can be delivered to a variety of terminals over dynamic network conditions in a scenario covered by the requirement of universal multimedia access (UMA). UMA enables terminals with limited communication processing, storage, and display capabilities to access rich multimedia content, as shown in Figure 3.14.

The UMA-related requirements addressed by the experiment are terminal capabilities, network characteristics, and user preferences all having an impact on a media resource delivery. The characteristics addressed are available bandwidth, screen size, media resource, scalability, and survivability. Each channel can be configured separately, and packets are forwarded to the appropriate channel using a packet filter and applying different limitations/delays to different traffic according to the specified filter rules. Dynamic variations can be applied to channel bandwidth, propagation delays, and packet loss. These can be either deterministic or statistical.

Text News Distribution to Web Sites

A news company creates raw textual news and distributes it to Web sites using batch-mode and real-time data feeds. The distribution infrastructure is based on technologies such as file transfer protocol (FTP) and Web data-transfer protocols. The news company specified that it is acceptable for site owners to perform certain editorial functions on the news, such as

- turning it into HTML
- annotations such as the addition of company *stock ticker symbols* and related hypertext links
- altering names to suit local convention.

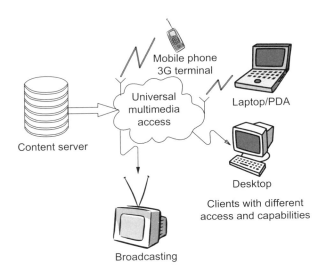

Figure 3.14 Universal multimedia access.

However, site owners are not allowed to

- change the meaning of the story
- paraphrase the story – it must appear in full or be properly quoted
- translate the story into a foreign language.

Video News Distribution

Video news footage is created and syndicated by Video News Company. The value chain consists of the rights header, the distributor, the broadcaster, into whose programming schedule the footage is included, and the consumer. For syndicated footage, there are strict rules regarding the scope of distribution defined by the right holders. Primarily, these will limit the geographic regions in which to create content for their domestic market and will wish to retain distribution rights in that region. However, exclusions may also include such rules as markets sharing the same language as the domestic market and may be defined by the geographies covered by trade associations. Creators of video footage may specify rules that vary from one application to another. Rules may differ depending on.

- the type of video content (sport, general news, fashion, entertainment, etc.);
- the type of broadcast program into which the news footage is eventually embedded.

E-Commerce and Channel Partner Distribution

An information provider has a number of fine-grained information services that can be incorporated by channel partner distributors into their own online offerings. The rules affecting the rights of the channel partner are encoded into component used contracts that are aggregated into a single contract via an online administration system. The rules governing the final consumption of information, via the channel partner, are similarly encoded into contract components to be integrated with components of the channel partner to form the contract with the consumer. An online contract/licensing interface is provided so that the full text of the contract does not need to be downloaded. This is particularly important for consumption on small devices, where only relevant portions of the licence are required at any one time. Rules are also required to specify how the channel partner may (or may not) aggregate this content with that of other suppliers.

Real-Time Financial Data Distribution

An information provider distributes real-time prices for stacks, bands, and other financial instruments via an online feed. Access to the feed itself is controlled and access to individual data items flowing within the feed is governed by fine-grained permission rules. The permission rules are defined in part by the suppliers of the original real-time information. Therefore, the final control is an amalgamation of external rules and rules added by the information provider. External rules differ on a source-by-source basis, so the ideal would be for the external agency to supply preencoded rules for inclusion into the final rule-set. Permission rules need to take into account metadata included within the data items flowing in the head. This may be a simple identifier within the data item or may by more complex metadata such as *typing* information to associate the data with particular products or services. The mapping process takes, as its input, the products and services to which the end user has subscribed and determines in real time which data elements can be viewed by the end user.

Picture Archive Access
A media company hosts and archives high-quality news pictures for download by media organizations for inclusion into online and print publications. Service access rules need to accommodate various levels of service package. The service packages control

- the number of images that can be downloaded in a particular time period
- the age of images that are accessible within the archive
- the pricing of images, which may include such complexity as subscription charge for the package (which includes a number of prepaid image downloads), a per-image charge, charges differentiated by the type, quality, and size of image, according to the intended publication usage.

3.6.5 User Requirements

A user is any entity that interacts in the MPEG-21 environment or makes use of a digital item. Such users include individuals, consumers communities, organizations, corporations, consortia, governments, and other standards bodies and initiatives around the world [28]. Users are defined by their role in a specific transaction, as shown in Figure 3.15.

From a technical point of view, MPEG-21 makes no distinction between a *content provider* and a *consumer*. Both are users. A single entity may use content in many ways (publish, deliver, consume, etc.). So all parties interacting within

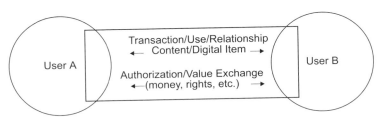

Figure 3.15 Users' relationship for a certain interaction [28]. (©2001 ISO/IEC.)

MPEG-21 are categorized as users equally. A user may assume specific or even unique rights and responsibilities according to his/her interaction with other users within MPEG-21.

At its most basic level, MPEG-21 provides a framework in which one user interacts with another user and the object of that interaction is a digital item commonly called content. Such interactions are creating content, providing content, archiving content, rating content, archiving and delivering content, aggregating content, delivering content, consuming content, and regulating content. The interactions between users using digital items may be described by seven core qualifiers as presented in Figure 3.16. These are Digital Item Declaration, Digital Item Identification and Description, Content Handling and Usage, Intellectual Property Management and Protection, Networks and Terminals, Content Representation, and Event Reporting.

Users represent a wide and diverse set of interests and no list of their requirements could ever be complete. Some common, broad user requirements for an MPEG-21 multimedia framework are

- the ease of understanding the terms of use
- the enforcement of business and usage rules through the value chain
- the support of regulations and rules and the incorporation of societal factors as necessary

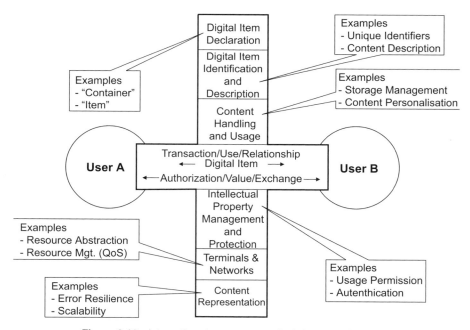

Figure 3.16 Interactions between users [28]. (©2001 ISO/IEC.)

- the provision of user protection, including reliability of service, stationary rights, liability and insurance for purchase, loss and damage and escrow arrangements to eliminate risks
- the management and protection of user privacy
- the personalization of content
- the ability to add metadata as content moves through the value chain
- the verification of the integrity of digital items and the provision of means to check when, by which method, and by which authority this integrity was verified
- the ability to track content and transactions
- the secure use of content and exchange of value
- the enabling quality and flexibility of interactive service
- the provision of metrics to communicate performance for each segment and function of MPEG-21 transactions, content interaction, and so on
- the enabling of interoperability of other multimedia frameworks with and into MPEG-21
- the support of existing standards outside of MPEG
- the ability to provide for, support, adopt, reference or integrate business processes and a library of common, standard intrabusiness processes.

From a technical perspective, these functionalities are expressed as requirements in each of the architectural elements: interoperability, transparency, robustness, integrity, scalability, extensibility, customization, event management, protection, rights management, standard metrics and interfaces, privacy, and interactivity.

3.7 IETF MULTIMEDIA INTERNET STANDARDS

The Internet Engineering Task Force (IETF) is a standardization body focused on the development of protocols used in IP-based networks. The IETF consists of many working groups and is managed by the Internet Engineering Steering Group (IESG), Internet Architecture Board (IAB), and Internet Society (ISOC). The IETF is an open international standardization body of network designers, operators, vendors, and researchers focused on the development of Internet standard protocols for use on the Internet and intranets [29]. One of the key strengths of the Internet is its global connectivity. For this connectivity, it is essential that all the hosts in the Internet interoperate with each other, understanding the common protocol at various levels [30]. In ref. [31] it is written that the Internet, a loosely organized international collaboration of autonomous, interconnected networks, supports host-to-host communication through voluntary adherence to open protocols and procedures defined by Internet Standards. The Internet standardization process of IETF under the legal umbrella of the Internet Society (ISOC), an international academic society, is the key to the success of GII over IP-based networks such as the Internet.

The IETF standards process involves several organizations: Internet Society (ISOC), Internet Architecture Board (IAB), Internet Engineering Society Group (IESG), Internet Assigned Number Authority (IANA), and IETF itself [21].

The IETF is an open international community of network designers, operators, vendors, and researchers concerned with the evolution of Internet architecture and smooth operation of the Internet. It is the principal body engaged in the development of new Internet Standard specifications.

The Internet Society represents an international organization concerned with the growth and evolution of the worldwide Internet, and the social, political, and technical issues that arise from its use.

The Internet Architecture Board is characterized by the Internet Society Trustees to provide oversight of the architecture of the Internet and its protocols. It appoints the IETF chair and is responsible for approving other IESG candidates put forward by the IETF nominating committee.

The Internet Engineering Steering Group is responsible for the actions associated with the proposition of the technical specification along the standards track, including the initial approval of new Working Groups and the final approval of specifications as Internet Standards.

The IANA is responsible for assigning the values of these protocol parameters for the Internet. IANA publishes tables of the currently assigned numbers and parameters in Request for Comments (RFC) entitled *Assigned Numbers*. IANA functions as the *top of the pyramid* for Internet address assignment, establishing policies for these functions.

To construct GII using IP-based protocols, it is necessary to have IP transmitted globally over any time layer technologies. That is, IP should be able to be transmitted over various access and back one link layer protocols. For the medium access network, IETF has IP over Cable Data Network (IPCDN) and IP over Vertical Blanking Interval (VBI). A new Working Group (WG) is being considered for IP over asynchronous digital subscriber loop (ADSL). Because cable and ADSL modems use the well-known Ethernet or ATM link-layer technologies, it is not necessary to newly create a protocol to use IP over them. It is still necessary to define a management information base (MIB) for the new modems. On the other hand, a satellite link is a new technology using IP protocols and several protocols must be developed to accommodate it. Satellite links have their own packet format, receive only sites, and long delay. There are IP over VBI, unidirectional link routing and TCP over satellite. For the backbone network, IP over ATM over synchronous optical network/synchronous digital hierarchy (SONET/SDH) and IP over Point-to-Point Protocol (PPP) over SONET/SDH are two major protocol stacks for which all the standardization effort of IETF has been completed. However, in a pure IP network, the overhead of the ATM header, the role of which is duplicated by that of IP, is not negligible. SONET/SDH itself has some header overhead. Moreover, high-bit-rate data link control (HDLC) framing of PPP is not very efficient, and the short period of the SONET/SDH scrambler makes PPP over SONET/SDH dangerous to network operators.

Because IETF has no formal membership, is open to any individual, and is operated with *rough consensus*, it is not very meaningful for other standardization bodies to exchange liaisons with IETF. While a liaison from IETF to another organization may significantly affect the organization, a liaison from the organization can act only as well as an ordinary participant. For the timely contribution to the Internet standard process, standardization organizations should send their members to IETF rather than spending time trying to exchange liaisons.

3.8 INDUSTRIAL FORA AND CONSORTIA

The environment for telecommunications is changing drastically and quickly, as well as both politically and technically. Deregulation of the telecommunication market is proceeding in several countries and is being scheduled or planned for in many other countries. This deregulation will affect the process of standardization among international organizations. The emerging multimedia technologies and GII require broader collaboration among not only *de jure* standardization organizations but also with the *de facto* industry standardization bodies. Multimedia communication is one of the most important features of the GII [22]. Cooperation and information sharing with standards development organizations and other communications industry organizations are important in achieving globally interoperable multimedia infrastructure. Information sharing and cooperation are essential to avoid parallel or overlapping activities and provide common solutions.

We will present the activities of industrial fora and consortia for multimedia, described from the viewpoint of the current status of standards related to multimedia communications such as ATM Forum, Telecommunications Consortium (IMTC), Digital Audio-Visual Council (DAVIC), MPEG-4 Industry Forum, Internet Streaming Media Alliance (ISMA), 3rd Generation Partnership Project (3GPP), 3rd Generation Partnership Project 2 (3GPP2), Digital Video Broadcasting/Advanced Television Systems Committee (DVB/ATSC).

ATM Forum Industry Cooperation

ATM Forum is an international nonprofit organization formed with the objective of accelerating the use of asynchronous transfer mode (ATM) products and services through a rapid convergence of interoperability specification. In addition, the Forum promotes industry cooperation and awareness [32]. Since its formation in 1991, the ATM Forum has generated very strong interest within communications in industry. Currently, the ATM Forum consists of approximately 150 member companies, and it remains open to any organization that is interested in accelerating the availability of ATM-based solutions. ATM Forum consists of a worldwide Technical Committee, a marketing awareness program such as Broadband Exchange, and the User Committee, through which ATM end users participate.

The ATM Forum Technical Committee works with other worldwide standards bodies selecting appropriate standards, resolving differences among standards, and recommending new standards when existing ones are absent or inappropriate [33].

The Technical Committee was created as one, single worldwide committee in order to promote a single set of specifications, thereby ensuring interoperability between all vendors as ATM products and services become available. The Technical Committee consists of a variety of working groups, which investigate different areas of ATM technology.

The ATM market awareness activities include [34]

- broadband exchange
- ATM roadmap.

The ATM Forum has recognized the need for networking technology to find the seamless connections that maximize value, identify strengths, and promote advances. The Forum hosts day-long seminars at each of its quarterly technical meetings for all interested parties from the networking and broadband industries to participate in these important opportunities to promote ideas and identify important new areas of technical collaboration.

As the ATM industry moves towards a new era of broadband networking, it becomes evident that a new direction is needed to have and to drive the necessary development work in broadband technologies. The roadmap identifies six key emerging market opportunities: three new networking architectures and three innovative application areas.

The User Committee, formed in 1993, consists of ATM end users. This group interacts regularly with the membership to ensure that ATM Forum technical specifications meet real-world end user needs. In this rapidly growing organization, topics of discussion are varied, including issues such as interoperability among vendors and migration of the installed base to ATM.

The ATM Forum communicates key results throughout development of new implementation agreements, issues, work play, and schedules with other organizations to accelerate problem resolution in all organizations. Organizations with which the ATM Forum cooperates are illustrated in Figure 3.17. The ATM Forum works in collaboration on a number of activities. Examples include the Frame Relay Forum, Asynchronous Digital Subscriber Line (ADSL) Forum, European Telecommunications Standards Institute (ETSI), and Internet Engineering Task Force (IETF) in developing and adapting certain IP scenarios to ATM, the ETSI BRAN (Broadband Radio Access Network) activity, and the ATM Forum. Wireless ATM (WATM) implementation agreement activities are not only being coordinated, but the work has been divided between the latter two organizations.

Because the ITU-T is promoting ATM standardization for BISDN public networks, the ATM Forum initially targeted the use of ATM in private networks. It has primarily promoted the development of common implementation specifications for ATM local area networks (LANs). The ATM Forum has close relations not only with the ITU-T but with other standardization organizations and fora, such as the American National Standards Institute (ANSI), ETSI, IETF, Frame Relay Forum (FRF) and Digital Audio-Visual Council (DAVIC). It contributed to standardization work by adopting existing standards and specifications, and through upstream activities. The development of basic implementation specifications at the ATM Forum

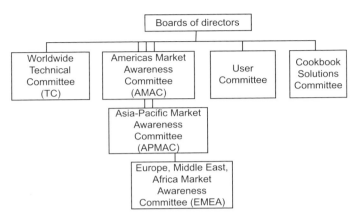

Figure 3.17 The ATM Forum organizations [32]. (©1998 IEEE.)

reached its peak in 1996. The Forum is now shifting its focus to developing implementation specifications and enlightening society about such applications.

The ATM Forum User–Network interface (UNI) Specification Version 4.0 for improved signaling was completed prior to the ITU-T standardization. It includes the available bit rate (ABR) specification in traffic management (TM) 4.0 and leaf-multipoint-connection. Broadband intercarrier interface (B-ICI) versions 2.0 and 2.1, including network service access point (NSAP) support, network control center interface (NCCI), and best-effort type, were completed. An interim local management interface (ILMI) on a UNI is improved. Private network node interface (PNNI) 1.0 was completed. It defines the routing protocols needed for on-demand connection setups. LAN emulation (LANE) was also completed; it enables customers to put their existing upper-layer protocols on an ATM network without having to change them. For example, a method for encapsulating constant bit rate (CBR) coded MPEG-2 signals into ATM adaptation layer type 5 (AAL5) was defined for video-on-demand (VoD) service.

ATM is a very good match for new high-frequency radio spectrum, which is becoming available for wireless communications. A short packet interface is required to accommodate the transmission error characteristics associated with the frequency band and matches the characteristics of ATM. ETSI BRAN and ATM Forum are collaborating in the development of standards and specifications for both private and public access scenarios. ETSI is responsible for developing radio-layer standards involving the physical layer, media access control (with QoS), data link control, and radio resource management, while the ATM Forum addresses *mobile* ATM protocol extensions involving hand off control, location management, routing considerations, traffic and QoS control, resource management, and wireless network management. In the future, vendors will use a combination of ETSI standards and ATM Forum specifications to develop the new generation of wireless ATM (WATM) product.

Telecommunication Information Network Architecture (TINA)

Telecommunication information network architecture (TINA) has been designed to develop an integrated service approach, taking into consideration whatever the information and/or telecommunications industries provide and the market requirements. It allows for dynamically changing structures with regard to stakeholders and their roles, domain borders, ownership, and price structures. The development of TINA was dramatically accelerated by the creation of the TINA Consortium (TINA-C) at the end of 1992 by around 40 leading companies of network features, telecom vendors, and IT vendors. TINA-C was created with the main goal of producing a set of architectural software specifications validated by various experiments. By the end of 1997, TINA-C had delivered a computing architecture, a service architecture, and a network architecture that are paving the future of telecom services using cooperative solutions in a competitive world. The Consortium has been a force in telecom and information technologies integration. It has issued a set of coherent, validated specifications that can be used as a whole, or as a set of partial technical solutions, when developing any kind of telecommunication and information services. At the completion of the first phase (1993–1997), TINA-C had reached its original objectives in terms of architectural framework, component specifications, and demonstration of feasibility. The second phase of TINA-C started in 1998 to facilitate the market-driven adoption of the TINA-C architecture [35].

For the telecommunications architecture people, TINA stands as a vision for telecommunications networks beyond intelligent networks; TINA also means integration of network services and network operations software on the same platform. For the computer software architect, TINA is a challenging opportunity to introduce the modern principles of computing, object orientation, and distributed computing to a real-time sensitive environment and an application that is truly global in scale. While TINA offers a vision of the infrastructure of the future, it also allows for gradual evolution and coexistence of the new with the present, allowing stockholders a spectrum of options ranging from taking leaps to taking small steps, dictated by their business plans. Another set of TINA goals comes from the regulatory perspective. It can quickly accommodate regulatory changes, particularly easing entry of new business entities and allowing changing relationships between different parts of the industry. Some of the specific objectives of TINA are:

- supporting clearly defined interfaces
- supporting flexible reference points
- uniform support of management
- support of new services
- intelligent mobility support
- scalability
- reusability of software (rapid introduction of services)

- separation of service from network transport
- technology independence.

Many of these objectives are no different from those for intellectual network, telecommunications management network (TMN), and others. The difference is that TINA is able to incorporate a new set of technologies to get closer to attaining the objectives.

The organization of TINA-C is shown in Figure 3.18. The General Forum is responsible for setting the strategic direction of the consortium, as well as for the administrative issues, and the election of Board Members. The technical Forum is responsible for technical issues. The Architecture Board (AB) is responsible for overall consistency of TINA specifications. The working groups meet more frequently that the other bodies and produce TINA specifications.

The TINA-C administrative office includes the Chief Executive Officer (CEO), Chief Technical Office (CTO), and a small support staff. Besides providing the operation infrastructure, the objective of this central office is to coordinate the technical projects among the member companies, and represent TINA-C in industry activities. TINA-C is an open consortium in the sense that its membership is open to any organization that has an interest in its activities. The final documents are posted on a public Web site, but meeting reports and documents in progress are available only to members.

The value and influence of TINA has been measured by the extent to which it provides for industry-implementable results to accelerate the availability of TINA-like products. Given that the TINA-C objective is to provide a common software architecture for the provisioning of telecommunications and information services, its achievement consists of sets of specifications for systems and components that allow meeting the objectives listed earlier. One can easily understand the goal of the TINA architectures directed to provisioning of any kind of services, running on a global scale, on different network technologies, allowing flexibility in combining any kind of media (voice, data, etc.) to multimedia services, and any kind of connectivity, allowing third-party connection setup and broadcasting as well as multiparty involvement. Because of that scope, the TINA-C initiative can be seen as a most valuable complement to GII activities. There is no preference for any particular service provided in the architectural area of TINA; since it is market-driven, its primary application areas are expected to be teleconferencing and computer-supported collaborative work, high-quality Internet information services, home shopping, video on demand (VoD), next-generation Universal Mobile Telecommunications Systems (UMTS), and virtual private network customization and management in a global environment.

TINA has been following the viewpoint separation defined by the Reference Model of Open Distributed Processing (RM-ODP) using its enterprise, information, computational, and engineering viewpoint specification rules. In short, in a TINA system:

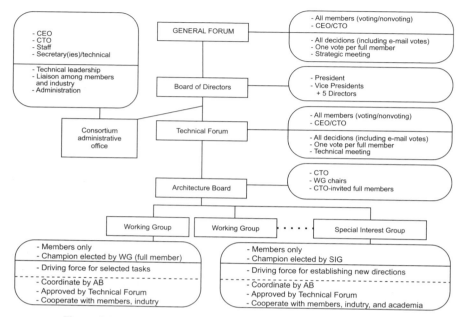

Figure 3.18 The TINA-C organizational structure [35]. (©1998 IEEE.)

- The enterprise viewpoint defines business roles and policies for the parties involved.
- The information viewpoint focuses on information-bearing entities and the relationships among them.
- The computational viewpoint models data types and functions in terms of interacting computational objects.

The engineering viewpoint describes distribution and deployment scenarios of objects allowing their execution and interaction based on a distributed processing environment (DPE), a connecting signaling network called a kernel transport network (KTN), and the computing and communication environment (NCCE) infrastructure.

The formal description techniques applied to the network specification process used in TINA are a simplified version of the guidelines for the definition on managed objects (GDMO) together with the generic relationship model (GRM) for information modeling, and for forthcoming work specification and description language (SDL) for formal behavior description. TINA-ODL can be viewed as a superset of the object management group's interface definitions language (OMG-IDL). The extensions are needed to meet the specific telecommunications requirement enhancements in terms of handling multiple interfaces, continuous information flow-related stream interfaces, QoS parameters, and others. TINA-ODL plays an important role in the current OMG component description language. Open distributed processing modeling concepts in TINA are shown in Figure 3.19.

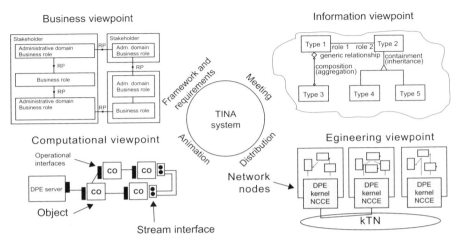

Figure 3.19 Open distributed processing modeling concepts in TINA [35]. (©1998 IEEE.)

The overall structure of a TINA system is depicted in Figure 3.20. The encapsulated service application and network resource component objects reside on top of a DPE, which hides the complexity of distribution and heterogeneity from the service developer. Thus, an abstraction from different data representations, or current position of the object or, if desired, a masking of failure detection and recovery actions is obtained.

Additional features of the basic DPE communication environment are provided by generically defined DPE services. Examples are services that provide trader and notification functions, performance monitoring, and transaction services. Any TINA application can make use of those services. The work performed in TINA was directed towards specifying requirements of DPE services that meet explicit telecommunication needs. A logical separated network, the kTN is designed to transport information between different DPE nodes. There are different implementations of DPE nodes available, one typically represented by common object request broker architecture (CORBA). The layer describing the NCCE contains operating systems, communication protocol stacks, and others. The transport network guarantees the handling of continuous information flows such as audio and video streams.

The TINA session model concept divides all processes involved in the provisioning of a service in a given time frame. Three main sessions have been identified, as shown Figure 3.21: the access session, the service session, and the communication session.

All activities in user–provider interactions to establish and maintain a service, including procedure for subscription, user profile settings, and authentication, are termed the access session, allowing also combining and/or conducting of several services at the same time.

The service session corresponds to the provisioning and usage of the service itself, handling the overall control and management of service and user interaction

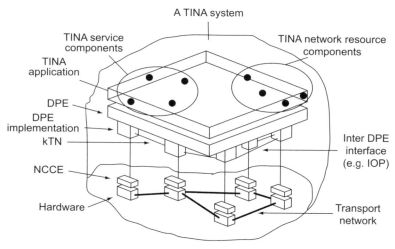

Figure 3.20 An overall structure of a TINA system [35]. (©1998 IEEE.)

during runtime. It can be subdivided into the user service session and the provider service session. The user service session is responsible for managing the involved user's activity and resource attributes. The provider service session provides service logic and functions needed for joining and inviting further participations.

The communication session provides the required network resources according to the quality of service requests of the service session. It provides an abstraction of the actual connectivity needed for running the service.

The TINA session model is applicable to a wide range of different business domain relationships. One example in the TINA service architecture is an extension of the user agent/provider agent concepts through peer agents in a computational

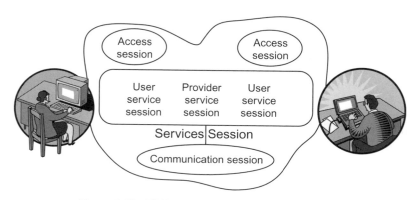

Figure 3.21 TINA session model [35]. (©1998 IEEE.)

model. Another is the extension of the TINA service session graph, representing different service session relationships in the information model. The most prominent area of applying the session concept is therefore TINA's service architecture [36]. It provides reusable components for service access and control, subscription, and accounting management. The service architecture supports a wide range of services, such as telecommunications and Internet services, and provides the means for customer-tailored services.

The TINA network modeling principles are based on the TINA Network Resources Information Model (NRIM), which contains specifications of the information elements, such as trail, links, and network termination point pools that represent the topological and connectivity structure in a network [36]. Control and management of network resources have been handled in a common manner, unlike the usual separation of control and management functions in existing networks. In terms of the session concept, the scope of network resource architecture describes the communication session. It allows for multipoint-to-multipoint connection and handles complex communication sessions based on multimedia service requirements.

One major goal of TINA-C has been to overcome, on the one hand, limitations of existing networks and service provisioning concepts, such as intelligent networks and traffic management network (TMN). Also, on the other hand the goal has been to protect the huge investment undertaken by telecommunications operators around the globe. A way out of these requirements is paved by migration and interworking scenarios from today's networks to TINA's architectural solutions. Depending on the interest, either to replace a whole system or gradually evolve to TINA, interworking units or adaptation units will be put in place. Figure 3.22 depicts a solution for interworking between IN and TINA by using a CORBA/SS7 gateway to bridge between the INAP and Inter-Orb Protocol (IOP), allowing exploitation of the full benefits of TINA's service architecture, in the shape of distributed IN components [37].

As a large and complex telecommunications and information system with global coverage, a TINA system can only be made possible by the cooperation of many service providers, network operators, and telecom and IT vendors in setting and observing international standards. A clear focus on making it more feasible for the industry to develop and use TINA-based products distinguishes the current activities from the first phase of the consortium, which was focused more on creating a conceptual/architectural underpinning for all subarchitectures of TINA. The projects of the current TINA activities reflect the product focus in the consortium goals. TINA specifications and architecture have been used in Object Management Group (OMG) for stream handling, in the Network Management Forum (NMF) for Distributed Processing Environment (DPE) introduction and Service Management in the ATM Forum for information modeling, and in the International Telecommunication Union – Telecommunication Standardization Sector (ITU-T) for Object Distributed Language (ODL) and business modeling.

Quality of Service in TINA

It has been recognized that QoS will be crucially important in multimedia broadband services as well as for its role in telecom service management. Free network service

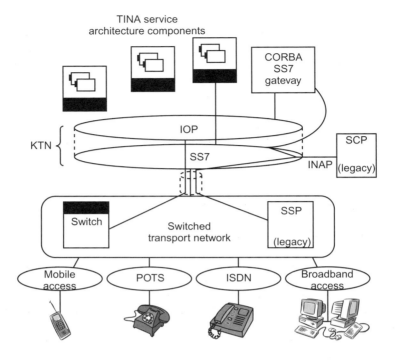

Figure 3.22 TINA interworking with existing networks [35]. (©1998 IEEE.)

does not guarantee availability of resources. Thus, it cannot guarantee QoS, which is not simply a technical matter. It can be properly understood only when the problem is viewed from both the enterprise and service point of view [38]. Figure 3.23 illustrates what we perceive as the four basic elements of QoS [39].

Service quality, which user applications expect from the network, needs to be understood in such a way that it is

- easy to express and understand from the user's point of view
- interpreted quantifiably and eventually mapped onto allocation of network resources.

Although service quality may not be explicitly stated, most applications assume a certain level of QoS from the network.

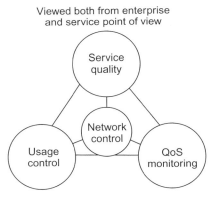

Figure 3.23 Four basic elements in QoS quartet [39]. (©1998 IEEE.)

Usage control is a mechanism to increase predictability of user behavior. Leaky bucket usage parameter control (UPC) is used at the edge of the network.

Service quality is translated into resource requirements, which are realized by QoS-oriented connection management protocols such as synchronous transfer mode user–network interface (UNI), and Resource Reservation Protocol (RSVP) [40, 41]. In combination with resource configuration, admission control, the multiplexing scheme, priority control, and network control guarantee the QoS of the connection.

A QoS guarantee cannot make sense unless network performance is monitored properly. Protocol level performance is relatively easy to monitor, but not necessarily by the kind of QoS an application may request or what the user perceives as QoS.

All four elements of QoS have to exist in any QoS-oriented network. Furthermore, they have to be integrated with a billing system based on sound economic principles. TINA serves as a hosting framework to which QoS elements can be added and upgraded, evolving itself towards high QoS oriented telecom applications.

Under a given pricing structure, the user is expected to behave most rationally. Suppose there is a set of QoS parameters that represent service quality. By assuming that the user has a hypothetical service quality function (SQF), the requirement would be stated as the following monotonic condition:

$$(\forall x)(\forall y) SQF(x) \leq SQF(y)$$
$$\Longrightarrow \text{Price}(x) \leq \text{Price}(y)$$

where x and y are two sets of QoS parameters, and Price() is a function representing the pricing structure. SQF is a vector, and the set of vectors are only partially ordered. This is due to the fact that different services qualities such as audio and video cannot be compared. Better video quality means better SQF, while all other service qualities are equal. The service quality function SQF should be simple, intuitively appearing and understandable, and yet expressive enough to describe a discernable difference of service quality, which may be reflected in the pricing structure. An SQF is a multidimensional function. Each dimension corresponds to an independent, discernable service quality such as audio or video. The value of each SQF dimension ranges from 0 to 1, where 1 should correspond to the best conceivable quality.

Example 3.3   This is an example of an SQF and its dimensions:

- *Audio*: notable audio service quality (e.g., CD, FM radio, 8 Kbps codec) should be mapped onto the scale.
- *Video*: notable video quality (e.g., NTSC, MPEG-2 352 × 240, HDTV) should be mapped onto the scale.
- *Response time*: an important metric for interactive games, but also a useful performance measure of operational interfaces. In contrast to audio and video, which are mainly transport network (TN) QoS, this metric may relate more to kernel TN (kTN) signaling network QoS.
- *Throughput*: this metric can be useful for file transfer protocol (FTP)-like service.

TINA makes a basic distinction between computational interfaces: operational interfaces, which correspond to management or control aspects of objects, and stream interfaces, which are ports of audio/visual multimedia traffic. In accordance with the distinction of interfaces, transport networks are categorized into two groups: the kTN is an extension of the signaling network, which is bound to operational interfaces, and the TN is bound to stream interfaces to carry multimedia traffic. Although both kTN and TN may be carried over a ubiquitous common transport protocol such as transmission control protocol (TCP) or user datagram protocol (UDP) on Internet protocol (IP), it should be noted that they present different requirements. The former emphasizes ubiquitous availability, reliability, real-time performance, and security. The latter emphasizes performance of the stream connection quality of service and billability. Naturally, our main focus is the binding of stream interfaces and stream flow connection (SFC) on TN, which are called stream binding.

Stream binding is a generic mechanism for multimedia, multiparty communication. As such, point-to-multipoint connection is an essential part of it, particularly for broadcasting or teleconferencing applications. The object management group (OMG) model is timed toward intranets or corporate virtual private networks, where the distinction between service provider and connectivity provider does not exist, or is not necessary.

International Multimedia Telecommunications Consortium (IMTC)

International Multimedia Telecommunications Consortium (IMTC) is an international community of companies working together to facilitate the availability of real-time, rich-media communications between people in multiple locations around the world. Rich media refers to converged communication sessions that incorporate voice and one-way (or two-way) data and one-way (or two-way) video. Members of this community include Internet application developers and service providers, teleconferencing hardware and software suppliers, telecommunications companies and equipment vendors, end users, educational institutions, government agencies, and nonprofit corporations [42]. The IMTC focuses on

- promoting standards that enable real-time, rich-media communications
- identifying utilization of multimedia products and services
- developing and submitting technology interoperability recommendation to official standards bodies such as European Telecommunication Standards Initiative (ETSI), Internet Engineering Task Force (IETF), Third-Generation Partnership Project (3GPP), International Organizations for Standardization (ISO), and International Telecommunications Union (ITU-T)
- initiating scheduled interoperability test sessions between suppliers of rich-media products and services
- educating and promoting the business and consumer communities on the benefits and implementation of multimedia products.

The IMTC identified interoperability as a key impediment to industry growth and initiated formal standards recommendations and test programs to improve interoperability of the products and services. The IMTC was formed in 1994 through a merger of the Consortium for Audio graphics Teleconferencing Standards (CATS) and the Multimedia Communications Community of Internet (MCCOI). CATS formed in 1993 focused on the T.120 standards suite for multipoint data conferencing. MCCOI also formed in 1993, focused on the ITU-T H.320 (ISDN) and H.324 (POTS) standards suites for multipoint video communications. In 1995 the IMTC merged with the Personal Conferencing Work Group (PCWG). The PCWG has also focused on multimedia teleconferencing standards as well as the needs of users of products and services in this category.

From the mid-1990s, as the rich-media communications concepts extended beyond traditional switched networks to include IP networks, the IMTC acted as an industry convergence point for voice, data, and video over IP. At that time, IMTC's efforts resulted in the promulgation of the ITU-T H.323 (packet-based video) standard, agreement on the initial voice over IP implementation, integrated interoperability tests involving the ITU-T T.120, H.320, H.324, H.323 standards and emerging IETF requests for comments such as session invitation protocol (SIP).

Today, the IMTC retains its focus on addressing rich-media deployment obstacles and interoperability in wide area and enterprise networks. The initiatives IMTC sponsors now include IP Version 6 (IPv6), 3GPP technology, packet switched streaming techniques, enterprise network address translation (NAT)/firewall traversal and wireless mobility. Its efforts enable service providers and vendors to create more compatible rich-media products, applications, and services, which in turn facilitate the widespread adoption of the offering by protecting end users' capital instruments and meeting usability expectations.

Together with the IPv6 Forum, IMTC has initiated joint activities focused on prolifering IPv6-based multimedia and conferencing applications, meeting the need for improved routing and security, addressing mobile (3GPP) issues, and defining enterprise firewalls, security, and management issues. More recently, IMTC has begun to highlight the interoperability of multimedia offerings for mobile phone users. These offerings are based on the 3GPP specifications for streaming and interacting with multimedia content on demand and in real time over 3G wireless networks.

Membership in the IMTC is by company. Current members include Internet application developers, teleconferencing hardware and software service suppliers, telecommunications companies, teleconferencing service providers and end users, educational institutions, government agencies, and nonprofit corporations.

MPEG-4 Industry Forum (M4IF)

The MPEG-4 Industry Forum is a not-for-profit organization with the goal to further the adoption of the MPEG-4 standard, by establishing MPEG-4 as an accepted and widely used standard among applications developers, service providers, content creators, and end users. The activities of the MPEG-4 Industry Forum (M4IF) generally start where MPEG stops. These include issues that MPEG cannot deal with, for example because of ISO rules, such as clearance of patents. A list of M4IF's current activities includes [43]:

- promoting the standard, and serving as a single point of information on MPEG-4 technology, products, and services
- carrying out interoperability tests, which lead to an ecosystem of interoperable products
- developing and establishing an MPEG-4 Certification program, which will come with the right to carry M4IF's logo
- organization of MPEG-4 exhibitions and tutorials (Geneva 2000, San Jose 2001 and 2002)
- organization of and participation in many trade show events – M4IF has shown floor presence together with some of its members
- initiating discussions leading to the potential establishment of patent pools outside of M4IF, which should grant a license to an unlimited number of applications throughout the world under conditions that are demonstrably free of any unfair competition
- studying licensing models for downloadable software decoders, such as Internet players.

The purpose of M4IF will be pursued by promoting MPEG-4, making available information on MPEG-4, making available MPEG-4 tools or giving information on where to obtain these, creating a single point for information about MPEG-4, and creating industrial focus around the usage of MPEG-4. The goals are realized through the open international collaboration of all interested parties, and are reasonably applied uniformly and openly. M4IF will contribute the results and its activities to appropriate format standards bodies if applicable.

MPEG-4 builds on the proven success of three fields [29, 33]:

- digital television
- interactive graphics applications (synthetic content)
- interactive multimedia (World Wide Web, distribution of and access to content).

MPEG-4 provides the standardized technological elements enabling the integration of the production, distribution, and content access paradigms of the three fields. M4IF is vital to the success of the MPEG-4 standard, since the work done by MPEG is necessary but not sufficient. In its attempt to get MPEG-4 widely adopted, M4IF picks up where MPEG stops. There are more than just technical issues to resolve. For example, M4IF does much marketing work and the recent public debate over the licensing terms for MPEG-4 visual profile was greatly assisted by the Forum's industry-wide public discussion on the announced schemes. In addition, M4IF has an advanced program of cross-vendor product interoperability testing.

The M4IF provides the industry with a means to work in a precompetitive, post-standardization environment to help building out the MPEG-4 ecosystem. Other activities of the forum include product certification, several working groups, access to MPEG committee members, and an annual conference.

Internet Streaming Media Alliance (ISMA)

The Internet Streaming Media Alliance (ISMA) creates a set of specifications for Internet Streaming. ISMA has chosen specific MPEG-4 Audio and Visual Profiles and levels and augmented these with IETF transport specifications to create cross-vendor interoperability for video on the Internet. The first ISMA-compliant implementations have started to emerge [44]. ISMA recommends streaming protocols for the Internet, including MPEG-4 Simple Visual and Advanced Simple profiles for video, MPEG-4 High Quality Audio Profile for audio, MPEG-4 file format for file storage, and real-time protocol (RTP) as well as real-time streaming protocol (RTSP) for streaming protocols and control.

Generally speaking, ISMA is a nonprofit corporation formed to provide a forum for the creation of specifications that define an interoperable implementation for streaming rich media (video, audio, and associated data) over IP networks. ISMA provides a forum for the creation and sponsorship of market and user education programs to accelerate the demand for products based on these specifications.

ISMA maintains relationships and liaison with educational institutions, government research institutes, other technology consortia, and other organizations that support and contribute to the development of relevant specifications and international standards. In developing specifications, ISMA utilizes relevant, established standards that exist and proposes additions or refinements as needed to relevant standards body efforts that are still in development. Just as the adoption of standards mark-up languages has fueled innovation and the explosive use of today's World Wide Web, the goal of Internet Streaming Media Alliance is to accomplish the same for the next move of rich Internet content, streaming video, and audio. In creating an interoperable approach for transporting and viewing streaming media, content creators, product developers, and service providers will have easier access to the expanding commercial and consumer markets for streaming media services. To date, the prohibitive costs associated with rolling out streaming video services that support all current, disparate formats has kept many potential service providers and other adopters from taking full advantage of existing market opportunities. The emerging class of Internet appliances stands to benefit from a single standard as these devices often cannot afford to have multiple streaming media players installed to view differently formatted video content from the Web.

Standards for many of the fundamental pieces needed for a streaming rich media over IP solution do exist. The ISMA adopts parts or all of those existing standards and contributes to those still in development in order to complete, publish, and promote a systemic, end-to-end specification that enables cross-platform and multi-vendor interoperability. The first specification from the ISMA defines an implementation agreement for streaming MPEG-4 video and audio over IP networks. The Alliance's ongoing work to animate the specifications includes adopting methods for digital rights management, reliable quality of service, and other relevant technologies.

Third Generation Partnership Project (3GPP/3GPP2)

The Third Generation Partnership Project (3GPP) is a collaboration agreement that was established in December 1998. The collaboration agreement brings together a number of telecommunications standards bodies, which are known as *Organization partners* [45].

3GPP defines standards for third-generation mobile networks and services starting from the global system for mobile (GSM)-based systems. In their wireless terminal specifications, 3GPP uses MPEG-4 simple visual profile for video, MPEG-4 file format for multimedia messaging and RTP, RTSP for streaming protocols and control.

The original scope of 3GPP was to produce globally applicable Technical Specifications and Technical Reports for a third-generation mobile system based on evolved GSM core networks and the radio access technologies that they support (Universal Terrestrial Radio Access (UTRA), both Frequency Division Duplex (FDD) and Time Division Duplex (TDD) modes). The scope was subsequently amended to include the maintenance and development of the global system for mobile communications, technical specifications, and technical reports including evolved radio access technologies and enhanced data rates for GSM evolution (EDGE). The discussions that led to the signaling of the 3GPP Agreement were recorded in a series of slides called the *Partnership project description*, which describes the basic principles and ideas on which the project is based. The Partnership Project Description has not been maintained since its first creation, but the principles of operation of the project still remain valid. In order to obtain a consolidated view of market requirements, a second category of partnership was created within the project, called *Market representation partners. Observer* status is also possible within 3GPP for those telecommunication standards bodies that have the potential to become organizational partners.

The Third Generation Partnership Project 2 (3GPP2):

- is a collaborative third-generation (3G) telecommunications specifications-setting project
- comprises North American and Asian interests developing global specifications for Cellular Radiotelecommunication Intersystem Operations network evolution to 3G and global specifications for the radio transmission technologies (RTTs).

3GPP2 defines standards for third-generation mobile networks and services starting from code division multiple access (CDMA) based systems and paying much attention to mobile multimedia. In the wireless terminal specification, 3GPP2 uses MPEG-4 simple visual profile for video, MPEG-4 file format for multimedia messaging and RTP, RTSP for streaming protocols and control.

3GPP2 was born out of the International Telecommunications Union's (ITU) International Mobile Telecommunications IMT-2000 initiative, covering high-speed, broadband, and IP-based mobile systems featuring network-to-network interconnection, feature/service transparency, global roaming, and seamless services independent of location. IMT-2000 is intended to bring high-quality mobile multimedia telecommunications to a worldwide mass market by achieving the goals of increasing the speed and ease of wireless communications, responding to the problems faced by the increased demand to pass data via telecommunications, and providing *anytime, anywhere* services.

In 1998, when serious discussions about working on the IMT-2000 initiative began, it became evident that the goals of globalization and convergence could not be accomplished efficiently using traditional standards-setting processes, often characterized as *too slow* given the speed with which technology was forging ahead. Bodies such as the Global Standards Collaboration (GSC) and Radio Standardization (RAST) helped to forge understanding of issues and work plans among all Participating Standards Organizations (PSOs).

Digital Video Broadcasting (DVB)

During 1991, broadcasters and consumer equipment manufacturers discussed how to form a concerted pan-European platform to develop digital terrestrial TV.

Towards the end of that year, broadcasters, consumer electronics manufacturers, and regulatory bodies came together to discuss the formation of a group that would oversee the development of digital television in Europe – the European Launching Group (ELG).

The ELG expanded to include the major European media interest groups, both public and private, the consumer electronics manufacturers, common carriers, and regulators. It drafted a Memorandum of Understanding (MoU) establishing the rules by which this new and challenging game of collective action would be played. The concept of the MoU was a departure into unexplored territory and means that commercial competitors needed to appreciate their common requirements and agendas. The MoU was signed by all ELG participants in September 1993, and the Launching Group became DVB (Digital Video Broadcasting). Around this time, the Working Group on Digital Television prepared a study on the prospects and possibilities for digital terrestrial television in Europe. The highly respected report introduced important new concepts, such as a proposal to allow several different consumer markets to be served at the same time (e.g., portable television and HDTV). In conjunction with this activity, change was coming to the European satellite broadcasting industry. It was becoming clear that the state-of-the art media access control (MAC) systems would have to give way to all-digital technology. DVB provided the forum for gathering all the major European television interests into one group. It promised to develop a complete digital television system based on a unified approach. It became clear that satellite and cable would deliver the first broadcast digital television services [46].

DVB presents the best, most practical solution for the consumer electronics industry, the broadcaster, and the viewer. The success of DVB depends upon the development of a coherent set of methods of bringing digital television into the home, by taking advantage of the special characteristics of a range of delivery media, including satellite, cable, and terrestrial, as well as satellite master antenna television (SMATV) and multipoint distribution systems (MDS). DVB also ensures that essential elements of the DVB standards are common to all delivery environments, providing benefits from economies of scale.

The DVB system provides a complete solution for digital television and data broadcasting across the range of delivery media. DVB adds to the MPEG transport stream multiplex, the necessary elements to bring digital television to the home through cable, satellite, and terrestrial broadcast systems. Also, the question often arises as to how DVB might help with the framework for the interactive television services of the future.

The DVB core system has the following philosophy:

- The system is designed as a container to carry flexible combinations of MPEG-2 video and audio, or other data.
- The system uses the common MPEG-2 transport stream (TS) multiplex.
- The system uses a common service information (SI) system giving details of the programs being broadcast.

- The system uses a common first-level Read–Solomon (RS) forward error-correction (FEC) system.
- Modulation and additional channel coding systems, if any, are chosen to meet the requirements of the different transmission media.
- A common scrambling system is available.
- A common conditional access interface is available.

The DVB family of standards includes

- DVB-S digital satellite system for use in the frequency bands up to 11/12 GHz
- DVB-C digital cable delivery system, compatible with DVB-S
- DVB-CS digital SMATV system, adopted from DVB-C and DVB-S, to serve community antenna installations
- DVB-MC digital multipoint distribution system, using microwave frequencies above about 10 GHz for direct distribution to viewers' homes – based on the DVB-C cable delivery system
- DVB-MS digital multipoint distribution system, using microwave frequencies above about 10 GHz for direct distribution to viewers' homes – based on the DVB-S satellite delivery system
- DVB-T digital terrestrial television system, designed for terrestrial 8 MHz (and 7 MHz) channels
- DVB-SI service information system for use by the DVB decoder to configure itself and to help the user to navigate DVB bit streams
- DVB-TXT DVB fixed-format teletext transport specification
- DVB-CI DVB common interface for use in conditional access and other applications
- DVB-IPI transport of the DVB services over IP-based networks.

DVB-S is a single carrier system. It can be modeled as a kind of onion. In the center, the onion's core is the payload, which is the useful bit rate. Surrounding this are a series of layers to make the signal less sensitive to errors and to arrange the payload in a form suitable for broadcasting. The video, audio, and other data are inserted into fixed-length MPEG transport stream packets. The packetized data constitutes the payload. The first step is to form the data into a regular structure by inverting synchronization bytes every eighth packet header. The next step in the processing is to randomize the contents. The following step is to add a Read–Solomon forward error-correction overhead to the packet data. This is a very efficient system, which adds less than 12 percent overhead to the signal. It is called the outer code. Next, convolutional interleaving is applied to the packet contents. Following this, a further error correction system is added, using a punctured convolutional code. This second error correction system, the inner code, can be adjusted, in the amount of overhead, to suit the needs of the service provider. Finally, the signal is used to modulate the satellite broadcast carrier using quadrature phase-shift keying (QPSK). In essence, between the multiplexing and the physical transmission, the system is tailored to the specific channel properties. The system is arranged to adopt to the error characteristics of the channel. Burst errors are randomized and two layers of forward error correction are added. The second level, or inner code, can be adjusted to suit the circumstances (power, bit rate available). There are thus two variables for the service provider: the total size of the onion and the thickness of the second error-correction outer skin. In each case, the receiver will discover the right combination to use by very rapid trial and error on the received signal. An appropriate combination of payload size and inner code can be chosen to suit the service operator's environment.

In DVB-C, the cable network system has the same core as the satellite system. The modulation system is based on quadrature amplitude modulation (QAM) rather then QPSK. No inner code forward error-correction is needed. The system is centered on 64-QAM, but lower-level systems, such as 16-QAM and 32-QAM, can also be used. In each case, the data capacity of the system is traded against robustness of the data. Higher-level systems, such as 128-QAM and 256-QAM, may also become possible, but their use will depend on the capacity of the cable network to cope with the reduced decoding margin.

DVB-MC is the digital multipoint distribution system, which uses microwave frequencies below approximately 10 GHz for direct distribution to viewers' homes. It is based on the DVB-C cable delivery system and will therefore enable a common receiver to be used for both cable transmission and this type of microwave transmission.

DVB-T system specifications, for the terrestrial broadcasting of digital television signals, were approved by the Steering Board in December 1995. The work was based on a set of user requirements produced by the Terrestrial Commercial Module of the DVB Project. DVB members contributed to the technical development of DVB-T through the DTTV-SA (Digital Terrestrial Television – Systems Aspects) of the Technical Module. As with the other DVB standards, MPEG-2 sound and vision coding forms the payload of DVB-T. Other elements of the specification include a transmission scheme based on orthogonal frequency division multiplexing (OFDM), which allows for the use of either 1705 carriers or 6817 carriers. Concatenated error correction is used. Also, Read–Solomon outer coding and outer convolutional interleaving are used, in common with the other DVB standards. The inner coding is the same as that used for DVB-S. Two-level hierarchical source coding is not used, since its benefits do not justify the extra receiver complexity involved. Because of the multipath immunity of OFDM, it is potentially possible to operate an overlapping network of transmitting stations with a single frequency. In the areas of overlap, the weaker of the two received signals is like an echo signal. However, if the two transmitters are far apart, the time delay between the two signals will be large and the system will therefore need a large guard interval.

DVB-SI provides the elements necessary for the development of the electronic program guides (EPG) that are likely to become a feature of the new digital television services. More elaborate EPGs may also be provided, perhaps as additional elements via a receiver interface. DVB-SI needs to describe the technical attributes of each service offered by an individual broadcaster. Other information, such as start time, the name of the service provider, and the classification of the event (sports, news, etc.), is given.

DVB systems are developed through consensus in the working groups of the Technical Module. Members of the groups are drawn from the general assembly of the project. Once standards have been published through ETSI, they are available at a nominal cost for anyone, worldwide. Because the standards are open, all the manufacturers making compliant systems are able to guarantee that their DVB equipment will work with other manufacturers' DVB equipment [46]. Owing to the use of MPEG-2 packets as *data containers* and the critical DVB service information surrounding and identifying those packets, DVB can deliver to the home almost anything that can be digitized, whether it is high definition TV, multiple channel standard definition TV, or even exciting new broadband multimedia data and interactive services. For each specifications, a set of user requirements is compiled by the Commercial Module. These are used as constraints on the specification. User requirements outline market parameters for a DVB system (price band, user function, etc.). The Technical Module then develops the specification, following these user requirements. The approval process within DVB requires that the Commercial Module supports the specification before it is finally approved by the Steering Board.

In contrast to earlier initiatives in Europe and the United States, the DVB Project works to strict commercial requirements established by the organizations that work every day to meet its needs. Working to tight timescales and strict market requirements means achieving a considerable economy of scale, which ensures that, in the transformation of the industry to digital, broadcasters, manufacturers, and, finally, the viewing public will benefit. Although DVB began as a European project, its membership has now spread around the globe, and reflecting this, the project members now aim at achieving world standards. Liaison takes place regularly with the ITU-R and ITU-T on the world standardization of systems developed under the DVB project.

Digital Audio Visual Council (DAVIC)

Since the inception the Digital Audio Visual Council (DAVIC), DVB members have understood the importance of working closely with it. DAVIC covers an extremely wide field, generally extending well outside the area of broadcasting, and DAVIC seeks to provide end-to-end interoperability for the use of digital images and sound across countries and between applications and services. DAVIC liaison officers have been appointed in DVB to coordinate the efforts of both groups. Harmonization with DAVIC has been achieved in many areas. The DVB Cable and Satellite Systems have been adopted by DAVIC. The DVB subtitling group has also reached broad agreement with DAVIC, after a good deal of discussion and a willingness to change its own draft specification. Other harmonization agreements have been reached in areas including interactive services, receiver interfaces, and service information, and work will continue to ensure the greatest possible commonality.

The general purpose of DAVIC is to standardize system architectures, interfaces between systems, and the communication protocols necessary for providing services, such as VoD and application systems. DAVIC is composed of a Board of Directors, a Management Committee, a Membership and Nomination Committee, a Finance and Audit Committee, a Strategic Planning Advisory Committee, and Technical Committees (TCs), by whom technical specifications are studied. The first DAVIC specification, DAVIC 1.0, was finalized in December 1995. In September 1996, the second DAVIC specification, DAVIC 1.1, was completed and published. The main items added by DAVIC 1.1 were:

- fundamental specifications for access from a set-top terminal to the Internet
- Java application programming interface (API) specifications for changing a *set-top box* into a *virtual machine* by downloading software
- transmission standards for terrestrial broadcasting by microwave.

DAVIC 1.2 was established in December 1996. It added:

- basic security for DAVIC 1.0 systems
- high-quality audio and video standards
- Internet access by high-speed video network
- ADSL ATM mapping.

DAVIC 1.3 was published in September 1997. It includes several technical points:

- DAVIC service and system management
- communication services (telephony, conferencing, and multiplayer gaming)
- multiple servers and services
- a still picture display control API for the set-top unit.

DAVIC Technical Committees are driven by the need to deliver specifications of the highest possible quality given the limitations of time and technology within a given deadline. A fundamental *rule* is that only one tool shall be specified for any functionality. This rule avoids unnecessary duplication of components that are difficult to change in the target applications. In making their decisions, the TCs must balance other criteria such as technical merit, cost, and wide usage. If necessary, a TC can adopt a duplicated tool or, where required, choose a technical solution not drawn from existing standards. The final DAVIC specification represents the agreed-upon deliberations, sometimes achieved by ballot, of the member companies across the entire communications and information delivery spectrum. The degree of international consensus is high because the proportion of DAVIC membership from Asia, Europe, and the Americas is roughly identical. In addition to participation by established DAVIC members, each DAVIC meeting can be attended by outside observers. The association DAVIC was closed according to its status after five years of activity and remains only active through its Web site. It represented all sectors of the audiovisual industry: manufacturing (computer, consumer electronics, and telecommunications equipment) and service broadcasting, telecommunications and cable television (CATV), as well as a number of government agencies and research organizations. DAVIC has been creating the industry standard for end-to-end interoperability of broadcast and interactive digital audiovisual information and of multimedia communications. DAVIC has had a vision of an audiovisual world where producers of multimedia content can reach the widest possible audience, where users are protected from obsolescence and have seamless access to information and communication carrier can offer effective transport, while manufacturers can provide hardware and software to support unrestricted production, and use of information.

Advanced Television Systems Committee (ATSC)

The Advanced Television Systems Committee (ATSC) is an international, nonprofit membership organization developing voluntary standards for the entire spectrum of advanced television systems. Specifically, ATSC is working to coordinate television standards among different communications media focusing on digital television, interactive systems, and broadband multimedia communications. ATSC is also developing digital television implementation strategies and presenting educational seminars on the ATSC standards [47]. ATSC was formed in 1992 by the member organizations of the Joint Committee on Inter Society Coordination (JCIC), the Electronic Industries Association (EIA), the Institute of Electrical and Electronic Engineers (IEEE), the National Cable Television Association (NCTA), and the Society of Motion Picture and Television Engineers (SMPTE). ATSC was incorporated in January 2002.

ATSC Digital TV Standards include digital high definition television (HDTV), standard definition television (SDTV), data broadcasting, multichannel surround-sound audio, and satellite direct-to-home broadcasting.

In December 1996, the United States Federal Communications Commission (FCC) adopted the major elements of the ATSC Digital Television (DTV) Standard (A/53). The ATSC DTV Standard has been adopted by the governments of Canada (in November 1997), South Korea (November 1997), Taiwan (May 1998), and Argentina (1998).

European Broadcasting Union (EBU)

European Broadcasting Union (EBU) is the largest professional association of national broadcasters in the world. Activities include operation of the Eurovision and Euroradio networks, coordination of news and sports programming, promotion of technical standardization legal advice, and the defense of public service broadcasting [48].

The Digital Strategy Group is a group of the EBU Administrative Council established to evaluate policy and strategy options for EBU members in the digital environment. The group shares experiences and examines whether those are common elements of policy in events such as new media. On-Line Service Group includes heads of multimedia and Internet Services in EBU members. It shares experiences in the development of Internet-based services and coordinates activities for major events that are offered via the Web. Forum is an annual event designed to bring to light new problems and new solutions in the area of new media. It primarily discusses economic and content aspects and is open to all EBU members.

REFERENCES

1. K. Asatani and S. Nogami, Trends in standardization on multimedia communications, *IEEE Comm. Magazine*, 35, 112–116 (1997).
2. K. Asatani and S. Nogami, Trends in standardization of telecommunications on GII, multimedia and other network technologies and services, *IEEE Comm. Magazine*, 34, 33–46 (1996).
3. M. H. Sherif, A framework for standardization in telecommunications and information technology, *IEEE Comm. Magazine*, 39, 94–100 (2001).
4. M. H. Sherif and D. K. Sparrell, Standards and innovations in telecommunications, *IEEE Comm. Magazine*, 30, 22–29 (1992).
5. G. Schroder and M. H. Sherif, The road to G.729, *IEEE Comm. Magazine*, 35, 48–54 (1997).
6. M. H. Sherif, Contribution towards a theory of standardization in telecommunications, *1st IEEE Conference on Standardization and Innovation in Information Technology* (SIIT'99), Aachen, Germany, September 15–17, 1999, pp. 143–148.
7. K. Jakobs, *Participation in Standardization Processes – Impact, Problems and Benefits*, Viewing Publishers, 2000.
8. J. Kipnis, Beating the system: abuses of the standards adoption process, *IEEE Comm. Magazine*, 38, 102–105 (2000).
9. Alliance for Telecommunications Industry Solutions, *1995 Annual Report of Committee T1 – Telecommunications*, March 1996.
10. ETSI, *Report of the Sixth Strategic Review Committee on European Information Infrastructure*, Sophia Antipolis, France, June 1995.
11. The Telecommunication Technology Committee, *Preparations for Promotion of NII of Japan and GII (Proposal for TTC Standardization Activities)*, Strategic Research and Planning Committee Information Infrastructure Task Group, November 1995.
12. N. B. Seitz and K. C. Glossbrenner, Performance standards for the GII, *IEEE Comm. Magazine*, 36, 116–121 (1998).
13. K. Asatani, F. Bigi, and P. A. Probst, Telecommunications standardization for the new millennium: ITU-T's strategies, *IEEE Comm. Magazine*, 39, 124–130 (2001).
14. ISO/IEC/ITU, *GII Standards Policy Development Meetings*, Geneva, June 1996.
15. TSG, ISO/IEC JTC1, *Standards Necessary to Define Interfaces for Application Portability (IAP) – Final Report*, Comp. Stds. and Interfaces 15, pp. 469–514, April 1991.
16. ISO/IEC Directives, Pt. 1, *Procedures for technical work*, available at http://www.iso.ch/dire/directives.html.

17. ISO/IEC and ITU-T, Guide for ITU-T and ISO/IEC JTC1 Cooperation, Annex A to Recomm. A.23, October 1996.
18. ISO/IEC and ITU-T, Guide for ITU-T and ISO/IEC JTC1 Cooperation, Annex K to ISO/IEC JTC1 Directives, December 1996.
19. *Mapping between ISO/IEC Standards and ITU-T Recommendations*, Jan. 1991, available at http://www.itu.int/itudoc/itu-t/com7.html.
20. L. Chiariglione, *MPEG and multimedia communication*, 1997, available at http://www.cselt.stet.it/ufv/leonardo/paper/mpeg.htm.
21. R. Hovey and S. Braduer, The organizations involved in the IETF standard process, RFC2028, October 1996.
22. M. Tatipamula and B. Khasnabish, *Multimedia Communications Networks Technologies and Services*, Artech House, Hingham, MA, 1998.
23. K. Asatani, Standardization of network technologies and services, *IEEE Comm. Magazine*, 32, 86–91 (1994).
24. ITU-T SG13, *IP Project Version 5*, May 2001.
25. ITU MEDIACOM2004, *Project description – Version 3.0*, March 2002.
26. ISO/IEC JTC1/SC29/WG11 Doc.N5231, October 2002.
27. ISO/IEC JTC1/SC29/WG11 Doc.N4512, December 2001.
28. ISO/IEC TR2100-1, 2001 (E), ISO/IEC JTC1/SC29/WG11 Doc.N4333, Information technology – Multimedia framework (MPEG-21) – Part 1: Vision, technologies and strategy, July 2001.
29. K. R. Rao, Z. S. Bojkovic, and D. A. Milovanovic, *Multimedia Communication Systems: Techniques, Standards and Networks*, Prentice Hall PTR, Upper Saddle River, NJ, 2002.
30. B. Carpenter, *Architectural Principles of the Internet*, RFC1958, June 1996.
31. S. Braduer, Ed., *The Internet Standards Process – Revision 3*, RFC2026, October 1996.
32. G. H. Dobrovski, The ATM Forum: developing implementation agreements, *IEEE Comm. Magazine*, 36, 121–125 (1998).
33. K. R. Rao and Z. S. Bojkovic, *Packet video communications over ATM networks*, Upper Saddle River, NJ: Prentice Hall PTR, 2000.
34. Available at http://www.atmforum.com/aboutforum/atmf.html.
35. Y. Inone, D. Guha, and H. Berndt, The TINA consortium, *IEEE Comm. Magazine*, 36, 130–136 (1998).
36. TINA-C, *Service architecture baseline*, v.5.0, June 1997, available at http://www.tinac.com.
37. EURESCOM P508D1, *Initial assessment of the options for evolving to TINA*, available at http://www.eurescom.de/public/deliverables/dfp.htm.
38. K. Werbach, Digital tornado: the Internet and Telecommunications Policy, OPP Working Paper No. 29, FCC, Washington, DC, March 1997.
39. T. Hamada et al., Service quality in TINA: Quality of service trading in open network architecture, *IEEE Comm. Magazine*, 36, 122–130 (1998).
40. ATM Forum, *ATM User–Network Interface (UNI) Signaling Specification, Version 4.0*, July 1996.
41. N. Venkatasubramanian et al., An integrated metric for video QoS, *Proc. ACM Multimedia Conference*, Seattle, WA, pp. 371-381, November 1997.

42. *IMTC by law*, rev. Feb. 2003, available at http://www.impte.org/imptebody.htm.
43. *MPEG-4 Industry Forum MP4*, Nov. 2002, available at http://www.m4if.org.
44. Available at http://www.isma.tv
45. Available at http://www.3gpp.org
46. Available at http://www.dvb.org
47. Available at http://www.atsc.org
48. Available at http://www.ebu.ch
49. ISO/IEC JTC1/SC29/WG11 Doc.N4991, *MPEG-21 Use case scenario document*, July 2002.

4

APPLICATIONS LAYER

The challenge of multimedia communications is to provide applications that integrate text, image, and video information and to do it in a way that presents ease of use and interactivity. We start with discussing ITU and MPEG applications from the integration, interactivity, storage, streaming, retrieval, and media distribution points of view. After that, digital broadcasting is described. Beginning from the convergence of telecommunications and broadcasting infrastructures, we continue with an overview of digital audio, radio broadcasting, digital video broadcasting, and interactive services in digital television infrastructure. This chapter also reviews mobile services and applications (types, business models, challenges, and abstracts in adopting mobile applications) as well as research activities, such as usability, user interfaces, mobile access to databases, and agent technologies. Finally, this chapter concludes with the current status of universal multimedia access technologies and investigates future directions in this area: content representation, adaptation and delivery, intellectual property and protection, mobile and wearable devices.

4.1 INTRODUCTION

The lists of Web sites maintained by standards, groups, fora, consortia, alliances, research institutes, private organizations, universities, industry, professional/technical societies, and so on, are quite extensive, and every effort is made to compile an

---

Introduction to Multimedia Communications, By K. R. Rao, Zoran S. Bojkovic, and Dragorad A. Milovanovic
Copyright © 2006 John Wiley & Sons, Inc.

148   APPLICATIONS LAYER

up-to-date list. The types of organizations producing standards, specifications, recommendations, or prestandards are

- *International standards bodies* such as ISO, IEC, and ITU-T – these develop standards and recommendations applicable throughout the world.
- *Regional standards bodies* such as ETSI – these produce standards applicable to large parts of the world.
- *National standards bodies* – these produce standards applicable to a single nation.
- *Prestandardization bodies such as IETF* – they produce *pre-standards* or *functional applications*, which are not standards in the legal sense, but are often used as the basis of standards that are later adopted by a regional or international standards organization.
- *Industry standards bodies* – these produce standards for the IT industry, some of which are later adopted by ISO.

ITU, a United Nations organization, is responsible for coordination of global telecom networks and services among governments and the private sector. Application areas in ITU GII Standardization projects are:

- home entertainment
- electronic commerce
- interbusiness processes
- personal communications
- public service communications
- publications.

Application aspects in GII include:

- medical informatics
- libraries
- electronic museums
- road transport informatics
- industrial multimedia communication
- ergonomics
- character set
- geographic information systems.

ITU GII services are:

- data interchange
- graphic

- audio
- data processing
- data management
- cost billing
- network control.

Services building blocks are:

- access methods
- addressing
- compression
- cost
- quotation
- data navigation
- data portability
- identification
- internationalization
- interoperability testing
- latency control
- nomadicity/mobility
- priority management
- privacy/ownership
- quality of service
- route selection
- search
- security.

The Internet Engineering Task Force (IETF) is a large, open, international community of network designers, operators, vendors, and researchers concerned with the evolution of the Internet architecture and smooth operation of the Internet. IETF applications areas are:

- Application configuration access protocol
- Content navigation
- Digital audio video (DAV) searching and locating
- Large-scale multicast applications
- Detailed revision/update of message standards
- Electronic data interchange – Internet integration
- Extension to file transfer protocol (FTP)
- Hyper text transfer protocol (HTTP)

- Instant messaging and presence protocol
- Internet fax
- Internet open trading protocol
- Internet routing protocol
- Calendaring and scheduling
- Mail and directory managements
- Message tracking protocol
- Network news transfer protocol (NTP) extensions
- Scheme registration
- Uniform resource registration procedures
- User article standard update
- WWW distributed authoring and versioning
- Web replication and caching
- Web version and configuration management.

As for multimedia applications over IP, we have

- IP telephony
- Media gateway control protocol
- Reliable multicast transport
- Performance implications of link characteristics
- Internet fax.

IETF transport WG is responsible for

- Audio/video transport
- Context and micromobility routing
- Differentiated services (diffserv)
- IP performance metrics
- IP storage
- IP telephony
- Integrated services over specific link layers
- Media gateway control
- Middlebox communication
- Multicast-address allocation
- Multiparty multimedia session control
- Network address translators
- Network file systems
- Performance implications of link characteristics

- Reliable multicast transport
- Reliable server pooling
- Resource reservation setup protocol
- Robust header compression
- Internet service
- Session initiation protocol
- Signaling transport
- Telephone number mapping.

The Advanced Television Systems Committee (ATSC) is an international organization that is establishing voluntary technical standards for advanced television systems. ATSC DigitalTV standards include digital high definition television (HDTV), standard definition television (SDTV), data broadcasting, multichannel surround-sound audio, and satellite direct-to-home broadcasting.

Digital Video Broadcasting (DVB) is a consortium of companies in the field of broadcasting, manufacturing, network operation, and regulatory matters that have come together to establish common international standards for the move from analog to digital broadcasting. For each specification, a set of user requirements is compiled by the Commercial module. These are used as constraints for a DVB system (price-band, user functions, etc.). The Technical module then develops the specification, following these user requirements. The approval process within DVB requires that the commercial module supports the specification before it is finally approved by the steering board.

The multimedia home platform (MHP) defines a generic interface between interactive digital applications and the terminals on which those applications execute. This interface decouples different providers' application from the specific hardware and software details of different MHP terminal implementations. It enables digital content providers to address all types of terminal, ranging from low-end to high-end set-top boxes, integrated digital TV sets, and multimedia PCs. The MHP extends the existing, successful DVB open standards for broadcast and interactive services in all transmission networks including satellite, cable, terrestrial, and microwave systems.

Resources for streaming media professionals include industry, service providers, infrastructure, and content providers. Technology for streaming media comprises network protocols, networks, media packing and formats, A/V compression, encryption, and watermarking.

Application-related projects are in support of a particular application area pertinent to the EII/GII. The number of applications of the EII is potentially limitless.

The ATM Forum is an international nonprofit organization formed with the objective of accelerating the use of asynchronous transfer mode (ATM) products and services through a rapid convergence of interoperability specifications. In addition the Forum promotes industry cooperation. The ATM Forum remains open to any organization that is interested in accelerating the availability of

ATM-based solutions. IT consists of a worldwide Technical committee, marketing programs such as Broadband exchange, and the User committee, through which ATM end users participate.

The 3rd Generation Partnership Project (3GPP) is a collaboration agreement that was established in 1998. The original scope of 3GPP was to produce globally applicable technical specifications and technical reports for a 3rd generation mobile system based on evolved GSM core networks and the radio access technologies that they support: Universal terrestrial radio access (UTRA), both frequency division duplex (FDD) and time division duplex (TDD) modes. The scope was subsequently amended to include the maintenance and development of the GSM. The dimensions that led to the signing of the 3GPP Agreement were recorded in a series of slides called the *Partnership Project Description*, which describes the basic principles and ideas on which the project is based. In order to obtain a consolidated view of market requirements, a second category of partnership was created within the project called *Market Representation Partners*. Observer status is also possible within 3GPP for those telecommunication standards bodies that have the potential to become Organizational partners, but which, for various reasons, have not yet done so. A permanent project support group called the *Mobile Competence Centre* (MCC) has been established to ensure the efficient day-to-day running of 3GPP.

ITU-T Study Group 16 has initiated the project MEDIACOM2004, which addresses the development of related standards. The intent of the document *Project Description* [1] is to review the project and how it relates to the ISO/IEC JTC1/SC29 MPEG-21 project, identifying areas of overlap and potential cooperation.

The scope of MPEG-21 is the integration of the critical technologies enabling transparent and augmented use of the critical multimedia resources across a wide range of networks and devices. The aim is to support functions such as content creation, content production, content distribution, content consumption and usage, content packing, intellectual property management and protection, user privacy, terminals and network resource abstraction, content representation, and event reporting. MPEG will elaborate the elements by defining the syntax and semantics of their characteristics, such as interfaces to the elements. MPEG-21 will also address the necessary framework functionality, such as protocol associated with the interfaces and mechanisms to provide a repository, composition conformance, and so on.

The project MEDIACOM2004 was established by Study Group 16 in its capacity as lead study group for multimedia studies in the ITU-T. The intent of the project is to generate a framework for multimedia communication studies for use within the ITU. The formation of this project has been endorsed by the World Telecommunication Standardization Assembly 2000.

The main objective of MEDIACOM2004 is to establish a framework for the harmonized and coordinated development of global multimedia communication standards. It is intended that this will be developed with the support of, and for use by, all relevant ITU-T and ITU-R study groups, and in close cooperation with other regional and international standards development organizations (SDOs).

4.1 INTRODUCTION

The MEDIACOM framework standardization areas are:

- Application design (ITU-T SG7,9,16, ITU-R)
- Middleware design (ITU-T SG7,9,16, SSG, ITU-R, JTC1, IETF, W3C)
- Network design (ITU-T SG13, 2,3,11,15, SSG, ITU-R).

The key framework study areas (FSA) identified for study within MEDIACOM2004 are [2]

- multimedia architecture
- multimedia applications and services
- interoperability of multimedia systems and services
- media coding
- quality of service and end-to-end performance in multimedia systems
- security of multimedia systems and services
- accessibility to multimedia systems and services.

These FSAs correspond to Questions B/16 through H/1 of the questions now under study by SG16. Table 4.1 shows the matrix of relationships between MEDIACOM2004 Framework Study Areas and MPEG-21 architectural elements and parts [1].

Although both projects involve the definition of a *multimedia framework*, in very general terms, the two projects can be differentiated. MPEG-21 addresses primarily the augmented use of multimedia resources across a wide range of networks and devices by all the users of the value chain, including end users, content creators, right holders, and so on. On the other hand, MEDIACOM2004 addresses primarily a framework for multimedia communications standardization.

With the growing ubiquity and mobility of multimedia-enabled services, universal multimedia access (UMA) is emerging as one of the important components for the next generation of multimedia applications. The basic concept underlying UMA is universal or seamless access to multimedia content, by automatic selection and adaptation of content based on the user environment [3]. Methods in this context may include selection among different pieces of content or many different variations of a single piece of content. Methods for adaptation include rate reduction, adaptive spatial and temporal sampling, quality reduction, summarization, personalization, and reediting of the multimedia content. The different relevant parameters in the user environment include device capabilities, available bandwidth, user preferences, usage context, as well as spatial and temporal awareness. UMA is particularly addressed by scalable or layered encoding, progressive data representation and object- or scene-based coding. The example is in MPEG-4 [4], which inherently provides different embedded quality levels of the same content. From the network perspective, UMA involves important concepts related to the growing variety of the communication channels, dynamic bandwidth variation and perceptual quality of

Table 4.1 The matrix of relationships between MEDIACOM2004 and MPEG-21 [1]

MEDIACOM 2004

	Digital Item Declaration	Digital Item Identification and Description	Content Handling and Usage	IPMP	Terminals and Networks	Content Representation	Event Reporting
Architecture	X						
Applications and services	X	X	X		X		X
Interoperability	X				X		
Coding				X	X	X	
Quality of service					X	X	
Security	X		X	X	X		
Accessibility					X		

MPEG-21

	Vision, Technologies and Strategy	Digital Item Declaration	Digital Item Identification and Description	IPMP Architecture	Rights Expression Language	Rights Data Dictionary	Digital Item Adaptation
Architecture	X	X					X
Applications and services	X	X		X	X		X
Interoperability	X		X				X
Coding	X			X			
Quality of service	X			X			
Security	X	X		X	X	X	
Accessibility	X				X		X

©2002 ISO/IEC.

service (QoS). UMA also involves different preferences of a user (recipients of the content) or a content publisher in choosing the form, the quality, or the personalization of the content. UMA promises an integration of these different perspectives into a new class of content adaptive applications that could allow users to access multimedia content without concern for specific coding formats, terminal capabilities, or network conditions [5, 6].

A new business and technology called *data broadcasting* is a concept whereby a combination of video, audio, software programs, streaming data, or other digital multimedia content is transmitted continuously to intelligent devices where it can be manipulated.

The broadcast concept means that although a return path is available, it is not required. Content is received without being requested. There are many services and applications that can be operated within a data broadcasting system. IT can be used for background routing of large e-mails, fast delivery of content to Internet service providers and for a new and very powerful way of videoconferencing in corporations. Another data broadcasting concept is that of a *channel*: a constantly updating media experience that combines rich broadcast with interaction. Such a *channel* need not be confined to single media. It can have streams of video in one part of a screen and interactive content of any sort in another part, and can provide the ultimate electronic entertainment. Data broadcasting can also provide a range of other experiences like virtual universities, powerful corporate presentations, and so on [7].

European Broadcasting Union (EBU) is the largest professional association of national broadcasters in the world, with active members in Europe, North Africa, and the Middle East, and associate members. Activities include operation of the Eurovision and Euroradio networks, coordination of news, promotion of technical standardization, legal advice, and the defense of public service broadcasting. At present, EBU technical activity is divided into three main areas: broadcasting systems (new systems and spectrum management), network systems, and production systems.

Each area is supervised by a Management Committee, which besides the priorities sets up projects, when necessary, for detailed studies. The Management Committee reports to the Technical Committee, but, within their spheres of competence, they decide EBU policy.

Since the creation in 1990 of the EBU/ETSI Joint Technical Committee (JTC), the European Broadcasting Union has been a full partner of the European Telecommunications Institute. Remember that ETSI is the body responsible for the establishment of voluntary European standards for telecommunications systems and technologies, including those intended for broadcast transmission and emission. The JTC was enlarged in 1995 to include the European Committee for Electrotechnical Standardization in matters relating to the standardization of radio and television receivers and related apparatus.

The Digital Video Broadcasting (DVB) system provides a complete solution for digital television and data broadcasting across the range of delivery media. The digital vision and sound-coding systems adopted for DVB use sophisticated compression techniques. The MPEG standard specifies a data stream syntax. The system designer is given a *toolbox* from which to make up systems incorporating greater or lesser degrees of sophistication. In this way, services avoid being over engineered, yet are able to respond fully to market requirements and are capable of evolution. MPEG Layer III is a digital compression system that takes advantage of the fact that a sound element will have a masking effect on lower-level sounds (or on noise) at nearby frequencies. This is used to facilitate the coding of the audio with low data rates. Sound elements that are present, but would not be layered even if reproduced faithfully, are not coded. The MPEG Layer II system can achieve a sound quality that is very close to compact disk. The system can be used for mono, stereo, or multilingual sound. Also, MPEG-2 audio can be used for surround sound [8]. In addition to the flexibility in source formats, MPEG-2 allows different profiles. Each profile offers a collection of compression tools that together make up the coding system. A different profile means that a different set of compression tools is available.

In order to develop technical recommendations for interactive multimedia and data broadcast services, the end user and service provider requirements have to be defined. These include aspects such as [9]

- fixed and mobile terminals
- forward and return data path requirements
- traffic patterns (randomly distributed, bursty, etc.)
- security and privacy protection
- transaction models
- reliability (accuracy and continuity of transmission, error detection and correction, etc.)
- location and identification of resources including content and downloadable applications.

Over the decades, the delivery alternatives for broadcast services have increased with the availability of new technologies. A detailed interactive and multimedia broadcast system model will be developed that will allow the identification of the various hardware, software, and protocol elements for which recommendations need to be developed. The forward and return channels will not necessarily use the same transmission medium. The work will therefore consider following delivery alternatives such as

- terrestrial over-the-air broadcast
- microwave and millimetric-based broadcast
- satellite-based broadcast
- wired systems (e.g., cable television, broadband networks, etc.)
- IP delivered webcasting

- nonreal-time systems for delivery of broadcast content (e.g., TV-Anytime, replay, Wireless home distribution network, WHDN)
- other relevant wireless systems.

Multimedia Home Platform (MHP) defines a generic interface between interactive digital applications and the terminals on which those applications execute. This interface decouples different providers' applications from the specific hardware and software details of different MHP terminal implementations. It enables digital content providers to address all types of terminals ranging from low-end to high-end set top boxes, integrated digital TV sets, and multimedia PCs. The MHP extends the existing, successful DVB open standards for broadcast and interactive services in all transmission networks including satellite, cable, terrestrial, and microwave systems [10].

While some interactive services may use existing broadcast allocated bands for the forward path, interactive services may require the allocation of new bands for the return path. Furthermore, in some instances where interactive multimedia broadcast systems may be deployed independently from sound and television broadcast services, new spectrum may be required for the forward path either in the existing broadcast frequency bands or outside these bands. These requirements have to be determined taking into consideration factors such as service traffic pattern [11] or forward and return signal coverage area.

It is expected that users may own different pieces of equipment to interface with to verify new applications for use with several services, including broadcasting. It is necessary to make provision for this, and the orderly development of such applications requires that the interoperability of equipment is ensured and that it meets the requirements of the application.

4.2 ITU APPLICATIONS

With the rapid growth of the Internet and mobile telecommunications and convergence of technology and services, standardization plays a more and more important role in making telecommunication accessible to all, thus hastening development and growth in all domains. ITU has its last Radio Assembly (RA-2000) and the World Telecommunication Standardization Assembly (WTSA-2000) in the year 2000. The year 2001 was busy for ITU-R and ITU-T. In 2001, the ITU-T took concrete measures without delay to implement the decision of WTSA. The ATM Access Point (AAP) process was efficiently launched at the end of January 2001. By the end of December 2001, there were about 190 new or revised technical recommendations approved by AAP with more than 60 percent approved in less than two months after the texts were identified as mature. ITU-T had several study group meetings held in different regions. A pioneer workshop on IP and MEDIA-COM2004 was held in April 2001. ITU-T strengthened its cooperation with IETF, Third Generation Partnership Projects (3GPPs) and other standard development organizations.

The issues of convergence were addressed by several study groups in ITU and by many SDOs outside ITU. Within ITU, ITU-T SG16 studies multimedia services and systems. ITU-T SG2 studies cable technologies, and ITU-R SG6 studies interactivity in multimedia services. Interactive multimedia systems can be defined as those systems that allow the delivery of multimedia content, which may include a combination of text, data, graphics, image, animation, voice, sound, and video, with which the user may interact, through broadcast services. Interactivity and multimedia have been under study in the ITU-R sector since at least 1996. In fact, in 1997, Study Group 11 adopted a Draft Recommendation that addresses guidance for the harmonious development of interactive television services. It covers the subject areas of interaction channels, interactive services, and transport mechanisms. This draft was approved by ITU member states in 1998 as Recommendation ITU-R BT.1369, dealing with basic principles for a worldwide common family of systems for the provision of interactive television services. Owing to the importance of the subject, in 1997, Study group 11 decided to establish Task Group 11/5, Internet television broadcasting.

The Recommendation Assembly (RA-2000) held in Istanbul decided to merge the former Study Group 10 (Broadcasting Services) and former Task Group 11/5, and JTG 10–11 were merged to form Working Party 6M (Interactivity and Multimedia).

Other ITU-T and ITU-R SGs study a wide range of issues of new services and functions. The leadership of the leading SGs dealing with multimedia convergence feel it is necessary to have a better understanding of the concept of multimedia services and they would like to harmonize their studies on multimedia services and technologies.

There is a need for standardized international multimedia applications and services that will fully meet evolving user needs and guarantee the compatibility of multimedia systems and terminals on a worldwide basis. Across different networks, SG16 has developed a general methodology for description of services as F.700 and applies it to F.700 series service recommendations. The user requirements are captured through specific application scenarios and a modular approach facilitates evolution and interoperability. We will study a consistent approach for various generic multimedia applications and services taking into consideration the increasing technical convergence of the telecommunication sectors, television and computer fields.

Convergence is an essential issue in the MEDIACOM2004 project. The convergence of information technology, telecommunications, radio communications, broadcasting, and the interactive computer environment coupled with the development of new networks (especially those based on IP technology and mobile networks) is fueling the roll-out of all types of new multimedia services. The convergence will address the types of new multimedia services. The convergence will make the prospect of the concept of total multimedia mobility become a reality in the near future. MEDIACOM2004 project will ensure that the architecture is applicable to various services, independently form the networks that support them. Generic functional descriptions of services will rely on a modular approach allowing reusability for different services and flexibility in the design of new services.

Figure 4.1 represents application of the reference model to the description of multimedia services. Although recommendations will be drafted with generic features, specific adaptations will be provided when necessary, so that the Recommendations can be applied to different networks.

Work and study items are as follows:

- Identify multimedia services and applications studied by the ITU and other bodies and produce a map of their interrelationship.
- Identify priorities to respond to Administration requirements and market demands.
- Identify missing and overlapped areas.
- Review the methodology for service definition to include technologies such as service development through scenarios and definition of *service capabilities* rather than full *services*.
- Harmonize development of multimedia applications and services with the help of network service.
- Ldentify the services and applications to be explored by SG16.
- Define the scope of these services and applications under SG16 responsibility.
- Define the requirements for these services and applications.

The cooperation and interaction activity will be led by ITU-T SG16. Other bodies will include ITU-T and ITU-R SGs, ISO, IETF, ETSI, MPEG, DVB, DAVIC, and TV Anytime. The H series of recommendation that are presently designed for fixed networks will also be extended to mobiles and satellites so that the user will be able to access all services in various environments. The design of gatekeepers, which has already started to be standardized in the H.323 family, will be extended so as to ensure transparency for users over the boundary between IP and non-IP networks for most applications.

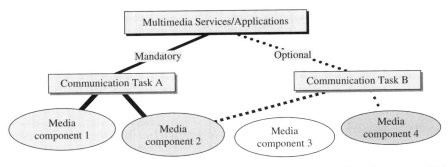

Figure 4.1 Application of the reference model to the description of multimedia services [12]. (©2001 ITU-T.)

Today, we observe convergence in the fields of information technology, telecommunications, radio communications, broadcasting, and the development of new networks, especially networks based on IP technology and mobile networks. Network-independent protocols such as the T.120 and T.130 services will be provided with extensions meeting the particular needs of different types of services and applications. Recommendations will be drafted with generic features and, when necessary, specific adaptations added so that they can be applied to different networks.

4.2.1 Multimedia Services and Systems

ITU-T Study Group 16 is the lead study group for ITU-T studies forming a defined program of work involving a number of study groups. This lead study group is responsible for the study of the appropriate core questions (Question 6/16, Multimedia Applications and Services; Question 1/16, Multimedia Systems, Terminals and Data Conferencing; Question 2/16, Multimedia Over Packet Networks Using H.323 Systems; Question 4/16, Video and Data Conferencing Using Internet Supported Services). In addition, in consultation with the relevant study groups and in collaboration, where appropriate, with other standards bodies, the lead study group has the responsibility to define and to maintain the overall framework, to coordinate, assign, and prioritize the studies to be done by the study groups, and to ensure the preparation of consistent, complete, and timely recommendations. In Resolution 2, Annex A, Part 2, World Telecommunication Standardization Assembly (WTSA) 200 appointed Study Group 16 as Lead Study Group on multimedia services, systems and terminals, e-business and e-commerce.

To conclude, ITU-T Study Group 16 is responsible for ITU-T Recommendations on multimedia service definition and multimedia systems, including the associated terminals, modules, protocols, and signal processing. ITU-T SG16 is active in all aspects of multimedia standardization, including

- multimedia terminals
- multimedia systems and protocols
- multimedia architecture
- conferencing
- multimedia quality of service
- interworking
- mobility
- security
- speech and audio coding
- video coding
- PSTN modems and interfaces
- data protocols
- terminals
- accessibility.

We are going now to present some examples of key technologies standardized in ITU-T Study Group 16.

Multimedia over IP. H.323 is the international standard and the market leader for IP telephony. H.323 networks in operation today are carrying hundreds of millions of minutes of voice traffic per month. H.323 has proven to be an extremely scalable solution that meets the needs of both service providers and enterprises. H.248 has been developed in close cooperation with the IETF and defines the protocols used by Media gateways, a vital component in VoIP networks.

Speech coding. G.728 is the standard for speech coding at 8 kbit/s with toll quality. It is now widely used in many multimedia applications. Annexes to the Recommendation exist to specify a low-complexity version, floating point versions, and silence suppression technique.

Video coding. Building on the success of H.262 and H.263, Study Group 16 is now working in conjunction with the MPEG committee of ISO/IEC to define the next generation of video coding technology in a new Joint Video Team toward new standard H.264.

PSTN modems. Study Group 16 is also responsible for the development of recommendations for voice band and PSTN modems. In 1998, the Recommendation V.90 was approved and has become the ubiquitous technology for Internet access, with annual sales of over 100 million products.

Multimedia Applications and Services

The objective is to develop a consistent approach for various generic multimedia applications and related services taking into consideration the increasing technical convergence of the telecommunications, television and service developed by SG 16.

ITU-T Study Group 16 has successfully established H.300 and T.120 series system recommendations largely for real-time video and data conferencing applications in various network environments. Based on these systems, we can configure other applications such as distance learning, telemedicine, and satellite offices where additional sets of specifications would be required to make different systems interoperable. Utilization of rich IP supported services together with the audiovisual functionalities would play an important role in these applications. Furthermore, there is a wide range of multimedia services other than multimedia conferencing that is dealt with inside and outside ITU. Across different networks, SG16 has developed a general methodology for description of services as F.700 and applied it to F.700 series service recommendations. The user requirements are captured through specific application scenarios and a modular approach facilitates evolution and interoperability.

As for study items, they are as follows:

- Identify multimedia services and applications studied by ITU and other bodies and produce a map of their interrelationship.
- Identify priorities to respond to the market demand.
- Identify missing and overlapped areas.
- Harmonize development of multimedia applications and services with those of network services.
- Apply the general service description methodology defined in F.700 to all the multimedia services.
- Extend this methodology to generic applications.
- Identify the services and applications to be explored by SG16.
- Define the scope of these services and applications under SG16 responsibility.
- Define the requirements for these services and applications under SG16 responsibility that contribute to development of the specification.

Multimedia Systems, Terminals, and Data Conferencing

Since the first set of recommendations for audiovisual communication systems for NISDN environments (H.320) were established in 1990, additional audiovisual communication systems have been developed including the H.324 series of recommendations for audiovisual communications over fixed and mobile (wireless) telephone networks and the H.310 series of recommendations for point-to-point and multipoint BISDN networks. For data sharing in point-to-point and multipoint environments, the T.120 series of recommendations has been developed. Following this, several enhancements with respect to multipoint communication, use of new audio and video coding, security features, and use of data conferencing and control continue to be developed in the form of new recommendations or revision of existing recommendations. To respond to the market needs, these enhancements may be included in the existing systems and remain competitive in the marketplace. Particular attention is concentrated on supporting of advanced coding technologies, interworking with other terminals accommodated in different networks, and enhancements to cover other services.

Study items include

- Improvements in quality aspects, audio quality, picture quality, delay, hypothetical reference connection and performance objectives, taking account the studies in relevant SGs.
- Enhancement by use of optional advanced audio and visual coding (e.g., H.264).
- Continued enhancements relating to error protection for use in error-prone environments such as mobile networks.

- Specifications necessary for accommodating new services other than conversational services, such as retrieval, messaging, distribution services applicable to the supported recommendations.
- Establishment of recommendations to support channel aggregation in mobile terminals.
- Enhancement of existing H-series recommendations.
- Possibility of new multimedia terminal system for all networks.

Multimedia Over Packet Networks Using H.323 Systems

ITU-T Study Group 16 has created in H.323 a widely used system of protocols for multimedia conferencing and video/Internet telephony over packet networks, including the Internet and LANs (local area networks). The work is currently focused on mobility, interactions, and stimulus-based call signaling in H.323 combined with network control of terminating call services.

As for study items, this question will cover the ongoing work in H.323, H.225.0, H.450.X, and H.332. Other items to be cover include

- alignment with goals related to MEDIACOM2004, IP, GII
- operation of the H.323 system over all kinds of physical layers (cable, xDSL, mobile, etc.)
- negotiating optimal transport mechanisms, such as H.223 or H.323 Annex C
- operation in the same fashion in both public and private networks
- support of accessibility
- system robustness
- possibility of new multimedia terminal system for all networks.

Video and Data Conferencing Using Internet Supported Services

ITU-T Study Group 16 has focused on the H.300 and T.120 series standardization of real-time audiovisual conversational services and systems, such as videoconferencing, videophone, and data conferencing, over various types of networks. As a platform of those systems, the personal computer (PC) is used more and more to implement not only control of functions but also audiovisual coding and presentation. The PC is also capable of accessing rich Internet supported services such as hypertext multimedia data retrieval, e-mail, and multimedia file transfer, which are becoming widely accepted to make business activities more efficient. The network infrastructure to support the video and data conferencing service is converging to the IP-based one that has been developed to provide Internet services. This Question addresses how to enhance video and data conferencing systems by use of Internet-supported services in business situations like conferencing, distance learning, and telemedicine. The study items include the following:

- Architecture to integrate video and conferencing functions with Internet-supported service functions.
- Protocols to implement the above integration.

- Mechanisms for synchronization between audiovisual and other service presentations.
- Multipoint aspects of the integrated systems.
- Verification tests for interoperability.
- Alignment with goals related to MEDIACOM2004, IP, and GII projects.

4.2.2 Integrated Broadband Cable Networks and Television and Sound Transmission

The area of responsibility for ITU-T Study Group 9 (SG9) is to prepare and maintain recommendations on

- Use of cable and hybrid networks, primarily designed for television and sound program delivery to the home, as integrated broadband networks to also carry voice or other time-critical services, video on demand, interactive services, and so on.
- Use of telecommunication systems for contribution, primary distribution and secondary distribution of television, sound programs, and similar data services.

Also, it is the lead study group on integrated broadband cable and television networks.

IP Cablecom is an effort organized in SG9 for the purpose of making efficient progress towards the development of a coordinated set of recommendations that will specify an architecture and a set of integrated protocol interfaces that operate as a system to enable the efficient delivery of time-critical interactive services over cable television networks using IP. The new set of recommendations will address

- Conditional access methods and practices for digital cable distribution to the home (Question 6/9).
- Cable television delivery of advanced multimedia digital services and applications that use IP and/or packet-based data (Question 12/9).
- Voice and video IP applications over cable television networks (Question 13/9).

The initial focus of work within IP Cablecom has been targeted to provide an integrated system for cable that can support a wide variety of time-critical interactive services within a single zone. For the purpose of the IP Cablecom architecture, a zone is defined as the set of devices (client, gateway, and others) that are under the control of a single supervisory function, which is referred to as a cable management server (CMS). Future work will consider the study of issues associated with communication between zones as well as issues associated with the support of intelligent client systems.

4.2 ITU APPLICATIONS 165

In their conversion to digital television, cable television systems in many countries are also provisioning very high-speed bidirectional data facilities to support, among other payloads, those utilizing IP. These facilities can also be used to supply other digital services to the home, based on packet data, exploiting the broadband capacity provided by hybrid fiber/coaxial (HFC) digital cable television systems and interconnecting local, and geographically distinct digital cable television systems through direct connections or managed backbones. The envisaged range of packet-based data services to be provided encompasses those services and applications that are based on the use of it. The technology considered for the delivery of these packet-based data services over the cable television infrastructure covers the use of the relevant transmission protocols, including IP and enhancements thereof. Many cable television operators are upgrading their facilities to provide two-way capability and are using this capability to provide high-speed IP data and other various multimedia services. These operators want to expand the capabilities of this delivery platform to include bidirectional voice communication and other time-critical services.

Based on Recommendation J.112 *Transmission Systems for Interactive Cable Television Services*, Voice/Video over IP services over IP-based cable television networks are expected to be available to the market. New recommendations for these applications are strongly required to meet the demand of new services in a timely manner.

According to the draft Recommendation J.160, the IP Cablecom architecture at a very high level has to connect with the following networks: HFC access network, managed IP network, and PSTN. System architecture should describe the specifications of the functional components and define the interfaces between these networks and IP-based cable television networks. The reference architecture for IP Cablecom is shown in Figure 4.2. The cable modem HFC access network provides high-speed,

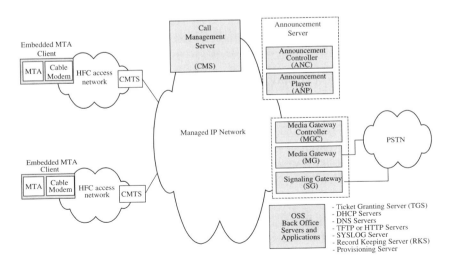

Figure 4.2 IP Cablecom reference architecture [13]. (©2002 ITU-T.)

reliable and secure transport between the customer premises and the cable headend. This access network may provide all cable modem capabilities including quality of service (QoS). The cable modem HFC access network includes the following functional components: the cable modem (CM), multimedia terminal adapter (MTA), and cable modem termination system (CMTS). The managed IP network serves several functions.

First, it provides interconnection between the basic IP Cablecom functional components responsible for signaling media, provisioning, and QoS establishment. In addition, the managed IP network provides IP connectivity between other managed IP and Cablecom modem HFC networks. The managed IP network includes the following functional components: call management server (CMS), announcement server (ANS), several operational support system (OOS) back-office servers, signaling gateways (SG), media gateway (MG), and media gateway controller (MGC). Both the signaling gateway (SG) and media gateway (MG) provide connectivity between the managed IP network and PSTN.

An IP Cablecom zone consists of the set of MTAs in one or more cable modem HFC access networks that are managed by a single functional CMS. Interfaces between functional components within a single zone are defined in the IP Cablecom specifications. Interfaces between zones have not been defined and will be addressed in future phases of the IP Cablecom architecture.

An IP Cablecom domain is made up of one or more IP Cablecom zones that are operated and managed by a single administrative entity. An IP Cablecom domain may also be referred to as an administrative domain. Interfaces between domains have not been defined in IP Cablecom and are for further study.

Table 4.2 lists the set of IP Cablecom Recommendations planned for development, in order to meet the urgent market requirements of cable operators, along with the current status of each document.

The Packet Cable multimedia specification, which was approved in June 2003, will enable operators to offer sophisticated applications, according to the organization and vendors [13].

4.2.3 Interactivity in Broadcasting

The objectives of ITU-R SG6 Working Party 6M are to develop recommendations for

- approaches and requirements specific to interactive multimedia production, coding, and multiplexing systems
- requirements specific to interactive multimedia services carried on radio communication broadcast systems
- access to webcasting and the use of the broadcast receiver as an interactive Web terminal [14].

In pursuance of those systems, Working Party 6M is

- Examining existing questions relating to interactive multimedia and data applications involving broadcast and broadcast-like platforms, as identified in Recommendation ITU-R BT.1369 [15]
- Assisting in the coordination of studies on transportation of content for interactive multimedia and data broadcasting systems within the ITV and with other international bodies.
- Contributing to studies on the impact on spectrum requirements of the development of interactive multimedia and data broadcast systems.
- Developing recommendations, which will facilitate the implementation of the transportation of content for interactive multimedia and data broadcast services.
- Making recommendations for common content format and application program interfaces (API) to achieve maximum commonality among all applications of multimedia, including broadcasting, where necessary coordinating with other relevant bodies, in particular ITU-T study groups.
- Making recommendations for technologies to facilitate commonality of conditional access systems and protection of program content.
- Producing reports and handbooks as required.

Owing to the nature of interactive multimedia and data broadcasting, the activities of the working party will be carried out in close cooperation with other groups, in order to support rather than duplicate work being carried out else where.

In order to develop technical recommendations for interactive multimedia and data broadcasting services, the end user and service provider requirements have to be defined. These include aspects such as

- fixed and mobile user terminals
- forward and return data path requirements
- traffic pattern (randomly distributed, bursty, etc)
- security and privacy protection transaction models
- reliability (accuracy and continuity of transmission, error detection and correction, etc)
- location and identification of resources including content and downloadable applications.

Over the decades, the delivery alternatives for broadcast services have increased with the availability of new technologies. A detailed Interactive and Multimedia Broadcast System model will be developed that will allow the identification of the various hardware, software, and protocol elements for which Recommendations need to be developed. The forward and return channel will not necessarily use the

Table 4.2 IP Cablecom Recommendations together with the recommendations scope [13]. (© 2002 ITU-T.)

IP Cablecom	Recommendation Name	Status	Recommendation Scope
J.160	Architecture Framework	Approved	Defines architecture framework for IP Cablecom networks including all major system components and network interfaces necessary for delivery of IP Cablecom services.
J.161	Audio/Video Codecs	Approved	Defines the audio and video codecs necessary to provide the highest quality and the most resource-efficient service delivery to the customer. Also specifies the performance required in client devices to support future IP Cablecom codecs and describes suggested methodology for optimal network support for codecs.
J.163	Dynamic Quality-of-Service	Approved	Defines the QoS architecture for the *Access* portion of the PacketCable network, provided to request applications on a per-flow basis. The access portion of the network is defined to be between the multimedia terminal adapter (MTA) and the cable modem termination system (CMTS). The method of QoS allocation over the backbone is unspecified in this document.
J.162	Network-Based Call Signaling	Approved	Defines a profile of the Media Gateway Control Protocol (MGCP) for IP Cablecom embedded clients, referred to as the network-based call signaling (NCS) protocol. MGCP is a call signaling protocol for use in a centralized call control architecture, and assumes relatively simple client devices.
J.164	Event Messages	Approved	Defines the concept of *event messages* used to collect usage for the purposes of billing within the IP Cablecom architecture.
J.165	Internet Signaling Transport Protocol (ISTP)	Approved	Defines the Internet Signaling Transport Protocol (ISTP) for IP Cablecom PSTN signaling gateways. ISTP is a protocol that provides a signaling interconnection service between the IP Cablecom network control elements (call management server and media gateway controller) and the PSTN C7 signaling network through the C7 signaling gateway.

J.168	MTA MIB	Approved	Defines the MIB module, which supplies the basic management objects for the MTA device.
J.169	NCS MIB	Approved	Defines the MIB module that supplies the basic management object for the NCS protocol.
J.166	MIBs Framework	Approved	Describes the framework in which IP Cablecom MIBs (management information base) are defined. It provides information on the management requirements of IP Cablecom-specified devices and functions, and how these requirements are supported in the MIB. It is intended to support and complement the actual MIB documents, which are issued separately.
J.167	MTA Device Provisioning	Approved	Defines the protocol mechanisms for provisioning of an IP Cablecom embedded-MTA device by a single provisioning and network management provider.
J.170	Security	Approved	Defines the security architecture, protocols, algorithms, associated functional requirements and any technological requirements that can provide for the security of the system for the IP Cablecom network.
J.171	PSTN Gateway Call Signaling	Approved	Defines a trunking gateway control protocol (TGCP) for use in a centralized call control architecture that assumes relatively simple endpoint devices. TGCP is designed to meet the protocol requirements for the media gateway controller to media gateway interface defined in the IP Cablecom architecture.
J.172	Management Event Mechanism	Approved	Defines the management event mechanism that IP Cablecom elements can use to report asynchronous events that indicate malfunction situations and notification about important nonfault situations.
J.173 (pls)	Embedded-MTA Device Specification	Approved	Specifies minimum device requirements for embedded multimedia terminal adapters in the areas of physical interfaces, power requirements, processing capabilities, and protocol support.
J.174	Interdomain Quality of Service	Approved	Defines an architectural model for end-to-end quality of service for IP Cablecom inter- and intradomain environments.

same transmission medium. Therefore, the following delivery alternatives must be considered:

- terrestrial over-the-air broadcast
- microwave and millimetric-based broadcast
- satellite-based broadcast
- wired systems (e.g., cable television, broadband networks, etc)
- IP delivered webcasting
- nonreal-time systems for the delivery of broadcast content (TV-Anytime, Wireless home distribution networks, WHDN)
- other relevant wireless systems.

Achievement of compatibility between multimedia systems developed by various service providers will require definition and development of international recommendations for items such as

- reservation engines
- execution engines
- API specifications
- security and transaction models.

Here, an *engine* is defined as a computing system performing a specific task.

While some interactive services may use existing broadcast allocated bands for the *forward path*, interactive services may require the allocation of new bands for the return path. Furthermore, in some instances where interactive multimedia broadcast systems may be developed independently from sound and television broadcast services, new spectrum may be required for the forward path, either in the existing broadcast frequency bands or outside these bands. These requirements have to be determined taking into consideration factors such as

- service traffic patterns
- forward and return signal coverage area.

Working party 6M will develop contributions to other ITU study groups, their working parties and task groups, responsible for spectrum allocation, spectrum planning, and interference and protection issues.

Initially it is expected that users may own different pieces of equipment to interface with a variety of new applications for use with several services, including broadcasting. Also, it is necessary to make provision for this. The orderly development of such applications requires that the interoperability of equipment is ensured and that it meets the requirements of the applications. However, it is anticipated that in the longer term we will see more towards greater commonality of equipment.

Working Party 6M will liaise closely with other ITU study groups and working parties as and when necessary, particularly with ITU-R SG8 and ITU-T SGs 9, 13, 15, and 16.

Interactive Multimedia

Interactive multimedia is the multimedia content with which the user can interact. Downloaded with broadcast material and stored in a receiver, the user interacts at the local level, selecting different camera angles, different drama scenarios, categories of goods for sale, electronic program guides, World Wide Web pages, and so on. A functional reference model for interactive multimedia is presented in Figure 4.3. Specific information requested by the user implies the need for a channel to contact the broadcaster, the return channel (sometimes called the interactive channel) and a dedicated channel for the broadcaster to provide the information – the forward interactive channel (sometimes called the download channel).

4.3 MPEG APPLICATIONS

The need for any standard comes from an essential requirement relevant for all applications involving communication between two or more parts. Interoperability is thus the requirement expressing the user's dream of exchanging any type of information without any technical barriers, in the simplest way. Without a standard way to perform some of the operations involved in the communication process and to structure the data exchanged, easy interoperability between terminals would be impossible. Having said that, it is clear that a standard shall specify the minimum number of tools needed to guarantee interoperability, to allow the incorporation of technical advances, and thus to increase the lifetime of the standard, as well as stimulate industrial technical competition. The experience of a standard also has

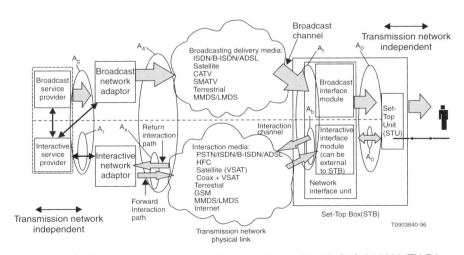

Figure 4.3 Functional reference model for interactive multimedia [14]. (©2003 ITU-R.)

important economic implications, because it allows the sharing of costs and investments and the acceleration of applications deployment.

Established in 1988, the Moving Picture Experts Group (MPEG) has developed digital audiovisual compression standards that have changed the way audiovisual content is produced by manifold industries, delivered through all sorts of distribution channels, and consumed by a variety of devices. MPEG is a standardization body composed of leading experts in the areas of video and audio compression and communication. Operating under the auspices of the International Organization for Standardization (ISO), the MPEG group specified a series of standards for video and audio compression. In addition to compression, MPEG also specifies technologies for system-level functions such as tools to multiplex and transport compressed audiovisual information to be played back in a synchronized manner. MPEG only specifies the decoding process for audiovisual information with the condition that the encoded stream be decodable by all compliant decoders.

When encoding a video, encoders can use any motion estimation technique to compute the motion vectors, but the motion vectors must be encoded as specified in the standard. MPEG-1 and MPEG-2 primarily address compression of audio and video, with parts of these standards addressing communication and transport issues. The next in the series of MPEG standards, MPEG-4, enables interactive presentation in addition to improved audiovisual compression. MPEG-4 also addresses synthetic audio and video compression. MPEG is continuing its work with MPEG-7, addressing content description and indexing. As for MPEG-21, it is addressing content management, protection, and transactions in the very early stages.

4.3.1 Multimedia PC

The MPEG-1 standard was the first in the series of standards developed by the MPEG committee. The MPEG-1 standard was intended for video coding at 1.2 Mbps and stereo audio coding at around 250 Kbps [16, 17], together resulting in bit rates compatible with those of a double-speed CD-ROM player. The typical frame size for MPEG-1 video is 352 × 240 at 30 frames per second (fps) noninterlaced. Larger frame sizes of up to 4095 × 4095 are also allowed, resulting in higher bit rate video streams.

MPEG-1, formally known as ISO/IEC 11172, is a standard in five parts. The first three parts are Systems, Video, and Audio. Two more parts complete the suite of MPEG-1 standards: *Conformance testing*, which specifies the methodology for verifying claims of conformance to the standard by manufacturers of equipment and produces bit streams, and *Software simulation*, a full C-language implementation of the MPEG-1 standard (encoder and decoder). Manifold have been the implementations of the MPEG-1 standard: from software implementation running on a PC in real time, to single board for PCs, to the so-called VideoCD, and so on [18, 19].

MPEG-1 Systems

The MPEG-1 Systems standard defines a packet structure for multiplexing coded audio and video data into one stream and keeping it synchronized. It thus supports multiplexing of multiple coded audio and video streams, where each stream is referred to as an elementary stream. The systems syntax includes data fields that allow synchronization of elementary streams and assists in parsing the multiplexed stream after random access, management of decoder buffers, and identification of timing of the coded program. Thus, the MPEG-1 Systems specifies the syntax to allow generation of systems bit streams and semantics for decoding these bit streams. The mechanism for generating the timing information from decoded data is provided by the system clock reference (SCR) fields. The presentation playback or display synchronization information is provided by presentation time stamps (PTS), which represent the intended time of presentation of decoded video pictures or audio frames.

To ensure guaranteed decoder buffer behavior, the MPEG Systems specifies the concepts of a systems target decoder (STD) and decoding time stamp (DTS). The DTS differs from the PTS only in the case of pictures, which require additional reordering delay during decoding process.

MPEG-1 Video

The MPEG-1 Video standard specifies the video bit stream syntax and the corresponding video decoding process. The MPEG-1 Video syntax supports three types of coded frames or pictures: intra (I) pictures coded separately by themselves, predictive (P) pictures coded with respect to the immediately previous I or P picture, and bidirectionally predictive (B) pictures, coded with respect to the immediately previous I or P picture as well as the immediately next P or I picture. In terms of coding order, P-pictures are causal, whereas B-pictures are noncausal and use two surrounding causally coded pictures for prediction. In terms of compression efficiency, I-pictures are the most expensive, P-pictures are less expensive than I-pictures, and B-pictures are the least expensive. However, because B-pictures are noncausal, they incur additional (reordering) delay.

Example 4.1 Figure 4.4 shows an example picture structure in MPEG-1 video coding that uses a pair of B-pictures between two reference (I or P) pictures. In MPEG-1 video coding,

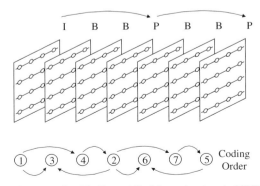

Figure 4.4 An example of I-, P- and B-picture structure in MPEG-1 coding.

an input video sequence is divided into units of groups of pictures (GOPs), where each GOP starts with an I-picture, while the rest of the GOP contain an arrangement of P-pictures and B-pictures. A GOP serves as a basic access unit, with the I-picture serving as the entry point to facilitate random access. For coding purposes, each picture is further divided into one or more *slices*. Slices are independently decodable entities that offer a mechanism for resynchronization and thus limit the propagation of errors. Each slice is composed of a number of macroblocks. Each macroblock is basically a 16×16 block of luminance (or alternatively, four 8×8 blocks) with corresponding chrominance blocks.

MPEG-1 video coding can exploit both the spatial and the temporal redundancies in video scenes [16]. Spatial redundancies are exploited by using block discrete cosine transform (DCT) coding of 8×8 pixel blocks, resulting in 8×8 blocks of DCT coefficients. These coefficients then undergo quantization, zig-zag scanning, and variable length coding. A nonlinear quantization matrix can be use for weighting of DCT coefficients prior to quantization. In this case we allow perceptually weighted quantization, in which perceptually irrelevant information can be easily discarded, further increasing the coding efficiency. Temporal redundancies are exploited by using block motion compensation to compensate for interframe motion of objects in a scene. This results in a significant reduction of interframe prediction error.

MPEG-1 Audio

The MPEG-1 Audio standards specifies the audio bit stream syntax and the corresponding audio decoding process. It is a generic standard that does not make any assumptions about the nature of the audio source, unlike some vocal-tract model coders that work well for speech only. MPEG-1 audio coding exploits the perceptual limitations of the human auditory system, and thus much of the compression comes from removal of perceptual irrelevant parts of the audio signal.

The MPEG-1 Audio standard consists of three layers. These layers also represent increasing complexity, delay, and coding efficiency. In terms of coding methods, these layers are related, because a higher layer includes the building blocks used for a lower layer. The sampling rates supported by MPEG-1 Audio are 32, 44.1, and 48 KHz. Several fixed bit rates in the range of 32 to 224 Kbit/s per channel can be used. In addition, Layer III supports variable bit rate coding. Good quality is possible with layer I above 128 Kbit/s, with layer II around 128 Kbit/s, and layer III around 64 Kbit/s. MPEG-1 Audio has four modes: mono, stereo, dual mono with two separate channels, and joint stereo. The optional joint stereo mode exploits interchannel redundancies.

MPEG-1 Constraints

Although the MPEG-1 Video standard allows the use of fairly large picture sizes, high frame rates and corresponding high bit rates, it does not necessarily imply that every MPEG-1 video decoder support these parameters. In fact, to keep decoder complexity reasonable while ensuring interoperability, an MPEG-1 video decoder need only conform to a set of constrained parameters that specify the largest horizontal size (720 pels/line), the largest vertical size (576 lines/frame), the maximum number of macroblocks per picture (396), the maximum number of macroblocks per second (396 × 25), the highest picture rate (30 frame/sec), the maximum bitrate (1.86 Mb/sec), and the largest decoder buffer size (376 × 832 bits).

To conclude, since MPEG-1 was intended for digital storage such as CD-ROM, the MPEG-1 System was not designed to be tolerant to bit errors. Furthermore, to keep the overhead small, the MPEG-1 System streams contain large variable length packets. To deliver MPEG-1 streams on the Internet, specialized mapping of MPEG-1 streams, for example, on the payload of real-time transport protocol (RTP) packets has been specified [20–22]. MEPG-1 was optimized for applications at above 1.5 Mbps on a reliable storage medium such as CD-ROM and as such is not suitable for broadcast quality applications. To address the requirements of broadcast television and high-quality applications, the MPEG committee began its work on MPEG-2 in 1992.

4.3.2 Digital TV and Storage Media

Digital video is replacing analog video in the consumer marketplace. A prime example is the introduction of digital television in both standard-definition and high-definition formats. Another example is the digital versatile disk (DVD) standard, which is replacing videocassettes as the preferred medium for watching movies. The MPEG-2 video standard has been one of the key technologies that enabled the acceptance of these new media.

The original target for the MPEG-2 standard was TV-resolution video and up to five-channel audio of very good quality at about 4 to 15 Mbit/s for applications such as digital broadcast TV and DVD. The standard has been deployed for a number of other applications including digital cable or satellite TV, video on ATM networks [23], and high-definition TV (HDTV). Also, MPEG-2 provides support to a number of technical features, the most important of which is support to content addressing, encryption, and copyright identification. For example, the MPEG-2 systems transport stream has been designed so that it can be used to carry a large number of television programs. For this reason it provides support to signal the content of the programs by means of tables that describe which program can be found where. This specification has been extended by regional initiatives to identify more features, such as the nature of the program, the scheduled time, the interval between starting times, and so on.

MPEG-2 Systems defines two special streams called encryption controlled message (ECM) and encryption management message (EMM) that carry information that can be used to decrypt information carried by the MPEG-2 transport stream if this has been encrypted. The encryption system itself is not specified by MPEG [19]. Also, MPEG-2 Systems provides support for the management of audiovisual works' copyright. This is done by managing the rights of that particular audiovisual work followed by a field that gives the identification number of the work, as assigned by the society. For instance, this information enables the monitoring of the flow of copyrighted work through a network.

Based on the target applications, a number of primary requirements were derived as follows [24]:

- random access or channel switching within a limited time, allowing frequent access points every half second
- capability for fast forward and fast reverse, enabling seek and play forward or backward at several times the nominal speed
- support for scalable coding to allow multiple simultaneous layers and to achieve backward compatibility with MPEG-1
- a system for synchronized playback and time-in or access of audiovisual data
- a defined subset of the standard to be implementable in practical real-time decoders at a reasonable cost in hardware.

Besides the preceding requirements, a number of other requirements also arose, such as support for a number of picture resolutions and formats (both interlaced and noninterlaced), a number of sampling structures for chrominance, robustness to errors, coding quality trade-off with coding delay, and the possibility of real-time modes at a reasonable cost for at least a defined subset of the standard.

Like its predecessor, the MPEG-2 standard has several parts [24]: Systems, Video, Audio, Conformance, Software, Digital Storage Media – Command and Control (DSM-CC), Advanced Audio Coding (AAC), real-time interface. We will now briefly discuss each of the four main components of the MPEG-2 standard.

MPEG-2 Systems

The MPEG-2 Systems was developed to improve error resilience and the ability to carry multiple programs simultaneously without requesting them to have a common time base. In addition, it was required that MPEG-2 Systems should support ATM networks [23]. The MPEG-2 Systems specification defines two types of streams: the program stream is similar to the MPEG-1 Systems stream but uses modified syntax and new functions to support advanced functionalities. The requirements of MPEG-2 program stream decoders are similar to those of MPEG-1 system stream decoders. Like MPEG-1 Systems decoders, program streams decoders employ long and variable-length packets. Such packets are well suited for software-based processing in error-free environments, such as when the compressed data are stored on a disk. The packet sizes are usually in the range of 1 to 2 Kbytes, chosen to model disk sector sizes (typically 2 Kbytes). However, packet size as large as 64 Kbytes is also supported. The program stream includes features not supported by MPEG-1 Systems, such as hooks for scrambling of data, alignment of different priorities to packets, information to assist alignment of elementary stream packets, indication of copyright, indication of fast forward, fast reverse, and other trick modes for storage devices, an optional field for network performance testing and an optional numbering of sequence of packets.

The transport stream offers robustness necessary for noisy channels as well as the ability to include multiple programs in a single stream. The transport stream uses fixed-length packets of size 188 bytes with a new header syntax. It is therefore more suited for hardware processing and for error-correction schemes. Thus, the transport stream is well suited for delivering compressed video and audio over error-prone channels such as coaxial cable television networks and satellite transponders. Furthermore, multiple programs with independent time bases can be multiplexed in one transport stream. In fact, the transport stream is designed to support many functions such as asynchronous multiplexing of programs, fast access to a desired program for channel hopping, multiplexing of programs with clocks unrelated to the transport clock, and correct synchronization of elementary streams for playback, to allow control of decoder buffers during start-up and playback for constant bit rate and variable bit rate programs, to be self-describing and to tolerate channel errors.

A basic structure that is common to the organization of both the program stream and the transport stream is the packetized elementary stream (PES) packet. PES packets are generated by packetizing the continuous stream of compressed data generated by video and audio encoders. Demultiplexers and a process similar to that for transport streams are followed.

Both types of MPEG-2 Systems, those using program stream multiplexing and those using transport stream multiplexing, are illustrated in Figure 4.5. An MPEG-2 system is capable of combining multiple sources of user data along with MPEG encoded audio and video. The audio and video streams are packetized to form audio and video PES packets, which are sent to either a program multiplexer or a transport multiplexer, resulting in a program stream or a transport stream.

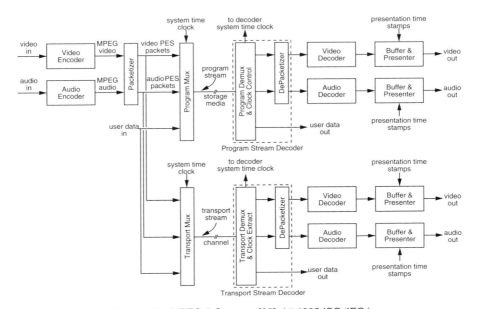

Figure 4.5 MPEG-2 Systems [25]. (©1995 ISO/IEC.)

Program streams are intended for error-free environments such as digital storage media (DSM), whereas transport streams are intended for noisy environments such as terrestrial broadcast channel.

Transport streams are decoded by the transport multiplexer (which includes a clock extraction mechanism), unpacketized by a depacketizer, and sent to audio and video decoders for audio and video decoding. The decoded signals are sent to a display device and its speaker at the appropriate time. Similarly, if program streams are employed, they are decoded by a program stream.

MPEG-2 Video

As in the case of MPEG-1, MPEG-2 does not standardize the video encoding process of the encoder. Only the bit stream syntax and the decoding process are standardized. MPEG-2 video coding is based on the block motion compensated DCT coding of MPEG-1. Coding is performed on pictures, where a picture can be a frame or field, because with interlaced video each frame consists of two fields separated in time. A sequence is divided into groups of pictures assuming frame coding. A frame may be coded as an intra (I) picture, a predictive (P) picture, or a bidirectionally predictive (B) picture. Thus, a group of pictures may contain an arrangement of I-, P-, and B-coded pictures. Each picture is further partitioned into slices, each slice into a sequence of macroblocks and each macroblock into four luminance blocks and corresponding chrominance blocks. The MPEG-2 video encoder consists of various components such as an inter/intraframe/field DCT encoder, a frame/field motion estimator and compensator, and the variable length encoder. The frame/field DCT encoder exploits spatial redundancies and the frame/field motion compensator exploits temporal redundancies in the interlaced video signal. The coded video bit stream is sent to the systems multiplexer, SysMux, which outputs either a transport or a program stream.

Figure 4.6 shows a simplified block diagram of an MPEG-2 video decoder that receives bit streams to be decoded from the MPEG-2 Systems demux. The MPEG-2 video decoder consists of a variable-length decoder, an

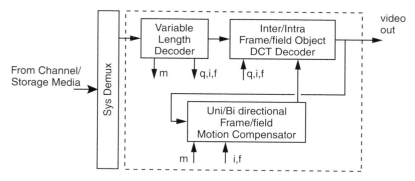

Figure 4.6 MPEG-2 video decoder.

inter/intraframe/field DCT decoder, and a uni/bidirectional frame/field motion compensator. After demultiplexing, the MPEG-2 video bit stream is sent to the variable-length decoder for decoding of motion vectors (m), quantization information (q), inter/intradecision (i), frame/field decision (f), and the data consisting of quantized DCT coefficient indices. The inter/intraframe/field DCT decoder uses the decoded DCT coefficient indices, the quantization information, the inter/intradecision and the frame/field information to dequantize the indices to yield DCT coefficients blocks and then inverse transform the blocks to recover decoded pixel blocks. The uni/bidirectional frame/field motion compensator, if the coding mode is inter, uses motion vectors and frame/field information to generate motion-compensated prediction of blocks, which are then added back to the corresponding decoded prediction error blocks output by the inter/intra frame/field DCT decoder to generate decoded blocks. If the coding mode is intra, no motion compensated prediction needs to be added to the output of the inter/intraframe/field DCT decoder. The resulting decoded pictures are the output on the line labeled video out.

In terms of interoperability with MPEG-1 video, the MPEG-2 video standard was required to satisfy two key elements: forward compatibility and backward compatibility. Because MEPG-2 video is a syntactic superset of MPEG-1 video, it is able to meet the requirement of forward compatibility, meaning that an MPEG-2 video decoder ought to be able to decode MEPG-1 video bit stream. The requirement of backward compatibility means that the subset of MPEG-2 bit streams should be decodable by existing MPEG-1 decoders. This is achieved via scalability. Scalability is the property that allows decoders of various complexities to be able to decode video of resolution or quality commensurate with their abilities from the same bit stream. A generalized codec structure for MPEG-2 scalable video coding is shown in Figure 4.7. This structure essentially allows only spatial and temporal resolution scalabilities.

The generalized codec supports two scalability layers, a lower layer, specifically refereed to as the base layer and a higher layer that provides enhancement of the base layer. Input video goes through a preprocessor and results in two video signals, one of which is input to the MPEG-1/MPEG-2 nonscalable video encoder and the other to the MPEG-2 enhancement video encoder. Depending on the specific type of

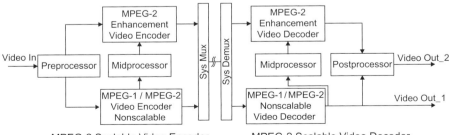

Figure 4.7 Generalized codec for MPEG-2 scalable video coding [25]. (©1995 ISO/IEC.)

scalability, some processing of decoded video format of the MPEG-1/MPEG-2 nonscalable video encoder may be needed in the midprocessor before it is used for prediction in the MPEG-2 enhancement video encoder. The two coded video bit streams, one from each encoder, are multiplexed in the SysMux. At the decoder end, the MPEG-2 SysDemux performs the inverse operation of unpacking from a single bit stream or two substreams, one corresponding to the lower layer and the other corresponding to the higher layer [26].

Two amendments to MPEG-2 Video took place after completion of the original standard. The first amendment was motivated by the needs of professional applications, and tested and verified the performance of the higher chrominance spatial resolution format called 4:2:2 format. The second amendment to MPEG-2 Video was motivated by the potential of applications in video games, education, and entertainment, and was involved in developing, testing, and verifying a solution for efficient coding of multiviewpoint signals, including at least the case of stereoscopic video. This involves exploiting correlation between different views of a scene.

MPEG-2 Audio

The MPEG-2 Audio standards are specified in Part 3, which specifies tools that are backward compatible with MPEG-1, and Part 7, which specifies standards for non-backward compatible (NBC) audio, now called advanced audio coding (AAC). Digital multichannel audio systems employ a combination of p front and q back channels. For example, we have three front channels (left, right, center) and two back channels (surround left and surround right), to create surreal and theater-like experiences. In addition, multichannel systems can be used to provide multilingual programs, audio augmentation for the visually impaired, enhanced audio for the hearing impaired, and so on. The MPEG-2 Audio standard addresses such applications [24, 27].

In Figure 4.8 a generalized coded structure illustrating MPEG-2 Multichannel Audio coding is presented. Multichannel Audio is shown undergoing conversion by the use of a matrix operation resulting in five converted signals. It consists of five signals, that is, left (L), center (C), right (R), left surround (Ls), and right surround (Rs). Two of the signals are encoded by an MPEG-1 audio encoder to provide compatibility with the MPEG-1 standard, while the remaining bit streams from the

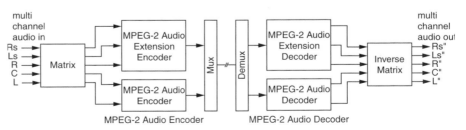

Figure 4.8 A generalized codec for MPEG-2 backward-compatible multichannel audio coding [27]. (©1995 ISO/IEC.)

two encoders are multiplexed in Mux for storage or transmission. At the decoder, an MPEG-1 audio decoder decodes the bit stream input to it by Sys demux and produces two decoded audio signals: the other three audio signals are decoded by an MPEG-2 audio extension decoder. The decoded audio signals are reconverted back to the original domain by using Inverse Matrix and they represent approximate values indicated by L, C, R, Ls, and Rs.

In Figure 4.9, a simplified reference model configuration of an advanced audio coding (AAC), that is, audio code, is shown. Multichannel audio undergoes transformation via time-to-frequency mapping, which output is subject to various operations such as joint channel coding, quantization, coding, and bit allocation. A psychoacoustic model is employed at the encoder and controls both the mapping and bit allocation operations. The output of the joint channel coding, quantization, coding, and bit allocation unit is input to a bit stream formatter that generates the bit stream for storage or transmission. At the decoder, an inverse operation is performed at what is called the bit stream unpacker, following which dequantization, decoding, and joint decoding occur. Finally, an inverse mapping is performed to transform the frequency domain signal to its time domain representation, resulting in reconstructed multichannel audio output.

MPEG-2 DSM-CC

Coded MPEG-2 bit stream may be stored on a variety of digital storage media (DSM), such as CD-ROM, magnetic type disks, DVDs, and others. This presents a problem for users when trying to access coded MPEG-2 data, because each DSM may have its own control command language, forcing the user to know many such languages. Moreover, the DSM may be either local to the user or at a remote location. When it is remote, a common mechanism for accessing various digital storage media over a network is needed. Otherwise the user has to be informed about the type of DSM, which may not be known or possible. The MPEG-2 DSM-CC is a set of generic control commands independent of the type of DSM that addresses these two problems. Thus, control commands are defined as a specific application protocol to allow a set of basic functions specific to MPEG bit streams. The resulting control commands do not depend on the type of DSM, or whether the DSM is local or remote, the network transmission protocol,

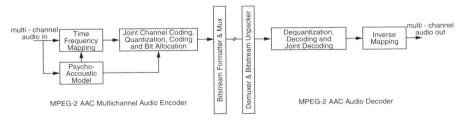

Figure 4.9 A generalized codec for MPEG-2 advanced audio coding (AAC) multichannel audio [27]. (©1997 ISO/IEC.)

or the operating system with which it is interfacing, and the bit streams, the MPEG-2 program streams, or the MPEG-2 transport streams.

The DSM control commands can generally be divided into two categories. The first category consists of a set of very basic operations such as the stream selection, play and store commands. Stream selection enables a request for a specific bit stream and a specific operation mode on that bit stream. Play enables playback of a selected bit stream of a specific speed, direction of play (to accommodate a number of trick modes such as fast forward and fast reverse), or other features such as pause, resume, step through, or stop. Store enables the recording of a bit stream on a DSM. The second category consists of a set of more advanced operations such as multiuser mode, session reservation, server capability information, directory information, and bit stream editing. In multiuser mode, more than one user is allowed access to the same server within a session. Server capability information allows the user to be notified of the capabilities of the server, such as playback, fast forward, fast reverse, slow motion, storage, demultiplex, and remultiplex. Directory information allows the user access to information about directory structure and specific attributes of a bit stream, such as type, sizes, bit rate, entry points for random access, program description, and others. Typically, not all this information may be available through the API. Bit stream editing allows creation of new bit streams by insertion or detection of portions of bit streams into others.

A simplified relationship of DSM-CC with Multimedia and Hypermedia Experts Group (MHEG) standard and scripting language has been developed. The MPEG standard is basically an interchange format for multimedia objects between applications. MPEG specifies a class set that can be used to specify objects containing monomedia information, relationships between objects, dynamic behavior between objects, and information to optimize real-time handling of objects. MPEG classes include the content class, composite class, link class, action class, script class, descriptor class, and result class. Applications may access DSM-CC either directly or through an MHEG layer: moreover, scripting languages may be supported through an MHEG layer (Fig. 4.10). The DSM-CC protocols also form a layer higher than the transport protocols layer. Examples of transport protocols are the TCP, UDP, the MPEG-2 program streams, and the MPEG-2 transport stream.

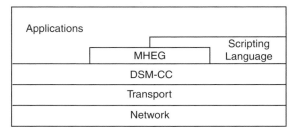

Figure 4.10 DSM-CC centric view of MHEG, scripting language and network.

The DSM-CC provides access for general applications, MHEG applications, and scripting languages to primitives for establishing or deleting network connections using user–network (U–N) primitives and communication between a client and a server across a network using user–user (U–U) primitives (Fig. 4.11). The U–U operations may use a remote procedure call (RPC) protocol. Both the U–U and the U–N operations may employ message passing in the form of exchanges of a sequence of codes, that is, the scenarios of U–N and U–U interaction. A client can connect to a server either directly via a network or through a resource manager located within the network. A client setup is typically expected to include a session gateway, which is a user–network interface point, and a library of DSM-CC routines, which is a user–user interface point. A server typically consists of a session gateway, which is a user–network interface point, and a service gateway, which is a user–user interface point. Depending on the requirements of an application, both user–user connection and user–network connection can be established.

MPEG-2 Profiling

Profiling is a mechanism by which a decoder, to be compliant with a standard, has to implement only a subset and further only certain parameters' combinations of the standard. The MPEG-2 Video standard extends the concept of constrained parameters by allowing a number of valid subsets of the standard organized into `Profiles` and `Levels`. A profile is a defined subset of the entire bit stream syntax of a standard. A level of a profile specifies the constraints on the allowable values for parameters in the bit stream. The Main profile is the most important profile of MPEG-2 Video. It supports nonscalable video syntax. Further, it consists of four levels: Low, Main, High1440, and High. Again, as the name suggests, the Main level refers to TV resolution, and High1440 level refers to two resolutions for HDTV. The low level refers to MPEG-1 constrained parameters.

To conclude, because of its flexibility and support for a wide range of applications, MPEG-2 has been successful. MPEG-2 has been universally adopted for high-quality audiovisual applications including digital broadcast TV, DVD, HDTV, and digital cinema.

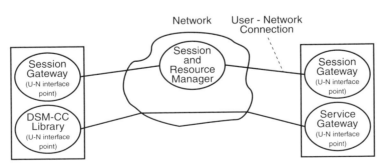

Figure 4.11 DSM-CC user–network and user–user interaction [25]. (©1995 ISO/IEC.)

4.3.3 Multimedia Conferencing, Streaming Media, and Interactive Broadcasting

The previous MPEG standards, MPEG-1 and MPEG-2, have found widespread use in the digital multimedia industries. Streaming video over the Internet as well as video CDs conform to the MPEG-1 standards. Digital broadcast TV, as well as DVDs, were driven by the emergence of MPEG-2. It is expected that the technologies developed in MPEG-4 will revolutionize the digital multimedia industry.

In multimedia conferencing applications, the terminal may be a multimedia PC (equipped with camera, microphone, and speakers), or it can be part of a high-end videoconferencing system. In fact, it can very well be the case that these two kinds of terminals are communicating in the same virtual environment. Indeed, the notion of shared communication space is the main idea of this application, and the use of MPEG-4 Systems to represent the shared data is the technical choice. This application highlights two key features of MPEG-4: dynamic scene representation and scalability.

In streaming content over the Internet, the client terminal is a multimedia terminal connected to the Internet. An example of such a terminal is a PC with multimedia features. The MPEG-4 based application may be received over the network from a remote server. The user of this application first enters a specific address and is connected to the server. As a result of establishing a connection to the server a *page* appears on the client terminal's screen. This page looks very much like an HTML page with Java applet-based animations. In fact, the page is not created as a result of receiving an HTML file but is actually an MPEG-4 application.

In interactive broadcast application, the MPEG-4 receiver may be the traditional home set-top box, a part of high-end home theater, and is connected to a high-bandwidth broadcast network at its input end. The receiver could also be a conventional multimedia terminal connected to the broadcast network. The key concept of MPEG-4 is *create once–access everywhere*, and the tools that give support to allow content creators and service providers to make their content available across the entire range of available delivery systems. In order to receive broadcast content, the user needs to be connected to the broadcast server and have managed to tune to the broadcast channel or program of his choice. The client terminal acquires the necessary scene description information for this channel. The necessary audio and visual streams, constituents of the program on the tuned channel, are acquired and the user is now able to watch and possibly interact with the program. A key feature is the compressed nature of the scene description information. The gain factor is indeed directly linked to the amount of content that you can deliver in such broadcast carousel. Another important feature is the built-in support of intellectual property management and protection in MPEG-4 Systems, allowing the intellectual property rights on the content to be respected. Finally, the definition of the transport of MPEG-4 content on top of MPEG-2 Systems will allow a smooth and backward compatible evolution from existing MEPG-2 broadcast applications to fully interactive television.

Traditionally, the visual component of multimedia presentations is a rectangular video frame. Even when text is overlaid on a video frame, the text and video are composed into a single rectangular frame before being compressed or transmitted. Advances in image and video encoding and representation techniques have made possible encoding and representation of audiovisual scenes with semantically meaningful objects [28, 29]. The traditionally rectangular video can now be coded and represented as a collection of arbitrarily shaped visual objects. The ability to create object-based scenes and presentations creates many possibilities for a new generation of applications and services. The MPEG-4 series of standards specify tools for such object-based audiovisual presentations [30].

The MPEG-4 standard consists of the following parts: Systems, Video, Audio, Conformance, Software, Delivery Multimedia Integration Framework (DMIF). We will briefly discuss the current status of the four main components (Systems, Video, Audio, and DMIF) of the MPEG-4 standard. Note that the MPEG-4 standard is aimed at Internet and intranet video, wireless video, interactive home shopping, video e-mail, home movies, virtual reality games, simulation and training, and media object databases. The MPEG-4 standard addresses the following functionalities [31].

- Content-based interactivity allows the ability to interact with important objects in a scene. The MPEG-4 standards extends the types of interaction typically available for synthetic objects to natural objects as well as hybrid (synthetic–natural) objects to enable new audiovisual applications. It also supports the spatial and temporal scalability of media objects.
- Universal accessibility means the ability to access audiovisual data over a diverse range of storage and transmission media. It is important that access be available to applications via wireless networks. Also, MPEG-4 provides tools for robust coding in error-prone environments at low bit rates. MPEG-4 is also developing tools to allow fine granularity media scalability for Internet applications.
- Improved compression allows an increase in efficiency of transmission or a decrease in the amount of storage required. Because of its object-oriented nature, MPEG-4 allows very flexible adaptations of degree of compression to the channel bandwidth or storage media capacity. The MPEG-4 coding tools, although generic, are still able to provide state-of-the-art compression, because optimization of MPEG-4 coding was performed on low-resolution content at low bit rates.

The conceptual architecture of MPEG-4 is depicted in Figure 4.12. It comprises three layers: the compression layer, the sync layer, and the delivery layer. The compression layer is media aware and delivery unaware; the sync layer is media unaware and delivery unaware; the delivery layer is media unaware and delivery aware. The compression layer performs media encoding and decoding into and from elementary streams. The sync layer manages elementary streams and their synchronization and hierarchical relations. The delivery layer ensures transparent access to content

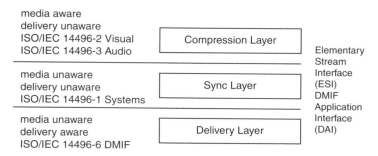

Figure 4.12 Various parts of MPEG-4.

irrespective of delivery technology. The boundary between the compression layer and the sync layer is called the elementary stream interface (ESI).

MPEG-4 Systems

The Systems part of MPEG-4 makes the standard radically different in the tools for object-based representation. The most significant difference in MPEG-4 Systems standards is the scene composition at the user terminal. The individual objects that make up a scene are transmitted as separate elementary streams and are composed upon reception according to the composition information delivered along with the media objects. The MPEG-4 Systems standard specifies tools to encode the composition of objects and interaction with objects. The standard itself is coded and has a general framework to support different codes. The MPEG-4 Systems standard also specifies a file format for storing object-based presentations and a Java framework called MPEG-J that allows manipulation of scene composition information using Java applets. These new tools make MPEG-4 Systems an application framework for interactive audiovisual content, blurring the line between applications and content. The MPEG-4 Systems version 1 represents the departure from the previous MPEG standards. The object-based nature of MPEG-4 necessitated a new approach to MPEG-4 Systems, although the traditional issues of multiplexing and synchronization are still quite important. For synchronization the challenge for MPEG-4 Systems was to provide a mechanism to handle a large number of streams, which result from the fact that a typical MPEG-4 scene may be composed of many objects. In addition the spatiotemporal positioning of these objects forming a scene (or scene description) is a new key component. Further, MPEG-4 Systems also had to deal with issues of user interactivity with the scene. Another item, added late during MPEG-4 Systems version 1 development, addresses the problem of management and protection of intellectual property related to media content. To conclude, the MPEG-4 Systems version 1 specification covers the following aspects [32]:

- terminal model for time and buffer management
- coded representation of scene description

- coded representation of metadata – object descriptors (and others)
- coded representation of audiovisual (AV) content information – object content information (OCI)
- an interface to intellectual property – Intellectual Property Management and Protection (IPMP)
- coded representation of sync information – SyncLayer (SL)
- multiplex of elementary streams to a single stream – FlexMux tools.

Version 2 of MPEG-4 Systems extends the version 1 specification. Additional capabilities are

- MPEG-4 file format – a file format for interchange
- MPEG-4 over IP and MPEG-4 over MPEG-2
- Scene description – application texture, advanced audio, message handling
- MPEG-J – Java-based flexible control of fixed MPEG-4 Systems.

MPEG-4 is an object-based standard for multimedia coding. It codes individual video objects and audio objects in the scene and delivers, in addition, a coded description of the scene. At the decoding end, the scene description and individual media objects are decoded, synchronized, and composed for presentation. Figure 4.13 shows the high-level architecture of an MPEG-4 terminal [32]. This architecture shows the MPEG-4 stream delivered over the network/storage medium via the delivery layer, which includes transport multiplex, TransMux, and an optional multiplex called FlexMux. Demultiplexed streams from the FlexMux leave via the delivery multimedia integration framework application interface (DAI) and enter the sync layer, resulting in the synchronization layer (SL), and packetized elementary streams that are ready to be decoded. The compression layer encapsulates the functions of the media, scene description, and object descriptor decoding, yielding individual decoded objects and related descriptors. The composition and rendering process uses the scene description and decoded media to compose and render the audiovisual scene and passes it to the presenter. A user can interact with the presentation of the scene, and the actions necessary as a result are sent back to the network/storage medium through the compression, sync, and delivery layers.

System Decoder Model

MPEG-4 defines a system decoder model (SDM) [33]. This is a conceptual model that allows precise definition of decoding events, composition events, and times at which these events occur. The MPEG-4 system decoder model exposes resources available at the receiving terminal and defines how they can be controlled by the sender or the content creator. The SDM is shown in Figure 4.14. After delivery at the multimedia integration framework (DMIF) application interface, we have the FlexMux buffer, which is a receiver buffer, that can store the FlexMux resources that are

APPLICATIONS LAYER

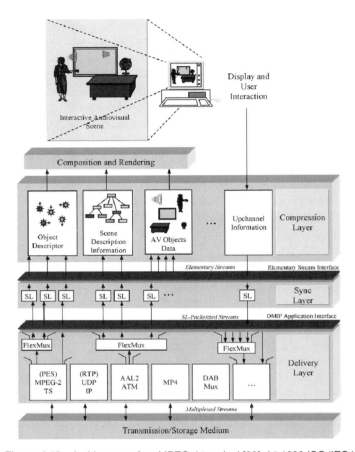

Figure 4.13 Architecture of an MPEG-4 terminal [33]. (©1999 ISO/IEC.)

used during a session. Further, the SDM is composed of a set of decoders (for the various audio or visual object types) provided with two types of buffers: decoding and composition. The decoding buffers are controlled by clock references and decoding time stamps. In MPEG-4, each individual object is assumed to have its own clock or object time base (OTB). Several objects may share the same clock. In addition, coded units of individual objects are associated with decoding time stamps (DTS). The composition buffers that are present at the decoder outputs form a second set of buffers. Their use is related to object persistence. In some situations, a content creator may want to reuse a particular object after it has been presented. By exposing a composition buffer, the content creator can control the lifetime of data in this buffer for later use. This feature may be particularly useful in low-bandwidth wireless environments.

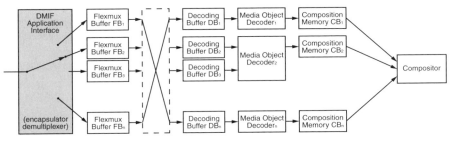

Figure 4.14 The MPEG-4 system decoder model [33]. (©1999 ISO/IEC.)

Scene Description

Scene description refers to the specification of the spatiotemporal positioning and behavior of individual objects [34]. It allows easy creation of compelling audiovisual content. The scene description is transmitted in a separate stream from the individual media objects. This allows one to change the scene description without operating on any of the consistent objects themselves.

The MPEG-4 scene description extends and parameterizes virtual reality modeling language (VRML), which is a textual language for describing the scene in three dimensions (3D). There are at least two main reasons for this. First, MEPG-4 needed the capability of scene description not only in three dimensions but also in two dimensions (2D). A textual language was not suitable for low-overhead transmission, and thus a parametric form binarizing VRML called Binary Format for Scenes (BIFS) was developed [35].

In VRML, nodes are the elements that can be grouped to organize the scene layout by creating a scene graph. In a scene graph the trunk is the highest hierarchical level, with branches representing children grouped around it. The characteristics of the present node are inherited by the child node. VRML supports a total of 54 nodes [36]. A row classification of nodes can be made on the basis of whether they are graphing nodes or leaf nodes. VRML nodes can be divided into two main categories: graphical nodes and nongraphical nodes. The graphical nodes can be divided into three subcategories: grouping nodes, geometry nodes, and attribute nodes. The graphical nodes are the nodes that are used to build the rendered scenes. The nongraphical nodes augmented the 3D scene by providing a means of adding dynamic effects such as sound, event triggering, and animation. The nongraphical nodes can also be divided into three subcategories with many nodes per subcategory: sound, event triggering, and animation. Each VRML node can have a number of fields that parameterize the node. Fields in VRML form the basis of the execution model. There are four types of fields: fields, event_In field, event_Out field, and exposed Field. The first field carries data, values that define characteristics of a node. The second event_In field, accepts incoming events that change its value to the value of the event itself (sink). The third event_Out field outputs its value as an event (source). The fourth, exposed_field, allows acceptance of a new value and can send out its value as an event (source and sink).

Binary Format for Scenes, although based on VRML, extends it in several directions. First, it provides a binary representation in VRML 2.0. This representation is much more efficient for storage and communication than straightforward binary representation of VRML text. Secondly, in recognition that BIFS needs to represent not only 3D scenes, but also normal 2D (audiovisual) scenes, it adds a number of 2D nodes, including 2D versions of several 3D nodes. Thirdly, it includes support for MPEG-4 specific media such as video, facial animation, and sound by adding new nodes. Fourthly, it improves the animation capabilities of VRML and further adds streaming capabilities. BIFS supports close to 100 nodes – one-half from VRML and one-half new nodes. It also specifies restrictions on semantics of several VRML nodes [37, 38].

Individual object data and scene description information are carried in separate elementary streams (ESs). As a result, BIFS media nodes need a mechanism to associate themselves with the ESs that carry their data (coded natural video object data, etc.). A direct mechanism would necessitate the inclusion of transport-related information in the scene description. An important requirement in MPEG-4 is transport independence [26, 39]. As a result, an indirect way was adopted, using object descriptors (ODs). Each media node is associated with an object identifier, which in turn uniquely identifies an OD. Within an OD, there is information on how many ESs are associated with this particular object and information describing each of those streams. The latter information includes the type of the stream, as well as how to locate it within the particular networking environment used. This approach simplifies remultiplexing (e.g., going through a wired–wireless interface), as there is only one entity that may need to be modified.

Multiplexing

MPEG-4 for delivery *supports* two major types of multiplex: the TransMux and FlexMux [30]. The TransMux is not specified by MPEG-4, but hooks are provided to enable any of the commonly used transport as needed by an application. Further, FlexMux, although specified by MPEG-4, is optional. This is a very simple design [40–42] intended for systems that may not provide native multiplexing services. An example is the data channel available in GSM cellular telephones. Its use, however, is entirely optional and does not affect the operation of the rest of the system. The FlexMux provides two modes of operation: *simple* and *muxcode* modes. Their key underlying concept in the design of the MPEG-4 multiplex is network independence. The MPEG-4 content may be delivered across a wide variety of channels, from very low bit rate wireless to high-speed ATM, and from broadcast systems to DVDs.

The next level of multiplexing in MPEG-4 is provided by the sync layer (SL), which is the basic conveyor of timing and framing information. The sync layer specifies a syntax for packetization of elementary streams into access units or parts thereof. Such a packet is called an SL packet. A sequence of such packets is called an SL-packetized stream (SPS). Access units are the only semantic entities that need to be preserved from end to end. Access units are used as the basic unit for synchronization. An SL packet consists of an SL packet header and an SL packet payload.

SL packets must be framed by a low-layer protocol, for example, the FlexMux tool. Packetization information is exchanged between an entity that generates an elementary stream and the sync layer. This relation is specified by a conceptual interface called the elementary stream interface (ESI).

MPEG-4 Visual

The envisaged applications of MPEG-4 Visual include mobile video phone, information access-game terminal, video e-mail and video answering machines, Internet multimedia, video catalogs, home shopping, virtual travel, surveillance, and networked video games.

With the phenomenal growth of the Internet and the World Wide Web, the interest in more interactivity with content provided by digital television is increasing. Additional text, still pictures, audio, or graphics that can be controlled by the end user can increase the entertainment value of certain programs or can provide valuable information that is unrelated to the current program but of interest to the viewer. Television station logos, customized advertising, and multiwindow screen formats that allow the display of sport statistics or stock quotes using data casting are prime examples of increased interactivity. Providing the capability to link and synchronize certain events with video will even improve the experience. By coding and representing not only frames of video but video objects as well, such new and exciting ways of representing content can provide completely new ways of television programming.

The enormous popularity of cellular phones and palm computers indicates the interest in mobile communications and computing. Using multimedia in these areas would enhance the end user's experience and improve the usability of these devices. Narrow bandwidth, limited computational capacity, and reliability of the transmission media are limitations that currently hamper widespread use of multimedia here. Providing improved error resilience, improved coding efficiency, and flexibility in assigning computational resources may bring this closer to reality.

Content creation is increasingly turning to virtual production techniques; an extension of the well-known technique is chroma keying. The scene and actors are recorded separately and can be mixed with additional computer-generated special effects. By coding video objects instead of frames and allowing access to the video objects, the scenes can be rendered with higher quality and with more flexibility. Television programs consisting of composited video objects and additional graphics and audio can then be transmitted directly to the end user, with the additional advantage of allowing the user to control the programming in a more sophisticated way.

The popularity of games on stand-alone game machines and on PCs clearly indicates the interest in user interaction. Most games are currently using 3D graphics, both for the environment and for the objects that are controlled by the players. Addition of video objects to these games would make the games even more realistic, and, using overlay techniques, the objects could be made more lifelike. Essential is the access of individual objects, and using standard-based technology would make it possible to personalize games by using personal video databases linked in real time into games.

Streaming video over the Internet is becoming more popular, using view tools as software plug-in for a Web browser. News updates and live music shows are some examples of streaming video. Here, bandwidth is limited because of the use of modems, and transmission reliability is an issue, as packet loss may occur. Increased error resilience and improved coding efficiency will improve the experience of streaming video. In addition, scalability of the bit stream, in terms of temporal and spatial resolutions but also in terms of video objects, under the control of the viewer, will further enhance the experience and also the use of streaming video.

The Visual part of MPEG-4 integrates a number of visual coding techniques from two major areas: natural video and synthetic visual [39, 43–46]. MPEG-4 Visual addresses a number of functionalities driven by applications, such as robustness against errors in Internet and wireless applications, low-bit-rate video coding for videoconferencing, high-quality video coding for home entertainment systems, object-based and scalable object-based video for flexible multimedia, and mesh and face coding for animation and synthetic modeling. Thus, MPEG-4 Visual integrates functionalities offered by MPEG-1 video, MPEG-2 video, object-based video, and synthetic visual. More precisely, the MPEG-4 Video version 1 specification covers the following aspects:

- *Natural video* – motion-compensated DCT coding of video objects.
- *Synthetic video tools* – mesh coding and face coding of wire frame objects.
- *Still texture decoding* – wavelet decoding of image texture objects.

A simplified high-level view of MPEG-4 Visual decoding is shown in Figure 4.15. The visual bit stream to be decoded is demultiplexed and variable length decoded into individual streams corresponding to objects and fed to one of four processes: face decoding, still texture decoding, mesh decoding, or video

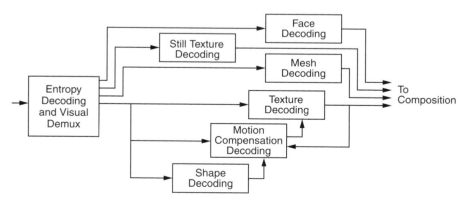

Figure 4.15 MPEG-4 Visual decoding.

decoding. The video decoding process further includes shape decoding, motion compensation decoding, and texture decoding. After decoding, the output of the face, still texture, mesh and video decoding process are sent for composition.

MPEG-4 includes significant improvements to the audio and visual coding. Even though MPEG-4 video can be operated anywhere from 5 kbps to 10 Mbps, it is not expected to replace MPEG-2 video. Beyond just compressing video frames, MPEG-4 Visual standard introduces new functionalities that include encoding arbitrarily shaped video objects, wavelet-based still image coding, and face and body animation tools [47].

Natural Video

In MPEG-4, each picture is considered as consisting of temporal instances of objects that undergo a variety of changes such as translations, rotations, scaling, and brightness and color variations. Moreover, new objects enter a scene and/or existing objects depart, leading to the presence of temporal instances of certain objects only in certain pictures. Sometimes, scene change occurs and thus the entire scene may either be reorganized or replaced by a new scene. Many MPEG-4 functionalities require access not only to an entire sequence of pictures, but also to an entire object and, further, not only to individual pictures but also to temporal instances of these objects within picture. The concept of video objects (VOs) and their temporal instances, video object planes (VOPs), is central to MPEG-4 video [48–51]. A VOP can be fully described by texture variations and shape representation. In natural scenes, VOPs are obtained by semiautomatic or automatic segmentation, and the resulting shape information can be represented as a binary shape mask. On the other hand, for hybrid natural and synthetic scenes generated by blue screen composition, shape information is represented by an 8-bit component, referred to as gray scale shape.

Example 4.2 Figure 4.16 shows a picture decomposed into a number of separate VOPs. The scene consists of two objects (head and shoulders and a logo) and the background. The objects are segmented by semiautomatic or automatic means and are referred to as VOP1 and VOP2,

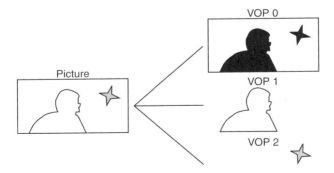

Figure 4.16 Semantic segmentation of picture into VOPs.

while the background without these objects is represented to as VOP0. In this manner, each picture in the sequence is segmented into VOPs. Thus, a segmented sequence contains a set of VOPs, a set of VOP1s, and a set of VOP2s. In other words, in this example, the segmented sequence consists of VO, VO1, and VO2. Individual VOs are encoded separately and multiplexed to form a bit stream the users can access and manipulate. Together with VOs, the encoder sends information about scene composition to indicate where and when VOPs of a VO are to be displayed. However, this information is optional and may be ignored at the decoder, which may use user-specified information about composition.

Example 4.3 An example of video object plane (VOP) structure in MPEG-4 video coding that uses a pair of B-VOPs between two references is shown in Figure 4.17. A pair of B-VOPs is represented, too. In MPEG-4 Video, an input sequence can be divided into groups of VOPs (GOVs). Each GOV starts with an I_VOP and the rest of the GOV contains an arrangement of P_VOPs and B_VOPs. For coding purpose, each VOP is divided into a number of macroblocks. Each macroblock is essentially a 16 × 16 block luminance (or alternatively, four 8 × 8 blocks) with corresponding chrominance blocks. An optional packet structure can be imposed on VOPs to provide more robustness in error-prone environments.

MPEG-4 video coding exploits the spatial and temporal redundancies. Spatial redundancies are exploited by block DCT coding, while temporal redundancies are exploited by motion compensation. In addition, MPEG-4 video needs to code the shape of each VOP. Shape coding in MPEG-4 also uses motion compensation for prediction. MPEG-4 video coding supports both noninterlaced video (as in MPEG-1 video coding) and interlaced video (MPEG-2 video coding).

A simplified block diagram showing systems demultiplex and the MPEG-4 video decoder is presented in Figure 4.18. The MPEG-4 video decoder receives bit streams to be decoded from the MPEG-4 Systems demux. The MPEG-4 video decoder consists of a variable length decoder, an inter/intraframe/field motion compensation. After demultiplexing, the MPEG-4 video bit stream is sent to the variable-length decoder for decoding of motion vector (m), quantizer information (q), inter/intradecision (i), frame/field decision (f), shape identifiers (s), and the data consisting of quantized DCT coefficient indices. The shape identifiers are decoded by the shape decoder. They may employ shape motion prediction using previous shape (ps) to generate the current shape (cs). The inter/intraframe/field DCT decoder uses the

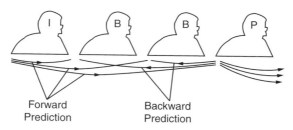

Figure 4.17 Video object plane (VOP) coding structure.

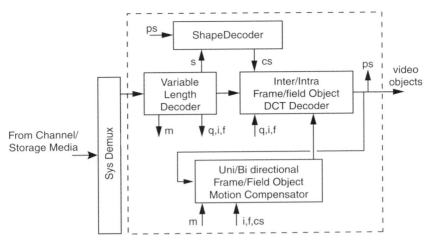

Figure 4.18 Systems demultiplex and the MPEG-4 video decoder [33]. (©1999 ISO/IEC.)

decoded DCT coefficient indices, the current shape, the quantizer information, the inter/intradecision, and the frame/field information to dequantize the indices to yield DCT coefficient blocks inside the object and then inverse transform the blocks to recover decoded pixel blocks. The uni/bidirectional frame/field motion compensator, if the coding mode is inter (based on the intra/interdecision), uses motion vectors, current shape, and frame/field information to generate motion-compensated prediction blocks that are then added back to the corresponding decoded prediction error blocks output by the inter/intraframe/field DCT decoder to generate decoded blocks. If the coding made is intra, no motion-compensated prediction needs to be added to the output of the inter/intraframe/field DCT decoder. The resulting decoded VOPs are output on the line labeled video objects.

MPEG-4 offers a generalized scalability framework supporting both temporal and spatial scalabilities [52, 53]. Scalable coding offers a means of scaling the decoder complexity of processors and/or when memory resources are limited and often time varying. Further, scalability also allows graceful degradation of quality when the bandwidth resources are limited and continually changing. It even allows increased resilience to errors under noisy channel conditions. Temporally scalable encoding offers decoders a means of increasing the temporal resolution of decoded video using decoded enhancement layer VOPs in conjunction with decoded base layer VOPs. On the other hand, spatial scalability encoding offers decoders a means of decoding and displaying either the base layer or the enhancement layer output.

Figure 4.19 shows a two-layer generalized codec structure for MPEG-4 scalability [52, 54]. Because MPEG-4 Video supports object-based scalability, the preprocessor is modified to program VO segmentation and generate two streams of VOPs

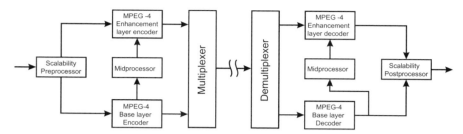

Figure 4.19 MPEG-4 video scalability decoder.

per VO by spatial or temporal preprocessing, depending on the scalability type. One such stream is the input to the lower layer encoder, that is, the MPEG-4 nonscalable video encoder. The higher layer encoder is identified as the MPEG-4 enhancement video encoder. The role of the midprocessor is the same as in MPEG-2, either to spatially upsample the lower layer VOPs or to let them pass through, in both cases to allow prediction of the enhancement layer VOPs. The two encoded bit streams are sent to MPEG-4 SystemsMux for multiplexing. The operation of the scalability decoder is essentially the inverse of that of the scalability encoder, just as in the case of MPEG-2. The decoded output of the base and enhancement layers is two streams of VOPs that are sent to the postprocessor, either to let the higher layer pass through or to be combined with the lower layer.

Synthetic Video

MPEG-4 is well suited for applications that incorporate natural and synthetic content. For example, a customized agent model can be defined for games or Web-based customer service applications. To this effect, MPEG-4 enables integration of face animation with multimedia communication and presentations and allows face animation over low-bit-rate communication channels, for point-to-point as well as multipoint connections with low delay. In many applications, the integration of face animation and text-to-speech synthesizer is of special interest. Simple animations of generic 2D objects can be used to enhance multimedia presentation or electronic multiplayer games. The 2D mesh tool in MPEG-4 can also be used to create special effects such as augmented reality video clips, in which natural video objects are overlaid by synthetic graphics or text, and to allow manipulation of video by the user.

The tools included in the synthetic video subpart of MPEG-4 Video are facial animation, mesh-based representation of general, natural, or synthetic visual objects, as well as still image texture [30].

Facial animation in MPEG-4 Video is supported via the facial animation parameters (FAPs) and facial definition parameters (FDPs), which are sets of parameters designed to allow animation of faces, reproducing expressions, emotions, and speech pronunciation, as well as defintion of facial shape and texture. The same set of FAPs, when applied to different facial models, results in reasonably similar expressions and speech pronunciation without the need to initialize or calibrate the model. On the other hand, the FDPs allow the definition of a precise facial shape and texture in the setup phase. If the FDPs are used in the setup phase, it is also possible to produce the movements of particular facial features precisely. Using a phoneme-to-FAP conversion, it is possible to control facial models accepting FAPs via text-to-speech (TTS) systems. This conversion is not standardized [55, 56]. The setup stage is necessary, not to create face animation, but to customize the face at the decoder. The FAP set contains two high-level parameters, visemes and expressions. A viseme is a visual correlate of phonemes. The viseme parameter allows viseme rendering without having to express them in terms of other parameters. Also it enhances the results of other parameters, ensuring the correct rendering of visemes. All the parameters involving translation movement are expressed in terms of the facial animation parameter units (FAPUs). These units are defined in order to allow interpretation of the FAPs on any facial model in a consistent way, producing reasonable results in terms of expression and speech pronunciation. The FDPs are used to customize the proprietary face model of the decoder to a particular face or to download a face model along with the information about how to animate it. The FDPs are normally transmitted once per session, followed by a stream of compressed FAPs. The FDP set is specified using the FDP node (in MPEG-4 Systems), which defines the face model to be used at the receiver [57–59].

The mesh-based representation of general, natural, or synthetic visual objects is useful for enabling a number of functions such as temporal rate conversion, content manipulation, animation, augmentation, and transfiguration (merging or replacing natural video with synthetic). MPEG-4 Visual includes a tool for triangular mesh-based representation of general-purpose objects. A visual object of interest, when it first appears in the scene, is tessellated into triangular patches resulting in a 2D triangular mesh. The vertices of the triangular patches forming the mesh are referred to as the node points. The node points of the initial mesh are then tracked as the VOP moves within the scene. The 2D motion of a video object can thus be compactly represented by the motion vectors of the node points in the mesh. Motion compensation can then be achieved by texture mapping the patches from VOP to VOP according to affine transforms. Coding of video texture or still texture of objects is performed by the texture coding tools of MPEG-4. Thus, efficient storage and transmission of the mesh representation of a moving object (dynamic mesh) require compression of its geometry and motion [59–65].

The still image texture is coded by the discrete wavelet transform (DWT). This texture is used for texture mapping or faces or object represented by mesh. The data can represent a rectangular or an arbitrary shaped VOP. Besides coding efficiency, an important requirement for coding texture map data is that the data should be coded in a manner facilitating continuous scalability, thus allowing many resolutions or qualities to be derived from the same coded bit stream. Although DCT-based coding is able to provide comparable coding efficiency as well as a few scalability layers, DWT-based coding offers flexibility in organization and number of scalability layers [60]. The basic steps of zero-tree wavelet-based coding scheme are as follows:

- decomposition of the texture using the discrete wavelet transform (DWT)
- quantization of the wavelet coefficients
- coding of the lowest frequency subband using a predictive scheme
- zero-tree scanning of the higher order subband wavelet coefficients
- entropy coding of the scanned quantized wavelet coefficients and the significant map.

MPEG-4 Audio

The Audio part of MPEG-4 integrates a number of audio coding techniques [66]. MPEG-4 Audio addresses a number of functionalities driven by applications, such as robustness against packet loss or change in transmission bit rates for Internet-phone systems, low-bit-rate coding, higher quality coding for music, improved text-to-speech (TTS), and object-based coding for musical orchestra synthesization. Audio scenes may be usefully described as the spatiotemporal combination of audio objects. An *audio object* is a single audio stream coded using one of the MPEG-4 Audio coding tools. Audio objects are related to each other by mixing, effects processing, switching and delaying, and may be spatialized to a particular 3D location [67]. More precisely, the MPEG-4 Audio version 1 specification covers the following aspects:

- *Low-bit-rate audio coding tools* – code excited linear predictive (CELP) coding and coding based on parametric representation (PARA).
- *High-quality audio coding tools* – time–frequency mapping techniques, advanced audio coding (AAC), and Twin VQ.
- *Synthetic audio tools* – text-to-speech (TTS) and structured audio.

Natural Audio

Natural audio coding in MPEG-4 includes low-bit-rate audio coding as well as high-quality audio coding tools [30, 66]. Figure 4.20 is a simplified block diagram for

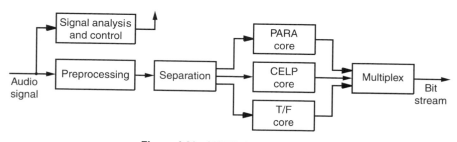

Figure 4.20 MPEG-4 and encoding.

integration of MPEG-4 natural audio coding tools. Only the encoding end is shown here. It consists of preprocessing, which facilitates separation of the audio signals into types of components to which a matching technique from among PARA, CELP, and T/F coding may be used. Signal analysis and control provide the bit rate assignment and quality parameters needed by the chosen coding technique [68].

The PARA coder core provides two sets of tools. The harmonic vector excitation coding (HVXC) tools allow coding of speech signals at 2 kbit/s. The individual line coding tools allow coding of nonspeech signals such as music at bit rates of 4 kbit/s and higher. Both sets of tools allow independent change of speed and pitch during the decoding and can be combined to handle a wider range of signals and bit rates.

The CELP coder is designed for speech coding at two different sampling frequencies: 8 and 16 kHz. The speech coders using the 8 kHz sampling rate are referred to as narrowband coders and those using the 16 kHz sampling rate as wideband coders. The CELP coder includes tools offering a variety of functions including bit rate control, bit rate scalability, speed control, complex scalability, and speech enhancement. By using the narrowband and wideband CELP coders, it is possible to span a wide range of bit rates (4 to 24 kbit/s). Real-time bit rate control in small steps can be provided. Many tools and processes have been designed to be commonly usable for both narrowband and wideband speech coders.

The T/F coder provides high-end audio coding and is based on MPEG-2 AAC coding. The MPEG-2 AAC is a state-of-the-art audio compression algorithm that provides compression superior to that provided by order algorithms. AAC is a transform coder and uses a filter bank with a finer frequency resolution that enables superior signal compression. AAC also uses a number of new tools such as temporal noise shaping, backward adaptive linear prediction, joint stereo coding techniques, and Huffman coding of quantized components, each of which provides additional audio compression capability. Furthermore, AAC supports a wide range of sampling rates and bit rates, from 1 to 48 audio channels, up to 15 low-frequency enhancement channels, multilanguage capability, and up to 15 embedded data streams. The MPEG-2 AAC provides five-channel audio coding capabilities.

Synthetic Audio

The text-to-speech (TTS) system synthesizes speech as its output when a text is provided as its input. In this case, the TTS changes the text into a string of phonetic symbols and the corresponding basic synthetic units are retrieved from the preprepared database. The MPEG-4 TTS not only can synthesize speech according to the input speech, but also executes several other functions:

- speech synthesis with the original prosody from the original speech
- synchronized speech synthesis with facial animation (FA) tools
- synchronized dubbing with moving pictures, not by recorded sound, but by text and some lip shape information

- functions such as stop, resume, forward, backward without breaking the prosody, even in the applications with facial animation (FA)/motion pictures (MP)
- ability of users to change the replaying speed, tone, volume, speaker's sex, and age.

The MPEG-4 TTS can be used for many languages because it adopts the concept of the language code, such as the country code for an international call. For MPEG-4 TTS, only the interface bit stream profiles are the subject of standardization [30, 66]. Because there are already many different types of TTS and each country has several of a few tens of different TTS synthesizing its own language, it is impossible to standardize all things related to TTS.

MPEG-4 Profiling

MPEG-4 Visual profiles are defined in terms of visual object types. There are six video profiles: Simple profile, Simple Scalable profile, Core profile, Main profile, N-bit profile, and Scalable Texture profile. There are two synthetic visual profiles: the Basic Animated Texture profile and Single Facial Animation profile. There is one hybrid profile, Hybrid profile, that combines video object types with synthetic visual object types.

MPEG-4 Audio consists of four profiles: Main profile, Scalable profile, Speech profile, and Synthetic profile. The Main profile and Scalable profile consist of four levels. The Speech profile consists of two levels, and the Synthetic profile consists of three levels.

MPEG-4 Systems also specifies a number of profiles. There are three types of profiles: Object Descriptor (OD) profile, Scene Graph profiles, and Graphics profiles. There is only one OD profile, called Core profile. There are four Scene Graph profiles – Audio profile, Simple 2D profile, Complete 2D profile, and Complete profile. There are three Graphics profiles – Simple 2D profile, Complete 2D profile, and Complete profile.

4.3.4 Media Description, Searching, and Retrieval

Multimedia search and retrieval applications include large-scale multimedia search engines on the Web, media asset management systems in corporations, audiovisual broadcast servers, and personal media servers for customers. Application requirements and user needs often depend on the context and the application scenarios. Professional users may want to find a specific piece of content from a large collection within a tight deadline, and leisure users may want to browse the clip art catalog to get a reasonable selection. Online users may want to filter through a massive amount of information to receive information pertaining to their interest only, whereas offline users may want to get informative summaries of selected content from a large repository.

To describe various types of multimedia information, the MPEG-7 standard has the objective of specifying a standard set of descriptors as well as description schemes for the structure of the descriptors and their relationships. This description will be associated with the content itself to allow fast and efficient searching for material of the user's interest.

In MPEG-7, the perceived advantages are the ease of fast and efficient identification of audiovisual content that is of interest to the user by

- allowing the same indexed material to be identified by many identifications (e.g., search and filtering) engines
- allowing the same identification engine to identify indexed material from many different sources.

There are many applications domains that would benefit from the MPEG-7 standard (education, journalism, entertainment, GIS, surveillance, biomedical applications, film, video and radio archives, audiovisual content production, multimedia portals). The MPEG-7 applications document organizes the example applications into three sets:

- *Pull applications*—applications mainly following a pull paradigm, notably storage and retrieval of audiovisual databases, delivery of pictures and video professional media production, commercial musical applications, sound effects libraries, historical speech databases, movie scene retrieval by memorable auditory events, and registration and retrieval of trademarks.
- *Push applications*—applications mainly following a push paradigm, notably user agent-driven media selection and filtering, personalized television services, intelligent multimedia presentations, and information access facilities for people with special needs.
- *Specialized professional applications*—applications that are particularly related to a specific professional environment, notably teleshopping, biomedical, remote sensing, educational, and surveillance applications.

MPEG-7 is conceptually an audiovisual information representation standard, but the representation satisfies very specific requirements. The Audio and Video parts provide standardized audio only and visual only descriptors, the Multimedia Description Schemes (MDS) part provides standardized description schemes involving both audio and video descriptors, the Description Definition Language (DDL) provides a standardized language to express description schemes, while the Systems part provides the necessary glue that enables the use of the standard in practical environments. Lastly, the Reference Software contains the huge contributions made by the community to develop the MPEG-7 Open Source code. MPEG-7 fulfills a key function in the forthcoming evolutionary steps of multimedia. So, much as MPEG-1, MPEG-2, and MPEG-4 provided the tools through which the current abundance of audiovisual content could happen, MPEG-7 will provide the means to navigate through this wealth of content [69]. Thus, MPEG-7 is the content representation standard for multimedia information search, filtering, management, and processing [70, 71]. The need for MPEG-7 grows because, although more and more multimedia information is available in compressed digital form and a number of search engines exist on the Internet, they do not incorporate special tools or features to search for audiovisual information, as much of the search is still aimed at textual documents. A goal of MPEG-7 is to enable search for multimedia on the Internet and improve the current situation caused by proprietary solutions by standardizing interfaces for description of multimedia content. MPEG-7 aims to address finding content of interest in *pull* applications, such as selection and filtering to extract content of interest within broadcast channels. It is expected that MPEG-7 will work not only with MPEG, but also with non-MPEG coded content.

MPEG-7 Tools

Early in development of MPEG-7, it was clear that a number of different tools were needed to achieve the standard's objectives. These tools are descriptors (the elements), description scheme (the structures), a description definition language (DDL) (for extending the predefined set of tools), and a number of systems and tools [72].

MPEG-7 descriptors describe various types of multimedia information. This description will be associated with the content itself, to allow fast and efficient searching for material of a user's interest. Audiovisual material that has MPEG-7 data associated with it can be indexed and searched. This material may include still pictures, graphics, 3D models, audio, speech, video, and information about how these elements are combined in a multimedia presentation. Special cases of these general data types may include facial expressions and personal characteristics.

Figure 4.21 shows a possible processing chain and scope of the MPEG-7 standard. Although MPEG-7 does not standardize the feature extraction, the MPEG-7 description is based on the output of feature extraction, and although it does not standardize the search engine, the resulting description is consumed by the search engine. Descriptions can vary according to the types of data, for example, color, musical harmony, textual name, and so on. Descriptions can also vary according to the application, for example, species, age, number of percussion instruments, information accuracy.

MPEG-7 Descriptors

The syntax and semantics of the descriptor provide a description of the feature. However, for fully representing a feature, one or more descriptors may often be

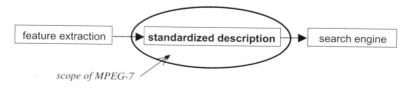

Figure 4.21 Scope of MPEG-7 standard.

needed. For example, for representing a color feature, one or more of the following descriptors may be used: the color histogram, the average of its frequency components, the motion field, and the textual description.

The description allows an evolution of the corresponding feature via the descriptor value. For descriptors, core experiments are needed to allow further evaluation and development of the few preselected proposals considered to be promising in the initial evaluation [73]. Several types of descriptors, such as color, texture, motion, and shape, are the subject of such core experiments and standardized test conditions (e.g., content, parameters, evaluation criteria). Each core experiment will be finalized. There are two core experiments on motion description. The first is related to motion activity and the second to motion trajectory. The motion activity experiment aims to classify the intensity or pace of the action in a segment of a video scene. For example, a segment of a video scene containing a goal scored in a soccer match may be considered as highly active, whereas a segment containing the subsequent interview with the player may be considered to be of low activity. The motion trajectory experiment aims to describe efficiently the trajectory of an object during its entire life span as well as the trajectory of multiple objects in segments of a video scene.

Two core experiments on shape descriptors are considered: the first related to simple nonrigid shapes and the second related to complex shape. The simple shape experiments evaluate the performance of competing proposals based on a number of criteria such as exact matching, similarity-based retrieval, and robust retrieval of small nonrigid deformations. The complex shapes experiment expects to evaluate the performance of competing proposals based on a number of criteria such as exact matching and similarity-based retrieval.

Examples for descriptors include a time code for representing duration, color moments and histograms for representing color, and a character string for representing a title.

MPEG-7 Description Schemes

A description scheme (DS) specifies the structure and semantics of the relationship between its components, which may be both descriptors and description schemes. Following the recommendations of the MPEG-7 evaluation process [73], a high-level framework common to all media description schemes and a specific framework on a generic visual description scheme is designed [74, 75].

The syntactic structure DS describes the physical entities and their relationship in the scene. It consists of zero or more occurrences of each of the segment DS, the region DS, and the DS describing the relation graph between the segments and regions. The segment DS describes the temporal relationship between segments (groups of frames) in the form of a segment tree in a scene. It consists of zero or more occurrences of the shot DS, the media DS, and the meta information DS. The region DS describes the spatial relationship between regions in the form of a region tree in a scene. It consists of zero or more occurrences of each of the geometry DS, the color/texture DS, the motion DS, the deformation DS, the media information DS, and the media DS.

The semantic structure DS describes the logical entities and their relationship in the scene. It consists of zero or more occurrences of an event DS, the object DS, and the DS describing the relation graph between the events and objects. The event DS contains zero or more occurrences of the events in the form of an event tree. The object DS contains zero or more occurrences of the objects in the form of an object tree.

The analytic/synthetic model DS describes cases that are neither completely syntactic nor completely semantic, but rather in between. The analytic model DS specifies the conceptual correspondence such as projection or registration of the underlying model with the image or video data. The synthetic animation DS consists of the animation stream defined by the model event DS, the animation object defined by the model object DS, and the DS describing the relation graph between the animation streams and objects.

The visualization DS contains a number of view description schemes to enable fast and effective browsing and visualization of the video program. The global media DS, global media information DS, and global meta information DS correspondingly provide information about the media content, file structure, and intellectual property rights. A simple example is a movie, temporally structured as scenes and shots, including some textual descriptors at the scene level, and color, motion, and audio amplitude descriptors at the shot level

MPEG-7 Description Definition Language (DDL)

The DDL is expected to be a standardized language used for defining MPEG-7 description schemes and descriptors. Many of the DDL proposals submitted for MPEG-7 evaluation were based on modifications of the extensible markup language (XML). The evaluation group recommended that the design of the MPEG-7 DDL to be based on XML be enhanced to satisfy MPEG-7 requirements [73]. The current status of the DDL is documented in reference [76]. The current list of DDL requirements comprises

- ability to compose a DS from multiple DDS
- platform and application independence
- support for primitive data types (e.g., text, integer, real, data, time, index)
- ability to describe composite data types (e.g., histograms, graphs)
- ability to relate descriptions to data of multiple media types
- support for distinct name spaces
- ability to reuse, extend, and inherit from existing DSs and descriptors
- capability to express spatial relations, temporal relations, structural relations, and conceptual relations
- ability to form links and/or references between one or several descriptions
- a mechanism for intellectual property information management and protection for DSs and descriptors.

MPEG-7 Systems

The basis of classification of MPEG-7 applications into the categories push, pull, and hybrid provides the capabilities needed for MPEG-7 Systems [73]. In push applications, besides the traditional role, MPEG-7 Systems has the main tasks of enabling multimedia data filtering. In pull applications, MPEG-7 Systems has the main task of enabling multimedia data browsing. A generic MPEG-7 system may have to enable both multimedia data filtering and browsing [77]. A typical model for MPEG-7 Systems is to support client–server interaction, multimedia DDL, multimedia data management, multimedia consumption, and aspects of multimedia data presentation. This role appears to be much wider than the role of MPEG-4 Systems.

4.3.5 Media Distribution and Consumption

MPEG-21's approach is to define a framework to support transactions that are interoperable and highly automated, specifically taking into account digital rights management (DRM) requirements and targeting multimedia access and delivery using heterogeneous networks and terminals.

Solutions with advanced multimedia functionality are becoming increasingly important as individuals are producing more and more digital media, not only for professional but also for personal use. All these *content providers* have many of the same concerns: management of content, protection of rights, protection from unauthorized access/modification, and protection of privacy of providers and consumers. New solutions are required to manage the access and delivery process of these different content types in an integrated and harmonized way, entirely transparent to the different users of multimedia services. The need for these solutions motivates the MPEG-21 Multimedia framework initiative that aims to enable transparent and augmented use of multimedia resources across a wide range of networks and devices.

MPEG-21 aims at defining a normative open framework for multimedia delivery and consumption for use by all the players in the delivery and consumption chain. This open framework will provide content creators and service providers with equal opportunities in the MPEG-21 enabled open market. This will also be to the benefit of the content consumers, providing them access to a large variety of content in an interoperable manner [78]. The MPEG-21 vision can thus be summarized as follows: to define a multimedia framework to enable transparent and augmented use of multimedia resources across a wide range of networks and devices used by different communities.

MPEG-21 is based on two essential concepts:

- the definition of a fundamental unit of distribution and transaction – the digital item (DI)
- the concept of users interacting with DIs.

The DIs can be considered the *what* of the multimedia framework (e.g., a video collection, a music album) and the users can be considered the *who* of the multimedia framework.

A digital item (DI) is a structured digital object with a standard representation, identification, and associated metadata within the MPEG-21 framework. The entity is the fundamental unit of distribution and transaction within this framework. MPEG-21 describes a set of abstract terms and concepts to form a useful model for defining DIs (digital item declaration, DID). Within this model, a DI is the digital representation of *some work* (the result of a creative process), and, as such, it is the unit that is acted upon within the model.

In MPEG-21, a user is any entity that interacts within the MPEG-21 environment and/or makes use of DIs. Such users include individuals, consumers, communities, organizations, corporations, consortia, and governments. Users are identified specifically by their relationship to another user for a given interaction. At its most basic level, MPEG-21 can be seen as providing a framework in which one user interacts with another user and the object of that interaction is a DI. Some such interactions are creating content, providing content, archiving content, rating content, enhancing and delivering content, aggregating content, retail selling of content, facilitating transactions that occur from any of the above, and regulating transactions that occur from any of the above.

MPEG-21 Parts

Part 1. Vision, Technologies, and Strategy. MPEG first produced a technical report to introduce the MPEG-21 Multimedia framework. The fundamental purposes of this technical report are

- To define a vision for a multimedia framework to enable transparent and augmented use of multimedia resources across a wide range of networks and devices to meet the needs of all users.
- To achieve the integration of components and standards to facilitate harmonization of technologies for the creation, management, transport, manipulation, distribution, and consumptions of DIs.
- To define a strategy for achieving a multimedia framework by the development of specifications and standards based on well-defined functional requirements through collaboration with other bodies.

Part 2. Digital Item Declaration (DID). A DID is a document that specifies the makeup, structure and organization of DI. As such, this technology contains three nominative clauses:

- *Model* – the DID model describes a set of abstract terms and concepts to form a useful model for defining DIs.
- *Representation* – this clause contains the normative description of the syntax and semantics of each of the DI declaration elements.
- *Schema* – this clause contains the normative XML schema comprising the entire grammar of the DID representation in XML.

Part 3. Digital Item Identification (DII). The DID can contain information about the item and/or parts thereof. The DII provides for a normative way to express how this identification can be expressed and associated with DIs, containers, components, and/or fragments thereof, by including them in a specific place in the DID. Digital items (DIs) and their parts within the MPEG-21 are identified by encapsulating uniform resource identifiers URIs [79] into the identifier element. Examples of likely identifiers include descriptive, control, revision tracking, and/or identifying information.

Part 4. Intellectual Property Management and Protection (IPMP). This part defines an interoperable framework for IPMP. This work builds further on IPMP work in MPEG-4, which is aimed at providing IPMP hooks. The project includes standardized ways of retrieving IPMP tools from remote locations and exchanging messages between IPMP tools and between these tools and the terminal. It also addresses authentication of IPMP tools and has provisions for integrating rights expressions according to the Rights Data Dictionary and the Rights Expression Language (REL).

Part 5. Rights Expression Language (REL). This is a machine-interpretable language intended to provide flexible, interoperable mechanisms to support transparent and augmented use of digital resources in publishing, distributing, and consuming of electronics books, broadcasting, digital movies, digital music, interactive games, computer software, and other creations in digital form in a way that protects digital content and honors the rights, conditions, and fees specified for digital contents. It is also intended to support exchange of sensitive or private digital content. The REL is also intended to provide flexible interoperable mechanisms to ensure personal data are processed in accordance with individual rights and to meet the requirement for users to be able to express their rights and interests in a way that addresses issues of privacy and use of personal data.

Part 6. Rights Data Dictionary (RDD). This is a dictionary of key terms that are required to describe rights of those who control DIs, including intellectual property rights and the permission they grant, that can be unambiguously expressed using a standard syntactic convention and can be applied across all domains in which rights and permissions need to be expressed. RDD comprises a set of clear, consistent, structured, integrated, and uniquely identified terms to support the MPEG-21 REL [80]. RDD specifies the structure and core of this dictionary and specifies how further terms may be defined under the governance of a registration authority. The RDD system is designed to support the mapping and transformation of metadata from the terminology of one namespace (or authority) into that of another namespace (or authority) in an automated or partially automated way, with minimum ambiguity or loss of semantic integrity.

MPEG-21 in a Universal Multimedia Access (UMA) Environment

Universal multimedia access (UMA) deals with the delivery of media resources under different network conditions, user preferences, and capabilities of terminal devices. The primary motivation of UMA is to enable terminals with potentially limited communication, processing, storage, and display capabilities to access rich-media resources [81, 82].

UMA presents the solution for wired and wireless systems to access the same media resource provider, each of them receiving media resources enabled for their systems capabilities. The UMA concept is presented in Figure 4.22. The UMA application suits the next generation mobile end wireless systems [83, 84]. UMA is a dynamic force behind the development of services in 3G systems in that UMA enabled services are what the end users will benefit from.

As far as access and distribution of media resources using heterogeneous terminals and networks is concerned, there is a good match between the goals of UMA and MPEG-21. A way to address the UMA requirements and additionally include the IPMP requirements is hence the use of DIs in the MPEG-21 multimedia framework. Within UMA, the following factors have been identified as having a main impact on streaming media: content availability, terminal capabilities, network characteristics, user preferences, and the natural environment of the user. The

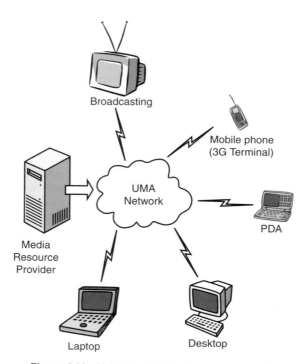

Figure 4.22 Universal multimedia access concept.

streaming media characteristics of these factors are available bandwidth, error characteristics, screen size, content scalability (spatial, temporal, and spectral), adaptivity, interactivity, synchronization, and multiplexing.

MPEG-21 Digital Item Adaptation

The central concept of MPEG-21 digital item adaptation (DIA) is that DIs are subject to a resource adaptation engine, as well as a description adaptation engine, which together produce the adapted DI. Illustration of DI adaptation is shown in Figure 4.23. DIA specifically addresses requirements with respect to the usage environment and to media resource adaptability.

DIA addresses requirements that are specific to the usage environment descriptions, where the emphasis is on identifying characteristics that should be described by the DI usage environment description. More specifically, MPEG-21 will support the following descriptions:

- the usage environment, at least in terms of terminal, network delivery, user, and natural environment capabilities (hardware platform, software platform, network characteristics)
- terminal capabilities, including acquisition properties, device type (encoder, decoder, gateway, router, camera) and profile, output properties, hardware properties (processor speed, power consumptions, memory architecture), software properties, system properties (processing modules, interconnection of components, configuration options, ability to scale resource quality, manage multiple resources), and IPMP-related capabilities
- network capabilities, including delay characteristics (end-to-end delay, one-way delay, delay variation), error characteristics (bit-error rate, packet loss,

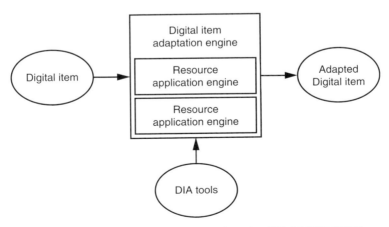

Figure 4.23 Illustration of digital item adaptation [83]. (©2003 IEEE)

burstiness) and bandwidth characteristics (amount of bandwidth, variation in bandwidth)
- delivery capabilities, including the types of transport protocols supported and the types of connections supported (broadcast, multicast, unicast)
- user characteristics including user preferences and demographic information
- natural environment characteristics, including location, type of location (indoor, outdoor, public place, home, office), available access networks in a given area, velocity of user or terminal
- service capabilities, including user roles and types of services
- interactions and relations among users, including the relation between various dimensions of the usage environment description.

Digital item adaptation (DIA) addresses requirements that are specific to media resource adaptability, including context representation format and resource-complexity descriptions. More specifically, MPEG-21 will support

- format descriptions that are independent of the actual content representation formats
- the description of scalable content representation formats
- descriptions that may be automatically extracted in a format-independent way from the resource – the descriptions will allow the generation of the media resource in a format-independent way
- the description of resources in terms of perceived quality and associated processing complexity; terminal QoS allows for resource scaling according to terminal capabilities: to scale the resource optimally, resource description information should be provided, so in this way, the terminal can better consider the trade-off between resource quality and the computational complexity that needs or is available
- the description of metadata in terms of perceived importance and associated processing complexity.

Additionally, MPEG-21 will provide descriptions of content representation formats at different syntactic layers for adapting resources and descriptions.

MPEG-21 and MEDIACOM2004

As previously stated, the intent of the project MEDIACOM2004 is to generate a framework for multimedia communications studies for use within the ITU and other standard development organizations (SDOs). The main objective of MEDIACOM2004 is to establish a framework for the harmonized and coordinated development of global multimedia communication standards. It is intended that this will be developed with the support of, and use by, all relevant ITU-T and ITU-R study groups, and in close cooperation with other regional and international SDOs.

The key framework standardization areas (FSAs) identified for study within MEDIACOM2004 are

- multimedia architecture
- multimedia applications and services
- interoperability of multimedia systems and services
- media coding
- QoS and end-to-end performance in multimedia systems
- security of multimedia systems and services
- accessibility to multimedia systems and services.

With respect to MPEG-21, MEDIACOM2004 can be considered to primarily address a framework for multimedia communications standardization, whereas MPEG-21 primarily addresses the augmented use of multimedia resources across a wide range of networks and devices by all the users of the value chain. In this case we arrive at the optimal collaboration between MPEG-21 and MEDIACOM2004.

To conclude, MPEG-21 offers solutions to support interoperable exchanging, accessing, consuming, trading, and otherwise manipulating digital items between users in an efficient, transparent, and interoperable way.

4.4 DIGITAL BROADCASTING

Media contains audio, video, and multimedia, in addition to still images, graphics, and computer animations. Mixtures of them occur in various combinations. Applications such as films, music, games, and so on, are most common today. The distribution of media can take place based on a great number of technical solutions.

There have been various advances in digital technology that have all come together to make digital terrestrial TV, as well as digital TV, in general, possible, including [85]

- The development of the MPEG-2 video compression standard, which takes advantages of similarities in the picture from one frame to another, as well as compressing the actual data in the same frame.
- In the case of digital terrestrial coded orthogonal frequency division multiplex (COFDM), it has only recently become viable to produce semiconductor devices fast and cheap enough to perform the fast Fourier transform (FFT) operation that is an essential aspect of orthogonal frequency division multiplex (COFDM).
- Even if the technology is available it has to be available at the right cost to consumers, to create the market. This has also only recently been possible with advances in the semiconductor manufacturing process.

Broadcasting means in general a distribution from one to many/all over a broadcasting network. Traditional broadcasting implies the use of a broadcasting network, which may be terrestrial, cable, or satellite. In 1992 an organization called Digital Video Broadcasting (DVB) was set up to set standards in the areas of digital television broadcasting in Europe. This voluntary group, made up of over 200 organizations, published the digital satellite (DVB-S) specification and digital cable (DVB-C) specification in 1994. The digital terrestrial (DVB-T) specification was then finalized in late 1996/early 1997. The Advanced Television Systems Committee (ATSC) has completed its data broadcast standard. This specification enables multimedia content providers to deliver applications using broadband channels of digital television. It will also enable the integration of databased information technologies and television audiovisual.

In future, new broadband networks such as fiber-copper or fixed broadband wireless access (BWA) solutions will also provide broadcasting services. The media distribution over such networks is called streaming media distribution or just streaming using point-to-point or multicasting IP solutions. Streaming media generally use two different technologies: real-time streaming and progressive streaming.

In real-time streaming media the system keeps the bandwidth and media signal matched to each other. This enables the user always to get the content in real time, but in some cases the bandwidth/quality may change during the transmission. Dedicated streaming media servers are required, including special real-time streaming protocols. For example, real-time multiple user video games can be regarded as real-time systems, but are generally not considered as such. Real-time streaming is suited for live events such as music and video, which in principle should never stop or pause when the transport is started. When transported over the Internet, this is an ideal case, but, in reality, periodical pauses may occur. In most cases, errors or last information cannot be retransmitted, which means that quality deterioration will occur when the network is congested or has some other transport problem. In universal mobile telecommunication system UMTS/3G, this can often be avoided by using a direct broadband link between the content service provider and the UMTS operator's controlled infrastructure. A quality real-time streaming system needs a fairly broad bandwidth and constant bandwidth/bit rate. Real-time streaming requires special streaming servers like Quick Time Streaming Server, Real Server, or Windows Media Server. For Internet application these servers are the most frequently used. They require a rather complicated setup and administration, but they give a fairly good control of media delivery. Distribution problems may occur with various filtering systems.

Progressive streaming means a progressive download for full content or only for a needed prebuffering. There is no matching needed to bandwidth/bit rate like in real-time streaming. If errors or missing packets occur, retransmission can be requested without notable errors. However, this is dependent on buffer size, the bit rates compared to the content and the error rate. As no special protocol is needed, the content can be stored in an http or FTP site. Progressive streaming is generally used when we need high picture quality and to download audio files. The data rate is not dependent on the transmission system, but has only an impact on the download time. Progressive streaming is also good for systems like TV Anytime and audio downloading, as well as other on-demand types of services.

The aim of this section is to clarify the terms and system capabilities for digital television/radio services, including both terrestrial and satellite components, and to identify models of cooperation and resulting technical issues. The commercial requirements of these applications can then be addressed. However, we shall start with the convergence of telecommunications and broadcast infrastructures.

4.4.1 Convergence of Telecommunications and Broadcast Infrastructure

Broadcasting deals with *one-to-many* transmissions and is suitable for distribution of audio/video/multimedia content. The channel is unidirectional. For interactive applications, a return channel can be provided using, for example, a telecommunication network. Conversely, telecommunications deal mainly with *one-to-one* connections. Typical applications are telephony, bidirectional exchange of data, and on-demand access to multimedia content. The communication channel is bidirectional and usually smaller in its bit rate. UMTS and DVB systems can become two elements in an integrated broadcast and telecommunication system. The UMTS component of such a cooperative system can be used for communications aimed at individual users, whereas the broadcasting component can become the point-to-multipoint path or even the unicast delivery extension [86]. To continue, we will deal with outlining the technical background of relevant mobile telecommunications and broadcast services.

Mobile Telecommunications

The second-generation mobile systems (e.g., GSM) have been designed to fit into the traditional telephony architecture. The basic versions typically implement a circuit-switched service, focused on voice traffic, and only offer data rates up to 14.4 kbit/s. GSM cells can be between 100 m and 35 km in diameter based on the environment and the expected traffic load. Higher data rates can be achieved employing 2.5G systems (e.g., Global Packet Radio System, GPRS), enabling up to about 2 Mbit/s. Packetized data communications is achieved with 3G technology (e.g., Universal Mobile Telecommunication System, UMTS).

UMTS is the third-generation popular cellular mobile system designed also to offer multimedia/Internet access to portable-mobile terminals. UMTS networks are characterized by small cells, especially in densely populated areas. The target is to carry high-capacity bidirectional multimedia services via radio. The main advantage of UMTS over GSM is the capability to deliver higher bit rate multimedia services (typically 384 kbit/s), such as Internet pages and video clips, to portable phones and other types of mobile terminals in an *always on* mode. Therefore, UMTS networks and terminals are able to deliver audio services and low-resolution video services to mobile terminals. Advanced modes of UMTS will also support restricted possibilities for multicast and cell-mode broadcast, similar to the current GSM Cell Broadcast, but with increased capacity. UMTS is a member of a family of interworking standards for indoor, vehicular, and pedestrian communications. Application on satellites is also feasible. This family is defined in the ITU's International Mobile Telecommunication Standard IMT-2000.

A comparison of mobile communication technology is given in Table 4.3, which summarizes the main characteristics of the mobile cellular radio systems. These parameters will be used to develop various applications and scenarios of network cooperation [87].

Broadcasting

A broadcasting system, combined with an interactive channel from a telecommunication system, proves the cheapest and most efficient solution for services that many users share. Today's market demands global access by multiple users and portable or mobile reception. The cellular phone industry exemplifies this demand. Portable and mobile reception can be offered, through a digital terrestrial radio system, or sometimes through satellite. Introducing multimedia in a broadcast system is not straightforward. Existing multimedia systems build on two-way communication links and error-free transmission. However, broadcasters cannot guarantee an error-free or noninterrupted channel. Systems used solely for broadcasting audio or TV were designed with these difficulties in mind. Adding services with multimedia in broadcasting networks demands new concepts that take radio channel characteristics into account [86].

The digital audio broadcast (DAB) system is designed to provide reliable, multiservice digital sound broadcasting for reception by mobile receivers, using a simple, nondirectional antenna. It can operate at any frequency up to 3 GHz for mobile reception (higher for fixed reception) and may be used on terrestrial, satellite, and cable broadcast networks. In addition to digital sound, DAB enables the transmission of data in a flexible, general-purpose multiplex. The system provides a signal that carries a multiplex of several digital services simultaneously. The system

Table 4.3 Comparison of mobile communications technologies

	Telecoms Cellular Systems		
	GSM	GPRS	UMTS
Spectrum bands	900 MHz and 1800 MHz	900 MHz and 1800 MHz	2000 MHz and 2500 MHz
Regulation	Telecom, licensed	Telecom, licensed	Telecom, licensed (unlicensed)
Max. throughput	14.4 kbit/s	115 kbit/s	384–2000 kbit/s
Typical throughput	9.6 kbit/s	30 kbit/s	30–300 kbit/s
Transfer mode	Circuit	Packet	Circuit/packet
Primary applications	Voice	Data	Voice and data
Mobility support	High	High	Low to high
Coverage	Wide	Wide	Low to high

bandwidth is about 1.5 MHz; the available bit rates for these services range from 0.6 to 1.7 Mbit/s

The DVB project is a consortium of around 300 companies from more than 35 countries, in the fields of broadcasting, manufacturing, network operation, and regulatory matters, that have come together to establish common international standards for the move from analog to digital broadcasting. For example, DVB-T is the standard for wireless terrestrial reception of DVB, either in stationary use (at home) or in mobile use. DVB-T uses the same VHF/UHF spectrum as analog (PAL, carrying 5 to 30 Mbit/s in 8 MHz bandwidth).

Application Scenario

Broadcast and telecommunication network cooperation refers to the joint usage of these two complementary technologies in order to provide new features that each technology individually cannot provide in a satisfactory manner. Such cooperation can also improve the efficiency of existing services through better utilization of spectrum and/or higher performance. An example of that concept is TV/Internet convergence in the *wired world*. Another example is the cooperation of UMTS and DVB-T [87], which gives additional benefits as shown in Table 4.4. The abilities of different networks (UMTS only, DVB-T only, and UMTS plus DVB-T) to support the different classes of applications are indicated. All DVB-T only services are distribution services without a return channel. Therefore, no personalized or real on-demand services are feasible. Also, both networks have their specific strengths with low overlaps. Cooperation of both networks has the potential (++) to deliver a better performance compared to either of the networks.

Table 4.4 Cooperation of UMTS and DVB-T

	UMTS	DVB	UMTS + DVB
Entertainment			
TV, radio programs	−	++	++
Audio, video on demand	0	−	0
Games, interactive TV	+++	−	++
General information			
News, weather, financial info	+	+	++
Travel, traffic, maps	+	+	++
Commercial info	+	+	++
Personalized information			
Web browsing, file transfer	+	−	++
Individual traffic info, navigation	+	−	++
Emergency, location-based services	++	−	++
Business and commerce			
Remote access, mobile office	++	−	++
E-mail, voice, unified messaging	++	−	++
E-commerce, e-banking	++	−	++

Services and applications can be categorized with respect to different criteria. We can classify these as follows:

- *The user view*: what does the user expect from an application, what are the performance targets in speed, QoS, and so on, in which environment will the application be used – portable, mobile, and so on.
- *The network view*: how can an application/service be implemented, which resources are required, what capabilities are available in a specific network technology.

The applications can be presented to the user via a data carousel, similar to teletext. The feature availability of large memories in a terminal enables the intermediate storage of this data, reducing considerably the delay of the data carousel. Other applications are dedicated to personalized information, such as real on-demand services, business-related applications, or remote medical diagnosis. These applications require a bidirectional communication link, which is not feasible with a pure distributing broadcast network.

The relevant network service modes for cooperation scenarios are distribution and retrieval. The term *network service mode* refers to the way a service is transported in a specific network. Distribution refers to the broadcasting of information from one source to multiple users, which is the normal mode of DVB networks. A special form of the distribution mode is *unicasting*, which devotes temporarily all or a part of the channel capacity to a single use, or *multicasting* to a user group. A broadcast network like DVB-T can implement a mixture of broadcasting, multicasting, and unicasting. Retrieval/interaction refers to the delivery of information on an individual basis, in response to a single user. The information flow is bidirectional, point-to-point, mostly asymmetrical, and nonreal-time. Most IP-based multimedia services Web-browsing falls into this category. UMTS is a typical network supporting mainly this mode.

Various network parameters have a critical influence in the manner the applications are provided on the networks. The most important ones are the bit rate, the data volume, the burstiness of traffic, the asymmetry and the duration of a session.

In addition, wireless services and applications enter into a second classification, from the user (and terminal) *mobility* point of view:

- portable use (fixed and pedestrian)
- mobile use (from 60 km/h to 130 km/h in cars, up to 300 km/h in trains)
- mobile use in maritime (30 km/h) and aeronautical (1000 km/h) applications.

4.4.2 Digital Radio Broadcasting

The problems with analog transmission are quite well understood among communication engineers. In the last 40 years, substantial research efforts have been spent on developing the basis of digital communication technologies, starting from Shannon's information theory and pulse-code modulation (PCM) to the theory of digital filtering and signal processing. In essence, digital communication allows incorporation of channel coding to ensure the fidelity of received digital representation of the source signal (i.e., the result of source coding) and regeneration of the signal without degradation. Today, not only the telephone network backbone, but most media signals are also represented, stored, and transmitted in digital forms. Examples are the compact disk (CD) for music, high-definition television (HDTV), and so on. These formats also provide improved quality, in terms of audio bandwidth and picture resolution, over the analog formats. The traditional terrestrial radio is the best communication and broadcasting service to become digital, at least in the North America region. The drive to all-digital radio broadcasting thus gained momentum in the late 1980s as CD music became ubiquitous and audio compression techniques demonstrated ever-increased efficiency due to the introduction of perceptual audio coding [88, 89].

In the early 1990s, progress towards the digital broadcast of audio programs took place along several directions. In Europe, the European Union (EU) attempted to unify broadcasting across national boundaries by supporting a development effort called EUREKA-147 [90, 91]. A standard carrying the name EUREKA-147 was adopted by the European Community in 1995. A number of countries have since announced plans to test and adopt the system for future digital broadcasting. The EUREKA-147 system was designed to operate in several frequency bands, most commonly in the L-bands (1500 MHz). Sometimes it is also referred to as *new-band* radio. These spectral bands were allocated by the EU and approved in 1992 by the World Administrative Radio Conference (WARC) for the new digital audio/radio service. Adoption of EUREKA-147 in the United States, although strongly supported by the Consumer Electronics Manufacturers Association (CEMA), was met with difficulty in spectrum allocation. The National Association of Broadcasters (NAB) in the United States favored a technology called in-band, on-channel (IBOC). This technology allows a station to smoothly migrate into digital broadcasting without having to seek a new operating license from the Federal Communications Commission (FCC) [92], or abruptly discontinuing its analog transmission. This is the case for both AM (510–1710 kHz) and FM (88–108 MHz) bands. The terrestrial U.S. digital audio/radio systems will first be introduced as hybrid IBOC systems where digital transmission is added to existing analog FM and analog AM. These systems will then evolve into all-digital IBOC systems where the analog signals are replaced by additional digital transmission. A recommendation for a world standard for digital audio/radio below 30 MHz has been recognized by the International Telecommunications Union (ITU) [93]. Part of this standard is being developed by Digital Radio Mondiale (DRM) [94, 95]. For the medium-wave AM band, the U.S. hybrid IBOC and all-digital IBOC are also part of the world standard.

Another push to digital audio broadcasting in North America came from proponents of direct satellite transmission. Direct satellite broadcast (DSB) for television has been in service since the early 1990s. However, it has not been extended to audio service, which, according to market studies, is quite attractive in mobile applications. Proponents of the plan convinced the FCC to realize two bands of spectrum, 12.5 MHz each, around 2.6 GHz (S-band) for such a satellite-based digital audio broadcast service. Subsequently, the allocated spectrum was mentioned in 1997 and two spectrum licenses, namely *Sirius* [96] and *XM* [97], set out to develop the systems, with a target broadcast launch date sometime in the later part of 2001. This is often referred to as satellite digital audio/radio services (SDARS) [98].

Service in SDARS is subscription based. A subscriber pays a monthly fee to receive the digitally protected broadcast signal. With the allocated spectrum, each broadcast company is able to provide about 100 channels of audio programs, some mostly music, other mostly voice-oriented talk. The two broadcasters employ different satellite technologies: one uses a geosynchronous system, while the other uses a geostationary system. These two systems require different signal relay plans in order to provide proper coverage for areas that may be blocked by terrain or buildings. A distinct feature of SDARS compared to terrestrial systems is that a listener can stay with a particular program throughout the entire North American region without having to switch channels due to the nature of the satellite coverage.

Global radio is a potential future provider of satellite digital radio in Europe. It is set for an early 2005 launch and aims at providing 200 channels of audio [99]. Three satellites in 24-hour elliptic orbit will be used. One main beam and seven spot beams over Europe are planned. Thus, local programming in separate languages is possible.

Finally, Internet and multimedia capabilities of personal computers (PCs), both in software and hardware, have given rise to an entirely new paradigm in radio broadcast, that is, the so-called *webcast* or *Internet radio*. Using media streaming technologies, instead of an electromagnetic (EM) wave receiver, a PC can download a *radio* or TV program from a server, that is, the *webcaster*, and allow the user to listen and watch without being limited by typical wireless constraints, such as control limits and spectrum availability. Proper streaming technologies coupled with efficient audio coding techniques, plus the virtually unlimited reach of the Internet, make webcast a new favorite of many listeners who *time* to stations that are a continent apart and are otherwise unreachable on traditional radio waves.

In short, digital audio/radio services, whether over terrestrial transmission, relayed by satellite, or in the form of media streaming via the Internet, are taking place at the turn of the century, after nearly 80 years of operation in analog modes. The advance is primarily due to the progress in digital audio coding and some key innovations in transmission technologies. In what follows, we will focus on the terrestrial and satellite systems as they represent the most profound depth of complexity and technical challenges. The advances in audio coding played a critical role in making digital audio broadcasting possible for the given current spectral and regulatory constraints. Hence, we will discuss the recent progress in audio coding.

Audio Coding

There are many advantages of using digital transmission and digital representations of audio, including an increased robustness to channel coding and the ability to regenerate signals without accumulating degradation. The digital nature of the links also increases the flexibility of the underlying audio format of the signals being broadcast. Digital representation allows the systems to transmit data, for example, stock quotes and messages, to use encryption algorithms and to manage access to subscription-based services at receivers.

The advantages of digital systems impose the corresponding requirements on the source coding algorithms used to encode the digital audio signals. They include

- the compression rate of the raw information
- the format represented in the compressed bit stream
- the algorithm's robustness to channel errors
- the audio quality, possibly as a function of the signal type (music or speech) and/or station
- the delay introduced by source coding
- the complexity of the source encoder and decoder.

Many of these considerations are specific to the broadcast environment and differ greatly from those of speech communication and storage/retrieval type applications. Source coding design represents a trade-off in multiple source-related factors and potentially a trade-off with a channel coding algorithm. For example, one can minimize source coder delay, but this may come at the expense of compression efficiency for a fixed quality level. Similarly, minimizing the delay introduced by channel coding can be at the expense of error-correction performance and the bit-error characteristics seen by source coders. Therefore, the choice of source coder is a compromise between the issues specific to the application, system, and even audio program.

The audio broadcast application differs from many other applications in that there are stringent requirements and expectations on both speech quality and music quality. This importance not only reflects requirements from the broadcasters and listeners themselves, but also the expectations of artists, talk radio hosts, and advertisers, who create the broadcast content. Selecting an appropriate coding technology often involves a balance between performance attributes as a function of signal type as well as the hardware requirements different technologies impose. There is no ideal technology satisfying all concerns at low bit rates. There are two main categories of coding technologies that are available to various digital audio broadcast applications, that is, speech and audio coding technologies. In general, speech coding technologies use model-based or waveform-based technologies that take advantage of the redundancies in the production and perception mechanism of speech [100].

Speech coding designs often achieve high speech quality at rates less than 1 bit per input sample, which is quite notable, especially at lower sampling rates of 8 and 16 kHz, where up to 21 of the 25 critical bands of hearing are covered in the acoustic bandwidth [101, 102].

Audio coders, in contrast, rely less on speech-specific attributes and more on the statistical redundancy common in many audio signals and general principles on the human auditory perception [103–105]. Common audio coder designs include transform or filter-bank signal decomposition combined with perception models and/or lossless coding techniques such as Huffman coding [106].

A general picture of the audio quality as a function of acoustic bandwidth is shown in Figure 4.24. A general summary of the application areas, bit rates, coding paradigms, and the nominal audio quality is shown in Figure 4.25. The overlap in speech and audio coding technologies is clearly visible.

Matching the trade-offs of the different paradigms to the source material, transmission channels, and hardware requirements is the challenge faced by the source coding technology in digital audio broadcasting systems. Some of these challenges have resulted in new advances in the area of speech and audio coding, including ideas on statistical multiplexing of multiple programs in a perceptually meaningful way using diversity in the source stream in both embedded and multidescriptive fashions, involving quality of audio coders on speech signals, and using multiple paradigms within a single coding structure [107].

The general scheme for audio coding is shown in Figure 4.26. General source coding algorithms maximize objective measures such as the SNR for the given bit rate.

Perceptual Audio Coding

Perceptual audio coders explore human perception with the aim of minimizing the perceived distortion for a given bit rate. Compression in a perceptual audio coder involves two processes: redundancy reduction and irrelevancy reduction. The filter bank of a perceptual audio coder yields a high degree of redundancy reduction due to the statistical nature of audio sources; for example, the energy of many audio sources is often concentrated in a few subbands of the entire signal bandwidth. The efficiency of the coder is further improved without impairing the audio quality by shaping the quantization noise according to perceptual considerations. This is the basis for irrelevancy reduction. One way irrelevancy reduction is achieved is by taking masking effects of the human auditory system into account. Masking describes the phenomenon in which one signal (in this case quantization noise)

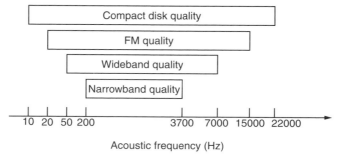

Figure 4.24 Acoustic bandwidth.

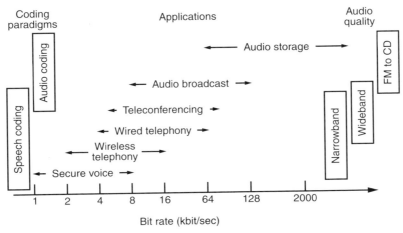

Figure 4.25 Application space versus bit rate, paradigms, and acoustic bandwidth.

becomes inaudible in the presence of another signal (in this case, the coded version of the input signal). Such masking happens in both the time and frequency domains. In the frequency domain, the level below which the masked signal becomes inaudible is termed the masked threshold. This threshold is a function of the masking signal and is often computed by considering the masking effect of each component of audio signal [108]. The masked threshold is computed by considering the masking effect of each spectral component of the audio signal as shown in Figure 4.27. During the encoding process, the spectral coefficients of the filter bank of a perceptual audio coder are grouped into coding bands. Each of these coding bands is quantized separately, such that the resulting quantization error is just below the unmasked threshold, as shown in Figure 4.28.

The structure of a generic perceptual audio encoder (monophonic) is shown in Figure 4.29. The input samples are converted into a subsampled spectral representation using a filter bank [109]. A perceptual model estimates the signal's masked threshold [108]. For each spectral coefficient this gives the maximum coding error that can be allowed in the audio signal while still maintaining perceptual transparent signal quality. The spectral values are quantized such that the error will be just below the masked threshold. Thus, the quantization noise is hidden by the respective transmitted signal. The resulting quantizer indices are coded with a lossless coder.

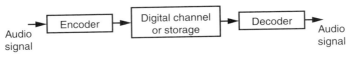

Figure 4.26 General block scheme of audio encoding/decoding system.

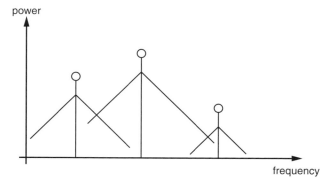

Figure 4.27 The masked threshold computed by considering the masking effect of each spectral component of the audio signal.

The coded spectral values and additional side information are packed into a bit stream and transmitted to the decoder or stored for future decoding.

The decoder shown in Figure 4.30 reverses this process. It contains three main functions. First, the bit stream is parsed, yielding the coded spectral values and side information. Secondly, the lossless decoding of the spectral indices is performed, resulting in the quantized spectral values. Finally, the spectral values are transformed back into the time domain.

Variable Bit Rate Coding and Constant Bit Rate Transmission

Nonstationary signals as audio signals have a varying amount of inherent perceptual entropy as a function of time. Variable bit rate compression techniques are therefore natural means of approaching the compression limit of audio signals, that is, perceptual entropy for transparent audio coding. However, most broadcasting applications

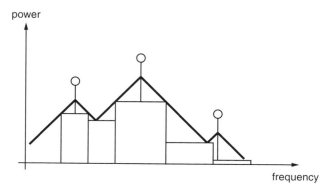

Figure 4.28 Coding bands of the spectral coefficients. Each coding band is quantized. The error is below the masked threshold.

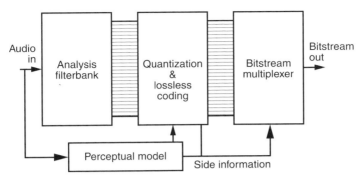

Figure 4.29 Perceptual audio encoder.

require a constant bit rate transmission. When a variable bit rate source coder is used together with a constant bit rate transmission channel, the output at the source coder needs to be buffered to absorb the variations in the bit rate. Figure 4.31 shows an audio encoder/decoder with a buffered bit stream to enable a constant bit rate transmission. In this scenario, k bits from the audio encoder are put into a first-in first-out (FIFO) buffer at a variable bit rate of $M[k]$ b per frame from the source coder. Bits are removed from the FIFO buffer at a constant bit rate of $R_d b$ per frame where R_d is equal to the rate of transmission channel. The number of data bits in the buffer after the processing of frame k, $l[k]$, can be expressed iteratively as

$$l[k] = l[k-1] + M[k] - R_d$$

Of course, some initial buffer level of $l[k]$ bits is assumed.

The buffer represents an interesting trade-off influencing the source coder design. The larger the buffer size, the more variations in bit rate can be absorbed and the less the impact is to the audio quality due to constant bit rate transmission. The size of the

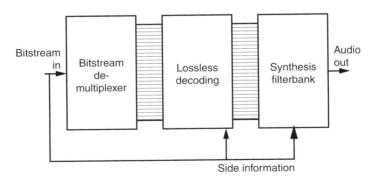

Figure 4.30 Perceptual audio decoder.

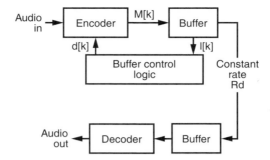

Figure 4.31 Audio encoder and decoder with a constant bit rate transmission channel [107]. (©2002 IEEE.)

buffer is restricted by constraints on timing delay and cost. In such a system, buffer control logic is necessary. The ultimate goal of the buffer control is to provide the best possible perceptual quality for a given buffer size restriction. To influence the encoding process and $M[k]$ in a perceptually meaningful way, the buffer control logic determines a level of quantization distortion in frame k through a perceptual criterion $d[k]$. The distortion criterion $d[k]$ determines how much noise is added above the masked threshold. If $d[k] = 0$, then frame k is encoded with coding distortion just below the masked threshold. If $d[k] > 0$, the coding distortion is allowed to exceed the masked threshold. In general, the larger the value of $d[k]$, the smaller the number of bits that will be required to encoded frame k. Therefore, the criterion $d[k]$ regulates the bit rate coming out of the source encoder. To select the required value of $d[k]$, many buffer control schemes for audio coders use two processing loops: the outer loop and the inner loop [110]. The outer loop determines for each frame k a bit rate $M_d[k]$ at which the frame should be encoded. The bit rate $M_d[k]$ is computed as a function of the buffer level $l[k-1]$ and the perceptual entropy or a related measure of the frame. The inner loop then iteratively re-encodes the frame at different levels of distortion $d[k]$ until the bit rate of the encoded frame $M[k]$ is sufficiently close to $M_d[k]$, keeping $d[k]$ to a minimum.

Joint Audio Coding

Satellite digital radio services broadcast a large number of radio programs (up to 100) simultaneously. A large number of programs and/or better audio quality of the programs (better performance) can be achieved if $N>1$ radio programs are encoded jointly with a shared bit stream. Thus, it is better if N channels share a common stream at NRd kbit/s than if each program is encoded individually, each with a single bit stream at Rd kbit/s. To achieve this, a buffer-control scheme for joint coding is used that dynamically allocates the channel capacity among the audio coders sharing the common bit stream. In Figure 4.32, it is shown how N audio encoders are connected to form a joint encoder with a joint bit stream. The bit rate of each joint frame $J[k]$ is the sum of the bit rates of the frames of the individual encoders $M_n[k]$

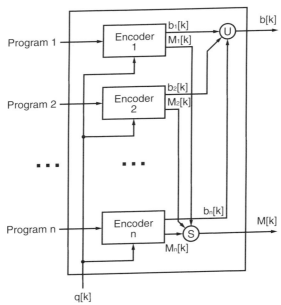

Figure 4.32 Audio coders connected to form a joint encoder, with a joint bit stream [107]. (©2002 IEEE.)

where $n \in \{1,2,\ldots,N\}$, that is,

$$J[k] = \sum_{n=1}^{N} M_n[k]$$

A distortion criterion $D[k]$ common to all encoders is used since it is simpler than dealing with a separate distortion criterion for each encoder. In addition, by having the same perceptual distortion criterion, the buffer control has the same average quality/bit rate impact on each audio encoder. It is also possible to consider different criteria for each encoder.

A buffered joint encoding scheme with a receiver is shown in Figure 4.33. The joint frames of the joint encoder are put into FIFO joint buffer. A buffer control scheme determines $D[k]$ such that the buffer level does not overflow. The bits in the joint buffer are transmitted to the receiver with a constant bit rate NR_d. Once a joint frame arrives at the receiver, the bits of the desired radio program P are extracted and placed into the decoder buffer by the program parser.

Example 4.4 Assume that the bit rates of the audio coders $M_n[k]$, where $n \in \{1,2,\ldots,N\}$ are independent random variables with means $m_i = m$ and $\sigma_n^2 = \sigma^2$. It then follows that the mean and variance of the joint bit rate $J[k]$ are Nm and $N\sigma^2$, respectively. Assume also that the

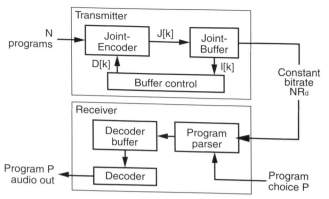

Figure 4.33 Buffered joint audio encoding scheme with a receiver [107]. (©2002 IEEE.)

average bit rate available for the N audio coders is NR_d. The standard deviation of the bit rate normalized by the desired bit rate R_d for one audio coder is σ/R_d, whereas the standard deviation of the joint encoder bit rate normalized by the total available bit rate NR_d is only $\sigma/\sqrt{N}R_d$.

A second important advantage of joint coding is that the different audio coders can operate at different average bit rates according to the individual demands of their audio inputs. The dependence of the perceived quality of the decoded audio on each channel's program material is greatly reduced.

Embedded and Multistream Audio Coding

In embedded and multidescriptive audio coding, the bit stream of the source is divided into a number of subsets that can be transmitted over independent channels. The subsets can be combined into various substreams, each of which can be independently decoded. Each subset is a substream that can be decoded independently. Multiple subsets can also be combined and decoded together to get higher quality. In the case of embedded coding, these subsets or layers have a hierarchy. The first layer, the *core* layer, is essential to all descriptions (subsequent layers of the bit stream) and can be used on its own to produce a decoded output. All other *enhancement* layers can be combined with the core and then decoded to produce output with increased quality. The enhancement layers may themselves be ordered even though, like multidescription coding, the layers may be combined in various ways. The division of the bit stream can be presented as in Figure 4.34. The bit stream is divided in such a way that the core bit stream provides basic audio quality, while bit streams 1–3 enhance the audio quality.

Core is the core part of the bit stream. It is self-sufficient and can be decoded independently of the other substreams.

- Enhancement Layer 1 consists of enhanced high-frequency spectral coefficients. This sub bit stream enhances the audio bandwidth of the core.

4.4 DIGITAL BROADCASTING   227

	LOW BANDWIDTH	HIGH BANDWIDTH
STEREO	2	3
MONO	CORE	1

Figure 4.34 The division of the bit stream.

- Enhancement Layer 2 consists of encoded left–right difference spectral coefficients. Given these, the core can be enhanced from mono to stereo.
- Enhancement Layer 3 consists of encoded high-frequency left–right difference spectral coefficients. Given 1 and 2 and this sub bit stream, the core is enhanced to high audio bandwidth stereo.

The core bitstream can be enhanced in different ways for better audio quality. For embedded audio coding, the substreams of Figure 4.34 can be used as shown in Figure 4.35. The core can be combined with the different substreams to enhance the audio quality.

For multistream audio coding, several independent bit streams are formed given the building blocks of Figure 4.34. Figure 4.36 shows two independent bit streams (core + 1, core + 2), which, when combined, yield enhanced audio quality (core + 1 + 2).

Another possibility for multimedia audio coding is encoding the audio signal using complementary quantizers and sending the information from each quantizer

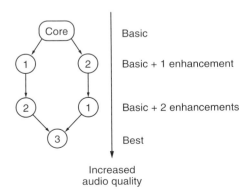

Figure 4.35 The core bit stream in embedded audio coding.

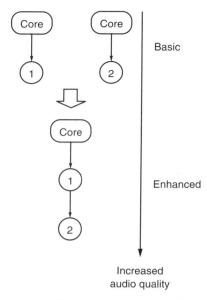

Figure 4.36 Multistream audio coding.

in different streams [111]. If information from both quantizers is received, then the quantizers are combined and the audio signal is decoded with less distortion.

Terrestrial Digital Audio Broadcasting

In North America, terrestrial radio commonly refers to broadcast in the FM band (88–108 MHz) and the AM band (510–1710 kHz). To circumvent the difficulty in allocating a new spectrum for digital audio broadcasting over terrestrial channels and to allow current analog radio stations to migrate into digital transmission without causing disruption in consumer adaptation, the National Association of Broadcasters (NAB) has been supporting the development of in-band on-channel (IBOC) technology. The support for IBOC technology is mostly based on a migration plan that the NAB deems sensible and acceptable. The migration to all-digital audio broadcasting will take two steps. The first step is to move from today's analog transmission to a hybrid system, which inserts digital signals along the two sidebands of the host analog signal. The second and final step is to totally replace the analog host signal with digital signals, which may carry additional services, as the market adapts gradually to the new system. Such a strategy with IBOC transition from analog to digital is depicted in Figure 4.37.

Invoking a digital radio system, one needs to set up the requirements of the system. The requirements for the IBOC system can be addressed along several dimensions, such as coverage, service quality, spectral efficiency, feature set, and compatibility.

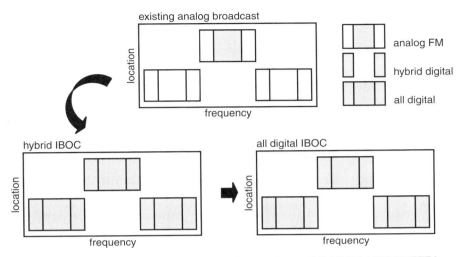

Figure 4.37 In-band on-channel transition from analog to digital [107]. (©2002 IEEE.)

The coverage of existing AM and FM stations, in reference to the contours limited by interference and by noise, shall not be compromised due to the digital signal in both hybrid and all-digital modes. The digital system must provide a service area that is at least equivalent to the host station's analog service area, while simultaneously providing suitable protection in cochannel and adjacent channel stations. Such a requirement ensures market stability in the service areas.

Audio quality in both hybrid and all-digital modes shall be significantly better than that of existing analog AM and FM modes. In fact, an original appeal in moving to digital systems was the improvement in audio quality, potentially to the level of CD quality in FM systems and to the level of analog FM quality in AM systems.

Spectral efficiency provided by IBOC shall be better that existing AM and FM bands in both hybrid and all-digital modes. Spectral efficiency refers to the ratio between the source signal bandwidth and the transmission signal bandwidth at given audio quality.

Both the hybrid and the all-digital modes shall support a subset of new features such as auxiliary data channel and an automated public safety infrastructure.

Deployment of IBOC in either hybrid or all-digital mode shall not impact existing analog stations or analog receivers. Insertion of digital signals will not create additional interface to the existing analog signal. The hybrid transmission mode will be backward compatible with current analog receivers already in use. The all-digital mode will be backward compatible with hybrid IBOC receivers. In short, the system will afford a smooth transition from analog to digital services.

All these requirements provide a design guideline in the development of the hybrid system and the eventual goal of an all-digital system.

Evolution of IBOC

In the early 1990s, in light of the development of the EUREKA-147 system in Europe, the Consumer Electronics Manufacturer's Association (CEMA) and proponents of EUREKA-147 urged the National Radio Systems Committee (NRSC) to consider a plan for digital audio services. Several systems were proposed, including the L-band EUREKA-147 system at two different bit rates and S-band satellite system, as in-band adjacent-channel (IBAC) system, and various IBOC systems. The key idea of an IBAC system is to find vacant channels in the current AM and FM bands for digital broadcasting.

Unsatisfactory performance of early digital audio radio systems was mostly due to the relatively high bit rates needed for audio coding. The lowest coding audio rate attempted in these systems was 128 kbit/s, which could not be supported by the digital transmission scheme.

As the spectral allocation issue became more prominent, the NAB in the mid-1990s started to focus on IBOC systems. In the mean time, advances in perceptual audio coding and orthogonal frequency division multiplexing (OFDM) or digital multitone technology for digital transmission have inspired new hope for the IBOC system. In 1996, audio coders like perceptual audio coders (PAC) and MPEG-2 advanced audio coders (AAC) were shown to be able to code stereo music at 96 kbit/s without causing audible degradation from original CD materials [106, 112]. These advances inspired a collaboration between two of the original proponents of the IBOC system, USA Digital Radio (USA DR) and Lucent Technologies, who joined forces to develop a working IBOC system in 1997.

USA DR and Lucent separated in 1999, although development efforts continued in each individual company. In 1999, Lucent Technologies formed Lucent Digital Radio (LDR) to signify its commitment to this particular technology area. LDR moved rapidly into a new system, with key advances such as multistream audio coding, which can be considered a new generation system.

During 1998 and 1999, the NRSC established Test and Evaluation Guideline documents to assist the technology proponents in self-testing programs so as to identify information that would be needed by NRSC to validate the ability of the new system. Digital Radio Express (DRE) was merged into USA DR in 1999, and in 2000, the two remaining proponents, USA DR and LDR, with somewhat different system designs, joined together to become a sole company called JBI-QUITY Digital Corporation [113]. Attributes of both systems have been combined and, in August 2001, test results were presented to the NRSC. Based on the evolution of these results, NRSC made a recommendation for approval of the FM system to the Federal Communications Commission (FCC) in November 2001 and the AM system in April 2001 [114]. Deployment of both AM and FM hybrid IBOC was scheduled for the 2002/2003 time frame.

Other Terrestrial Systems

The largest deployed terrestrial digital audio radio service system is EUREKA-147 [90, 115–117]. This system was the outcome of a large European consortium activity in the early 1980s. The project was carried out in the context of the

EUREKA series of research projects, and project 147 began in 1986 to develop a digital audio broadcasting system. The system specification was finalized in 1994 and was adopted as a worldwide ITU-R standard in 1994 and as an ETSI standard in 1997. The system is operational in many Western European Countries and Canada, as well as in several Asian countries and Australia. EUREKA-147 is different from IBOC in many ways. Rather than using existing AM and FM bands, it assumes newly allocated bands. To obtain efficient frequency use, several programs are multiplexed and transmitted on a single carrier. Such an ensemble has a transmission bandwidth of 1.536 MHz. Using OFDM modulation (differential quadrature phase-shift keying QPSK for each carrier), the gross capacity of this ensemble is about 2.3 Mbit/s. Varying levels of error protection can be selected, resulting in net bit rates of 0.6–1.8 Mbit/s. Error protection levels can be set for individual programs within an ensemble. Its audio compression scheme relies on MPEG 1, 2 Layer II, which requires 128–192 kbit/s for stereo audio broadcasts. It supports both 48 kHz and 24 kHz sampling frequencies and bit rates from 32 to 384 kbit/s in mono, stereo, and dual-channel mode. Its basic frame size is 24 ms. Besides audio, EUREKA-147 supports program associated data and generic data. The latter is organized in 24 ms logical frames with a data rate of n times 8 kbit/s. The system has been designed for mobile reception over a wide range of frequencies (30 MHz and 3 GHz). Owing to its robust design against multifading, it is possible to operate in a so-called single frequency network (SFN) mode, where several (geographically separated) transmitters all broadcast the same ensemble at the same frequency. This allows robust coverage of a large area. Another advantage of an SFN is that it provides a very power-efficient network compared to (analog) FM for the same coverage efficiency. The need for multiplexing requires careful coordination among content providers and collective responsibility for the transmitter infrastructure. This approach has been found quite feasible in Europe, where broadcasting in general has been organized at a national level and is intended for national coverage. This is in contrast to the United States where local programming is preferred. Although established for sound delivery, its most successful applications rely on it as a robust high-speed wireless data delivery service. Recent proposals of combining EUREKA-147 with general packet radio system (GPRS) have indicated that this is a viable commercial option.

The Digital Radio Mondiale (DRM) consortium has developed a system for digital broadcasting at frequencies below 30 MHz [94, 95]. This system has already been recognized by the ITU in a draft recommendation [93]. The DRM system has been developed based on some key requirements; that is

- The audio quality must be improved over that achieved by analog AM
- The DRM signal must fit within the present channel arrangements in the AM bands
- The DRM signal should support operation of an SFN.

APPLICATIONS LAYER

The capacity available for audio within a single 9 or 10 kHz (U.S.) AM channel is limited. Audio coding rates from as low as 10 kbit/s up to mid-20 kbit/s have been proposed. For the lower rates speech coders can be used, while for the higher rates MPEG-2 AAC with spectral band replication (SBR) is used [95]. Data and audio bit streams are multiplexed. The transmission system is based on OFDM, which provides the need for adaptive equalization. Constellation sizes varying from 16-QAM (4 bit/s/Hz) to 64-QAM (6 bit/s/Hz) have been proposed. The channel coding used in the system is multilevel coding. A variety of OFDM configurations, system bandwidths, and data rates have been included in the standard.

In-Band On-Channel (IBOC) FM Systems

Digital broadcasting in the FM band inside the Federal Communications Commission (FCC) emission mask can take place in a so-called hybrid IBOC system where the digital information is transmitted at a lower power level (typically 25 dB lower) than the analog host FM signal. This digital transmission is achieved in subbands on both sides of the analog host signal. The composite signal is typically 400 kHz wide with the FM carrier in the middle. The digital subbands are about 70 kHz wide at the upper and lower edges of the composite signal. The basic FM power spectrum is shown in Figure 4.38.

One current design proposal for hybrid IBOC FM systems uses a single 96 kbit/s perceptual audio coding (PAC) source stream duplicated for transmission over two sidebands using OFDM modulation [106, 118, 119].

To ensure graceful degradation in the presence of one-sided first adjacent channel interference, an alternative system uses multistream transmission on the two sidebands combined with multidescriptive audio coding [120]. Further robustness to this type of interference is obtained by introducing a bit error sensitivity classifier in the audio coding algorithm and by transmitting bits in separate classes with different channel codes and different frequency bands [121].

In-Band On-Channel (IBOC) AM Systems

The AM systems differ from FM systems in many aspects, particularly in terms of the nature of interference due to the modulation scheme. For the FM systems, the

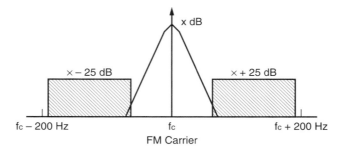

Figure 4.38 Basic FM power spectrum [107]. (©2002 IEEE.)

digital and analog signals are transmitted without overlapping in frequencies. In the AM systems, simultaneous transmission of analog and digital signals in the same frequency is not only possible but, because of linear analog modulation, it is also necessary because of the severe bandwidth limitations in the AM bands.

The AM channels are very different and less well understood for digital transmission to mobile receivers. First, daytime and nighttime conditions are very different. During daytime conditions, fairly good stable channels with interference slowly increasing and decreasing in certain bands are obtained when driving in the coverage area. The stable interference is caused by cochannel and adjacent channel interference from other AM or IBOC-AM stations. Impulsive noise should also be taken into account in the signal design. Doppler plays a small role in AM transmission in contrast to the FM case. During nighttime, the AM channels can change rapidly due to skyway interference.

One proposal to transmit the digital audio signal on top of the analog AM signal consists of using a 30 kHz transmission bandwidth as shown in Figure 4.39. The digital data are transmitted through bands A–C. In this case, some severe second adjacent interference may occur in certain coverage areas and data transmitted in bands A or C can be lost completely. For the FM case, the digital audio bit stream may be duplicated and transmitted on both sides of the analog host to provide a robust solution to this problem. However, in the AM case, there is not enough bandwidth to transmit a duplicated bit stream. To provide a solution to this problem, a more robust strategy is built on embedded/multidescriptive audio coding and separate channel coding/modulation in several frequency bands. The audio decoder has the capability of blending down to a lower bit rate, when a certain subband in the hybrid IBOC-AM signal is subjected to severe interference. The audio quality of this lower bit rate audio signal is better than that of analog AM. Thus, a graceful degradation is achieved along with a higher degree of robustness to certain channel conditions.

Similar to FM systems, the modem proposed for a hybrid IBOC-AM system is typically an OFDM modem. The modulation scheme proposed for daytime transmission is quadrature amplitude modulation (QAM) using 16-QAM, 32-QAM, or

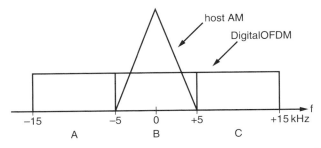

Figure 4.39 Power spectrum of a 30 kHz hybrid IBOC-AM system. The digital data are transmitted in bands A–C [107]. (©2002 IEEE.)

64-QAM. For nighttime transmission, since the channel can change very rapidly due to skyway interference, 8-PSK (phase-shift keying) modulation is proposed. The bandwidth and transmission power are extremely limited in hybrid IBOC-AM systems. To protect the digital audio bit stream, bandwidth-efficient forward error-correction (FEC) schemes and coded modulation schemes have to be designed.

Satellite Digital Audio Radio Services

A satellite provides a large coverage to many receivers, and transmission delay is not an issue in broadcasting. Most broadcast use of satellites has been limited to television services. In addition, these services mainly provide signals to stationary rather than mobile receivers [122]. The basic satellite broadcasting configuration is presented in Figure 4.40. It consists of a central programming and production facility, which transmits (uplinks) the broadcast signals to a satellite. The satellite takes these signals and shifts the up-link frequency to the proper down-link frequency, applies power amplification, and directs the signal to its designed footprint/service area. The signals are received by both spatio-varying and mobile receivers within this area and are processed to retrieve audio baseband signals. To make uninterrupted broadcast reception possible, it is necessary to maintain a line of sight (LOS) with the satellite. Depending on the elevation angle between the service area and the

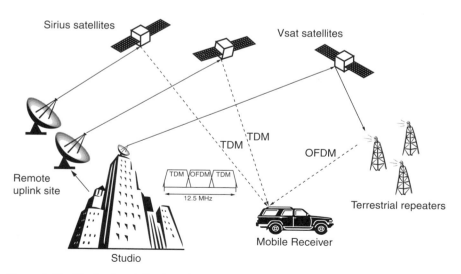

Figure 4.40 The basic satellite broadcasting configuration to match the Sirius Satellite Radio system [107]. (©2002 IEEE.)

satellite, this might be difficult to guarantee, especially for mobile receivers. To make these systems more reliable for mobile users, one option is to provide additional transmission diversity using terrestrial repeaters. Fortunately, the situation where this is needed the most coincides with important market areas and is therefore economically feasible. If the use of terrestrial repeaters is not an option or can only be used sparsely, another solution for countries or regions at high altitudes would be the use of elliptical (or geosynchronous) orbits. This option requires more satellites and a switching scheme between the satellites to make sure active transmissions are coming from the satellite having the highest elevation angle with respect to the service area. To make these systems more robust, it is common to introduce time diversity. To do this, the same program is broadcast from different sources, for example, two different satellites or one satellite and one terrestrial.

Example 4.5 Suppose that one of the channels is delayed with respect to the other, for example by 4 s. Referring to the nondelayed channel as the late channel, at the receiver the early channel is delayed by the same amount to time align the two programs. Listening is done on the late channel end if the transmission is interrupted by blockage, and the listener is switched to the stored early channel. The concept of time diversity is illustrated in Figure 4.41. By delaying stream 1 during transmission, the receiver has two versions available. The content will be impaired at different time instants.

The first SDARS system was Digital Satellite Radio (DSR) and was operational from 1989 to 1999. A more recent system is the Astra Digital Radio (ADR) system. This system provides digital radio services on its geostationary TV broadcasting satellites. It covers central Europe and uses stationary receivers [123]. Also, it uses MPEG Audio Layer I at 192 kbit/s for stereo signals. Forward error-correction (FEC) is based on a punctured convolution code with code rate 3/4 resulting in a 256 kbit/s gross bit rate per channel. Transmissions are done in the 11 GHz range. Owing to path losses at these frequencies, the antennas need dishes with diameters between 0.5 and 1.2 m.

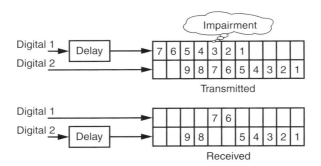

Figure 4.41 Concept of time diversity.

Another more recent SDARS system is Worldspace, which provides digital radio services to developing countries using three geostationary satellites [124]. The system operates in the L-band (1452–1492 MHz) and consists of three geostationary satellites:

- AfriStar (21°), covering Africa and the Middle East
- AsiaStar (105°), covering China, India, Japan
- AmeriStar (95°), covering Central and South America.

Each satellite has three spot beams. For each spot, two time division multiplex (TDM) streams are delivered. Each TDM stream contains 96 so-called prime rate channels (PRC). Each PRC transmits at 16 kbit/s. For a typical high-quality stereo signal the signal is transmitted at 128 kbit/s using MPEG Audio Layer III, thereby occupying 8PRC. At these rates, each spot can deliver $2 \times 12 = 24$ audio channels. Services can be multimedia (audio, image, and moving image and data) and can be individually encrypted for pay-per-use. Most receivers are used in the stationary mode (home and portables). By using two broadcast channels per program and introducing time diversity, it is possible to make the reception more robust for mobile receivers, although, most likely, additional terrestrial repeaters are needed.

In the United States, no frequency allocation exists for SDARS in the L-band. In 1994, the Federal Communications Commission (FCC) allocated the 2310–2360 MHz (S-band) for SDARS. In 1996, Congress mandated that the 3210–3220 MHz and 2345–2360 MHz portions of this band should be auctioned for wireless terrestrial communications services (WCS). The remaining bands were auctioned for SDARS services to Sirius Satellite Radio, Inc., based in New York, and American Mobile Radio Corporation (now XM Satellite Radio, Inc.) based in Washington, DC. Both services are similar. For example, for a monthly fee of about $10/month, they provide approximately 50 music channels and 50 talk radio channels. The quality and audio bandwidth of the music channels is somewhere between analog FM and CD quality (stereo), while the talk radio programs approach the quality of mono FM. Most of the programming is commercial-free and can be received anywhere in the continental United States. This wide coverage allows for the broadcast of programs that in local markets only have a small target audience, but nationwide would reach a much larger audience. This diversity could be one of the attractions of SDARS. The nationwide coverage and commercial-free nature are other appealing factors.

To allow for flexible program allocation, it is necessary to have a control channel. This control channel is also used for controlling access. Since the service is a subscription-based service, all information has been encrypted and mechanisms are in place to control access. Although the main purpose of SDARS is the delivery of audio content, the systems provide, in essence, a bit pipe to the mobile user. This allows for delivery of data services. Both XM and Sirius are looking into reselling some of their capacities for such services.

Status and Systems Evolution

XM started their limited-area services in October 2001, while Sirius started limited-area commercial services in February 2002. At this time, both services cover the continental United States. The main market is mobile receivers in cars. Many car manufacturers have committed to 2003 models with limits in receivers and antennas. It is expected that there will be a strong interest for these antennas. XM has a business model based on a combination of channels with and without commercials, while Sirius has committed to only commercial-free content. However, it remains to be seen how good the coverage is for each of these services. Interrupted services for music distribution are not well received by end users and could quickly reduce the excitement for SDARS. On the other hand, the ability to receive a large selection of programs continuously throughout the United States has a very strong value proposition that will change radio forever.

The success of broadcasting depends on the availability of affordable receivers that can be used in many scenarios. Most digital radios require a significant amount of digital processing and require special-purpose VLSI. Economics will push down prices to make radios affordable to anyone. EUREKA-147 receivers, for example, have quickly become available and have come down in price significantly. One potential hurdle for after-market satellite receivers is the need for an external antenna, which requires additional wiring and installation. New cars that are equipped, as standard, with satellite radio most likely will become one of the driving forces for acceptance of this format.

Further cost reduction is expected by providing radios that integrate the various broadcasting formats. Since the standards are quite different in terms of frequency band, modulation schemes, and compression formats, it could be that the only way to provide multiple standards is by use of programmable platform. From a service perspective, a consumer would likely subscribe to only view, integration with IBOC is more likely, since this will eventually replace the current freely available analog FM and AM services. Note that some techniques may see initial deployment only to be phased out later, while new techniques may be introduced as the service develops. The reasons for such dynamics in the design are both technical (as a result of hardware and performance considerations) as well as nontechnical (due to mergers and market pressures). The dynamics also reflect the flexibility inherent in the digital nature of the design.

4.4.3 Digital Video Broadcasting

Digital Video Broadcasting (DVB) was originally created in 1993 to provide a set of open and common technical mechanisms by which digital television broadcasts can be delivered to consumers, recognizing a range of television-based businesses. Some years later, much of that original objective has been achieved and the DVB, as well as the world in which it operates, has advanced. The range of contents that DVB systems have to handle has expanded beyond television. That content is increasingly being carried on nonboard networks and to non-TV devices. The connection between the transfer of data and the consumption of the services is becoming less tightly coupled, due to storage and load networking. As a result, DVB membership is expanding at an accelerating rate with many organizations joining from well outside the original charter, extending the scope both geographically as well as beyond the core broadcast markets.

The DVB family of specifications covers all aspects of DVB, starting from a principle of being a complete solution for all delivery media, and to make all these media as transparent as possible. For content providers, DVB systems compromise a system of *pipes* all seamlessly connected, eventually delivering MPEG-2 bit streams to the home. The bit streams may have been routed through satellite (DVB-S), cable (DVB-C), and/or terrestrial (DVB-T) networks on the way. At the end of this process, they can be easily decoded.

With its advanced work on the Multimedia Home Platform (MHP), which includes defining a common application programming interface (API) for receivers of the future that will be capable of dealing with many kinds of service and interactivity models, DVB is providing a technical solution to the new, exciting, and uncharted territories of digital media convergence.

Using MPEG-2 as a stream of *data containers* labeled with full addressing and processing instructions (in the form of DVB service information), DVB provides the receiver at the other end of the *pipe* with the location, network, broadcast, and program information it needs to jump from one element of the multiplex to another and between multiplexes without missing a beat, shifting gear to decode HDTV, SDTV, data, and automatically setting the optimum system parameters.

DVB is a transmission scheme based on the MPEG-2 video compression/transmission scheme. DVB provides superior picture quality with the opportunity to view pictures in standard format for wide screen (16:9) format, along with mono, stereo, or surround sound. It also allows a range of new features and services including subtitling, multiple audio tracks, interactive content, and multimedia content, where, for instance, programs may be linked to World Wide Web material.

Satellite transmission has led the way in delivering digital TV to viewers. A typical satellite channel has 36 MHz bandwidth, which may support transmission at up to 35–40 Mbps using QPSK modulation. The video, audio, control data and user data are all formed into fixed-sized MPEG-2 transport packets. The complete coding process may be summarized by

- inverting every eighth synchronization byte
- scrambling the contents of each packet
- Reed–Solomon coding at 8 percent overhead
- interleaved convolutional coding (the level of coding ranges from $\frac{1}{2}$ to $\frac{7}{8}$ depending on the intended application)
- modulation using QPSK of the resulting bit stream.

The MPEG-2 coding and compression system was chosen after an analysis and comparison of potential alternatives. Unlike other compression tools that claim to provide greater degrees of compression for given quality, but which are as yet unproven for a wide range of program material or over different broadcasting systems, MPEG-2 is tried and tested. It has been repeatedly shown to be capable of providing excellent quality pictures at bit rates that are practical for the services that will be required. From a commercial point of view, the adoption of MPEG-2, an existing, driven standard, was advantageous, because it allowed DVB to concentrate its efforts on finding ways of carrying the already well-specified MPEG-2 data packets through a range of different transmission media, including satellite, cable, terrestrial, and so on. DVB can effectively be regarded as the bridge between broadcasters and the home, over which MPEG-2 data packets can be carried. Another important reason for choosing MPEG-2 was that it includes techniques for the inclusion of program-specific information (PSI) for the configuration of decoders. DVB extended these techniques to provide a complete service information (SI) capability, enabling receivers to automatically time to particular services, and to decode a mixture of services and service components, including television, sound, and data. SI also allows services to be grouped into categories with relevant schedule information, making it possible to provide user-friendly program guides. Another important consideration was the design of the decoder while still retaining the flexibility to make quality improvements at the encoding of the chain.

DVB has cooperated closely with the Digital Audio Visual Council (DAVIC), whose mission includes the whole range of multimedia transmissions, and many DVB systems have been accepted within the DAVIC standard. MPEG-2 is a video, audio, and data coding scheme, which can be applied to a range of applications beyond broadcasting, and many multimedia features may, in time, be available from DVB services. DVB systems can naturally carry any or all of the items utilized for multimedia presentations, including text, still pictures, graphics, and different types of moving images, and allow multimedia extensions to be added as they appear. It is important to recognize that the key element of DVB is broadcasting. Therefore, DVB members have been focusing on the broadcast market for immediate commercial feature.

The first reason for a broadcaster to select DVB is that DVB makes much better use of available bandwidth. For satellite direct-to-home (DTH) broadcasters this advantage is clear. Where a satellite transponder used to carry one analog channel, DVB can offer up to 18 digital channels. DVB-T for terrestrial digital broadcasting offers broadband two-way interactivity. Today, the production, contribution, and distribution of television is almost entirely digital. The last step, transmission to the end user, is still analog. DVB brings this last step into the digital age.

The DVB system has the capability to utilize a return path between the set-top decoder and the broadcaster. This can be used by a subscriber management system. It requires a modem and the telephone network, or cable TV return path, or even a small satellite up-link. This return path can be used for audience participation, such as voting, games playing, teleshopping, and telebanking, and also for delivering messages to the decoder. DVB already offers a kind of interactivity without the need for a return path, simply by the breadth of program choice available, for example, multiple sports events and near video on demand.

DVB Interoperabilities

DVB is interoperable in two key senses. The first is between different delivery media, where a maximum commonality between systems allows problem-free transport of programs across the barriers between different delivery media. For example, one of the key reasons for the successful implementation of DVB-T receivers is that they share many of the same components used in DVB-S and DVB-C receivers made by the same manufacturers and installed in their millions of receivers across the world. This is known as cross-medium interoperability. Secondly, all DVB equipment interoperates, which means that equipment from many different manufacturers can be connected together and will work. This allows the content provider, network operator, and service provider to freely choose the best added value equipment implementations from various manufacturers all the way through the broadcast chain. This is known as cross-manufacturers interoperability.

DVB's work is carried out by members from all parts of the broadcasting value-chain and lately even members of the computer industry have become active in the working groups of the project. This is a reflection of the natural process of convergence, which is a consequence of the widespread digitization of all media. From DVB-T to DVB-S to DVB-C, the DVB system has been very widely adopted. While we do not suggest that DVB-T be adopted anywhere where it has not yet been adopted, the commonality between all the DVB systems is an undeniable fact.

MPEG-2 and DVB

Data is already being sent over DVB networks using the MPEG-2 transport stream. A variety of proprietary encoding schemes are being used. Data transmission may be simplex or full duplex (using an interaction channel for the return) and may be unicast (point-to-point), multicast (one to many), or broadcast. Typical configuration for providing direct-to-home (DTH) Internet delivery using DVB is shown in Figure 4.42.

In an effort to standardize services, the DVB specification suggests data may be sent using one of five profiles [125]:

- *Data piping* – where discrete pieces of data are delivered using containers to the destination.
- *Data streaming* – where the data takes the form of a continuous stream that may be *asynchronous* (i.e., without timing, as for Internet packed data), *synchronous* (i.e., tied to a fixed rate transmission clock, as for emulation of a synchronous communication link), or *synchronized* (i.e., tied via time stamps to the decoder clock).
- *Multiprotocol encapsulation* (MPE) – the technique is based on digital storage media – command control (DSM-CC) and is intended for providing LAN emulation to exchange packet data.
- *Data carousel* – scheme for assembling data sets into a buffer that are played out cyclic manner (periodic transmission). The data sets may be of any format or type and the data are sent using fixed sized DSM-CC sections.

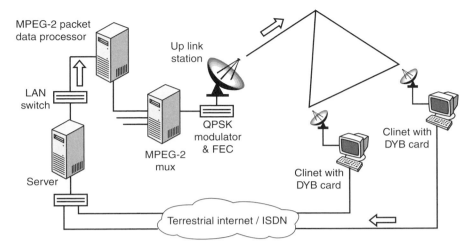

Figure 4.42 Typical configuration for providing DTH Internet delivery using DVB.

- *Object carousels* – resemble data carousels, but primarily intended for data broadcast services. The data sets are defined by the DVB network independent protocol specification.

The MPEG-2 standards define how to format the various component parts of a multimedia program. They also define how these components are combined into a single synchronous transmission bit stream. The process of combining the streams is known as multiplexing. The multiplexed stream may be transmitted over a variety of links. Standards/products are available for [86]:

- radio frequency links
- digital broadcast satellite links
- cable TV networks
- standard terrestrial communication links
- microwave line of sight links (wireless)
- digital subscriber links
- packet/cell links (ATM, IP, IPv6, Ethernet).

Many of these formats are being standardized by the DVB Project.

DVB Project
The DVB Project is recognized throughout the world as a success in standardization. Major elements of the success are a result of the active involvement in the television value chain. They work in the business environment rather than in standardization bodies only, and decided that commercial requirements had to follow technical specifications. Early involvement of requirements and standardization bodies has

proven to be essential for success within the geographical area Europe, and created the basis for DVB's success in spreading out the DVB specifications into *de facto* standards in many parts of the world outside Europe. DVB started with broadcasting and gradually moved to the specification of return channels in the telecom domain, and finally to the specifications for interactive and data broadcasting in the software domain.

Over its first years of existence, DVB has provided robust and reliable technology that has facilitated the rollout of digital TV to a large number of markets across the world. However, the growth and advance of DVB presents the Project with many new policy and management challenges.

- First, the expanding range of technological and commercial opportunities can put pressure on the Project to develop new technology in an uncoordinated and disjointed fashion.
- Secondly, as DVB membership becomes increasingly broad, stresses and strain may develop as result of the divergent aims and agendas of the various constituencies, which may even be operating under different regulatory regimes.
- Lastly, the importance of liaison functions and the need for speed will increase since other bodies, and even individual companies, may address parts of the expanding scope of DVB.

The main drivers of DVB's future work will be to ensure that content is able to move from providers across the various global, local, and in-home networks to consumer devices, including the aspects of local storage and retransmission towards other devices. The main drivers for this DVB development work will be the businesses that exploit the content value chain from content creation to consumption by the consumer.

DVB's vision for the future will be to build a content that combines the stability and interoperability of the world of broadcast with innovation and multiplicity of services in the world of the Internet.

Therefore, DVB's strategy for the future will be

- the delivery of content through networks to consumer devices in a transparent and interoperable fashion
- the continuing development of the technology of broadcast networks such as cable, satellite, terrestrial, and so on
- the commercial exploitation of the movement and consumption of content in a secure manner
- a robust framework for new, interactive services.

In the face of these developments and challenges, the members of the DVB Project have decided that the time is now ripe to renew the Project with a fresh vision and strategy for the future.

DVB 2.0

The core of DVB's new mission is to provide the tools and mechanisms to facilitate interoperability and interworking among different networks, devices, and systems to allow content and content-based services to be passed through the value chain to the consumer. However, each element in the value chain has significant diversity and in many areas this diversity is increasing. So, providing a comprehensive set of tools and mechanisms for every permutation of content, network, and so on, is a massive task.

DVB 2.0 will exist in a world where the converging technologies and markets of broadcasting, telecommunications, and information technology will blur if not eradicate the traditional boundaries among these once separate fields. The main focus of DVB 2.0 will be to ensure that content is able to move from providers across the various global, local, and in-home networks to consumer devices including the aspects of local storage and retransmission towards other devices. The main drivers for this DVB development work will be the businesses that export the content value chain from content creation to consumption by the consumer. In the rapidly developing technology of the converged world, much of the innovation surrounding content creation, distribution, and consumption will come from the Internet and from IP-based industries.

DVB is a market-driven project, but it needs to become especially focused on the major business requirements of its members in deciding the areas of work it should be tackling. The DVB 2.0 needs to agree a framework of Strategic Work Areas that the project should focus on in delivering its new mission. The Strategic Work Areas will need to be consistent with one another and together they should from a coherent package of work for the DVB 2.0. The maintenance of existing DVB specifications should be included as a Strategic Work Area in its own right.

The Strategic Work Areas are as follows:

- maintenance of existing DVB specifications
- the completion and maintenance of the multimedia home platform (MHP)
- storage of content in consumer devices and networks accessed by the consumer
- the scalability of services and content to enable delivery to different devices via a variety of different networks
- delivery of content through broadband IP networks
- securing – transactional mechanisms, copy management, authentication mechanisms, access control
- the alliance of broadcast systems and services with 3rd and 4th generation mobile telecommunication networks and services
- home networking, both wired and wireless.

With the adoption of Strategic Work Areas, it may be beneficial for the bulk of the detailed work on commercial requirements and technical solutions to be concentrated with specialist Strategic Work Area groups. This will

- ensure that work is tightly focused on the task in hand
- concentrate expertise within the project on appropriate areas and tasks
- speed up decisions at the working level
- reduce the possibility that requirements or solutions expand in an uncontrolled fashion.

Once the specific Strategic Work Area groups are satisfied that they have completed their tasks, then the results would be submitted to the wider DVB forum to ensure that it fits within DVB's wider mission and strategy as a whole.

The DVB 2.0 mission is focused on providing tools and mechanisms to enable content-based business to exploit market opportunities for providing materials and services to consumers. However, there are a number of closely related business areas that use technology similar to DVB 2.0, but which lie outside the DVB 2.0 newly defined areas of interest. DVB 2.0 should be able to use and exploit the technology used in such business areas, but it should not itself get directly involved in developing such technology unless it is of direct relevance to one of the Strategic Work Areas defined by DVB 2.0. Areas that are considered to be beyond DVB's 2.0 remit under its new mission include

- consumer to consumer services, such as voice and video telephony systems
- authoring systems for content creation
- noncontent-based business to business services
- noncompression algorithms.

The success of DVB 2.0 and its broad acceptance beyond Europe, can be largely attributed to the robustness, cohesiveness, and quality of its resulting specifications.

Commercial Strategy for the Future of DVB

DVB was formed to drive the future of TV broadcasting and to respond to the commercial opportunities created by the new delivery infrastructure. This was driven by market demand and the realization of commercial value, particularly through pay TV, and advertising that supported public service broadcasting. The answer at the time was encapsulated by the word *digital*, which brought promises on high quality and economic multichannel programming. However, *digital* is much more than a means to offer more choices. It brings about great flexibility, but it also adds the complexities of managing the underlying stability of the consumer television experience. There is an accelerating requirement to maintain and improve the consumer offering in the face of diverse and competitive market development.

DVB has established a strong record for delivering specifications together with the consensus support for those specifications to become the prevailing worldwide open standards. There is an internal driver from within DVB's membership to use the powerful working methods of DVB to address new and difficult issues. Further success here reduces business risk for manufacturers, service providers, and content owners. It also benefits consumers by removing uncertainty and accelerating the adoption of new mass-market services at lowest possible cost.

The first wave of Internet adoption was based on PCs as the preferred client device for business and home users, with e-mail, Web browsing, music downloading, and e-commerce as the key applications. The second wave of the Internet's adoption is likely to be accelerated by devices other than computers being used for Internet applications: interactive TVs, personal digital assistants (PDAs), mobile devices connected to GSM, GPRS, and UMTS networks, and domestic appliances with IP connectivity. Also, it should be mentioned that some additional services such as *broadcast quality*, streamed and live video, which provide a substitute or complement to digital TV broadcasting over traditional networks, should be taken into account. Internet technology linked to digital broadcasting and new consumer devices offer the essential building blocks for the second wave of Internet adoption. DVB has a key part to play in building links to related standard bodies, creating a shared vision for the market and delivering an agreed set of supporting specifications and other tolls for use by all players in *new media* value chains.

The new services will be implemented using combinations to create hybrid solutions. For example the United State's interactive TV service uses satellite data-broadcast technology linked to dial-up two-way interactivity over telephone modems and sophisticated client/server applications based on Oracle and OpenTV technology. Similar combinations of technology are likely to drive early adoption of broadcast and interactive services in cars and other forms of transport. Current DVB services are likely to be transported over Internet backbone networks and viewed on devices not primarily seen as TV receivers.

From a commercial perspective, all that leads to a view that DVB should evolve its approach for the next phase of its work. An interdisciplinary way of working will be necessary to bring together the diverse skills and new market perspectives required for success. There is a need to look at the whole picture as an *end-to-end* system, and not just a small part of a value chain or service delivery network in isolation. The issues of scalability and interactivity require particular attention.

New functions in the business chain are emerging. Metadata creation, management, and aggregation are examples. Now that digital TV services are well established in many marketplaces, often using bandwidth side-by-side in the same delivery mechanisms sometimes requires the transition to a purely digital solution, phasing out analog services and receivers.

DVB activities are often linked to software products and computer industry solutions. In the usual broadcast and telecom industry approach, technical solutions are established first. Competition then occurs in the speed of implementation, cost, quality, and the addition of extra features around an open system core. DVB needs to be adaptable and also to be proactive on open standards. Being open and timely is important but not sufficient when there are innovative proprietary solutions available. DVB's output needs to be done well and produced on time for the market.

Using the Internet to carry DVB services requires network on QoS management, together with security issues and geographic rights management. Existing broadcast networks often have implicit limits on their geographic reach, thus resolving issues of rights to deliver outside the intended market or territory. The Internet offers a truly global network where the absence of territorial restrictions is a powerful benefit to many users.

E-business requires a dependable communications environment that has an open network for easy access and yet contains the vulnerabilities through a pervasive approach to security. Instead of being layered onto networks, security capabilities should migrate into the core fabric of all network infrastructures and systems architectures. Data protection legislation is a key factor that needs to be well understood and the necessary protection must be considered at the requirements capture stage and in the systems design phase.

In the new DVB environment, there are content formats based on MPEG for DVB broadcasting and a variety of open and proprietary content formats used over open IP networks. While these two groups of formats will be used for many years, the relevance of IP transport streams will grow and some of these content formats will be changing over time. A critical work area is that of streaming media, and its positioning with respect to broadcast media. On the other hand, customers' terminal devices are required to decode different incompatible video streams, each of which is able to deliver TV equivalent image quality at data rates of around 2 Mbit/s. In order to do so, terminal devices will have to store many tens of MBytes of software decoders.

Interoperability is a continuing cornerstone of DVB's work and must remain a priority. Regularity aspects of DVB's work have always been carefully addressed. This will continue to be a vital activity, becoming increasingly complex in the convergent world of broadcasting, IP networking, and information and communications technology.

Architectural Framework for the Future DVB

In the following, an overview of existing and emerging areas within the DVB territory is presented. Trends of future development are identified and the areas in which DVB should get involved are proposed.

The potential of coaxial networks can be exploited either by using the existing techniques for broadcasting digital signals (DVB-C) or by offering digital communication services within a bandwidth of up to 1 GHz. Based on the available standards, it appears to be possible to offer virtually every service the private cable customers may wish to receive. For professional users who require communication services at very high data rates, optimized solutions still would need to be developed. The existing transmission standards may be replaced by something completely new, but such a move does not seem to be appropriate in the coming five to ten years. Research continues in this area. This may lead to a transmission standard that no longer differentiates between broadcast and private communication services but rather allocates the required transmission capacity to services and customers as momentarily required. Activities for DVB include observation of market and technology developments, maintenance of existing standards, and cooperation with `CableLabs`.

DVB-Satellite (DVB-S) is the world standard. There are two developments in the world of satellites that may have an impact on DVB. One is the ever closer orbital positioning of satellites that may require modifications of the DVB-S standard in order to make it more resilient against interference. The other is the development of regenerative satellites with on-board signal processing. Activities for DVB are observation of market and technology developments, maintenance of existing standards, development of DVB, return channel satellite (DVB-RCS) version for regenerative satellites, and possible upgrading of DVB-S for higher capacity and/or higher interference resilience.

DVB-Terrestrial (DVB-T) is in the process of being implemented in many countries of the world. It has been recently predicted that DVB-T will be the only world standard by 2006. Activities for DVB can be presented as observation of market and technology developments, maintenance of existing standards, verification of the usefulness of a possible DVB-T specification both technically and commercially, and contact with wireless network activities.

Wireless local loop technologies in the 2 GHz to 5 GHz bands may offer new low-cost ways of delivering broadband IP access in rural areas and DVB should track such developments and respond as necessary. Activities for DVB are observation of market and technology developments as well as maintenance of existing standards.

For fast Internet connections, xDSL is being rolled out in several countries. For example, in Germany, the customer demand is so high that delivery of xDSL equipment has become a real problem, causing delays in the connection of private homes of more than three months. xDSL will be used for delivering audio and video content into homes via streaming media. xDSL is a technology for only the last few miles. It is a key element of IP broadband networks. As for activities for DVB, they are expected in the area of DVB content via IP networks.

After decades of talking about Fiber To The Home (FTTH), it has started to be deployed in certain countries (for example, in parts of Germany and in Sweden). It is not clear yet if and when this deployment will become widespread. Observation of market and technology developments are activities for DVB.

Wireless Local Area Networks (WLAN) are typically used for business-to-business applications and are therefore not regarded as part of the DVB territory. As for DVB activities, they include observation of market and technology developments as well as maintenance of existing standards.

The classical mobile telecommunications networks today play a role in DVB because GSM can be used as a return channel for DVB services. Telephony services are increasingly being delivered via mobile phones. GPRS and third generation mobile telecommunications networks will technically be able to offer various multimedia services. Connection to the Internet is possible via such networks at moderate data rates. This is a very interesting area that is currently being addressed.

The variety of home terminal devices is growing and will continue to do so. In addition to PCs, set-top boxes, and the TV receivers we are already familiar with, equipment will be made available that includes more and more of the capabilities of state-of-the-art PCs, but without the look and feel of a PC. Next generation set-top boxes will incorporate hard disk storage, high computing power, DVD drives, and so on. The TV plus concept will become a reality in the marketplace. It will allow viewers the use of services that today can only be used via a PC or notebook. As a consequence, various new sorts of services will be possible that directly affect the business of DVB members. Digital recorders will become more widespread in consumer's homes. Modifications may be required for some DVB specifications and implementation guidelines to better accommodate digital recording.

The TV Anytime project is developing metadata specification that can be broadcast alongside normal program content, and can be used by a home disk recorder to select material for recording and for search and retrieval of content on the recorder's hard disk. Activities for DVB include cooperation with the TV Anytime forum and integration of TV Anytime proposals in the DVB specifications. Then, we have digital recording and its impact on DVB specifications and implementations guidelines. The work on the development of the multimedia home platform to cater for the hard disk environment needs to be continued.

Home Local Network standards already exist based on twisted-pair wires and/or optical fiber. Wireless connections based, for example, on IEEE802.11 are a reality within offices. In-home networks, especially the yet to be deployed wireless networks for residential homes, should interconnect terminal equipment seamlessly. Activities for DVB are observation of market and technology developments, maintenance of existing standards, as well as the development of wireless in-home network technology.

In the DVB environment there are mainly two groups of protocol and content formats, namely the MPEG-2 based transport stream, over, for example, DVB-C, DVB-S, DVB-T for TV and radio broadcasting and the IP family of protocols for packet-switched open network services. While these two groups of formats can be expected to be used for many years, the relevance of IP over transport streams will grow and the content formats on the top of IP will be changing over time. A critical question in this context is the commercial importance of streaming media and their positioning with respect to broadcast media. As for activities for DVB, we can say that the exploration of technology trends has started via the group called Internet Protocol Infrastructures (IPI).

Finally, there is a requirement to protect intellectual property made available via digital broadcast and via telecommunications networks from use and abuse by honest consumers in a stricter fashion than today. Digital signals can be cloned and thus the need to prevent unauthorized use of audiovisual content has become acute. There are several methods by which intellectual property can be protected, and the main two are by broadcasting copy protection signaling, and by watermarking the audiovisual content.

4.4.4 DVB/ATSC System

The ATSC digital television standard describes a system designed to transmit high-quality video and audio as well as data over a single 6 MHz channel. The system can deliver, reliably, about 19 Mbps of throughput in a 6 MHz terrestrial broadcasting channel and about 38 Mbps of throughput in a 6 MHz cable television channel. This means that encoding a video source whose resolution can be as high as five times that of conventional television resolution requires a bit rate reduction by a factor of 50 or higher. The objective is to maximize the information passed through the data channel by minimizing the amount of data required to represent the video image sequence and its associated audio. We want to represent the video, audio, and data sources with as few bits as possible while preserving the level of quality required for the given application.

DVB System

Owing to the use of MPEG-2 packets as *data containers* and the DVB service information surrounding and identifying those packets, DVB can deliver to the home almost anything that can be digitized, whether it is high-definition TV, multiple channel standard definition TV, or broadband multimedia data and interactive services. DVB has taken the transport stream (TS) structure from the digital information to be modulated and broadcast. The content being broadcast is completely independent of the DVB physical channel used as well as independent of the modulation and coding used by each physical channel. This allows for a great degree of flexibility and interoperability among the different broadcasting systems.

Baseline System. The DVB video coding is based on the MPEG-2 specification and uses the various profiles and levels that allows for different scan rates, different aspect ratios including $3:4$, $16:9$, or $2.21:1$; selectable resolution including standard definition TV (SDTV). DVB also accommodates transmission of $4:2:2$ TV format for contribution applications.

The DVB audio coding is also based on MPEG-2 specification allowing for multiple audio channels that are compatible with MPEG-1.

Data. There are different modes for various types of data, from teletext, subtitling or other vertical blanking interval data to the several formats specified for data broadcasting such as data piping, asynchronous data streaming, synchronous and synchronized data streaming, multiprotocol encapsulation, data carousels and object carousels.

System. The specification for service information (SI) is the glue for interoperability among different services and delivery systems, cable, terrestrial, microwave, or satellite.

Interactivity. Interactivity is accomplished by using network-independent protocols for DVB interactive services through the various standards defined by DVB for return channels including cable, terrestrial, and satellite.

Interfacing. Interfaces for the transport of MPEG transport stream through conventional telecommunication systems such as pleosynchronous digital hierarchy (PDH) or synchronous digital hierarchy (SDH) have been also defined. Also, means to convey asynchronous transfer mode (ATM) type of data is considered under guidelines for handling ATM signals in DVB systems.

Definition for interfacing networks at home are also standardized by the DVB to take into account access to the home as well as digital local network at home with the documents Home Access Network (HAN) and In-Home Digital Network (IHDN) and Home Local Network (HLN). The interfaces for integrated receiver decoder (IRD) are defined in the Common Interface specification for conditional access and other digital video broadcasting decoder applications as well as all other interfaces for IRDs.

Conditional Access. The conditional access (CA) is supported by the so-called common scrambling algorithm and there are standards for conditional access and for simulscrypt system and transport stream (TS).

Cable DVB-C. DVB-C describes the modulation scheme used in cable systems. As the cable channel has a better stability and signal-to-noise ratio, this system does not specify the use of the inner forward error-correction (FEC) system. Considering that the linearity can be kept well under control, the modulation system allows for varying amplitudes. Consequently a higher order of modulation is feasible, combining phase and amplitude modulations called quadrature amplitude modulation (QAM). The constellation mapping can be chosen from 16-QAM, 32-QAM, 64-QAM, 128-QAM, or 256-QAM. The channel coding includes a randomizer for energy dispersal followed by an outer FEC based on a shortened Reed–Solomon (RS) block code. A convolutional interleaver protects the signal against burst errors. A baseband shaping filter is used prior to modulation. The available useful bit rates depend on the modulation order selected and the bandwidth of the channel. For channel spacing of 8 MHz, the useful bit rate may range from 25 Mbps (16-QAM) to about 50 Mbps (256-QAM).

Terrestrial DVB-T. DVB-T takes account of variations in channel conditions caused by multipath reception, fast fading due to movement, and variations caused by meteorological conditions. In order to solve multipath reception, a guard interval is included with each symbol in such a manner that the length of the guard can be chosen to be $\frac{1}{4}, \frac{1}{8}, \frac{1}{16}$, or $\frac{1}{32}$ of the useful symbol duration. This guard interval for each symbol is an exact copy of the corresponding best part of the useful portion of the symbol. For this mechanism to be effective, the guard interval has to have an absolute length large enough to cancel any of the expected delays of multipath or multireception. This implies making the useful part of the symbol much longer that those used for DVB-C or DVB-S. The solution is to use many narrow band carriers, each modulated at a relatively slow symbol rate, to convey the high-speed digital modulation. A practical way of achieving such a system is using the orthogonal frequency division multiplex (OFDM) modulation system. The coding part of the system is identical to that used by the satellite system. The modulation part is where the major difference exists. Two modes have been selected, one called 2K, which consists in modulating 1705 carriers with a symbol length of 224 μs, and the other called the 8K mode, with 6817 carriers with symbol length of 896 μs. Depending on the coverage area along with the geographical environment and the use or not of small or large single frequency network (SFN) the network planner may choose the use of 2K or 8K modes and the length of the guard interval to be used from $\frac{1}{4}$, $\frac{1}{8}$, $\frac{1}{16}$, or $\frac{1}{32}$ of the useful symbol duration. The linearity is not a great constraint, and with the terrestrial system it is also possible to choose higher QAM modulation orders: 4-QAM, 16-QAM, or 64-QAM.

A hierarchical system has also been defined for the DVB-T, so it is possible to broadcast two completely independent transport streams (TS) with different degrees of prediction. This is accomplished by means of selecting 16-QAM or 64-QAM in a way that the high-priority TS maps its bits as QPSK onto the two most significant bits of the selected constellation, while the low-priority TS maps its bits as the two (for 16-QAM) or the four (for 64-QAM) least significant bits. Both the hierarchical mode and the normal mode have an mapping process of bits into constellation symbols for each carrier, such that an interleaving system is used to further protect the data from the spectral notches that may appear when high echoes or multireception is present. The overall system is called coded orthogonal frequency division multiplex (COFDM) due to the use of an inner convolutional FEC system. It provides many different useful bit rates depending on the code rate selected (CR), the guard interval selected (GI), the constellation selected, the channelization used (6 MHz, 7 MHz, or 8 MHz) and whether or not the hierarchical mode is used.

The guard interval protects against the interference produced by multipath reception; that is, it protects against reception of delayed versions of the main signal. It also protects from signals coming from nearby transmitters if they broadcast the same information in the same channel with time and frequency synchronization. Then, it is possible to cover a large area that is meant to receive the same program with multiple synchronized transmitters forming a single frequency network (SFN). The DVB-T system may be well suited to multiple access by keeping the packet identification (PID) system as used in the DVB baseline system. As for security, it may be achieved by the current common scrambling algorithm. The DVB-T is very well suited to portability and mobility of either the transmitters or the receivers, and is very appropriate to provide wideband mobile communications for multimedia applications.

Satellite DVB-S. DVB-S is a digital broadcasting system that uses QPSK as modulation scheme. Although this is a basic modulation system that is not highly bandwidth efficient, it is very resilient against the nonlinearity inherent in the satellite transponders, which typically work at saturation or near saturation. The channel coding includes a randomizer for energy dispersal followed by an outer forward error-correction (FEC) based on a shortened Reed–Solomon (RS) block code. A convolutional interleaver protects the signal against burst errors. Owing to the variability of the characteristics of the transmission channel, a second FEC based on a convolutional-punctured code is added for further protection. This inner protection code rate can be selected from five values ($\frac{1}{2}$, $\frac{2}{3}$, $\frac{3}{4}$, $\frac{5}{6}$, or $\frac{7}{8}$) to provide a flexible degree of protection traded from the useful data capacity. A base-band shaping filter is used prior to modulation. Finally, the modulation scheme used is conventional

Gray-coded QPSK modulation with absolute mapping. The available bit rates depend on the bandwidth of the transponder.

Microwave Multipoint Video Distribution Systems (MVDSs). The three microwave systems DVB-MS, DVB-MC, and DVB-MT are closely related to the three basic systems for satellite, cable, and terrestrial, respectively.

The DVB-MS is intended to provide digital TV distribution by multipoint video distribution system (MVDS) operating in the frequency band of 40.5 to 42.5 GHz, which has been harmonized within the European conference of Post and Telecommunication Administrations under recommendation T/R 52-01. However, the systems are applicable to other frequency bands above 10 GHz.

The DVB-MC is intended to provide digital TV distribution by microwave multipoint distribution systems (MMDs) operating below 10 GHz. The DVB-MT is intended to provide digital TV distribution by microwave multipoint distribution systems operating at any frequency above or below 10 GHz. It offers an alternative to DVB-MC, providing the advantages of OFDM for the microwave range of frequencies.

Digital Satellite News Gathering DVB-DSNG. DVB-DSNG, while having all the baseline, coding, and modulation parameters described in DVB-S, adds optional modulation modes as 8QPSK and 16-QAM with the inner FEC coding using a *pragmatic* trellis coding principle in order to fulfill special applications requirements.

Digital Satellite Transmission Channel. A simple generic model of a digital satellite transmission channel is shown in Figure 4.43. It is comprised of several basic building blocks, which include baseband processing and channel adaptation in the transmitter and the complementary function in the receiver. Central to the model is, of course, the satellite transmission channel. In a practical system the transmitter elements would not necessarily be implemented at the same location. Channel adaptation would most likely be done at the transmit satellite Earth station, while the baseband processing would be performed at a point close to the program source.

The service content can be any mixture of video, audio, and data. Delivery can be to a conventional television set via a set-top box or even to a personal computer.

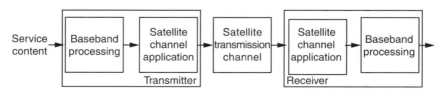

Figure 4.43 Generic model of the digital satellite transmission channel.

DVB Baseband Processing

The established MPEG-2 standard was adopted in DVB for the source coding of audio and video information and for multiplexing a number of source data streams and ancillary information into a single data stream suitable for transmission. Consequently, many of the parameters, fields, and syntax used in a DVB baseband processor are specified in the relevant MPEG-2 standards. Figure 4.44 is a conceptual block diagram of a DVB (MPEG-2) baseband and processor. Note that DVB baseband processors do not necessary physically implement all of the blocks. The input to the processor consists of a number of program sources. Each program source comprises any mixture of raw data and uncompressed video and audio, where the data can be, for example, teletext and/or subtitling information and graphics information such as logos. Each of the video, audio, and program-related data inputs is encoded and formatted into a packetized elementary stream (PES). Thus each PES is a digitally encoded component of a program.

The simplest type of service is a radio program that consists of a single audio elementary stream. A traditional television broadcast would comprise three elementary streams: one carrying coded video, one carrying coded stereo audio, and one carrying teletext.

Following packetization, the various elementary streams of program are multiplexed with packetized elementary streams from other programs to form a transport stream (TS).

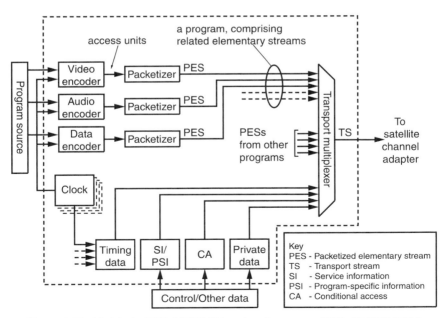

Figure 4.44 Model of a DVB (MPEG-2) baseband processor (126). (©1999 DVB.)

Each of the packetized elementary streams can carry timing information, or *time stamps*, to ensure that related elementary streams, for example, video and audio, are replayed in synchronism in the decoder. Programs can each have different reference clocks, or can share a common clock. Samples of each *program clock*, called program clock references (PCRs), are inserted into the transport stream to enable the decoder to synchronize its clock to the one in the multiplexer. Once synchronized, the decoder can correctly interpret the time stamps and determine the appropriate time to decode and present the associated information to the user.

Additional data is inserted into the transport stream, which includes program-specific information (PSI), service information (SI), conditional access (CA) data, and private data, that is, a data stream whose content is not specified by MPEG. The output of the transport multiplexer is a single data stream that is suitable for transmission or storage. It may be of fixed or variable data rate and may contain fixed or variable data rate elementary streams.

Certain nonvideo and nonaudio data are processed before being input to the transport multiplexer, whereas other data, such as the service information, are input directly to the transport multiplexer. The reason for this is that some data need to be synchronized with the video and audio content of a specific program. This data is encoded into program elementary stream (PES) packets in order to take advantage of the in-band time stamps and synchronization facilities of the PES. Two specific examples of this type of data are teletext and subtitling. Standard methods exist for transporting this information in DVB bit streams [126].

Teletext. The transport mechanism is intended to support the transcoding of the teletext data into the vertical blanking interval (VBI) of analog video, such that it is compatible with existing TV receivers equipped with teletext decoders. It is specified to be capable of transmitting subtitles with accurate timing with respect to the video. Teletext data are conveyed in PES packets as additional data. The PES packets are in turn carried by transport stream packets. The packet identifier (PID) of a teletext stream associated with a particular service (program) is identified in the service information or, more specifically, in the program map table (PMT) of the PSI for that service. A service may include more than one teletext data stream, provided that each stream can be uniquely identified.

Subtitling. The ETSI standard ETS300743 specifies the method by which subtitles, logos, and other graphical elements may be efficiently coded and carried in DVB (MPEG-2) bit streams. Subtitling data is carried in PES packets, which are in turn carried by transport packets. All the data of a subtitle stream is carried with a single PID. A single transport packet stream can carry several different streams of subtitles. The different subtitle streams could be subtitles in different languages for a common program, or could be subtitles in different programs (provided that the programs share a common program clock reference). Different subtitling streams are identified within a transport packet stream by their page identifiers. The presentation time stamp (PTS) in the PES packet provides the necessary timing information for the subtitling data.

4.4 DIGITAL BROADCASTING

Packetized Elementary Streams. The packetized elementary stream (PES) is designed for real-time data such as video and audio and provides all the necessary mechanisms for the transport and synchronization of video and audio elementary streams. The formation of a video PES is illustrated in Figure 4.45, while a summary of MPEG-2 levels is presented in Table 4.5.

Generally speaking, MPEG-2 encoders for broadcast applications operate at Main Profile, Main Level, and are capable of encoding 4 : 3 format television at quality comparable to that with existing PAL and SECAM standards, at bit rates from 1.5 Mbps to around 15 Mbps.

The input to the MPEG-2 encoder is assumed to be at Main Level. From Table 4.5 this means that the luminance component is sampled at 720 times per line. Given that there are 576 active lines per frame, each picture (frame) consists of 720×576 luminance samples. The two color difference signals are each sampled at half this rate, producing the same amount of data per frame for the chrominance information. Consequently, since all samples are represented with 8 bits, each picture in the uncompressed video input represents approximately 830 kbytes of data.

Each picture in its uncompressed form is referred to as a presentation unit. The coder compresses each presentation unit to give a coded picture. A coded picture is known as an *access unit*.

The PES packets consists of a payload and a header. The payload of consecutive packets is filled with data bytes taken sequentially from the video elementary stream. The start of a new access unit need not coincide with the start of a PES packet payload, but may occur anywhere within the payload. Multiple access units may be carried in the payload of a single PES packet. Access units can also be split across PES packet boundaries. The PES packets may be of variable length, the length being indicated in the packet header. The header of each PES packet includes information that distinguishes PES packets belonging to one elementary stream from those of another within the same program. Up to 32 audio elementary

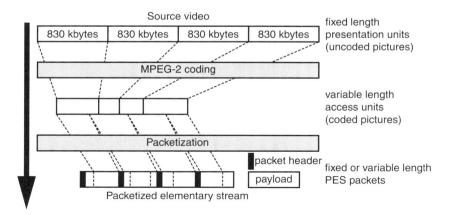

Figure 4.45 Construction of video packetized elementary stream.

256 APPLICATIONS LAYER

Table 4.5 Summary of MPEG-2 levels

MPEG-2 Level	Picture Resolution Samples/Line : Lines/Frame (25 frames/s)	Max. Bit Rate (Mbit/s)	Equivalent Quality (approx.)
Low	352 : 288	4	VHS
Main	720 : 576	15	PAL/SECAM
High1440	1440 : 1152	60	EDTV/HDTV
High	1920 : 1152	80	HDTV

streams and 16 video elementary streams can be uniquely identified within a single program. The header also provides other information such as the length of the PES packet and whether or not the payload is scrambled. The PES packet header periodically contains timing information that is necessary to ensure correct synchronization between related elementary streams in the decoder. These are the *time stamps*. A time stamp must occur at least every 0.7 seconds in a video or audio packetized elementary stream.

DVB utilizes the transport stream to convey information relating to multiple programs.

Transport Stream Multiplex. The transport stream protocol provides mechanisms for the fragmentation and reassembly of higher-level protocols and for error detection.

The transport stream is intended for broadcasting systems where error resilience is one of the most important properties. It supports multiple programs with independent time-bases, carries an extensible description of the contents of the programs in the multiplex, and supports remultiplexing and scrambling operations.

A transport stream multiplex always consists entirely of short, fixed-length transport packets. A transport packet is always 188 bytes long and comprises a header, followed by adaptation field or a payload, or both. The mapping of packets into transport packets is illustrated in Figure 4.46.

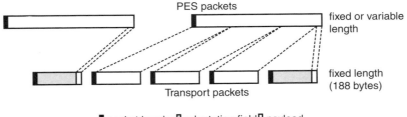

Figure 4.46 Dividing a packetized elementary stream (PES) packet into a number of transport packets.

The transport packets are usually shorter in length than the PES packets of an elementary stream. Consequently, each PES packet is divided among the payload parts of a number of transport packets. The start of the PES packet must occur at the start of a transport packet payload and a transport packet may only carry data taken from one PES packet. Since a PES packet is unlikely to exactly fill the payload of an integer number of transport packets, the payload of the last transport packet will be left partially empty. This excess capacity is deliberately wasted by including an adaptation field of the appropriate length in this transport packet.

All the packetized elementary streams that are to be multiplexed together are converted to transport packets. The resultant transport packets are output sequentially in order to form an MPEG-2 transport stream, so long as the chronological order of packets belonging to the same elementary stream is preserved. Transport packets carrying other information, such as service information or conditional access data, are also added to the multiplex, together with *null* packets that fill up any spare multiplex capacity.

A single transport stream can carry many different programs, each comprising many packetized elementary streams. A 13-bit packet identifier (PID) field is used in the header of each transport packet to distinguish transport packets containing data from one elementary stream. The MPEG-2 syntax allows for up to 8175 uniquely identifiable data streams to be accommodated in a single transport stream. The PID is used in the first stage of transport stream demultiplexing. MPEG-2 demultiplexers usually pipe all information carried in transport packets with a video PID directly to the video decoder chip. Similarly, all information carried in transport packets with an audio PID is routed directly to the audio decoder chip. The transport packet header also includes a continuity counter, which is incremented for successive transport packets belonging to the same elementary stream. This enables a decoder to detect the loss of transport packet and to take the necessary action to conceal any areas that this loss would otherwise produce.

Service information (SI) data forms a part of the DVB transport stream. It is included for two main reasons: so that an integrated receiver decoder (IRD) can automatically configure itself for the selected service and so that the user can be provided with information to assist in selection of service and/or events within the bit stream. SI data for automatic configuration of the IRD is mostly specified within MPEG-2 as program-specific information (PSI). Additional SI data is specified within DVB to complement the PSI by providing data to aid the automatic timing of IRDs and additional information intended for display to the user. This is usually referred to as DVB-SI.

Program-Specific Information (MPEG-2). The program-specific information (PSI) data consists of tables: the program association table (PAT), the program map table (PMT), and the conditional access table (CAT). The PAT and PMT are mandatory and must be contained in every DVB-compatible data stream. The CAT must be contained in the data stream if a conditional access (CA) system is employed. The PSI tables are used to physically describe the content of an MPEG-2 transport stream. Each packet of a transport stream is tagged with a PID value to indicate to which elementary stream its payload belongs. In order to determine which elementary streams belong to a particular program, the IRD needs additional information that relates PID values of the elementary streams to programs. This information is related in the PSI. The PAT, PMT, and CAT of the PSI provide information only for the multiplex in which they are contained. They do not give any information on programs carried in other transport streams, even if these are delivered by the same physical network.

Conditional Access. The term conditional access (CA) is used to describe systems that control access to programs and services. CA systems consists of several blocks, including

- the mechanism to scramble the program or service and to convey the necessary CA data to the receiver
- the subscriber management system (SMS), in which all customer data are stored
- the subscriber authorization system (SAS), which encrypts and delivers code words to enable the program or service to be correctly descrambled.

Neither the SMS nor the SAS have been standardized by DVB.

Multicrypt and Simulcrypt. There are two basic ways to allow viewers to access programs that have been processed by different CA systems. These are referred to as *multicrypt* and *simulcrypt*. In order to receive pay-TV services or other services with controlled access, the integrated receiver decoder (IRD) needs to be equipped with a CA module. The function of this module is to receive and interpret the CA information contained in a transport stream and to descramble those programs that have been authorized. The multicrypt approach requires the CA module in the IRD to be standardized to a certain extent, in particular through the provision of a common interface. With simulcrypt, the CA module can be entirely proprietary and can be embedded within the IRD so that physical access to the unit is difficult.

In multicrypt applications, the descrambling of programs controlled by different CA systems requires insertion of the appropriate CA module into the common interface. The CA module is also physically more accessible than is the case with simulcrypt, which can give raise to security concerns. However, multicrypt has two important benefits. First, it allows service to operate entirely independently, with no need to share CA information and no need to enter into a commercial agreement with other service providers who are addressing the same market with different CA systems. Secondly, the architecture of the IRD is more *open*. Consequently, the IRD is less likely to become obsolete as services evolve or if the viewer wishes to subscribe to a different package controlled by a different CA system.

In simulcrypt applications, the viewer is able to use a single CA system built into his IRD to watch all programs, irrespective of the CA system used to scramble each program. This has some security benefits with respect to multicrypt because the CA module is not usually physically accessible to the user. Furthermore, there is potentially some cost savings in the manufacture of the IRD since the common interface need not be provided, although the cost of implementing this interface has fallen considerably and continues to fall. In contrast to multicrypt, the simulcrypt approach requires CA system operators to establish commercial agreements. It also implies dedicated IRD hardware that cannot be upgraded or adopted for use with the CA systems.

Satellite Channel Adaptation. The DVB-S system adopts the baseband digital signals at the output of the MPEG-2 transport multiplexer to the satellite channel characteristics. The processes involved are illustrated in Figure 4.47, which is in fact the functional block diagram of the DVB-S system. The processes are

- transport multiplex adaptation and randomization for energy dispersal
- outer coding (i.e., Reed–Solomon)
- convolutional interleaving
- inner coding (i.e., punctured convolutional code), baseband shaping for modulation
- modulation.

The complementary functions are implemented in reverse order in the receiver.

The input to the satellite channel adaptor is an MPEG-2 transport stream and consequently it consists of a series of fixed-length packets 188 bytes long.

The transport stream data are randomized so that the RF spectrum approximates that generated by truly random data (i.e., it is noise-like and does not contain significant spectral line components).

Randomization is implemented by adding a pseudorandom binary sequence (PRBS) to the transport stream data using the exclusive-OR function. It is performed over sequential groups of eight bytes. The PRBS is restarted at the beginning of each new group.

The sync byte of the transport packet in a group of eight packets is bitwise inverted to provide an initialization signal for the descrambler in the receiver.

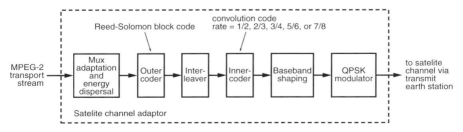

Figure 4.47 Functional block diagram of the DVB-S system.

This process is referred to as transport multiplex adaptation. The MPEG-2 sync byte is the first byte of a transport packet's header.

The sync bytes or the other seven transport packets are not randomized in order to aid other synchronization functions. All remaining transport packet data are randomized, including the rest of the transport packet headers.

The randomization process is active even when there is no input signal so that transmission of an unmodulated carrier is avoided.

Error Correction Coding. The three blocks labeled outer coder, interleaver, and inner coder presented in Figure 4.47 together provide the error correction mechanism. The outer code is a Reed–Solomon (RS) block code, and the inner code is a convolutional code. In the receiver, a Viterbi decoder precedes the RS decoder. The function of the Viterbi decoder is to decode the inner (convolutional) code. The function of the RS code is to correct any residual error appearing in the output of the Viterbi decoder. Uncorrected errors at the output of the Viterbi decoder tend to occur in bursts. However, the RS code is only effective if errors occur randomly, rather than in groups. Consequently, to obtain an effective concatenated error correction scheme, the output of the Viterbi decoder needs to be randomized. Randomization is performed by the de-interleaver in the receiver and the interleaver in the transmitter.

ATSC Digital Television

A basic block diagram representation of the ATSC digital television system is shown in Figure 4.48. This is the ITU-R digital terrestrial television broadcasting model [127], adopted by the International Telecommunication Union, Recommendation factor (ITU-R) Task Group 11/3 (Digital Terrestrial Television Broadcasting). According to this model, the digital television system can be seen to consist of three subsystems [128]: source coding and compression, service multiplex and transport, and RF/transmission.

Source coding and compression refers to the bit rate reduction methods, also known as data compression, appropriate for application to the video, audio, and ancillary digital data streams. The term *ancillary data* includes control data, conditional access control data, and data associated with the program audio and video services. Ancillary data can also refer to independent program services. The coder minimizes the number of bits needed to represent the audio and video information. The digital television system employs the MPEG-2 video stream syntax for the coding of video and the Digital Audio Compression Standard (AC-3) for coding audio.

Service multiplex and transport refers to the means of dividing the digital data stream into packets of information, the means of uniquely identifying each packet or packet type, and the appropriate methods of multiplexing video data stream packets, audio data stream packets, and ancillary data stream packets into a single data stream. In developing the transport mechanism, interoperability among digital media, such as terrestrial broadcasting, cable distribution, satellite distribution, recording media, and computer interfaces, was a prime consideration. The digital

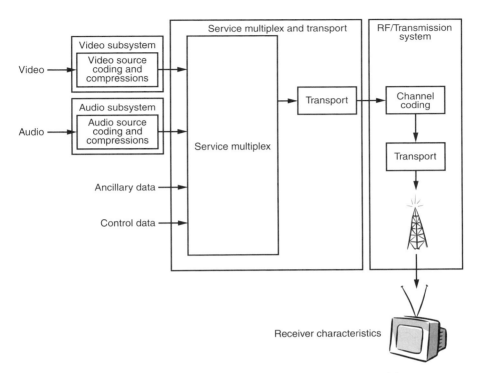

Figure 4.48 ITU-R digital terrestrial television broadcasting model.

television system employs the MPEG-2 transport stream syntax for the packetization and multiplexing of video, audio, and data signals for digital broadcasting systems. The MPEG-2 transport stream syntax was developed for applications where channel bandwidth or recording media capacity is limited and the requirement for an efficient transport mechanism is paramount. It was designed also to facilitate interoperability with the ATM transport mechanism [127].

RF/transmission refers to channel coding and modulation. The channel coder takes the data bit stream and adds additional information that can be used by the receiver to reconstruct the data from the received signal, which, due to transmission impairments, may not accurately represent the transmitted signal. The modulation (or physical layer) uses the digital data stream information to modulate the transmitted signal. The modulation subsystem offers two modes: a terrestrial mode (8VSB), and a high data rate mode (16VSB).

Video Subsystem. The video compression system takes in an analog video service signal and outputs a compressed digital signal that contains information that can be decoded to produce an approximate version of the original image sequence. The goal is for the reconstructed approximation to be imperceptibly different from the original for most viewers, for most images, and for most of the time. In order to approach such fidelity, the algorithms are flexible, allowing for frequent adaptive changes in the algorithm depending on scene content, history of the processing, estimates of image complexity, and perceptibility of distortions introduced by the compression.

Figure 4.49 shows the video subsystem within the digital television system. Signals presented to the system are digitized and sent to the encoder for compression, and the compressed data then are transmitted over a communications channel. On being received, the possibly error-corrupted compressed signal is decompressed in the decoder, and reconstructed for display.

Audio Subsystem. As illustrated in Figure 4.50, the audio subsystem comprises the audio encoding/decoding function and resides between the audio inputs/outputs and the transport subsystem. The audio encoder(s) is (are) responsible for generating the audio elementary stream(s), which are encoded representations of the baseband audio input signals. The flexibility of the transport system allows multiple audio elementary streams to be delivered to the receiver. At the receiver, the transport subsystem allows multiple audio elementary streams to be delivered to the receiver. At the receiver, the transport subsystem is responsible for selecting which audio streams(s) to deliver to the audio subsystem. The audio subsystem is responsible for decoding the audio elementary stream(s) back into baseband audio.

An audio program source is encoded by a digital television audio encoder. The output of the audio encoder is a string of bits that represent the audio source, and is referred to as an audio elementary stream. The transport subsystem packetizes the audio data into PES packets, which are then further packetized into transport packets. The transmission subsystem converts the transport packets into a modulated

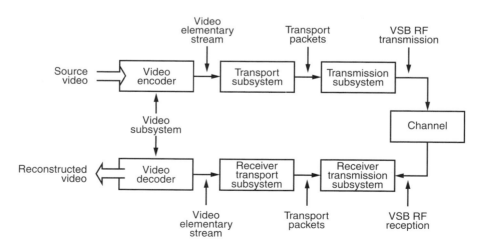

Figure 4.49 Video subsystem within the digital television system.

4.4 DIGITAL BROADCASTING

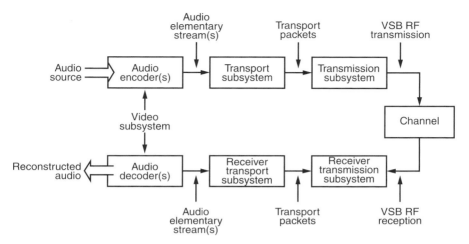

Figure 4.50 Audio subsystem within the digital television system.

RF signal for transmission to the receiver. At the receiver, the received signal is demodulated by the receiver transmission subsystem. The receiver transport subsystem converts the received audio packets back into an audio elementary stream, which is decoded by the digital television audio decoder. The partitioning shown is conceptual, while practical implementation may differ. For example, the transport processing may be broken into two blocks: one to perform PES packetization, and the second to perform transport packetization. Alternatively, some of the transport functionality may be included in either the audio coder or the transmission subsystem.

Transport Systems and Service Multiplex. The transport system standard is based on MPEG-2 and is given in ISO/IEC 13818-1 as constrained and extended for the digital television standard. The transport system employs the fixed-length, transport stream packetization approach defined by MPEG. This approach to the transport layer is well suited to the needs of terrestrial broadcast and cable television transmission of digital television. The use of moderately long fixed-length packets matches well with the needs and technologies for error protection in both terrestrial broadcast and cable television distribution environments. At the same time, it provides great flexibility to accommodate the initial needs of the service to multiplex video, audio, and data, while providing a well-defined path to add additional services in the failure in a fully backward-compatible manner. By basing the transport layer on MPEG-2, maximum interoperability with other media and standards is maintained.

A transport layer based on a fixed-length packetization approach offers a great deal of flexibility and some significant advantages when attempting to multiplex data related to several applications on a single bit stream. These are dynamic capacity allocation, scalability, extensibility, robustness, cost-effective receiver implementations, and MPEG-2 compatibility.

The use of fixed-length packet offers complete flexibility to allocate channel capacity among video, audio, and auxiliary data services. The use of a packet identifier (PID) in the packet header as a means of bit stream identification makes it possible to have a mix of video, audio, and auxiliary data, which is flexible and which need not be specified in advance. The entire channel capacity can be reallocated in bursts for data delivery. This capability could be used to distribute decryption keys to a large number of receivers during the seconds preceding a popular pay-per-view program.

The transport format is scalable in the sense that the availability of a larger bandwidth channel may also be exploited by adding more elementary bit streams at the input of the multiplexer, or even multiplexing these elementary bit streams at the second multiplexing stage with the original bit stream. This is a valuable feature for network distribution, and also serves interoperability with cable plants capable of delivering a higher data rate within a 6 MHz channel.

Because there will be possibilities for future services that we cannot anticipate today, it is extremely important that the transport architecture provide open-ended extensibility of services. New elementary bit streams can be handled at the transport layer without hardware modification by assigning new packet identifiers at the transmitter and filtering these new PIDs in the bit stream at the receiver. Backward compatibility is assured when new bit streams are introduced into the transport system, as existing decoders will automatically ignore new PIDs. This capability could possibly be used to compatibly introduce *1000-line progressive formats* or 3D HTV, by sending augmentation data along with the basic signal.

Another fundamental advantage of the fixed-length packetization approach is that the fixed-length packet can form the basis for handling errors that occur during transmission. Error correction and detection processing may be synchronized to the packet structure so that one deals at the decoder with units of packets when handling data loss due to transmission impairments. After detecting errors during transmission, one recovers the data bit stream from the first good packet. Recovery of synchronization within each application is also aided by the transport packet header information. Without this approach, recovery of synchronization in the bit stream would have been completely dependent on the properties of each elementary bit stream.

A transport system based on fixed length enables simple decoder bit stream demultiplex architecture. The decoder does not need detailed knowledge of the multiplexing strategy or the source bit rate characteristics to extract individual elementary bit streams at the demultiplexer. All the receiver needs to know is the identity of the packet, which is transmitted in each packet header at fixed and known locations in the bit stream. The only important timing information is for bit-level and packet-level synchronization.

The transport system is based on the MPEG-2 system specification. An additional advantage of MPEG-2 compatibility is interoperability with other MPEG-2 applications. The MPEG-2 format is likely to be used in a number of other applications, including storage of compressed bit streams, computer networking, and non-HDTV delivery systems. MPEG-2 transport system compatibility implies that digital television transport bit streams may be handled directly in these other applications. We ignore for the moment the issue of bandwidth and processing speed.

Transport Subsystem. Organization of functionality in a transmitter–receiver pair for a single digital television program together with the location of the transport subsystem in the overall system is illustrated in Figure 4.51. The transport resides between the application (video, audio) encoding/decoding function and the transmission subsystem. At its lowest layer, the encoder transport subsystem is responsible for formatting the encoded bits and multiplexing the different components of the program for transmission. At the receiver it is responsible for recovering the bit streams for the individual application decoders and for the corresponding error signaling. At a higher layer, multiplexing and demultiplexing of multiple programs within a single bit stream can be achieved with an additional system-level

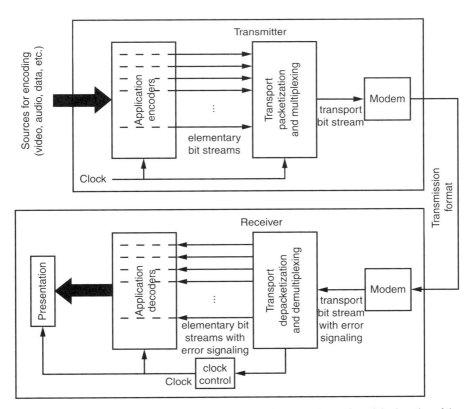

Figure 4.51 Organization of a digital television transmitter–receiver pair and the location of the transport subsystem [127]. (©1994 ITU-R.)

multiplexing or demultiplexing stage, before the modem in the transmitter and after the modem in the receiver. The transport subsystem also incorporates higher level functionality related to identification of applications and synchronization of the receiver. The data transport mechanism is based on the use of fixed-length packets that are identified by headers. Each header identifies a particular application bit stream (also called an elementary bit stream), which forms the payload of the packet. Applications supported include video, audio, data, program system control information, and so on. The elementary bit streams for video and audio are themselves wrapped in a variable-length packet structure called the packetized elementary stream (PES) before transport processing. The PES layer provides functionality for identification, and synchronization of decoding and presentation of the individual application.

It should be noted that the PES layer is not required for all applications. Its use is required for both video and audio in the Digital Television Standard.

Moving up one level in the description of the general organization of the bit streams, elementary bit streams sharing a common time base are multiplexed, along with a control data stream, into *programs*. Note that a program in the digital television system is analogous to a channel in the NTSC system in that it contains all the video, audio, and other information required to make up a complete television program. These programs and an overall system control data stream are then asynchronously multiplexed to form a multiplexed system.

At this level, the transport is also quite flexible in two aspects:

- It permits programs to be defined as any combination of elementary bit streams; specifically, the same elementary bit stream can be present in more than one program (e.g., two different video bit streams with the same audio bit stream); a program can be formed by combining a basic elementary bit stream and a supplementary elementary bit stream (e.g., bit streams for scaleable decoders); programs can be tailored for specific needs (e.g., regional selection of language for broadcast of secondary audio); and so on.
- Flexibility at the systems layer allows different programs to be multiplexed into the system as desired, and allows the system to be reconfigured easily when required. The procedure for extraction of separate programs from within a system is also simple and well-defined. The transport system provides other features that are useful for both normal decoder operation and for the special features required in broadcast and cable applications. These include decoder synchronization, conditional access, and local program insertion. The transport bit stream definition directly addresses issues relating to the storage and playback programs. It is a fundamental requirement for creating programs in advance, storing them, and playing them back at the desired time. The programs are stored in the same format in which they are transmitted, that is, as a transport bit stream.

The transport bit stream consists of fixed-length packets with a fixed and a variable component to the header field as illustrated in Figure 4.52. Each packet consists

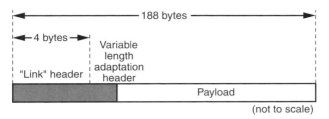

Figure 4.52 Transport packet format.

of 188 bytes and is constructed in accordance with the MPEG-2 transport syntax and semantics. The choice of this packet size was motivated by a few factors. The packet needs to be large enough so that the overhead due to the transport headers does not become a significant portion of the total data carried. It should not be so large that the probability of packet error becomes significant under standard operating conditions (due to inefficient error correction). It is also desirable to have packet lengths consistent with the block sizes of typical, block-oriented, error correction methods, so that packets may be synchronized to error correction blocks, and the physical layer of the system can aid the packet-level synchronization process in the decoder. Another reason for the particular packet length selection is interoperability with the ATM format. The general philosophy is to transmit a single MPEG-2 transport packet in four ATM cells. There are, in general, several approaches to achieve this functionality.

The contents of each packet are identified by the packet headers. The packet header structure is layered and may be described as a combination of a fixed-length link layer and a variable-length adaptation layer. Each layer serves a different function similar to the link and transport layer functions in the OSI layered model of a communications system. In the digital television system, this link and adaptation-level functionality are used directly for the terrestrial broadcast link on which the MPEG-2 transport bit stream is transmitted. However, in a different communications system, for example, ATM, the MPEG-2 headers would not play a role in implementing a protocol layer in the overall transmission system. The MPEG-2 headers would be carried as part of the payload in such a case and would continue to serve as identifiers for the contents of the data stream.

Main and Associated Services. Multiple audio services are provided by multiple elementary streams. Each elementary stream is conveyed by the transport multiplex with a unique PID. There are a number of audio service types that may (individually) be coded into each elementary stream. Each elementary stream is tagged as to its service type using the `bsmode` bit field. There are two types of *main service* and six types of *associated service*. Each associated service may be tagged (in the AC-3 audio descriptor in the transport `PSI` data) as being associated with one or more main audio services. Each AC-3 elementary stream may also be tagged with a language code.

Associated services may contain complete program mixes, or may contain only a single program element. Associated services that are complete mixes may be decoded and used as is. Table 4.6 indicates service types. In general, a complete audio program (what is presented to the listener over the set of loudspeakers) may consist of a main audio service, an associated audio service that is a complete mix, or a main audio service combined with one associated audio service. The capability to simultaneously decode one main service and one associated service is required in order to form a complete audio program on certain service combinations. This capability may not exist in some receivers.

Ancillary Data Services. The digital television system affords the opportunity to augment the basic television and audio service with ancillary digital data services. The flexibility of the MPEG-2 transport layer employed in the new digital services allows it to be easily introduced at any time. The ancillary services envisioned are a mix of associated audio services and textual services. The associated audio services are simply transported as separable audio streams.

At the option of broadcasters, an interactive program guide database may be transmitted in the transport stream. The program guide database contains information relating to the programs currently being transmitted and information for programs that will be transmitted at future times. It also contains time and control information to facilitate navigation and as such it allows suitably equipped receivers to build an interactive screen grid of program information.

At the option of broadcasters, certain system information may be transmitted in the transport stream. In general, system information contains information on the locations of available transport streams and on how to acquire these streams and their constituent services. System information is especially useful in networks where the operator has control over more than one transport multiplex. To allow for interoperability between broadcast and other transmission environments, the system information mechanisms identified for broadcast applications are a compatible subset of a protocol.

In order to facilitate simple acquisition of the system information by a receiver, a specific transport PID is reserved for its transmission.

Table 4.6 Service types [127]

bsmode	Type of Service
000 (0)	Main audio service: complete main (CM)
001 (1)	Main audio service: music and effects (ME)
010 (2)	Associated service: visually impaired (VI)
011 (3)	Associated service: hearing impaired (HI)
100 (4)	Associated service: dialog (D)
101 (5)	Associated service: commentary (C)
110 (6)	Associated service: emergency (E)
111 (7)	Associated service: voice-over (VO)

©1994 ITU-R.

Compatibility with Other Transport Systems. The transport system interoperates with two of the most important alternative transport systems. It is syntactically identical with the MPEG-2 transport stream definition, with the Digital Television Standard being a subset of the MPEG-2 specification. Digital Television Standard completely specifies the special constraints and requirements of the subset. The transport system also has a high degree of interoperability, with the ATM definition being finalized for broadband ISDN. Furthermore, as several of the cable television and direct broadcast satellite (DBS) systems currently in use or being designed employ MPEG-2 transport layer syntax, the degree of interoperability with such deployed systems should be quite high (possibly requiring a translation if the cable television or DBS system deploys a slightly incompatible MPEG-2 variant).

Interoperability has two aspects. The first is syntactic and refers only to the coded representation of the digital television information. The second relates to the delivery of the bit stream in real time. This aspect of interoperability is beyond the scope of this discussion, but it should be noted that to guarantee interoperability with a digital television receiver conforming to the Standard, the output bit stream of the alternative transport system must have the proper real-time characteristics.

DVB Services over IP-Based Networks

A wide range of specifications will be built upon the basic architecture document in order to define the usage and implementation of IP-based DVB services. The DVB-IP architecture is applicable to all systems and service implementations, using integrating receiver decoders, TV sets and multimedia PCs, as well as clusters of such devices connected to home networks [129]. The prime target for standardization by DVB is the home network and devices to enable high-volume, low-cost equipment. The suite of standards should be complete from layer 1 up to and including the application layer. The principle of one tool for a function should be employed to simplify the architecture and to keep cost low. In order to describe the complex system that is necessary for the delivering of DVB services over IP-based networks, the following three sections describe the inherent functionality from different points of view: the Layer Model, the Home Reference Model, and Modules for the Home Network Elements.

Layer Model. A diagram of high-level reference model for DVB services on IP is shown in the Layer Model of Figure 4.53. This model is intended to show the domains relevant to DVB services on IP, with the basic peer-to-peer information flows at the different layers. A management plane is included for management and control purposes.

The four communicating domains are briefly described as follows:

- *Content provider*: the entity who owns or is licensed to sell content or content assets. Although the service provider is the primary source for the client at home, a direct logical information flow may be set up between content provider and home client, for example, for rights management and protection. This flow is shown in the layered model.

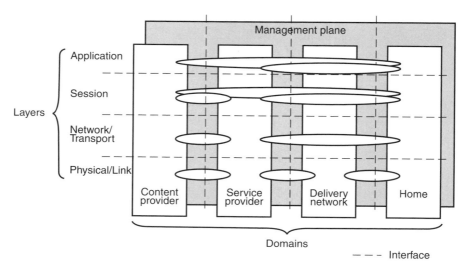

Figure 4.53 Layer Model of DVB services on IP.

- *Service provider*: the entity providing a service to the client. Different types of service provider may be relevant for DVB services on IP, for example, simple Internet Service Providers (ISPs) and Content Service Providers (CSPs). In the context of DVB services on IP, the CSP acquires/licenses content from content providers and packages this into a service. In this sense, the service provider is not necessarily transparent to the application and content information flow.
- *Delivery network*: the entity connecting clients and service providers. The delivery system is usually composed of access networks and core or backbone networks, using a variety of network technologies. The delivery network is transparent to the IP traffic, although there may be timing and packet loss issues relevant for A/V content streamed on IP.
- *Home*: the domain where the A/V services are consumed. In the home, a single terminal may be used for service consumption, but also a network of terminals and related devices may be present for this purpose.

As mentioned above, the service provider entity covers various kinds of service provider types, especially broadband ISPs and CSPs. It should be noted that although we treat these two business roles separately, a single company can very well act in both roles. In such a case, the end user can be offered a single subscription covering both the ISP and the CSP service offerings.

The broadband ISP typically provides the following services and functions (non-exhaustive list):

- *Addressing services.* For residential customers, the ISP typically provides the IP addresses and associated configuration to the delivery network gateway and/or to the home network end devices by a dynamic mechanism.
- *Authentication* and *authorization* of the Internet (or IP) access service subscriber. The authentication may be explicit, typically based on PPP or a Web login, or it may be implicit, for example, based on some link layer ID that can be tied to the subscriber. It is noted that authentication is important for the ISP to be able to fulfill its obligations (like handling abuse), but it may also serve as the basis for various kinds of service differentiation. Such differentiation may include different access bandwidth rates, the right to receive/source prioritized traffic and value added services.
- *Naming services*, which provide for the translation between symbolic names and IP addresses, implemented by DNS.
- *IP connectivity.* This is the basic service and provides broadband access to the Internet. However, the IP connectivity can also cover a QoS enabled access to various sources of content, logically parallel to a best effort Internet access. It is noted that downstream IP multicast needs to be supported to carry IP-based TV services, although the access service may impose certain restrictions on the ranges of multicast addresses accessible. The address ranges may be determined by the access rights of the specific subscription.
- Termination of the *session control* protocol for multicast-based services.
- *Accounting* of the services associated with the IP access subscription.
- *Value added services.* This service category can involve services like a network-based firewall, network-based storage for the end user, caching services, and e-mail.

It is noted that an ISP may extend its service offering by the category *value added services* above. Such services may then also include content-related services like VOD. In such a case, the ISP party also acts as a CSP and may then make use of a common user authentication for all services; that is, the end user does not necessarily need to have different user accounts per service, but a single user account can cover the IP access service as well as a VOD service. However, the ISP and the CSP can also be separate parts, each one having a separate authentication process of the associated users. In this context we will assume the latter, since that can be regarded as the general case.

In the layer model, each communication protocol layer presents only a service interface to the layer above, hiding its own implementation details. This has the advantage of reducing complexity and cost and enables easy layer changes and improvements. The following comprise a brief description of the protocol layers:

- The physical layer of a communication link describes the scheme that constitutes a bit on the media, the bit synchronization, the media, and the connector shape and size. This level is not aware of packets or frames used for multiplexing.

- The link layer handles the media access control, with addressing and arbitration to enable end-systems to receive only the packets addressed to them. It may also provide error control, flow control, and retransmission of corrupted or lost packets. This layer is very dependent on the physical media and may contain specific control and linking mechanisms.
- The IP network layer enables an elementary end-to-end logical link by providing routing, packet forwarding, segmentation and reassembly. Only formatting of packets with an IP header and IP source and destination address is needed for implementation of the end-to-end communication link. Because IP hardly makes any assumption about the link layer, it can be transported using almost any link layer technology.
- The transport layer uses the IP layer to create a flow-controlled, error-controlled end-to-end communication link, protected against corruption, duplication, and packet loss. The measure of protection and the processing required for it needs to be weighed against the needs of the service. The transport layer may also provide multiplexing of services on a single IP link, using, for example, port numbers. Popular IP transport protocols are UDP and TCP. UDP provides multiplexing, but no flow or error control, while TCP provides error detection and control, flow control, and multiplexing.
- The session layer sets up or closes the connections needed to start or terminate an application.
- The application layer consists of two sublayers: the API and application sublayers. They provide end-to-end application control and command. For DVB services, this layer is specified as MHP.

Home Reference Model

The architecture of the DVB-IPI network must support the following (nonexhaustive) list of possible scenarios [130]:

- A home network can be simultaneously connected to multiple and heterogeneous delivery networks. As an example, in a typical scenario, ADSL and DVB-S are both available at the home. Load balancing may be possible among the different delivery networks in order to optimize the utilization and throughput of the networks and to minimize the delay.
- End users can choose the service provider. As an example, the ISPs and the CSPs may be independent of each other.
- Different end users in the same home network can select different service providers.
- Access to the content is independent of the underlying hardware. As an example, terminals with different capabilities (e.g., CPU power, display size, storage capacity) may be allowed to access the same content through the use of transcoding resources, or through the use of device-specific resources.
- Roaming of end users between delivery networks should be possible. As an example, the personal environment of a (SOHO) user stored on a home server

should be accessible from different external locations. Adequate security aspects need to be taken into account.

The Home Reference Model based on these scenarios for the DVB-IPI home network is described in Figure 4.54. It consists of the Home domain of the layer model. Furthermore, it shows the interconnection with the Delivery Network domain. This home reference model shows the elements that can be present in the home and their mutual relations. Based on the fact that this is just a reference model, elements can be removed or added as long as the connection between a home network end device and the delivery network is still possible. The collection of all these home network elements forms the Home Network (HN). The elements present in the Home Reference Model are described as follows:

- *Delivery network gateway* (DNG): the device that is connected to one or multiple delivery networks and one or multiple home network segments. It contains one or more connecting components so that it can interconnect the delivery network(s) with the home network segment(s) on any of the OSI layers. This means that it can be a so-called "null" device, a wire interconnecting the networks on OSI layer 1, that it can function as a bridge or router interconnecting different link layer technologies, or that it can act as a gateway also providing functionality on the OSI layer 4 and above.
- *Home network segment* (HNS): this element consists of a single link layer technology and provides a layer 2 connection between home network end devices and/or connecting components. The connecting components are not part of a

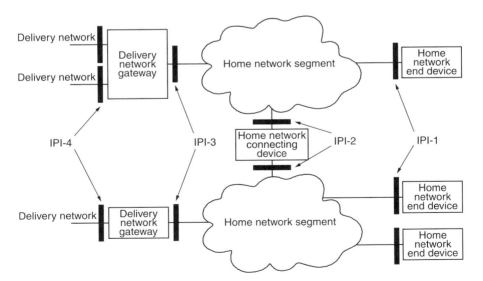

Figure 4.54 Home Reference Model [130]. (©2002 DVB.)

home network segment. So, each home network segment is connected to another home network segment via a connecting component. The separation of a home network into home network segments does not imply that each segment needs to be an "IP-subnet". A home network segment can be wired or wireless.

- *Home network connecting device* (HNCD): this device, which contains one or more connecting components, connects two or more home network segments with one another and functions as a bridge, router, or gateway.
- *Home network end device* (HNED): the device that is connected to a home network and that typically terminates the IP-based information flow (sender or receiver side). This does not imply that this home network end device needs to be the end point of the non-IP-based information flow. So, it can still serve as an application level gateway to other non-IP-based network technologies.

Modules for Home Network Elements

The functionality of the various elements identified in the Home Reference Model can be described with modules that take into account the layer structure Layer Model. These modules contain protocols for specific required functionalities. Once defined, they can be reused where needed. The modules will be addressed in the various DVB-IPI specifications.

Services

There are several ways to describe services that will be available on DVB over the IP technology framework. Here, services will be divided into categories depending on the *users' service perception*; the starting point will be concrete services. Then, each service category will be described to better understand underlying technological implications. The analysis will point out what kind of technological framework is capable of implementing the services: if interactive channels are needed or not, what delivery methods and formats will be used. DVB services, or services related to DVB, as the *New Mission* of DVB states, will be considered.

In this description, intended for discussion, we used the following guidelines:

- user requirements and needs that services aim to satisfy
- main functionalities provided (audio and video delivery, download, file transfer, etc.)
- technology (the most suitable network or device, etc.).

Services are divided into categories to aggregate similar functionalities. The categories we found are the following:

- *entertainment services* – general leisure and spare time services (e.g., sports news, online gaming)

- *general information services* – main purpose of these services is to provide users with up-to-date information
- *educational services* – they provide educational content, distance learning functionalities
- *messaging services* – users can use them to exchange information, like mail messages;
- *communication services* – they provide real-time connection among users, like video conferencing (video telephony, voice over IP)
- *service information* – these services provide to users information about other services; they make it easy for the user to connect to a service and to use it.

In what follows, we will present service description.

Entertainment services include PayTV (audio/video), video on demand (audio/video), music (audio), pictures download, games, and have the following features:

- Short description – general leisure and spare time services, for example, sport news, online gaming.
- Interaction – in their basic version, these services require no interaction. Video on demand requires interaction.
- Games can be divided into different types. Some games require a network for interaction with other players. Other games (e.g., gambling) require an interaction with the provider.
- More evolved scenarios include the possibility of interaction for the user, requiring a full bidirectional connection with the provider and the possibility to download useful software, presentations, and so on; service related.
- Possible delivery method – in their basic version, they are push, point-to-multipoint services. Point-to-point connection is needed for video on demand and some types of games.
- Considering IP encapsulated DVB, at the Network Layer, IP multicast is suitable to convey PayTV, music, pictures download, games. IP unicast is suitable for video (and audio) on demand.
- Possible format – MPEG-2 audio and video. Other formats can be jpeg/png for pictures, java applications for games.
- End devices – these services are well suited to STBs and PCs. Powerful handheld devices can also be a target.

General information services are advertising, sports news, entertainment news, emergency information, general news, travel information, and stock exchange information, and have the following characteristics:

- Short description – the main purpose of these services is to provide users with up-to-date information on different subjects (sports, entertainment, and travel). Advertising is also considered an information service here.

- Interaction – in their basic version, these services require no interaction.
- Possible delivery method – in their basic version, they are push, point-to-multipoint services. They essentially need one-way transport functionalities.
- Considering IP encapsulated DVB, IP multicast is suitable to convey them.
- Possible format – MPEG-2 audio and video for advertising, sport news, entertainment news, general news, emergency information.
- Travel information can be integrated with text information.
- Stock exchange is text based (e.g., html or xml pages).
- End devices – these services are well suited to STBs and PCs. Powerful handheld devices could also be a target.

Educational services include computer/STB based training, and distance learning, and have the following attributes:

- Short description – they provide educational content, distance learning functionalities, distinct from general information services because of their specificity.
- Interaction – in their basic version, these services require no interaction (distance learning without interaction).
- More evolved scenarios include the possibility of interaction for the user, requiring a full bidirectional connection with the provider and the possibility to download useful software, presentations, and so on. In this case the consumer machine (PC or STB) has to support all the involved formats.
- Possible delivery method – in their basic version, they are push, point-to-multipoint services. They can be conveyed on both one-way and bidirectional channels.
- Considering IP encapsulated DVB, IP multicast is suitable to convey them. Interactivity is obtained by adding a bidirectional channel (e.g., TCP based).
- Possible format – MPEG-2 audio and video. Other formats for support material download (jpeg/png/gif for pictures, java applications, etc).
- End devices – these services are well suited to STBs and PCs. Handheld devices can be considered for very basic services.

Messaging services include e-mail and multimedia messaging:

- Short description – users can use them to exchange information, like mail messages.
- Messages could be as simple as chat transactions and as complex as video messages. Messages can be addressed to a single user or to the listeners of a program.
- Interaction – these services require interaction.
- Possible delivery method – they need bidirectional channels. Messaging can be both a point-to-point and a point-to-multipoint service.

- Considering IP encapsulated DVB, at the Network Layer, IP unicast is convenient for e-mail. For messaging addressed to many users, IP multicast is convenient for forward channel, IP unicast for interaction with the provider.
- Possible format – text, html. Other supported formats can be jpeg/png/gif for pictures, video clips, and attachments in general.
- End devices – these services are well suited to STBs and PCs. Handheld devices are well suited for these services.

Communication services are video conferencing, video telephony, and voice over IP, and have the following characteristics:

- Short description – they provide real-time connection among users, like video conferencing. Content is generated by one or several users and it is sent to other users in real time.
- Interaction – these services require *real-time* interaction.
- Possible delivery method – they need bidirectional channels; they can be both point-to-point and point-to-multipoint services.
- Considering IP encapsulated DVB, at the Network Layer, IP multicast is convenient to forward the streams to the subscribers. IP unicast or multicast can be used to send the streams to the provider.
- Possible format – mainly MPEG-2 audio and video. Text, html enriches the communication when convenient and useful. The service offers to the user the possibility of sending support material (jpeg/png/gif pictures, documents, applications, etc).
- End devices – these services are well suited to STBs and PCs. Handheld devices are well suited for the basic functionalities of these services.

Service information includes the Electronic Program Guide (with detailed content description), service discovery, and selection.

- Short description – these services provide to the users information about other services, and make it easy for the user to connect to a service and to use it.
- Interaction – in their basic version, these services require no interaction.
- Traditional DVB EPG and SI information is a pure push service. Service discovery, selection and description for DVB services over IP can be a push or a pull service.
- Possible delivery method – push or pull services.
- They can be conveyed on both one-way and bidirectional channels.
- Considering IP encapsulated DVB, at the Network Layer, IP multicast is suitable to convey service information.

- Possible format – DVB specified format for traditional DVB EPG. XML, SDP specified format (and other formats?) for service discovery, selection and description over IP.
- End devices – these services are well suited to STBs, PCs, and handheld devices.

Requirements for Service Authentication, Authorization, and Accounting

Service providers need methods for authentication, authorization, and subsequently billing of users for using their networks and services. The following lists some requirements for authentication, authorization, and accounting (AAA) in DVB services on IP networks.

The technologies used should

- be based on standards used within the Internet community, such as the AAA Framework described in RFC 2904 and RFC 2905
- offer authentication, authorization, and accounting on a per service basis
- support multiple services running simultaneously on/over the same customer premises equipment (CPE)
- allow the use of different accounting policies, for example, one for video-audio, one for Internet access (e.g., Web-browsing), and another for IP-telephony
- allow each service provider to do its own authentication, authorization, and accounting
- prevent CPE, which does not contain AAA functionality, to enter the broadband access network
- be secure, that is, users cannot circumvent authentication, authorization, and accounting, for example, by "hacking" the CPE.

4.4.5 DVB-MHP Application and Interactive Television

The DVB project has already defined a number of open standards for digital video broadcasting, which enable broadcast and interactive services over all transmission networks including satellite, cable, terrestrial, and microwave systems. The later activity is the harmonization of specifications for multimedia home platform (MHP) to ensure key features such as application interoperability, application download, scalability, and upgradability. MHP defines a generic interface between interactivity digital applications and the terminals on which those applications execute. It enables digital content providers to address all types of terminals, ranging from low-end to high-end set-top boxes, integrated digital TV sets, and multimedia PCs. The MHP extends the existing, successful open standards for broadcast and interactive services in all transmission networks.

MHP encompasses the peripherals and the interconnection of multimedia equipment via the in-line digital network. The MHP solution covers the whole set of technologies that are necessary to implement digital interactive multimedia in the home

– including protocols, common application programming interface (API) languages, interfaces, and recommendations. At the beginning of 1996, the UNITEL universal set-top box project was launched by the Program of the European Commission. The main aim of this project was to raise awareness of the benefits of developing a common platform for user-transparent access to the widest range of multimedia services. Promising progress has since been achieved towards the harmonization of what is now widely called the MHP. The MHP Launched Group was born from the UNITEL initiative in order to open the project to external parties via joint meetings. Key representatives of the High Level Strategy Group took part in this group, and this collaboration eventually led to the transfer of these activities to the DVB project. Two DVB workgroups were subsequently set up [131]:

- A commercially-oriented group, DVB-MHP, to define the user and market requirements for enhanced and interactive broadcasting in the local cluster (including Internet access).
- A technical group, DVB-TAM (Technical issues Associated with MHP), to work on the specification of the DVB application programming interface (API).

Different reference models have been defined for each MHP system currently in use. UNITEL used object-modeling tools to define the application classes and functionalities that would ultimately identify the hardware and software resources required for an MHP system. With this system, users would be able to access

- enhanced broadcasting services
- interactive broadcasting services
- Internet services.

This model offers system modularity through the use of key interfaces. These interfaces will be able to maintain the stability of MHP systems as they evolve – both in terms of hardware and software. Backward compatibility will be supported to the largest possible extent, for example, by using scalable applications.

The reference model consists of five layers [132]:

- application (content, script) and media (audio, video, subtitle) components
- pipes and streams
- the API and native navigation (selective functions)
- platform/system software or middleware, including the interactive engine, the run-time engine (RTE) or virtual machine, the application manager, and so on
- hardware and software resources and associated software.

As a market-led organization, the DVB project started MHP-related work with the definition of function and commercial user requirements. This was done by a specialized ad hoc group of the Commercial Module of the project. The key elements of the agreed set of requirements cover

- interoperability
- scalability, extensibility
- upgradability
- separation of data and applications
- support for CA systems
- open standards.

Starting from these commercial requirements, the DVB Technical Module and its ad hoc group on Technical Aspects of the MHP (TAM) developed an MHP reference model and technical requirements. After a long selection process, fundamental decisions for the MHP architecture were taken in 1996. The whole process in the DVB Project is summarized in Table 4.7. For the MHP specification release 1, a final working draft was approved by the DVB Technical Module in January 2000. First implementations of the DVB MHP are ongoing and early MHP prototypes were shown in Berlin, in September 1999.

The main target of this platform harmonization is the step forward from today's *vertical markets*, where each service provider operates his own platform with head-end, conditional access (CA) system, transmission system, and set-top boxes (STBs), to future *horizontal markets*, where each of these functions has well-defined interfaces and enables multiple implementations. In the horizontal market scenario, any digital content provider can then address any type of terminal ranging from low-end to high-end STBs, integrated TV sets, or multimedia PCs. Possible services include PayTV, free-to-air TV, electronic program guides, stock tickers, and the Internet. For the MHP in a horizontal market, a key element is the API, which provides a platform-independent interface among applications from different providers and the manufacturer-specific hardware and software implementation.

Table 4.7 Steps in the working process of the DVB project [133]

No.	Process Step
1	Commercial requirements
2	Technical requirements
3	Requests for detailed specifications (RDS)
4	Framework (basic architecture)
5	First Working Draft
6	Working Draft
7	Draft standard
8	International standard

MHP Layers

The DVB MHP is primarily built for digital broadcasting scenarios. It receives content via a DVB broadcast channel and can use an additional interaction channel. The usual display is a TV screen and the MHP is operated via a remote control. Figure 4.55 illustrates the three basic layers of the MHP. Resources like MPEG processing, I/O devices, CPU, and graphics processing provide access to available interfaces. The system software includes an application manager and support for basic transport protocols. It also implements a virtual machine, which decouples the manufacturer-specific hardware and software from the standardized APIs. The applications only access the platform via these APIs.

Elements of the DVB-MHP

An overview of the DVB-MHP solution is given in Figure 4.56. The core elements are the general architecture, transport protocols, content formats, a Java Virtual Machine (Java VM) and DVB-J APIs, application lifecycle and signaling, security, and the detailed profile definition. As a later step, further specifications for minimum platform capabilities, conformance, and authoring guidelines may follow.

DVB specifications are already available for transmission of broadcast and interactive data. To ensure interoperability, it is only necessary to select for both cases at least one mandatory protocol stock on which applications can always rely. Mandatory broadcast protocols are shown in Table 4.8, while mandatory interactive protocols are presented in Table 4.9.

The selection of formats for different content types is an important issue for interoperability. The following formats are available for an MHP (the support for particular formats is profile-dependent):

- static formats (PNG, JPEG, MPEG-2 I(P) frames, MPEG-1/2 audio clips, text)
- streaming formats (MPEG-2 video, MPEG-1/2 audio, subtitles)
- resident fonts (in particular for bandwidth-limited networks, such as terrestrial, it is important that one or more font sets are always available on the MHP)

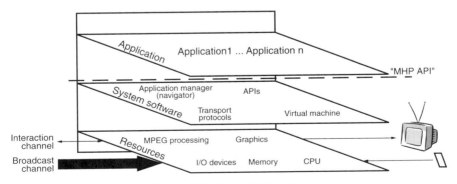

Figure 4.55 Basic MHP layers.

APPLICATIONS LAYER

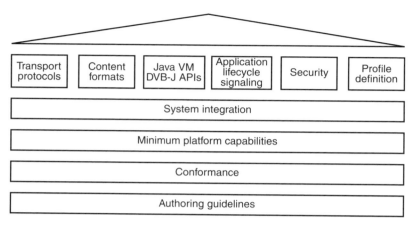

Figure 4.56 Elements of the DVB-MHP.

- downloadable fonts (in addition to resident fonts, a service provider has the possibility to download his own font set to exactly control the look and feel of his application; this includes a format and locator specification)
- HTML, including the evolution of XML, is considered as the declarative content format for use in the MHP, in particular for the Internet Access profile. The first release of the MHP includes the basic hooks for HTML applications signaling and lifecycle. Work is ongoing for the detailed specification of these HTML elements.

DVB-MHP uses a virtual machine (VM) concept, which provides a common interface to different hardware and software implementation. The VM is based on the Java specification from Sun Microsystems, which owns the trademark Java and potentially holds further intellectual property rights (IPR) on technical elements. The basic elements of the language were created with the expectation that any system running Java programs would have certain user input and output devices (keyboard, pointers, microphones, display screens, audio speakers) and data communications parts for networking. Accordingly, the language expects that the system will deal with those devices in detail, and will need only some general instructions.

Table 4.8 Mandatory broadcast protocols [133]

Applications
DVB Object carousels
DSM-CC Object carousels
DSM-CC Data carousels
MPEG-2 Sections
MPEG-2 Transport Stream

4.4 DIGITAL BROADCASTING

Table 4.9 Mandatory interactive protocols [133]

Applications	
TCP	UDP
IP	
Network specific	

The DVB-specific platform has been named DVB-J. In addition, the DVB project decided on an MHP declaration that defines the process for multiple independent implementations of the specification and assures the acceptance of international standards organizations.

Figure 4.57 shows the structure of the DVB-J platform. The system software consists of all implementation-specific parts like the real-time operating system, drivers, and firmware. The application manager includes a *navigator*, which enables a neutral access to all services. The implementation-specific system software and navigator allow manufacturers to differentiate themselves – besides different hardware characteristics – with specific user interfaces and features, which are considered important for the successful development of an open horizontal market.

The envisaged horizontal market implies special requirements for application scenarios, as no vertical market operator controls the operation. A state machine for applications has been defined, which considers the application states `loaded`, `paused`, `active`, `destroyed` and their respective behaviors, as shown in Figure 4.58. The system supports start of applications by the user and also-start applications, which are, for example, started when the user selects a service.

The handling of multiple concurrently running applications is a special case in the horizontal market scenario. Those applications have to share the resources of the MHP. Some resources may only be available once (e.g., the screen), while others can be shared among multiple applications (e.g., memory). The MHP considers the special case of cooperative applications coming from one service provider.

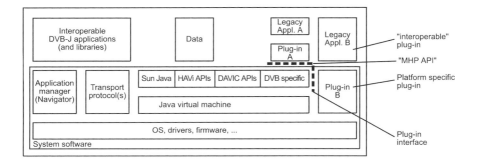

Figure 4.57 The structure of the DVB-J platform.

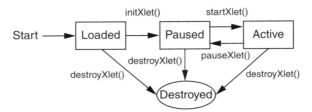

Figure 4.58 Lifecycle for DVB-J applications.

In addition to available DVB service information (SI) specification, a specific application signaling was defined. It can provide special information like location of the application(s), associated data, required MHP profile, required resources, auto-start features, and related cooperative applications.

As there are already many DVB systems with different APIs in operation, support for such existing systems is essential to enable the migration to a future common API. Therefore, the DVB-J platform provides support for plug-ins to implement legacy APIs.

Plug-ins can be implemented in DVB-J (type A in Fig. 4.57). Such an interoperable plug-in can be used on all DVB-J platforms. Another way is direct implementation on the system software (type B), which requires a platform-specific plug-in. Further specification of supporting APIs and the lifecycle is foreseen for a next revision of the specification.

Before loading and starting download applications, it is important to authenticate the source (content provider) and to check the integrity of the application. Otherwise, an application can behave in an undesired manner (e.g., print to screen) or even allocate resources (e.g., generate call-charges using the modem).

There are techniques available to add electronic signatures to applications, which can be validated using certificates stored in the MHP. Such certificates contain only public keys with the advantage that no secret information is required in the MHP. The MHP specification defines the technical means for the application authentication. The MHP specification also includes techniques for secure communication over the interaction channel (return channel).

The wide range of possible MHP applications cannot be covered with a single MHP variant. Therefore, different MHP profiles have been defined that combine a set of functionalities. In this way, it is ensured that applications developed for a certain profile will run on MHPs of that profile. Currently, different profiles are considered for the applications areas *enhanced broadcasting*, *interactive broadcasting*, and *Internet access*. Only the first two are defined in the first release of the MHP specification.

The definition of the DVB-MHP has involved a number of leading companies from consumer electronics, the IT industry, software houses, and service providers, who contribute from different perspectives and with different objectives to the process. In this constellation, it is not always simple to find solutions, acceptable to all, which are not merely compromises, but also good technical solutions. Owing to these circumstances, the process to develop the DVB-MHP has needed more time than originally expected. However, this cooperation has perhaps provided the unique opportunity to prepare the way for the broad introduction of digital TV with open standards.

Applications

The predictable environment described by the reference model will readily allow applications to be authored and tested. Compliance with the reference model will ensure that applications execute properly, independent of the precise MHP implementation. The integrity of the look, feel, and functionalities of each application will have to be ensured, and the design of the original application provider must be preserved – irrespective of the platform implementation. It should be possible to design scalable applications that maintain compatibility across a range of receiver implementations.

DVB-TAM (technical issues Associated with MHP) defines an application as a functional implementation of an interactive service that is realized as software modules. An application can also be seen as a set of organized functions that request activation of MHP hardware and software resources.

Procedural applications, based on low-level functions and primitives, are used when very strong optimization is required at the host level (e.g., to minimize the platform footprint and maximize the use for the transmission resources). Procedural applications are generally platform-dependent and, hence, each one must be verified on the different host platforms.

Declarative applications use high-level functions and primitives. This allows us to define a platform-independent reference model that can verify whether such applications comply in terms of cross-platform compatibility and performance accuracy.

In reality, applications are neither fully declarative nor fully procedural. As an example, declarative applications can make use of procedural enhancements to improve their performance. This allows us to reduce the size of the application and to reduce its execution time by using routines written in executable code. Platform independence is ensured by relying on embedded RTEs, virtual machines, or other interactive engines. It is more difficult to achieve compliance of the compiled code routine libraries for different platforms, if they are not taken into account at the time of the platform design.

Applications are identified and signaled to indicate their availability, and an appropriate mode of access is presented to the user. Applications are launched automatically or by request. The application presentation can be nominal or down-sized (if scalable), thus maximizing the use of the available resources. Application management encompasses interruptions, failures, priority modes, and dynamic resource allocation. The application must release the system resources it has used, when quitting.

Application script and content are grouped together in application objects, which are converted into DSM-CC carousel objects. DSM-CC has been standardized by MPEG for the retrieval and transport of MPEG streams, and has been adopted by

DVB. DSM-CC UU is the interface that allows us to extract DSM-CC carousel objects from the broadcast stream, or via an interactive access to a remote server.

DSM-CC carousel objects allow one or more application objects to be carried in one module of the data carousel. Objects can be arranged in modules, in order to optimize the performance and use of memory. DSM-CC also includes compression tools to format the application objects and carousel modules, and mechanisms to ensure the secure downloading of the carousel objects.

DVB-TAM has defined an API as a set of high-level functions, data structures, and protocols that represent a standard interface for platform-independent application software. It uses object-oriented languages and it enhances the flexibility and reusability of the platform functionalities.

An application describes a set of objects according to the definition of high-level APIs. It defines the interface (via the interactive engine), among the applications, and the software and hardware resources of the host. The primitives that are embedded in the application objects are interpreted, and the resources that are requested by the corresponding declarative and procedural functions are activated. The interpreter is an executable code.

UNITEL identified the following API requirements:

- *openness* (it should be specified in such a way that it can be used in the implementation of other interfaces)
- *abstraction* (it should not expose its implementation – it should also hide aspects of the underlying software and hardware)
- *evolution* (it should be flexible and easily extendible)
- *scalability* (it should be hardware-independent in order to take advantage of future improvements in hardware and of the characteristics of different hardware implementations; the API itself can be updated or complemented by, for example, adding new libraries – e.g., procedural extensions – using download mechanisms).

According to the application format, low-level and/or high-level APIs will be used to deal, respectively, with procedural and declarative functions:

- Low-level APIs are more procedural and tend to access low-level procedural functions. The API interprets the application function or primitive, but also knows how to activate the resources.
- High-level APIs are more declarative. The higher the level of abstraction declaration (i.e., the hiding of the system implementation), the stronger is the system independence. The API interprets the application function or primitive, but does not need to know that the corresponding resources will be activated.

The specification of an open API should lead to the embedding of this hardware-independent facility within DVB receivers.

DVB-MHP has stated that API should

- support applications that are locally stored as well as those that are downloaded in either real time or nonreal time)
- preserve the *look and feel* of the application
- enable access to databases (e.g., DVB-SI)
- allow room for competition among implementers.

An open and evolutionary (modular, portable, flexible, extendible) API is vital for the implementation of platforms in an unfragmented horizontal market. This will allow different content and service providers to share different implementations of compliant platforms.

The application launch function, and the application and presentation control function, provide the facilities to run an application. The application code may already be resident in the STU or it may be obtained via a session to a remote server. After loading, the application is launched and execution is transferred to the new code.

It is the application manager's responsibility to

- check the code and data integrity
- synchronize the commands and information
- adapt the presentation graphics format to suite the platform display
- obtain and dispose of the system resources
- manage the error signaling and exceptions
- initiate and terminate any new sessions
- allow the sharing of variables and contents
- conclude in an orderly and clean fashion.

DVB has the following security requirements (although the security model itself has not yet been defined).

- The API should be accompanied by a system that incorporates a common security model for the applications and data. It should enable full compatibility among the signals transmitted by the different broadcasters and content providers.
- The security model should include a description of the procedures and entities that must be put into place to support the associated secret management issues. It should be independent of CA systems. The MHP API should give access to CA functions, if and when required.

Among the important security aspects to be addressed are (1) machine protection against abusive requests for system resources (e.g., excessive demands on memory) and (2) protection against nonauthorized access to data (e.g., private data).

288 APPLICATIONS LAYER

Multimedia Car Platform

The multimedia car platform (MCP) project is based on the results and achievements of the two predecessor projects MEMO (Multimedia Environments for Mobiles) and MOTIVATE (Mobile Television and Innovative Receivers).

The major streams of the MCP projects are

- service definition and implementation based on user cases and service requirements for the car driver and passenger
- specification for an open multimedia platform in the car integrating communication, entertainment, and navigation
- implementation of the first multimedia car terminal in the world
- specification of the architecture of a hybrid network, including service handover and interoperability among different networks.

The MEMO system architecture and protocol are used as the starting point for the work in MCP. However, while MEMO only provided specifications for DAB as a broadcast radio interface, MCP will also provide for DVB-T. The MCP network reference model is shown in Figure 4.59. MCP started work in January 2000. In a very short time it attracted major interests of the car industry. One of the most important European manufacturers has become involved in MCP. MCP will encourage convergence among telecommunication, broadcasting, and media, which at present have partly prohibitive cross-regulations.

MCP will actively promote changes in regulations in Europe, namely to allow dynamic usage of time and frequency for data in broadcasting networks.

Migration Process and Future Operational Issues

Migration is primarily the process by which a population of receivers based on proprietary software systems are all converted to a population of MHP receivers that use the common DVB-MHP system and, particularly, the API. According to

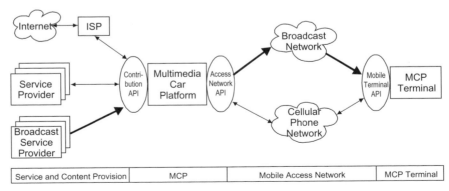

Figure 4.59 Multimedia car platform reference model.

DVB-MHP, the migration process will be initiated when service providers have begun to offer services in a format that is compatible with the MHP solution.

DVB receivers already make use of a large number of common elements, including the modulation and multiplexing schemes, MPEG-2 audio and video, the DSM-CC UU interface and protocols, the Common Interface (for conditional access and other uses), and the DVB-SI.

Nevertheless, a number of elements differ between implementations:

- the mechanisms that combine application script and code, data, and content into application objects
- compression tools
- the format of procedural functions
- libraries (e.g., procedural extensions, graphics)
- data carousels or other cyclic data delivery mechanisms
- downloading procedures and tools
- memory allocation and management (e.g., application queues and garbage collection)
- interactivity
- the formats of the variables
- security procedures.

DVB requests that, in multiprovider/multiapplication environments, the MHP solutions should be based on the separation of data. This enables different applications to use these data (if in a common data format), particularly as different applications can be implemented to accomplish the same task. It will be possible to reserve part of the data for specific applications.

At the system level, migration should be considered carefully in order to achieve the largest possible use of the DVB-TAM API. This will help to maintain receiver portability and mobility, particularly for digital terrestrial broadcasting where the limited number of programs is another reason to support solutions that favor a horizontal retail market. The consumer will probably not invest in several receivers if the added content value is limited.

Migration will not be easy. It will require substantial effort and collaboration, for example, to maintain backward compatibility with currently deployed platforms.

The wide use of a common API will raise new operational issues. There will be significant changes in the modes of operation of the service providers who, currently, target well-defined and proven receiving platforms. In order to accommodate different implementations of platforms, all using a common API, we will have to follow certain generic guidelines:

- applications will have to be downloadable and should not rely on persistent storage when the application is not active

- common libraries (procedural extensions, graphics etc.) and resident programs should be embedded in order to limit the size of the application
- application, data, and declarative interfaces should be organized in accordance with common generic schemes
- the same data carousel object format should be used, and the same mechanisms should be applied for the delivery of these objects over the streams being broadcast
- common compression schemes should be adopted
- similar start-up and closing application procedures should be used
- the amount of reinscriptable Flash memory that is available should be defined.

MHP platforms will evolve and will be able to support more complex and, hopefully, scalable applications. These may require further API extensions. Future evolution should certainly aim at increasing the degree of commonality of these system elements and procedures. This should contribute towards increasing the cost-effectiveness of the system. It should also help to increase the lifetime of the equipment.

DVB and Interactivity

Throughout its brief history, television has evolved continuously to enhance the reality of the viewing experience. The transition from analog to digital and from passive to interactive is the latest step in a long series of improvements to the system. Just as with the change from black-and-white to color, the evolution to interactivity will require systemwide changes, affecting program creation, storage, and broadcasting. The end user will not see a big quality improvement, but he will have access to a large number of programs. For example, in the United Kingdom, British Sky Broadcasting launched its interactive services – SkyDigital – in October 1998. This was a bouquet of more than 150 programs. This significant number of programs requires a new type of interface between the user and his TV set. An electric program guide (EPG) allows the user to select a specific program from a list of broadcasting programs. The digital receiver integrates a computer to manage this EPG and its interface to the end user. The same computer can be used in versatile ways for other functions. In particular, it allows access to a large number of interactive services such as home banking, home shopping, information-on-demand, interactive subscriber services, interactive advertising, educational programs, play-along games, community services, and access to online services. These services add value to the network offering and allow the network to significantly differentiate its digital TV offering from its analog TV offering.

The key element for interactive services is the software operating system managing the computer embedded in the digital decoder. The OpenTV operating system provides a powerful and coherent application program interface (API) to implement the interactive services. The OpenTV API enables content developers to develop a single version of their interactive applications. These can be used on many different networks and decoders, facilitating the development of interactive services with minimal increments. It provides advantages to the content developers, service providers, network operators, receiver manufacturers, and interactive TV consumers.

A key technology that is seldom discussed is the authoring systems – the technology that will allow content producers to make interactive television. OpenTV has developed *OpenAuthor*, a system dedicated to the creation of OpenTV applications, targeted at the nontechnical community. This allows a broadcaster to create an interactive TV service using media-rich dynamic content that is of similar or higher quality than hypertext markup language (HTML) authoring tools.

The environment is aware of the specific audio/video/graphic hardware of the interactive TV set-top box. It can combine content delivery on push broadcast links as well as pull point-to-points links. Extensive use is made of MPEG compression to deliver graphics that can be processed instantaneously by the set-top box hardware, making interactive satellite services a reality.

Interactive Television

The superior quality of digital television, the wide acceptance of standards, and the ability of digital television to provide more functions mean that the future of television is undoubtedly digital.

The storage and transmission of digital data is only really viable when it is compressed. During post-production a piece of content may be compressed and decompressed many times – great care needs to be taken to ensure that the quality does not suffer as a result of this.

However, once data is in its final form it only needs to be compressed once at a central location before being placed in the storage hierarchy ready for transmission. If the data is transmitted in its compressed digital form all the way to the receiving television then decompression has to be done in every single household. Set-top boxes capable of performing this decompression are still rather expensive.

Meanwhile, many systems are being tested where the decompression is carried out at some central point so that even though the data is stored digitally and may be transmitted part of the way in digital form, it is decompressed and converted to analog form for the final stage of the transmission. It therefore arrives at the user's television in a form that can be used immediately, thus avoiding the cost of the more expensive set-top boxes.

Systems like this normally have a limited number of decompressors and therefore the number of users that can be serviced simultaneously is limited. Such systems may be suitable for certain types of services in hotels and businesses and for limited home trials of interactive television, but full interactive services are almost certain to require full decompression capabilities in the home.

Interactive television has actually been with us for a long time. If you use a teletext service, you interactively request what data you wish to view. However, in such a system the user is not actually requesting data to be transmitted, but is selectively viewing information that is already broadcast.

A similar technique for video is called Near Video on Demand. For example, a satellite can be using 250 channels to broadcast the top 10 movies, each one being transmitted 50 times, with start times staggered by five minutes. When you select a movie, the system determines on which channel the movie is about to start and sets your television to that channel – you will see the beginning of the movie in less than five minutes. If you pause the movie, when you return the system switches channels so that you do not miss any of it.

The benefit of these systems is that they do not require a return path. In order to achieve full video on demand (where the user can select ANY movie to start NOW), or interactive services, then you need a return path and a much more sophisticated control system.

Let us look now at each of the components of a full system in much more detail.

Content. Vast amounts of material exist already in the form of old films and television shows. Control of these is of great importance, hence the vast number of mergers, alliances, joint ventures, and so on, that which are currently occurring among the media companies who own the material and the communications and computer organizations that can help them get into the home. Equally important is the new wave of interactive development, which is required to provide the infrastructure and all the new services that will become possible as the use of interactive television grows. Components of the infrastructure, such as the navigation systems, will involve sophisticated use of all the available media to provide, for example, entertaining but useable menu systems for the user. New authoring tools (such as the joint IBM/Apple ScriptX project from Kaleida) and a new breed of graphic designers and programmers are necessary to produce these systems. Similar tools and skills will be required to produce program content where the program is not just a simple film or television show, but a new concept such as interactive home shopping. As with many of the components of interactive television, versions of these facilities exist to run trials, but much more work on standards and tools is required before they will be ready for mass production for a mass market.

Compression Capabilities. Consumers and businesses are already using many techniques for compressing digital video. The key compression standard at present is the ISO/MPEG standard [26].

Storage Hierarchy and Control System. The term *video server* has been coined to describe the central system on which the data is stored. In a simple system this could be anything from a single PC up to a large mainframe capable of serving many thousands of concurrent interactive sessions. But a sophisticated system may well involve massive memories for holding critical components and highly popular videos, massive high-speed data banks for data that is currently being used, and archival storage as automated tape libraries for older or less popular material. The volumes involved are many hundreds of terabytes. The server is essentially a fast switch with direct access to all this storage; the number and type of processors involved will depend on the anticipated load and the service level required. Furthermore these systems may be split across multiple locations and copies of popular material may be held at many sites. Managing this vast *video jukebox in the sky* will require extremely sophisticated software and access techniques. Data will need to be shared and exchanged between the local sites and the center and between the various storage hierarchies. There is also the challenge of providing an isochronous data stream. This is a data stream that must be provided at a specific rate. If the rate falls too low, the recipient will not get a smooth video reception. Multiple users accessing different parts of the same video on a single disk offer a significant challenge, which is being addressed by techniques such as fragmenting or replicating each video across multiple drives. In addition to all the software needed to optimize the storage, maintenance, and access to these enormous video databases, there is also the requirement for a control system to service the user requests and interface to the storage hierarchy, the transmission system, and the subscriber management system. Simple versions of these systems exist today to handle the many trials currently being carried out. This is one of the main areas that need extensive development in order to sustain the mass market rather than a limited trial.

Transmission System. For video, the transmission must not only be at a sufficiently high bit rate, but it must also be delivered isochronously. There are a number of different methods of transmission: twisted pair (e.g., copper telephone wires), coaxial cable, fiber optic cable, satellite, microwave, and so on. Different systems have different limitations, so let us consider each in turn.

- *Satellite.* This is really a wireless broadcast medium, so although it will handle many interesting new applications in the future, there are some interesting challenges for fully interactive systems where every user requires a different signal to be sent to his television. Another key problem is the latency – the delay inherent in satellite-based systems; extensive work is being done in this area with new methods of packing the data to endeavor to reduce the delays during an interactive conversation. Satellite transmission is also (as far as domestic use is concerned) a one-way system, providing no possibility of a return path. However, there is a very simple return path available: the ordinary telephone – just dial up the center and use your touchtone telephone to communicate with the control system.
- *Twisted pair.* This is the most common existing wired system, as it is present in millions of telephone lines going to houses. It is also the most limited in its bandwidth. However, the upgrading of the backbone networks has meant that the limitations are often in only the last mile as the wire enters the house, and improvements in technology enable the necessary speeds to be transmitted into the home and provide a low-speed (64 kbps) return path suitable for most interactive applications. ADSL-2 should allow MPEG-2 transmission at up to 6 Mbps. Twisted pair is not fast enough to allow the analog transmissions that can take advantage of cheap set-top boxes, but since the future is undoubtedly digital, this is only a short-term limitation. A more likely limiting factor on the development is the much higher speed available on coax and fiber-optic cables – there may be insufficient interest in twisted pair on a worldwide basis to ensure its success.

- *Coaxial and fiber-optic cable.* A coax can provide 100 channels, each of which is effectively a 36 Mbps pipe; these can be further broken down into 12×3 Mbps MPEG-2 digital television channels, thus giving a total of 1200 channels (plus spare capacity for control and management) as opposed to one on a twisted pair. (There are many variations in this calculation, but all indicate an enormous number of channels.) Likewise, a fiber-optic cable can provide up to 1,500,000 times the capacity of twisted pair. However, the current analog cable systems are used mainly for broadcasting signals at present. To become fully interactive requires that they introduce switching systems to enable true one-to-one transmissions to take place.
- *Return path.* As we have seen, most of the systems are already capable of providing a return path, and the telephone system is always available as an alternative. The return path for most interactive television applications does not need to very fast (64 kbps is quite adequate) since it is transmitting short bursts of control information rather than the much larger media components that are traveling to the home along the main path. This is obviously not sufficient for two-way video applications, such as video conferencing, which requires full video stream capability in both directions.
- *Set-top box.* The future television will require more intelligence than the current dumb domestic television. This is currently being solved by adding a *black box* known as the set-top box. There are many varieties being used in the current trials. New ones being developed now are incorporating very significant computing power so that within a couple of years your television will probably be the most powerful PC in the house. Unfortunately, set-top-boxes with full MPEG decompression capabilities will be a little expensive initially; hopefully, standardization and mass production will get the price down. To overcome the initial inertia caused by the high entry cost, there are bound to be many promotional schemes for renting or even giving them to you just so that you can get access to all the chargeable services they provide. It is unlikely that the television and PC will converge entirely in the next few years as there are many fundamental differences in how they are presently used:
  - The television and PC are often used by different people for different purposes in different parts of the house.
  - Large screens for PC use are still much more expensive than their television counterparts; many people will be unwilling to replace their television screen with a smaller PC screen or pay the extra cost involved in a large PC screen.
  - The television will remain a mainly passive experience for many, whereas the PC is normally highly interactive.
  - The television must be easy to use; the PC, which is often task rather than entertainment oriented, can have much more complex interfaces as the user is much more motivated to learn them.

- The television needs to be controllable with a simple remote control that can be operated from a comfortable chair at a distance; the PC has many complex interface devices: keyboard, mouse, joystick and so on, which are likely to involve a good work surface and provision for wiring.
- Most importantly, people sit back in a comfortable chair to watch television – tests have shown that they will not move closer. This drastically limits the amount of text that you can show on a television screen. A PC, on the other hand, is expected to be used up close and to display vast amounts of detailed information.
- Further complications arise when there are multiple televisions in a house – will we need multiple set-top boxes, or will they be sophisticated enough to handle multiple requests and data streams simultaneously? Initial interactive television systems are likely to have a very tight integration among the set-top box, the remote control, the navigation system, and the control system. The standards do not yet exist to control this, so everyone's system will be different. Hopefully, as the technology matures, standards will be developed so that the set-top box becomes a standard commodity item that can be integrated into the television itself.
- *Remote control and navigation system.* The navigation system, which allows a user to select a service and then specify what particular aspect of that service he requires, is a critical part of a successful system. For example, the user must be able, with a simple remote control, to select that he wishes to view a film, then search by many different criteria until finding the particular film he wants. Analysis shows that to surf through 500 channels to find out what is on would take about 43 minutes – by which time the program you have decided to watch is probably finished. Surveys show that the average consumer with access to 50 channels only actually uses about seven of them. An intelligent set-top box could of course help with this, since you could program it with your preferences and it could recommend what you should watch. Without facilities like this and excellent navigation systems for interactive services, there will be a plethora of failures.
- *Subscriber management.* Many television systems are paid for by advertisers. When you can view what you want, and program your set-top box to skip all the advertisements, then how many advertisements will you watch? Who is going to pay for all these new services? Will you need to subscribe or register? Will you pay per use? What do people actually want on their television? How much are they willing to pay for it? Who do they pay? How do all the people who have contributed to what you see get their share? If the quality is not sufficient (movie watchers will compare it to VHS, games players to 16-bit consoles, etc.), if the price is not right, if the system is not useable, then it would not succeed. We need advances in scrambling and conditional access systems to ensure that people are paying for what they get. The registration, usage, billing, and administrative systems behind the future interactive television services may well develop into a major industry in their own right [135].

Many trials are in progress, and have been for some time now. Some of the current market research indicates that the demand for some of the possible services is minimal and that some of these services are still years away and very expensive. What is currently being tested will almost certainly not be what will become available on the mass market. Despite the business risks, there have been vigorous investments by the content owners and also by the technology and telecommunications industries.

4.4.6 Interactive Services in Digital Television Infrastructure

Digital television (DTV) technology appears commercially today in hybrid digital–analog systems, such as digital satellite and cable systems, and it serves as the default delivery mechanism for high-definition TV (HDTV). All digital sets, such as HDTV, can display higher resolution of digital format and do not require additional external conversion equipment. The digital format for TV broadcast transport has several key advantages over analog transport. For service operators, the key benefit is the high transport efficiency – digital compression packs five or more times as many channels in a given distribution–network bandwidth. This, in turn, increases the operator's revenue potential by delivering more content and pay-per-view events, including near video-on-demand (NVOD) movies with multiple, closely spaced start times. End users have a larger program selection with spaced start times. End users have a larger program selection with CD-quality audio and better picture quality potential, even when viewed in a hybrid setup with analog TV sets [136].

Figure 4.60 shows an example of a satellite-based DTV broadcast system. However, most of the discussion that follows is independent of the physical distribution network and applies to cable and over-the-air digital transmission as well.

Within the constraints of their service subscriptions, individual users can watch any channel by tuning into the appropriate program within the broadcast multiplex. The individual set-top box that resides on the user's premises handles the channel timing.

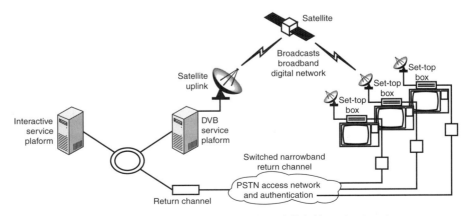

Figure 4.60 Components of a satellite-based digital broadcast system.

A digital video broadcast network distributes audio and video streams to subscribers using a transport protocol. In standard-based implementations, the MPEG-2 transport stream carries digital data over the broadcast network. The MPEG-2 transport structure may contain multiple video and audio channels, as well as private data. MPEG packet identifiers (PIDs) uniquely identify all program components. A typical digital set-top box contains a control microprocessor and memory, a network interface and tuner and demodulator for cable and satellite, a transport stream demultiplexer, and MPEG audio and video decoders. A set-top box also implements user interface components such as the remote-control input and on-screen graphical display capability for output, which are used for controls and for the electronic programming guide. Where appropriate, the conditional access keys distributed to authorized subscribers are used to decrypt the encrypted content in the set-top boxes. Similar functionality can also be packetized into PC-compatible interface cards to let PCs receive digital video, audio, and data from the broadcast network.

From the data delivery point of view, the DTV infrastructure provides a broadband digital distribution network, data transport protocols, and digital terminals (set-top decoders) on the user's premises. As such, it provides a powerful platform for delivering information and data services that not only enrich but fundamentally transform the television viewing experience. DTV systems always provide a one-way broadcast path for distributing digital video. Optionally, a return communication link can be provided to allow the *upstream* flow of data from users to the service center. The return channel often supports the impulse buying of pay-per-view and NVOD events. The return channel is usually implemented via a narrowband communication link such as a public switched telephone network (PSTN), or an integrated services digital network (ISDN). Cable systems with two-way enabled plants can implement a return channel over the cable infrastructure. Since both cable and satellite DTV systems use the same data transport mechanism and protocols – the MPEG-2 transport stream – the physical nature of the underlying distribution network is transparent to data services.

Data broadband technology enables DTV service providers to enhance their customers' television viewing experience by providing a wide range of interactive services as an incremental add-on to the DTV broadcast infrastructure. Depending on the underlying digital video broadcast system infrastructure, the following classes of interactive services are possible:

- broadcast-only interactive services
- broadcast with a batch return channel
- broadcast with an online return channel.

Table 4.10 summarizes types of broadcast-based services a DTV infrastructure with different return channel options can deploy. Note that user-level capability and interactivity is a function of the network and connectivity infrastructure. An important challenge facing DTV service designers lies in dividing data services that operate in the most common, broadcast-only environment and scale-up in

APPLICATIONS LAYER

Table 4.10 Interactive data services as a function of underlying infrastructure capability [136]

	Digital Broadcast (One to Many) Downstream			Point-to-Point	
	One-Way Plant (Satellite or Cable)			Two-Way Plant (Cable Only)	
	No Return	Polled Return (PSTN)	Real-Time Return (PSTN)	Cable Return (Real-Time)	
Network	Broadcast One-way	Broadcast Polled return	Broadcast Phone return, dial-up	Broadcast Two-way HFC	Switched Two-way FTTC, ATM
Interactivity	Local	One-way (user response)	Two-way	Two-way	Two-way
User level function	Browse (view interactively)	Browse + batch transaction	Browse + real-time transactions	Browse + real-time transactions	Full service

HFC, hybrid fiber-coax; FTTC, fiber to the curb; ATM, asynchronous transfer mode.
©1998 IEEE.

user-level functions with the increased capability of the underlying DTV infrastructure, when available.

Interactive Broadcast Data Services

Interactive broadcast data services can be broadly categorized as follows:

- *Interactive data broadcast (IDB) services.* Primarily data-only channels, with optional background audio. When a user selects such a service, a data-only screen displays. The user may use hot-spot or hyperline-style mechanisms.
- *Interactive video broadcast (IVB) services.* Combine video and data channels to provide an enhanced TV viewing experience. Service delivery and user experience of combined video-data services can be further categorized in terms of their temporal relationship.
- *Unsynchronized video and data.* Video and data may be typically related or unrelated. Common examples include a simple interactive icon overlay, a partial data overlay such as a ticket display, or a data screen with a video (broadcast) insert.
- *Synchronized video and data.* In this mode, video and data are both typically related and authorized to be synchronized at playback.

Table 4.11 shows a sampling of interactive services.

Table 4.11 Classification of sample services [136]

Service	Broadcast (No Return Channel)	Polled Return Channel	Real-Time Return
Primary user-level service capability	Browse (local interactivity)		
Broadcast Video Services			
Broadcast video	Tune	Tune	Tune
Electronic program guide	View, tune to selection	View, tune to selection	View, tune to selection
Impulse PPV, NVOD	View (order by phone)	View, order (smart card log)	View, order
Interactive Data Broadcast Services			
Information services	Browse (data carousel)	Browse and acknowledge	Browse and request
Games	Download and play	Download and play (delayed comparison of scores)	Download and play, real-time comparison of scores, multiplayer
Home shopping	Browse (order by phone)	Browse, order (delayed confirmation)	Browse, order
Interactive Video Broadcast Services			
Enhanced program information	Additional information broadcast, synchronized on current video program	Additional information broadcast, with delayed requests	Fully interactive, ability to request additional information
Interactive advertising	Browse (service information)	Browse, order coupons, brochures, goods (delayed confirmation)	Browse, order on line
Play-along programming	Play along, keeps local score	Play along, keeps local score, delayed comparison of scores	Play along, local and networked scoring in real time
Online Services			
E-mail, forums	Receipt only	Receipt with delayed reply, post	Full interactive receive and replay
Internet access	Intranet broadcast		Full service
VOD	Not supported	Not supported	Fully interactive

© 1998 IEEE.

300 APPLICATIONS LAYER

Many DTV systems use an intermittent batch return channel for billing of impulse pay-per-view (IPPV) events. In this approach, the return channel is usually implemented using a telephone line with a low-speed, dial-up modem. At specified intervals, the control billing system polls individual set-top boxes to retrieve the accumulated information. Optionally, the set-top box might dial up the central system when its local event storage almost fills up or the IPPV viewing credit nearly runs out.

Data Carousel Concept

To support low-end set-top boxes with small amounts of application memory and no additional storage capacity, the data streams are broadcast cyclically on the DTV network. In effect, the DTV network acts as a large serial disk for storage. This approach gives rise to what is known as the data carousel. The approach allows clients with a local catching capability to find the necessary data and code on the network at any time, with the worst-case access latency equal to the carousel cycle duration. Figure 4.61 shows the basic layout of a data carousel. The carousel data stream consists of an indexing and naming mechanism to locate objects within the data carousel, application code to download to the receiver when the user tunes into the carousel data channel, and application data objects that the user terminal retrieves at run time in response to the user's interactive requests.

A DTV network provides an ideal platform for distributing data broadcasts in a carousel fashion. The interactive data services (carousel) are multiplexed in the MPEG-2 transport stream. From the management and distribution points of view, data service files can be handled in exactly the same manner as any other DTV stored content (such as NVOD). This lowers system acquisition and operation costs for interactive services because the familiar service center equipment and procedures for TV signal delivery, such as the scheduler and NVOD server, also distribute data services.

For more detailed information on these topics, references [137–139] are recommended as research papers dealing with some theoretical concepts of data

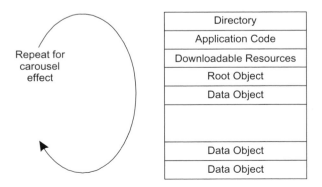

Figure 4.61 Structural layout of a data carousel.

broadcast. A detailed treatment of DTV components and infrastructure can be found in reference [140]. Reference [141] deals with user interaction issues.

DVB with Return Channel via Satellite (DVB-RCS)
Increased interactivity is a general tendency for telecommunication services today, an aspect that is also reflected in the more traditional distribution services like radio and television. Customers want to choose, sort, order, store, and manipulate what they receive on their terminals, and ideally also interact from the same terminals. The distribution network becomes an asymmetric interactive network, with a possible evolution towards fully symmetric communication. This convergence between communication and broadcasting leads to an evolution from broadcasting to multicasting or point-to-multipoint communication, where the difference lies in the possibility to offer contents/services designed for individuals or groups of people with restricted access and billing. This evolution will also have consequences for satellite communications, certainly the most broadcast-oriented medium of all.

There are several ways to design a return channel for satellite multicast services, and many believe terrestrial return channels to be the most cost-effective and practical. Commonly proposed terrestrial return channels are PSTN, ISDN, or GSM. However, there is large worldwide interest in a definition of a return channel via satellite, and there are several reasons for that. First, as mentioned above, the *normal* consumer does not want to be bothered by technical setups with interconnections between TV, PC, and telephone. A solution where all the technical equipment is concentrated within one box, and without the fear of blocked telephone lines and so on, will certainly be appealing for many people. Another reason to choose satellite services is the increased traffic in the terrestrial networks, which often results in blocking or reduced quality of service. The instantly available capacity on a satellite link can with efficient resource allocation for instance be set to 2 Mb/s. A 100 Mbyte file will need about 7 minutes for transfer over satellite, whereas the time required over a 64 kb/s terrestrial line will be about $3\frac{1}{2}$ hours. Finally, there is the advantage, both for users and operators, that both channels are on the same medium. This enables a better control with QoS and network management. The terrestrial infrastructure is often not controlled by the same operator as that for satellite, and this is certainly the case when national borders are crossed.

Owing to the recognized need for a specification of a return channel via satellite, the DVB-TM (Digital Video Broadcasting–Technical Module) created an ad hoc group in early 1999, called the DVB-RCS (DVB–Return Channel via Satellite). The DVB project itself was launched in 1993, under the auspices of the EBU (European Broadcasting Union). The motivation was to promote a common, standard, European platform for digital TV broadcasting, and the idea was supported by all players: broadcasters, operators, standardization bodies, media groups, and industry [142].

Figure 4.62 gives the general DVB return channel reference model. In this model, the interactive network is depicted as independent from the forward channel. Very often, however, the forward interaction channel, or forward signaling channel, is integrated in the forward transport stream (TS). This is also the case in the DVB-RCS specification, where the forward signaling is part of the DVB-STS. Figure 4.63

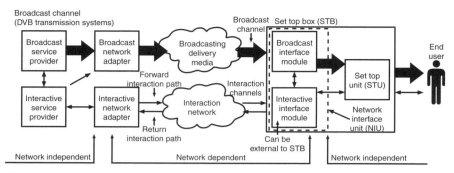

Figure 4.62 General reference model of DVB for interactive networks [143]. (©2000 DVB.)

shows a simplified diagram of a network architecture. Actually, the DVB-RCS reference model is far more complex than this; however, the wish to indicate all possible network realizations may obscure the simplicity of the concept. Usually, several RCSTs will be connected to the interactive satellite network, consisting also of the satellite, an Earth station, and a network control center (NCC). In Figure 4.63, the Earth station antenna acts both as a feeder for the forward path and as a gateway for the return path. The NCC shall handle the synchronization, give correction messages to the terminals, and allocate resources.

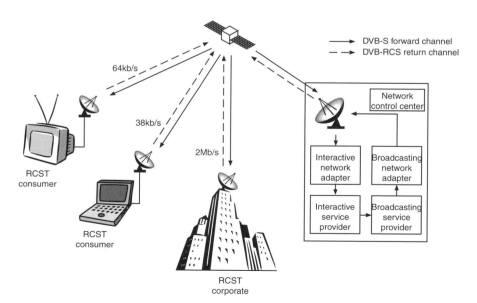

Figure 4.63 Simplified diagram network architecture for DVB-RCS systems [143]. (©2000 DVB.)

One of the main challenges for the DVB-RCS group has been to specify inexpensive terminals. The CR indicates ex-factory prices on the order of 1000 Euros, 3000 Euros, and 50,000 Euros for Consumer, Prosumer, and Corporate terminals respectively. Cost limitations will also imply EIRP limitations and possible use of suboptimal components such as nonlinear amplifiers.

As soon as interactive services are considered, the delay becomes a matter of concern with implications on several levels ranging from synchronization, log-on algorithms, to the delay perceived by the user after having made a request. This aspect highlights the need for efficient transport mechanisms, a need to be balanced against the contradicting need for flexibility. The NCC is in charge of network control, which will include several RCSTs, but may also include several satellites, feeders, gateways, and even several networks. The RCST network to manage is a multipoint-to-point structure, far more complex to administrate than the opposite, the point-to-multipoint.

The NCC is thus in charge of the control of every RCST in the network as well as the network as a whole. A terminal will log on after having received general information by *listening* to the forward link. The information given there is on the status of the network, and, most important, the forward link provides the network clock reference (NCR). When the RCST has obtained synchronization with the NCR, it will use one of the designated slots (indicated in the forward channel) for log-on request in a slotted-aloha manner. If the terminal is successful with the request, the NCC will forward various tables containing general network and terminal-specific information. The specific information is about necessary frequency, timing, and power level corrections to be performed by the terminal before transmission starts. These tables will also indicate the resources allocated for the terminal, and it is possible to ask for different services or increased capacity during transmission. The NCC has the possibility, with certain intervals, to correct the transmission parameters of the RCST, and if something goes wrong during transmission, the NCC shall also have the possibility to force the log-off of the RCST. The continuous signaling from the NCC is provided according to MPEG-2 [144, 145].

The DVB-RCS specification is also restricted to the indoor unit. The outdoor unit (the radio frequency part) will be specified by ETSI in [146]. The DVB-RCS physical layer contains specification of time slots and frames organized in superframes. The sequencing is controlled by means of the NCR, the access method is MF-TDMA (multiple frequency time division multiple access). Otherwise, the specification contains energy dispersion, two types of channel codes (concatenated Reed–Solomon/convolutional coding and turbo-codes), prefix emplacement, Nyquist filtering, and QPSK modulation, most of which is well known from the DVB-S specification.

Many satellite operators have shown their interest in the return channel via satellite technology, and concrete plans for operation of such services in the near future exist. The consumer market has been evaluated to have a potential market of some millions just in Europe, and as soon as higher volumes of terminals are produced, reasonable prices for the consumer market will be reached.

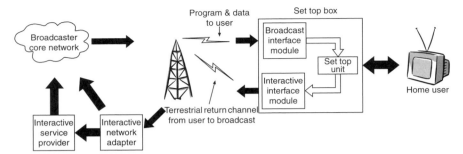

Figure 4.64   DVB-RCT network.

DVB Terrestrial Return Channel System (DVB-RCT)

The terrestrial return channel system (DVB-RCT) system is able to provide interactive services for terrestrial digital TV, using the existing infrastructure already used to broadcast DVB-T services. The terrestrial return channel system (DVB-RCT) is based on in-band (IB) downstream signaling. Accordingly, the forward information path data are embedded into the MPEG-2 TS packet, themselves carried in the DVB-T broadcast channel. The forward information path is made up of MPEG-2 TS packets having a specific PID and carrying the medium access control (MAC) management data. The return interaction path is mainly made up of ATM cells mapped onto physical bursts. ATM cells include application data messages and medium access control management data. The MAC messages control the access of the RCTTs to the shared medium [147].

The interactive system consists of a forward interaction channel (downstream), which is based upon an MPEG-2 transport stream conveyed to the user via a DVB-T compliant terrestrial broadcast network, and a return interaction channel based on a VHF/UHF transmission (upstream). A typical DVB-RCT system is illustrated in Figure 4.64.

The downstream transmission from the base station (INA) to the RCTTs (NIUs) provides synchronization and information to all RCTTs. That allows RCTTs to synchronously access the network and then to transmit upstream synchronized information to the base station. RCTTs can use the same antenna used for reception of the broadcast channel signal. The return channel signal may either be transmitted directly to a base station co-located with the broadcast transmitter site, or to a base station included in a cellular network of base stations.

To allow access by multiple users, the VHF/UHF radio frequency return channel is partitioned both in the frequency and time domains, using frequency division (FD) and time division (TD). A global synchronization signal, required for the correct operation of the upstream demodulator at the base station, is transmitted to all users via global DVB-T timing signals. Time synchronization signals are conveyed to all users through the broadcast channel, either within the MPEG2 Transport Stream or via global DVB-T timing signals. More precisely, the DVB-RCT frequency synchronization is derived from the broadcast DVB-T signal while the time synchronization results from the use of MAC management packets are conveyed through the broadcast channel.

The DVB-RCT system follows the following rules:

- Each authorized RCTT transmits one or several low bit rate modulated carriers towards the base station (INA).
- The carriers are frequency-locked and power ranged and the timing of the modulation is synchronized by the base station (INA).
- On the INA side, the upstream signal is demodulated, using an FFT process, just like the one performed in a DVB-T receiver.

For correct operation of the demodulator at the base station, the carriers modulated by each RCTT shall be synchronized both in the frequency and time domains. The frequency tolerance for any carrier produced by an RCTT, in regard to its nominal value, depends on the transmission mode used (i.e., the intercarrier spacing).

Video-On-Demand (VoD) Systems in Broadcast Environment

With the advances in digital video technology, video-on-demand (VoD) has not been a commercial success because its technology is still very expensive. Thus, many researchers have made enormous efforts on the design of a VoD system in order to reduce the operational cost. Currently, there are mainly two directions by which to provide a cost-effective VoD service. The first is to use multicasting/broadcasting techniques to share the system resources. The second is to use a proxy server to minimize the backbone network transmission cost.

For the last few years, different multicasting/broadcasting protocols have been developed to improve the efficiency of a VoD system. Staggered broadcasting [148] was the simplest broadcasting protocol proposed in the early days. Since the staggered broadcasting scheme suffered from long start-up latency, some efficient broadcasting protocols such as *skyscraper* [149], *fast data* [149] and *harmonic* [150] were then proposed. In skyscraper and fast data broadcasting, a video is divided into increasing sizes of segments that are transmitted into logical channels of the same bandwidth so as to reduce the start-up delay. In harmonic broadcasting, instead of using channels with the same capacity, the video is divided into equal-sized segments and the system broadcasts the video segments into logical channels of decreasing bandwidth. In these broadcasting schemes, customers are required to receive data from several channels simultaneously and a buffer should be installed in each receiver. A review of different data broadcasting techniques for VoD services can be found in reference [151]. The results showed that efficient broadcasting protocols can support a nearly true VoD service and the waiting time can be reduced to less than 15 seconds.

In order to support true (zero-delay) VoD service in the multicast/broadcast environment, *patching* and *hierarchical stream merging* (HSM) protocols were also developed to serve each customer immediately as the request is made. The idea of patching is that a client first downloads data on two channels simultaneously.

One is called a regular channel, which is used to serve a batch of customers. The other is called a patching channel, and provides the leading portion of video so that the customer can be served without waiting for the next regular channel. For the HSM scheme, the customers hierarchically merge with the existing multicast groups so that the bandwidth requirement can be further reduced compared with the patching protocol. The basic rule of the HSM scheme is that a customer always listens to the closest earlier stream that is still active. Once he/she has caught up with the closest stream, he/she merges into it and his/her own stream is terminated.

In addition to *data sharing*, hierarchical network architectures [152] have also been exploited to provide cost saving, as well as increased quality of service to end users in a VoD system. In reference [152] a queuing model is developed with the two-tier architecture to decide which video and how many copies have to be maintained at each distributed server. In this architecture, there are a couple of metropolitan video servers, each connected to a number of local video servers. When a customer requests a video, he/she is generally served by a local server. If the local server is too congested or the requested video is not stored in it, the customer will then be served by a metropolitan server. Instead of storing the video programs as an entity in the local servers, a server caching scheme [153] in a distributed system was proposed in which the local server precaches a portion of video to serve the local customers. It is clearly seen that there is a trade-off between the limited backbone bandwidth and the cost of the local servers. Thus, a cost model for the server caching was developed to address the problem of how to minimize the cost.

However, most of the work mentioned above mainly focused on minimizing the system resources under the homogeneous environment. Recently, some people have considered how the video data can be efficiently delivered under a heterogeneous environment such as the Internet. They proposed to use the layered video streams [154] to flexibly provide different qualities of videos by transmitting a different number of layers according to the available bandwidth between the server and customers. To serve the low-bandwidth environment, the system simply transmits the base layer of the video to provide the basic video quality, which is acceptable for playback. If the customers have broadband service, the quality can be increased by receiving more numbers of enhancement layers of video.

4.4.7 DVB and Internet

Data broadcasting will produce global changes to the way media is created, distributed, and used. *Data broadcast* is a concept whereby a combination of video, audio software programs, streaming data, or other digital/multimedia content is transmitted continuously to intelligent devices such as PCs, digital set-top boxes, and hand-held devices, where it can be manipulated. The broadcast concept means that although a return path may be available, it is not required. Content is received without being requested.

There are many services and applications that can be operated within a data broadcasting system. It can be used for background routing of large e-mails, fast delivery of content to Internet service providers, and for a new and very powerful way of videoconferencing in corporation. Another data broadcasting concept is that of a channel: a constantly updating media experience that combines rich broadcast with interaction. Such a channel need not be confined to single media; it can have streams of video in one part of a screen and interactive content of any sort in other parts, and can provide the ultimate electronic entertainment. Data broadcasting can also provide a range of other experiences, like virtual universities, powerful corporate presentations, and so on.

Telecommunication is a diverse business, but it can be divided into two very basic concepts:

- point-to-point communication
- point-to-multipoint communication.

Point-to-point communication is converging around the Internet with its TCP/IP protocols. TCP/IP has a number of advantages, but it has one problem. It is slow – principally for three reasons:

- *The acknowledgement process.* Every time you send an information package, a message needs to go back through the network confirming that it has been received by the individual.
- *Strain on server.* When one million people want the same information, the server has to send it out one million times and read all the package acknowledgements at the same time.
- *Strain on networks.* What is sent to John is not the same as what is being sent to Paul and Mary. So they cannot share the same infrastructure. Three users generally means three times the capacity requirements.

From technical point of view, data broadcaster can be positioned as the technology merges broadcasting with the Internet. Figure 4.65 shows the convergence of telecommunications infrastructures, broadcasting, and the Internet. The phenomenal drive of the Internet and electronic commerce, and the introduction of digital television (and higher bandwidth mobile phones) are processes in the market that prepare the way for ultimate convergence.

Data broadcasting will typically provide online data into a computer or another IP device at more than 100 times (and up to several thousand times) the speed of a fast Internet connection, which means that it creates a strategic inflection point for the Internet business as well as for the broadcasting business. This strategic inflection point dramatically changes some of the commercial rules, since it provides new opportunities for operating companies and to provide a powerful direct communication line between companies and consumers. The phenomenon will also provide strategic inflection for media companies and broadband network providers who are already directly involved in the distribution of bits and bytes. So from a commercial point of view, we can position data broadcast as the phenomenon that enables players in the value chains to provide fast and rich communication to their users anywhere, anytime, and anyhow.

308 APPLICATIONS LAYER

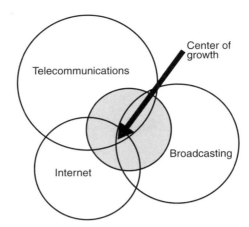

Figure 4.65 The convergence of telecommunication infrastructures, broadcasting, and the Internet.

IP Multicast

Any piece of information of any type to be sent out by any media can have three potential types of audience:

- exclusively one receiver – unicast
- a selected group of receivers – multicast
- all the potential receivers – broadcast.

Unicast is what you would call point-to-point communication. It would in physical mail be equivalent to sending a letter to your friend in Chicago. Multicast is point-to-multipoint communication. It would be like direct mail ("send the same to this specific list of addresses: xxxx"). Broadcast is also point-to-multipoint communication. It would be like when you send the same brochure to everyone in a given community ("send to all in area code xxx"). Multicast is, in other words, a restricted form of broadcast. Unicast on the other hand, is not broadcast.

There are similar choices in IP communications. Information can be sent on a point-to-point basis, but it is also possible to adopt a point-to-multipoint approach (IP addressing reserves a range of addresses to IP multicasting 224.0.0.0–239.255.255.255). Point-to-multipoint IP multicast addresses do not belong to particular terminals, but to a group of them, called an IP Multicast Group. IP multicast groups are dynamic, meaning that any user can join and leave the multicast group at any time.

With IP multicast, a single packet can have multiple destinations and is not split up until the last possible moment. This means that it can pass through several routers before it needs to be divided to reach its final destinations. Or it can bypass all routers, which means that it is never split at all. Both scenarios lead to much more efficient transmission and also ensure that packets reach multiple destinations at roughly the same time. IGMP (Internet Group Management Protocol) is a protocol used to dynamically build routing information on how to distribute multicast data. IGMP messages are transmitted in the opposite direction to the desired multicast path.

IP multicast can be used over any one-way infrastructure (such as television satellite), but there are also developments under way that make it feasible to use it over the Internet's two-way infrastructures. While the Internet as a whole is not currently multicast enabled (i.e., not all routers support IP multicast routing protocols) multicast-enabled subnet islands have been connected together through *tunnels* to form the Mbone (multicast backbone). Tunnels simply encapsulate IP multicast packets inside TCP and send point-to-point to the end of the tunnel where the multicast packet is de-encapsulated. Mbone enables distribution and access to real-time interactive multimedia (data, graphics, audio, video, etc.).

Another interesting development is the reliable multicast protocols. Examples of this are LGMP (Local Group Multicast Protocol) and RMTP (Reliable Multicast Transport Protocol).

Traditional mail is delivered to houses, flats, office buildings, and mansions. The final destination for Internet content is today almost entirely PCs. However, there will soon be a multitude of other devices that can receive digital content. The IP anywhere era started with PCs, but the next devices will be set-top boxes (as digital television is rolled out), and mobile devices (especially with the commercial rollout of 3G technology, which will enable broadband), and finally a multitude of household utilities. This provides a key challenge for the media industry, since content often will flow through numerous different topologies from the time it leaves the content provider until it reaches the final user. The complexity is growing considerably.

Audio/Video Streaming

A/V streaming in a data broadcasting environment is somewhere between what we know from those media. In terms of viewer experience it is very much radio and television. The picture can be clear, sharp, and big. It is updated at 30 frames/s, like television. Sound can be CD/DVD quality, or even surround sound. And it arrives live – there is no download time previous to viewing. However, from a technical aspect, it is not terribly different from the way it is done on the Internet. The content is not analog; it is divided into small digital packages, each of which has a header and a payload (content).

A/V streaming in a data broadcasting environment is

- best effort
- time-critical
- immediate view.

A/V streaming implies that a prerecorded file or a live event is broadcast. Unlike package delivery or cached content delivery, the file does not have to be completely received before it is played. However, contrary to cached content or package delivery, streamed content is not cached on the receiver side, but is displayed as it is received (which has a key advantage, since it might otherwise take up huge amounts of hard drive space). But this means that the user has no control over when to start the streaming. He can only choose when to view the streaming file and this may or may not be in the middle of the broadcast, as we know from radio and television. This solution is near-on-demand scheduling.

A/V streaming is extremely suitable for distribution of real-time information to a large audience. If A/V streaming is just considered as a stand-alone service, it does not offer any value over a business television or closed captioning system. However, A/V streaming combines with other delivery services and provides real added value.

Example 4.6 *Corporate announcement.* A/V streaming can be used as part of a broadcasting corporate announcement. A video viewer shows the president of a company announcing the launch of a new product, while in a separate browser window a slide show highlighting the major points of the announcement is shown. Or a chief financial officer announces quarterly results of a company to analysts and institutional investors – perhaps via a network operated by a major investment bank for its key customers. While the A/V stream is used to distribute this particular information there can be simultaneous background transmission of earning reports, financial statements, and press releases to the viewer so they can browse at their leisure. This background transmission would then use cached content delivery as the format. Synchronization of media streams and other content types can be achieved by using SMIL (Synchronized Multimedia Interaction language) W3 standard.

Example 4.7 *Daily stock market updates.* A/V streaming is also used for regular events. While the stock exchange is open, a financial institution broadcasts hourly financial status and trends, and at the same time broadcast press releases of companies mentioned during the broadcast as cached content. Analysis of these companies is delivered as a reliable broadcast and can be bought by the end user.

The typical architecture for the reception of the A/V stream is one where A/V packets pass through the subscription filter and dispatch like any other delivery type. They are subsequently passed up to a viewer that is either stand-alone or embedded in a Web browser. Using this architecture allows the imposition of subscription control on a stream, while still providing all the advantages of HTML embedding of the viewer. Embedding the viewer inside a Web browser allows for support of multimedia integration of different media types into one user experience. Architecture like the one presented allows integration with any streaming technology, as the stream player does not see a difference between a direct stream and a stream received through a subscription filter and dispatcher.

In contrast to cached content delivery and package delivery, the stream is only available when it is broadcast. This makes it important that a user is alerted when a stream starts. This alert functionality is best performed by the broadcast guide, which may alert the user before the stream starts. If the user does not acknowledge that he is ready to receive the broadcast, the filter will automatically discard the stream.

Three different scheduling and playout methods allow covering of the A/V streaming requirements.

Streaming from File. Audio and video are here encoded and stored in a file. The file is then scheduled for later broadcast and uploaded to the operator of the distribution network. At the scheduled broadcast time, the playout begins from the media file stored at the broadcaster's location. This scheduling and playout method is particularly useful when a media event has been prerecorded some time before the broadcast is scheduled. The upload does not need to happen at the same time as the booking, but needs to happen before playout. This independent playout does not require much bandwidth for the upload as long as the content is available at the broadcaster's location before the playout starts. For very large content the upload can even occur offline by sending CDs or DVDs to the broadcaster via physical mail. However, streaming from file is a preferred method for prerecorded content.

Streaming from File with Propagation. If content is prerecorded, but there is no time to upload the content to the broadcaster, the *streaming from file with propagation* method can be used. In this method, the schedule is created and the bandwidth for the streaming broadcast is booked. Because of the booking, the broadcasters assign the content provider a TCP/IP address and port to which to send the content. At the given playout time the broadcaster accepts a connection on this TCP/IP port and replicates the stream for broadcast, encapsulating it with the correct subscription information. The advantage of this method is that the content does not need to be ready in advance of the scheduled playout time and that the broadcaster does not need to have sufficient storage for the complete A/V file. The disadvantage of this method is that we do not have any buffering at the broadcaster location, making it necessary to have guaranteed bandwidth available from the content provider to the broadcaster.

Live Event Streaming. Live event streaming is, as the name says, a vehicle for broadcasting streams covering live events. The broadcast is scheduled exactly as in the file propagation method. A video camera at the location of the event captures the event, and an encoder converts the video stream into an MPEG stream. At the time of the broadcast, this stream is accepted on a TCP/IP port at the broadcaster's location (assuming that the system is IP based). The stream is then wrapped into subscription packages and replicated onto the broadcast stream. The advantage of this is that the content is not stored anywhere and is directly broadcast. As in the file streaming with propagation methods, the lack of buffering at the broadcast site requires guaranteed bandwidth to be available from the content provider to the broadcaster.

Scheduling streaming content for playout adds some unique considerations to the booking and layout system. The first one is a phenomenon called *audio and video thinning*. The issues are:

- An A/V stream is recorded with a certain frame rate and resolution resulting in unique bandwidth requirements.
- If the recorded bandwidth is larger than the scheduled playout bandwidth, or if the playout system needs to reduce bandwidth because of resource constraints or for economic reasons, we need to automatically change the bandwidth.

This process is called *thinning*. Video thinning can be achieved by reducing frame rate or changing resolution. Audio thinning can be achieved by reducing data rate or by switching from stereo to mono sound. When booking for an A/V stream event the content provider can set hints for the playout system to determine what thinning method to use if or when required.

Another issue is what is called *live feed preemption*. When scheduling for cached content delivery, the scheduling system needs any two of the following three parameters to calculate the third:

- data size
- bandwidth
- broadcast elapsed time.

This is very simple. However, A/V streams are not scheduled based on size, but based on length of playout. If we do not book bandwidth for the appropriate time, the playout might be stopped when the scheduled time is over. Content providers need to make conscious decisions if it is acceptable that a stream is preempted or alternatively book bandwidth for extra time. A situation like this happens frequently when live events run longer than expected.

A booking and scheduling system may accept last-minute booking for in-progress streams if bandwidth is still available [155].

Data Streaming

Data streaming or real-time data is a time-critical, best effort, and viewed-when-broadcast type of delivery. Bandwidth for the streaming data is scheduling with the broadcaster, but the data are always broadcast as soon as they are received from the content provider. So how does this differ from the basic concepts of A/V streaming?

A/V streaming is continuous streaming. Picture frames are sent at constant intervals. Data streaming, on the other hand, is discrete streaming. Data items do not arrive at constant time intervals, but need to be rebroadcast on a first-in first-out principle. Data streaming applications cannot afford human intervention to reformat and redistribute, as this would destroy their real-time nature.

Example 4.8 Imagine this: you are trading in the stock market. Not once or twice a year, but often, and with real money. You spend several hours a day tracing what goes on there. You want to know about how prices move, how much is traded, at which bid processing time, and so on. You want to be able to track selected prices as charts, or to link their updates into your portfolio (which you run on a spreadsheet). And you want all the information in absolutely real time, so that you know the precise market prices when you execute a trading order. So you need a system with data streaming.

Linking Broadcasting Services to the Internet

We have now looked at five different basic broadcasting formats:

- cached content
- package delivery
- A/V streaming
- streaming data
- broadcasting guide data.

All these enable a content distributor to broadcast digital content to the end user. One good question is how this can be combined with the Internet. How do the two media merge?

One of the strongest applications of data broadcasting is to use it as building blocks for three-layered hybrid (or hyper) medium, where the three layers are *pure broadcast*, *walled garden*, and *the Internet*, as shown in Figure 4.66.

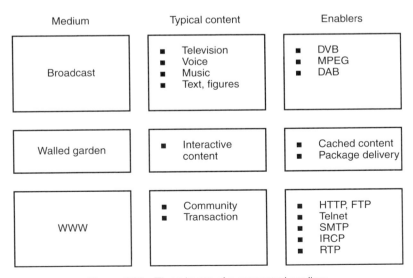

Figure 4.66 Three layers of a converged medium.

The Broadcast Layer. What you should experience at the highest level in this combined medium is something that is very close to television (which is, after all, a great commercial hit: there are at least 1.4 billions units installed). This means a layer that is, most of the time, dominated by streaming video, streaming data, and CD quality sound. However, there should be a direct connection from this layer into the two other layers.

Walled Garden Layer. The walled garden layer should contain interactive content that has been broadcast to the hard drive. The user should be able to explore this content, which may be very media-rich, and which will be instantly accessible. There should be direct connections from areas in this walled garden into the Internet.

Internet Layer. Access from the broadcast layer and the walled garden layers into the Internet should happen in two different ways: initiated by the user or initiated by the broadcast content (more about that later).

The basic navigation method in such a system will be based on the user clicking on buttons or links, entering URLs, or launching Web searchers. Broadcast content, walled garden content, and Internet content should often be available alongside one another and simultaneously within a single interface. Users should thus not bother about the technologies behind it; what matters is that you get a single, seamless medium that combines the speed, drama, and emotion of broadcast with communities, diversity, and transactions of the Internet.

Multimedia Content over Broadcast Network

The conventional way of looking at digital broadcasting is to recognize it as a different way to deliver television to homes. We know that with digital compression, four or more times as many standard definition channels can be delivered within one analog channel. There has long been discussion about placing Internet services on digital TV, but that resurrects lean back/lean forward debate about interactive and passive users. However, somewhere between the huge consumer appeal of Internet services and fast multimedia delivery must be a real business opportunity. Great as the Internet is, we know its limitations. Streaming remains a very limited market, and when the masses want the same information, the system cannot cope.

Today, multimedia, rather than just pure text, is available via IP-based services. Placing this onto DVB-T – the so-called IP over DVB or datacasting – opens up new awareness. The huge bandwidth and the ability of DVB-T to give reliable data delivery as well as fixed operations are both unique opportunities for multimedia. Return channels can be provided for mass consumption, as IP over DVB makes a lot of sense.

IP Datacast Forum (IPDC)

The IP Datacast (IPDC) Forum was launched in 2001, with the aim to promote and explore the capabilities of IP-based services over digital networks, and includes companies representing most aspects of the broadcast and multimedia industries.

The Forum members share a common vision of using DVB and DAB broadcasting networks to deliver IP-based audio, video, data, graphics, and other broadband information directly to the user, whether at home, at work, or on the move. The potential applications range from entertainment to information and business services.

More specifically, the Forum aims to

- Identify and resolve key business issues affecting datacast services.
- Support and promote members to participate in work enabling interoperability, ensuring security of content.
- Encourage and facilitate pilot projects.
- Represent and promote the interests of IP datacasting to create a harmonized regulatory environment across Europe.

The Forum, therefore, has three key areas of interest: business, technical, and regulatory. In order to promote relevant and appropriate study, the Forum is proposing the use of two reference service models:

- datacasting in cooperation with cellular telephony, and
- datacasting to devices with local storage.

By focusing on two tangible service types, it is anticipated that the study will quickly identify areas where existing applications, protocols, and standards fail to meet the anticipated service requirements.

The business area of this Forum has the objective of identifying viable business models using the principles of IP datacasting. Although it is generally understood that combining the economy and ubiquity of broadcasting with the vitality and dynamics of the Internet is a good idea, this does not necessarily mean that it represents sound business sense. To determine whether this is the case, one of the first tasks undertaken by the Forum has been to study the potential business cases for IPDC and present the results in an overview business opportunity document. One of the principal outcomes from the group was the visualization of a generic business model, shown in Figure 4.67.

Working from a foundation of simple assumptions,

- consumers want content and are willing to pay for it
- Internet technology is clearly a winning platform
- the majority of content is intended for mass consumption
- many consumers like to share experiences born by content
- broadcast platforms are very efficient for one-to-many distribution.

316 APPLICATIONS LAYER

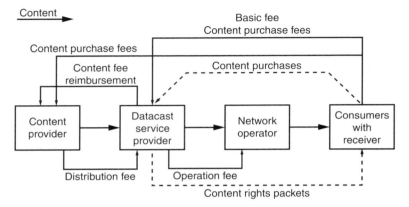

Figure 4.67 Visualization of a generic business model.

The business study group identified the business models, players, and consumption models that can form the basis of viable services. The primary conclusion from the study was that there are five business cases worthy of further study.

The first of these is a business-to-business case – a business IP delivery service within an enclosed environment. This case has the simplest business model and revenue flows. The remaining four areas are all business-to-consumer models but with differentiation between the consumption environment, which may be fixed, portable, mobile-automotive, and mobile-transportation.

The other key output from the business study was to set commercial requirements for the forthcoming technical studies. The resultant document will drive the work of the technical group in commercially viable directions.

One of the common issues noted across the potential IPDC members at their very first gathering was that of regulation. Those members that had attempted to implement trial services had found that regulatory bodies across Europe have difficulty in understanding the needs and requirements of datacasting. Although it is based upon broadcast technologies, datacasting cannot be regarded as a broadcast service in the traditional sense of sound and pictures intended for real-time linear consumption. Conversely, it is equally misplaced in the world of telecoms, the traditional home of data communications technologies, as it lacks the switched-circuit privacy and security of telephony.

To date, the prime output from the Forum's technical group has been a document summarizing the current technological landscape for IP datacast services. Like the business group, the technical group also felt that it needed a generic visualization of a datacast system on which it could focus its thoughts. This reference technical model is shown in Figure 4.68.

The key purpose of the model is to identify the domains and interfaces that constitute an overall generic system. The content domain refers to broadcasters, publishing houses, and media centers that create, retain, and distribute multimedia content.

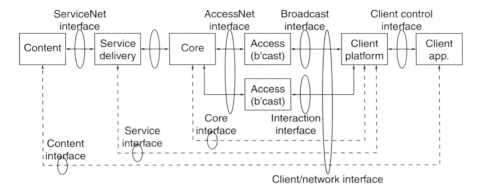

Figure 4.68 Reference technical model.

The service domain is where that content is aggregated into branded services, and the core domain then provides distribution to the base stations and transmitters that complete the last-mile delivery to the user.

Many of the key transport interfaces, denoted by solid lines, already exist, but may require enhancement to cope with specific requirements raised by the datacast model. Note that there are two parallel client access last-mile physical interfaces: broadcast (DVB/DAB) and bidirectional (GSM/3G telephony). It is anticipated that a combination of broadcast and one-to-one technologies will provide the optimum functionality for both mass and personalized consumption. The higher layer and compound interface, shown as dotted lines, largely comprise existing standards and protocols.

The IPDC Forum has set some clear objectives. Early work has resulted in valuable foundation documentation that will pave the way for future study work. With the ultimate goals of clear standards, an industry consensus upon implementation, and a harmonized regulatory approach, the Forum sees a healthy future for IP datacasting.

Datacasting

The convergence of digital content and communication networks enables a new service, datacasting. Any digital content such as graphics, video, audio – and including normally IP-delivered items such as software and Web pages – can be distributed by digital broadcasting via satellite, cable, or terrestrial transmission. Users can have various kinds of receivers residing in living rooms, cars, or even in their pockets.

The widely adopted DVB and DAB services provide excellent platforms to carry datacasting services to consumers. This one-to-many delivery means that, for popular content, the delivery costs per receiver can be a fraction of the traditional two-way Internet access. A variety of such convergence options has been studied in EU research programs.

DVB-T makes a huge transmission capacity available – from 10 Mb/s to more than 20 Mb/s per 8 MHz TV channel – that can be used for datacasting. It supports mobile operations, so massive audiences can be reached anywhere, anytime, in a cost-effective way. This offers a new platform for broadband services and contributed to universal access.

Scheduled content delivery is the basic service of datacasting, with digital multimedia content distributed over broadband networks. This resembles traditional TV broadcasting, but it also can distribute Web pages, application software, music files, a multimedia information carousel of local events, and so on.

Datacasting over DVB-T can provide a huge choice of content and services. Thus, a sophisticated electronic service guide (ESG) is needed so that users can easily find and filter contents of interest. Also, content providers require mechanisms for content protection and payment. Most of the required technology exists for these needs, and various standards organizations – specifically DVB, IEC, and IETF – are working to agree on standards for datacasting as a service platform.

Figure 4.69 outlines the generic business functions needed to implement a datacast service. The business system creates numerous business models and various role combinations for actual companies. The (generic) role of content providers is to create attractive content. Content aggregators take the financial responsibility of selling content through various channels. The datacast service operator makes content available for subscribers via a service that incorporates adequate content protection and payment solutions.

Broadcast network operators allocate distribution capacity to datacast service operators according to the ESG and manage the overall timing of the service mix and usage of capacity. They are the wholesalers of coverage and capacity to distribute broadcasting services – for example, selling capacity both for TV broadcasting and datacasting.

Telecom service operators provide two-way access for interactive services – at least to offer connections between users' terminals and Web sites hosting e-commerce, decryption key delivery, or other online services related to datacast content. However, telecom service operators can also offer user authentication, advanced and

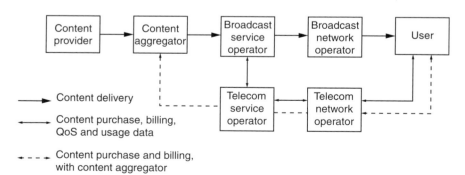

Figure 4.69 Generic business functions to implement a datacast service.

cost-efficient billing, and customer relationship management for the use of datacasting services. They can act as distributors of the datacasting service for their own customer bases.

Some services that can be offered with datacasting include

- audio streams of music according to a specific genre, such as contemporary Argentinean tango, together with related Web pages
- audio streams of local radio stations targeting various ethnic or demographic groups
- repeating carousels, for stockmarket information, local event guides, timetable and traffic information – even for maps and navigation support
- downloading of files containing music, movies, games, or other application software
- public or commercial announcements from various authorities or companies as video clips or Web pages.

Datacast's huge data capacity means that dozens of the above services can be delivered simultaneously. To manage this, it must be easy for consumers to find the content information they want to receive and view live or cache for later consumption.

The ESG should not simply list content available and times for each service provider; the services and content should be hierarchically categorized. Consumers can browse this guide to find services that match their particular interests.

An example of a hierarchical datacast service guide is presented in Figure 4.70. The ESG information can be used to set personal profiles to control content reception. For example, it can immediately show traffic announcements and download video highlights of a soccer match for later viewing while playing music from the music channel. The enjoyment of datacast services and content depends on the capabilities of the user's terminal. Three service segments exist:

- Service for PCs can exploit their ample processing power, large memory for downloads and various available applications. It is not only games, Web

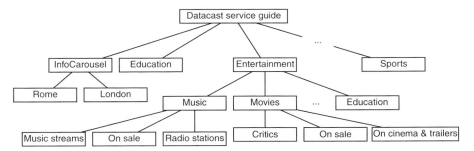

Figure 4.70 Example of hierarchical datacast service guide.

pages and PC software that are ideal for this environment, but also education-related services fit well.
- Television or entertainment-centric datacasting services offer powerful means to distribute multimedia content to homes. Instead of paper, CDs, or cassettes, datacasting delivers electronic newspapers, music, or movies. Public services would fit in this segment. The multimedia home platform (MHP) will also support datacasting services for television-centric consumption.
- Mobile datacasting will open one-way delivery of broadband multimedia anywhere, anytime offering great possibilities for innovative services and products. Entertainment and information, as well as local and community services, all will attract mobile users.

Adding a two-way, interactive communication channel allows further enhancement – for instance, enabling the purchase of rights to use protected content, Internet browsing prompted by datacast Web pages, or the automatic filtering of local traffic announcements based on phone cell position. Generally, it also supports applications requiring significant download capacity, but only a narrow upload connection. Making multimedia datacasting a viable business for content and service providers requires content protection and payment solutions. Protection can be anything from preventing unauthorized reception to full copyright protection. Payment solutions should not only support subscription fees, but also small content providers requiring payment schemes covering just a single content item. For consumers, datacasting allows more choices of access to multimedia content, while for the media industry, datacasting opens digital channels that can support safe and well-managed distribution of branded, high-value content. For governments and public services – for instance, education – datacasting provides cost-efficient public access.

To make datacast a reality requires, in many cases, changes in regulations because broadcasting is tied to radio and television services. This restricts datacast services, significantly limiting the opportunities that digital broadcast technology can offer. Also, further standardization is needed to enable a competitive market for terminals and service infrastructure and, above all, services.

Internet Broadcast Based on IP Simulcast

A technique called IP simulcast for transmitting data over the Internet from a sender simultaneously to multiple receivers is described in reference [156]. IP simulcast shows significant advantages over existing techniques, including IP unicast and IP multicast. For example, it resolves all the issues and problems involved in implementing the IP multicast. Similar to IP multicast, IP simulcast reduces the server (or sender) overhead by distributing the load to each client (receiver). Each receiver becomes a repeater, which rebroadcasts its received content to two child receivers (repeaters), forming a broadcast pyramid, as illustrated in Figure 4.71.

This method significantly reduces the needed network bandwidth for the server-sender, because the server sends just one copy of the data, which the receivers-repeaters then rebroadcast. Thus, the cost of service provision is borne by the receivers (rather than the sender), who have typically paid for the fixed, often unused, bandwidth. In this way, the IP simulcast concept provides broadcast functionality at a lower cost than multicast.

Unlike IP multicast, which requires special routers for its implementation and several other features, IP simulcast does not have any special requirements for its implementation. The number of clients in the IP simulcast pyramids grows as a binary tree. A one-tree-level pyramid has two clients, a two-level pyramid has six clients, and so on. The number of clients in the nth level equals 2^n. For example, for a broadcast system with 10 levels, the number of clients in the last level is $2^{10} = 1024$, and the total number of clients in the pyramid is $1024 + 1022 = 2046$. For the IP simulcast pyramid, consisting of 16 levels (131,000 nodes), the end-to-end delay to the nodes at the bottom of the pyramid equals 3 to 4 seconds.

The repeater-receiver performs conventional client functions, including error recovery and detection of the lost connection. Consequently, unlike IP multicast, IP simulcast provides guaranteed delivery of packets. IP multicast services make no provision for error recovery. The lost packets must be either ignored or recovered from the server at the cost of increased server bandwidth. IP simulcast uses a radically different model of digital broadcast, referred to as the *repeater-server* model. In this model, the server manages and controls the interconnection of repeaters. While the server may resemble a conventional server, the repeater contains server functions in addition to conventional client functions. In essence, each repeater not only plays the data stream back to its audience, but also transmits the data stream to two other repeaters.

As Figure 4.71 illustrates, IP simulcast builds on the new repeater-server model. The server sends the data only to two repeater-receivers, and then the packets are rebroadcast by each level of repeaters to the next level. This process builds a

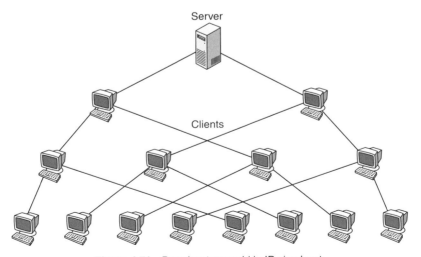

Figure 4.71 Broadcast pyramid in IP simulcast.

pyramid network that the server manages and controls. In addition, to assure a reliable data transmission, retransmission of lost packets or packets with errors is requested through secondary feeds (the dashed lines in Fig. 4.72).

The server functions include the following features.

- *Digitization of the program source.* A typical source program might include analog audio and video. These analog program sources are digitized into streams of time-varying data.
- *Synchronization of the digital source.* Streams of time-varying data may come from various sources such as digitization of analog sources, stored compressed data on disk, and digital data from animation programs, authoring programs, or other sources. Source programs can be interrupted, overlaid, or otherwise synchronized with advertising spots, and they can be scheduled throughout the day. The various sources of digital data must be synchronized and time-stamped for playback.
- *Compression of the source.* Each stream of time-varying digital data can be compressed to reduce its size and transmission time. The compression technique is a trade-off among various factors including the compression ratio, perceived quality, complexity of compression and decompression, scalability, and noise immunity.
- *Collection of the compressed source into transmission packets.* IP transmission is a packed-based protocol. The data is collected into IP packets in preparation for transmission. Compressed data can be represented by several alternative packetization schemes to adapt to different transmission lines or computers

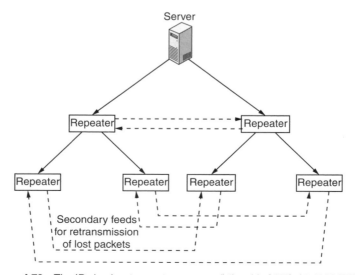

Figure 4.72 The IP simulcast repeater-server relationship [156]. (©1998 IEEE.)

of different power. Each of these packetization schemes can be used to feed alternate pyramid repeaters.
- *Transmission of compressed source transmission packets.* Two feeds are supported, each to be received and retransmitted by its destination repeater.
- *Connection of repeaters.* Each repeater sends a request to the server asking to be serviced with the transmission stream. The server responds by selecting an available repeater to serve as the requesting repeater's source. The transmission stream is then fed to the requesting repeater. The server also selects a secondary feed for the requesting repeater. Error retransmission occurs over this secondary feed.
- *Collection of statistics.* The server monitors the construction and breaking of connections. Each repeater-client has responsibility for collecting the transmitted data streams and playing them back to its audience. The repeater-client's functions include:
  - *Establishment of connections.* The repeater-client issues a connection request to the server. The server will establish an individual connection to the repeater-client.
  - *Reconnection.* The client must determine if a connection is broken and then attempt reconnection.
  - *Retransmission requests.* Requests are issued to the repeater-client's secondary feed to request retransmission of missing packets.
  - *Error recovery.* In cases where a packet cannot be recovered, the repeater-client must perform some recovery action (play silence, replay the last packet, degrade quality, and so on).
  - *Decompression of received data stream.* The received data is decompressed in anticipation of playback.
  - *Playback of data streams.* The decompressed data plays back to the repeater-client's audience.
  - *Synchronization with the server.* The playback rate must match the server's capture rate to avoid overflow or starvation of the repeater-client's buffers. The repeater-client must adapt to the small differences in playback rate that may exist.

The repeater-transmitter performs some conventional server functions, including:

- *Transmission of compressed source transmission packets.* The system supports secondary feeds, each received and retransmitted by its destination repeater.
- *Retransmission of error packets.* Each repeater-transmitter supports a secondary feed. On requests, a missed packet is retransmitted to the secondary feed's destination.

The broadcast system subdivides into fractional streams for transmission purposes. The system organizes repeaters for each fractional stream into a binary stream, propagating the fractional stream through all repeaters. Each repeater collects fractional streams into a single stream, which causes a superposition of the binary tree into a single *bush* that represents the transmission of the full system. The topology of the superposition is chosen so that the two levels of the fractional tree become separated by one-half the width of the stage in the tree. This topology ensures that no repeater starves due to the failure of a single feeding repeater.

Each repeater collects packets into a buffer, which helps compensate jitter delays. Additional buffering performs error recovery. After the jitter and error delay have played out, the received packets are broadcast to the next level. When a repeater-receiver wants to leave the broadcast, it issues a disconnection request to the server. If the queue of the repeaters waiting for connection is not empty, a repeater is selected from the queues, and the server issues fractional feed requests to the parents of the terminating repeater. On the other hand, if the repeater connection queue is empty, the oldest node on the bottom stage serves as the replacement node.

IP simulcast requires a little more bandwidth than the traditional IP multicast, but much less than IP unicast. Compared to the reliable IP multicast, IP simulcast requires about the same network bandwidth. Owing to its simplicity, easy implementation, efficiency, and inexpensive initial cost for the server and network bandwidth, IP simulcast proves ideal for many current and potential webcast-based applications. Other potential applications include distance learning, electronic software distribution (including software updates), real-time broadcasting of critical data (like stock prices), database replication and file transfer, videoconferencing, and many others.

Integrated Multimedia Mailing System

Current work that supports audio and video in e-mail user agents generally falls into one of four categories:

- *Capture and playback tools with mailing capability.* Several applications, separate from the e-mail user agent, record and/or play back audio and video messages. These applications create multimedia files that require an existing e-mail system to transport the messages as MIME attachments.
- *Streaming tools.* ImageMind's Video Express e-mail supports the composition of multimedia messages with streaming technology for delivery. This system records audio and video messages and sends them to a central message store. The receiver uses an existing text-based mailing system to receive an executable attachment, which initiates the streaming process for the playback of messages.
- *Presentation tools.* Several researchers have developed applications for the synchronized presentation of various multimedia forms. MHEGAM integrates an MHEG (Multimedia and Hypermedia experts Group) editor into the mailing systems to let a user compose and transfer multimedia messages consisting of prerecorded components. The MHEG editor allows the creation of hypermedia

documents with proper synchronization among audio, video, graphics, images, and text.
- *Integrated capture and mailing tools without playback.* NetActivity integrates recording and transport capabilities into one application. However, while it lets users record and send QuickTime format video messages, it does not receive messages. An existing e-mail application is necessary to receive the message, in addition to a QuickTime movie player to view the attachment.

`VistaMail` is an example of a fully integrated multimedia mailing system, where messages represent a fundamental shift in how people perceive e-mail messages. This system lets users easily create and send new audio and video components. The system treats these components as part of a single unified message rather than as additions to the text position.

`VistaMail` message represents a fundamental shift in how people perceive e-mail messages. Video and audio are no longer considered special additions to a text e-mail message. Instead, the combination of text, audio, and video constitutes the e-mail message. People often find it difficult to gauge the tone of a text-only e-mail message. Many people have used emoticons (smiles) to convey their emotions. Audio and video can convey true emotions accurately without emoticons. The integrated nature of `VistaMail` brings a more personal touch to e-mail messages.

In contrast to the systems above, `VistaMail` fully integrates the capability to record, play back, send, and receive audio and video messages into a single system. It also includes the standard features found in most text-based e-mail applications.

The `VistaMail` system architecture consists of user agents, message stores, and the message transfer system (MTS). We can divide the user agents on the sender and receiver sides into functional components with different responsibilities as shown in Figure 4.73.

On the sender side these components consists of the

- capture component
- integration component
- sending component.

On the receiver side, the components consists of

- display component
- assembly component
- retrieval component.

The Sender Side. The capture component receives direct input from the user. It primarily captures the media components that make up the e-mail message in real time. For example, if a message contains audio and video, the capture component must communicate directly with the microphone and digital camera, synchronizing

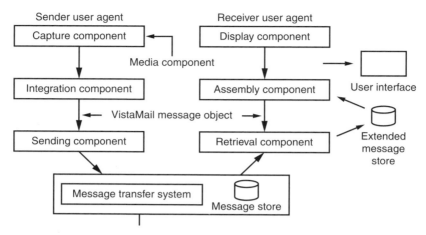

Figure 4.73 The `VistaMail` system architecture [157]. (©1998 IEEE.)

the recording of the audio and video streams. Once the capture component has captured all the message components, it passes these components down to the integration component. The integration component integrates the separate message components and adds important header information to create a single `VistaMail` message object. This message object is finally passed down to the sending components, which establish communication with the standard MTS.

The Receiver Side. After the MTS deposits the `VistaMail` message object into the message store on the receiver side, the retrieval component extracts the object from the message store. The retrieval component then disassembles the message object back into the header information, text component, and multimedia components. The system then places multimedia components into an extended message store (EMS), separate from the standard message store of the MTS, which still contains the text components.

The header information, which includes references to the message's specific audio and video components in the EMS, is then passed to the assembly component. Using these header references to obtain the multimedia components in the EMS, the assembly component – when directed by the user – assembles the separate message components and passes them to the display component, which then plays the message in a synchronized session.

Once the `VistaMail` message object arrives at the receiver side, the retrieval component disassembles it into its different components and places them in the standard and extended messages stores as shown in Figure 4.74. When the sender finishes the message, the transfer system sends it to the receiver. Once stored in the message store on the receiver side, the message may be played back to the receiver in a synchronized session. This is represented in Figure 4.75 which illustrates the complete process.

There are two different implementations of the message transfer system (MTS). The first implementation (Fig. 4.76a) sends the video and audio components using a new protocol. While the text component still transmits through SMTP, the system sends the video and audio components to a special media server located in the same network as the recipient's mail server. Rather than having the video and audio components extracted from the standard message store by the assembly component, this server deposits these multimedia components directly into the EMS. In addition to resolving the capacity problems of the message stores, this means VistaMail no longer needs to perform the potentially costly task of decoding the multimedia components. The server implementation simply opens up a connection and transfers the separate files. A more sophisticated server may make specific accommodations for the media types, such as laying out the audio and video sequentially on disk. While this approach improves performance, the use of a nonstandard protocol can hinder the acceptance of the system.

The second implementation (Fig. 4.76b) sends the VistaMail multimedia components as MIME attachments via SMTP as a text e-mail message. This implementation lets VistaMail read and send messages to other e-mail systems. The drawback to this implementation, however, is that the accumulated VistaMail messages on the receiver side may exceed the capacity or specified limits of the message stores. Because the video and audio components of a VistaMail message may be quite large, several VistaMail messages may rapidly fill or exceed a disk quota for the message stores before the assembly component can separate these multimedia components. In a multiuser setting, for example, this setup may be problematic for a system administrator in charge of allocating disk space for users.

Figure 4.76 shows alternative message transfer systems with (a) the use of a separate media server for delivering media components directly to the EMS and (b)

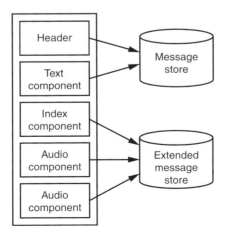

Figure 4.74 Receiver side of the VistaMail system [154]. (©1998 IEEE.)

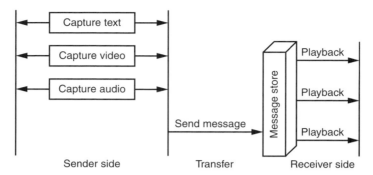

Figure 4.75 Message on the receiver side [157]. (©1998 IEEE.)

using standard protocols (SMTP and MIME) to deliver all the components of the `VistaMail` message object. The dotted arrows from the message store and EMS represent the extraction process that the assembly component performs. Standard protocols in the actual implementation are used. MTA is the message transfer agent.

To conclude, the acceptance of a multimedia e-mail system depends on

- perceptual quality of received multimedia e-mail messages (control of resources to achieve bounded jitter and skews)
- network technology progress to increase bandwidth to the home environment
- disk technology progress to allow storage of large multimedia messages
- privacy when composing and receiving audio-video messages
- easy use of the integrated system.

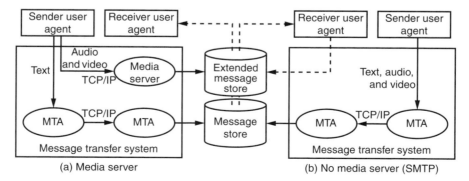

Figure 4.76 Message transfer systems for `VistaMail` [157]. (©1998 IEEE.)

The increase in the number of systems becoming available to facilitate composition and playback of audio and video messages illustrates the increased interest in multimedia e-mail.

4.5 MOBILE SERVICES AND APPLICATIONS

The world of mobile communication is a phenomenon. The major advances in wireless technology have elevated mobile communications to a critical position in the modern organization. Consequently, research and innovation in mobile communications have acquired new levels of significance in academia, industry, and other communities. There is a growing demand to provide a thorough understanding of mobile communications, along with its foundation, to meet industrial and academic need.

As wireless technologies evolve, the coming mobile revolution will cause dramatic and fundamental changes to the world. This revolution has already begun and is gaining momentum. The revolution will impact numerous facets of our daily lives and the way business is conducted. It will provide important data in real time to assists decision-makers, exert great influence on communications between businesses and their customers, and transform the way we live our lives.

New services in 3G will require new business models. Interdependent market players need to identify their roles before presenting user-friendly services with underlying complex technology and billing principles.

Advances in wireless technology and mobile devices give rise to a new kind of e-commerce – mobile commerce. Mobile commerce transactions are conducted via mobile devices using wireless telecommunication networks and other wired e-commerce technologies. Mobile commerce (also increasingly known as m-commerce or mobile e-commerce) enables wireless information exchanges and business transactions. Mobile commerce means different things to different people. To customers, it represents convenience, while merchants associate it with a huge earning potential, and service providers view it as a large unexplored market [158].

The rapid growth of the cell phone market makes people very excited about the future of mobile commerce. Mobile commerce was usually viewed as a mobile extension of traditional Internet-based e-commerce [159–162].

M-commerce should not be understood simply as a new distribution channel, a mobile Internet, or a substitute for PCs. Internet-based e-commerce applications that are successful on PCs may not necessarily meet similar accomplishments in m-commerce [163]. M-commerce should emphasize the unique features that Internet-based e-commerce may not have, such as communications, personal touch, location-related and time-critical services. To unleash the value of m-commerce, a radical shift in thinking is required [164].

This perception suggests that we go beyond the existing Internet-based e-commerce business model and search for models more appropriate for m-commerce.

4.5.1 Mobile Services

Wireless and mobile networking have presented an entirely new way for companies to better serve their customers. Mobile services will enable users to make purchases, request services, access news and information, and pay bills, using mobile communication devices such as PDAs, laptops, and cellular phones. Another mobile service area that has much potential is the mobile government. Mobile devices can enhance the e-government concept by allowing constituents to access government services through mobile devices.

The essence of mobile services revolves around the idea of reaching customers and partners regardless of their location. It is about delivering the right information to the right place at the right time. This flexibility of mobile services is made possible by the convergence of the Internet and wireless technologies. Some of the key drivers for mobile services are the following:

- *Mobility.* Mobility is the primary advantage of mobile services. Users can get any information they want, whenever they want, regardless of their location, through Internet-enabled mobile devices. Mobile services fulfill the need for real-time information and for communication anytime. Through their Internet-enabled mobile devices, mobile users may be engaged in activities, such as meeting people or traveling, while conducting transactions or receiving information.
- *Reachability.* Through mobile devices, business/government entities are able to reach customers/constituents anywhere, anytime. With a mobile terminal, a user can be in touch with and accessible to other people anywhere anytime. The user can also limit his/her *reachability* to particular persons or at particular times.
- *Localization.* The knowledge of a user's physical location at a particular moment also adds significant value to mobile services. With location information available, many location-based services can be provided. For example, with the knowledge of a user's location, the mobile application can alert him/her quickly when his or her friend or colleague is nearby. It can also help the user locate the nearest restaurant or ATM.
- *Personalization.* Although enormous amounts of information, services, and applications are currently available on the Internet, not all information is relevant to all users. Mobile services can be personalized to filter information and to provide services in ways appropriate to a tailored user.

The growth of mobile services will depend on the development and deployment of enabling technologies. These technologies include, but are not limited to, network technologies, service technologies, mobile middlewares, mobile commerce terminals, mobile location technologies, mobile personalization technologies, and content delivery and format. Presented below are some of the major technologies that are making mobile services a reality.

4.5 MOBILE SERVICES AND APPLICATIONS

- GSM – operating in the 900 MHz and the 1800 MHz (1900 MHz in the United States) frequency band, GSM (global system for mobile communication) is the prevailing mobile standard in Europe and most of the Asia-Pacific region. It also serves as the basis for other network technologies such as HSCSD (high-speed circuit-switched data) and GPRS (general packet radio service). The wide adoption of the GSM standard makes it economically feasible to develop innovative mobile applications and services.
- SMA – *short message service* (SMS) enables the sending and receiving of text messages to and from mobile phones. Currently, up to 160 alphanumeric characters can be exchanged in each SMS message. Widely used in Europe, SMS messages are mainly voicemail notification and simple person-to-person messaging. It also provides mobile information services, such as news, stock quotes, sports, and weather. SMS chat is the latest feature and is growing in popularity.
- WAP – *wireless application protocol* (WAP) is an open and global standard for mobile solutions, designed specifically to deliver Web information to mobile terminals. As an end-to-end application protocol, it attempts to provide solutions to the challenges in developing mobile applications, such as connecting mobile terminals to the Internet and making mobile terminals become capable of communicating with other devices over a wireless network. It also permits the design of interactive and real-time mobile services.
- UMTS – *universal mobile telecommunications system* (UMTS), the so-called 3G technology, aims to offer higher bandwidth, and packet-based transmission of text, voice, video, and multimedia, which are needed to support data-intensive applications. Once UMTS is fully implemented, computer and phone users can be constantly connected to the Internet and have access to a consistent set of services worldwide. Integrating the functions of a whole range of different equipment, a 3G mobile phone can be used as a phone, a computer, a television, a paper, a video conferencing center, a newspaper, a diary, and a credit card.
- Fourth-generation technologies – although 3G technologies are just emerging, research has commenced on 4G technologies. These research initiatives encompass a variety of radio interfaces and even an entirely new wireless access infrastructure. Better modulation methods and smart antenna technology are two of the main research areas that will enable fourth-generation wireless systems to outperform third-generation wireless networks.
- Bluetooth – named after a tenth-century Danish king who conquered Scandinavia, Bluetooth is a low-power radio technology for communication and data exchange. Using a single chip with built-in radio transmission circuitry, Bluetooth is an inexpensive short-range wireless standard supporting local area networks (LANs). It was developed to replace the cables and infrared links within a ten-meter diameter. Bluetooth can be used to link electronic devices, such as PCs, printers, mobile devices, and PDA, to wireless data networks.

332 APPLICATIONS LAYER

- GPS – *global positioning system* (GPS) is a system of satellites orbiting the Earth. Because the satellites are continuously broadcasting their own position and direction, GPS receivers can calculate the exact geographic location with great accuracy. Originally developed in the United States for military use, GPS is now also used for civilian purpose. For example, GPS is used in car navigation systems.
- XML – *extensible markup language* (XML) is a meta-language, designed to communicate the meaning of data through self-describing mechanisms. It tags data and puts content into context, thereby enabling content providers to encode semantics into their documents. For XML-compliant information systems, data can be exchanged directly even between organizations with different operating systems and data models, as long as the organizations agree on the meaning of the data that is exchanged. XML is heavily used in mobile applications development.

4.5.2 Types of Mobile Services

Mobility creates many new opportunities for the business world to embrace. Meanwhile, the current applications can be streamlined by integrating mobility. Adding mobility to services will create anytime/anywhere information access, facilitate information sharing, and provide location-based and presence-based services. Based on the mobile infrastructure and mobile technologies, mobile services can be classified into various categories (Fig. 4.77).

- mobile B2B
- mobile B2C
- mobile C2C
- mobile government.

Mobile B2B includes mobile transactions within and among organizations. Mobility integrated into organizational operations has the potential to make information flow more efficient, to coordinate operations within the extended enterprise,

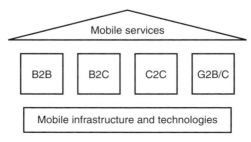

Figure 4.77 Mobile service applications.

and thus improve supply chain management. For example, mobile e-mail and the Internet enable instant data information exchange within organizations and among business partners, and facilitate information sharing and interactivity within the supply chain network to help business cope with complex business environments. With a laptop, a GSM modem and a connected mobile phone, corporate users can dial into the corporate network's mail server and stay in touch with their organizations while on the move. Together with mobile e-mail, the mobile B2B services enable sales professionals and customer care forces to track order status, access marketing information, check customer feedback, report problems, consult with technicians, and identify locations anytime and anywhere. In addition, field technicians can use mobile devices to communicate logistics, machine status, customer information, and order and billing information.

Mobile services can also be used to manage logistics and work flow, and streamline inventory and distribution control. For example, Bluetooth devices are ideal for inventory control. Instead of users making manual connection, Bluetooth devices connect and communicate spontaneously. Instantly linked to the system and easy to handle, Bluetooth devices provide fast data sharing and quick check, and enable the supply chain members to share stock data.

Mobile B2C transactions are retailing transactions with individual shoppers, known as customers, or consumers. Mobile services can provide customers automated, unassisted operations directly from mobile terminals. For example, United Parcel Services (UPS) has begun to use wireless devices to track shipments, and its customers can determine the estimated delivery time with a PDA or mobile phone. Mobility also promises business units, such as financial institutions, powerful channels to reach out to their customers through always-on, always-with-you mobile devices. Mobile commerce can increase customer satisfaction by pushing information to mobile users, and putting them in a better-informed position. Listed below are some applications in mobile B2C.

- *Mobile shopping* – mobility extends the ability of customers to make transactions across time and location, and thus provides them with personalized and immediate opportunities to purchase. Customers can make decisions on the spot and do not have to go to alternate sources. Mobile shopping is now available in many wireless Web sites.
- *Mobile financial services* – mobile financial services open a new service channel for financial institutions. They are also a key commercial driver for the mobile commercial market in Europe and beyond. Financial services, such as mobile banking and mobile brokering, are available at various wireless Internet Web sites.
- *Mobile ticketing* – ticket master online-citysearch (TMCS) offers a service that allows users to purchase tickets using mobile communication devices. The service is reported to give cellular phone users access to tickets, local news, and information. The advantage of mobile ticketing is obvious: customers do not have to go in person to a ticket booth, or to call an agency or an outlet. It is

clearly more convenient to select or book tickets for movies, theatres, operas, and concerts directly from a mobile device.
- *Mobile news, sports, and other information* – the range of information that can be provided wirelessly is unlimited. Along with news and sports information, many other types of information are now available from wireless devices.
- *Mobile advertising* – since the location of a mobile device can be determined precisely, the stores around the mobile device user can transmit user-specific information, such as current sales or specials, and alert the user to similar upcoming events. Wireless coupons enable advertisers to deliver geographically targeted and time-sensitive messages to willing consumers directly, with promotional offers virtually anytime and anywhere. Mobile advertising will increase marketing efficiency and allow direct offers suited to user profiles or stated user preferences.
- *Mobile entertainment* – mobile entertainment may take on many forms, such as mobile gaming, mobile music, and mobile video. It presents mobile users new ways to entertain themselves on the move [165].

In the mobile C2C category, consumers sell directly to other consumers via mobile devices. In the near future, individuals will be able to use their mobile handsets to seek potential buyers for their properties, autos, and so on, interact and negotiate with them, and conclude transactions. For example, buyers and sellers can interact outside a football stadium regarding tickets for a game. Individuals can also advertise personal services and sell knowledge and expertise on the mobile Internet.

Mobile government is the extension of e-government as the Internet goes wireless. It enables people to access government departments and organizations through Web sites on the Internet, regardless of physical location or time of the day, via their mobile handsets. It has the potential to make viable and highly productive connections between government and constituents (G2C), government and business (G2B), government and employee (G2E), and government and government (G2G). In the near future it is expected that mobile constituents will be able to pay taxes, order birth certificates, renew vehicle licenses, reserve campsites, and so forth, via their mobile devices. In a mobile government age, constituents can expect more convenience, responsiveness, and personalized services from their governments.

4.5.3 Business Models

Content providers can be classified in numerous ways, and a simple content value chain includes content creation, content aggregation, to content distribution. This chain is being broken down into separate functions as new value chains are being established [166].

The *content producer* creates content. This content can either be professional quality published content that requires equipment and experienced staff, or material recorded by a user. A *content publisher* obtains the right to use and distribute

4.5 MOBILE SERVICES AND APPLICATIONS

content. Content rights may be restricted to particular geographic regions or forms of delivery. A common type of player is the *content aggregator*, acquiring content from multiple sources and packing it for service providers or end users. The aggregator may also add a brand of its own to the package. TV broadcasters, movie studios, and music producers have traditionally dominated the content publishing area.

To generalize business models, the content producers, like production companies, sell exclusive rights to content aggregators. These rights often have geographical and time limits. Production companies hereby collect revenues directly from content aggregators. This means also that content aggregators, often television companies, do not have rights to the same content. The aggregators, however, collect revenues from advertising or licenses. The interface with content distributors handles rights differently, as the aggregators want to reach as large a public as possible. As a consequence, distributors, often cable television companies, can distribute content from all aggregators, and collect revenues through user subscriptions.

Figure 4.78 illustrates the established media provider value chain, from content producer to content distributor. This reflects principles and is a simplified illustration of reality, where no obvious lines can be drawn between various market players. Some market players can cover the entire value chain. It is also shown how content aggregates towards the end of the value chain. This implies that rights to content can get more expensive in this direction, as aggregators collect revenues from advertisers or licenses. However, as financial capital would also be added from advertisers, the content does not have to be more costly to the final customer.

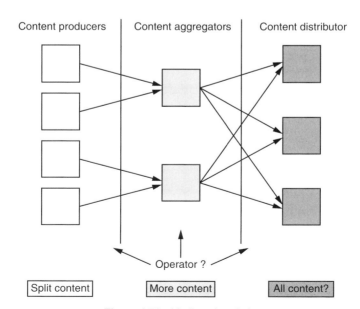

Figure 4.78 Media value chain.

A business model is the method by which a firm builds and uses its resources to offer its customers a better value chain than its competitors. There are many different definitions of business models [167]. To achieve m-commerce success, people are looking for innovative *killer applications* or extensions of existing e-commerce applications in a mobile environment. However, it is not the application but the business model behind the application that really determines the success. Key components of m-commerce business models are presented in Table 4.12.

Scope is essentially about the market segments, geographic areas, or demographics to which the proposed value should be offered. In the mobile environment, the large population of cellphone users are the really big users of m-commerce. A single applications would not work for everybody and there will be a whole set of niche applications that are relevant to each target audience. Here we segment the market into three categories: mobility, countries/cultures, and demographics [168, 169].

For people sitting in an office or at home, PCs connected to the Internet will serve them much better than mobile devices. The targeted users of m-commerce must be those who are on the move, going out, traveling, service delivering, and so on. Examples are police, fire fighters, traffic enforcement, health inspectors, building inspectors, transportation inspectors, fire inspectors, management–employees, and so on. This market segment can be very large in comparison with the *stationary* customers.

Countries can be segmented by means of product diffusion patterns and response elasticities. Beyond national boundaries, a special regional culture may play a crucial role in product adoption. While Internet-based e-commerce naturally has a global focus, m-commerce has to target a specific country or

Table 4.12 Key components of m-commerce business models

Scope of Market	Value Proposition	Revenue Sources	Roles in Value Chain
Mobility	Ubiquitous communication	Basic communication charges	Mobile communication providers
Countries/ cultures	Emergency and time-critical information services	Value-added charges	Equipment and device suppliers
Demographic	Location-sensitive services	Sales of mobile communication equipment and devices	Portal providers
	Pocket e-wallet	Sales of mobile applications	Application developers
	Portable entertainment	Advertising	Value-added service providers
	Improving productivity of mobile workforce		

region. For instance, in Japan and Europe, customers pay for every landline call, whether local or long distance. This not only means that mobile phones are more convenient, but that they are also only marginally more expensive [170, 171].

Demographics include measurable characteristics of people such as age, gender, income, education, and occupation. Demographics have a great influence on the success of specific m-commerce applications. M-commerce players must explore the needs of each demographic group and decide which type of service fits into which group. Generally, young people like *hanging around* and keeping in touch with their friends and family; traveling business people need to connect to their home office; elderly people have their concerns about immediate medical assistance [170].

A critical part of business model analysis is the determination of the sources of revenues. Here we identify several revenue sources for m-commerce: basic communication charges, value-added service charges, sales of mobile communication equipment and devices, fees of mobile application development, and advertising [167].

A subscription is required for using mobile communications or other services such as Internet access for a specific period of time (usually a contract of one or two years). Large numbers of subscriptions represent one of the major revenues for wireless operators. This stream of revenues is maintained by various discount packages/plans provided by carriers, particularly targeted at different consumer groups according to their patterns/preferences for mobile communications.

Users may subscribe to specific services such as headlines news. Services can also be charged for, based on each use, such as travel assistance services, financial transactions, bill payments, or home burglary notification service. Since mobile communications are paid for by customers, any value-added service will increase network traffic and therefore generate extra revenues for network operators and products/service providers.

Owing to the rapid expansion of and high investment in building mobile communication infrastructure, selling mobile communication equipment and handheld devices is a big source of revenue for equipment providers such as Ericsson and Nokia. Additionally, some m-commerce applications may require the use of specific handheld devices. Sales of handheld devices therefore are the major revenue stream for m-commerce players.

Unlike Internet-based applications (usually Web-based), m-commerce applications are most often network-specific and even device-dependent. Owing to the capacity limitations of mobile handhelds and incompatible wireless network protocols, it usually takes many steps (i.e., gateways) to connect different systems in m-commerce applications. The development of m-commerce applications therefore is more complex and expensive. This fact makes the fees for development and deployment of mobile applications one major revenue source for m-commerce applications developers [172].

Advertising is a major source of revenue for Internet-based e-commerce, but a minor source for m-commerce, due to the high cost of delivery through mobile communications. For m-commerce, advertising is mainly carried out in two ways: subscribers are paid to listen to advertisements, or subscribers are attracted by adding extra value, particularly through targeting advertisements to their on-the-spot needs. Carriers also believe that this will boost mobile usage and increase their market share. In Western Europe, up to 97 percent of mobile subscribers who have tried a sponsored service have accepted the offer of *paid advertisement* [173, 174].

A related issue is the billing system for revenue collection. Unlike individual billing by different e-commerce services, revenue collection in mobile commerce requires a convergent billing system, which must be capable of covering the customer bases as it expands, and handling complex and innovative new products and raising structures. The precise and fair allocation of revenues between value chain parties of m-commerce also depends on a scalable and flexible billing system such as airtime-based, user patterns-based, specific application-based, location-based, and so on. It is not surprising that wireless network carriers have been eager to upgrade or replace their existing and often inadequate billing infrastructure over the last few years.

In brief, when we identify revenues for m-commerce, we should shift our thinking from traditional Internet-based e-commerce in the following ways:

- from a resource abundant, free service dominating *socialist* system to a scarcity/ownership-based *capitalist* system
- from flat-rate Internet access charges to traffic-based communication charges
- from advertising-centric to subscription-centric
- from transaction cost reduction to improved internal business productivity
- from simple individual billing system to complex multibilling systems.

4.5.4 Value Chain Structure and Coordination

The goal of delivering seamless m-commerce applications can only be achieved by a group of companies that jointly provide related products/services. This group is called an m-commerce value chain, in which all parties concerned interrelate with one another to create value for mobile users that use the service, and share the revenue.

Consumer online services demand that diverse needs must be combined to create and deliver value. Cooperation, collaboration, and consolidation have been the key watchwords, as arrangements are struck between computer hardware and software, entertainment, creative content, news distribution, and financial services have seized opportunities by aligning competencies and assets via mergers and acquisitions, resulting in a major consolidation of information-based industries [175].

4.5 MOBILE SERVICES AND APPLICATIONS

There are three coordinated value chains:

- *services* – where various categories of content providers play an important role
- *infrastructure* – which is dominated by the makers and the operators of mobile telecommunications
- *devices* – with cell phones, where makers and retailers of all phones control the markets.

A basic model on value chains adapted to m-commerce was defined in reference [176]. Here, m-commerce is defined as any transaction with a monetary value, either direct or indirect, that is conducted over a wireless telecommunication network. The model consists of six core processes in two main areas:

- content
- infrastructure and services.

In the content area, the three core processes are content creation, content packing, and market making.

- *Content creation* – the focus in this value space is on creating digital material such as audio, video, and textual information. For example, digital news feeds and real-time stock information are available from *Reuters*.
- *Content packing* – in this box, we are likely to see digitizing, formatting, editing, customizing, and the use of software to combine and package content. This can include, for example, packing *Reuters* stock information, the online version of the business and financial newspaper.
- *Market making* – marketing and selling content is the primary role of mobile portals. This includes program development, service delivery, and customer care. A mobile Web portal provides a one-stop shop for a large number of services.

In the infrastructure and service area, the three core processes are mobile transport, mobile services and delivery support, and mobile interface and applications.

- *Mobile transport* – this is the basic network involved in communications, including transportation, transmission, and switching for voice and data.
- *Mobile services and delivery support* – this involves, for example, the infrastructure in connecting to the Internet, security, the server platform, and payment systems. Standards such as i-mode are key building blocks towards enabling the delivery of Internet services via mobile handsets.
- *Mobile interface and applications* – this process centers on integrating the infrastructure and systems with user hardware, software, and communications. This includes the user interface, navigation, and application development, as well as the authoring tools.

Business models are often more important than technical solutions! The end user perspective and the commercial perspective are more important than the technical possibilities. The reason is argued throughout the thesis as follows: mobile services will not succeed because they are technically doable; they will only be successful if they add value to the user. However, the technical architecture of a mobile service has great impact on the business model.

The actual business model for mobile services and the revenue split depend on what the content provider has to offer and what strategy the operator has.

4.5.5 Challenges and Obstacles in Adopting Mobile Applications

Current technical limitations of mobile devices and wireless communications, coupled with business concerns, complicate the development of mobile applications. In this section, we elaborate on the challenges and obstacles in adopting mobile applications: strategy changes, investment risk, mobile devices limitations, incompatible networks, competing Web languages, security concerns, and trust.

- *Strategy changes* – to stay competitive and realize genuine productivity benefits from mobile services, many organizational processes need to be redesigned to fully integrate mobile services. They will have to make fundamental changes in organizational behavior, develop new business models, and eliminate the inefficiencies of the old organizational structures. The process of rethinking and redesigning is a demanding task. For example, implementing mobile government is more than developing a Web site on the mobile Internet. Actually, the problem involves rethinking and reengineering the way government conducts business. Unlike traditional government, which is organized around agencies and bureaucracies, mobile government in the information age will be deployed around the needs of its citizens. This entails rethinking how government should be organized from the perspective of its citizens, and reengineering how government can better perform its functions according to the needs of its citizens, rather than the requirements of bureaucracies.
- *Investment risk* – a major problem faced by mobile services is the huge investment required for their implementation and operation. Engineering massive organizational and system changes to strategically reposition an organization is complicated as well as expensive. For example, a company will have to build a mobile infrastructure and invest money in mobile devices. But implementing the mobile technology itself does not guarantee that an implementing organization will reap any benefits from mobile services. Expertise in a field other than technology is also a prerequisite for successful applications of mobile services. How can organizations obtain a payoff from their investment in wireless technology? Understanding the costs and benefits of mobile commerce is difficult, particularly when the technology is changing and evolving at a rapid pace.

- *Mobile devices limitations* – current wireless devices include phones, handheld or palm-sized computers, laptops, and vehicle-mounted interfaces. While mobile terminals demonstrate a greater extent of mobility and flexibility, they are inferior, in several aspects, to personal computers or laptops. The screen is small and the display resolution is low. The small and multifunction keypad complicates user input. Because of the need to be physically small and light, these input and output mechanisms impede the development of user-friendly interfaces and graphical applications for mobile devices. Mobile handsets are also limited in computational power, memory and disk capacity, battery life, and surfability. These drawbacks in mobile devices do not support complex applications and transactions, and consequently limit usage of mobile services in complicated business environments. However, mobile communication technologies and devices are advancing in leaps and bounds, and many of the existing limitations will disappear.
- *Incompatible networks* – multiple, complex, and competing protocols exist in today's cellular network standards. As previously mentioned, GSM is a single standard used by the network operators in Europe and the Pacific Asian region. But TDMA (time-division multiple access) and CDMA (code division multiple access) are widely used in the United States. These different standards have resulted in the global incompatibility of cellular handsets. The network incompatibility poses problems for companies in communicating with their customers.
- *Competing Web languages* – in addition to incompatible networks, there are a number of competing Web languages. Newer mobile phones will incorporate WAP and its WML. The fact that incompatible standards are utilized in mobile devices today makes the process of creating successful m-commerce applications even more difficult. The need for standardization of Web languages appears extremely urgent. The mobile communications within organizations and the interactions among organizations and their customers will not see significant improvements until the issue of competing Web languages is addressed.
- *Security concerns* – compared to its wired counterpart, wireless communications are more vulnerable. Although most wireless data networks today provide a reasonable level of encryption and security, the technology does not ensure transmission security in the network infrastructure. Data can be lost due to mobile terminal malfunctions. Worse yet, these terminals can be stolen and ongoing transactions can be altered. In short, the mobility enjoyed in m-commerce also raises many more challenging security issues. Serious consideration must be given to the issue of security in developing mobile applications.
- *Trust* – in each transaction, each party involved needs to be able to authenticate its counterparts, to make sure that received messages are not tampered with, to keep the communication content confidential, and to believe that the received messages came from the correct senders. Owing to the inherent vulnerability of the mobile environment, users in mobile commerce are more concerned

about security issues involved with mobile transactions. U.S. consumers are not ready to buy mobile services. They first need to be assured that their financial information is secure, and that wireless transactions are safe. The mass adoption of mobile commerce will not be realized until users begin to trust mobile services [177].

4.5.6 Research Activities

Research plays a vital role in solving problems in current mobile applications. Thus, we will present the research that should be carried out to address the challenges facing the field.

- *Identify killer applications* – for mobile services to succeed, one or more killer applications must be developed to compel individuals to purchase and use mobile devices in their daily and commercial activities. The killer application(s) for mobile services should take full advantage of mobility, provide services directly relevant to the needs of mobile users, and benefit users in immediacy and efficiency.
- *Enhance usability of mobile devices* – currently, the usability of mobile devices is poor due to the various limitations of mobile terminals. Future mobile devices are expected to be smaller and more wearable, and they will possess larger processing capability and storage capacity. Screens for cellular phones will be made larger, making them easier to read and more visually appealing. Meanwhile, offline methods that require no direct connection of mobile devices to the network can help to minimize the technical limitations. Future mobile devices will also support Bluetooth technology, allowing them to access nearby appliances such as vending machines and televisions using very low-cost, short-range moderate-bandwidth connections. With such capabilities, mobile devices will support a combination of different communication connections to provide a variety of mobile services.
- *Design user-friendly interface* – unlike the wired computing environment where large screens are available, mobile applications have to operate on small and often wearable mobile devices that can only include small screens. Researchers are now developing voice-based and pen-based interactions to replace the present keyboard and mouse interactions. Pen-based interaction on touch screens may replace the mouse; voice-based interaction may be used for activation and control of functions such as voice dialing. Some studies on the user interface for mobile devices have been reported in the workshop series on human computer interaction with mobile devices [178].
- *Consolidate network infrastructure* – bandwidth and coverage are major issues for network infrastructure. The former allows more data to be exchanged between servers and mobile devices, thus supporting multimedia content delivery. The latter minimizes the complications of connection losses when a mobile device moves beyond a network boundary or crosses from one network to

another. These two issues directly affect the quality of mobile data transfer, and therefore are critical to the further development and future deployment of mobile applications [179].

- *Develop content delivery and a format for mobile commerce* – at present, much attention has been given to providing visual access to Web content. As a result, WML and compact HTML (cHTML) are now widely used. Voice access can also be employed to enable Web content to be displayed on mobile devices. VoiceXML is a new markup language for creating voice–user interfaces to Web applications or voice capabilities; it is important to study how a combined voice, screen, and keyboard (or button) access to the Web can be realized by integrating the features in VoiceXML with wireless markup language [180].
- *Improve mobile access to databases* – to allow users to run applications on their mobile devices without having to maintain constant connection with the servers and pay expensive connection fees, at least part of the database systems must be able to reside on the mobile devices. It will be necessary for these mobile database systems to require little memory and to be able to transfer their data to the centralized database systems. In some cases, a mobile database system may only manage a portion of a large central database, pulling in additional data on demand and pushing back data that is not required. In a mobile environment where users are constantly on the move and few computing resources are available, query processing and data recovery capabilities for these mobile database systems will have to be further improved.
- *Explore agent technologies* – the relatively high cost of connection time and data exchange for mobile devices discourages the adoption of mobile applications by cost-sensitive organizations. Agent technologies can alleviate this problem. Mobile users can contact agents to look for products and services, to locate merchants, to negotiate prices, and to make payments. All of these activities can be performed without having the mobile devices constantly connected to the network. In an agent-based mobile commerce framework, agents can be envisioned as merchants, consumer, and other brokering services, interacting with one another to enable electronic transactions.
- *Address security issues* – research on how to improve security in mobile commerce must be carried out, due to the vulnerability of mobile devices and wireless networks. To meet security requirements such as authentication, integrity, confidentiality, message authentication, and nonrepudiation in mobile commerce, additional security software and features (e.g., certificates, private, and public keys) will have to be installed on mobile devices. However, due to the limited computing capability of mobile devices, at some point it might be necessary to establish additional servers to store information, perform security checking, and conduct electronic payments on behalf of mobile devices [181].

Mobile services have the potential to significantly and positively impact the way we do our business. Although there remain a number of technical, regulatory, and social challenges to overcome, mobile communications and mobile devices will continue to develop and incorporate additional functionalities in the coming years. Undoubtedly, mobile services will be developed with the advancement in mobile communication technologies.

4.6 UNIVERSAL MULTIMEDIA ACCESS

The previous years have seen a variety of trends and developments in the area of communications and thus multimedia access. In particular, we are going to see delivery of all types of data for all types of users in all types of conditions. This section discusses the current status of universal multimedia access (UMA) technologies and investigates future directions in this area.

Key developments and trends from the last few years have set the scene for ubiquitous multimedia consumption. In summary, these are

- wireless communications and mobility
- standardized multimedia content
- interactive versus passive consumption
- the Internet and the World Wide Web (WWW).

An important step in the progression of ubiquitous multimedia has been the accomplishment of (almost) full mobility. With the explosion of mobile networks and terminals, users can now realize being constantly online. Originally, only voice communication was feasible, but limited Internet access and video telephony are now possible. Already, phone usage is reaching saturation levels in several countries. Moreover, mobile terminals are becoming increasingly sophisticated, especially with the emerging third-generation mobile technology. They are offering additional services, notably multimedia-centered services.

Another vital feature of human interface with the environment has also found its way into communications applications and devices: interactivity. While the consumption of audio and video has been passive for many decades, the Internet adventure and the diffusion of games have shown the importance of a deeper, interactive relationship between the user and multimedia content. This has generated an expectation of multimedia beyond passive television viewing. Interactivity also invaded the media with hundreds of cable TV channels today providing interactive capabilities for hyperlinking, voting, chatting, and the like. In general, interactivity provides the capability of tuning the content according to the user's needs and wishes, in short, personalizing the contents to offer more individual and private experiences.

The way users think today about multimedia content is strongly determined by the Internet. It is a part of everyday life because of broad content availability, simple yet powerful interaction with the content, and easy access. Integration into other familiar devices and technology is the next, even more challenging development stage.

The strength of the Internet is that it provides versatile bidirectional interactivity and allows peer-to-peer communication. This personalized experience marks a jump from broadcasting to narrowcasting, where the same content is distributed to everybody, but is tuned to a consumer's personal context. As yet this is limited to WWW pages, but soon it will extend to full multimedia content.

While there is a wealth of audio and visual data on the Internet today, it is increasingly difficult to find the information we need (in spite of the significant advances in terms of search engines). The gap between our expectation of information and the delivered information increases daily, at least for those less familiar with search engines. Description, structure, and management of information are becoming critical. One solution is offered by portals, the Web virtual libraries and shopping centers, which provide users gateways to organized and specific multimedia domains. This growing body of structured information (even if that structure is still limited) increasingly needs to be accessed from a diverse set of networks and terminals.

The variety of delivery mechanisms to those terminals is also growing; currently, these include satellite, radio broadcasting, cable, mobile, and copper using xDSL. At the end of the distribution path are the users, with different devices, preferences, locations, environments, needs, and possibly disabilities. Figure 4.79 shows how different terminals access rich multimedia content through different networks.

The notion is that UMA represents associated technologies enabling any content to be available anytime, anywhere, even if after adaptation. This may require that content be transcoded from, for example, one bit rate or format to another or transcoded across modalities, for example, text to speech. UMA concentrates on altering the content to meet the limitations of a user's terminal or network. *Universal* applies here to the user location (anywhere) and time (anytime), but also to the content to be accessed (anything), even if that requires some adaptation to occur. UMA requires a general understanding of personalization involving not only the user's need and

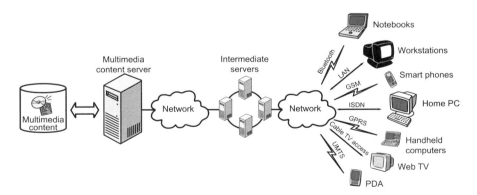

Figure 4.79 Different terminals' access reaching multimedia content.

preferences, but also the capabilities of the user's environment (e.g., the network characteristics; the terminal where the content will be presented; and the natural environment where a user is located, such as the location, temperature, and altitude).

Technologies that will allow a UMA system to be structured are just starting to appear. Among the most relevant are adaptation tools that process content to fit the characteristics of the consumption environment. These adaptation tools have to consider individual data types (e.g., video or music) as well as structured content, such as portals and MPEG-21 digital items [83].

Thus, adaptation extends from individual multimedia objects to the presentation of multiple, structured elements. Content and usage environment or context descriptions (both metadata) are central to content adaptation since they provide information that can control a suitable adaptation to be or being standardized; examples are content and usage environment description, delivery protocols, and right expression mechanisms. Today, UMA service deployment is limited not only by network and terminal bottleneck, but also by the lack of standard technologies that allow some services to hit mass markets at acceptable prices, for example, mobile video streaming [83].

While today's UMA technologies are offering adaptation to a terminal, tomorrow's technologies will provide users with adapted and informative universal multimedia experiences (UMEs). This is the critical difference between UMA and UMEs: the latter clearly acknowledge that the end point of universal multimedia consumption is the user and not the terminal. With this, mass media becomes mass customization.

Universal multimedia experience (UME) is the notation that a user should have an equivalent, informative experience anytime, anywhere. Typically, such an experience will consist of multiple forms of multimedia content. Each will be adapted as in UMA, but rather than to limits of equipment, to limits that ensure the user has a worthwhile and informative experience. Thus, the user is central and the terminal and network are purely vehicles of the constituent content.

The primary difference between accessing the content (UMA) and ensuring a consistent user experience (UME) is a shift in focus from data delivery to the terminal to experience delivery to the users themselves. The key aspects are to ensure that our systems can deliver a meaningful experience that will deliver information to the user and, in turn, allow the user to build knowledge. Human knowledge evolves as result of millions of individual experiences at different times and locations and in various contexts; it is this past that a user carries and adds to while consuming multimedia experiences. A profitable multimedia experience is thus one that takes the end user beyond the simple content presentation to information and knowledge; this may involve personal communication, entertainment, or education. The more powerful the experience is, the higher its value in terms of resulting knowledge for the task.

4.6.1 Existing and Emerging Technologies

Since there are many UME-relevant technologies and it is not possible to address all of them, particular emphasis will be given to signal-processing-related technologies.

While some of the technologies considered are already well established, providing a significant technological basis for the development and deployment of UME applications (such as content scalability and transcoding) there are still vital technologies missing for a complete UME system vision. Many of these technologies are directly related to particular usage environments. While multimedia adaptation for improved experiences is typically thought of in the context of more constrained environments (e.g., mobile terminals and networks), it is also possible that the content has to be adapted to more sophisticated environments, for example, with three-dimensional (3D) capabilities. Whether the adaptation processing is to be performed at the server, at the terminal, or partially at both, is something that may have to be determined case by case, depending on such criteria as computational power, bandwidth, interfacing conditions, and privacy issues.

Scalable coding represents the original data such that certain subsets of the total coded bit stream still provide a useful representation of the original data if decoded separately. There is currently a significant number of scalable coding schemes available, each with different characteristics in terms of the coding technology, the domain in which they provide scalability (e.g., spatial, temporal, quality), the granularity of the scalability provided, and the efficiency/cost of providing scalability. MPEG-4 is the content representation standard where the widest range of scalability mechanisms is available, notably in terms of data types, granularities, and scalability domains.

Transcoding describes the conversion of a coded (image, video, 3D, audio, or speech) bit stream to another bit stream representing the same information in a different coding format or with the same format but with reduced quality (less bit rate), spatial resolution, frame rate (for video), or sampling rate (for audio). The fundamental issue in transcoding is to achieve these outcomes without requiring the complete decoding and reencoding of the content [4].

With the explosion of multimedia content, it soon became evident that there was a need for efficient content description (with so-called metadata) to achieve more efficient content management, retrieval and filtering. Content description is also essential for effective usage environment adaptation if systems are to avoid computationally expensive content analysis at adaptation time. This is particularly critical when content has to be adopted in real time. Rapid navigation and browsing capabilities are prime requirements of user interaction with structure content, and data abstraction techniques enable such facilities through the provision of different types of summaries.

4.6.2 Content Representation

While there will be a necessity for future content representation tools, it is increasingly desirable to keep formats backward compatible and cross compatible, for example, with the use of common file formats for MPEG-4 and JPEG2000. This ensures that storage and transport of content can be handled uniformly while maintaining the different purposes of actual content streams. Further, there are interesting proposals that explicitly separate the content from the bit stream format. In particular, several techniques are suggested for describing a bit stream in terms of XML metadata and manipulating it using XML tools. The idea behind this is that future systems can be implemented quickly with just the provision of high-level mapping between one bit stream format and a second. This opens the possibility of broader flexibility in content representation and transcoding, while maintaining standard solutions [86].

Currently, the work to provide more efficient fine granularity video scalability solutions is still ongoing. For example, there is work under development based on MPEG-4 Parts 2 and 10 (video coding solutions); the latter is also known as advanced video coding (AVC) or ITU-T H.264. MPEG is also studying wavelet-based video coding scalability, notably considering the relevance that scalable coding assumes in the context of the MPEG-21 multimedia framework.

Fine granular scalability for audio remains a significant challenge for researchers. Traditional audio and speech coders (e.g., ITU G.728, ITU G.729) are bit rate centric. For instance, audio and speech coders are separated by model dependence, but in each group certain algorithms dominate for certain bit rates and signal bandwidths. This is partly due to the perceptual audio effects exploited by various coders and the necessity of tuning these to bit rates. This has proved to be a significant obstacle in the development of a single algorithm that operates in limited bandwidth and quality ranges, for example, the adaptive multirate (AMR) coders. One new trend in audio coding is the shift from low-rate compression focus to a perceptual quality focus. In particular, lossless audio coding and scalability from lossy to lossless are now considered important areas. Coupled with the move to higher quality, a new focus is the delivery of multichannel audio and the scalability of such experiences to, for example, users with headphones [5].

4.6.3 User Environment Description

For the user to obtain the best possible multimedia experience, it is essential that the content is adapted taking into account as many relevant usage environment parameters as possible. Typical relevant usage environment dimensions are the network, terminal, natural environment, and user; each of these dimensions may be characterized by multiple parameters.

Standard usage environment description tools like content description tools are needed to ensure a high degree of interoperability among terminals and severs. MPEG-21 digital item adaptation (DIA) is a relevant development in this area [83]. In part, it targets the specification of standard description tools for the usage environment.

Since multimedia experiences are centered in the user, the incorporation of the user into the adaptation and delivery processes needs to be at a deeper level than today's typical approach. As an example, the MPEG-21 IDA specification only considers a narrow set of user preferences for multimedia selection and adaptation. With the future emergence of continuously wearable devices (e.g., with capability of measuring several human parameters such as blood pressure, cardiac rhythm, and temperature), it is possible to take a much wider variety of data into account during the multimedia adaptation process. The result will be to determine the user's current characteristics, for example, it terms of mental and physical moods or circumstances.

Besides the selection and standard description of the relevant features, the provision of the capabilities in question would require a (nonstandard) solution to map the physiological sets into psychological dispositions or moods with which certain types of multimedia content are associated. This would likely involve complex psychophysiological studies so as to enrich the user/human dimension that UMEs require.

One of the big problems with context description is the large set of descriptors that could be generated. It is also clear that many descriptors will only be relevant to a small subset of applications and situations. One possible solution is intelligent agent technologies that can cope with a large, diverse set of context inputs, and determine those that are relevant and thus simplify adaptation processing.

4.6.4 Content Adaptation

A central role in adapting multimedia content to different usage environments is played by content adaptation. Content adaptation using a UMA engine is shown in Figure 4.80. Content adaptation may be performed at various locations in the delivery chain or even distributed across several nodes. Moreover, it may be performed in real time or offline, providing a set of variations from which to select at adaptation time.

While adaptation through transcoding is well known, crossmodal adaptation is a more challenging process that may range from a simple key-frame extraction process (to transform video into images) to more complex conversions such as text-to-speech, speech-to-text, or even video-to-text or speech. A video-to-speech cross-modal conversion may be as simple as using the information in the textual annotation field of the associated metadata stream (e.g., MPEG-7 stream describing MPEG-2 or MPEG-4 content) or as complex as analyzing the content, recognizing some object (e.g., for which specific models are available), and synthesizing some speech based on the recognized objects.

The adaptation of structures content such as portals or MPEG-21 digital items may be divided into two stages: first, filtering the content components to give a

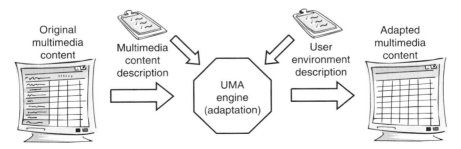

Figure 4.80 An example of content adaptation.

set suitable for the consumption environment and, second, after a certain content component is selected, adapting that component in terms of transcoding or crossmodal conversion.

4.6.5 Intellectual Property and Protection

A key factor in the delivery of multimedia content at present is an increasing desire by intellectual property owners to establish, control, and protect their rights. A number of schemes for protecting music and video content have been used, most notably, the protection of DVD content and, more recently, attempts by the music industry to protect CDs from being copied. Such techniques have generally proved to be limited in their level of protection, and new technologies continue to be introduced. In practice, it seems likely that it is the platforms themselves that must be altered to incorporate mechanisms that protect content from copying.

Currently, various bodies are creating standards (an example is the work in MPEG on intellectual property management and protection, a rights expression language, and a data dictionary that will provide a framework for digital rights management). The enforcement technologies are likely to be a significant area of research and development for the foreseeable future. For UMEs, the expression of rights to access and alter content metadata, perform (or prevent) certain adaptations, and the enforcement of those rights will be critical if content providers are to be willing to distribute material into new user environments. Further, for UMEs, intellectual property management and protection will be an issue not only for the content but also for usage environment information. Usage environment information can reveal personal information such as location and user's state of health; it is unlikely that users would be happy to have such information freely available on a network.

4.6.6 Presentation Conditions

In general, universal multimedia involves adaptation of high-quality/functionality content to reduced functionality usage environments, such as mobile terminals. This is a growing trend, as presentation devices become smaller, thinner, and lighter. There are more sophisticated content presentation conditions and devices that may also require adaptation processing to achieve an improved user experience. Presentation devices may be very broad in their capabilities, ranging form small mobile devices to sophisticated immersive rooms. For example, it is possible to imagine a multimedia device that is simply a pair of semitransparent (see through) glasses capable of overlaying stereoscopic visual data, both natural and synthetic; this example illustrates future human-information wearable interfaces that are both network- and signal- (image, video, and 3D) processing enabled.

Presentation conditions and devices also determine the type and level of interactivity provided to the user. While simple, portable devices mostly have limited interaction capabilities, more sophisticated systems such as immersive environments may provide very powerful interaction mechanisms, such as tactile gloves to manipulate objects in a virtual world. The presentation interface itself may also be adapted based on user preferences and skills; today, game interfaces can change depending on the user's skill level, but, in the future, multimedia experience may adapt to a consumer's age, past experience, and desired outcomes.

Three-dimensional audiovisual environments provide an interesting example of advances in presentation. In terms of visual information, this may imply the adaptation of available 3D content to specific types of 3D or stereoscopic displays or even the processing of two-dimensional (2D) content to provide 3D-like visual sensations. The same 3D content may also have to be adapted to rather simple consumption conditions, such as a personal digital assistant (PDA), targeting the provision of 3D navigational capabilities, or even 3D sensory impressions using glasses-based technologies [182].

In terms of audio information, users are now expected and experiencing near-cinema-grade sound in their living rooms. Already, amplifiers offer several modes of different *acoustic environments*, such as concert, stadium, theater, and studio. This is just the beginning of the possibilities, and it is quite feasible to adapt the audio delivery to a room's acoustics to maximize a user's experience and to provide elaborate manipulation of audio objects and streams in space. Just as with video, the consumption of audio, particularly in 3D (either simulated using stereo systems or actual multispeaker systems), is likely to become less passive and more interactive in the near future [5]. To achieve this, sophisticated adaptation algorithms will be a vital part of ensuring a quality user experience.

4.6.7 Mobile and Wearable Devices

A key aspect of the desire of users for multimedia content delivery has been the significant uptake in mobile devices both in the cellular phone and PDA markets. While it remains unclear whether the enhanced phone or wireless networked PDA will prevail, the desire for the consumption of content on such devices is clear. Various companies are offering proprietary low-complexity codec solutions for the current limited processing power devices, and several have MPEG-4 decoders available. In future, a multimedia experience may be provided to a user using multiple mobile terminals (all carried by the user) or even nonpersonal terminals (e.g., the larger screen in a coffee shop).

Beyond simple mobile devices, wearable devices are a new area of research. The prospect of *suits* of high-capability processing and sensor devices offers significant possibilities for future multimedia delivery. The vision of wearable computers is to move computing from being a *primary task* to a situation where the computer is permanently available and augments another activity that the user is performing. For example, when a user walks down a street or rides a bicycle, sensors can make spatiotemporal context information available to the computer nodes to alter user interfaces and multimedia delivery. The embodiment of computers and humans into a *user* removes the typical sharp boundaries between the user and terminal and can have a profound effect on multimedia delivery.

Coupled with the increase in mobile terminals, today's passive networks that route traffic from source to destination are evolving. Increasingly, some level of processing is being offered in switches, and the extension of this concept to active or programmable network nodes provides a platform for adaptations that will be required to deliver UMEs. This may be a solution to the limited processing power available in small, mobile devices since a user (or, more likely, his agent) can request a network service provider to perform certain processing (e.g., transcoding) on his behalf. In practice, this is an extension of the current transcoding of HTML WWW content to wireless applications protocol (WAP) content in some mobile networks. The potential of programmable networks is much greater, however; the programmable network node (equipped with suitable software) can be the assembly point for a UME. In such a case, an agent with suitable usage environment information would perform the necessary constituent content adaptations before delivering the UME as a set of, for example, low-bandwidth content streams.

4.6.8 Content Delivery

The traditional content delivery model is one where users access content from a set of servers. The *master* of the content is thus the administrator of the server and, in general, he controls who has access to the content. An alternative vision of content delivery is one where every use is interconnected and any user can act as a server of content for any other user.

Peer-to-peer networking has been a long-term form of communication; face-to-face conversation, telegraphy, telephony, and the postal system are all examples. It has thus been natural that peer-to-peer has quickly grown in electronic form on the Internet and in cellular systems. At first, e-mail was the prime example, but now we see instant messaging, the file-sharing networks and, among cellular-users, short message services (SMS) growing rapidly. The latter is already being improved with multimedia messaging services as users demand, and improved peer-to-peer transfer is commonplace. This will require changes in the way we consider rights and security of access to content for users.

The impact of a mixed peer-to-peer and server-based architecture is that adaptation will be required both in professional delivery from servers as well as among individual users. Since the latter may not have the processing capability or software to deliver complex adaptations, a new breed of network services may emerge to provide content adaptation to peer-to-peer users. This further emphasizes the necessity of transparency of the users' wishes to exchange content as easily as they can. Hence, the use of intelligent agents to ensure consistent user experiences will be a logical and necessary step.

REFERENCES

1. MEDIACOM2004, *Project Description*, March 2002.
2. ISO/IEC JTC1/SC29/WG11, MPEG02/M8221, *Updated Description of the Relationship Between the MPEG-21 and MEDIACOM 2004 Projects*, March 2002.

3. A. Perkis et al., Universal multimedia access from wired and wireless systems, *Transactions on Circuits, Systems and Signal Processing, Special Issue on Multimedia Communications*, Birkhauser, Boston, MA, 20(3), 387–402 (2001).
4. A. Vetro, Ch. Christopoulos, and H. Sun, Video transcoding architectures and techniques: an overview, *IEEE Signal Processing Magazine*, 20, 18–29 (2003).
5. K. Homayounfar, Rate adaptive speech coding for universal multimedia access, *IEEE Signal Processing Magazine*, 20, 30–39 (2003).
6. P. vanBeek et al., Metadata-driven multimedia access, *IEEE Signal Processing Magazine*, 20, 40–52 (2003).
7. L. Twvede, P. Pircher, and J. Bodenkamp, *Data Broadcasting*, Wiley, Hoboken, NJ, 2001.
8. Available at http://www.dvb.org
9. L. Eronen, User centered research for interactive television, *Proc. European Conference on Interactive Television*, pp. 5–12, Brighton, UK, April 2003.
10. Available at http://www.mhp.org
11. Sh. Lamont, Case study: successful adoption of a user-centered design approach during the development of interactive television applications, *Proc. European Conference on Interactive Television*, pp. 13–18, Brighton, UK, April 2003.
12. MEDIACOM2004, A framework for multimedia standardization, *Project Description*, 2001.
13. ITU SG9, IPCablecom, *Project Description*, 2002.
14. ITU-T, *Scope of Study Group 6 – Broadcasting Service*, October 2003.
15. Available at http://www.itu.int/itu-r/asp/scope
16. ISO/IEC SC29/WG11 MPEG 11172-2, *Generic Coding of Moving Pictures and Associated Audio (MPEG-1 Video)*, November 1991.
17. ISO/IEC SC29/WG11 MPEG 11172-3, *Generic Coding of Moving Pictures and Associated Audio (MPEG-1 Audio)*, November 1991.
18. ISO/IEC SC29/WG11 MPEG 11172-1, *Generic Coding of Moving Pictures and Associated Audio (MPEG-1 Systems)*, November 1991.
19. L. Chiariglione, MPEG and multimedia communications, *IEEE Trans. CSVT*, 7, 5–18 (1997).
20. IETF Networks Working Group, *RTP Payload Format for MPEG1/MPEG2 Video*, RFCT 2050, IETF, 1998.
21. IETF Networks Working Group, *RTP: A Transport Protocol for Real-Time Applications*, IETF, July 1999.
22. IETF Networks Working Group, RFC2343, *RTP Payload Format for Bundled MPEG*, IETF, May 1998.
23. K. R. Rao and Z. S. Bojkovic, *Packet Video Communications Over ATM Networks*, Prentice Hall PTR, Upper Saddle River, NJ, 2000.
24. A. Puri and T. Chen (Eds.), *Multimedia Systems, Standards and Networks*, Marcel Dekker, New York, NY, 2000.
25. ISO/IEC SC29/WG11 MPEG-2 System Group 13818-1, *Information Technology – Generic Coding of Moving Pictures and Associated Audio: Part 1 Systems*, 1995.

26. A. Puri, Video coding using the MPEG-2 compression standard, *Proc. SPIE Visual Communications and Image Processing*, 1199, 1701–1713 (1993).
27. ISO/IEC SC29/WG11 MPEG-2 13818-3, *Information Technology – Generic Coding of Moving Pictures and Associated Audio: Part 3 Audio*, 1995.
28. ISO/IEC SC29/WG11 MPEG-4 14386-1, *Information Technology – Generic Coding of Moving Pictures and Associated Audio: Part 1 Systems*, April 1999.
29. ISO/IEC SC29/WG11, MPEG-2 13818-2, *Information Technology – Generic Coding of Moving Pictures and Associated Audio: Part 2 Video*, 1995.
30. A. Puri and A. Eleftheriodis, MPEG-4: a multimedia coding standard support mobile applications, *ACM Mobile Networks and Application Journal*, 3, 5–32 (1998).
31. ISO/IEC SC29/WG11, *MPEG-4 Requirements Group Doc. N2456*, December 1998.
32. ISO/IEC SC29/WG11, *MPEG-4 FDIS 14496-1 Doc. N2501, Information Technology – Generic Coding of Moving Pictures and Associated Audio: Part 1 Systems*, November 1998.
33. ISO/IEC SC29/WG11, *MPEG-4 System Group 14496-1 Doc. N2739, Part 1 Systems/PDAM1*, March 1999.
34. O. Auro et al., MPEG-4 systems: overview, *Signal Processing: Image Communications*, 15, 281–298 (2000).
35. E. D. Scheirer, R. Vaananen, and J. Huopaniem, AudioBIFS: describing audio scenes with the MPEG-4 multimedia standard, *IEEE Trans. Multimedia*, 1, 237–250 (1999).
36. ISO/IEC 14772, *The Virtual Reality Modeling Language*, 1997, available at http://www.vrml.org/specifications/VRML97.
37. ISO/IEC SC29/WG11 MPEG-4 Doc. N1162, *Report on the Ad Hoc Group on Evolution of Tools and Algorithms of Video Submissions for MPEG*, January 1996.
38. Z. Bojkovic and D. Milovanovic, Audiovisual integration in multimedia communications based on MPEG-4 facial animation, *Circuits, Systems and Signal Processing*, 20, 311–339 (2001).
39. ISO/IEC SC29/WG11 MPEG-4 Doc.N2459, *Overview of the MPEG-4 Standard*, October 1998.
40. W. G. Gardner, Reverberation algorithms, in M. Khars and K. Brandenburg, Eds., *Applications of Digital Signal Processing to Audio and Acoustics*, Kluwer, New York, NY, 1998.
41. J. P. Jullien, Structured model for the representation and the control of room acoustical quality, *Proc. Int. Conf. Acoustics*, Trondheim, Norway, pp. 517–520, 1995.
42. J. M. Jot, Efficient models for reverberation and distance rendering in computer music and virtual audio reality, *Proc. Int. Conf. Computer Music*, Greece, pp. 236–243, 1997.
43. ISO/IEC SC29/WG11 MPEG-4 FDIS 14496-2 Doc.N2502, *Information Technology – Generic Coding of Moving Pictures and Associated Audio: Part 2 Visual*, November 1998.
44. ITU-T Experts Group on Very Low Bitrate Visual Telephony, *ITU-T Recommendation H.263: Video Coding for Low Bit Rate Communication*, December 1995.
45. F. I. Parker and K. Waters, *Computer Facial Animation*, AK Paters, 1996.
46. T. Sikovar and L. Chiariglione, The MPEG-4 video standard and its potential for future multimedia applications, *Proc. IEEE ISCAS*, Hong Kong, pp. 326–329, June 1997.
47. H. Kalva, *Delivery MPEG-4 Based Audio-Visual Services*, Kluwer, Boston, MA, 2001.

48. ISO/IEC SC29/WG11 MPEG-4, Video verification model editing committee, Doc. N2552, *The MPEG-4 Video Verification Model 12.0*, December 1998.
49. R. V. Cox et al., On the application of multimedia processing to communications, *Proc. IEEE*, 86, 755–824 (1998).
50. T. Sikora, The MPEG-4 video standard verification model, *IEEE Trans. CSVT*, 7, 19–31 (1997).
51. C. leBuhan et al., Shape representation and coding of visual objects in multimedia applications: an overview, *Ann. Telecomm.*, 53, 164–178 (1998).
52. ISO/IEC SC29/WG11 MPEG-4, Video verification model editing committee, Doc. N1796, *The MPEG-4 Video Verification Model 8.0*, July 1997.
53. J. L. Mitchell et al., *MPEG Video Compression Standard*, Chapman and Hall, New York, NY, 1996.
54. ISO/IEC SC29/WG11 MPEG-4, Video verification model editing committee, Doc. N1869, *The MPEG-4 Video Verification Model 9.0*, May 1998.
55. G. A. Abrantes and F. Pereira, MPEG-4 facial animation technology: survey, implementation and results, *IEEE Trans. CSVT*, 9, 290–305 (1999).
56. F. Lavagelto and R. Pockay, The facial animation engine: toward a high-level interface for the design of MPEG-4 compliant animated faces, *IEEE Trans. CSVT*, 9, 277–289 (1999).
57. L. Chen, J. Ostermann, and T. Huang, Adaptation of a generic 3D human face model to 3D range data, *First Workshop on Multimedia Signal Processing*, pp. 274–279, Princeton, NJ, June 1998.
58. J. Ostermann, Animation of synthetic faces in MPEG-4, *Computer Animation*, 49–52 (June 1998).
59. H. Tao et al., Compression of facial animation parameters for transmission of talking heads, *IEEE Trans. CSVT*, 9, 264–276 (1999).
60. A. M. Tekalp, *Digital Video Processing*, Prentice Hall, Englewood Cliffs, NJ, 1995.
61. Y. Altmibasak and A. M. Tekalp, Occlusion-adaptive, content-based mesh design and forward tracking, *IEEE Trans. Image Processing*, 6, 1270–1280 (1997).
62. Y. Altmibasak and A. M. Tekalp, Closed-form connectivity preserving solutions for motion compensation using 2D meshes, *IEEE Trans Image Processing*, 6, 1255–1269 (1997).
63. Y. Wang and O. Lee, Active mesh – a feature seeking and tracking image sequence representation scheme, *IEEE Trans Image Processing*, 3, 610–624 (1994).
64. C. Tokles et al., Tracking motion and intensity variations using hierarchical 2D mesh modeling, *Graphical Models Image Processing*, 58, 553–573 (1996).
65. P. J. L. vanBeek et al., Hierarchical 2D mesh representation, tracking and compression for object-based video, *IEEE Trans CSVT*, 9, 353–369 (1999).
66. ISO/IEC SC29/WG11 MPEG-4 Audio Group, FDIS 14496-3 Doc.N2503, *Information Technology – Generic Coding of Moving Pictures and Associated Audio: Part 3 Audio*, November 1998.
67. ISO/IEC SC29/WG11 MPEG-4 Doc.N2725, *Overview of the MPEG-4 Standard*, 1999.
68. K. Brandenburg et al., MPEG-4 natural audio coding, *Signal Processing: Image Communications*, 15, 423–443 (2000).

69. R. S. Manjunath, P. Salembier, and T. Sikora (Eds.), *Introduction to MPEG-7, Multimedia Content Description Interface*, Wiley, Hoboke, NJ, 2002.
70. *The MPEG Home Page*, available at http://www.mpeg.org.
71. R. Koenen and F. Pereira, MPEG-7: a standardized description of audiovisual content, *Signal Processing: Image Communications*, 16, 5–13 (2000).
72. ISO/IEC SC29/WG11 Doc.N4320, *MPEG-7 Requirements*, July 2001.
73. ISO/IEC SC29/WG11 MPEG-7 Requirements Doc.N4582, *Report of the ad hoc Group on Evaluation Logistic*, March 1999.
74. ISO/IEC SC29/WG11 MPEG-7 Requirements Doc.N2732, *Description Schemes*, March 1999.
75. ISO/IEC SC29/WG11 MPEG-7 Video Group Doc.N2694, *Generic Visual Description Scheme*, March 1999.
76. ISO/IEC SC29/WG11 MPEG-7 Requirements Doc.N2731, *DDL Development*, March 1999.
77. ISO/IEC SC29/WG11 MPEG-7 Doc.N4546, *Input for MPEG-7 Systems Work*, March 1999.
78. ISO/IEC SC29/WG11 MPEG-21 Doc.N4041, *MPEG-21 Overview*, March 2001.
79. IETF RFC2396, *Uniform Resource Identifiers (URI): Generic Syntax*, 1998.
80. Available at http://www.itsc.ipsj.or.jp/sc29/29w42911.htm
81. A. Permis et al., Universal multimedia access from wired and wireless systems, *Transactions on Circuits, Systems and Signal Process, Special Issue on Multimedia Communications*, Birkhauser, Boston, 20, 387–402 (2001).
82. R. Mohan, J. R. Smith, and C. S. Li, Adapting multimedia internet content for universal access, *IEEE Trans Multimedia*, 1, 104–114 (1999).
83. J. Bomans, J. Gelissen, and A. Perkis, MPEG-21: the 21st century multimedia framework, *IEEE Signal Processing Magazine*, 20, 53–62 (2003).
84. 3GPP, *Packet-Switched Streaming Service*, TS26.35, February 2002.
85. M. Massel, *Digital Television: DVB-T, COFDM and ATSC8-VSB*, Digital TV Books Asia, 1999/2000.
86. K. R. Rao, Z. S. Bojkovic, and D. A. Milovanovic, *Multimedia Communication Systems: Techniques, Standards and Networks*, Prentice Hall PTR, Upper Saddle River, NJ, 2003.
87. Ad hoc Group DVB-UMTS, R. Lueder, *Summary of Report No.1*, February 2002.
88. J. D. Johnston, Transform coding of audio signals using perceptual noise criteria, *IEEE J. Select. Areas Comm.*, 6, 314–323 (1987).
89. K. H. Braundenburg and G. Stoll, ISO MPEG-1 Audio: a generic standard for coding of high-quality digital audio, *J. Audio. Eng. Soc.*, 42, 780–792 (1994).
90. W. Hoeng and T. Lanterbach, *Digital Audio Broadcasting Principles and Applications*, Wiley, New York, NY, 2001.
91. EUREKA 147, *Digital Audio Broadcasting System*, April 2002, available at www.eurekadab.org.
92. Federal Communication Commission, August 2002, Washington, DC, available at www.fcc.org.
93. Draft Recommendation ITU-R BS [Doc6/63], *System for Digital Sound Broadcasting Bands Below 30 MHz*, ITU-R, September 2001.

94. *Digital Radio Mondiale*, June 2002, available at www.xmradio.com.
95. J. Stott, *Digital Radio Mondiale (DRM) – Key Technical Features*, Technical Report, March 2001.
96. Sirius Satellite Radio, available at www.siriusradio.com.
97. XM Satellite Radio, Washington, DC available at www.xmradio.com.
98. D. Layer, Digital radio takes to the road, *IEEE Spectrum*, 38, 40–46 (2001).
99. Global Radio, Luxembourg, available at www.globalradio.com.
100. W. Bkleijn and K. K. Paliwal (Eds.), *Speech Coding and Synthesis*, Elsevier, New York, NY, 1995.
101. B. Scharf, *Foundations of Modern Auditory Theory*, Academic Press, New York, NY, 1970.
102. J. D. Johnston, Transform coding of audio signals using perceptual noise criteria, *IEEE J. Select. Areas Comm.*, 6, 314–323 (1987).
103. T. Painter and A. Spanias, Perceptual coding of digital audio, *Proc. IEEE*, 88, 117–119 (2000).
104. ISO/IEC JTC1/SC29/WG11, FDIS 14496-3, *Coding of Audio and Visual Objects, Part 3: Audio*, October 1998.
105. ATSC, *Digital Audio Compression Standard (AC-3)* available at www.atsc.org/standards/a52
106. D. Sincha et al., The perceptual audio coder (PAC), in V. Madisetti and D. B. Williams, Eds.; *The Digital Signal Processing Handbook*, CRC/IEEE, Boca Raton, FL, 1997.
107. Ch. Faller et al., Technical advances in digital audio radio broadcasting, *Proc. IEEE*, 90, 1303–1333 (2002).
108. J. H. Hall, Auditory psychophysics for coding applications, in V. Madisetti and D. B. Williams, Eds.; *The Digital Signal Processing Handbook*, CRC/IEEE, Boca Raton, FL, 1997.
109. J. P. Princen and A. B. Bradley, Analysis and synthesis filter bank design based on time domain aliasing cancellation, *IEEE Trans Acoust. Speech, Signal Processing*, ASSP-34, 277–284 (1986).
110. F. R. Jean et al., Two-stage bit allocation algorithm for stereo audio coder, *Proc. Inst. Elect. Eng. – Vis. Image Signal Process.*, 143, 331–336 (1996).
111. R. Arean, J. Kapcevic, and V. K. Goyal, Multiple description perceptual audio coding with correlated transforms, *IEEE Trans Speech Audio Processing*, 8, 140–145 (2000).
112. M. Bosi et al., ISO/IEC MPEG-2 Advanced audio coding, *J. Audio Eng. Soc.*, 45, 789–814 (1997).
113. IBIQUITY Digital Corporation, Warren, NJ, available at www.ibiquity.com.
114. NAB National Radio Systems Committee, Washington, DC, available at www.nab.org/scitech/nrsc.asp.
115. EUREKA 147, *Digital Audio Broadcasting System*, available at www.eurekadab.org, April 2002.
116. ETSI ETS300401, *Radio Broadcasting Systems: DAB to Mobile, Portable and Fixed Receivers*, May 1997.
117. WorldDab, available at www.worlddab.org.

118. J. D. Johnson et al., AT&T perceptual audio coding (PAC), in N. Glishrist and C. Grewin, Eds.; *Collected Papers on Digital Audio Bit-Rate Reduction*, AES, 1996, pp. 73–82.
119. N. S. Jayant and E. Y. Chen, Audio compression: technology and applications, *AT&T Tech. J.*, 74, 23–34 (1995).
120. C. E. W. Sundberg et al., Multistream hybrid in-band-on-channel systems for digital audio broadcasting in the FM band, *IEEE Trans Broadcast*, 45, 410–417 (1999).
121. D. Singh and C. E. W. Sundberg, Unequal error protection (UEP) for perceptual audio coders, *IEEE ICASSP*, 2423–2426, April 1999.
122. W. Pritchard and M. Ogata, Satellite direct broadcast, *Proc. IEEE*, 78, 1116–1140 (1990).
123. *Astra*, Betzdorf, Luxemburg, available at www.worldspace.com.
124. *Worldspace*, Washington, DC, available at www.worldspace.com.
125. ETS300 421 DVB-Satellite (1999), ETS300 429 DVB-Cable (1999), ETS300 744 DVB-Terrestrial (1999).
126. DVB-Eutelsat Technical Guide, *Overview of DVB*, Annex B, June 1999.
127. ITU-R Task Group11/3, *Report of the Second Meeting of ITU-R Task Group 11/3*, ITU, Geneva, January 1994.
128. ITU-R Document TG 11/3-2, *Outline of Work for Task Group 11/3, Digital Terrestrial Television Broadcasting*, ITU, June 1992.
129. TM2456 *Commercial Requirements for Multimedia Services over Broadband IP in a DVB Context* (CM255r4), March 2001.
130. DVB Doc.A071, *Architectural Framework for the Delivery of DVB-Services Over IP-Based Networks*, February 2002.
131. G. Lutteke, *DVB MHP: Concept and Impact, MHP Cable Workshop*, Munich, Germany, January 2002.
132. J. van der Meer and C. M. Huiyer, *Interoperability Between Different Interactive Engines for Digital Television: Problems and Solutions*, Philips, Eindhoven, Netherlands, June 1997.
133. C. Voyt, The DVB MHP specification – a guided tour, *World Broadcast Engineering*, 4–8 (2000).
134. L. Eronen, User centered research for interactive television, in *Proc. European Conference on Interactive Television: From Viewers to Actors*, EuroITV, pp. 5–12, April 2003.
135. M. Milenkovic, Delivering interactive services via a digital TV infrastructure, *IEEE Multimedia*, 34–43 (1998).
136. A. Acharya, M. Franklin, and S. Zennim, Prefetching from a broadcast disk, *Proc. Int. Conf. Data Engineering*, pp. 276–285, Los Alamitos, CA, 1996.
137. T. E. Browen et al., The data cycle architecture, *Comm. ACM*, 30, 71–81 (1992).
138. T. Imielinski, S. Viswanathan, and B. R. Badrinath, Data on air: organizations and access, *IEEE Trans. Knowledge Data Eng.*, 9, 353–372 (1997).
139. D. Minole, *Video Dialtone Technology*, McGraw-Hill, New York, NY, 1995.
140. J. Nilsen, *Usability Engineering*, Academic Press, Chestmant Hill, MA, 1993.
141. Available at http://www.dvb.org

142. V. Paxal, *DVB with Return Channel via Satellite*, DVB-Telenor, AS050400, 2000, pp. 1–5.
143. ETSI EN301 958, *DVB-RCT Interaction Channel for Terrestrial Networks*, 2003.
144. ETSI EN300 421, *Digital Broadcasting Systems for Television, Sound and Data Services: Framing Structure, Channel Coding and Modulation for 11/12 GHz Satellite Services*, 1994.
145. ETSI Draft DEN SES-00, *Satellite Interactive Terminals (SIT) and Satellite User Terminals (SUT) Transmitting Towards Satellites in Geostationary Orbit in the 29.5 GHz to 30 Ghz Frequency Bands*, May 1999.
146. ETSI EN301 958, 2002.
147. J. W. Wong, Broadcast delivery, *Proc. IEEE*, 76, 1566–1577 (1998).
148. L. John and L. Tseng, Fast data broadcasting and receiving scheme for popular video services, *IEEE Trans. Broadcasting*, 44, 100–105 (1998).
149. L. John and L. Tseng, Harmonic broadcasting protocols for video-on-demand service, *IEEE Trans. Broadcasting*, 43, 268–271 (1997).
150. K. A. Hua, Y. Cai, and S. Shen, Patching: a multicast technique for video-on-demand services, *Proc. ACM International Conference on Multimedia*, pp. 191–200, 1998.
151. O. K. Li, W. J. Liao, X. X. Oui, and W.-M. Wong, Performance model of interactive video-on-demand systems, *IEEE J. Selected Areas in Comm.*, 14, 1099–1109 (1996).
152. G. S. H. Chan and F. Tobagim, Distributed servers architecture for networked video services, *IEEE/ACM Trans. Networking*, 9, 125–136 (2001).
153. C. Guo et al., Optimal caching algorithm based on dynamic programming, *Proc. of SPIE International Symposium on Convergence of IT and Communications (ITCom01)*, 4519, 285–295 (2001).
154. L. Tvede, P. Pircher, and J. Bodenkamp, *Data Broadcasting: Merging Digital Broadcasting With the Internet*, Wiley, Hoboken, NJ, 2001.
155. B. Furth, R. Westwater, and J. Ice, Multimedia broadcasting over the Internet: Part I, *IEEE Multimedia*, 5, 78–82 (1998).
156. Ch. Hess, D. Lin, and K. Nahrstedt, VistaMail: an integrated multimedia mailing system, *IEEE Multimedia*, 5, 13–23 (1998).
157. K. Sian, E. Lim, and Z. Shen, Mobile commerce: promises, challenges and research agenda, *Journal of Database Management*, 12, 3–10 (2001).
158. L. Liebman, Whose client is it, anyway?, *Communications News*, 38, 76 (2001).
159. M. Donegan, The m-commerce challenges, *Telecommunications*, 34, 58 (2000).
160. A. K. Ghosh and T. M. Swaminathan, Software security and privacy risks in mobile e-commerce, *Communications of the ACM*, 44, 51–57 (2000).
161. E. Schwartz, Mobile commerce takes off, *InfoWorld*, 22, 1–32 (2000).
162. J. Clarke III, Engineering value propositions for m-commerce, *Journal of Business Strategies*, 18, 133–148 (2001).
163. N. Nohria and M. Leestma, A moving target: the mobile commerce customer, *MIT Sloan Management Review*, 42, 104 (2001).
164. K. Sian and Z. Shen, Mobile communications and mobile services, *Int. J. Mobile Communications*, 1, 3–14 (2003).

165. B. W. Wirtz, Reconfiguration of value chains in converging media and communication market, *Long Range Planning*, 34(4), Elsevier Science, pp. 489–506, August 2001.
166. A. Afman and L. C. Tuci, *Internet Business Models and Strategies*, McGraw Hill, New York, NY, 2001.
167. T. Sweeney, Tech obstacles keep wireless e-retailers grounded, *Information Week*, 837, 72–76 (2001).
168. C. Schnederman, Are you billing in real-time, *Telecommunications*, 35, 71–72 (2001).
169. V. Kumar and A. Nagpal, Segmenting global markets: look before you leap, *Marketing Research*, 13, 8–13 (2001).
170. J. Rohwer, Today Tokyo, tomorrow the world, *Fortune*, 142, 140–152 (2000).
171. C. Goldman, The m-commerce horizon, *Wireless Review*, 17, 14 (2000).
172. E. Turban et al., *Electronic Commerce 2002: A Management Perspective*, Prentice Hall, Upper Saddle River, NJ, 2002.
173. B. Schulz, The m-commerce fallacy, Network World, 18, 77–82 (2001).
174. C. Sclenter and M. J. Shaw, A strategic framework for developing electronic commerce, *IEEE Internet Computing*, 1, 20–28 (1997).
175. S. J. Barnes, The mobile commerce value chain: analysis and future developments, *International Journal of Information Management*, Elsevier, 22, 91–108 (2002).
176. K. Sian and Z. Shen, Building consumer trust in mobile commerce, *Communications of the ACM*, 46, 1 (2003).
177. C. Johnson, *GIST Technical Report G98-1, Proc. of Workshop on Human Computer Interaction with Mobile Devices*, Glasgow, Scotland, 1998.
178. U. Varshney, R. J. Vetter, and R. Kalakota, Mobile e-commerce: a new frontiers, *IEEE Computer*, 33, 32–38 (2000).
179. *Voice XML 1.0 Specification*, March 2000, available at http://www.voicexml.org.
180. D. V. Thanh, Security issues in mobile commerce, in *Proc. International Conference on Electronic Commerce and Web Technologies*, pp. 412–425, EC-Web 2000, London, 2000.
181. F. Pereira and J. Bernett, Universal multimedia experiences for tomorrow, *IEEE Signal Processing Magazine*, 20, 63–73 (2003).

BIBLIOGRAPHY

182. V. Bhaskaran and K. Konstantinides, *Image and Video Compression Standards: Algorithms and Architectures*, Kluwer, Norwell, MA, 1997.
183. A. Bovik (Ed.), *Handbook of Image and Video Processing 2nd Ed.*, Academic Press, Orlando, FL, 2000.
184. V. Castelli and L. D. Bergman, *Image Databases: Search and Retrieval of Digital Imagery*, Wiley, Hoboken, NJ, 2002.
185. J. Chen, U. V. Koc, and K. J. R. Liu, *Design of Digital Video Coding Systems*, Marcel Dekker, New York, NY, 2001.
186. N. Freed and N. Borenstein, *Multipurpose Internet Mail Extension (MIME), Part One: Format of Internet Message Bodies*, RFC2045, November 1996.

187. N. Freed and N. Borenstein, *Multipurpose Internet Mail Extension (MIME), Part Two: Media Types*, RFC2046, November 1996.
188. N. Freed and N. Borenstein, *Multipurpose Internet Mail Extension (MIME), Part Three: Message Extension for Non-ASCII Text*, RFC2047, November 1996.
189. N. Freed and N. Borenstein, *Multipurpose Internet Mail Extension (MIME), Part Five: Conformance Criteria and Examples*, RFC2049, November 1996.
190. N. Freed and N. Borenstein, *Multipurpose Internet Mail Extension (MIME), Part Four: Registration Procedures*, RFC2048, November 1996.
191. M. Ghanbari, *Standard Codecs: Image Compression to Advanced Video Coding*, IEE, London, UK, 2003.
192. M. Ghanbari, *Video Coding: An Introduction to Standard Codecs*, IEE, London, UK, 1999.
193. J. D. Gibson et al., *Multimedia Compression: Applications and Standards*, Morgan Kaufmann, San Francisco, CA, 1997.
194. L. Guan, S. Y. King, and J. Larsen (Eds.), *Multimedia Image and Video Processing*, CRC Press, Boca Raton, FL, 2000.
195. A. Hanjalic, *Content-Based Analysis of Digital Video*, Kluwer, Boston, MA, 2004.
196. A. Hanjalic, G. C. Langelaar, P. M. B. van Roosmalen, J. Biemond, and R. L. Lagendijk, *Image and Video Databases: Restoration, Watermarking and Retrieval*, Elsevier, 2000.
197. B. Haskell, A. Puri, and A. Netravali, *Digital Video: An Introduction to MPEG-2*, Chapman & Hall, 1996.
198. K. Hayler, The IP Datacast Forum, *Broadcast Engineering*, A special supplement, pp. 4–5, August 2002.
199. W. J. Heng and K. N. Ngan, *Digital Video Transition Analysis and Detection*, World Scientific, River Edge, NJ, 2003.
200. ISO 10021 and ITU-TS X.420, *Message Handling Systems*, Geneve, October 1991.
201. ISO/IEC SC29/WG11 MPEG-7 Doc.N3752, *Overview of the MPEG-7 Standard*, October 2000.
202. J. Kamarainess, Datacasting: a concrete examples of convergence, *Broadcast Engineering, A Special Supplement*, pp. 10–11, August 2002.
203. M. K. Mandal, *Multimedia Signals and Systems*, Kluwer, Norwell, MA, 2002.
204. B. S. Manjunath, P. Salembier, and T. Sikora, *Introduction to MPEG-7*, Wiley, Hoboken, NJ, 2002.
205. MEDIACOM2004 Project, available at http://www.itu.int/itu-t/studygroups/com16/mediacom2004/index.html.
206. I. Pandzic and R. Forchheimer, *MPEG-4 Facial Animation*, Wiley, Hoboken, NJ, 2002.
207. K. K. Parhi and T. Nishitani (Eds.), *Digital Signal Processing for Multimedia Systems*, Marcel Dekker, New York, NY, 1999.
208. W. B. Pennebaker and J. L. Mitchell, *JPEG Still Image Data Compression Standard*, Van Nostrand Reinhold, 1993.
209. W. B. Pennebaker, J. L. Mitchel, C. Fogg, and D. LeGall, *MPEG Digital Video Compression Standard*, Chapman & Hall, 1997.
210. F. Pereira and T. Ebrahimi (Eds.), *The MPEG-4 Book*, IMSC Press, 2002.
211. J. B. Postel, *Simple Mail Transfer Protocol*, RFC821, August 1982.

212. A. Puri and T. Chen (Eds.), *Multimedia Systems, Standards and Networks*, Marcel Dekker, New York, NY, 2000.
213. K. R. Rao and J. J. Hwang, *Techniques and Standards for Image, Video and Audio Coding*, Prentice Hall, Upper Saddle River, NJ, 1996.
214. I. E. G. Richardson, *H.264 and MPEG-4 Video Compression*, Wiley, Hoboken, NJ, 2003.
215. I. E. G. Richardson, *Video Codec Design*, Wiley, Hoboken, NJ, 2002.
216. M. J. Riley and I. E. G. Richardson, *Digital Video Communications*, Artech House, 1997.
217. A. Sadka, *Compressed Video Communications*, Wiley, Hoboken, NJ, 2002.
218. E. Stare et al., *The Multimedia Car Platform, European Commission Project*, January 2000.
219. M. T. Sun and A. R. Reibman, *Compressed Video Over Networks*, Marcel Dekker, New York, NY, 2000.
220. D. S. Taubman and W. M. Marcellin, *JPEG2000: Image Compression Fundamentals, Standards and Practices*, Kluwer, Norwell, MA, 2002.
221. Y. Wang, J. Ostermann, and Y. Q. Zhang, *Video Processing and Communications*, Prentice-Hall, Upper Saddle River, NJ, 2002.
222. J. Watkinson, *The MPEG Handbook: MPEG-1, MPEG-2, MPEG-4*, Focal Press, Oxford, UK, 2001.
223. C. Weinshenk, Multimedia goes beyond voice, *Communications Technology*, 20, 14–15 (2003).
224. P. C. Yip and K. R. Rao, *The Transform and Data Compression Handbook*, CRC Press, Boca Raton, FL, 2001.
225. Y. Yuan and J. J. Zhang, Towards an appropriate business model for m-commerce, *Int. J. Mobile Communications*, 1, 35–56 (2003).

5

MIDDLEWARE LAYER

Much of the complexity and cost of building networked applications can be alleviated by the use of highly flexible, efficient, and secure middleware, which is a systems software that resides between the applications and the underlying operating systems and networks. It provides reusable services that can be composed, configured, and deployed to create multimedia applications rapidly and robustly. This chapter is organized as follows. First we present middleware for multimedia. Next we describe media coding. In particular we analyze MPEG-4 multimedia content representation, core compression technologies, transcoding architectures and techniques, as well as multimedia implementation. We continue with media streaming. Owing to the explosive growth and great success of the Internet, as well as increasing demand for multimedia services, streaming media over the Internet has drawn tremendous attention from both academia and industry. After that we will also review infrastructures for multimedia content management, watermarking frameworks and technologies, digital right management systems, and security infrastructures for content distribution, as well as copyright protection. Finally, this chapter concludes with middleware technologies for multimedia networks.

5.1 INTRODUCTION

The term *middleware* has acquired numerous meanings that would allow it to be just about any piece of software that sits between systems. In the strict sense, middleware

---

Introduction to Multimedia Communications, By K. R. Rao, Zoran S. Bojkovic,
and Dragorad A. Milovanovic
Copyright © 2006 John Wiley & Sons, Inc.

is transport software that is used to move information from one program to one or more other programs, shielding the developers from dependences on communication protocols, operating systems, and hardware platforms. Middleware provides the plumbing necessary for applications to exchange data, regardless of the environment in which they are running. Transactions, data broadcasting, packages, and extensible markup language (XML) often ride on middleware in the enterprise. The concept of middleware dates back to the 1980s when companies wanted one package to move data between mainframes, databases, and user terminals. Modern middleware extends this concept to the widespread distribution of data in real time across a remarkable variety of services, clients, and sites. Middleware as used in this sense means that data is sent between systems in messages, which are similar to data packets on the network. These messages have headers, which indicate the distinction and payloads of varying sizes and formats that contain the actual data. Message-oriented middleware (MOM) originally appeared in the form of message queues.

When a message was sent to another system, it was stored in a message queue on the destination system. Whenever the destination system needed the data, it looked in the queue for the message. If it was there, the message was retrieved; if not, the system would wait until the data arrived in the queue. This approach proved reliable, but slow. It is still used today in many transaction-oriented environments, where security of transactions and integrity of message delivery are a high priority. A second model, called publish and subscribe (or pub/sub), evolved form the need to deliver messages in real time, especially to a large number of clients. In the pub/sub model, clients register for certain kinds of messages in which they are interested, and a server sends the clients those messages in real time. The emphasis of the pub/sub model is to send data from one server to many clients as fast as possible. Typical applications might be stockbrokers needing the latest process on certain bonds or equities. These processes typically are sent in real time to all brokers who subscribe to this information.

5.2 MIDDLEWARE FOR MULTIMEDIA

The goal of the International Telecommunication Union (ITU) is to play a *catalytic* role in facilitating the development of truly global information superhighways (http://www.GlobalCollaboration.org). There are four main ways this will be done: by developing standards that will enable the different networks that are likely to make up the global information infrastructure such as global mobile personal communications satellite (GMPCS) services and helping manage their use; by providing policy and technical assistance to developing countries in partnership with the private sector and other international organizations; and by providing a forum where government and industry can discuss policies and strategies for making the vision of the GII a reality. Looking at these four main ways, it may be concluded that they correspond quite well with ITU's core activities, although carried out to a different degree and in a different manner, but nevertheless promoting the evolution of GII as an entire system.

5.2.1 GII Standardization Projects

Focusing on standardization, the ITU can be seen as a GII facilitator by providing global interconnectivity and interoperability through its standards. Global specifications are universally seen as necessary for timely successful GII. Such standards can achieve applicability and interoperability, and support for cultural diversity. In addition to the global developments under way by consortia and industry related to GII, various national and regional organizations are concentrating on developing their own particular national and regional information infrastructure. For the development of GII, the ITU is ideally placed to be the integrator, linking developing and developed countries.

A society can be considered to consist of a socioeconomic part and an industrial part (with which standards bodies are concerned), each with a number of specific roles. Figure 5.1 illustrates a number of specific roles in which the information society operates [1]. In fact, the information society operates in the roles of and between two parts: the socioeconomic part and the industry part.

The socioeconomic part is

- establishing lifestyles, courtesy, and customs (e.g., teleshopping practices, etiquette for a virtual society)
- creating culture and arts (e.g., multimedia books, teleorchestration)
- building regulations, laws, and provisions (e.g., medical law in telemedicine).

The industry part is

- creating applications and services (teleshopping, video services)
- building information networks (e.g., telecommunications networks, CATV, satellites)
- manufacturing facilities (e.g., terminals, transmission systems).

The development of the GII is expected to lead to increasing and improved person-to-person communications and to future, and as yet unknown, businesses and interpersonal applications. This move to an *information society* in which individuals

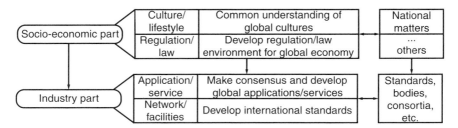

Figure 5.1 Configuration of an information society and its roles [1]. (©1998 IEEE.)

have secure global access to all kinds of information and services, and which recognizes and meets cultural diversities and sensitivities, is expected to be as far-reaching in its social and economic impact as the move from the agrarian society to the industrial age.

Real and very significant standardization challenges are posed by each of the three component terms encompassed by the GII.

- *Global* – global standards are required for the information and infrastructure components of the GII. The globalization of business, the ease of information access, and the ease of personal mobility ensure that there is no longer any particular national or regional way of doing business.
- *Information* – the purpose of the global infrastructure is to enable users to globally manage the creation, storage, delivery, and use of information. Adequate global standards for the representation of and secure context-specific access to or exchange of information independent of the location of the information provider and information user are needed to realize the benefits of the GII.
- *Infrastructure* – the technological convergence and interconnection of telecommunications equipment, computers, and much of consumer electronics has led to new demands on the communications infrastructure by information providers and users.

A key objective of the GII standardization work is to avoid overlap of work and for the various bodies to collaborate on preparation of the necessary standards. The ITU-T has agreements with a number of fora and consortia, as well as with other standards bodies, for the exchange of information, recognition of results, and referencing of their standards in ITU-T Recommendations. A particularly important area for collaboration is that of the evolution of the Internet and its relation to GII. Experts in the various ITU-T Study Groups involved in GII work are increasingly becoming involved in the work of the Internet Engineering Task Force (IETF), which will help ensure that the necessary convergence is achieved.

GII is also one of the areas identified by the Global Standards Collaboration among the national, regional, and international standards bodies in support of the work of the ITU. These meeting provide a framework for the exchange of information, establishing of objectives to accelerate the process of global telecommunications standards development, and promotion of interconnectivity and interoperability.

Table 5.1 lists the ITU GII standardization projects. They have been grouped into the major themes of framework aspects (F projects), network aspects (N projects), middleware aspects (M projects), applications aspects (A projects), and Internet-related aspects (I projects).

Telecommunication networks are currently providing voice and data services worldwide with a high level of reliability and defined QoS and are based on different network technologies with interworking among them. Extension of the networks to include broadband capabilities is based on the ATM technology. ATM is also being

Table 5.1   List of ITU GII standardization projects [1]

Number	Name of Project
F.1	Principles and framework for GII
F.2	Scenarios and key interfaces for GII
F.3	Information appliance
F.4	End-to-end interoperability
N.1	Architecture and layer 1 aspects of wideband/broadband access infrastructures for GII
N.2.1	Signalling and control aspects of wideband/broadband access interfaces for GII
N.2.2	Signalling and control aspects of wideband/broadband network element to network element interface for GII
N.3	Network interworking for the GII
N.4	Access to and interworking with IP-based networks
N.5.1	*Intelligent mobility* for the GII, IMT2000 (former FPLMTS)
N.5.2	*Intelligent mobility* for the GII, Global mobility
N.6	Harmonization of B-ISDN signalling protocols and their interfaces to public broadband networks
N.7	Enhanced network intelligence for the GII
N.8	Quality of service and network performance
N.9	Addressing for the GII
M.1	Network-oriented middleware and network operating systems for GII
M.2	APIs harmonized with network capabilities
M.3	Technical framework for electronic commerce
M.4	Middleware for multimedia
M.5	System management
M.6	Security (end-to-end)
M.7	High-level naming
M.8	Object-oriented environments
M.9	Operating environment and user interfaces
A.1	Medical informatics
A.2	Libraries
A.3	Electronic museums
A.4	Road transport informatics
A.5	Electronic purse
A.6	Industrial multimedia communication
A.7	Ergonomics
A.8	Character set
A.9	Geographic information systems
I.1	Overview of Internet-related issues
I.2	Multimedia over Internet protocols
I.3	Interworking between Web services and PSTN/ISDN

©1998 IEEE.

enhanced to provide not only for connection-oriented network services, but also to meet the requirements of connectionless network capabilities and services supported by these capabilities.

Networks based on Internet protocols provide a platform that allows users connected to different network infrastructures to have a common set of applications and to exchange data with a unified QoS. The IP protocol suite is evolving to include voice, data, and video applications with defined QoS. These convergence trends in networking technology are illustrated in Figure 5.2. Additionally, terrestrial radio, cable, and satellite networks are providing local broadcast entertainment services and are also evolving to provide interactive voice, and video services.

As a basis for formal standardization process, a GII functional model has been defined. This is an abstract description of a system and is developed in such a way that it is independent of any implementation of the system. Its purposes are

- To allow freedom in methods of implementation without affecting the operation of the overall system.
- To allow large-scale functional integration inside one equipment or software module while retaining a manageable and scalable description of the equipment or software module.
- To allow the dynamic creation of services that can be tailored to the needs of users.

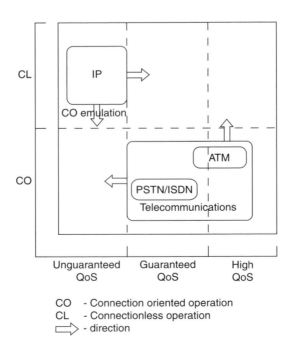

Figure 5.2 Convergence trends in network technology [1]. (©1988 IEEE.)

The use of functional models is finding more widespread use in both the telecommunications and IT industries, and there are many examples of functional models and functional modeling methodologies, including

- the information and computational viewpoints of the open distributed processing reference model
- the use of service-independent building blocks in intelligent networks
- the layered description of the synchronous digital hierarchy interface description and the use of functional blocks in the equipment specifications
- the extensive use of layers of application programming interfaces in operating and software support system architecture.

The basic types of functions and basic types of logical interface in the GII and their likely relationship to various roles in the GII are as follows.

- *Applications functions* – the logical entities of applications (usually implemented in software and normally called *applications objects*).
- *Middleware functions* – the logical entities of middleware (also usually implemented in software and normally called *middleware objects*), including
  - *Service control function* – the middleware functions that are the logical entities that allow building services from service components and the associated resources, and control the interaction of the user with the service (these functions are sometimes associated with session control);
  - *Management functions* – the middleware functions that are the logical entities that perform the management of all the other functions.
- *Baseware functions* – the logical entities that allow application and middleware functions to run, communicate with other functions by interfacing with network functions, and interface to users (functions normally associated with operating systems or virtual machines such as Java virtual machine). These include
- *Network functions* – the baseware functions that are the logical entities that support communication among separated locations in the GII and include transport functions and control functions;
- *Processing and storage functions* – the baseware functions that are the logical entities that execute middleware and application components and store information;
- *Human–computer interfacing functions* – the baseware functions that are the logical entities that allow application components to present information to and gain input from a human user.

The relationship among functions and the role in which they perform their functions is determined by the needs of the role. For example, a communications and networking of information role will require many middleware functions in order to have the resources to be able to offer the range of service components associated with this role. However, while a generic communications role will need to have many network functions in order to be able to offer the service components associated with the role, it may also have some middleware functions in order to provide a slightly wider range of service. In this way the description of each role has some flexibility, and then the functionality among different roles may overlap.

5.2.2 EII Standardization Projects

Information technology and the use of IP-based networks and applications (e.g., electronic commerce) have become critical factors in the development of telecommunication networks [2]. Data traffic is growing at more than ten times the rate of voice traffic, and it is estimated that in the near future, data will account for 80 percent of all traffic carried by telecommunications networks. Therefore, with this rapid change, the past concept of telephone networks, which also carry data, will be replaced by the concept of data networks that also carry voice. In this regard, seamless interworking between IP-based networks and telecommunications networks and interoperability of their respective applications/services is essential to meet the burgeoning business requirements placed on modern communications networks.

Another important trend in telecommunications networks is the emergence of mobile networks and a significant increase of customers subscribing to them. The work on the 3rd generation mobile networks performed by ITU under the name of the IMT-2000 has, as its main features, an increased data speed of 384 kbps up to 2 Mbps, a global roaming capability, and the virtual home environment (enabling users to move seamlessly between fixed and mobile networks). These features will provide an additional infrastructure for the IP-based network services with fast, ubiquitous access through the global roaming capability.

Dramatic changes have also been taking place in the satellite industry that are driven by the growth of wireless and Internet traffic. Many new systems have been and are being planned, designed, and deployed to offers services such as integrated voice/data/video, Internet access, wireless access, and so on. Services offered by satellite networks are also moving quickly towards the end users by providing them direct high-speed access.

One of the goals of EII standardization is to allow many components to form the information infrastructure, like telecommunications networks based on a variety of different technologies and operated by competing organizations, computing platforms using different hardware and software and distributed processing facilities. Remember that the functions involved in interactions regarding the information infrastructure are: application involving and handling functions, application supporting platforms, basic and enhanced telecommunications functions, as well as distributed information processing and storage functions. Also, application invoking and handling functions, as well as application supporting platforms, in combination support the structural roles of the enterprise model.

The following standardization and specifications highways have been identified using the enterprise model: network-oriented standardization/specification highway, applications-oriented standardization specification highway, and architecture-oriented standardization/specification highway. Each highway has been refined in terms of specific standardization projects. For middleware the standardization projects are as follows:

- technical framework for electronic commerce
- middleware for multimedia
- browsing and searching
- systems management
- security
- basic services to support distributed application
- fundamental processing services in end user systems
- middleware for human–computer interfaces
- high-level naming
- object-oriented environments
- structured data files
- graphical representation of data elements.

The middleware-related projects relate to the general-purpose computing support for EII/GII applications. The standards converted by the projects are of general utility and are not devoted to any particular application. There are many bodies worldwide that are involved in producing standards for this area.

5.2.3 ITU-T SG16 Work Program

The rapidly developing world of the multimedia-related technologies and standards requires a framework to develop standards for applications, services, and systems that can respond to users' requirements in terms of mobility, ease of use, flexibility of systems, and end-to-end interoperability with specific quality requirements.

Consider

- the continuing trend in digitization
- the rapid growth of digital networks and, in particular, the Internet
- the increasing computational power in personal computers
- the convergence of various technologies including communications, broadcasting, information technology, and home electronics
- that multimedia topics are discussed in ITU-T SGs (e.g., 2, 7, 9, 12, 13, 16, and SSG), ITU-R SGs, international and regional SDOs, and also in external organizations
- emerging new communication services and applications that are resulting from the growth of the Internet and wireless technologies

- that with the emergence of high-speed, high-quality networks, society will request real-time multimedia communications as an extension of existing monomedia systems
- that the end-to-end performance of multimedia systems and services defined in the ITU and elsewhere should be dimensioned
- that it will be necessary to study the interfaces in the Information Appliance environment as consumer devices increasingly perform multimedia functions.

ITU-T SG16, the lead SG for multimedia, has initiated a project MEDIA-COM2004 (Multimedia Communication 2004) [3]. Within the framework of the MEDIACOM2004, SG16 is responsible for [4]

- developing a framework for multimedia services and systems that must be, as far as possible, independent from the underlying infrastructure
- developing appropriate services- and systems-related standards for applications in multimedia communications
- defining the interface with the relevant work areas assigned to other ITU-T and ITU-R SGs involved
- liaison and coordination with the relevant bodies outside the ITU
- ensuring that end-to-end interoperability is accommodated either by full compatibility among systems or by specification of the appropriate gateways
- maintaining a database of all multimedia standardization activities.

5.2.4 MEDIACOM2004 Middleware Design

As was pointed out, the goal of the MEDIACOM2004 project is to establish a framework for multimedia standardization [4]. This framework supports the harmonized and coordinated development of global multimedia communications standards across developing organizations/bodies as shown in Figure 5.3. The Middleware 2004 project is divided into a number of main framework study areas (FSAs) as shown in Table 5.2, each of which has an associated study question in SG16.

The roles of ITU-T/ITU-R SGs and other relevant bodies are shown in Tables 5.3 to 5.6. It consists of four parts regarding analysis of multimedia (MM) standardization roles in draft form.

An information communication service or application can be roughly divided into real-time and nonreal-time communications. The former is a stream type communication, for example, telephone, fax, and audiovisual communication, over the PSTN, ISDN, and so on. The latter is a burst type of communication, for example, on IP networks [5]. Typically, services and applications over circuit-switched networks have been developed by different ITU-T SGs, such as SG7 (Data Transfer), SG8 (Fascimile), and SG16 (Audiovisual Multimedia). On the other hand, in IP-based networks, the applications have typically been developed by the IETF or the World Wide Web consortia (W3C). However, there are exceptions; for example, at present, H.323 is the key Recommendation for real-time multimedia applications

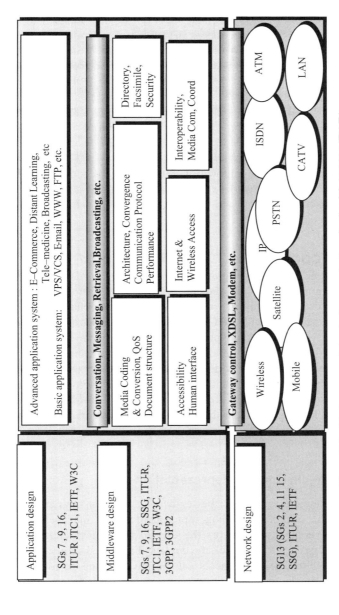

Figure 5.3 Study areas and developing organizations/bodies [3]. (©2001 ITU-T.)

Table 5.2 Framework for study areas [3]

Q. No	FSAs	Related Document
B	Architecture	Annex B
C	Applications and Services	Annex C
D	Interoperability	Annex D
E	Coding	Annex E
F	QoS and Performance	Annex F
G	Security	Annex G
H	Accessibility	Annex H

©2001 ITU-T.

over IP-based networks. The interworking of applications and services between the circuit-switched and IP networks can be realized through gateways (e.g., IP telephony, etc.).

5.3 MEDIA CODING

In media streaming, audio-video sequences are encoded offline and stored in a server. A user may access the server over a constant bit rate channel. For example, in streaming video we have video-on-demand, archived video news, and noninteractive distance learning. Before the playback, part of the video bit tream is preloaded in the scheduled time [6]. For these streaming video applications, since the video is encoded offline and the future video frames are available to the encoder, a more sophisticated bit-allocation scheme can be used to achieve better video quality. During the encoding process for streaming video, two requirements need to be considered: the preloading time that the video viewers have to wait and the physical buffer size at the receiver (decoder side).

5.3.1 Multimedia Content Representation

A general multimedia system can be segmented into four physical and/or logical components: the content, including representation and management, the server that delivers the content, the network that carriers the content, and the user terminal that plays and operates to playback the content. Figure 5.4 shows components of a multimedia system. The components can be physical and/or logical, with well-defined interfaces among them. There are three such interfaces, with each of the components possibly containing more interfaces.

Designing a system as a set of subsystems connected at published interfaces is necessary for the development of systems with such a broad scope. This also allows multiple vendors and service providers to provide components and services.

Table 5.3 Analysis of multimedia standardization roles (Part 1/4) [3]

	SG2	SG3	SG4	SG5	SG6	SG7	SG9	SG10	SG11	SG12	SG13	SG15	SG16	SSG-IMT2000	ITU-R
E-commerce						S							L		
Distant learning													M		
Telemedicine													M		X
TV service							M								
Web-casting													M		
MBone													M		
E-mail															
WWW															
Conversation conference		S											L		
Messaging						M									
Retrieval													M		
Broadcasting directory							L								X
Security	X	S				M							X		
Facsimile						L							L		
Communication protocols			X			X	X		X		X	X	X		
User gateway control								S					L		
Media coding and conversion										S			X		
Internet access									S		X	X	X		
Wireless access							X		S			X	M		X
Interoperability									S		X		L		
MediaCom coordination											S		L		

L, Lead SG; M, Main part of the work; S, Support or collaboration; X, Work on the subject.
©2001 ITU-T.

Table 5.4 Analysis of multimedia standardization roles (Part 2/4) [3].

	SG2	SG3	SG4	SG5	SG6	SG7	SG9	SG10	SG11	SG12	SG13	SG15	SG16	SSG-IMT2000	ITU-R
Network							S								
Mobile	S	S	S					S	L		S		S		X
WLAN			S					S	S				X		
Satellite			S					S	S						X
IP							X				L		X		
PSTN		S	S								M		L		
XDSL		S										M	S		
ISDN		S	S					S	S	S	L		X		
ATM		S	S						S		L		X		
CATV							M					M	X		
LAN							X								

L, Lead SG; **M**, Main part of the work; **S**, Support or collaboration; **X**, Work on the subject.
©2001 ITU-T.

Table 5.5 Analysis of multimedia standardization roles (Part 3/4) [3]

	ISO/IEC JTC1	ETSI	T1	IETF	ATMF	IMTC	IEEE	TIA	3GPP	3GPP2
			Application Design							
E-commerce				X						
Distant learning										
Telemedicine										
TV service										
Web-casting										
MBone				M						
E-mail				M						
WWW				M						
			Middleware Design (Services, Architecture, Media Coding, Protocols)							
Conversation conference	X	X		X		M				
Messaging	X			X						
Retrieval	X			X						
Broadcasting	X	X								
Directory				X						
Security	X	X		X		X		X		
Facsimile				X				X		
Communication protocols	M	X	X	X	X	X		X	X	X
User gateway control								X	X	
Media coding and representation	M							X	X	X
Internet access		X		M						
Wireless access		X						X		
Interoperability		X	X		M	M		X		
MEDIACOM coordination										

M, Main part of the work; X, Work on the subject.

©2001 ITU-T.

Table 5.6 Analysis of multimedia standardization roles (Part 4/4) [3]

	ISO/IEC JTC1	ETSI	T1	IETF	ATMF	IMTC	IEEE	TIA	3GPP	3GPP2
	Network Design (Architecture, Platform)									
Mobile	X	X					X	X	X	
Wireless LAN	X					X				
Satellite	X						X			
IP	X			M			X			S
PSTN (Modem)	X						X			
XDSL	X						X			
ISDN	X	X								
ATM	X	X			M		S			
CATV	X									
LAN	X		S				M	S		

M, Main part of the work; X, Work on the subject.
©2001 ITU-T.

The content creation process is a creation process akin to creating TV programming. As the amount of content on a system grows, it becomes difficult to manage the content in terms of searching, editing, compiling, and delivering. Content representation techniques should take these factors into consideration. With broadband access becoming common, service providers will be able to deliver high-quality audiovisual content to PCs. The major consideration for content creators is the ability to deliver and playback content on multiple platforms (TVs and PCs). Content representation techniques should allow delivery of alternative representations of content based on terminal resources.

Object-based representation of content allows reuse of objects in creating new presentations. As the amount of content grows, the ability to search through the objects and locating the right one becomes difficult. The MPEG-4 Systems layer includes a meta data stream called the *object content information* (OCI) stream, which can be used by content management systems for object location and retrieval. Content management should include tools that allow users to search for content based on the visual or textual description of objects [7].

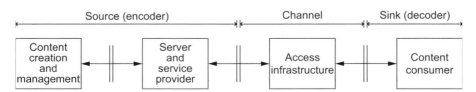

Figure 5.4 Components of an end-to-end multimedia system.

The future of object-based presentation results in content with complexity that varies with the structure of the presentation. The complexity of object-based presentations cannot be characterized just by bit rate. The complexity of a presentation depends on a number of objects in presentation, the type of objects, the dynamics of object addition and deletion, and user interactivity.

One important consideration during content creation is the suitability of content for delivery over networks with different capacities and to terminals with different computing, display, and storage resources. During the content creation process, authors should specify alternative representations for objects so that the server can deliver appropriate objects based on the feedback from the terminals and the networks. Determining the alternative representations for presentations dynamically is a difficult problem, especially when considering both terminal and network resources.

The basic function of the server is to manage sessions between the server and clients and to provide available services. The server is also responsible for publishing the content and services available for clients and to provide support for user interaction. The main consideration in server design is scalability. Also, a distributed server can deliver content to a range of clients to which it delivers. With object-based services, servers have to manage multiple object delivery in a single presentation. The complexity of a server increases when interactivity is supported, especially for object-based services [8].

When an object is added to a presentation as a result of user interaction, the server has to locate the object, retrieve it, and then deliver it to the client. The ability of servers to perform these functions efficiently depends on the object scheduling techniques and also on the underlying content representation format. For simple cases that do not include interactivity, objects can be efficiently delivered using techniques such as direct streaming that precompute transport layer packet headers [9]. These techniques may not be as useful when the design consideration is the server's ability to support content and service discovery. With object-based representation and functionality provided by a framework such as MPEG-4 Systems, the distinction between content and applications is blurring! The content can include instructions on how to respond to user interaction. Services will be able to support different applications just by supporting interactive object-based content.

The access network between a server and a client carries the encoded content to the end user. The configuration of the access network depends on the operating environment. The configuration of the access network varies with available end user connectivity. Some common access networks include cable, digital subscriber line (DSL), local multipoint distribution service (LMDS), and multichannel multipoint distribution service (MMDS). The main problem in delivering audiovisual content over access networks is resource reservation and renegotiation. In networks such as the Internet that do not yet support resource reservations, techniques such as dynamic rate shaping can be used to reduce the object bit rate, based on the state of the network [10]. Even in networks supporting QoS, techniques such as media filtering become necessary in negotiating and renegotiating QoS under resource constraints [11]. With object-based representation of audiovisual content, adapting content to meet network constraints can be done more efficiently. An example is structuring presentations to content objects with priorities proportional to their significance in the presentation.

Delivering object-based services may involve establishing and tearing down connections as objects in presentations are added and deleted. MPEG-4 specifies a delivery framework called DMIF that allows efficient object communication. The DMIF framework also allows development of applications without regard to the underlying network. Not all networks are suitable for carrying real-time multimedia traffic. The properties of data networks and their suitability to carry multimedia traffic are summarized in references [12, 13].

The content consumer, usually a user terminal with an attached monitor, is the last component of the multimedia delivery chain. These end user devices are connected to access networks and may have an upstream connection for server interaction and signaling. The digital TV infrastructure uses an MPEG-2 transport stream for delivering multiplexed audio, video, and images. For digital TV, the terminals would include MPEG audio and video decoders to playback the content. With object-based audiovisual representation, the presentations can contain many different media types and it is impractical to have a terminal with hardware decoders for all the possible media types. Terminals supporting object-based presentations would have to include software decoders and even programmable processors for efficient decoding. The buffer models have to change to accommodate multiple objects in presentations. With object-based multimedia content and the support for user interaction, user terminals are becoming more and more complex. Innovative architectures and representation techniques are needed to enable sophisticated applications.

5.3.2 Core Compression Technologies

As a follow on from the increasing role of video in the rapidly evolving world of multimedia systems, an important evolution in the concept of audiovisual information is taking place. While for a very long time video processing dealt exclusively with fixed-rate sequences of rectangular-shaped images, interest has been recently moving toward a more flexible concept in which the subject of the processing and encoding operations is a set of visual elements organized in both time and space in a flexible and arbitrarily complex way.

With the increasing demand for digital media, there has been significant interest in the deployment of image and video communication services. The World Wide Web is becoming a pervasive means for disseminating multimedia data (including image, video, text, and audio), and mobile communications are emerging from primarily voice-based services toward data services. As communication channels are limited in bandwidth, compression techniques are employed to reduce the amount of information to be transmitted. Image/video compression applies predictive coding schemes to remove spatial, temporal, or spatiotemporal redundancy among pixels. While bit rate reduction is achieved, a strong data dependence is created. An image relies on the integrity of the pixels it is predicted from for correct decoding.

If the pixels are corrupted with errors, the image cannot be decoded correctly. When the channel deteriorates because of noise, interference, and so on, the picture quality can suffer abruptly. Therefore, robust transmission of compressed multimedia data over noisy or unreliable communication channels has become an increasingly important application requirement.

Much effort has been invested to build error resilience into the compressed bit stream and ensure improving the quality of image/video transmission in noisy or error-prone environments [14]. The methods developed have reached a stage of maturity for standardization. Two international organizations have been heavily involved in the standardization of image, audio, and video coding methodologies, namely ISO/IEC and ITU-T. ITU-T Video Coding Experts Group (VCEG) develops international standards for advanced moving image coding methods appropriate for conversational and nonconversational audio/video applications. It caters essentially to real-time video applications. ISO/IEC Moving Picture Experts Group (MPEG) develops international standards for compression and coding, decompression, processing, representation of moving pictures, images, audio, and their combinations. It caters essentially to video storage, broadcast video, video streaming (video over Internet/DSL/wireless) applications [15].

The ISO MPEG-4 international standardization effort supports the advanced concept of visual information, covering the generic coding of audiovisual information for a wide range of rates and applications. It is well known that in MPEG-4, visual information is organized on the basis of the video object (VO) concept, which represents a time-varying visual entity with arbitrary shape that can be individually manipulated and combined with other similar entities to produce a scene. The information associated to a VO is represented as a set of video object layers (VOLs), each including the information corresponding to a given level of temporal or spatial resolution, so that scalable transmission and storage is allowed. Each VOL is considered as a sequence of video object planes (VOPs), which represent the information associated to given temporal instants and which substitute the traditional video frames. Within each VOP, shape and texture are encoded as different information items, following, in both cases, a predictive scheme. Shape is first coded in either a lossless or a lossy way, and then the texture is coded following a variation of the hybrid predictive-transform coding scheme and making use of shape information. Multiobject (MO) coding, involves the distribution of the bit rate among the different values that the user or the application assigns to the different objects. For instance, objects can be considered equally important so that the only interest is to obtain the best possible global quality [16]. But the user may prefer to ensure that the most relevant objects are always coded with quality above a minimum, as the speaker in the case of a video-conference system with speaker and background as different objects, or a video object showing an intruder in a surveillance application. Rate control (RC), understood as the regulation of the coder in order to optimize its operation imposed by the application and communications environments, incorporates in the MO case the bit allocation functionality, becoming a key element of the coding system.

ITU-T has been working on a video coding standard called H.26L since 1997. In August 1998, the first test model was ready and was demonstrated at MPEG's open call for technology in July 2001. In late 2001, ISO/IEC MPEG and ITU-T VCEG decided on a joint venture towards enhancing standard video coding performance – specifically in the areas where bandwidth and/or storage capacity are limited. This joint team of both standard organizations is called the Joint Video Team (JVT). The standard thus formed is called H.264/MPEG-4, and is presently referred to as JVT/H.26L/Advanced Video Coding (AVC) [17].

The JVT has had several meetings to date – at Fairfax, Virginia, in May 2002, Klagenfurt, Austria, in July 2002, Geneva, Switzerland, in October 2002, Awaji Island, Japan, in December 2002, and Pattaya, Thailand, in March 2003. The meeting in Japan in December 2002 marked the completion of the design and draft for the standard. Since the meeting in May 2002, the technical specifications are almost frozen and in May 2003 both organizations gave their final approval. Table 5.7 shows the image and video coding standards along with the year they were standardized and their major applications.

H.264/AVC (MPEG-4)

The video coding standards to date have not been able to address all the needs required by varying bit rates of different applications and at the same time meeting the quality requirements. H.264/AVC applications for video content include, but are not limited to, the following [18]:

- CATV – cable TV on optical networks, copper, and so on
- DBS – direct broadcast satellite video services
- DSL – digital subscriber line video services
- DTTB – digital terrestrial television broadcasting
- ISM – interactive storage media (optical disks, etc.)
- MMM – multimedia mailing
- MSPN – multimedia services over packet networks
- RTC – real-time conversational services (videoconferencing, videophone, etc.)
- RVS – remote video surveillance
- SSM – serial storage media (digital VTR, etc.).

Requirements for H.264/AVC arise from the various video applications that it aims at supporting, like video streaming, video conferencing, over fixed and wireless networks and over different transport protocols. H.264/AVC features thus aim at meeting the requirements evolving from such applications. The following lists the important requirements and how H.264/AVC meets them.

Robust (Error Resilient) Video Transmission

One of the key problems faced by previous standards is their layered nature, which results in less robust video transmission in packet lossy environments [19]. Previous standards contained header information about the slice/picture/GOP/sequence that was coded at the start of each slice/picture/GOP/sequence. The loss of packet

5.3 MEDIA CODING

Table 5.7 List of image and video coding standards

Standard	Main Applications	Year
JPEG	Image	1992
JPEG Extensions	Image	1996
JPEG LS Part I/Part II	Image	1998/1999
JBIG I	Fax	1995
MRC (Mixed Raster Content)	Color fax, Internet fax	1998
JBIG2	Fax	2000
H.261	Video conferencing	1990
H.262 (MPEG-2)	DTV, SDTV	1995
H.263 (Baseline)	Videophone	1998
H.262+ (Profile 3)		1999
H.263 + + (Profile 5)		2000
H.26L	VLBR video	2002
MPEG-1	Video CD	1992
MPEG-2	(Generic) DTV, SDTV, HDTV, DVD	1995
MPEG-4 Version 2	Interactive video (synthetic and natural)	1999
		2000
MPEG-7	Multimedia content description interface	2001
MPEG-21	Multimedia framework	2002
H.264/MPEG-4 (Part 10)	Advanced video coding	2003
Fidelity range extensions	HD-DVD, BD-DVD, DVB, D-Cinema	2004

containing this header information would make the data dependent on this header useless. H.264/AVC overcame this shortcoming by making the packets transmitted synchronously in a real-time multimedia environment self-contained. That is, each packet can be reconstructed without depending on the information from other packets. All information at higher layers is system-dependent, but not content-dependent, and is conveyed asynchronously. Parameters that change very frequently are added to the slice layer. All other parameters are collected in a `Parameter Set`. The H.264/AVC standard specifies a method to convey parameter sets in a special network abstraction layer (NAL) unit type. NAL is described in the next section. Different logical channels or out-of-band control protocols may be used to convey parameter sets from the coder to the decoder. In-band parameter set information and the out-of-band control protocol should not be used in combination. The parameter set concept is shown in Figure 5.5.

Network Friendliness
Previous video coding standards, such as H.261, MPEG-1, MPEG-2, and H.263 [87], were mainly designed for special applications and transport protocols, usually in a circuit-switched, bit-stream oriented environment. JVT experts realized the growing importance of packet-based data over fixed and wireless networks right at the beginning, and designed the video codec from that perspective. Common

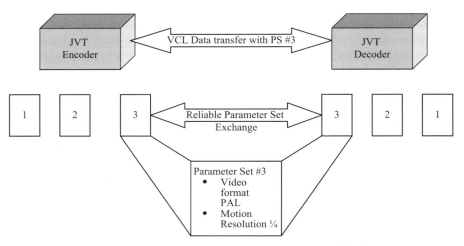

Figure 5.5 Parameter set concept [19]. (©2002 IEEE.)

test cases for such transmission include fixed Internet conversational services as well as packet-switched conversational services and packet-switched streaming services over 3G mobile networks. These IP networks usually employ IP on the network layer, UDP at the transport layer, and RTP at the application layer. IP and UDP offer an unreliable datagram service, while RTP makes the transport of media possible. Sequence numbers are used to restore the out-of-order IP packets. RTP payload does not add to the bit stream, but specifies how the media information should be interpreted. The standardization process of both JVT codec and RTP payload specifications for H.264/AVC is still an ongoing issue, but the goal of designing a simple coding scheme should be achieved [19, 20]. The JVT coding network environment is shown in Figure 5.6.

Support for Different Bit Rates, Buffer Sizes, and Start-Up Delays of the Buffer

In many video applications, the peak bit rate varies according to the network path and also fluctuates in time according to network conditions. In addition, the video bit streams are delivered to a variety of devices with different buffer capabilities. A flexible decoder buffer model, as suggested for H.264/AVC, would support this wide variety of video application conditions – bit rates, buffer sizes, and start-up delays of the buffer. The H.264/AVC Video coding standard requires that the bit stream to be transmitted should be decodable by a hypothetical reference decoder (HRD) without an underflow or overflow of the reference buffer. This aspect of the decoder is analyzed in [21] in order to support different bit rates. A typical reference buffer is shown in Figure 5.7.

The HRD coded picture buffer represents a means to communicate how the bit rate is controlled in the process of compression. The HRD contains coded picture

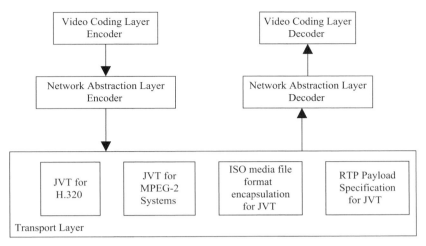

Figure 5.6 JVT coding in the network environment [19]. (©2002 IEEE.)

buffers (CPB) through which compressed data flows with a precisely specified arrival and removal timing, as shown in Figure 5.8 (HRD buffer verifiers). An HRD may be designed for variable or constant bit rate operation, and for low-delay or delay-tolerant behavior.

The HRD contains the following buffers, as shown in Figure 5.8:

- One or more coded picture buffers (CPB), each of which is either variable bit rate (VBR) or constant bit rate (CBR).
- One decoded picture buffer (DPB) attached to the output of one of the CPBs.

The multiple CPBs exist because a bit stream may conform to multiple CPBs. Considering the operation of a single CPB, data associated with coded pictures flow into the CPB according to a specified arrival schedule. Each coded picture is removed instantaneously and decoded by the decoder at CPB removal times. Decoded pictures are placed in the DPB at the CPB removal time. Finally, pictures are removed from the DPB towards the end of the DPB output time and the time that they are no longer needed as reference for decoding. The primary conditions are that the CPB neither underflow nor overflow and that the DPB does not overflow.

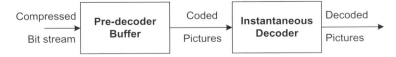

Figure 5.7 A hypothetical reference buffer.

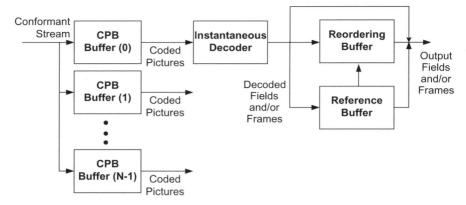

Figure 5.8 Hypothetical reference decoder (HRD) buffer verifiers [22]. (©2002 ITU-T.)

Improved Prediction

Earlier standards used a maximum of two prediction signals, one from a past frame and the other from a future frame in the bidirectional mode (B-picture), as shown in Figure 5.9. H.264/MPEG-4 allows multiple reference frames for prediction. A maximum of five reference frames may be used for prediction. Although this increases the complexity of the encoder, the encoder remains simple and the prediction is significantly improved. A multiple reference case of prediction signals from two past frames is illustrated in Figure 5.10.

Figure 5.11 shows the graph for luminance PSNR versus bit rate for the CIF video sequence Mobile and Calendar (30 fps) compressed with H.264/AVC. The graph shows that the multiple reference prediction always gives a better performance – improvement of around 0.5 dB for low bit rates of the order of 0.5 Mbps and around 1 dB for higher bit rates of the order of 2.5 Mbps.

Improved Fractional Accuracy

Fractional pel values add significantly to the accuracy of the reconstructed image. These are imaginary pel positions assumed to be stationed between physical pels. Their values are evaluated using interpolation filters. Previous standards have incorporated half-pel and quarter-pel accuracies. H.264/AVC improves prediction

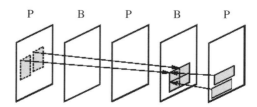

Figure 5.9 A bidirectional mode with one past and one future macroblock prediction signal.

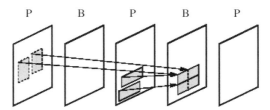

Figure 5.10 Multiple reference mode, which allows a linear combination of two past macroblock prediction signals.

capability by incorporating quarter-pel accuracies. This would increase the coding efficiency at high bit rates and at high video resolutions [22]. Integer and fractional position of pixels in a block are presented in Figure 5.12.

Significant Data Compression
Earlier standards implemented quantization steps with constant increments. H.264/AVC includes a scalar quantizer with step sizes that increase at the rate of 12.5 percent. Chrominance values have finer quantizer steps. This provides significant data compression [22].

Better Coding Efficiency
H.264/AVC uses UVLC (universal variable length coding), CAVLC (context-based variable length coding) and CABAC (context-based adaptive binary arithmetic coding) to provide efficient entropy coding. CABAC provides as good as 2 : 1 coding gain over MPEG-2.

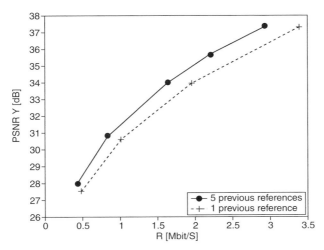

Figure 5.11 Luminance PSNR versus bit rate for the CIF video sequence Mobile and Calendar (30 fps) compressed with H.264/MPEG-4 Part 10 [23]. (©IEEE 2002.)

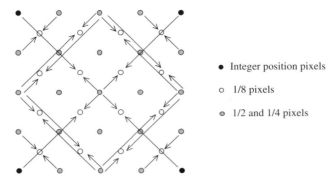

Figure 5.12 Integer and fractional position of pixels in a block.

Overlay Coding Technique
Faded transitions are such that the pictures of two scenes are laid on top of each other in a semitransparent manner, and the transparency of the pictures at the top gradually changes in the transition period. Motion compensation is not a powerful enough method to represent the changes between pictures in the transition during a faded scene. H.264/AVC utilizes an overlay coding technique that provides over 50 percent bit rate savings in both cross-fades and through-fades compared to earlier techniques [24].

Better Video Quality
H.264/AVC employs blocks of different sizes and shapes, higher resolution fractional-pel motion estimation, and multiple reference frame selection. These provide better motion estimation/compensation. On the transform front, H.264/AVC uses integer-based transform, which approximates the DCT used in other standards besides eliminating the mismatch problem in its inverse transform. Hence, the video received is of very good quality compared to the previous standards [25].

Group Capabilities
H.264/AVC grouped its capabilities into profiles and levels – Baseline, Main, and Extended profile [18]. A profile is a subset of the entire bit stream of syntax that is specified by the International Standard. Within each profile there are a number of levels designed for a wide range of applications, bit rates, resolutions, qualities, and services. A *level* has a specified set of constraints imposed on parameters in a bit stream. It is easier to design a decoder if the profile, level, and hence the capabilities are known in advance. The three profiles and their proposed areas of applications are shown in Table 5.8.

Figure 5.13 represents overlapping features of H.264/AVC MPEG-4 Part 10 profiles. The Baseline profile is intended to be the common profile supported by all implementations. Reference [26] suggests that the Baseline profile and its levels would probably be royalty-free for all implementations. A new amendment to

Table 5.8 H.264/MPEG-4 profiles and their major application areas

Profile	Applications
Baseline	Video conferencing, video telephony
Main	Broadcast video
Extended	Streaming media
High profiles	HD-DVD, BD-DVD, DVB, D-Cinema

H.264/AVC called Fidelity Range Extensions (FRExt Amendment 1) was added in August 2004. The *FRExt* project produced a suite of four new profiles called High profiles [27]. All of these profiles support all features of the prior Main profile, and additionally support an adaptive transform blocksize and perceptual quantization scaling matrices.

Basic Architecture of the Standard

Conceptually, H.264/AVC consists of two layers – the video coding layer (VCL) and the network abstraction layer (NAL). VCL is the core coding layer, which concentrates on attaining maximum coding efficiency. NAL abstracts the VCL data in terms of the details required by the transport layer and to carry this data over a variety of networks. The VCL layer takes care of the coding of transform coefficients and motion estimation/compensation information. NAL provides the header information about the VCL format, in a manner that is appropriate for conveyance by the transport layers or storage media. A NAL unit (NALU) defines a generic format for use in both packet-based and bit-streaming systems. The format for both the systems is the same except that the NAL unit in a bit-stream system can be preceded by a start code. The basic block diagram of a H.264/AVC coder is shown in Figure 5.14.

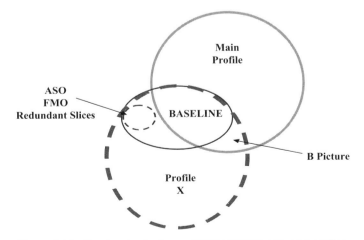

Figure 5.13 Overlapping features of H.264/MPEG-4 Part 10 profiles.

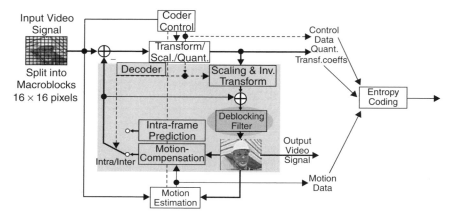

Figure 5.14 Basic block diagram of H.264/MPEG-4 Part 10 coder.

Common video formats used are CIF and QCIF, as shown in Figure 5.15. The luminance and chrominance blocks follow the 4:2:0 format. A picture is divided into macroblocks of 16 × 16 luma samples with two associated 8 × 8 chroma samples. A set of macroblocks forms a slice. Each macroblock belongs to exactly one slice. The minimum number of slices for a picture is one.

A sequence-GOP-picture-slice-macroblock-block is shown in Figure 5.16. Each macroblock (16 × 16) can be partitioned into blocks with sizes 16 × 16, 16 × 8, 8 × 16, and 8 × 8. An 8 × 8 block can further be subpartitioned into subblocks with sizes 8 × 8, 8 × 4, 4 × 8, and 4×4 as shown in Figure 5.17. Each block is motion compensated using a separate motion vector. The numbering of the motion vectors for the different blocks depends on the inter mode. For each block, the horizontal component comes first followed by the vertical component.

A coded block pattern (CBP) contains information of which 8 × 8 blocks – luminance and chrominance – contain transform coefficients. Notice that an 8 × 8 block contains four 4 × 4 blocks, meaning that the statement "8 × 8 block contains coefficients" means that "one or more of the four 4 × 4 blocks contain coefficients." The four least significant bits of CBP contain information on which of the four 8 × 8 luminance blocks in a macroblock contain nonzero coefficients. A 0 in position n of CBP means that the corresponding 8 × 8 block has no coefficients, whereas a 1 means that the 8 × 8 block has at least one nonzero coefficient.

For chrominance, three possibilities are defined in [28] as

nc = 0: No chrominance coefficients at all.

nc = 1: There are nonzero 2 × 2 transform coefficients. All chroma AC coefficients = 0. No EOB for chrominance AC coefficients is sent.

nc = 2: There may be 2 × 2 nonzero coefficients. At least one AC coefficient is nonzero. Ten end of blocks (EOB) (2 for DC coefficients and 2 × 4 = 8 for the eight 4 × 4 blocks).

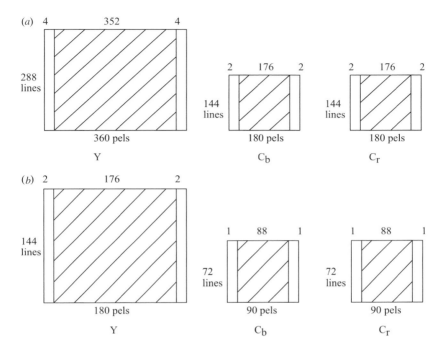

Figure 5.15 (*a*) Common intermediate format (CIF), and (*b*) quadrature common intermediate format (QCIF).

Statistics of CBP values in the case of Intra and Inter are different and hence different codewords are used.

Data partitioning may or may not be used with a slice. When data partitioning is not used, the coded slices start with a slice header and are followed by the entropy-coded symbols of the macroblock data for all the macroblocks of the slice in raster scan order. When data partitioning is used, the macroblock data of a slice is

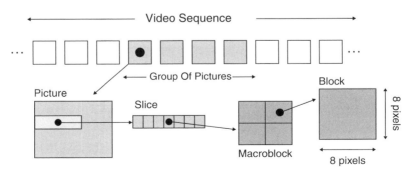

Figure 5.16 Sequence-GOP-picture-slice-macroblock-block.

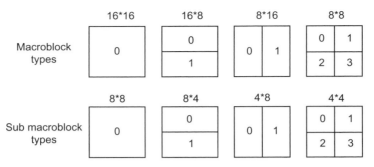

Figure 5.17 Macroblock subdivision type and block order [22]. (©2002 ITU-T.)

partitioned in up to three partitions – header information; intra coded block pattern (CBP) and coefficients; and inter coded block pattern and coefficients. The order of the macroblocks in the transmitted bit stream depends on the macroblock allocation map (MAP). The macroblock allocation map consists of an array of numbers, one for each coded macroblock indicating the slice group to which the coded macroblock belongs [22].

The H.264/AVC coder involves motion estimation and compensation in order to best provide a prediction for a block. The transform coefficients of a block are scaled and quantized and later entropy coded for transmission. Before entropy coding, the transform coefficients are reconstructed. A deblocking filter is applied to all reconstructed macroblocks of a picture and then stored for reference. These stored frames may then be used later for motion estimation and compensation. Reference [29] indicates that 52 QP (quantization parameter) values can be used, from -12 to $+39$. There is an increase in step size of about 12 percent from one QP to the next and there is no *dead zone* in the quantization process. The quantizer value can be changed at the macroblock level. Different QP values are used for luma and chroma.

Coding a picture on a slice basis [17] involves the decision as to whether it is to be Inter or Intra coded; coded with or without motion compensation. Intra-predictive coding indicates entropy coding of the transform coefficients of the blocks, without prediction from any reference frame, in estimating the coefficient values. Inter-predictive coding indicates using reference frames stored in the buffer to predict the values of the current frame. Motion compensation [233, 234] may be used when the motion compensated macroblock follows the mean square error (MSE) limit given by

$$\mathrm{MSE} = \frac{1}{256} \sum_{m=0}^{15} \sum_{n=0}^{15} [x_0(m,n) - x_{mc}(m,n)]^2$$

where $x_0(m, n)$ denotes original macroblock and $x_{mc}(m, n)$ denotes the motion compensated macroblock. The mean square error along with the variance of the input (VAR input) macroblock may be used in deciding whether Intra or Inter mode is used. Intra mode is used when the input variance is smaller and the MSE is greater. Please refer to reference [6] for further details. Each block is motion compensated using a separate motion vector.

Intraprediction

Intraprediction it is referred to as an Intra (I) slice. It uses only transform coding and the neighboring blocks of the same frame to predict block values. Intraprediction is performed on 16×16 luma and 4×4 luma blocks. No intraprediction is performed on 8×8 luma blocks.

Intra Prediction for 4 × 4 Luma Block

A 4×4 luma block contains samples as shown in Figure 5.18. There are nine intra prediction modes, as shown in Table 5.9. Modes 0, 1, 3, 4, 5, 6, 7, and 8 are directional prediction modes as indicated in Figure 5.19. Mode 2 is *DC prediction*.

Example 5.1. To illustrate intraprediction, let us consider two modes.

Mode 0: *Vertical prediction*. This mode shall be used only if A, B, C, D are available. The prediction in this mode shall be as follows:

a, e, i, m are predicted by A,

b, f, j, n are predicted by B,

c, g, k, o are predicted by C,

d, h, l, p are predicted by D.

Mode 2: *DC prediction has the following rules*. If all samples A, B, C, D, I, J, K, L are available, all samples are predicted by $(A + B + C + D + I + J + K + L + 4) \gg 3$. If A, B, C, and D are not available and I, J, K, and L are available, all samples shall be predicted by $(I + J + K + L + 2) \gg 2$. If I, J, K, and L are not available and A, B, C, and D are available, all samples shall be predicted by $(A + B + C + D + 2) \gg 2$. If all eight samples are not available, the prediction for all luma samples in the 4×4 block shall be 128. A block may therefore always be predicted in this mode.

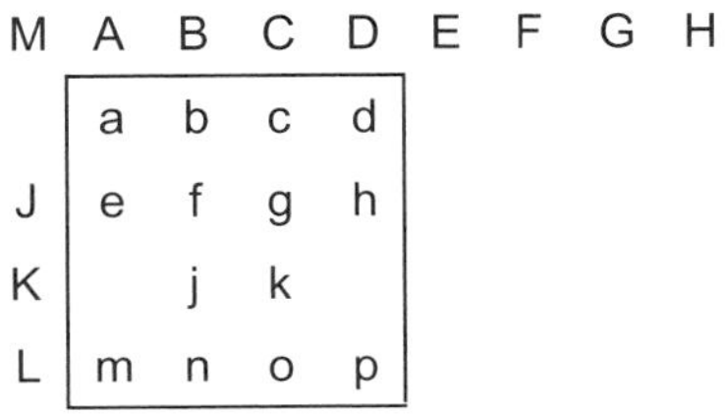

	M	A	B	C	D	E	F	G	H
		a	b	c	d				
	J	e	f	g	h				
	K		j	k					
	L	m	n	o	p				

Figure 5.18 Identification of samples used for intraspatial prediction [22]. (©2002 ITU-T.)

Table 5.9 Intraprediction modes for 4 × 4 luma block

Mode Name
Vertical
Horizontal
DC
Diagonal down/left
Diagonal down/right
Vertical–right
Horizontal–down
Vertical–left
Horizontal–up

Intraprediction for Chroma

Intraprediction for chroma blocks supports only one mode. An 8 × 8 chroma macroblock consists of four 4 × 4 blocks A, B, C, D as shown in Figure 5.20.

Example 5.2. There are four prediction cases depending upon whether S0, S1, S2 or S3 are inside or outside. For example, if all are inside:

$A = (S0 + S2 + 4)/8$
$B = (S1 + 2)/4$
$C = (S3 + 2)/4$
$D = (S1 + S3 + 4)/8$

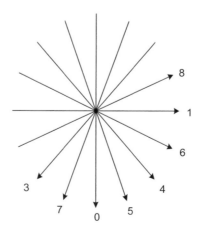

Figure 5.19 Intraprediction directions [22]. (©2002 ITU-T.)

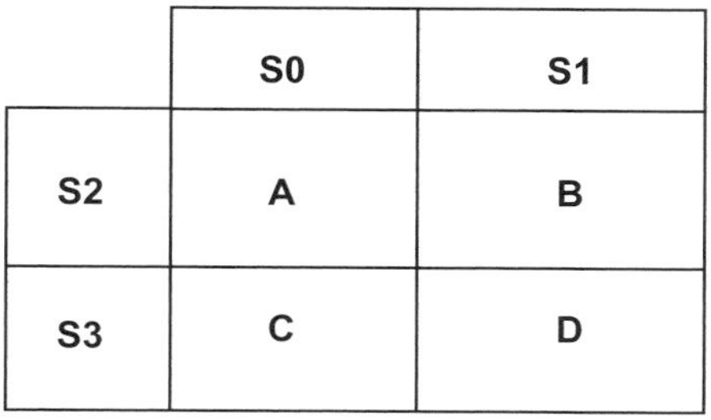

	S0	S1
S2	A	B
S3	C	D

Figure 5.20 Intraprediction for chroma macroblock.

Interprediction

Interprediction may be performed on 16×16, 16×8, 8×16, 8×8 blocks and 4×8, 8×4 and 4×4 subblocks. There are different types of interprediction pictures depending on the reference pictures used in the prediction. The P, B picture prediction concept is shown in Figure 5.21. As in previous standards, a Prediction (P) picture uses a previous encoded P or I picture for prediction. H.264/AVC allows single or multiple reference frames, a maximum of five reference frames. In case of multiple reference frames, interpolation filters are used to predict the transform coefficients. It is coded using forward motion compensation.

As in previous standards, a B picture uses both past and previous pictures as reference. Thus it is called Bidirectional prediction. It uses both forward and backward motion compensation. However, unlike previous standards, a B picture can utilize a B, P, or I picture for prediction. There are five prediction types supported by B picture. They are forward, backward, bipredictive, direct, and intra prediction modes. Forward prediction indicates that the prediction signal is formed from a reference picture in the forward reference frame buffer. A picture from the backward reference frame buffer is used for backward prediction. In both direct and bipredictive modes, prediction signal is formed by weighted average of a forward and backward prediction signal. The only difference is that the bipredictive mode has separate encoded reference frame parameters and motion vectors for forward, and backward, whereas in the direct mode, the reference frame parameters, forward, and backward motion vectors for the prediction signals are derived from motion vectors used in the macroblock of the picture [22]. An illustration of B picture prediction is given in Figure 5.22.

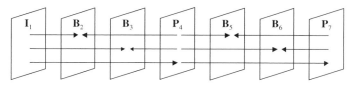

Figure 5.21 P, B picture prediction concept.

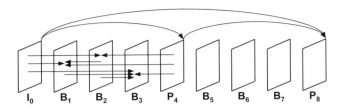

Figure 5.22 Illustration of B picture prediction.

It should be noted that the order of displaying these pictures is different from the order in which they are coded. An I picture is coded first. Then the P picture dependent on this I picture's prediction signal is coded next. Lastly, the B picture is predicted from the already coded P and I pictures.

Switched (S) is a new picture introduced for H.264/AVC. It is used in the Extended profile. Switched (S) pictures are of two types: Switched I (SI)–picture and Switched P (SP)–picture. SP coding makes use of temporal redundancy through motion-compensated inter prediction from previously decoded reference pictures, using at most one motion vector and reference picture index to predict the sample values of each block, encoded such that it can be reconstructed identically to another SP slice or SI slice. SI slice makes use of spatial prediction and can identically reconstruct to another SP slice or SI slice. The inclusion of these pictures in a bit stream enables efficient switching between bit streams with similar content encoding at different bit rates, as well as random access and fast playback modes. Macroblock modes for SP-picture are similar to P-picture, whereas SI-picture has 26 macroblock modes. Figure 5.23 represents S picture [22]. SP-pictures are much smaller (bit rate) than I-pictures.

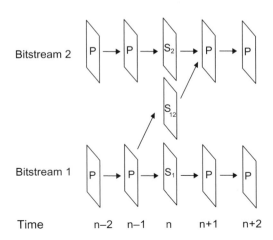

Figure 5.23 S picture [22]. (©2002 ITU-T.)

Deblocking Filter

A conditional filtering is applied to all block edges of a slice, except edges at the boundary of the picture and any edges for which the deblocking filter is disabled as specified by disable_deblocking_filter_idc. A deblocking filter is a mandatory part of the standard. This filtering is done on a macroblock basis, with macroblocks being processed in raster-scan order throughout the picture. For luma, as the first step, the 16 samples of the four vertical edges of the 4×4 raster are filtered beginning with the left edge, as shown on the left-hand side of Figure 5.24. Filtering of the four horizontal edges (vertical filtering) follows in the same manner, beginning with the top edge. The same ordering applies for chroma filtering, with the exception that two edges of eight samples each are filtered in each direction [22].

For each boundary between neighboring 4×4 luma blocks, a boundary strength Bs is assigned. Every block boundary of a chroma block corresponds to a specific boundary of a luma block. Bs values for chroma are not calculated, but simply copied from the corresponding luma Bs. If $Bs = 0$, filtering is skipped for that particular edge. In all other cases, filtering is dependent on the local sample properties and the value of Bs for this particular boundary segment.

For each edge, if any sample of the neighboring macroblocks is coded using intra macroblock prediction mode, a relatively strong filtering ($Bs = 3$) is applied. A special procedure with even stronger filtering may be applied on macroblock boundaries with both macroblocks coded using Intra macroblock prediction mode ($Bs = 4$). If neither of the neighboring macroblocks are coded using Intra macroblock prediction mode and at least one of them contains nonzero transform coefficient levels, a medium filtering strength ($Bs = 2$) is used. If none of the previous conditions is satisfied, filtering takes place with $Bs = 1$ if the following condition is satisfied:

- Any corresponding pair of motion vectors from the two neighboring macroblocks is referencing the same picture and either component of this pair has

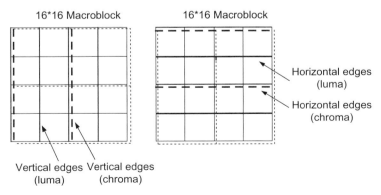

Figure 5.24 Boundaries in a macroblock to be filtered (luma boundaries shown with solid lines and chroma boundaries shown with dotted lines).

a difference of more than one sample. Otherwise filtering is skipped for that particular edge (Bs = 0).

Reference Frames

Reference frames are stored in the frame buffer. It contains short-term and long-term frames that may be used for macroblock prediction. With respect to the current frame, the frames before and after the current frame in the display sequence order are called the *reverse* reference frame and *forward* reference frame, respectively. These frames are also classified as *short-term* frame and *long-term* frame.

Memory management is required to take care of marking some stored frames as "unused" and deciding which frames to delete from the buffer for efficient memory management. There are two types of buffer management modes: adaptive buffering mode and sliding window buffering mode. Indices are allotted to frames in the buffer. Each picture type has a default index order for frames. Figure 5.25 gives an idea of storage of frames in the buffer for P picture.

Motion Vector/Estimation/Compensation

The macroblock to be encoded is compared with blocks from previous frames that are stored in the buffer. The process of finding the best match within the search window is called motion estimation. This best match is found from the reference frames stored in the buffer. The selected frames for prediction are indicated by the reference index associated with the frame index in the buffer. The process of encoding, based on the values predicted from the best match, is called *motion compensation* [30]. A block diagram emphasizing motion compensation is presented in Figure 5.26.

A motion vector is defined as a vector drawn from the reference. No vector component prediction can take place across macroblock boundaries that do not belong to the same slice. For the purpose of motion vector component prediction, macroblocks that do not belong to the same slice are treated as outside the picture. This motion vector is then encoded and transmitted. Motion compensation is applied to all blocks in a macroblock. Accuracy of half-pel and quarter-pel resolutions may be used during motion compensation.

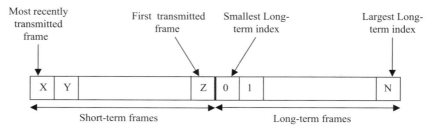

Figure 5.25 Example of short-term and long-term frame storage in the buffer.

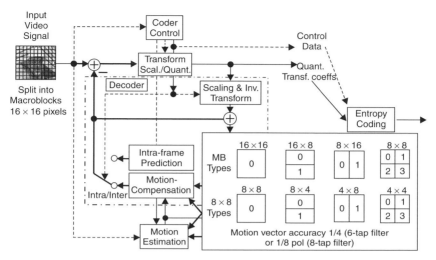

Figure 5.26 Block diagram emphasizing motion compensation [30]. (©2002 IEEE.)

Entropy Coding

Different types of encoding schemes are followed by H.264/AVC. Universal variable length coding (UVLC) is the default entropy coding. Reference [22] provides a table of code numbers and codewords that can be used. Zig-zag scan is used to read the transform coefficients as shown in Figure 5.27. Also, alternate scan is used [31].

The context-based adaptive variable length coding (CAVLC) method is used for decoding transform coefficients. The following coding elements are used:

- If there are nonzero coefficients, it is typically observed that there is a string of coefficients at the highest frequencies that are $+1/-1$. A common parameter Num-Trail is used that contains the number of coefficients as well as the number of *trailing 1s* (T1s). For T1s, only the sign has to be decoded.
- For coefficients other than the T1s, Level information is decoded.
- Lastly, the Run information is decided. Since the number of coefficients is already known, this limits possible values for Run.

Context-based adaptive binary arithmetic coding (CABAC) provides large bit-rate reduction compared to UVLC-based entropy coding. It involves constructing context models to predict the current symbol under consideration. The nonbinary symbols are converted to binary using binary decisions called bins and these bins are then encoded using binary arithmetic coding.

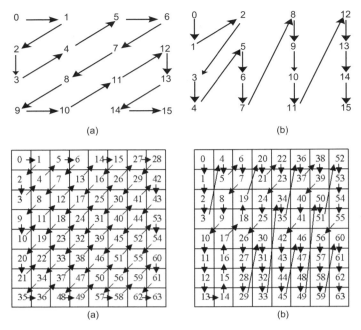

Figure 5.27 (a) Zig-zag scan and (b) alternate scan [31]. (©2002 ITU-T.)

Decoding Process

The decoder follows the reverse process of the coder (Figure 5.28). A macroblock or subpartition is decoded in the following order:

- Parsing of syntax elements using VLC/CAVLC or CABAC
- Interprediction or intraprediction
- Transform coefficient decoding
- Deblocking filter.

Profiles and Levels

Taking into account that all the users may not require all the features provided by H.264/AVC, Profiles and Levels have been introduced [32]. It specifies a set of algorithmic features and limits, which shall be supported by all decoders conforming to that profile. The encoders are not required to make use of any particular set of features supported in a profile. For any given profile, Levels generally correspond to processing power and memory capability on a codec. Each level may support a different picture size – QCIF, CIF, ITU-R 601 (SDTV), HDTV, S-HDTV, D-Cinema, and data rate varying from a few tens of kilobits per second (kbps) to hundreds of megabits per second (Mbps). The noninteger level numbers are referred as *intermediate levels*. All levels have the same status, but note that some applications may choose to use only the integer-numbered levels.

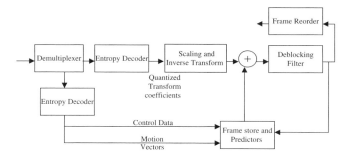

Figure 5.28 H.264/MPEG-4 decoder [30]. (©2002 IEEE.)

The profile ID for baseline profile is 66. It supports video conferencing and video telephony applications. The decoders supported by this profile support the following features [32]:

- I and P slice types
- deblocking filter
- pictures with field picture flag equal to 0
- pictures with alternate scan flag equal to 0
- pictures with macroblock adaptive frame field flag equal to 0
- zig-zag scan
- $\frac{1}{4}$-sample interprediction
- tree-structured motion segmentation down to 4×4 block size
- VLC-based entropy coding
- arbitrary slice order (ASO) – the decoding order of slices within a picture may not follow the constraint that the first macroblock in the slice is monotonically increasing within the NAL unit stream for a picture
- flexible macroblock ordering (FMO) (number of slice groups -1) <8
- the macroblocks may not necessarily be in the raster scan order; the MAP assigns MBs to a slice group and a maximum of eight slice groups are allowed
- 4 : 2 : 0 chroma format
- redundant slices – these belong to the redundant coded picture; this picture is not used for decoding unless the primary coded picture is missing or corrupted.

Main Profile
The profile ID for main profile is 77. It supports the broadcast video application. The decoders supported by this profile support the following features [32]:

- B slice type
- CABAC
- adaptive biprediction (weighted prediction)

- all features included in the baseline except
  - arbitrary slice order (ASO): in Main profile, the decoding order of slices within a picture follows the constraint that the first macroblock in the slice is monotonically increasing within the NAL unit stream for a picture
  - flexible macroblock ordering (FMO): in Main profile (number of slice groups $-1) < 0$
  - redundant slices
  - pictures with field picture flag equal to 1
  - pictures with macroblock adaptive frame field flag equal to 1
  - capable of decoding bit streams confirming to baseline profile if the following additional sequence parameter set constraints are obeyed
    - more_than_one_slice_group_allowed_flag is equal to 0
    - arbitrary_slice_order_allowed_flag is equal to 0
    - redundant_slices_allowed_flag is equal to 0.

Extended Profile
The profile ID for extended profile is 88. The decoders supported by this profile support the following features [32]:

- B slice type
- SP and SI slice types
- data partitioning slices
- adaptive biprediction (weighted prediction)
- all features included in the Baseline profile
- pictures with field picture flag equal to 1
- pictures with macroblock adaptive frame field flag equal to 1.

All video decoders supporting the Extended profile shall also support the Baseline profile.

Context-Based Adaptive Binary Arithmetic Coding (CABAC)
A general block diagram for CABAC is shown in Figure 5.29. The following scheme is used by CABAC [22]:

- Context models are created based on the neighboring symbols referred to as context modeling. The initialization of context models is explained in the following subsection.
- Nonbinary symbols are mapped into a sequence of binary decisions called bins.
- For each bin, a context variable is defined by an equation of prior transmitted symbols. The possible numerical values of a context variable are called contexts and each context has a probability distribution associated with it.

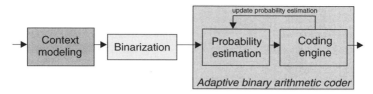

Figure 5.29 Schematic block diagram for CABAC.

- The bins are then encoded with adaptive binary arithmetic coding. After coding of each bin, the probability model is updated using the values of encoded bins.

Fractional Pel Accuracy

H.264/MPEG-4 supports one-quarter and one-eighth pel accuracy and the fractional sample accuracy is indicated by a parameter called motion resolution in H.264/AVC. If motion resolution has value 0, quarter sample resolution with a 6-tap filter is applied to luma samples in the block. If motion resolution value is 1, one-eighth sample interpolation with 8-tap filter is used. The interpolation is equivalent to upsampling of the frame. Figure 5.30 shows the interpolation process for the two motion vector accuracies of $\frac{1}{4}$- and $\frac{1}{8}$-pel accuracies [33]. In case of $\frac{1}{4}$-pel motion vector (MV) accuracy, the frame has to be upsampled by a factor of 4, but by a factor of 8 for $\frac{1}{8}$-pel MV accuracy. A combination of filters may be used to upsample by the required factor.

One-eighth-pel accuracy for motion estimation/compensation has been subsequently dropped as the extra complexity did not justify the savings in bit rate.

Chroma Sample Interpolation

Motion-compensated chroma prediction values at fractional sample positions shall be obtained using the equation

$$v = [(s - d^x)(s - d^y)A + d^x(s - d^y)B + (s - d^x)d^y C + d^x d^y D + s^2/2]/s^2$$

where A, B, C, and D are the integer position reference picture samples surrounding the fractional sample location; d^x and d^y are the fractional parts of the sample position in units of one-eighth samples for quarter sample interpolation or one-sixteenth samples for one-eighth sample interpolation; and s is 8 for quarter sample

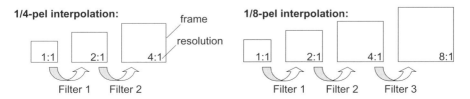

Figure 5.30 Fractional pel accuracy.

interpolation and s is 16 for one-eighth sample interpolation. The relationships between the variables in the equation and reference picture positions are illustrated in Figure 5.31.

4 × 4 Integer Transform

The previous video coding standards relied on discrete cosine transform (DCT) that provided the transformation but produced inverse transform mismatch problems. H.264/AVC uses an integer transform with a similar coding gain as a 4 × 4 DCT. It is multiplier-free, involves additions, shifts in 16-bit arithmetic, thus minimizing computational complexity, especially for low-end processors. The transformation of input pixels $X = \{x_{00} \ldots x_{33}\}$ to output coefficients $Y = \{y_{00} \ldots y_{33}\}$ is defined by [22]

$$Y = \begin{bmatrix} 1 & 1 & 1 & 1 \\ 2 & 1 & -1 & -2 \\ 1 & -1 & -1 & 1 \\ 1 & -2 & 2 & -1 \end{bmatrix} \begin{bmatrix} x_{00} & x_{01} & x_{02} & x_{03} \\ x_{10} & x_{11} & x_{12} & x_{13} \\ x_{20} & x_{21} & x_{22} & x_{23} \\ x_{30} & x_{31} & x_{32} & x_{33} \end{bmatrix} \begin{bmatrix} 1 & 2 & 1 & 1 \\ 1 & 1 & -1 & -2 \\ 1 & -1 & -1 & 2 \\ 1 & -2 & 1 & -1 \end{bmatrix}$$

In the 16 × 16 Intra prediction mode each 4 × 4 residual block is first transformed as above.

The 16 luma DC coefficients (Figure 5.32) of 16 (4 × 4) blocks, are then transformed using the *Walsh–Hadamard* transform:

$$Y_D = \left(\begin{bmatrix} 1 & 1 & 1 & 1 \\ 1 & 1 & -1 & -1 \\ 1 & -1 & -1 & 1 \\ 1 & -1 & 1 & -1 \end{bmatrix} \begin{bmatrix} x_{D00} & x_{D01} & x_{D02} & x_{D03} \\ x_{D10} & x_{D11} & x_{D12} & x_{D13} \\ x_{D20} & x_{D21} & x_{D22} & x_{D23} \\ x_{D30} & x_{D31} & x_{D32} & x_{D33} \end{bmatrix} \begin{bmatrix} 1 & 1 & 1 & 1 \\ 1 & 1 & -1 & -1 \\ 1 & -1 & -1 & 1 \\ 1 & -1 & 1 & -1 \end{bmatrix} \right) //2$$

Chroma DC coefficients (Figures 5.33 and 5.34) of four 4 × 4 blocks of each chroma

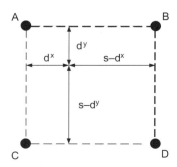

Figure 5.31 Fractional sample position dependent variables in chroma interpolation and surrounding integer position samples A, B, C, and D.

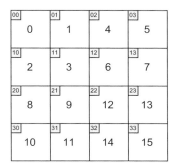

Figure 5.32 Assignment of indices.

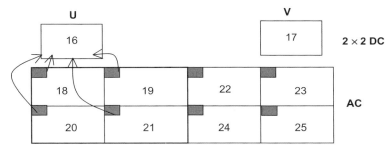

Figure 5.33 8 × 8 MB – four 4 × 4 blocks for each chroma component.

component are transformed using the *Walsh–Hadamard* transform:

$$Y = \begin{bmatrix} 1 & 1 \\ 1 & -1 \end{bmatrix} \begin{bmatrix} DC_{00} & DC_{01} \\ DC_{10} & DC_{11} \end{bmatrix} \begin{bmatrix} 1 & 1 \\ 1 & -1 \end{bmatrix}$$

Multiplication by two can be performed either through additions or through left shifts, so that no actual multiplication operations are necessary. Thus, the transform is multiplier-free.

For input pixels with 9-bit dynamic range (because they are residuals from 8-bit pixel data), the transform coefficients are guaranteed to fit within 16 bits, even when the second transform for DC coefficients is used. Thus, all transform operations can

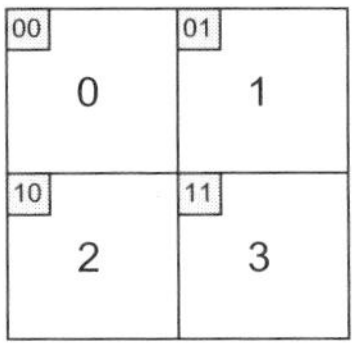

Figure 5.34 Assignment of indices for a chroma block corresponding to a 4 × 4 luma block.

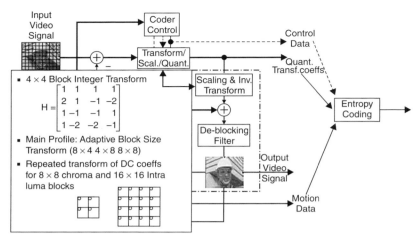

Figure 5.35 Block diagram emphasizing adaptive block size transform [30]. (©2002 IEEE.)

be computed in 16-bit arithmetic. In fact, the maximum dynamic range of the transform coefficients fills a range of only 15.2 bits; this small headroom can be used to support a variety of different quantization strategies, which are outside the scope of this specification.

The inverse transformation of normalized coefficients $Y' = \{y'_{00} \ldots y'_{33}\}$ to output pixels X' is defined by

$$X' = \begin{bmatrix} 1 & 1 & 1 & \frac{1}{2} \\ 1 & \frac{1}{2} & -1 & -1 \\ 1 & -\frac{1}{2} & -1 & 1 \\ 1 & -1 & 1 & -\frac{1}{2} \end{bmatrix} \begin{bmatrix} y'_{00} & y'_{01} & y'_{02} & y'_{03} \\ y'_{10} & y'_{11} & y'_{12} & y'_{13} \\ y'_{20} & y'_{21} & y'_{22} & y'_{23} \\ y'_{30} & y'_{31} & y'_{32} & y'_{33} \end{bmatrix} \begin{bmatrix} 1 & 1 & 1 & 1 \\ 1 & \frac{1}{2} & -\frac{1}{2} & -1 \\ 1 & -1 & -1 & 1 \\ \frac{1}{2} & -1 & 1 & -\frac{1}{2} \end{bmatrix}$$

Multiplications by $\frac{1}{2}$ are actually performed via right shifts, so that the inverse transform is also multiplier-free. The small errors introduced by the right shifts are compensated by a larger dynamic range for the data at the input of the inverse transform.

The transform and inverse transform matrices have orthogonal basis functions. Unlike the DCT, however, the basis functions do not have the same norm. Therefore, for the inverse transform to recover the original pixels, appropriate normalization factors must be applied to the transform coefficients before quantization and after dequantization. By the above exact definition of the inverse transform, the same operations will be performed on coder and decoder sides. Thus we avoid the usual problem of inverse transform mismatch (Figure 5.35).

Comparative Studies

MPEG-2, H.263, MPEG-4, and H.264/AVC are based on similar concepts and the main differences involved are the prediction signal, the block sizes used for transform coding, and the entropy coding [34]. H.264/AVC has some specific additional features that distinguish it from other standards. H.264/AVC uses a more flexible motion compensation model supporting various rectangular partitions in each macroblock. Previous standards allowed only square-sized partitions in a macroblock. Multiple reference pictures also help in better prediction, although the complexity is increased. Moreover, quarter-pixel accuracies provides high spatial accuracy. Table 5.10 provides a comparison of MPEG-1, MPEG-2, MPEG-4, and H.264/AVC.

H.264/AVC has very good features, such as multiple reference frames, CABAC, different profiles, and is suitable for applications like video streaming and video conferencing [35]. Error resilience is achieved by using parameter sets, which can be either transmitted in-band or out-of-band. It is more flexible compared to the previous standards and this enables improved coding efficiency. However, it should be noted that this is at the expense of added complexity to the coder/decoder. Also, it is not backward compatible to the previous standards. The level of complexity of the decoder is reduced by designing it specifically for a profile and a level. H.264/AVC provides three profiles and levels within them. In all, H.264/AVC seems to have a combination for good video storage and real-time video applications [36].

Fine Granularity Scalability (FGS) Video Coding Techniques

The objective of video coding has traditionally been to optimize video quality at a given bit rate. Owing to the network applications, such as Internet streaming video, the objective is somewhat changed. This change is necessary because network video has introduced a new system configuration, and the network channel capacity varies over a wide range depending on the type of connections and the network traffic at any given time.

In a traditional communication system, the encoder compresses the input video signal into a bit rate that is less than, and close to, the channel capacity, and the decoder reconstructs the video signal using all the bits received from the channel. In such a model, two basic assumptions are made. The first is that the encoder knows the channel capacity. The second is that the decoder is able to decode all the bits received from the channel fast enough to reconstruct the video. These two basic assumptions are challenged in Internet streaming video applications. First of all, due to the video server used between the encoder and the channel, as shown in Figure 5.36, plus the varying channel capacity, the encoder no longer knows the channel capacity and does not known at which bit rate the video quality should be optimized. Secondly, more and more applications use a software video client/decoder that has to share the computational resources with other operations on the user terminal. The video decoder may not be able to decode all the bits received from the channel fast enough for reconstruction of the video signal. Therefore, the objective of video coding for Internet streaming video is changed to optimizing the video quality over a given bit rate instead of at a given bit rate. The bit stream should be partially decodable at any bit rate within the bit rate range to reconstruct a video signal with optimized quality at that bit rate.

Table 5.10 Comparison of standards MPEG-1, MPEG-2, MPEG-4 Visual, and H.264/AVC

Feature/Standard	MPEG-1	MPEG-2	MPEG-4 Visual	H.264/MPEG-4 Part 10
Macroblock size	16 × 16	16 × 16 (frame mode) 16 × 8 (field mode)	16 × 16	16 × 16
Block size	8 × 8	8 × 8	16 × 16, 8 × 8, 16 × 8 [21]	8 × 8, 8 × 16, 16 × 8, 16 × 16, 4 × 8, 8 × 4, 4 × 4
Transform	DCT	DCT	DCT/Wavelet transform	4 × 4 integer transform
Transform size	8 × 8	8 × 8	8 × 8	4 × 4
Quantization step size	Increases with constant increment	Increases with constant increment	Vector quantization used	Step sizes that increase at the rate of 12.5%
Entropy coding	VLC	VLC (different VLC tables for intra and intermodes)	VLC	VLC, CAVLC, and CABAC
Motion estimation and compensation	Yes	Yes	Yes	Yes, more flexible, up to 16 MVs per MB
Pel accuracy	Integer $\frac{1}{2}$-pel	Integer $\frac{1}{2}$-pel	Integer $\frac{1}{2}$-pel $\frac{1}{4}$-pel	Integer $\frac{1}{2}$-pel $\frac{1}{4}$-pel
Profiles	No	Five profiles Several levels within a profile	Eight profiles Several levels within a profile	Three profiles Several levels within a profile
Reference frame	Yes One frame	Yes One frame	Yes One frame	Yes Multiple frames (as many as five frames allowed)
Picture types	I, P, B, D	I, P, B	I, P, B	I, P, B, SI, SP
Playback and random access	Yes	Yes	Yes	Yes
Error robustness	Synchronization and concealment [20]	Data partitioning, redundancy, FEC for important packet transmission [20]	Synchronization, data partitioning, header extension, reversible VLCs	Deals with packet loss and bit errors in error-prone wireless networks
Transmission rate	Up to 1.5 Mbps	2–15 Mbps	64 kbps ~ 2 Mbps	64 kbps–150 Mbps
Encoder complexity	Low	Medium	Medium	High

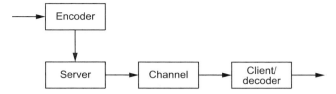

Figure 5.36 System configuration for Internet streaming video.

Before presenting FGS technique in MPEG-4, other scalable video coding techniques (SNR, temporal, spatial) will be briefly reviewed. More detailed descriptions can be found in references [37–42]. Signal-to-noise ratio (SNR) scalability is a technique to code a video sequence into two layers at the same frame rate and the same spatial resolution but different quantization accuracies. Temporal scalability is a technique to code a video sequence into two layers at the same spatial resolution, but at different frame rates. The base layer is coded at a lower frame rate. The enhancement layer provides the missing frames to form a video with a higher frame rate. Coding efficiency of temporal scalability is high and very close to nonscalable coding. Spatial scalability is a technique to code a video sequence into two layers at the same frame rate, but different spatial resolutions. The reconstructed base-layer picture is upsampled to form the prediction for the high-resolution picture in the enhancement layer.

FGS has been identified in MPEG-4 as a desired functionality, especially for streaming video applications. It is well known that wavelet coding based on zerotree arithmetic coding can also achieve the FGS functionality [43–47]. Initially, three tpes of techniques were proposed for FGS in MPEG-4, namely, bit-plane coding of the DCT coefficients [48], wavelet coding of image residue [49], and matching pursuit coding of image residue [50]. After some core experiments, bit-plane coding of the DCT coefficients was chosen due to its comparable coding efficiency and implementation simplicity [51].

The basic idea of FGS is to code a video sequence into a base layer and an enhancement layer. The base layer uses nonscalable coding to reach the lower bound of the bit-rate range. The enhancement layer is to code the difference between the original picture and the reconstructed picture using bit-plane coding of the DCT coefficients.

Figure 5.37 shows the FGS encoder, while Figure 5.38 represents the decoder structure. The bit stream of the FGS enhancement layer may be truncated into any number of bits per picture after encoding is completed. The decoder should be able to reconstruct an enhancement video from the base layer and the truncated enhancement-layer bit stream. The enhancement-layer video quality is proportional to the number of bits decoded by the decoder for each picture. The FGS decoder structure is the one standardized in the Amendment to MPEG-4.

To evaluate the coding efficiency of the FGS technique, extensive experiments have been performed to compare FGS with multiplayer SNR scalability, nonscalable coding, and simulcast. The test conditions are set up to cover a wide range of

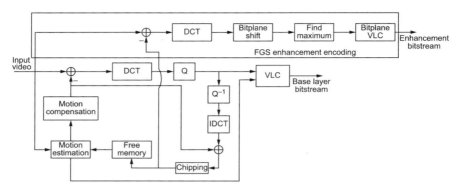

Figure 5.37 Fine granularity scalability (FGS) encoder structure [52]. (©2001 IEEE.)

sequences, bit rates, frame rates, and spatial resolutions. The number of bits for each FGS frame is truncated according to the channel bit rate and frame rate as follows:

```
bits/frame = bit rate/frame rate
```

To further improve visual quality of FGS enhancement video, two advanced features are included: frequency weighting and selective enhancement. Frequency weighting uses different weighting for different frequency components, so that more bits of the visually important frequency components are put in the bit stream ahead of that of other frequency components.

Selective enhancement uses different weighting for different spatial locations of a frame so that more bit-planes of some part of a frame are put in the bit stream ahead of those of other parts of the frame. Another advanced feature is to include resynchronization markers in the enhancement layer to make the FGS bit stream more

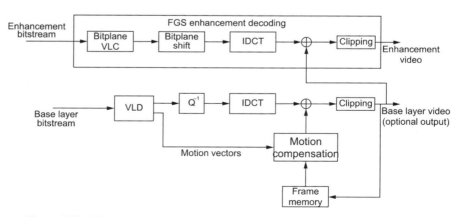

Figure 5.38 Fine granularity scalability (FGS) decoder structure [52]. (©2001 IEEE.)

error resilient for wireless applications where random burst error may occur. Yet another advanced feature is a combination of FGS with temporal scalability to increase the bit-rate range a scalable bit stream covers.

There are many advantages of using FGS for Internet streaming video applications: it allows separation of encoding and transmission, the server can transmit enhancement layer at any bit rate without transcoding, it enables video broadcast on the Internet to reach a large audience, and it provides a solution to the video server overload problem. It is expected that the Amendment of MPEG-4 will be used for many applications. Of course, there are many network-related issues to be addressed in order to take full advantage of FGS video coding.

H.264/AVC Over IP

H.264 is the denomination of ITU-T's most recent video codec Recommendation, which is also known as ISO/IEC 14496-10 or as MPEG-4 Advanced Video Codec (AVC). It is a product of the Joint Video Team (JVT) consisting of MPEG and the ITU's Video Coding Experts Group. H.264 consists of a video coding layer (VCL) and a network adaptation layer (NAL). The VCL consists of the core compression engine, and comprises syntactical levels commonly known as the block, macroblock, and slice levels. It is designed to be as network-independent as possible [35]. The VCL contains several coding tools that enhance the error resilience of the compressed video stream [53].

The NAL adapts the bit strings generated by the VCL to various network and multiplex environments. It covers all syntactical levels above the slice level. In particular, it includes mechanisms for

- The representation of the data that is required to decode individual slices (data that reside in picture and sequence headers in previous video compression standards).
- The start code emulation prevention.
- The support of supplementary enhancement information (SEI).
- The framing of the bit strings that represent coded slices for use over a bit-oriented network.

As a result of this effort, it has been shown that NAL design specified in the Recommendation is appropriate for the adaptation of H.264 over RTP/UDP/IP, H.324/M, MPEG-2 transport, and H.320. An integration into the MPEG-4 system framework is also well on its way to standardization.

The main motivation for introducing the NAL, and its separation from the VCL, is twofold. First, the Recommendation defines an interface between the signal processing technology of the VCL, and the transport-oriented mechanisms of the NAL. This allows for a clean design of a VCL implementation – probably on a different processor platform than the NAL. Secondly, both the VCL and NAL are designed in such a way that in heterogeneous transport environments, no source-based transcoding is necessary. In other words, gateways never need to reconstruct and re-encode a

VCL bit stream because of different network environments. This holds true, of course, only if the VCL of the encoder has provisioned the stream for the to-be-expected, or measured, end-to-end transport characteristics. It is, for example, the VCL responsibility to segment the bit stream into slices appropriate for the network in use, to use sufficient nonpredictively coded information to cope with erasures, and so forth.

Using IP as a transport, three major applications can currently be identified:

- conversational applications, such as video telephony and video conferencing
- the download of complete, pre-encoded video streams
- IP-based streaming.

IP-based streaming is a technology that, with respect to its delay characteristics, is somewhere in the middle between download and conversational applications.

There is no generally accepted definition for the term *streaming*. Most people associate it with a transmission service that allows the start of video playback before the whole video bit stream has been transmitted, with an initial delay of only a few seconds, and in a near real-time fashion. The video stream is either prerecorded and transmitted on demand, or a live session is compressed in real time – often in more than one representation with different bit rates – and sent over one or more multicast channels to a multitude of users. Owing to the relaxed delay constraints when compared to conversational services, some high-delay video coding tools, such as bipredicted slices, can be used. However, under normal conditions, streaming services use unreliable transmission protocols, making error control in the source and/or the channel coding a necessity. The encoder has only limited – if any – knowledge of the network conditions and has to adapt the error resilience tools to a level that most users would find acceptable. Streaming video is sent from a single server, but may be distributed in a point-to-point, multipoint, or even broadcast fashion. The group size determines the possibility of the use of feedback-based transport and coding tools.

IP networks can currently be found in two flavors: unmanaged IP networks, with the Internet as the most prominent example, and managed IP networks such as the wide-area networks of some long-distance telephone companies. An emerging third category can be addressed: wireless IP network based on the third-generation mobile networks.

H.264/AVC in Wireless Environment

The primary goals of H.264/AVC are improved coding efficiency and improved network adaptation. The syntax of H.264/AVC typically permits a significant reduction in bit rate compared to all previous standards such as ITU-T Recommendation H.263 and ISO/IEC JTC1 MPEG-4 Visual at the same quality level [54, 55].

The demand for fast and location-independent access to multimedia services offered on today's Internet is steadily increasing. Hence, most current and future cellular networks, like GSM-GPRS, UMTS, or CDMA-2000, contain a variety of packet-oriented transmission models allowing transport of practically any type of IP-based traffic to and from mobile terminals [237], thus providing users with a simple and flexible transport interface. The third-generation partnership project (3GPP) has selected several multimedia codecs for inclusion into its multimedia specifications. To provide a basic video service in the first release of 3G wireless systems, the well-established and almost identical Baseline H.263 and the MPEG-4 Visual simple profile have been integrated. The choice was based [56] on the manageable complexity of the encoding and decoding process, as well as on the maturity and simplicity of the design.

However, owing to the likely business models in emerging wireless systems in which the end user's costs are proportional to the transmitted data volume and also due to limited resources bandwidth and transmission power, compression efficiency is the main target for wireless video and multimedia applications. This makes H.264/AVC coding an attractive candidate for all wireless applications. However, to allow transmission in different environments, not only is coding efficiency relevant, but also seamless and easy integration of the coded video into all current and possible future protocol and multiplex architectures. In addition, for conversational applications the video codec's support of enhanced error-resilience features is of major importance.

Video transmission for mobile terminals is likely to be a major application in emerging 3G systems and may be a key factor in their success. The video-capable display on mobile devices paves the road to several new applications. Three major service categories were identified in the H.264/AVC standardization process [57]:

- circuit-switched and packet-switched conversational services (PCS) for video telephony and conferencing
- live or prerecorded video packet-switched streaming services (PSS)
- video in multimedia messaging services (MMS).

The transmission requirements for the three identified applications can be distinguished with respect to requested data rate, the maximum allowed end-to-end delay, and the maximum delay jitter. This results in different system architectures for each of these applications. A simplified illustration is provided in Figure 5.39. As MMS does not include any real-time constraints, encoding, transport, and decoding are completely separated. The recorded video signal is encoded offline and

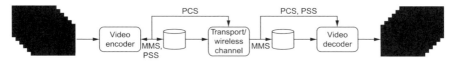

Figure 5.39 Wireless video application MMS, PSS, and PCS: differentiation by real time of offline processing for encoding, transmission, and decoding [57]. (©2003 IEEE.)

locally stored. The transmission is started using the stored signal at any time. The decoding process at the receiver is in general not started until the completion of the download. In PPS applications, the user typically requests precoded sequences, which are stored at a server. Whereas encoding and transmission are separated, decoding and display are started during transmission to minimize the initial delay and memory usage in mobile devices. Finally, in conversation services, the end-to-end delay has to be minimized to avoid any perceptual disturbances and to maintain synchronicity of audio and video. Therefore, encoding, transmission, and decoding are performed simultaneously in real time and, moreover, in both directions. These different ancillary conditions permit and require different strategies in encoding, transport, and decoding, as well as in the underlying network and control architecture.

In general, the available bandwidth and therefore the bit rates over the radio link are limited, and the costs for a user are expected to be proportional to the reserved bit rate or the number of transmitted bits over the radio link. Thus, low bit rates are likely to be typical, and compression efficiency is the main requirement for a video coding standard to be successful in a mobile environment. This makes H.264/AVC a prime candidate for use in wireless systems, because of its superior compression efficiency [55].

3G wireless transmission usually consists of two different bearer types, dedicated and shared channels. Whereas in dedicated channels one user gets assigned a fixed data rate for the entire transmission interval, shared channels allow a dynamic bit rate allocation similar to ATM and GSM GPRS. High-speed downlink packet access (HSDPA) will be an extension of the shared channel concept on the air interface. Except for MMS, all streaming and conversational applications are assumed to use dedicated channels in the initial phase of 3G wireless systems due to their almost constant bit rate behaviors. In modern system designs, an application can request one of many different QoS classes. QoS classes contain parameters like maximum error rate, maximum delay, and a guaranteed maximum data rate. Furthermore according to reference [58], applications are usually divided into different service classes: conversational, streaming, interactive, and background traffic. Characteristics and typical examples are shown in Table 5.11.

The H.264/AVC standard in the transport environment is shown in Figure 5.40. Both the VCL and NAL are parts of the H.264/AVC standard. The VCL specifies an efficient representation for the coded video signal. The NAL or H.264/AVC defines the interface between the video codec itself and the outside world. It operates on NAL units, which give support for the packet-based approach of most existing networks. At the NAL decoder interface, it is assumed that the NAL units are delivered in decoding order and that packets are either received correctly, are lost, or an error flag in the NAL unit header can be raised if the payload contains bit errors. The latter feature is not part of the standard as the flag can be used for different purposes. However, it provides a way to signal an error indication through the entire network. Additionally, interface specifications are required for different transport protocols that will be specified by the responsible standardization bodies.

5.3 MEDIA CODING

Table 5.11 QoS service classes in packet radio systems

Traffic Class	Fundamental Characteristics	Typical Examples
Conversational	Preserve time relation among information entities of the stream Conversational pattern (stringent and low delay)	Streaming voice and video telephony, video conferencing
Streaming	Preserve time relation (variation) among information entities of the stream	Multimedia (video, audio, etc.)
Interactive	Request response pattern Preserve data integrity	Web browsing, network games
Background	Destination is not expecting the data within a certain time Preserve data integrity	Background download of e-mails, files, and so on

Although the design of the H.264/AVC codec basically follows the hybrid design (motion compensation with lossy coding of residual signal) of prior video coding standards such as MPEG-2, H.263, and MPEG-4 Visual, it contains many new features that enable it to achieve a significant improvement in terms of compression efficiency. This is the main reason why H.264/AVC will be very attractive for use in wireless environments with the costly resource bit rate. The main features for significantly increased coding efficiency are multiframe motion-compensation, generalized B-pictures concepts, quarter-pel motion accuracy, intracoding utilizing prediction in the spatial domain, in-loop deblocking filters, and efficient entropy-coding methods [35].

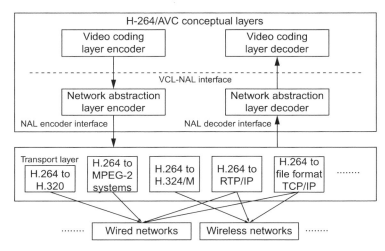

Figure 5.40 H.264/AVC standard in transport environment [57]. (©2003 IEEE.)

The normative part of a video coding standard in general only consists of the appropriate definition of the order and semantics of the syntax elements and the decoding of error-free bit streams. This allows a significant flexibility at the encoder, which can, on the one hand, be exploited for pure compression efficiency, and on the other hand, several included features in the standard can be selected by the encoder for other purposes such as error resilience, random access, and so on. A typical encoder, with the main encoding options, is shown in Figure 5.41. The encoding options relevant for wireless transmission are now highlighted. The recorded video data are preprocessed by appropriate spatial and temporal preprocessing such that the data rates and displays in a wireless environment are well matched. For the quantization of transform coefficients, H.264 coding uses scalar quantization. The quantizers are arranged in such a way that there is an increase of approximately 12.5 percent from one quantization parameter (QP) to the next. The quantized transform coefficients are converted into coding symbols and all syntax elements of a macroblock (MB), including the coding symbols, are conveyed by entropy coding methods. An MB can always be coded in one of several intra modes with and without prediction, as well as various efficient inter modes. Each motion-compensated mode corresponds to a specific partition of the MB into fixed-size blocks used for motion description, and up to 16 motion vectors may be transmitted for MB. In addition, for each MB, a different reference frame can be selected. Finally, an NAL unit in single-slice mode consists of the coded MBs of an entire frame or subset of frame.

In addition to pure compression efficiency features, additional tools for different purposes have been included in H.264/AVC. We will highlight these features with application to wireless video transmission. Because of the strict separation of encoding, transmission, and decoding, the main issue for MMS is compression efficiency.

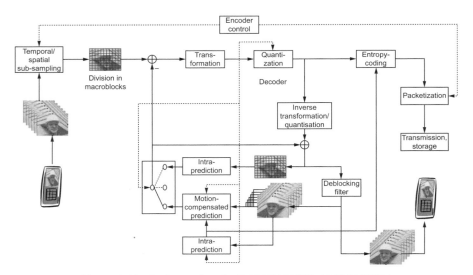

Figure 5.41 Coding options for H.264/AVC [57]. (©2003 IEEE.)

Other helpful features include the insertion of regular intra frames with instantaneous decoder refresh (IDR) for random access and fast forward. The rate control is typically applied such that video quality is almost constant over the sequence, regardless of the scene complexity, except for constraints from the hypothetical reference decoder (HRD). If time, memory capabilities, and battery power permit, several encoding passes for optimized rate distortion (RD) performance would even be possible. On the link layer, reliable transmission strategies known for data transmission such as file download are used.

Owing to online transmission and decoding, streaming applications involve more technical challenges than MMS. Usually, pre-encoded data are requested by the user, which inherently does not allow an adaptation to the transmission conditions such as bit rate or error rate in the encoding process. However, the receiver usually buffers the received data and starts playback after a few seconds. Once starting playback, a continuous presentation of the sequence should be guaranteed. As wireless channels usually show ergodic behavior within a window of a few seconds, reliable transmission schemes can be applied on the link layer, especially when the channel is known at the transmitter or retransmissions for erroneous link layer packets can be used as, for example, in the acknowledged mode. Slow variance due to distance, shadowing, or varying multiuser topology in the supported cell with renewed resource allocation, transform the wireless channel into a slowly varying variable-bit-rate channel. With an appropriate setting of the initial delay and receiver buffer, a certain quality of service can be guaranteed [59].

Furthermore, general channel-adaptive streaming technologies, which allow reacting to variable-bit-rate channels, have gained significant interest recently. These techniques can be grouped into three different categories. First, adaptive media playout is a new technique that allows a streaming media client, without the involvement of the server, to control the rate at which data are consumed by the playout process. Therefore, the probability of decoder buffer underflows and overflows can be reduced, but still noticeable artifacts in the displayed video occur [60]. A second technology for a streaming media system is proposed, which makes decisions that govern how to allocate transmission resources among packets. A flexible framework to allow RD optimized packet scheduling is also provided. Finally, it is shown that this RD-optimized transmission can be supported, if media streams are pre-encoded with appropriate packet dependencies, possibly adapted to the channel (channel-adaptive packet dependency control) [61].

The latter techniques are supported by H.264/AVC by various means. As the streaming server is, in general, aware of the current channel bit rate, the transmitter can decide to send several pre-encoded versions of the same content, taking into account the expected channel behavior. If the channel rate fluctuates only within a small range, frame dropping of nonreference frames may be sufficient, resulting in well-known temporal scalability. Switching of versions can be applied at I-frames that are also indicated as instantaneous decoder refresh (IDR) pictures to compensate for large-scale variations of the channel rate. In addition, H.264/AVC supports efficient version switching with the introduction of synchronization-predictive (SP) pictures (Fig. 5.24) [62]. Note that quality scalable video coding methods such as

MPEG-4 FGS [63] are not supported by H.264/AVC and such extensions of H.264/AVC are currently not planned. To conclude, H.264/AVC promises some significant advances of the state of the art of standardized video coding in mobile applications. In addition to excellent coding efficiency, the design of H.264/AVC also takes into account network adaptation, providing large flexibility for its use in wireless applications.

5.3.3 Transcoding Architectures and Technologies

Generally speaking, transcoding [235] can be defined as the conversion of one coded signal to another. While this definition can be interpreted quite broadly, it should be noted that research on video transcoding is usually very focused. As the number of networks, types of devices, and content representation formats increase, interoperability among different systems and different networks is becoming more important. Thus, devices such as gateways, multipoint control units, and servers must be developed to provide a seamless interaction between content creation and consumption. Transcoding of video content is one key technology to make this possible [64].

In the earliest work on transcoding, the majority of interest focused on reducing the bit rate to meet an available channel capacity. Additionally, researchers investigated conversions between constant bit rate (CBR) stream and variable bit rate (VBR) streams to facilitate more efficient transport of video. As time has moved on and mobile devices with limited display and processing power become a reality, temporal resolution reduction has also been studied. Furthermore, with the introduction of packet radio services over mobile access networks, error resilience video transcoding has gained a significant amount of attention lately, where the aim is to increase the resilience of the original bit stream to transmission errors. Some of these common transcoding operations are illustrated in Figure 5.42. Original

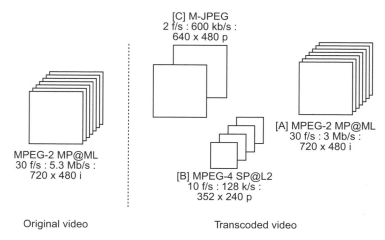

Figure 5.42 Common video transcoding operations [64]. (©2003 IEEE.)

video is encoded in an MPEG-2 format (MP@ML) at 5.3 Mbits. The input resolution is 720 × 480 interlaced, and the temporal rate is 30 frames per second. In case (A), original video is transcoded to a reduced bit rate of 3 Mbit/s. In (B), original video is transcoded to an MPEG-4 format (SP@L2) at 128 kbit/s. The output resolution is 352 × 240 progressive and temporal rate is 10 frames per second. Finally, (C) represents original video, which is transcoded to a Motion-JPEG (M-JPEG) sequence of images at a temporal rate of 2 frames per second, bit rate of 600 kbit/s, and output resolution of 640 × 480 p.

It is always possible to use a cascaded pixel-domain approach that decodes the original, performs the appropriate intermediate processing, if any, and fully re-encodes the processed signal subject to any new constraints.

In transmitting a video bit stream over a heterogeneous network, the diversity of channel capacities in different transmission media often gives rise to problems. More specifically, when connecting two transmission media, the channel capacity of the outgoing channel may be less than that of the incoming channel so that bit rate reduction is necessary before sending the video bit stream over the lower bit rate channel. In the application of video on demand (VoD), the server needs to distribute the same encoded video stream to several users through channels with different capacities, so the encoded video stream needs to be transcoded to specific bit rates for each outgoing channel. This problem also occurs in multipoint video conferencing, where the bundle of multiple video bit streams may exceed the capacity of the transmission channel and may require a bit rate reduction [65].

The simplest way to implement transcoding is to concatenate a decoder and an encoder. The decoder decompresses the bit stream, which was encoded at a bit rate R_1, and then the encodes this reconstructed video at a lower bit rate R_2. This strategy is relatively inefficient because very high computational complexity is required. Because there are common parts in both the decoder and the encoder, simplifying this cascaded architecture to reduce the complexity is possible.

We will concentrate on the transcoding of block-based video coding schemes that use hybrid DCT and motion compensation. In such schemes, the frames of the video sequence are divided into macroblocks (MBs), where each MB consists of a luminance block (e.g., of size 16 × 16, or alternatively, four 8 × 8 × blocks) along with corresponding chrominance blocks (e.g., 8 × 8 C_b and 8 × 8 C_r).

In reference [66], an architecture is introduced by assuming that the motion vectors decoded from the bit streams can be reused by the encoder.

Based on the principle that the DCT and the inverse DCT are linear operations, the number of DCT and IDCT modules was reduced from 3 to 2, and the number of frame memories from 2 to 1. Moreover, the common MC modules in the decoder and encoder are merged. This architecture results in transcoders with much reduced complexity.

To perform the MC, the macroblocks in the DCT domain need to be transformed to the pixel domain. The MC macroblocks in the pixel domain also need to be transformed back to the DCT domain for further processing. In the simplified architecture, the most time-consuming parts are in the DCT and IDCT modules, so it is highly desirable if these modules can be avoided in this architecture. Using the

DCT domain interpolation algorithm developed for transform domain video composition, the transcoding can be performed without the DCT and IDCT modules.

To speed up the transcoding operation, a video transcoder usually reuses the decoded motion vectors when re-encoding the video sequences at a lower bit rate. In many situations, frame-skipping may occur in the transcoding process. When the frame-skipping occurs in the transcoder the motion vectors from the incoming bit stream are not directly applicable because the motion estimation of the current frame is no longer based on the immediately previous frame. The motion vectors referring the current frame to its previous nonskipped frame need to be obtained. To reduce the computational complexity of obtaining these motion vectors, a combination of bilinear interpolation and search range reduction methods can be used. The bilinear interpolation roughly estimates the motion vectors from the incoming motion vectors. After the bilinear interpolation, motion estimation can be performed in a smaller search range.

For video transport over a narrow channel, usually even a good bit rate control strategy cannot allocate sufficient bits to each frame, specifically when the frames are very complex or the objects in the frame are very active. In these cases, frame skipping is a good strategy to both conform to the limitation of the channel bandwidth and maintain the image quality at an acceptable level. Actually, low bit rate video coding standards (such as H.261, H.263) allow frames to be skipped. If the frames are randomly skipped, some important frames may be discarded, resulting in an abrupt motion. This approach can be applied to video transcoders for the bit rate reduction [64].

Video transcoding is also used for video combining in a multipoint control unit (MCU). The MCU is a central server that controls multiple incoming video streams to each conference participant in a multipoint conference. To combine the multipoint incoming video streams, the MCU can perform the multiplexing function in the coded domain, but this simple scheme is not flexible and cannot take advantage of image characteristics to improve the quality of the combined images. With a transcoding approach for video combining, it is possible to achieve better video quality by allocating bits in a better way [68].

Bit Rate Reduction

The objective of bit rate reduction is to reduce the bit rate while maintaining low complexity and achieving the highest quality possible. Applications requiring this type of conversion include broadcast and Internet streaming. Ideally, the quality of the reduced rate bit stream should have the quality of a bit stream directly generated with the reduced rate. The most straightforward way to achieve this is to decode the video bit stream and fully re-encode the reconstructed signal at the new rate. This approach is illustrated in Figure 5.43, which represents transcoding in cascaded pixel-domain. The best performance can be achieved by calculating new motion vectors and mode decisions for every MB at the new rate [69]. However, significant complexity saving can be achieved, while still maintaining acceptable quality, by reusing information contained in the original incoming bit streams and also considering simplified architectures [70, 71].

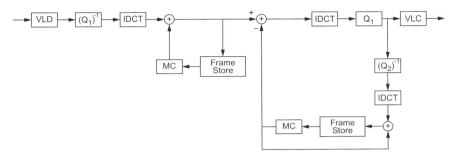

Figure 5.43 Cascaded pixel-domain transcoding architecture for bit rate reduction [64]. (©2003 IEEE.)

Transcoding Architectures

Figure 5.44 shows simplified transcoding architectures for bit rate reduction with an open-loop system in (*a*) and a closed-loop system in (*b*). In the open-loop system, the bit stream is variable-length decoded (VLD) to extract the variable-length code words corresponding to the quantized DCT coefficients, as well as MB data corresponding to the motion vectors and other MB-level information. In this scheme, the quantized coefficients are inverse quantized and then simply requantized to

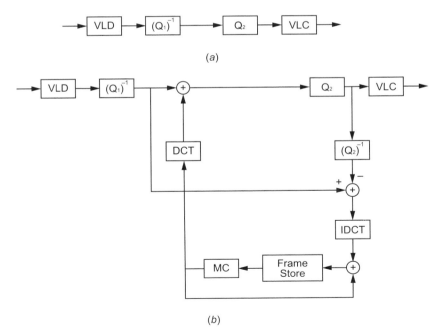

Figure 5.44 Simplified transcoding architectures for bit rate reduction: (*a*) open-loop and (*b*) closed-loop [64]. (©2003 IEEE.)

satisfy the new output bit rate. Finally, the requantized coefficients and stored MB-level information are variable length coded (VLC). An alternative open-loop scheme, which is less complex, directly cuts high-frequency data from each MB [69]. As MBs are processed, code words corresponding to high-frequency coefficients are eliminated as needed so that the target bit rate is met. Regardless of the technologies used to achieve the reduced rate, open-loop systems are relatively simple since a frame memory is not required and there is no need for an inverse IDCT. In terms of quality, better coding efficiency can be obtained by the requantization approach since the variable-length codes that are used for the requantized data will be more efficient. However, open-loop architectures are subject to drift.

In general, the reason for drift is the loss of high-frequency information. Beginning with the I-frame, which is a reference for the next P-frame, high-frequency information is discarded by the transcoder to meet the new target bit rate. Incoming residual blocks are also subject to this loss. When a decoder receives this transcoded bit stream, it will decode the I-frame with reduced quality and stores it in memory. When it is time to decode the next P-frame, the degraded I-frame is used as a predictive component and added to a degraded residual component. Considering that the purpose of the residual is to accurately represent the difference between the original signal and the motion-compensated prediction and now both the residual and predictive components are different than what was originally derived by the encoder, errors would be introduced in the reconstructed frame. This error is a result of the mismatch between the predictive and residual components. As time goes on, this mismatch progressively increases, resulting in the reconstructed frames becoming severely degraded.

The closed-loop system architecture aims to eliminate the mismatch between predictive and residual components by approximating the cascaded decoder–encoder architecture [72]. The main difference in structure between the cascaded pixel-domain architecture and this simplified scheme is that reconstruction in the cascaded pixel-domain architecture is performed in the spatial domain, thereby requiring two reconstruction loops with DCT and two IDCTs. On the other hand, in the simplified structure that is shown in Figure 5.44b, only one reconstruction loop is required with one DCT and one IDCT. In this structure, some arithmetic inaccuracy is introduced due to the nonlinear nature in which the reconstruction loops are combined. However, it has been found the approximation has little effect on the quality. With the exception of this slight inaccuracy, this architecture is mathematically equivalent to a cascaded decoder–encoder approach [72].

The described closed-loop architecture provides an effective transcoding structure in which the MB reconstruction is performed in the DCT domain. The closed-loop architecture provides an effective transcoding structure in which the MB reconstruction is performed in the DCT domain. However, since the memory stresses spatial domain pixels, the additional DCT/IDCT is still needed. This can be avoided, however, by utilizing the compressed-domain methods for MC [73]. In this way, it is possible to reconstruct reference frames without decoding to the spatial domain; several architectures describing this reconstruction process in the compressed domain have been proposed [74]. It was found that decoding completely in the compressed domain could yield equivalent quality to spatial-domain decoding [75]. However, this was achieved with floating-point matrix multiplication and proved to be quite costly. In [74] this computation was simplified by approximating the floating-point elements by power-of-two fractions so that shift operations could be used, and in reference [76], simplifications have been achieved through matrix decomposition techniques.

Regardless of the simplification applied, once the reconstruction has been accomplished in the compressed domain, one can easily requantize the drift-free blocks and VLC the quantized data to yield the desired bit stream. In reference [74], the bit reallocation has been accomplished using the Lagrangian multiplier method. In this formulation, sets of quantizer steps are found for a group of MBs so that the average distortion caused by transcoding error is minimized.

In reference [77], further simplifications of the DCT-based MC process were achieved by exploiting the fact that the stored DCT coefficients in the transcoder are mainly concentrated in low-frequency areas. Therefore, only a few low-frequency coefficients are significant and an accurate approximation to the MC process that uses all coefficients can be made.

In reduced resolution transcoding, drift errors are caused by many factors, such as requantization, motion vector truncation, and downsampling. Such errors can only propagate through intercoded blocks. By converting some percentage of interceded blocks to intracoded blocks, drift propagation can be controlled. In the past, the concept of intrarefresh has successfully been applied to error-resilience coding schemes, and it has been found that the same principle is also very useful for reducing the drift in a transcoder [78].

The intrarefresh architecture for spatial resolution reduction is illustrated in Figure 5.45. In this scheme, output MBs are subject to a DCT-domain downconversion, requantization, and variable-length coding. Output MBs are either derived directly from the input bit stream, that is, after variable-length decoding and inverse quantization, or retrieved from the frame store and subject to a DCT. Output blocks that originate from the frame store are independent of other data, hence coded as intrablocks; there is no picture drift associated with these blocks.

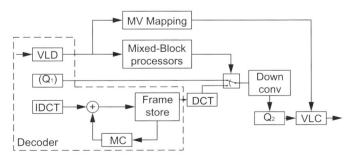

Figure 5.45 Intrarefresh architecture for reduced spatial resolution transcoding [64]. (©2003 IEEE.)

The decision to code an intrablock from the frame store depends on the MB coding modes and pictures statistics. In the first case, based on the coding mode, an output MB is converted when the possibility of a mixed block is detected. In the second case, based on the picture statistics, the motion vector and residual data are used to detect blocks that are likely to contribute to a larger drift error. For this case, picture quality can be maintained by employing an intracoded block in its place. Of course, the increase in the number of intrablocks must be compensated by the rate control by adjusting the quantization parameters so that the target rate can accurately be met. This is needed since intrablocks usually require more bits to code. Further details on rate control can be found in reference [79].

Reducing the temporal resolution of a video bit stream is a technique that may be used to reduce the bit rate requirements imposed by a network, to maintain higher quality of coded frames, or to satisfy processing limitations imposed by a terminal. For instance, a mobile terminal equipped with a 266 MHz general-purpose processor may only be capable of decoding and displaying 10 f/s. In another instance, the terminal may simply wish to conserve its battery life at the cost of receiving fewer frames. In both of these instances, one should keep in mind the dependencies that exist, such as the particular coding format, the given spatial resolution, power consumption properties, as well as the efficiency of the implementation. Also, when it comes to processing requirements, there are trade-offs that can be made between spatial and temporal resolutions.

As discussed earlier, motion vectors from the original bit stream are typically reused in bit rate reduction and spatial resolution reduction transcoders to speed up the re-encoding process. In the case of spatial resolution reduction, the input motion vectors are mapped to the lower spatial resolution. For temporal resolution reduction, we are faced with a similar problem in that it is necessary to estimate the motion vectors from the current frame to the previous nonskipped frame that will serve as a reference frame in the receiver. The general problem is illustrated in Figure 5.46. Since frame $(n-1)$ is dropped, a new motion vector to predict frame (n) from frame $(n-2)$ is estimated.

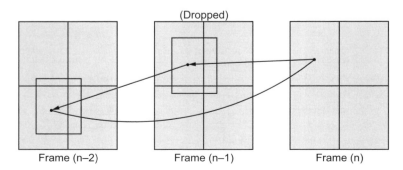

Figure 5.46 Motion vector re-estimation.

Transmitting video over wireless channels requires taking into account the conditions in which the video will be transmitted. In general, wireless channels have low bandwidth and higher error rate than wired channels. Error-resilience transcoding for video over wireless channels is needed in this case and has been studied in references [80, 81].

In reference [80], the authors present a method that is built on three steps. First, they use a transcoder that injects spatial and temporal resiliences into an encoded bit stream where the amount of resilience into an encoded bit stream is tailored to the content of the video and the prevailing error conditions, as characterized by bit error rate. The transcoder increases the spatial resilience by reducing the number of blocks per slice and increases the temporal resilience by increasing the proportion of intrablocks that are transmitted at each frame. Since the bit rate increases due to the error resilience, the transcoder achieves the (same) input bit rate at the output by dropping less significant coefficients as it increases resilience. Secondly, they derive analytical models that characterize how corruption (visual distortion) propagates in a video that is compressed using motion-compensated encoding and subjected to bit errors. Thirdly, they use rate distortion theory to compute the optimal allocation of bit rate between spatial resilience, temporal resilience, and source rate. Furthermore, they use the analytical models to generate the resilience rate-distortion functions that are used to compute the optimal resilience. The transcoder then injects this optimal resilience and improves video quality in the presence of errors, while maintaining the same input bit rate.

In reference [81], the authors propose error-resilience video transcoding for Internet-work communications using a general packet radio services (GPRS) mobile access network. The error-resilience transcoding takes place in a proxy, which provides the necessary output rate with the required amount of robustness. Here we use two error-resilience coding schemes: adaptive intrafresh (AIF) and feedback control signaling (FCS). The schemes can work independently or combined. Since both AIF and FCS increase bit rate, a simple bit rate regulation mechanism is needed that adapts the quantization parameters accordingly. The system uses two primary control feedback mechanisms. First, feedback signals that contain information related to the output channel conditions, such as bit error rate, delay, lost/received packets, and so on, based on the received feedback, AIF, and/or FCS can be used to insert the necessary robustness to the transcoded data. For example, in the case of increased bit error conditions, AIF is used as the major resilience block to stop the potential error accumulation effects resulting from transmission errors; for example, high motion areas are transcoded to intracoded MBs, which do not require MC. The second control feedback mechanism comprises adaptive rate transcoding.

This requires a feedback signaling method for the control of the output bit rate from the video transcoder. In this way, the signaling is originated from the output video frame buffer within the network-monitoring module, which continuously monitors the flow conditions. In the case of underflow, a signal is returned to the transcoder for an increase in bit rate. In the case of overflow, the signal indicates to the transcoder that it should decrease the bit rate. This is a relatively straightforward rate-controlling scheme for congestion control. Experiments showed superior transcoding performance over the error-prone GPRS channels to the nonresilient video.

Looking to the future of video transcoding, there are still many topics that require further study. One problem is finding an optimal transcoding strategy. Given several transcoding operations that would satisfy given constraints, a means for deciding the best one in a dynamic way has yet to be determined.

In the work to construct utility functions, features are first extracted from video, then machine learning and classification techniques are used to estimate the subjective/objective quality of the video coded according to the transcoding operation. Overall, further study is needed towards a complete algorithm that can measure and compare quality across spatiotemporal scales, possibly taking into account subjective factors, and account for a wide range of potential constraints (e.g., terminal, network, and user characteristics). Another topic is the transcoding of encrypted bit streams. The problems associated with the transcoding of encrypted bit streams include breaches in security by decrypting and reencrypting within the network, as well as computational issues. These problems have been circumvented with a secure scalable streaming format [82] that combines scalable coding techniques with a progressive encryption technique. However, handling this for nonscalable video and streams encrypted with traditional encryption techniques is still an open issue.

Multipoint Video Bridging

Multipoint videoconferencing is a natural extension of point-to-point video conferencing. With the rapid growth of video conferencing, the need for multipoint video conferencing is also growing [83, 84]. Multipoint video technology involves networking, video combining, and presentation of multiple coded video signals. The same technology can also be used for distance learning, remote collaboration, and video surveillance involving multiple sites. For a multiple video conference over a wide-area network, the conference participants are connected to an MCU in a central office. A video in the MCU combines the multiple coded digital video streams from the conference participants into a coded video bit stream that conforms to the syntax of the video coding standard and sends it back to the conference participants for decoding and presentation. The application scenario of four persons participating in a four-point video conference is shown in Figure 5.47. Although in the following, we will use the four-point video conference as an example, the discussion can be easily generalized to multiple video conference involving more sites.

One approach used in the video combiner to combine the multiple coded digital video streams is the coded domain combining approach. In this approach, the video combiner modifies the headers of the individual coded video bit streams from the conference participants, multiplexes the bit streams, and generates new headers to produce a valid code and combines video bitstream confirming to the video coding standard. For example, a code domain combiner can modify the headers of four coded QCIF (176 × 144 pixels) video sequences and concatenate them into a CIF (352 × 288 pixels) coded video sequence. Using the coded domain combining approach, since the combiner only needs to perform the multiplexing and header

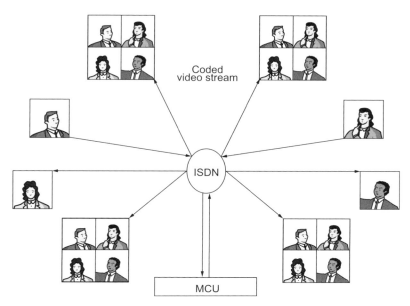

Figure 5.47 Four-point video conference over a wide-area network.

modifications functions in concatenating the video bit streams, the required processing is minimal. Also, because it does not need to decode and re-encode the video streams, it does not introduce quality degradation. However, the coded domain combining approach offers limited flexibility for users to manipulate the video bit streams.

Another solution is to use the transcoding approach. When using the transcoding approach for video combining, the video combiner decodes each coded video stream, combines the decoded video streams, and re-encodes the combined video at the transmission channel rate. Figure 5.48 shows a block diagram of video combining for multipoint video conferencing using the transcoding approach.

The transcoding approach requires more computations: it must decode the individual video stream and encode the combined video signal. The video quality using the transcoding approach will also potentially suffer from the double-encoding process because the video from each user needs to be decoded, combined, and then re-encoded, which introduces additional degradation.

Since the coded domain combiner simply concatenates bit streams, the combined bit rate will be the sum of the bit rates of all the individual bit streams. For example, in a four-point video conference where the bit rate of each participant's encoder is 32 kbps, the combined video bit rate will be about 128 kbps. Thus, for each conference participant, the input and output video bit rates are highly asymmetrical. For a four-point video conference over a wide-area network such as an integrated services digital network (ISDN), where about 128 kbps is available in the 2B channels, each user can encode his/her video at only about 32 kbps, so that the combined video can fit within 2B channels.

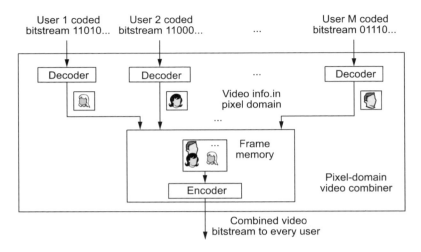

Figure 5.48 Video combining using the transcoding approach.

Active people need higher bit rates to produce good quality video, while inactive people only need lower bit rates to produce good quality video. With coded domain combining video, which allocates equal bandwidth to all participants, the bandwidth is usually too limited for an active person but is more than enough for an inactive person. It is difficult to dynamically allocate extra bandwidth to active people using the coded domain combining approach.

The transcoding approach allows each conference participant to encode the video using the full bandwidth of the transmission channel [85, 86]. Using the previous example, each participant will encode his/her video at 128 kbps instead of 32 kbps. Since each user can use the full bandwidth, and in the re-encoding process the rate-control strategies in video coding algorithm will dynamically allocate more bits to the active frames or areas, the overall quality can be more uniform compared to the coded domain combining approach where the quality of the active person may be much worse than that of the inactive people. The result is that the transcoding approach can provide a much more uniform video quality among conference participants. With complexity reduction techniques, more flexibility in video manipulation and better video quality, transcoding is a very practical approach for video combining in a multipoint video conference over a symmetrical wide-area network.

5.3.4 Multimedia Implementation

In the last two decades, progress in multimedia processing, such as digital image and video compression, has proceeded at an astounding pace. Image and video coding standards especially, developed by the Joint Picture Experts Group (JPEG) and Moving Picture Experts Group (MPEG) in the framework of the ISO/IEC, have enabled many successful digital image and video applications [87]. These applications range from digital camera, mobile communication devices, DVD and multimedia CDs on desktop and laptop computers, terrestrial and satellite digital TV, interactive digital cable television, to digital satellite networks. So, as there are methods on how best to achieve multimedia coding performance, there are also efficient techniques to process coded multimedia contents.

Multimedia processing consists of the capture, storage, manipulation, and transmission of multimedia objects such as text, handwritten data, audio objects, still images, 2/3D graphics, animation, and full-motion video. A number of implementation strategies have been proposed for processing multimedia data. These approaches can be broadly classified based on the evolution of processing architectures and on the functionality of the processors. So as to provide media processing solutions to different consumer markets, designers have combined some of the classical features from both the functional and evolution-based classifications, resulting in many hybrid solutions.

In recent years, improvements in microprocessor architecture and performance have enabled general-purpose computer clips to process compressed digital image and video in real time. Initially, this was limited only to workstations and later to high-end PCs. Now, similar functionality is available on processor clips approaching a consumer electronic price. The compelling reason for considering a computer-based television receiver or set-top box (STB) is flexibility. This flexibility is demanded by applications such as digital TV broadcasting, broadband Internet streaming, personal video recording (PVR), and so on. Thus, the same hardware resources used for watching TV can be used to implement a videotelephone or a Web browser that is capable of viewing a streaming audiovisual clip from the Internet. An application-specific processor (ASIC) only decoder is unlikely to implement these functions, because these applications require different types of source-coding standards (H.263, MPEG-4, Real-audio).

As TVs migrate from being passive broadcast audiovisual receivers towards evolving to interactive network computers, there will be a need to run large software applications in both headend and home terminals. These programs provide the interactive services because network bandwidth is still too limited and expensive for remotely implementing the rich graphical user interfaces (GUIs) needed to sell these services. These applications will need platform independence, large protected virtual address spaces, voluminous local persistent storage, the graphics rendering power to play video games, and networking capability to access streaming audiovisual contents from the server. A media-processor-based terminal architecture can provide a cost-effective (based on commodity pricing) execution environment for such applications because silicon resources, for the most part, are not dedicated to specific tasks such as video decoding or graphics rendering.

Video processing on programmable platforms offers significant advantages over dedicated hardware such as shorter development cycles, flexibility, and upgradability. Although the emergence of powerful media processors is making video processing on programmable platforms closer to reality, practical media-processor-based systems benefit from complexity-scalable video processing algorithms that can trade off video quality against computational resources while ensuring real-time performance.

Besides the applications derived from TV, portable telephone devices and palmtops are already migrating to true multimedia terminals, including still pictures and video. On the one hand, the variability of the many different standards requires flexible implementations likely achievable by general-purpose processors. On the other hand, the increasing complexity of the processing and the additional constraint of low-power devices requires a difficult search for appropriate trade-offs between flexibility and hardware-assisted coprocessing.

In this mobile/portable environment, another interesting approach for efficient implementation is to find the best trade-off between image quality (final quality of service), compression performance (available bandwidth), and necessary processor cycles (cost and autonomy). Thus, the difficulty of the problem increases because it is also necessary to explore different image processing technologies and find the ones that can better be implemented versus the desired trade-off range.

For the above reasons, the development of efficient media processors or general-purpose processors with instruction set architectural (ISA) extensions, with or without the assistance of hardware coprocessing, remains the core technological challenge for the next few years of the multimedia era.

One of the key functions in a video-compression encoder is motion estimation. For video coding standards such as MPEG and H.263, motion estimation is one of the most processing demanding functions at the encoder side. Techniques for reducing the complexity of motion estimation are very important for low-cost implementation of video encoders.

A fast motion estimation algorithm using only binary representation is desirable for both embedded system software optimization and hardware implementation with parallel architectures [88]. In a multimedia-embedded system, the video encoding module contains several major components including DCT, IDCT, motion estimation (ME), motion compensation, quantization, inverse quantization, bit rate control, and variable-length coding (VLC), where the most computationally expensive part is the motion estimation. Generally, ME takes around 50 percent of the total computational power for an optimized system. Many fast-search algorithms have been proposed [89–97]. For example, the all-binary ME algorithm features low computational complexity, reduced data bandwidth, simple software optimization, and pipelined hardware architectures. ABME uses all-binary representation for all layers [88].

With the growing popularity of battery-operated video devices such as portable video phones, hand-held DVD players, and mobile video conferencing units, reducing energy consumption of video encoders becomes a primary design requirement for many applications. A new architectural technique is proposed to reduce energy dissipation of frame memory through adaptive compression [98].

Fine granularity scalability (FGS) is the latest video compression technique provided by the MPEG-4 standard. By taking advantage of bit-plane coding of DCT residues, the compressed bit stream can be truncated at any location to support the fine rate scalability in the enhancement layer. However, both frame buffer scanning several times in bit-plane decoding and frame duplication simultaneously for base- and enhancement-layer decoding make FGS decoding more complex for implementation.

Paper [99] proposes a corresponding pair of efficient streaming schedule and pipeline decoding architectures to deal with the mentioned problems. The proposed method can be applied to the case of streaming stored FGS videos and can benefit FGS-related applications.

Texture coding based on discrete wavelet transform (DWT) is playing a leading role for its higher performance in terms of signal analysis, multiresolution features, and improved compression compared to existing methods such as DCT-based compression schemes adopted in the old JPEG standard. This success is testified by the fact that the wavelet transform has now been adopted by MPEG-4 for still-texture coding and by JPEG2000. Indeed, superior performance at low bit rates and transmission of data according to client display resolution are particularly interesting for mobile applications. The wavelet transform shows better results because it is intrinsically well suited to nonstationary signal analysis, such as images. Although it is a rather simple transform, DWT implementations may lead to critical requirements in terms of memory size and bandwidth, possibly yielding costly implementations. Thus, efficient implementations must be investigated to fit different system scenarios. In other words, the goal is to find different architectures, each of them specifically optimized for any specific system requirement in terms of complexity and memory bandwidth.

To facilitate MPEG-1 and MPEG-2 video compression, many graphics coprocessors provide the accelerators to the key function blocks, such as inverse DCT and motion compensation, in compression algorithms for real-time video decoding. The MPEG-4 multimedia coding standard supports object-based coding and manipulation of natural video and synthetic graphics objects. Therefore, it is desirable to use the graphics coprocessors to accelerate decoding of arbitrary-shaped MPEG-4 video objects as well [100]. It is found that boundary macroblock padding, which is an essential processing step in decoding arbitrarily shaped video objects, could not be efficiently accelerated on the graphics coprocessors due to its complexity. Although such a padding can be implemented by the host processor, the frame data processed on the graphics coprocessor need to be transferred to the host processor for padding. In addition, the padded data on the host processor need to be sent back to the graphics coprocessor to be used as a reference for subsequent frames. To avoid this overhead, there are two approaches of boundary macroblock padding. In the first approach, the boundary macroblock padding is partitioned into two tasks, one that the host processor can perform without the overhead of data transfers, and the second approach, in which two new instructions are specified and an algorithm is proposed for the next-generation graphics coprocessors or media-processors, which gives a performance improvement of up to a factor of nine compared to that with the PentiumIII [100].

5.4 MEDIA STREAMING

Advances in computers, networking, and communications have created new distribution channel and business opportunities for the dissemination of multimedia content. Streaming audio and video over networks such as the Internet, local area wireless networks, home networks, and commercial cellular phone systems has become a reality and it is likely that streaming media will become a mainstream means of communication. Despite some initial commercial success, streaming media still faces challenging technical issues, including quality of service (QoS) and cost-effectiveness. For example, deployments of multimedia services over 2.5G and 3G wireless networks have presented significant problems for real-time servers and clients in terms of high variability of network throughput and packet loss due to network buffer overflows and noisy channels. New streaming architectures such as pear-to-pear (P2P) networks and wireless ad hoc networks have also raised many interesting research challenges. This section is intended to address some of the principal technical challenges for streaming media by presenting a collection of the most recent advances in research and development.

5.4.1 MPEG-4 Delivery Framework

The framework is a model that hides to its upper layers the details of the technology being used to access the multimedia content. It supports virtual and known communication scenarios (e.g., stored files, remotely retrieved files, interactive retrieval from a real-time streaming server, multicast, broadcast, and interpersonal communication). The delivery framework provides, in ISO/OSI terms, a session layer service. This is further referred to as the delivery multimedia integration framework (DMIF) layer, and the modules making use of it are referred to as DMIF users. The DMIF layer manages sessions (associated to overall MPEG-4 presentations) and channels (associated to individual MPEG-4 elementary streams) and allows for the transmission of both user data and commands. The data transmission part is often referred to in the open literature as the user plane, while the management side is referred to as the control plane. The term DMIF, for instance, is used to indicate an implementation of the delivery layer for a specific delivery technology [101].

In the DMIF context, the different protocol stack options are generally named transport multiplexer (TransMux). Specific instances of a TransMux, such as a user datagram protocol (UDP), are named TransMux channels [102]. Within a TransMux channel, several streams can be further multiplexed, and MPEG-4 specifies a suitable multiplexing tool, the FlexMux. The need for an additional multiplexing stage (the FlexMux) derives from the wide variety of potential MPEG-4 applications, in which even huge numbers of MPEG-4 elementary streams (ESs) can be used at once. This is somewhat a new requirement specific to MPEG-4; in the IP world, for example, the real-time transport protocol (RTP) that is often used for streaming applications normally carries one stream per socket. However, in order to more effectively support the whole spectrum of potential MPEG-4 applications, the usage of the FlexMux in combination with RTP is being considered jointly between IETF and MPEG [103].

Figure 5.49 shows some of the possible stacks that can be used within the delivery framework to provide access to MPEG-4 content. Reading the figure as it applies for the transmitter side: ESs are first packetized and packets are equipped with

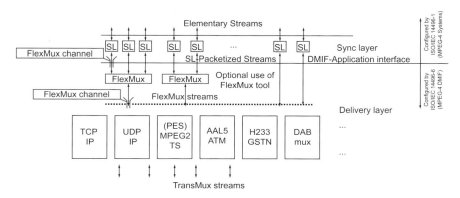

Figure 5.49 User plane in an MPEG-4 terminal [104]. (©2000 ISO/IEC.)

information necessary for synchronization (timing) – SL packets. Within the context of MPEG-4 Systems, the Sync Layer (SL) syntax is used for this purpose. Then, the packets are passed through the delivery application interface (DAI). They possibly get multiplexed by the MPEG-4 FlexMux tool, and finally they enter one of the various possible TransMuxes.

In order to control the flow of the ESs, commands such as PLAY, PAUSE, RESUME, and related parameters needed to be conveyed as well. Such commands are considered by DMIF as user commands, associated to channels. Such commands are opaquely managed (i.e., not interpreted by DMIF and just evaluated at the peer entity). This allows the stream control protocol(s) to evolve independently from DMIF. When real-time streaming protocol (RTSP) is used as the actual control protocol, the separation between use commands and signaling messages vanishes as RTSP deals with both channel setup and stream control. This separation is also void, for example, when directly accessing a file.

The delivery framework also is prepared for QoS management. Each request for creating a new channel might have associated certain QoS parameters, and a simple but generic model for monitoring QoS performance has been introduced as well. The infrastructure for QoS handling does not include, however, generic support for QoS negotiation or modification.

Of course, not all the features modeled in the delivery framework are meaningful for all scenarios. For example, it makes little sense to consider QoS when reading content from local files. Still, an application making use of the DMIF service as a whole need not be further concerned with the details of the actually involved scenario.

The approach of making the application running on top of DMIF totally unaware of the delivery stack details works well with MPEG-4. Multimedia presentations can be repurposed with minimal intervention. Repurposing means here that a certain multimedia content can be generated in different forms to suit specific scenarios, for example, a set of files to be locally consumed, or broadcast/multicast, or even interactively served from a remote real-time streaming application. Combinations of these scenarios are also enabled within a single presentation.

The delivery application service (DAI) represents the boundary between the session layer service offered by the delivery framework and the application making use of it, thus defining the functions offered by DMIF. In ISO/OSI terms, it corresponds to a Session Service Access Point.

The entity that uses the service provided by DMIF is termed the DMIF user and is hidden from the details of the technology used to access the multimedia content.

The DAI comprises a simple set of primitives that are defined in the standard in their semantic only. Actual implementation of the DAI needs to assign a precise syntax to each function and related parameters, as well as to extend the set of primitives to include initialization, reset, statistics monitoring, and any other housekeeping function.

DAI primitives can be categorized into five families, and analyzed as follows:

- service primitives (create or destroy a service)
- channel primitives (create or destroy channels)
- QoS monitoring primitives (set up and control QoS monitoring functions)
- user command primitives (carry user commands)
- data primitives (carry the actual media content).

In general, all the primitives being presented have two different (although similar) signatures (i.e., variations with different sets of parameters): one for communication from DMIF user to the DMIF layer and another for the communication in the reverse direction. The second is distinguished by the *Callback* suffix and, in a retrieval application, applies only to the remote peer. Moreover, each primitive presents both IN and OUT parameters, meaning that IN parameters are provided when the primitive is called, whereas OUT parameters are made available when the primitive returns. Of course, a specific implementation may choose to use nonblocking calls and to return the OUT parameters through an asynchronous callback. This is the case for the implementation provided in the MPEG-4 reference software.

The MPEG-4 delivery framework is intended to support a variety of communication scenarios while presenting a single, uniform interface to the DMIF used. It is then up to the specific DMIF instance to map the DAI primitives into appropriate actions to access the requested content. In general, each DMIF instance will deal with very specific protocols and technologies, such as the broadcast of MPEG-2 transport streams, or communication with a remote peer. In the latter case, however, a significant number of options in the selection of control plane protocol exists. This variety justifies the attempt to define a further level of commonality among the various options, making the final mapping to the actual bits on the wire a little more focused. Delivery applications interface (DAI) and delivery multimedia integration network interface (DNI) in the DMIF framework architecture are shown in Figure 5.50 [104].

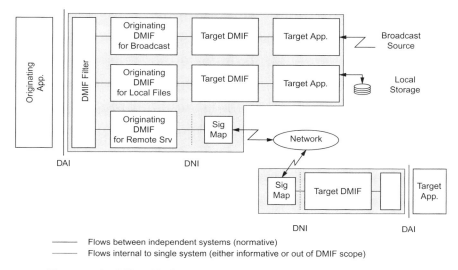

Figure 5.50 DAI and DNI in the DMIF architecture [104]. (©2000 ISO/IEC.)

The DNI captures a few generic concepts that are potentially common to peer-to-peer control protocols, for example, the usage of a reduced number of network resources (such as sockets) into which several channels would be multiplexed, and the ability to discriminate between a peer-to-peer relation (network session) and different services possibly activated within that single session (services).

DNI follows a model that helps determine the correct information to be delivered between peers but by no means defines the bits on the wire. If the concepts of sharing a TransMux channel among several streams or of the separation between network session and services are meaningless in some context, that is fine, and does not contradict the DMIF model as a whole.

The mapping between DAI and DNI primitives has been specified in reference [104]. As a consequence, the actual mapping between the DAI and a concrete protocol can be determined as the concatenation of the mappings between the DAI and DNI and between DNI and the selected protocol. The first mapping determines how to split the service creation process into two elementary steps, and how multiple channels managed at the DAI level can be multiplexed into one TransMux channel (by means of the FlexMux tool). The second protocol-specific mapping is usually straightforward and consists of placing the semantic information exposed at the DNI in concrete bits in the messages being sent on the wire.

In general, the DNI captures the information elements that need to be exchanged between peers, regardless of the actual control protocol being used.

DNI primitives can be categorized into five families, analyzed in the following:

- service primitives (create or destroy a session)
- channel primitives (create or destroy service)

- Transmux primitives (create or destroy a TransMux channel carrying one or more streams)
- channel primitives (create or destroy a FlexMux channel carrying a single stream)
- user command primitives (carry user commands).

In general, all the primitives being presented have two different but similar signatures, one for each communication direction. The *Callback* suffix indicates primitives that are issued by the lower layer. Different from the DAI, for the DNI the signatures of both the normal and the associated *Callback* primitives are identical. As for the DAI, each primitive presents both IN and OUT parameters, meaning that IN parameters are provided when the primitive is called, whereas OUT parameters are made available when the primitive returns. As for the DAI, the actual implementation may choose to use nonblocking calls, and to return the OUT parameters through an asynchronous callback. Also, some primitives use a *loop*( ) construct within the parameter list. This indicates that multiple tuples of those parameters can be exposed at once (e.g., in an array).

5.4.2 Streaming Video Over the Internet

Recent advances in computing technology, compression technology, high-bandwidth storage devices, and high-speed networks have made it feasible to provide real-time multimedia services over the Internet. Real-time multimedia, as the name implies, has timing constraints. For example, audio and video data must be played out continuously. If the data do not arrive in time, the playout process will pause, which is annoying to human ears and eyes.

Real-time transport of live video or stored video is the predominant part of real-time multimedia. Here, we are concerned with video streaming, which refers to real-time transmission of stored video. There are two modes for transmission of stored video over the Itnernet, namely the download mode and the streaming mode (i.e., video streaming). In the download mode, a user downloads the entire video file and then plays back the video file. However, full file transfer in the download mode usually suffers long and perhaps unacceptable transfer time. In contrast, in the streaming mode, the video contents are being received and decoded. Owing to its real-time nature, video streaming typically has bandwidth, delay, and loss requirements. However, the current best-effort Internet does not offer any quality of service (QoS) guarantees to streaming video over the Internet. In addition, for multicast, it is difficult to efficiently support multicast video while providing service flexibility to meet a wide range of QoS requirements from the users. Thus, designing mechanisms and protocols for Internet streaming video poses many challenges [105]. To address these challenges, extensive research has been conducted. To introduce the necessary background and give the reader a complete picture of this field, we cover some key areas of streaming video, such as video compression, application layer QoS control, continuous media distribution services, streaming servers, media

5.4 MEDIA STREAMING

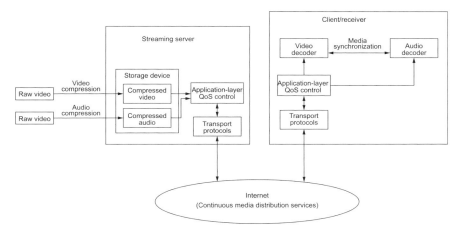

Figure 5.51 Basic building blocks in the architecture for video streaming [105]. (©2001 IEEE.)

synchronization mechanisms, and protocols for streaming media [106]. The relations among the basic building blocks are illustrated in Figure 5.51. Raw video and audio data are precompressed, by video and audio compression algorithms, and then saved in storage devices. It can be seen that the areas are closely related and they are coherent constituents of the video streaming architecture. We will briefly describe the areas. Before that it must be pointed out that upon the client's requests, a streaming server retrieves compressed video/audio data from storage devices and the application-layer QoS control module adapts the video/audio bit streams according to the network status and QoS requirements. After the adaptation, the transport protocols packetize the compressed bit streams and send the video/audio packets to the Internet. Packets may be dropped or experience excessive delay inside the Internet due to congestion. To improve the quality of video/audio transmission, continuous media distribution service (e.g., caching) is deployed in the Internet. For packets that are successfully delivered to the receiver, they first pass through the transport layers and are then processed by the application layer before being decoded at the video/audio decoder. To achieve synchronization between video and audio presentations, media synchronization mechanisms are required.

Video Compression
Since raw video consumes a large amount of bandwidth, compression is usually employed to achieve transmission efficiency. In this section, we discuss various compression approaches and requirements imposed by streaming applications on the video encoder and decoder.

In essence, video compression schemes can be classified into two approaches: scalable and nonscalable video coding. We will show the encoder and decoder in intramode and only use DCT. Intramode coding refers to coding video macroblocks without any reference to previously coded data. Since scalable video is capable of gracefully coping with the bandwidth fluctuations in the Internet [107], we are primarily concerned with scalable video coding techniques.

A nonscalable video encoder shown in Figure 5.52a generates one compressed bit stream, while a scalable video decoder is presented in Figure 5.52b. In contrast a scalable video encoder compresses a raw video sequence into multiple substreams as represented in Figure 5.53a. One of the compressed substreams is the base substream, which can be independently decoded and provides coarse visual quality. Other compressed substreams are enhanced substreams, which can only be decoded together with the base substream and can provide better visual quality. The complete bit stream (i.e., the combination of all substreams) provides the highest quality. An SNR scalable encoder as well as scalable decoder are shown in Figure 5.53.

The scalabilities of quality, image sizes, or frame rates are called SNR, spatial, or temporal scalabilities, respectively. These three scalabilities form the basic mechanisms, such as spatiotemporal scalability [108]. To provide more flexibility in meeting different demands of streaming (e.g., different access link bandwidths and different latency requirements), the fine granularity scalability (FGS) coding mechanism is proposed in MPEG-4 [109, 110]. An FGS encoder and FGS decoder are shown in Figure 5.54.

The FGS encoder compresses a raw video sequence into two substreams, that is, a base layer bit stream and an enhancement layer bit stream. Different from an SNR-scalable encoder, an FGS encoder uses bit-plane coding representing the enhancement stream. Bit-plane coding uses embedded representations. Bit planes of enhancement DCT coefficients are shown in Figure 5.55. With bit-plane coding, an FGS encoder is capable of achieving continuous rate control for the enhancement stream. This is because the enhancement bit stream can be truncated anywhere to achieve the target bit rate.

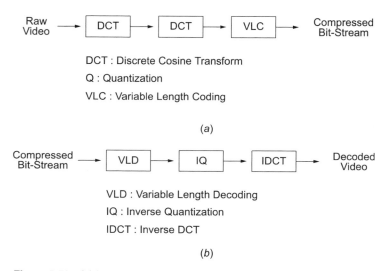

Figure 5.52 (a) Nonscalable video encoder, (b) nonscalable video decoder.

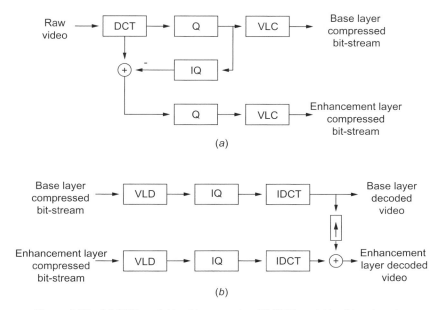

Figure 5.53 (a) SNR-scalable video encoder, (b) SNR-scalable video decoder.

Example 5.3. A DCT coefficient can be represented by 7 bits (i.e., its value ranges from 0 to 127). There are 64 DCT coefficients. Each DCT coefficient has a most significant bit (MSB). All the MSB from the 64 DCT coefficients form Bitplane 0 (Figure 5.55). Similarly, all the second most significant bits form Bitplane 1.

A variation of FGS is progressive fine granularity scalability (PFGS) [111]. PFGS shares the good features of FGS, such as fine granularity bit rate scalability and error resilience. Unlike FGS, which only has two layers, PFGS could have more then two layers. The essential difference between FGS and PFGS is that FGS only uses the base layer as a reference to reduce prediction error, resulting in higher coding efficiency.

Various Requirements Imposed by Streaming Applications
In what follows we will describe various requirements imposed by streaming allocations on the video encoder and decoder. Also, we will briefly discuss some techniques that address these requirements.

Bandwidth. To achieve acceptable perceptual quality, a streaming application typically has minimum bandwidth requirement. However, the current Internet does not provide bandwidth reservation to support this requirement. In addition, it is desirable for video streaming applications to employ congestion control to avoid congestion, which happens when the network is heavily loaded. For video streaming, congestion

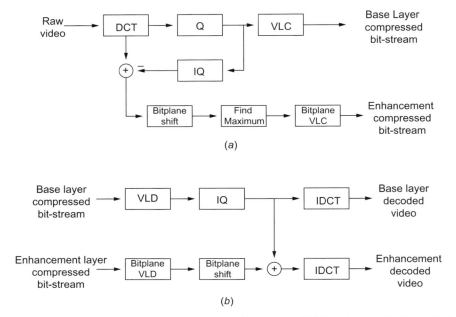

Figure 5.54 (a) Fine granularity scalability (FGS) encoder, (b) FGS decoder [105]. (©2001 IEEE.)

control takes the form of rate control, that is, adapting the sending rate to the available bandwidth in the network. Compared with nonscalable video, scalable video is more adaptable to the varying available bandwidth in the network.

Delay. Streaming video requires bounded end-to-end delay so that packets can arrive at the receiver in time to be decoded and displayed. If a video packet does

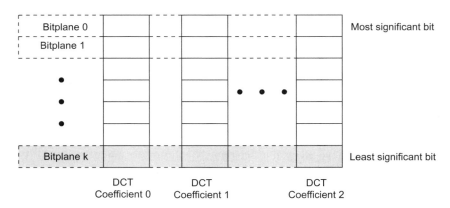

Figure 5.55 Bitplanes of enhancement DCT coefficients [105]. (©2001 IEEE.)

not arrive in time, the playout process will pause, which is annoying to human eyes. A video packet that arrives beyond its delay bound (e.g., its playout time) is useless and can be regarded as lost. Since the Internet introduces time-varying delay, to provide continuous playout, a buffer at the receiver is usually introduced before decoding [112].

Loss. Packet loss is inevitable in the Internet and can damage pictures, which is displeasing to human eyes. Thus, it is desirable that a video stream be robust to packet loss. Multiple description coding is such a compression technique to deal with packet loss [113].

Video Cassette Recorder (VCR) like Functions. Some streaming applications require VCR-like functions such as stop, pause/resume, fast forward, fast backward, and random access. A dual bit stream least-cost scheme to efficiently provide VCR-like functionality for MPEG video streaming is proposed in reference [105].

Decoding Complexity. Some devices such as cellular phones and personal digital assistants (PDAs) require low power consumption. Therefore, streaming video applications running on these devices must be simple. In particular, low decoding complexity is desirable.

We here present the application-layer QoS control mechanisms, which adapt the video bit streams according to the network status and QoS requirements.

Application-Layer QoS Control

The objective of application-layer QoS control is to avoid congestion and maximize video quality in the presence of packet loss. To cope with varying network conditions, and different presentation quality requested by the users, various application-layer QoS control techniques have been proposed [114, 115]. The application-layer QoS control techniques include congestion control and error control. These techniques are employed by the end systems and do not require any QoS support from the network.

Congestion Control. Bursty loss and excessive delay have a devasting effect on video presentation quality, and they are usually caused by network congestion. Thus, congestion control mechanisms at end systems are necessary to help reduce packet loss and delay. Typically, for streaming video, congestion control takes the form of rate control. This is a technique used to determine the sending rate of video traffic based on the estimated available bandwidth in the network. Rate control attempts to minimize the possibility of network congestion by matching the rate of the video stream to the available network bandwidth. Existing rate control schemes can be classified into three categories: source-based, receiver-based, and hybrid rate control.

Under source-based rate control, the sender is responsible for adapting the video transmission rate. Feedback is employed by source-based rate control mechanisms.

Based upon the feedback information about the network, the sender can regulate the rate of the video stream. The service-based rate control can be applied to both unicast [116] and multicast [117]. Figure 5.56 represents unicast and multicast video distribution.

For unicast video, existing source-based rate control mechanisms follow two approaches: probe-based and model-based. The probe-based approach is based on probing experiments. In particular, the source probes for the available network bandwidth by adjusting the sending rate in a way that could maintain the packet loss ratio p below a certain threshold P_{th}. There are two ways to adjust the sending rate: (1) additive increase and multiplicative decrease, and (2) multiplicative increase and multiplicative decrease [118].

The model-based approach is based on a throughput model of a transmission control protocol (TCP) connection. Specifically, the throughput of a TCP connection can be characterized by the following formula:

$$\lambda = \frac{1.22 + MTU}{RTTx\sqrt{p}} \tag{5.1}$$

where λ is throughput of a TCP connection, MTU (maximum transit unit) is the packet size used by the connection, RTT is the round-trip time for the connection, and p is the packet loss ratio experienced by the connection [119].

Under the model-based rate control, equation (5.1) is used to determine the sending rate of the video stream. Thus, the video connection can avoid congestion in a

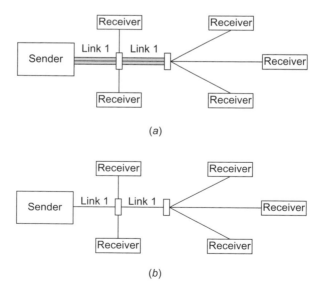

Figure 5.56 (a) Unicast video distribution using multiple point-to-point connections, (b) multicast video distribution using point-to-multipoint transmission.

way similar to that of TCP and it can compete fairly with TCP flows. For this reason, the model-based rate control is also called *TCP-friendly* rate control.

For multicast, under the source-based rate control, the sender uses a single channel to transport video to the receivers. Such multicast is called *single-channel* multicast. For single-channel multicast, only the probe-based rate control can be employed.

Single-channel multicast is efficient since all the receivers share one channel. However, single-channel multicast is unable to provide flexible services to meet the different demands from receivers with various access link bandwidths. In contrast, if multicast video were to be delivered through individual unicast streams, the bandwidth efficiency is low, but the service could be differentiated since each receiver can negotiate the parameters of the services with the source.

Under the receiver-based rate control, the receiver regulates the receiving rate of video streams by adding/dropping channels, while the sender does not participate in rate control. Typically, receiver-based rate control is used in multicast scalable video, where there are several layers in the scalable video and each layer corresponds to one channel in the multicast tree [105].

Similar to the source-based rate control, the existing receiver-based rate-control mechanisms follow two approaches: probe-based and model-based. The basic probe-based rate control consists of two parts [105]:

- When no congestion is detected, a receiver probes for the available bandwidth by joining a layer/channel, resulting in an increase of its receiving rate. If no congestion is detected after the joining, the joining experiment is successful. Otherwise, the receiver drops the newly added layer.
- When congestion is detected, a receiver drops a layer (i.e., leaves a channel), resulting in a reduction of its receiving rate.

Unlike the probe-based approach, which implicitly estimates the available network bandwidth through probing experiments, the model-based approach uses explicit estimation for the available network bandwidth.

Under the hybrid rate control the receiver regulates the receiving rate of video streams by adding/dropping channels, while the sender also adjusts the transmission rate of each channel based on feedback from the receivers. Examples of hybrid rate control include the destination set grouping and layered multicast scheme [120].

Architecture for source-based rate control is shown in Figure 5.57. An associated technique with rate control is rate shaping. The objective of rate shaping is to match rate of a precompressed video bit stream to the target rate constraints. A rate shaper (or filter), which performs rate shaping, is required for source-based rate control. This is because the stored video may be precompressed at a certain rate, which may not match the available bandwidth in the network. Many types of filters can be used, such as codec filter, frame-dropping filter, layer-dropping filter, frequency filter, and requantization filter [121].

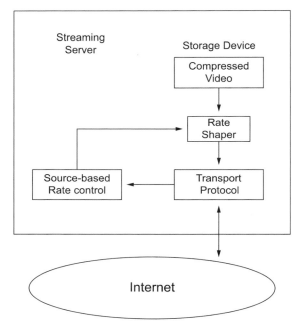

Figure 5.57 Architecture for source-based rate control [105]. (©2001 IEEE.)

In some applications, the purpose of congestion control is to avoid congestion. On the other hand, packet loss is inevitable in the Internet and may have significant impact on perceptual quality. This prompts the need to design mechanisms to maximize video presentation quality in the presence of packet loss. Error control is such a mechanism, and will be presented next.

Error Control. In the Internet, packets may be dropped due to congestion at routers, they may be misrouted, or they may reach the destination with such a long delay as to be considered useless or lost. Packet loss may severely degrade the visual presentation quality. To enhance the video quality in presence of packet loss, error-control mechanisms have been proposed.

For certain types of data (such as text), packet loss is intolerable while delay is acceptable. When a packet is lost, there are two ways to recover the packet: the corrupted data must be corrected by traditional forward error correction (FEC), that is, channel coding, or the packet must be retransmitted. On the other hand, for real-time video, some visual quality degradation is often acceptable while delay must be bounded. This feature of real-time video introduces many new error-control mechanisms, which are applicable to video applications but not applicable to traditional data such as text. In essence, the error-control mechanisms for video applications can be classified into four types, namely, FEC, retransmission, error resilience, and error concealment. FEC, retransmission, and error resilience are performed at both the source and the receiver side, while error concealment is carried out only at the receiver side.

The principle of FEC is to add redundant information so that the original message can be reconstructed in the presence of packet loss. Based on the kind of redundant information to be added, we classify existing FEC schemes into three categories: channel coding, source coding-based FEC, and joint source/channel coding [106].

For Internet applications, channel coding is typically used in terms of block codes. Specifically, a video stream is first chopped into segments, each of which is packetized into k packets; then, for each segment, a block code is applied to the k packets to generate an n-packet block, where $n > k$. To perfectly recover a segment, a user only needs to receive k packets in the n-packet block [122].

Source coding-based FEC (SFEC) is a variant of FEC for Internet video [123]. Like channel coding, SFEC also adds redundant information to recover from loss. For example, the nth packet contains the nth group of blocks (GOB) and redundant information about the $(n-1)$th GOB, which is a compressed version of the $(n-1)$th GOB with larger quantizer.

Joint source/channel coding is an approach to optimal rate allocation between source coding and channel coding [106].

Delay-Constrained Retransmission. Retransmission is usually dismissed as a method to recover lost packets in real-time video since a retransmitted packet may miss its playout time. However, if the one-way trip time is short with respect to the maximum allowable delay, a retransmission-based approach (called delay-constrained retransmission) is a viable option for error control.

For unicast, the receivers can perform the following delay-constrained retransmission scheme. When the receiver detects the loss of packet N, if $[T_c + RTT + D_a < T_d(N)]$ send the request for packet N to the sender, where T_c is current time, RTT is estimated round-trip time, D_a is a slack term, $T_d(N)$ is time when packet N is scheduled for display.

The slack time D_a may include tolerance of error in estimating RTT, the sender's response time, and the receiver's decoding delay. The timing diagram for receiver-based control is shown in Figure 5.58, where D_a is only the receiver's decoding

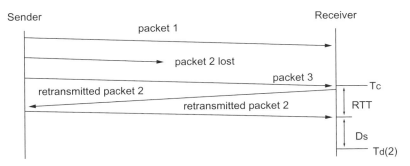

Figure 5.58 Timing diagram for receiver-based control [105]. (©2001 IEEE.)

delay. It is clear that the objective of the delay-constrained retransmission is to suppress requests of retransmission that will not arrive in time for display.

Error-Resilient Encoding. The objective of error-resilient encoding is to enhance robustness of compressed video to packet loss. The standardized error-resilient encoding schemes include resynchronization marking, data partitioning, and data recovery [124]. However, resynchronization marking, data partitioning, and data recovery are targeted at error-prone environments like wireless channels and may not be applicable to the Internet environment. For video transmission over the Internet, the boundary of a packet already provides a synchronization point in the variable-length coded bit stream at the receiver side. On the other hand, since a packet loss may cause the loss of all the motion data and its associated shape/texture data, mechanisms such as resynchronization, marking, data partitioning, and data recovery may not be useful for Internet video applications. Therefore, we will not present the standardized error-resilient tools. Instead, we present multiple description coding (MDC), which is promising for robust Internet video transmission [125].

With MDC, a raw video sequence is compressed into multiple streams (referred to as descriptions) as follows: each description provides acceptable visual quality; more combined descriptions provide a better visual quality. The advantages of MDC are:

- *Robustness to loss* – even if a receiver gets only one description (other descriptions being lost), it can still reconstruct video with acceptable quality.
- *Enhanced quality* – if a receiver gets multiple descriptions, it can combine them together to produce a better reconstruction than that produced from any one of them.

However, the advantages come with a cost. To make each description provide acceptable visual quality, each description must carry sufficient information about the original video. This will reduce the compression efficiency compared to conventional single description coding (SDC). In addition, although more description combinations provide a better visual quality, a certain degree of correlation between the multiple description has to be embedded in each description, resulting in further reduction of the compression efficiency. Further investigation is needed to find a good trade-off between compression efficiency and reconstruction quality from any one description.

Error Concealment. Error-resilient encoding is executed by the source to enhance robustness of compressed video before packet loss actually happens (this is called preventive approach). On the other hand, error concealment is performed by the receiver when packet loss has already occurred (this is called reactive approach). Specifically, error concealment is employed by the receiver to conceal the lost data and make the presentation less displeasing to human eyes.

The are two basic approaches for error concealment: spatial and temporal interpolations. In spatial interpolation, missing pixel values are reconstructed using neighboring spatial information. In temporal interpolation, the lost data are reconstructed from data in the previous frames. Typically, spatial interpolation is used to reconstruct the missing data in intracoded frames, while temporal interpolation is used to reconstruct the missing data in intercoded frames [126]. If the network is able to support QoS for video streaming, the performance can be further enhanced.

Continuous Media Distribution Services

In order to provide quality multimedia presentations, adequate support from the network is critical. This is because network support can reduce transport delay and packet loss ratio. Streaming video and audio are classified as continuous media because they consist of a sequence of media quanta (such as audio samples or video frames), which convey meaningful information only when presented in time. Built on top of the Internet (IP protocol), continuous media distribution services are designed with the aim of providing QoS and achieving efficiency for streaming video/audio over the best-effort Internet. Continuous media distribution services include network filtering, application-level multicast, and content replication.

Network Filtering. As a congestion control technique, network filtering aims to maximize video quality during network congestion. The filter at the video server can adapt the rate of video streams according to the network congestion status. Figure 5.59 illustrates an example of placing filters in the network. The nodes labeled R denote routers that have no knowledge of the format of the media streams and may randomly discard packets. The filter nodes receive the client's requests and adapt the stream sent by the server accordingly. This solution allows the service provider to place filters on the nodes that connect to network bottlenecks. Furthermore, multiple filters can be placed along the path from a server to a client.

To illustrate the operations of filters, a system model is depicted in Figure 5.60. The model consists of the server, the client, at least one filter, and two virtual

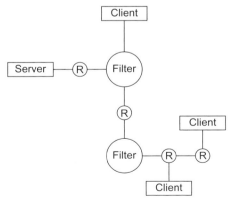

Figure 5.59 Filters placed inside the network.

Figure 5.60 Systems model of network filtering.

channels between them. Of the virtual channels, one is for control and the other is for data. The same channels exist between any pair of filters. The control channel is bidirectional, which can be realized by TCP connections. The model shown allows the client to communicate with only one host (the last filter), which will either forward the requests or act upon them. The operations of a filter on the data plane include: (1) receiving video stream from server or previous filter and (2) sending video client or next filter at the target rate. The operations of a filter on the control plane include: (1) receiving requests from client or next filter, (2) acting upon requests, and (3) forwarding the requests to its previous filter.

Typically, frame-dropping filters are used as network filters. The receiver can change the bandwidth of the media stream by sending requests to the filter to increase or decrease the frame dropping rate. To facilitate decisions on whether the filter should increase or decrease the bandwidth, the receiver continuously measures the packet loss ratio p. Based on the packet loss ratio, a rate-control mechanism can be designed as follows. If the packet loss ratio is higher that a threshold α, the client will ask the filter to increase the frame dropping rate. If the packet loss ratio is less than another threshold β ($\beta < \alpha$), the receiver will ask the filter to reduce the frame dropping rate [105].

The advantage of using frame-dropping filters inside the network include the following:

- *Improved video quality.* For example, when a video stream flows from an upstream link with larger available bandwidth to a downstream link with smaller available bandwidth, use of a frame-dropping filter at the connection point) between the upstream link and the downstream link can help improve the video quality. This is because the filter understands the format of the media stream and can drop packets in a way that gracefully degrades the stream's quality instead of corrupting the flow outright.
- *Bandwidth efficiency.* This is because the filtering can help to save network resources by discarding those frames that are late.

Application-Level Multicast. The Internet's original design, while well suited for point-to-point applications like e-mail, file transfer, and Web browsing, fails to effectively support large-scale content delivery like streaming-media multicast. In an attempt to address this shortcoming, a technology called IP multicast was proposed. As an extension to IP layer, IP multicast is capable of providing efficient multipoint packet delivery. To be specific, the efficiency is achieved by having one and only one copy of the original IP packet (sent by the multicast source) be transported along any physical path in the IP multicast tree. However, with a decade of research and development, there are still many barriers in deploying IP multicast. These problems include scalability, network management, deployment and support for higher layer functionality (e.g., error flow and congestion control) [127].

Application-level multicast is aimed at building a multicast service on top of the Internet. It enables independent content delivery service providers (CSPs), Internet service providers (ISPs), or enterprises to build their own Internet multicast networks and interconnect them into larger, worldwide media multicast networks. That is, the media multicast network can support peering relationships at the application level or streaming-media/content layer, where content backbones interconnect service providers. Hence, much as the Internet is built from an interconnection of networks enabled through IP-level peering relationships among ISPs, the media multicast networks can be built from an interconnection of content-distribution networks enabled through application-level peering relationships among various sorts of service providers, namely, traditional ISPs, CSPs, and applications service providers (ASPs).

The advantage of the application-level multicast is that it breaks barriers such as scalability, network management, and support for congestion control, which have prevented Internet service providers from establishing IP multicast peering arrangements.

Content Replication. An important technique for improving scalability of the media delivery system is content media replication. The content replication takes two forms: mirroring and caching, which are deployed by the content delivery service provider (CSP) and Internet service provider (ISP). Both mirroring and caching seek to place content closer to the clients and both share the following advantages:

- reduced bandwidth consumption on network links
- reduced load on streaming servers
- reduced latency for clients
- increased availability.

Mirroring is to place copies of the original multimedia files on the other machines scattered around the Internet. That is, the original multimedia files are stored on the main server while copies of the original multimedia files are placed on duplicate servers. In this way, clients can retrieve multimedia data from the nearest duplicate server, which gives the clients the best performance (e.g., lowest latency). Mirroring has some disadvantages. Currently, mechanisms for establishing a dedicated mirror on an existing server, while cheaper, is still an ad hoc and administratively complex process. Finally, there is no standard way to make scripts and server setup easily transferable from one server to another.

Caching, which is based on the belief that different clients will load many of the same contents, makes local copies of contents that the clients retrieve. Typically, clients in a single organization retrieve all contents from a single local machine, called a cache. The cache retrieves a video file from the origin server, storing a copy locally and then passing it on to the client who requests it. If a client asks for a video file that the cache has already stored, the cache will return the local copy rather that going all the way to the origin server where the video file resides. In addition, cache sharing and cache hierarchies allow each cache to access files stored at other caches so that the load on the origin server can be reduced and network bottlenecks can be alleviated [128].

Streaming Servers

Streaming servers play a key role in providing streaming services. To offer quality streaming services, streaming servers are required to process multimedia data under timing constraints in order to prevent artifacts (e.g., jerkiness in video motion and pops in audio) during playback at the clients. In addition, streaming servers also need to support video cassette recorder (VCR) like control operations, such as stop, pause/resume, fast forward, and fast backward. Streaming servers have to retrieve media components in a synchronous fashion. For example, retrieving a lecture presentation requires synchronizing video and audio with lecture slides. A streaming server consists of the following three subsystems: communicator (e.g., transport protocol), operating system, and storage system.

- The communicator involves the application layer and transport protocols implemented on the server. Through a communicator the clients can communicate with a server and retrieve multimedia contents in a continuous and synchronous manner.
- The operating system, different from traditional operating systems, needs to satisfy real-time requirements for streaming applications.
- The storage system for streaming services has to support continuous media storage and retrieval.

Media Synchronization

Media synchronization is a major feature that distinguishes multimedia applications from other traditional data applications. With media synchronization mechanisms, the application at the receiver side can present various media streams in the same way as they were originally captured. An example of media synchronization is that the movements of a speaker's lips match the played-out audio.

A major feature that distinguishes multimedia applications from other traditional data applications is the integration of various media streams that must be presented in a synchronized fashion. For example, in distance learning, the presentation of slides should be synchronized with the commenting audio stream. Otherwise, the current slide being displayed on the screen may not correspond to the lecturer's explanation heard by the students, which is annoying. With media synchronization, the application at the receiver side can present the media in the same way as they

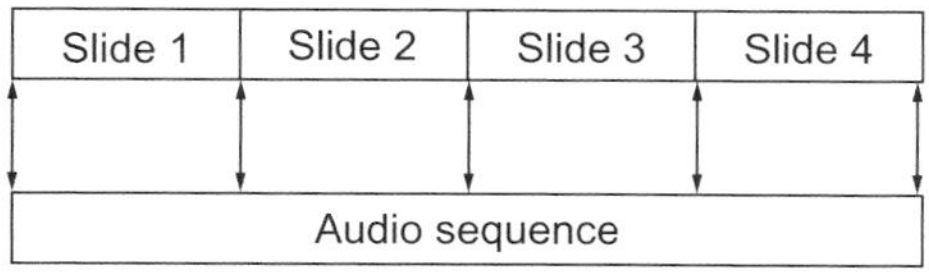

Slide 1	Slide 2	Slide 3	Slide 4
	Audio sequence		

Figure 5.61 Synchronization between the slides and the commenting audio stream.

were originally captured. Synchronization between the slides and the commenting audio stream is shown in Figure 5.61.

Media synchronization refers to maintaining the temporal relationships within one data stream and among various media streams. There are three levels of synchronization, namely, intrastream, interstream, and interobject synchronization. The three levels of synchronization correspond to three semantic layers of multimedia data as follows [129]:

Intrastream Synchronization. The lowest layer of continuous media or time-dependent data (such as video and audio) is the media layer. The unit of the media layer is a logical data unit such as a video/audio frame, which adheres to strict temporal constraints to ensure acceptable user perception at playback. Synchronization at this layer is referred to as intrastream synchronization, which maintains the continuity of logical data units. Without intrastream synchronization, the presentation of the stream may be interrupted by pauses or gaps.

Interstream Synchronization. The second layer of time-dependent data is the stream layer. The unit of the stream layer is a whole stream. Synchronization at this layer is referred to as interstream synchronization, which maintains temporal relationships among different continuous media. Without interstream synchronization, skew between the streams may become intolerable. For example, users could be annoyed if they notice that the movements of the lips of a speaker do not correspond to the presented audio.

Interobject Synchronization. The highest layer of the multimedia document is the object layer, which integrates streams and time-independent data such as text and still images. Synchronization at this layer is referred to as interobject synchronization. The objective of interobject synchronization is to start and stop the presentation of the time-independent data within a tolerable time interval, if some previously defined points of the presentation of a time-dependent media object are reached. Without interobject synchronization, for example, the audience of a slide show could be annoyed if the audio is commenting on one slide while another slide being presented.

The essential part of any media synchronization mechanism is the specifications of the temporal relations within a medium and between the media. The temporal relations can be specified either automatically or manually. In the case of audio/video recording and playback, the relations are specified automatically by the recording device. In the case of presentations that are composed of independently captured or otherwise created media, the temporal relations have to be specified manually (with human support). The manual specification can be illustrated by the design of a slide show: the designer selects the appropriated slides, creates an audio object, and defines the units of the audio stream where the slides have to be presented.

The methods that are used to specify the temporal relations include interval-based, axes-based, control flow-based, and event-based specifications. A widely used specification method for continuous media is axes-based specifications or time-stamping: at the source, a stream is time-stamped to keep temporal information within the stream and with respect to other streams; at the destination, the application presents the streams according to their temporal relation [130].

Besides specifying the temporal relations, it is desirable that synchronization be supported by each component on the transport path. For example, the servers store large amounts of data in such way that retrieval is quick and efficient to reduce delay; the network provides sufficient bandwidth, and delay and jitter introduced by the network are tolerable to the multimedia applications; the operating systems and the applications provide real-time data processing (e.g., retrieval, resynchronization, and display). However, real-time support from the network is not available in the current Internet. Hence, most synchronization mechanisms are implemented at the end systems. The synchronization mechanisms can be either preventive or corrective [131].

Preventive mechanisms are designed to minimize synchronization errors as data is transported from the server to the user. In other words, preventive mechanisms attempt to minimize latencies and jitters. These mechanisms involve disk-reading scheduling algorithms, network transport protocols, operating systems, and synchronization schedulers. Disk-reading scheduling is the process of organizing and coordinating the retrieval of data from the storage devices. Network transport protocols provide means for maintaining synchronization during data transmission over the Internet.

Corrective mechanisms are designed to recover synchronization in the presence of synchronization errors. Synchronization errors are unavoidable, since the Internet introduces random delay, which destroys the continuity of the media stream by incurring gaps and jitters during data transmission. Therefore, certain compensations (i.e., corrective mechanisms) at the receiver are necessary when synchronization errors occur. An example of corrective mechanisms is the stream synchronization protocol (SSP). In SSP, the concept of an *intentional delay* is used by the various streams in order to adjust their presentation time to recover from network delay variations. The operations of SSP are described as follows. At the client side, units that control and monitor the client end of the data connections compare the real arrival times of data with the ones predicted by the presentation schedule and notify the scheduler of any discrepancies. These discrepancies are then compensated by the scheduler, which delays the display of data that are *ahead* of other data, allowing the late data to *catch up*. To conclude, media synchronization is one of the key issues in the design of media streaming services.

Protocols for Streaming Media

Protocols are designed and standardized for communication between clients and streaming servers. Protocols for streaming media provide such services as network addressing, transport, and session control. According to their functionalities, the protocols can be classified into three categories: network-layer protocol such as Internet protocol (IP), transport protocol such as user datagram protocol (UDP), and session control protocol such as real-time streaming protocol (RTSP).

Network-layer protocol provides basic network service support such as network addressing. The IP serves as the network-layer protocol for Internet video streaming.

Transport protocol provides end-to-end network transport functions for streaming applications. Transport protocols include UDP, TCP, real-time transport protocol (RTP), and real-time control protocol (RTCP). UDP and TCP are lower-layer transport protocols while RTP and RTCP [132] are upper-layer transport protocols, which are implemented on top of UDP/TCP. Protocol stacks for media streaming are shown in Figure 5.62.

Session control protocol defines the messages and procedures to control the delivery of the multimedia data during an established session. The RTSP and the session initiation protocol (SIP) are such session control protocols [133, 134].

To illustrate the relationship among the three types of protocols, we depict the protocol stacks for media streaming. For the data plane, at the sending side, the compressed video/audio data is retrieved and packetized at the RTP layer. The RTP-packetized streams provide timing and synchronization information, as well as sequence numbers. The RTP-packetized streams are then passed to the UDP/TCP layer and the IP layer. The resulting IP packets are transported over the Internet. At the receiver side, the media streams are processed in the reversed manner before

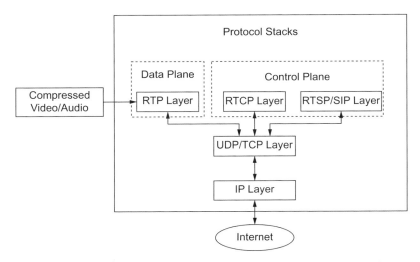

Figure 5.62 Protocol stacks for media streaming [105]. (©2001 IEEE.)

their presentations. For the control plane, RTCP packets and RTSP packets are multiplexed at the UDP/TCP layer and move to the IP layer for transmission over the Internet.

In what follows, we will discuss transport protocols for streaming media. Then, we will describe control protocols, that is, real-time streaming protocol (RTSP) and session initiation protocol (SIP).

Transport Protocols. The transport protocol family for media streaming includes UDP, TCP, RTP, and RTCP protocols [135]. UDP and TCP provide basic transport functions, while RTP and RTCP run on top of UDP/TCP. UDP and TCP protocols support such functions as multiplexing, error control, congestion control, or flow control. These functions can be briefly described as follows. First, UDP and TCP can multiplex data streams for different applications running on the same machine with the same IP address. Secondly, for the purpose of error control, TCP and most UDP implementations employ the checksum to detect bit errors. If single or multiple bit errors are detected in the incoming packet, the TCP/UDP layer discards the packet so that the upper layer (e.g., RTP) will not receive the corrupted packet. On the other hand, different from UDP, TCP uses retransmission to recover lost packets. Therefore, TCP provides reliable transmission while UDP does not. Thirdly, TCP employs congestion control to avoid sending too much traffic, which may cause network congestion. This is another feature that distinguishes TCP from UDP. Lastly, TCP employs flow control to prevent the receiver buffer from overflowing while UDP does not have any flow control mechanisms.

Since TCP retransmission introduces delays that are not acceptable for streaming applications with stringent delay requirements, UDP is typically employed as the transport protocol for video streams. In addition, since UDP does not guarantee packet delivery, the receiver needs to rely on the upper layer (i.e., RTP) to detect packet loss.

RTP is an Internet standard protocol designed to provide end-to-end transport functions for supporting real-time applications. RTCP is a companion protocol with RTP and is designed to provide QoS feedback to the participants of an RTP session. In other words, RTP is a data transfer protocol while RTCP is a control protocol.

RTP does not guarantee QoS or reliable delivery, but rather provides the following functions in support of media streaming:

- *Time-stamping*. RTP provides time-stamping to synchronize different media streams. Note that RTP itself is not responsible for the synchronization, which is left to the applications.
- *Sequence numbering*. Since packets arriving at the receiver may be out of sequence (UDP does not deliver packets in sequence), RTP employs sequence numbering to place the incoming RTP packets in the correct order. The sequence number is also used for packet loss detection.
- *Payload type identification*. The type of the payload contained in an RTP packet is indicated by an RTP-header field called payload type identifier. The receiver

interprets the contents of the packet based on the payload type identifier. Certain common payload types such as MPEG-1/2 audio and video have been assigned payload type numbers. For other payloads, this assignment can be done with session control protocols [136].
- *Source identification.* The source of each RTP packet is identified by an RTP-header field called Synchronization SouRCe identifier (SSRC), which provides a means for the receiver to distinguish different sources.

RTCP is the control protocol designed to work in conjunction with RTP. In an RTP session, participants periodically send RTCP packets to convey feedback on quality of data delivery and information of membership. RTCP essentially provides the following services.

- *QoS feedback.* This is the primary function of RTCP. RTCP provides feedback to an application regarding the quality of data distribution. The feedback is in the form of sender reports (sent by the source) and receiver reports (sent by the receiver). The reports can contain information on the quality of reception such as: (1) fraction of the lost RTP packets, since the last report; (2) cumulative number of lost packets, since the beginning of reception; (3) packet inter-arrival jitter; and (4) delay since receiving the last sender's report. The control information is useful to the senders, the receivers, and third-party monitors. Based on the feedback, the sender can adjust its transmission rate, the receivers can determine whether congestion is local, regional, or global, and the network manager can evaluate the network performance for multicast distribution.
- *Participant identification.* A source can be identified by the SSRC field in the RTP header. Unfortunately, the SSRC identifier is not convenient for the human user. To remedy this problem, the RTCP provides human-friendly mechanisms for source identification. Specifically, the RTCO SDES (source description) packet contains textual information called canonical names as globally unique identifiers of the session participants. It may include a user's name, telephone number, e-mail address, and other information.
- *Control packet scaling.* To scale that the RTCP controls packet transmission with the number of participants, a control mechanism is designed as follows. The control mechanism keeps the total control packets to 5 percent of the total session bandwidth. Among the control packets, 25 percent are allocated to the sender reports and 75 percent to the receiver reports. To prevent control packet starvation, at least one control packet is sent within 5 s at the sender or receiver.
- *Inter media synchronization.* The RTCP sender reports contain an indication of real time and the corresponding RTP time-stamp. This can be used in inter-media synchronization like lip synchronization in video.
- *Minimal session control information.* This optional functionality can be used for transporting session information such as names of the participants.

Session Control Protocols (RTSP and SIP). The RTSP is a session control protocol for streaming media over the Internet. One of the main functions of RTSP is to support VCR-like control operations such as stop, pause/resume, fast forward, and fast backward. In addition, RTSP also provides means for choosing delivery channels (e.g., UDP, multicast UDP, or TCP), and delivery mechanisms based upon RTP. RTSP works for multicast as well as unicast.

Another main function of RTSP is to establish and control streams of continuous audio and video media between the media server and the clients. Specifically, RTSP provides the following operations.

- *Media retrieval.* The client can request a presentation description, and ask the server to set up a session to send the requested media data.
- *Adding media to an existing session.* The server or the client can notify each other about any additional media becoming available to the established session.

In RTSP, each presentation and media stream is identified by an RTSP universal resource locator (URLS). The overall presentation and the properties of the media are defined in a presentation description file, which may include the encoding, language, RTSP URLs, destination address, port and other parameters. The presentation description file can be obtained by the client using HTTP, e-mail, or other means.

SIP is another session control protocol. Similar to RTSP, SIP can also create and terminate sessions with one or more participants. Unlike RTSP, SIP supports user mobility by proxying and redirecting requests to the user's current location.

To summarize, RTSP and SIP are designed to initiate and direct delivery of streaming media data from media servers. RTP is a transport protocol for streaming media data while RTCP is a protocol for monitoring delivery of RTP packets. UDP and TCP are lower-layer transport protocols for RTP/RTCP/RTSP/SIP packets and IP provides a common platform for delivering UDP/TCP packets over the Internet. The combination of these protocols provides a complete streaming service over the Internet.

Video streaming is an important component of many Internet multimedia applications, such as distance learning, digital libraries, home shopping, and video-on-demand. The best-effort nature of the current Internet poses many challenges to the design of streaming video systems. Our objective is to give the reader a perspective on the range of options available and the associated trade-offs among performance, functionality, and complexity.

5.4.3 Challenges for Transporting Real-Time Video Over the Internet

Transporting video over the Internet is an important component of many multimedia applications. Lack of QoS support in the current Internet, and the heterogeneity of the networks and end systems pose many challenging problems for designing video delivery systems. Four problems for video delivery systems can be identified:

bandwidth, delay, loss, and heterogeneity. Two general approaches address these problems: the network-centric approach and the end system-based approach [106]. Over the past several years extensive research based on the end system-based approach has been conducted and various solutions have been proposed. A holistic approach was taken from both transport and compression perspectives. A framework for transporting real-time Internet video includes two components, namely, congestion control and error control. It is well known that congestion control consists of rate control, rate-adaptive coding, and rate shaping. Error control consists of forward error correction (FEC), retransmission, error resilience and error concealment. As shown in Table 5.12, the approaches in the design space can be classified along two dimensions: the transport perspective and the compression perspective.

There are three mechanisms for congestion control: rate control, rate adaptive video encoding, and rate shaping. On the other hand, rate schemes can be classified into three categories: source-based, receiver-based, and hybrid. As shown in Table 5.13, rate control schemes can follow either the model-based approach or probe-based approach. Source-based rate control is primarily targeted at unicast and can follow either the model-based approach or the probe-based approach.

There have been extensive efforts on the combined transport approach and compression approach [137]. The synergy of transport and compression can provide better solutions in the design of video delivery systems.

Table 5.12 Taxonomy of the design space

		Transport	Compression
Congestion control	Rate control	Source-based Receiver-based Hybrid	
	Rate adaptive encoding		Altering quantizer Altering frame rate
	Rate shaping	Selective frame discard	Dynamic rate shaping
Error control	FEC	Channel coding	SFEC
		Joint channel/source coding	
	Delay-constrained retransmission	Sender-based control Receiver-based control Hybrid control	
	Error resilience		Optimal mode selection Multiple description coding
	Error concealment		EC-1, EC-2, EC-3

Table 5.13 Rate control

		Model-based	Probe-based
Rate control	Source-based	Unicast	Unicast/Multicast
	Receiver-based	Multicast	Multicast
	Hybrid		Multicast

Under the end-to-end approach, three factors are identified to have impact on the video presentation quality at the receiver:

- the source behavior (e.g., quantization and packetization)
- the path characteristics
- the receiver behavior (e.g., error concealment (EC)).

Figure 5.63 represents factors that have impact on the video presentation quality, that is, source behavior, path characteristics, and receiver behavior. By taking into consideration the network congestion status and receiver behavior, the end-to-end approach is capable of offering superior performance over the classical approach for Internet video applications. A promising future research direction is to combine the end system-based control techniques with QoS support from the network.

Different from the case in circuit-switched networks, in packet-switched networks, flows are statistically multiplexed onto physical links and no flow is isolated. To achieve high statistical multiplexing gain or high resource utilization in the network, occasional violations of hard QoS guarantees (called statistical QoS) are allowed. For example, the delay of 95 percent packets is within the delay bound while 5 percent packets are not guaranteed to have bounded delays. The percentage (e.g., 95 percent) is in an average sense. In other words, a certain flow may have only 10 percent packets arriving within the delay bound while the average for all flows is

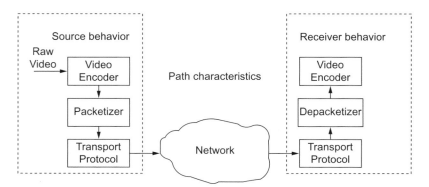

Figure 5.63 Factors that have impact on video presentation quality: source behavior, path characteristics, and receiver behavior [137]. (©2000 IEEE.)

95 percent. The statistical QoS service only guarantees the average performance, rather than the performance for each flow. In this case, if the end system-based control is employed for each video stream, higher presentation quality can be achieved since the end system-based control is capable of adapting to short-term violations.

As a final note, we would like to point out that each scheme has a trade-off between cost/complexity and performance. Designers can choose a scheme in the design space that meets the specific cost/performance objectives.

5.4.4 End-to-End Architecture for Transporting MPEG-4 Video Over the Internet

With the success of the Internet and flexibility of MPEG-4, transporting MPEG-4 video over the Internet is expected to be an important component of many multimedia applications in the near future. Video applications typically have delay and loss requirements, which cannot be adequately supported by the current Internet. Thus, it is a challenging problem to design an efficient MPEG-4 video delivery system that can maximize perceptual quality while achieving high resource utilization.

MPEG-4 builds on elements from several successful technologies, such as digital video, computer graphics, and the World Wide Web, with the aim of providing powerful tools in the production, distribution, and display of multimedia contents. With the flexibility and efficiency provided by coding a new form of visual data called visual object (VO), it is foreseen that MPEG-4 will be capable of addressing interactive content-based video services, as well as conventional storage and transmission video. Internet video applications typically have unique delay and loss requirements that differ from other data types. Furthermore, the traffic load condition over the Internet varies drastically over time, which is detrimental to video transmission. Thus, it is a major challenge to design an efficient video delivery system that can both maximize users' perceived quality of service (QoS) while achieving high resource utilization in the Internet.

Figure 5.64 shows an end-to-end architecture for transporting MPEG-4 video over the Internet. The architecture is applicable to both precompressed video and live video.

If the source is a precompressed video, the bit rate of the stream can be matched to the rate enforced by a feedback control protocol through dynamic rate shaping [138] or selective frame discarding [136]. If the source is a live video, it is used in the MPEG-4 rate adaptation algorithm to control the output rate of the encoder. On the sender side raw bit stream of live video is encoded by an adaptive MPEG-4 encoder. After this stage, the compressed video bit stream is first packetized at the sync layer (SL) and then passed through the RTP/UDP/IP layers before entering the Internet.

Packets may be dropped at a router/switch (due to congestion) or at the destination (due to excess delay). For packets that are successfully delivered to the destination, they first pass through the RTP/UDP/IP layers in reverse order before being decoded at the MPEG-4 decoder.

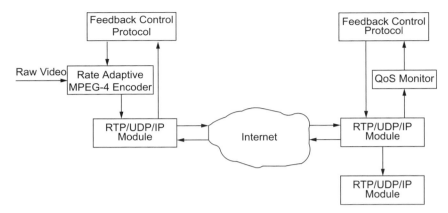

Figure 5.64 An end-to-end architecture for transporting MPEG-4 video [116]. (©2000 IEEE.)

Under the architecture, a QoS monitor is kept at the receiver side to infer network congestion status based on the behavior of the arriving packets, for example, packet loss and delay. Such information is used in the feedback-control protocol, which is sent back to the source. Based on such feedback information, the sender estimates the available network bandwidth and controls the output rate of the MPEG-4 encoder.

Figure 5.65 shows the protocol stack for transporting MPEG-4 video. The right half shows the processing stages at an end system. At the sending side, the compression layer compresses the visual information and generates elementary streams (ESs), which contain the coded representation of the VOs. The ESs are packetized as SL-packetized streams at the SL. The SL-packetized streams are multiplexed into FlexMux stream at the TransMux Layer, which is then passed to the transport protocol stacks composed of RTP, UDP, and IP. The resulting IP packets are transported over the Internet. At the receiver side, the video stream is processed in the reversed manner before its presentation. The left half shows the data format of each layer.

Figure 5.66 shows the structure of MPEG-4 video encoder. Raw video stream is first segmented into video objects, then encoded by individual VO encoder. The encoded VO bit streams are packetized before being multiplexed by the stream multiplex interface. The resulting FlexMux stream is passed to the RTP/UDP/IP module.

The structure of an MPEG-4 video decoder is shown in Figure 5.67. Packets from RTP/UDP/IP are transferred to a stream demultiplex interface and FlexMux buffer. The packets are demultiplexed and put into corresponding decoding buffers. The error concealment component will duplicate the previous VOP when packet loss is detected. The VO decoders decode the data in the decoding buffer and produce composition units (CUs), which are then put into composition memories to be consumed by the compositor.

To conclude, the MPEG-4 video standard has the potential of offering interactive content-based video services by using VO-based coding. Transporting MPEG-4

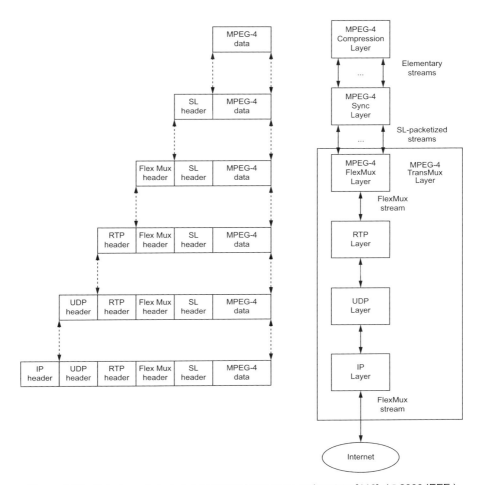

Figure 5.65 Data format at each processing layer at an end system [116]. (©2000 IEEE.)

video is foreseen to be an important component of many multimedia applications. On the other hand, since the current Internet lacks QoS support, there remain many challenging problems in transporting MPEG-4 video with satisfactory video quality. For example, one issue is packet loss control and recovery associated with transporting MPEG-4 video. Another issue that needs to be addressed is the support of multicast for Internet video.

5.4.5 Broadband Access

The demand for broadband access has grown steadily as users experience the convenience of high-speed response combined with *always on* connectivity. There are a

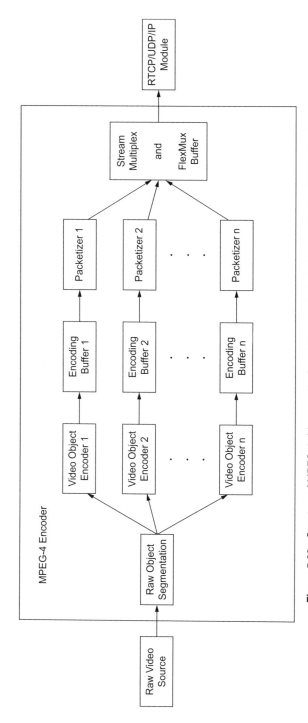

Figure 5.66 Structure of MPEG-4 video encoder for transporting over the Internet [116]. (©2000 IEEE.)

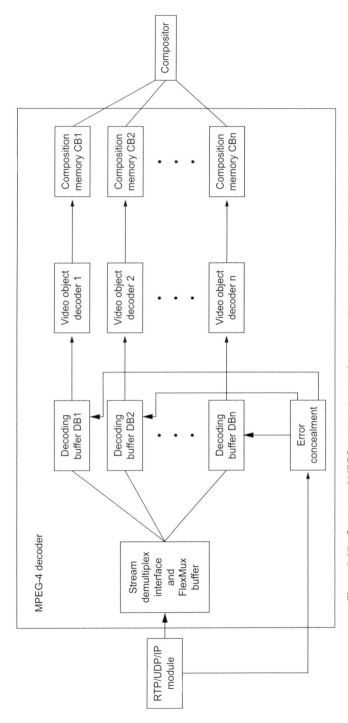

Figure 5.67 Structure of MPEG-4 video decoder for transporting over the Internet [116]. (©2000 IEEE.)

significant variety of different technologies to deliver broadband services. This variety is due in a large part to a corresponding variety of customer access scenarios. In urban metropolitan areas, fixed infrastructure is available and has been upgraded to support high-speed Internet services. In rural and remote areas Internet users are less likely to be able to obtain broadband access due to a combination of technological and economic factors.

Passive Optical Networks

For locations where subscribers are located in a reasonably dense topology and require very high-speed broadband access, a fiber medium is typically the ideal option. Passive optical networks (PONs) have evolved as a technology that is optimized for this application. Passive optical networks seek to reduce the cost of the fiber access infrastructure by a combination of minimizing (or eliminating) the need for active electronic systems in the distribution portion of the access network, reducing the total number of fiber interface ports for the access system, and reducing the total amount of fiber. PONs accomplish these goals by using a fiber optic tree with passive optical splitters in the distribution point to provide the branching to individual subscribers. A single optical interface port at the host terminal (either in the central office or a remote terminal location) can then provide a common optical interface (port) to a number of different subscribers instead of having a separate optical interface to communicate with each subscriber. The PON system currently favored by carriers uses either 155 or 622 Mbit/s in the downstream direction and 155 Mbit/s in the upstream direction to access 16–32 subscribers on a single PON tree. The standard for this system was developed through the Full Service Access Network (FSAN) forum and published by ITU-T. Although there has been a steady decline in the cost of optical components, this PON system still is not typically economical for access to home users. Home and small business users so far still lack the application to justify paying for this amount of bandwidth. On the other hand, for business users this bandwidth is often too little, which again restricts the applications for this PON system [139].

The traditional star topology of residential access does not allow true aggregation per neighborhood before reaching the user–network boundary, since the traffic needs to be commissioned (including service level agreement, SLA), checked for compliance (policed), charged, and monitored per customer, leading to the unavoidable cost burden of any retail sales situation. Thus, the user–network interfaces must stay near the customer. PON technology relieves this problem, allowing the traffic from different customers to be multiplexed using time-division multiple access (TDMA) under the arbitration of the medium access control (MAC) protocol in the upstream direction. So while the interfaces remain in the optical network unit (ONU) at the customer curb, building, or even home, the control is exercised by the optical line termination (OLT) at the root of the tree residing in the protected environment of the operator's central office. From there, the MAC protocol controls and polices the rates per customer, achieving significant cost reduction due to aggregation into a shared medium apart from that resulting from less fiber and the sharing of one optical transceiver for all customers on the network side (in the OLT). Thus, the MAC protocol as executor of TDMA in the upstream of the PON is of prime importance for cost-effectiveness, fairness, traffic profile control, and quality of service (QoS) guarantees [140–142].

Wireless Access

The potential of using fixed wireless for broadband access has been widely discussed [143, 144] and is realized in some urban and specialized applications using local multipoint distribution systems (LMDS) [145]. While considerable attention has been given to utilizing wireless in urban and metro areas, and isolated but densely populated villages, the prospect of serving rural areas with low population densities has received only limited attention, and typically for providing narrowband services [146]. Advances in wireless technology and system-level applications for data communications, now embodied in the WLAN 802.11 standards and commercialized by numerous suppliers as WiFi, open new possibilities for broadband fixed wireless access. Public use of WiFi is emerging in hot spots deployed in hotels, airports, coffee shops, and other public places [147]. The potential for using WiFi to extend Internet access to less densely populated and isolated areas is being explored by several entrepreneurial service providers. These companies typically utilize WiFi for last mile access and some form of radio link for backhaul as well. The proliferation of WiFi has resulted in significant reductions in equipment costs, with the majority of new laptop computers now being shipped with WiFi adapters built in.

Digital Subscriber Line

Digital subscriber line (DSL) systems are built on the existing twisted-pair telephone local loops. Hardware implementations with efficient signal processing together with advanced modulation techniques make possible data transfer at rates up to several megabits per second while preserving usual voice connectivity. Asymmetric DSL (ADSL) and very high speed DSL (VDSL) have emerged as the most popular among the family of DSL technologies. ADSL is well suited for increasing the transfer capabilities of the copper local loop. VDSL aims to be a last mile solution implemented in the local loop with deep fiber penetration.

Since the early 1990s, several standardization organizations have been participating in the development of hardware and protocols that are assembled into the ADSL architecture. It is enough to mention the ADSL Forum, ATM Forum, IETF, ITU, ANSI, and ETSI.

DSL multiservice access technology is capable of supporting more than one service (e.g., video, voice, and data) simultaneously over a shared access link. It is flexible enough to meet the high expectations set for both quality and quantity of services. Its multiple service delivery capability allows service providers to cost-effectively improve their local loop infrastructures and extend the economic and management benefits to the customer [148]. Being a scalable technology, DSL provides many advantages over other broadband technologies in network configurations. It is an efficient way to provide multimedia services. The idea of using physical layer transportation for voice has been around for a while. However, the potential for transporting both data and pulse code modulated (PCM) voice over the asymmetric DSL (ADSL) physical layer simultaneously has not been implemented yet. The uniqueness of the channelized voice over DSL (CVoDSL) solution lies in the fact that it transports voice over the physical layer, which eliminates voice packetization.

The voice bandwidth can be dynamically allocated in CVoDSL, so when voice lines are not in use, the bandwidth can be utilized for data transfer. CVoDSL coexists with plain old telephone service (POTS) by using a frequency well above the POTS band, which means analog dialup and fax modem can be used at the same time. The CVoDSL method intends to replace existing VoDSL while offering superior voice quality, bandwidth efficiency, and architectural flexibility [149].

With the rapid growth of Internet access and voice/data-centric communications, many access technologies have been developed to meet the stringent demand of high-speed data transmission and bridge the video bandwidth gap between ever-increasing high data rate core networks and bandwidth-hungry end user networks. To make efficient utilization of the limited bandwidth of existing access routes and cope with the adverse channel environment, many standards have been proposed for a variety of broadband access systems over different access environments (twisted pairs, coaxial cables, optical fibers, and fixed or mobile wireless). In the design and implementation domain of those systems, many research issues arise. First, multistandards coexist in many access environments. In addition, multimode or rate-adaptive designs are encountered frequently, even with one single standard.

Market demands, such as increasing Internet access, play a large role in advancing new technologies. Customers desire cost-effective, immediate, always available access to the Internet. ADSL technology utilizes the existing cooper twisted pair to fulfill these demands and more.

Over the past few years, various types of DSL have been introduced for different markets. These technologies are driven and supported by the industry, including service providers and vendors. Table 5.14 shows various DSL technologies and their data rate characteristics. ADSL has been chosen for CVoDSL transportation because of its popularity among small business and residential customers for asymmetric bandwidth characteristics.

Table 5.14 DSL techniques [149]

DSL Types	Rates	
	Upstream	Downstream
ADSL (30 kHz–1.104 MHz)	100–800 kb/s	1–8 Mb/s
HDSL	Fixed 784, 1544, 2048 kb/s	Fixed 784, 1544, 2048 kb/s
IDSL	128 kb/s	128 kb/s
SDSL	64–200 kb/s	800–2000 kb/s
VDSL	Up to 25 Mb/s	Up to 25 Mb/s

©2002 IEEE.

ADSL is characterized by a different line rate from the service provider to the customer than that from the customer to the service provider. Each ADSL logical data channel can comprise of four possible simplex and three duplex channels, which are known as bearer channels. The speed of the logical channel can, in fact, vary but must be a multiple of 32 kb/s. The downstream data channel can be a combination of simplex and duplex channels, but only duplex can be used for the upstream channel. Using PCM would allow up to 12 simultaneous voice channels over the upstream bandwidth [150].

The feasibility of VoDSL technology brought a spurt into the DSL market in terms of revenue generation. So far the DSL Forum has recognized only two VoDSL technologies. One of the methods is known as multiservice broadband network (MBN). It supports new generation voice networks based on soft switching and VoDSL standards developed by the IETF and ITU-T. The second VoDSL technology is broadband loop emulation services (BLES). It uses an asynchronous transfer mode (ATM) loop emulation method based on ATM Forum work. These methods have already been implemented in the DSL network. There are many disadvantages in these methods. Some of their drawbacks are inefficient use of bandwidth, network complexity, undesirable end-to-end delays, and poor voice quality.

Channelized voice over DSL operates in environments that integrate voice channels with traditional data services at the T interface. CVoDSL implementation requires support of CVoDSL Internet working function (IWF), dynamic rate repartitioning (DRR), signaling, transporting, and encoding [149].

Cellular Radio Networks

Convergence of wideband wireless access and Internet will be the next wave in the information industry, and it obviously becomes a focus of global investments. Fueled by such emerging technologies as all-IP direct signaling, super digital processing, smart antenna transceiver, broadband reconfigurable core, as well as converged wireless interfaces, the wireless system is taking on a more and more important role in Internet development. As communications evolve to this convergence, a new architecture will be required to support high-data-rate connection from 2 to over 100 Mb/s with various required qualities of service (QoS) based on the new spectrum requirement as well as the coexistence of the current spectrum for wideband wireless. To meet these critical demands, improvements in the wireless physical layer (modulation, diversity, coding, etc.) and link layer (access control, bandwidth allocation, etc.) are necessary.

In recent years, Internet technology has emerged as the major driving force behind new developments in the area of telecommunications networks. The volume of packet data traffic has increased at extreme rates. In order to meet these changing traffic patterns, more and more network operators adapt their strategies and plan to migrate to IP-based backbone networks. Clearly, the Internet will dominate our daily life in the future much more than today [151].

Meanwhile, mobile networks face a similar trend of exponential traffic increase and growing importance to users. In some countries, such as Korea, the number of mobile subscriptions has recently exceeded the number of fixed lines. This tremendous success was not expected in the 1980s, when today's second-generation mobile communication systems were designed.

The combination of both developments, the growth of the Internet and the success of mobile networks, suggests that the next trend will be an increasing demand for mobile access to Internet applications. It is therefore increasingly important that mobile radio networks support these applications in an efficient manner. Thus, mobile radio systems currently under development include support for packet data services. The most widely deployed standard for second-generation mobile radio networks is the Global System for Mobile Communications (GSM). Networks based on this standard are extended with the General Packet Radio Service (GPRS)/ Universal Mobile Telecommunication Systems (UMTS), which provides data rates up to 384 kb/s (2 Mb/s). The enhanced data rate for GSM evolution (EDGE) based version of GPRS that used eight-phase shift keying (PSK) can deliver 384 kb/s [152–154].

In brief, GPRS can be described as a service providing optimized access to the Internet, while reusing to a large degree existing GSM infrastructure. Advanced mobile terminals using multiple slots will offer more convenient and faster Internet access than today's technology. The GPRS concept allows volume-oriented charging, which permits users to have cheap and permanent connections to the Internet.

The demand for broadband Internet access is continually growing. Announced delays in the deployment of third-generation (3G) high-speed wireless networks as well as slow progress in satisfying demands for wired solutions such as DSL and cable modems place high expectations on alternative last mile technologies such as fixed wireless access (FWA). The aim of such access systems is to provide wireless high-speed Internet access, and in relevant markets, voice services, to fixed or nomadic/home offices (SOHO) located within reach of an access point or base transceiver station (BTS). Mainstream Internet applications are targeted such as Web browsing and e-mail, but also more demanding services such as real-time conferencing and/or voice. To maintain reasonably low costs as well as good penetration of the radio signals, mass market FWA systems typically use the sub-5 GHz band, examples of which are the so-called multipoint multichannel distribution services (MMDS) band (2.5–2.7 GHz) in the United States and the 3.5 GHz in international markets. The subscriber unit, sometimes referred to as customer premises equipment (CPE), is currently typically installed on a rooftop and communicates wirelessly to BTS several miles away. However, as line-of-sight (LOS) requirements are to be mitigated in the future, the CPE may be installed on the outside wall of the house or placed inside on a desktop. Future broadband access systems can also be envisioned to support portability and serve stationary lightweight unit users located anywhere within the coverage area. In any case, ubiquitous FWA networks must be able to cope with widely varying terrain features (flat, hilly, varying tree densities), urbanization levels, and user densities (rural, suburban, dense urban).

A careful radio design, and clever exploitation of the subscriber access unit's stationarity while in use and the (limited) directionality of the unit's antenna can allow for an order of magnitude improvement over other advanced mobile digital wireless systems such as 3G, in terms of data rates, access quality/reliability, and spectrum efficiency. In fact, upcoming FWA systems should offer performances comparable to wired technologies such as xDSL and cable modems to be truly attractive. Advantages of FWA include rapid deployment, high scalability, lower maintenance and upgrade costs compared to cable and DSL, and granular investments to match market growth. Nevertheless, a number of important issues including spectrum efficiency, network scalability, easily installable CPE antennas, and above all reliable non-LOS (NLOS) operation, need to be resolved before FWA can penetrate the market successfully [155].

5.4.6 Quality of Service Framework

Quality of service (QoS) has been a frequently used term and hot research topic. However, IP networks and wireless cellular systems have been looking at QoS provisioning from rather different perspectives. In IP and computer networks, researchers mainly concentrated on those techniques that could enable migrating traditional best effort Internet service with no guarantee on delay or throughput or even reliable delivery of packets into a more predictable architecture. Differentiated services (`DiffServ`), integrated services (`IntServ`), and various combinations of the two techniques have tried fulfilling QoS in the Internet by defining different architectures and reservation techniques.

While research on these and other alternative techniques is still ongoing, due to the large-scale architecture of the global Internet and the diversity in Internet nodes, routers, and hosts, Internet QoS still has a long way to go to become a reality. One implication issue in providing QoS over the Internet, arising from the fact that QoS should be maintained on an end-to-end basis, is that without a mutual commitment from all intermediate links and networks, QoS cannot be guaranteed. Such commitment is even difficult for a small private network, not to mention for the widespread Internet where each packet may travel on a different path from other packets of the same message.

For cellular systems, service quality has traditionally been guaranteed for certain measures, including call dropping probability, call blocking probability, security, and, more importantly for voice communications, delay limits. With the introducing of data services, cellular providers have tried to accommodate a certain level of service guarantee to data transactions, mainly those that used the traditional circuit-switched networks to deliver data packets.

Different from circuit-switched networks, packet-switched networks and their current commonly used IP networks do not dedicate any resources to message delivery unless a certain mechanism such as a reservation protocol is implemented. To guarantee QoS over integrated IP-cellular, therefore, it is necessary to look at the requirements of applications running over both systems.

The main application of the cellular system is voice communications, which require strict delay and delay variation constraints. Delivering voice packets over an IP network by means of techniques such as voice over IP then requires mechanisms that guarantee such delay requirements. On the other hand, for IP networks, with data applications such as e-mail, Web, and file transfer, other QoS measures such as reliable and error-free data transfer and high throughput are more critical than the delay requirement. The combination of different QoS requirements for different applications makes the problem even more challenging.

Modern cellular systems such as the UMTS have approached this issue by introducing a few traffic classes and handling them with different priorities. A similar approach has been taken in IP and computer networks by giving different levels and priorities and classifications to different traffic types and labeling them by using a few bits at the IP header. While the similarity between IP and wireless cellular initiatives in guaranteeing QoS on an end-to-end basis is more or less over this stage, we need to find more harmonized architectures to support QoS in future generations of wireless IP networks.

Therefore, in order to solve the issues of QoS provisioning in integrated IP-cellular systems (i.e., the next-generation mobile communications network), we must not only look at the individual techniques that satisfy particular service quality indicators, but more than that, look at architectural improvement that melts IP well inside cellular systems and vice versa. By architectural improvements we mean that all traffic, regardless of its originating applications, real-time or nonreal-time, voice or data, should be treated in a common way. Such improvements will make it easier to classify packets and service them based on their quality requirements.

Quality of Service Signaling for IP-Based Mobile Networks

Next-generation mobile networks will support heterogeneous radio access and will offer seamless services between different wireless access technologies. For different scenarios and applications, various radio access systems will complement one another. For instance, wireless LANs will offer high-speed data services with restricted mobility support for hot spots, while some cellular networks can provide real-time services. To support heterogeneous networks, mobile networks are moving toward *all-IP* networks, based on Internet protocols [156].

In heterogeneous, overlapping networks, a handover to more suitable access points offering more capabilities may be needed to enable additional services. For instance, when passing by a wireless hot spot, one can perform a handover to its access point for a short period of time to facilitate some demanding service, such as download of bulk data or video conferencing. However, in many cases, the availability of resources at the potential access point is not known before handover is performed. For QoS, this means that the resources at the new access point should be allocated before attaching to the network. This is often called *anticipated* or *planned* handover. This kind of handover mainly offers two advantages. First, it reduces handover latency, because most signaling to set up resources in the new path is carried out in advance. Secondly, it avoids unsuccessful handovers or unnecessary periods of QoS degradation, because handovers should only be performed if the resources are actually available.

Our QoS signaling architecture integrates resource management with mobility and location management. *Mobility management* protocols like MobileIP ensure that a mobile device is reachable by a home address, although the local IP address may change during handover. Further, micromobility protocols aim to reduce handover disruptions by handling handover signaling and packet forwarding locally if possible.

In addition to QoS signaling, *resource management* has to take care of admission control, allotment, and release of requested resources. To ensure QoS on the data path, there are several techniques such as differentiated services [157] or integrated services. Integrated services is based on per-flow resource reservation. Differentiated services (`DiffServ`) is a recent approach defined by IETF. Instead of manipulating the per-flow state at each router in a network, QoS preferences or guarantees are assigned to traffic aggregates, which are composed at the network edges [158]. This requires the marking of packets in a special field in the IP header, the DS field. While `IntServ` defined Resource Reservation Protocol (RSVP) [158] as a related end-to-end signaling protocol, the `DiffServ` standard lacks a control plane.

The main requirements for a QoS signaling architecture in future mobile IP-based networks are:

- independence of a particular QoS technique for provisioning of QoS on the data path (e.g., `IntServ` and `DiffServ`)
- independence of specific radio access technologies
- networking with different mobility concepts, including micromobility, for seamless handovers
- support for interdomain handovers, when a mobile node (MN) changes its point of attachment to a network that is administered by another organization.

In the following, we compare the main QoS approaches for their applicability in IP-based mobile networks. The many approaches to QoS in IP networks can be classified along the following two criteria.

The first criterion is whether signaling messages follow in the data path or not, often called *on-path* vs. *off-path* signaling. The RSVP protocol is the Internet standard for on-path QoS signaling and is used for other signaling purposes as well. However, RSVP is not very well suited to future mobile networks. The IETF is currently working on an on-path general-purpose signaling protocol. The advantage of on-path signaling is that failures affect both signaling and data path, and may be handled locally. For off-path signaling, resource requests are sent to a dedicated entity, which is then responsible for admission control and QoS setup along the data path.

Secondly, resources can be managed centrally by one entity (in each network domain) or decentrally in each router. For local resource control, each router manages the resources of the outgoing links. QoS signaling for anticipated handover between domains is shown in Figure 5.68.

In the centralized QoS architecture, a domain resource manager (DRM) (also called a bandwidth broker) handles the resources for one domain. The DRM maintains an up-to-date image of resources and reservations in its domain. The DRM may request resources from DRM in adjacent domains in order to provide end-to-end

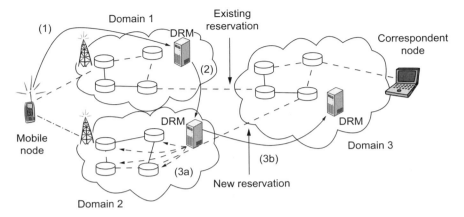

Figure 5.68 QoS signaling for anticipated handover between domains [156]. (©2004 IEEE.)

reservations. The central approach is flexible with respect to different QoS models (e.g., `IntServ` or `DiffServ`).

Typically, on-path signaling is used with local resource management, as in `IntServ` with RSVP. Central resource management suits off-path signaling, but can also be used with on-path signaling if the central resource manager is contacted by routers using additional protocols. Since a central approach has a single control point, the integration with different mobility schemes and location management is more flexible. DRM can determine the route and reserve resources for a new access point within its domain or by contracting a neighboring DRM. This is illustrated in three steps. After an anticipated handover request (STEP 1), the old DRM requests resources in a domain (STEP 2). After a successful reservation (STEPS 3a and 3b), the handover can take place. Note that off-path signaling is needed to reserve resources in the new domain. In this way, a handover only takes place if resource is available. It is important that the resource reservation may also be triggered by some network service, such as a mobility scheme that uses location prediction to find the suitable access point.

End-to-End QoS Issues in the IP Wired and Wireless Environment

QoS is still a topic that attracts a lot of attention in both the wired and wireless worlds. Regarding the former, the maturity of the integrated and differentiated services frameworks has led to a solidification of the research effort. The main IETF activity in the field is the definition of an end-to-end QoS signaling protocol within the context of the Next Steps in Signaling charter. On the other hand, in the wireless world there is still substantial research activity, primarily stemming from its intrinsic characteristics (i.e., the involvement of the radio interface and the implication imposed by mobility).

Taking for granted that different networks or domains are free to follow any QoS model, the focus is turned to the enabling factor for establishing an end-to-end QoS path that consists of different provisioning mechanisms in each domain that must, however, expose similar packet forwarding treatment. In order to accomplish this, we need an end-to-end QoS signaling protocol that can be interpreted in each domain, as well as a universal way to describe the end-to-end QoS requirements. Since wireless access networks have come on the scene, the end-to-end QoS requirements must be able to be translated not only to IP-based (Layer 3) service-level semantics for each QoS model, but also to those in Layer 2 [159].

When defining end-to-end QoS resource reservation and management mechanisms, it is imperative that certain requirements are taken into account since the applicability of such a solution will span several administrative boundaries, each with its own policies, resource control architectures, and traffic engineering mechanisms. For that reason, a *common language* (signaling protocol) is needed for communicating end to end, the QoS requirements of user traffic, while at the same time respecting the individualities of the autonomous operation of each traversed domain.

QoS signaling capabilities are needed to extend the provisioning of QoS in IP-based networks from a static model towards a dynamic one. They will provide the means for communicating the QoS requirements to network entities, and therefore establish the desired QoS level to the end-to-end path. Their role is to carry the information specified in the corresponding QoS framework, which can be understood and interpreted in a uniform way by all networks constituting the end-to-end path. To enable standardization of the carried information, and therefore an interoperable solution across different autonomous networks, it is indispensable to separate the signaling protocol from the carried information. In this way, control information can be carried by any signaling protocol, while being understood by any autonomous system.

When speaking of end-to-end QoS signaling solutions, where large transit networks are traversed by a great amount of traffic, scalability and performance issues should be taken into careful consideration. The number of signaling messages exchanged for the establishment and maintenance of reservations, reservation control, and packet forwarding state should be kept at a minimum to allow signaling protocols to scale with the ever-increasing traffic load. Although this coarse granularity comes at the expense of providing hard QoS guarantees to traffic flows, it allows for scalability and efficiency. QoS signaling mechanisms improve the trade-off between quality of guarantee and efficiency of the network, and help to provide differentiated delivery services for individual flows or aggregates, network provisioning, and admission control.

Another crucial issue in QoS signaling in mobile networks is the interaction between the signaling and mobility protocols. A totally independent operation can lead to ambiguities and even interoperability problems. Loose integration would define the interactions between the two protocols (e.g., how the mobility protocol triggers the transfer of signaling messages), while tighter integration would consider a single protocol carrying both mobility and network state information.

Individual QoS in Cellular Networks
In its general meaning, QoS refers to the ability of a telecommunications systems to provide an appropriate transport service to deliver various types of communications

traffic to different users satisfactorily. Sometimes it can be difficult to define the exact technical parameters required to ensure such delivery, especially due to the fact that perception of service quality may differ from one user to another.

Thus, it may be observed that whatever global QoS management concept is realized in a network, by definition it can never produce results that would guarantee the same level of satisfaction to each and every user. This becomes even more complicated in a cellular network, where the interface between the network and users, realized via radio connection, is not stationary. This nonstationarity nature of connectivity in cellular networks is partly a result of the circumstances common to any kind of telecommunication network (bandwidth overloading, its randomness over time), but also due to the inherent mobility features of cellular networks: the unexpected and ever changing physical location of mobile users.

The nonstationary nature of the radio network interface means that the chances of successful communication (establishing and completing call) depend on the physical location of a user relative to the serving network node, which typically will be the closest base station. It is obvious that if a user terminal will be located within an optimal distance and in favorable radio visibility conditions, it would greatly increase chances of successful communication with high QoS. On the contrary, being located near the edge of the coverage area (cell) makes communication more difficult and resource demanding.

It is clear that there might be different contributions having an impact on individual QoS (iQoS), some radio interface related (e.g., coverage issues), some depending on the user terminal itself (e.g., dropped calls due to weak battery, unable to sustain the required cell duration), or even users' personal behavior [160].

Even in the single cell, radio link characteristics differ in different parts of the cell, more so within the whole cellular network. A mobile user experiences different link conditions while moving in the natural environment within or between the cells. The median signal level (the long-term characterization of radio link budget for a particular point) may easily vary by some 10–20 dB just within the very short distances of a few or tens of meters.

All this supports a basic assumption of regular differences in link conditions that will be observed by different users within the same cell, especially if they are located at different spots for longer periods (at home, at work). This tendency and relationship between the dominant location of users and received QoS do not appear to be sufficiently addressed by QoS monitoring in current networks [161]. This shows a need to establish some means of monitoring and recording the iQoS actually experienced. This recording should be done for the whole duration of a call from the attempt to set up until it is terminated (naturally by the user or as a dropped call). Recorded data should then be collected and analyzed, first within a user terminal itself, then in some summarized form provided to the user and uploaded to the network management and control center as appropriate.

From a use psychology point of view, it may seem that if a user were able to know the QoS he/she has received, this would in itself increase his/her overall satisfaction with the service because of the feeling of being in control. This satisfaction of knowing what is happening and subsequent feeling of not being cheated may be further reinforced by, for example, making the measured iQoS one of the parameters in a billing system. Thus, users might be charged by operators not only by the amount (minutes/bytes) of received services, but also in accordance with the actual iQoS: charges should be made proportionate to real user satisfaction [162].

Availability of objective information on iQoS would also allow the practice of service-level agreements between network operator and users, which require individualizing the QoS to the service utilization of each single user. This is an essential part of modern relationships between users and service providers, but up to now it was not possible to realize in mobile cellular networks due to the described uncertainty of the actual QoS provided to every individual user.

5.4.7 Security of Multimedia Systems

In the mathematical model of Shannon, *the fundamental problem of communication is that of reproducing at one point either exactly or approximately a message selected at another point.* The pioneering papers of Shannon describe also cryptographic coding for ensuring confidentiality. Efficient, confidential, authentic, and reliable coding of digital sources is nowadays an issue for which a number of solutions exist both in multimedia communications and content delivery. However, a media is not always used in its digital representation. For example, in the case of image communication, the source is at the endpoint transformed into light through a display mechanism. The ultimate question arises: how can one ensure security of an image transformed into light? One solution would be to link intimately the image with a mark tracing the path of its distribution. Such a mark would be a trace of the process indicating the validity or integrity of a visual content but also revealing the source of illegal manipulation or redistribution. The ideal mark should be invisible and not erasable. These are the well-known requirements for watermarking. Some security requirements are passive (unremovable and invisible traces in the visual contents), and some more active (watermarks, copy control, etc.).

In recent years, advances in digital technologies have created significant changes in the way we reproduce, distribute, and market intellectual property (IP). Digital media can now be exploited by the IP owners to develop new and innovative business models for their products and services. The lowered cost of reproduction, storage, and distribution, however, also invites much motivation for large-scale commercial infringement. In a world where piracy is a growing potential threat, three complementary resources are in reserve to defend the rights of the IP owners: technology, legislation, and business models [163].

IP is created as a result of intellectual activities in the industrial, scientific, literary, and artistic fields. It is divided into two general categories:

- *industrial property*, which includes inventions (patents), trademarks, industrial designs, and geographic indications of source
- *copyright*, which includes literary and artistic works such as novels, poems and plays, films, musical works, artistic works such as drawings, paintings, photographs, and sculptures, and architectural designs.

Because of its high economic value, copyrighted entertainment content needs to be protected as long as the customer demand is present in the market.

Creation includes all the activities in developing a new product such as a movie, a TV program, a book or song. In making a movie, for example, the studio, producer, director, and actors all work together to create an IP that is fixed on a 35 mm film for commercialization. Depending on the release window, several distribution channels exist, ranging from theatrical performances to duplication on magnetic or optical media, to broadcasting off-the-air. Whatever the form of presentation, the consumer should have the necessary equipment and/or the rights to receive and consume the requested product.

The need for commodity protection has not changed over the years. Every commercial item needs some type of protection until it is introduced to the relevant market for consumption.

End-to-end security is the most critical requirement for the creation of new digital markets where copyrighted entertainment is a major product. After a brief overview of copyright and copyright industries, we will examine the technological, legal, and business solutions that help maintain the incentive to supply the lifeblood of the markets.

Important Aspects of the Copyright

Although copyright means *right to copy*, the term is now used to cover a number of exclusive rights granted to the authors for the protection of their work. The owner of copyright is given exclusive right to do and to authorize others to do any of the following [164]:

- to reproduce the copyrighted work in copies or phonorecord
- to prepare derivative works based upon the copyrighted work
- to distribute copies or phonorecords of the copyrighted work to the public by sale or other transfer of ownership, or by rental, lease, or lending
- to display the copyrighted work publicly, in the case of literary, musical, dramatic, and choreographic works, pantomimes, and pictorial, graphic, or sculptural works, including the individual images of motion pictures and other audiovisual works
- to perform the copyrighted work publicly by means of a digital audio transmission, in the case of sound recordings.

It is interesting to note that copyright is secured as soon as the work is created by the author in some fixed form. No action, including publication and registration, is needed in the Copyrights Office.

Publication is the distribution of copies or phonorecords of a work to the public by sale or other transfer of ownership, or by rental, lease, or leading. *Registration* is a legal process to create a public record of the basic facts of a particular copyright. Although neither publication nor registration is a requirement for protection, they provide certain advantages to the copyright owner.

The copyright law has different clauses for the protection of published and unpublished works. All unpublished works are subject to protection, regardless of the nationality or domicile of the author. The published works are protected if certain conditions are met regarding the type of work, citizenship, residence, and publication date and place.

International copyright laws do not exist for the protection of works throughout the entire world. The national laws of individual countries may include different measures to prevent unauthorized use of copyrighted works. Fortunately, many countries offer protection to foreign works under certain conditions through membership in international treaties and conventions.

Conditional Access Systems

In an end-to-end protection system, a fundamental problem is to determine whether the consumer is authorized to access the requested content. The traditional concept of controlling physical access to places (e.g., cities, buildings, rooms, highways) has been extended to the digital world in order to deal with information in binary form. A familiar example is the access-control mechanism used in computer operating systems to manage data, programs, and other system resources. Such systems can be effective in *bounded* communities (e.g., a corporation or a college campus) where the emphasis is placed on the original access to information rather than how the information is used once it is in the possession of the user. In contrast, the conditional access systems for digital entertainment content in *open* communities need to provide reliable services for long periods of time (up to several decades) and be capable of controlling the use of content after access [165].

We will look at two approaches for restricting access to content. The first approach has been used by satellite/terrestrial broadcasters and cable operators in the last few decades (for both analog and digital contents). The second approach has been adopted by the developers of emerging technologies for protecting Internet content.

A conditional access (CA) system [166–169], allows access to services based on payment or other requirements such as identification, authorization, authentication, registration, or a combination of these. Using satellite, terrestrial, or cable transmission, the service providers deliver different types of multimedia content ranging from the access programs to services such as PayTV, Pay-per-View, and Video-on-Demand.

CA systems are developed by companies, commonly called the CA providers, that specialize in the protection of audio/visual signals and secure processing environments. A typical architecture of a CA system and its major components are shown in Figure 5.69.

The common activities in this general model are as follows:

1. Digital content (called an *event* or *program*) is compressed to minimize bandwidth requirements. MPEG-2 is a well-known industry standard for coding audio/video streams. Other MPEG variations (MPEG-4, MPEG-7, and MPEG-21) are being considered for new applications.

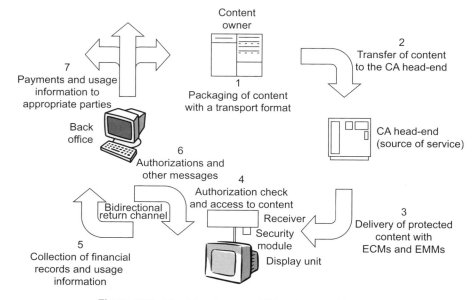

Figure 5.69 Conditional access (CA) systems architecture.

2. The program is sent to the CA head-end to be protected and packaged with entitlements indicating the access conditions.
3. The audio/video stream is scrambled and multiplexed with the entitlement message. In the context of CA systems, scrambling is the process of content encryption. This term is inherited from the analog protection systems where the analog video was manipulated using methods such as line shuffling. It is now being used to distinguish the process from the protection of descrambling keys. There are two types of entitlement messages associated with each program [170]. The entitlement control messages (ECMs) carry the decryption keys (called *control words*) and a short description of the program (number, title, date, time, price, rating, etc.) while the entitlement management messages (EMMs) specify the authorization levels related to services. In most CA systems, the EMMs can also be sent via other means such as telephone networks. The services are usually encrypted using a symmetric cipher such as data encryption standard (DES) or any other public domain or private algorithm. The lifetime and the length of the scrambling keys are two important system parameters. For security reasons, the protection of the ECMs is often privately defined by the CA providers, but public-key cryptography and one-way functions are useful tools for securing key delivery.
4. If the customer has received authorization to watch the protected program, the audio/video stream is descrambled by the receiver (also called *decoder*), and sent to the display unit for viewing. A removable security module (e.g., a

smart card) provides a safe environment for the processing of ECMs, EMMs, and other sensitive functions such as user authorization and temporary storage or purchase records. As for the program, it may come directly from the head-end or a local storage device. Protection of local storage (such as a hard disk) is a current research area.
5. The back office is an essential component of every CA system, handling billings and payments, transmission of EMMs, and interactive TV applications. A one-to-one link is established between the back office and the decoder (or removable security module, if it exists) using a *return channel*, which is basically a telephone connection via a modem. As with other details of the CA system, the security of this channel may be privately defined by the CA providers. At certain times, the back office collects the purchase history and other usage information for processing.
6. Authorization (e.g., EMMs) and other messages (system and security updates, etc.) are delivered to the customer's receiver.
7. Payments and usage information are sent to the appropriate parties (content providers, service operators, CA providers, etc.).

In today's CA systems, the security module is assigned the critical task of recovering the descrambling keys. These keys are then passed to the receiver for decrypting audio/video streams. The workload is therefore shared between the security module and its host. More recently, two separate standards have evolved to remove all the security functionality from navigation devices. In the United States, the National Renewable Security Standard (RSS) defines a renewable and replaceable security element for use in consumer electronics devices such as digital set-top boxes and digital TVs. In Europe, the Digital Video Broadcasting (DVB) project has specified a standard for a common interface (CI) between a host device and a security module [171].

The CA systems currently in operation support several purchase methods including subscription, pay-per-view, and impulsive pay-per-view. Other modules are also being considered to provide more user convenience and to facilitate payments. One such model use may be a *cash card* to store credits, which may be obtained from authorized dealers or ATM-like machines.

The DVB organization, a consortium of companies for establishing common international standards for digital broadcasting, has envisaged two basic CA approaches: simulcrypt and multicrypt.

- *Simulcrypt*. Each program is transmitted with the entitlement messages for multiple CA systems, enabling different CA decoders to receive and correctly descramble the program.
- *Multicrypt*. Each decoder is built with a common interface for multiple CA systems. Security modules from different CA system operators can be plugged into different slots in the same time decoder to allow switching among CA systems.

These architectures can be used for satellite, cable, and terrestrial transmission of digital television. The Advanced Television Systems Committee (ATSC) has recently adopted the Simulcrypt approach [172].

Copy Protection in Home Networks

A digital home network (DHN) is a cluster of digital audio/video devices including set-top boxes, TVs, DVD players, and general-purpose computing devices such as personal computers [173]. Figure 5.70 shows a typical home network with several sources of content feeding the audio/video devices.

The problem of content protection in home networks has the following dimensions:

- protection of content across digital interfaces
- protection of content on storage media
- management of rights associated with content.

This problem, it is believed, turns out to be the most difficult problem to solve for a number of technical, legal, and economic reasons.

- Private CA systems are, by definition, proprietary, and can be defined and operated using the strictest possible security means and methods. In comparison, the protection systems needed for the devices and interfaces in home networks have to be developed with a consensus among the stakeholders, making the determination of requirements very difficult.
- Renewability of a protection system needs to be properly defined and implemented. Important parameters are the cost, user convenience, and liabilities resulting from copyright infringements.

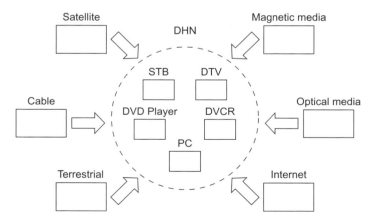

Figure 5.70 Digital home network (DHM) with its several content services.

- The new copyright legislation introduces controversial prohibitions subject to different interpretations.
- The payer of the bill for the cost of protection in home networks is unclear. Several business models are under consideration.

Two groups of technologies are believed to be useful in designing technical solutions: encryption-based and watermark-based. As each group presents strengths and weaknesses, some are of the opinion that both types of solutions should be implemented to increase the robustness to possible attacks.

Encryption and watermarking each provide a different *line of defense* in protecting content. The former, the *first line of defense*, makes the content unintelligible through a reversible mathematical transformation based on a secret key. The latter, the *second line of defense*, inserts data directly into the content at the expense of imperceptible degradation in quality. The theoretical level of security provided by encryption depends on the cipher strength and key length. Other factors such as tamper-resistant hardware or software also play an important role in implementations. Watermarking has several useful applications that dictate how and where the watermark is placed. For content-protection purposes, the watermark should be embedded in such a way that it is imperceptible to the human eye and robust against attacks. Its real value comes from the fact that the information it represents remains with the content in both analog and digital domains.

Two notable projects addressing security in home networking environments can be mentioned.

- The Video Electronics Standards Association (VESA) is an international nonprofit organization that develops and supports industrywide interface standards for the PC and other computing environments. The VESA and Consumer Electronics Association (CEA) have recently entered a memo of understanding that allowed CEA to assume all administration of the VESA Home Network Committee. The joint VESA/CEA committee, called R7.4, is now discussing different types of data security [174].
- The Universal Plug and Play (UPnP) Forum is an industry initiative designed to enable easy and robust connectivity among standalone devices and PCs from many different vendors. Although UPnP version 1 is moving towards completion, it does not specify countermeasures for security threats but instead relies on the protocols on which the UPnP architecture is built. However, because these measures are not sufficient to address all security-related scenarios, a working group has been formed within the Forum to investigate potential security enhancements [175].

Streaming Content Security

Video or audio streaming, or the real-time delivery of content over a wired or wireless data network, is the underlying technology behind many applications including video-on-demand and the delivery of educational and entertainment contents. In many applications, particularly those involving entertainment content, security issues such as conditional access and copy protection must be addressed. Since the content (particular the video) will often be compressed using a scalable compression technique and transported over a lossy packet network using the Internet protocol (IP), the security measures must not only be compatible with the compression technique and data transport but also be robust to errors as well. The underlying problems here are that a user may not receive the entire stream due to errors, the network transport (e.g., IP Multicast), or scalable compression (e.g., if MPEG-4 FGS is used). Some of the issues can be resolved by developing error-resilient security techniques (encryption and watermarking) and protocols (beyond the DRM protocols now being used) [176].

A fundamental problem in content streaming is that the network errors can cause the bit stream to be desynchronized from the receiver. This introduces a serious problem for encryption and watermarking.

- In the case of block encryption, any losses in the cipher text will result in the sequence not being properly decrypted. The use of stream ciphers eliminates this problem, but synchronization still needs to be maintained. Even if these problems are addressed, the receiver may still not *see* the entire bit stream by design because scalable compression may have been used. Here, the receiver only receives or uses the part of the bit stream that it needs.
- A similar problem exists with watermarking, where the receiver must process the bit stream to extract the watermark. If part of the bit stream is missing, then the watermark detector will be desynchronized and may not be able to detect the presence of the watermark.

5.5 INFRASTRUCTURE FOR MULTIMEDIA CONTENT DISTRIBUTION

Producing multimedia content today is easier than ever before. Using digital cameras, personal computers and the Internet, virtually every individual in the world is a potential content producer, capable of creating content that can be easily distributed and published. The same technologies allow content that would in the past remain inaccessible to be made available online.

MPEG-7 fulfils a key function in the forthcoming evolutionary steps of multimedia. As much as MPEG-1, MPEG-2, and MPEG-4 provided the tools through which the current abundance of audiovisual content could happen, MPEG-7 provides the means to navigate through this wealth of content. The *metadata* initiatives have all been developed to serve the specific needs of one business environment. This is the age in which content companies are by no means constrained by their own traditional delivery mechanisms. In the same way, content consumers are no longer tied to a single source of content. This is the real value of MPEG-7 [177].

The goal of the MPEG-7 standard is to allow interoperable searching, indexing, filtering, and access of audiovisual content by enabling interoperability among devices and applications that deal with audiovisual content description. MPEG-7 describes specific features of audiovisual content, as well as information related to audiovisual content management. MPEG-7 descriptions take two possible forms:

- a textual XML form suitable for editing, searching, and filtering
- a binary form suitable for storage, transmission, and streaming delivery.

Remember that this standard specifies four types of normative elements: descriptors, description schemes (DSs), a description definition language (DDL), and coding schemes [178].

MPEG-7 multimedia description schemes (MDSs) are metadata structures for describing and annotating audiovisual content. The description schemes (DSs) provide a standardized way of describing in extensible markup language (XML) the important concepts related to audio/video content description and content management in order to facilitate searching, indexing, filtering, and access. The DSs are defined using the MPEG-7 description definition language, which is based on the XML schema language, and are instantiated as documents or streams. The resulting descriptions can be expressed in a textual form (i.e., human readable XML for editing, searching, filtering) or compressed binary form (i.e., for storage or transmission) [179].

In the new MPEG-7 community that was gradually built up, it was not clear what the difference was between what was needed in an algorithm or an implementation and what was required in a standard. It was also not clear what were the interfaces to which the standard made reference. Even less clear was what characterized an MPEG-7 *encoder* and which were exact functions that were left for the *encoder optimization* and which were the subject of the standardization because they made reference to the MPEG-7 *decoder*. The demarcation between Audio or Visual Descriptors and Description Schemes seemed impossible to find. The difficulties highlighted not withstanding, MPEG-7 has turned out to be a very solid and effective standard. The Audio and Video parts provide standardized *audio only* and *visual only* descriptions, the Multimedia Description Scheme (MDS) part provides standardized description schemes involving both audio and visual description, the Description Definition Language (DDL) provides a standardized language to express description schemes, and the Systems part provides the necessary glue that enables the use of the standard in practical environments.

In comparison with other available emerging solutions for multimedia description, MPEG-7 can be characterized by [180]:

- its generality, related to its capability to consistently describe content from many application domains
- the integration of low-level and high-level descriptors into a single architecture, allowing the combination of the power of both types of descriptors

- its object-based data model, providing the capability to independently describe individual objects with a scene
- its extensibility, provided by the description definition language (DDL), which allows users to augment MPEG-7 to suit their own specific needs and the standard to keep evolving, and integrating novel description tools.

5.5.1 Content Description

MPEG-7 provides description schemes (DSs) for description of the structure and semantics of audio/video content. The structural tools describe the structure of the audio/video content in terms of video segments, frames, still and moving regions, and audio segments. The semantic tools describe the object, events, and motions from the real world that are captured by the audio/video content.

Structural Aspects of Content

The *Segment* DS describes the result of a spatial, temporal, or spatiotemporal partitioning of the audio/video content. The *Segment* DS can describe a recursive or hierarchical decomposition of the audio/video content into segments that form a segment tree. The *Segment Relation* DS describes additional spatiotemporal relationships among segments.

The *Segment* DS forms the base abstract type of the different specialized segment types: audio segments, video segments, AV segments, moving regions, and still regions. As a result, a segment may have spatial and/or temporal properties. For example, the *AudioSegment* DS can describe a temporal audio segment corresponding to a temporal period of an audio sequence. The *VideoSegment* DS can describe a set of frames of a video sequence. The *AudioVisualSegment* DS can describe a combination of audio and visual information such as a video with synchronized audio. The *StillRegions* DS can describe a spatial segment or region of an image or a frame in video. Finally, *MovingRegion* DS can describe a spatiotemporal segment or moving region of a video sequence.

There exist also a set of specialized segments for a specific type of AV content. For example, the *MosaicSegment* DS is specialized type of *StillRegion*. It describes a mosaic or panoramic view of a video segment constructed by aligning together and warping the frames of a *VideoSegment* upon each other using a common spatial reference system. The *VideoText* and the *InkSegment* DSs are two subclasses of the *MovingRegion* DS. The *VideoText* DS describes a region of video content corresponding to text or captions. This includes superimposed text as well as text appearing in a scene. The *InkSegment* DS describes a segment of an electronic ink document created by a pen-based system or an electronic whiteboard.

Since the *Segment* DS is abstract, it cannot be instantiated on its own. However, the *Segment* DS contains elements and attributes that are common to the different segment types. Among the common properties of segments is information related to creation, usage, media location, and text annotation.

5.5 INFRASTRUCTURE FOR MULTIMEDIA CONTENT DISTRIBUTION

The *Segment* DS can be used to describe segments that are not necessarily connected, but composed of several nonconnected components. Connectivity refers here to spatially connected if it is a group of connected pixels. A spatiotemporal segment (*MovingRegion*) is said to be spatially and temporally connected if the temporal segment where it is instantiated is temporally connected and if each one of its temporal instantiations in frames is spatially connected (note that this is not classical connectivity in 3D space).

Figure 5.71 illustrates several examples of temporal or spatial segments and their connectivity. Figure 5.71a and b illustrate a temporal and a spatial segment composed of a single connected component. Figure 5.71c and d illustrate a temporal and a spatial segment composed of three connected components.

Figure 5.72 shows examples of connected and nonconnected moving regions. In this last case, the segment is not connected because it is not instantiated in all frames and, furthermore, it involves several spatial connected components in some of the frames.

Note that, in all cases, the Descriptors and DSs attached to the segment are global to the union of the connected components building the segment. At this level, it is not possible to describe individually the connected components of the segment. If connected components have to be described individually, then the segment has to be decomposed into various subsegments corresponding to its individual connected components.

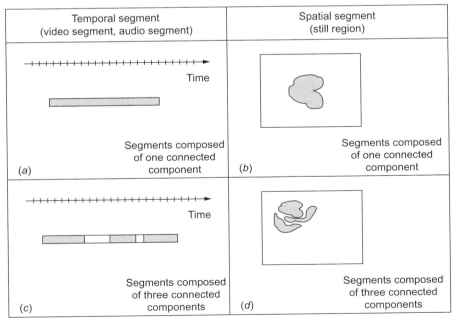

Figure 5.71 Examples of segments: (*a*), (*b*) segments composed of one connected component; (*c*), (*d*) segments composed of three connected components [177]. (©2002 Wiley.)

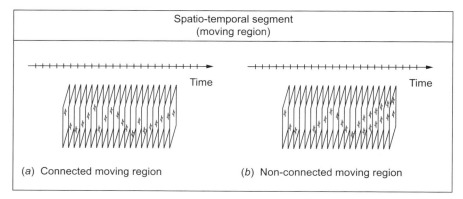

Figure 5.72 Examples of connected (a) and nonconnected moving region (b) [177]. (©2002 Wiley.)

The *Segment* DS is recursive; that is, it may be subdivided into subsegments, and thus may form a hierarchy (tree). The resulting segment tree is used to describe the media source, the temporal and/or spatial structure of the AV content. For example, a video program may be temporally segmented into various levels of scenes, shots, and microsegments; a table of contents may thus be generated based on this structure. Similar strategies can be used for spatial and spatiotemporal segments.

A segment may also be decomposed into various media sources such as various audio tracks or viewpoints from several cameras. The hierarchical decomposition is useful to design efficient search strategies (global search to local search). It also allows the description to be scalable: a segment may be described by its direct set of descriptors and DSs, but it may also be described by the union of the descriptors and DSs that are related to its subsegments. Note that a segment may be subdivided into subsegments of different types; for example, a video segment may be decomposed in moving regions that are themselves decomposed in still regions.

As is done in a spatiotemporal space, the decomposition is described by a set of attributes defining the type of subdivision: temporal, spatial, or spatiotemporal. Moreover, the spatial and temporal subdivisions may leave gaps and overlaps among the subsegments. Several examples of decompositions are described for temporal segments in Figure 5.73, where (a) and (b) describe two examples of decompositions without gaps or overlaps. In both cases, the union of the children corresponds exactly to the temporal extension of the parent, even if the parent is itself nonconnected (see case b). Case (c) shows an example of decomposition without gaps but no overlaps. Finally, case (d) illustrates a more complex case where the parent is composed of two connected components and its decomposition creates three children: the first one is itself composed of two connected components, whereas the two remaining children are composed of a single connected component. The decomposition allows gaps and overlaps. Note that, in any case, the decomposition implies that union of the spatiotemporal space defined by the children segments is

5.5 INFRASTRUCTURE FOR MULTIMEDIA CONTENT DISTRIBUTION

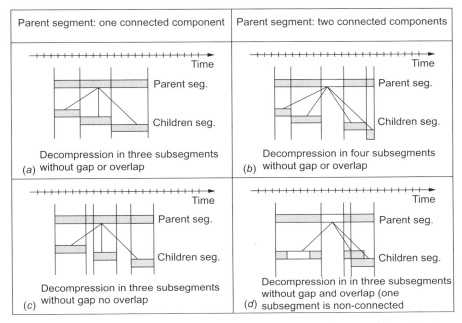

Figure 5.73 Examples of segment decomposition: (a), (b) segment decompositions without gap or overlap; (c), (d) segment decompositions with gap and overlap [177]. (©2002 Wiley.)

included in the spatiotemporal space defined by their ancestor segment (children are contained in their ancestors).

As described above, any segment may be described by creation information, usage information, media information, and textual annotation. However, specific features depending on the segment type are also allowed. These specific features are reported in Table 5.15. Most of the descriptions corresponding to these features can be extracted automatically from the original content. For this purpose, a large number of tools have been reported in the literature [177]. The institution of the

Table 5.15 Features for segment description [177]

Feature	Video Segment	Still Region	Moving Region	Audio Segment
Time	X		X	X
Shape		X	X	
Color	X	X	X	
Texture		X		
Motion	X		X	
Camera motion	X			
Audio features			X	X

©2002 Wiley.

488 MIDDLEWARE LAYER

decomposition involved in the *Segment* DS can be viewed as a hierarchical segmentation problem where elementary entities (region, video segment, and so forth) have to be defined and structured by inclusion relationship within a tree.

The semantic base DS describes narrative worlds and semantic entities in a narrative world. In addition, a number of specialized DSs are derived from the generic SemanticBase DS, which describes specific types of semantic entities, such as narrative worlds, objects, agent objects, events, places, and time, as follows. The Semantic DS describes narrative worlds that are depicted by or related to the AV content. It may also be used to describe a template for AV content. In practice, the Semantic DS is intended to encapsulate the description of a narrative world. The Object Ds describes a perceivable or abstract object. A perceivable object is an entity that exists; that is, it has temporal and spatial extent, in a narrative world.

An example of conceptual aspects description is illustrated in Figure 5.74. The narrative world involves *Tom Daniels* playing piano and his tutor. The event is characterized by a semantic time description *7–8 PM on the 14th of October 1998* and a semantic place *Carnegie Hall*. The description involves one event, to play, and four objects *piano, Tom Daniels, his tutor*, and *the abstract notion of musicians*. The last three objects belong to the class of Agent.

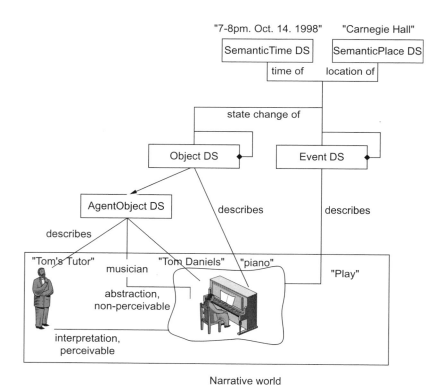

Figure 5.74 Example of conceptual aspect description [177]. (©2002 Wiley.)

5.5.2 Multimedia Content Management

Description schemes (DSs) describe the following information

- creation and production
- media coding, storage, and file format
- content message.

MPEG-7 provides DSs for AV content management. Many of the components of the content management DSs are optional. The instantiation of the optimal components is often decided in view of the specific multimedia application.

The *Creation Information* describes the creation and classification of the AV content and other related materials. The creation information provides a title (which may itself be textual or another piece of AV content), textual annotation, and information such as creators, creation locations, and dates. The classification information describes how the AV material is classified into categories such as genre, subject, purpose, language, and so forth. It also provides review and guidance information such as age classification, parental guidance, and subjective review. Finally, the related material information describes whether there exist other AV materials that are related to the content being described.

The *Media Information* describes the storage media such as the format, compression, and coding of the AV content. The media information DS identifies the master media, which is the original source from which different instances of the AV content are produced. The instances of the AV content are referred to as media profiles, which are versions of the master obtained perhaps by using different encodings, or storage and delivery formats. Each media profile is described individually in terms of the encoding parameters, storage media information, and location.

The *Usage Information* describes the usage information related to the AV content, such as usage rights, usage records, and financial information. The rights information is not explicitly included in the MPEG-7 description; instead, links are provided to the rights holders and other information related to rights management and protection. The rights DS provides these references in the form of unique identifiers that are under management by external authorities. The underlying strategy is to enable MPEG-7 descriptions to provide access to current rights owner information without dealing with information and negotiation directly. The usage record DS and available DSs provide information related to the use of the content such as broadcasting, on-demand delivery, CD sales, and so forth. Finally, the financial DS provides information related to the cost of production and the income resulting from content use. The usage information is typically dynamic in that it is subject to change during the lifetime of the AV content.

5.5.3 Multimedia Authentication Technologies

In most cases, a human will not be able to judge whether a multimedia signal is authentic by perceptual inspection. Ideally, the integrity and authenticity of multimedia data are ascertained by a system without access to information external to the challenged multimedia data itself, for example, common-knowledge, original multimedia signals. Multimedia authentication (MA) fulfills such a purpose. For example, when applied to the image, an authentication system would indicate the subsequent tampering and, in some cases, the security of modification and corresponding locations within the image. MA has been actively studied in the past and is finding an increasing number of critical applications in medicine, defense, commerce, industry, and the like. MA inherits many characteristics of many generic data authentications such as integrity verification, origination verification, nonrepudiation, and security [181]. However, MA has a few unique features that render generic data authentication algorithms well studied in cryptography inadequate or undesirable. Unlike other data, a multimedia signal can be represented equivalently in different forms or format; for example, an image represented in JPEG format that is subsequently converted to the GIF format carries exactly the same visual information. MA seeks to authenticate the multimedia content instead of its specific binary representation [182].

Multimedia signals typically contain a great amount of data. Many applications allow or even require certain processing, such as near-transparent compression, to be applied to multimedia without affecting its authenticity due to high redundancy and perceptual irrelevancy present in the signal. MA should be able to discriminate malicious manipulations from admissible manipulations. Other desirable features for MA include localization of tampered regions and indication of tamper severity and characteristics, so that the untampered portion can still be used and the altered content can be analyzed to determine if the semantic meaning is preserved and if the alteration is recoverable. Another desirable feature is to determine authenticity of a received segment of a signal, especially for an audiovisual signal, which typically has a long play time. In applications where the multimedia authenticator is generated and verified by different parties, it is desirable that knowledge for verification cannot deduce the secret to generate the authenticator. It is worth noting that some of these requirements are mutually competitive. A reasonable compromise is always necessary in the design of an authentication system.

A flowchart of a general multimedia authentication system is shown in Figure 5.75. A multimedia signal X along with an authentication key K_a is input into an authenticator T, which is then either tagged or embedded to the signal X and resulting output Y. In the verification stage, a received segment or whole signal Y' is input into a verification system along with the verification key K_v, where the

Figure 5.75 General multimedia authentication system flowchart diagram [183]. (©2004 IEEE.)

tagged or embedded authenticator T is extracted and compared with the authenticator calculated from the received signal to determine if Y' is authentic. In the watermarking case, a verification system may extract the watermark from Y' and compare with some *a priori* knowledge to make a decision. Some authentication systems may also give more information such as tamper locations and/or severity, and so on, when a received signal is determined not to be authentic. In a practical system, the verification key or the *a priori* knowledge used for verification should be content agnostic, that is, independent of the multimedia content, either the original or the challenged. This requirement rules out the possibility of using the original signal at the verification stage.

MA can be classified according to integrity criteria into hard (or complete) and soft authentication. Hard authentication rejects any modification to multimedia content. The only manipulation accepted by the hard authentication is lossless compression or format conversion that preserves visual pixel values or audio samples. This is similar to the classical authentication except that those lossless operations are also rejected by the classical authentication. Soft authentication passes certain content modifying, called incidental or admissible manipulations, and rejects all the rest, called malicious manipulation. Soft authentication can be further divided into quality-based authentication, which rejects any manipulations that lower the perceptual quality below an acceptable level, and content-based authentication, which rejects any manipulations that change the semantic meaning of the content. Classification of acceptable and unacceptable manipulations depends on a specific application. Soft authentication typically measures distortion in some metric between a feature vector from the received signal and the corresponding vector from the original signal and compares with the preset threshold to make decision on the challenged signal's authenticity. There is typically no sharp boundary between authentic and inauthentic signals. In many applications, it is often difficult to distinguish distortion caused by an incidental manipulation from that caused by a malicious manipulation. This intrinsic fuzziness makes the soft manipulation design challenging and, likely, ad hoc in most cases. Many soft authentication systems give a confidence (or a degree) of authentication instead of binary outputs.

In designing any practical authentication system, threat models and attacks need to be considered. While it is impossible or impractical to design a system to resist all forms of possible attacks, a good authentication system should be designed to survive common operations designed to reduce their effectiveness.

Some of the common attacks are covered in the following paragraphs.

Undetected modification. High redundancy and strong correlation in multimedia may be exploited to (maliciously) modify an authenticated media without being detected. Ill-defined distinction between incidental and malicious manipulations for soft authentication aggravates the problem. Tamper localization may enable an attacker to mount a successful attack by swapping components within the same signal or among different signals.

Authenticator transfer. The same high redundancy and strong correlation of multimedia may also be exploited to forge a valid authenticator for an arbitrary media signal for available authenticated signals. A famous mark transfer attack is the vector quantization attack proposed by Holliman and Memon to watermark-based authentication algorithms [184].

Information leakage. Large amounts of data in a multimedia signal or structures in the underlying secret information may be exploited to deduce the secret information or key, or to dramatically reduce the search space. Once the key is deduced, the attacker can then forge a valid authenticator to an arbitrary signal. The authentication verifier may also be exploited by an attacker to achieve the same goal.

Hard Authentication

Hard authentication rejects any modification to a multimedia signal. Most proposed algorithms are based on fragile watermarking, so the authenticator is embedded into the signal to be authenticated to simplify bookkeeping and maintenance of authenticators. In fragile watermarking, the inserted watermark is so weak that any manipulation to the multimedia content disturbs its integrity. Tampered parts of the multimedia signal may be located by checking the presence and integrity of the local fragile watermark. We will describe some major hard authentication schemes with image authentication. These technologies can be applied to other media types with minor modifications. For example, the frame index can be used in generating a video authenticator to avoid a frame reordering attack.

Single Pixel/Sample Authentication. A simple approach referred to as the *Yeung–Mintzer* scheme, which enables single pixel authentication, is shown in Figure 5.76 for grayscale images. The watermarked image I_w is generated by disturbing the original image I to enforce the relationship $L(i,j) = f(I_w(i,j))$, where L is a secret logo and f is a secret binary function. A simple way to generate the binary function f is to flip a coin for each possible pixel value. Illustrative pixel and logo values are also shown in the figure. Error diffusion is employed to reduce perceptual

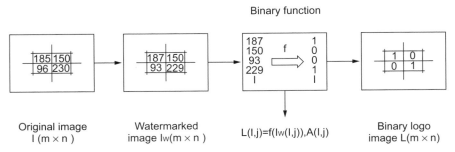

Figure 5.76 Yeung–Mintzer fragile watermarking scheme for grayscale images [183]. (©2004 IEEE.)

artifacts from the disturbance. Tampered pixels are found by examining visually or against the original logo the resulting binary image obtained by applying f to the challenged image. This scheme can be easily extended to color images and other multimedia signals.

This approach can locate a tampered pixel, but only half of modified pixels on average can be detected since each pixel is individually mapped to a binary value. The scheme's security depends critically on the secrecy of the underlying logo.

Block Authentication. Another approach is to partition a multimedia signal into disjoint parts: a signature part and an embedding part. The signature part captures all the significant information of the signal. An authenticator such as a message authentication code (MAC) or a digital signature is generated from the content of the signature part and is then embedded into the embedding part. One of the first fragile watermarking techniques was to insert key-dependent check sums of the seven most significant bits into the least significant bits (LSBs) of pseudorandomly selected pixels. If two blocks have identical logos, they can be swapped without detection. This can be avoided by using random logos.

Lossless Watermarking for Authentication. All the previously described watermarking-based schemes introduce small and irreversible distortion to signals to be authenticated. It is often desirable to design an authentication system that incurs no distortion to underlying signals like the classical data authentication, yet the authenticator is still embedded into the signal for easy storage and maintenance of authenticators. This can be achieved with recently proposed lossless watermarking schemes [185]. One approach is to use spatially additive, signal-independent robust watermarking to embed signal authentication data using a reversible modular addition. The watermark has to be robust enough to survive the reversible addition in the watermarking process so that for an unmodified watermarked signal, the authentication data can be correctly recovered and the original watermark can be subsequently regenerated and removed from the watermarked signal to recover the original signal. The amount of authentication data is typically constrained by the limited embedding capacity of the underlying robust watermark. Another approach is to losslessly compress some perceptually insignificant signal component, such as the LSB bit plane, for an image so the original component can be replaced by its compressed version appended with the authentication data for the signal. This scheme works only for signals whose components can be compressed to leave enough bits for authentication data. The lossless watermarking schemes can authenticate an image as a whole. They cannot indicate tamper locations once modification occurs. The perceptual quality of the watermarked signal needs to be considered to avoid annoying watermarking artifacts, even though watermarking can be reversed for authentic signals. Some signals may not be able to be authenticated by a lossless watermarking scheme if there is not enough available space to embed the authentication data.

Soft Authentication

In many applications, a multimedia signal may be processed after its generation. For example, a video clip may be transcoded to match the targeted devices. If hard authentication is used in these applications, any intermediate stage that performs legitimate processing on the multimedia signal will have to first verify authenticity of the signal to be processed and then authenticate the processed signal. This means that both authenticator generation and verification secrets have to be shared with these intermediate stages. Because of high correlation and perceptual irrelevancy in multimedia signals, some modifications, such as high-quality lossy compression, do not generate detectable perceptual distortion to human end users. Multimedia signals that are modified yet retain their original perceptual quality or semantic meaning are desired to be considered as authentic in many applications.

We will describe two types of authentication algorithms: those that accept only manipulations preserving the perceptual quality (quality-based authentication) and those that accept manipulations that preserve multimedia's semantic meaning (content-based authentication).

Quality-Based Authentication. This algorithm uses a quantization-based watermarking scheme to embed a pseudorandom pattern into a signal to check integrity and measure distortion [186]. Figure 5.77 shows a variant of the scheme, where a set of image blocks are pseudorandomly selected to embed data for distortion measurement.

Each selected block n is first transformed into the DCT domain, and then each $F_{i,j}^n$ is modified as $F_{i,j}^n \rightarrow F_{i,j}^{n,W} = M_{i,j}^n \{ \lfloor F_{i,j}^n / M_{i,j}^n \rfloor + r_{i,j}^n \text{sign}(F_{i,j}^n) \}$, where $r_{i,j}^n$ is a key-based random number in the interval $(0,1)$, $M_{i,j}^n$ is the masking value for the frequency bin, $\lfloor . \rfloor$ round towards 0, and $\text{sign}(x)$ is 1 or -1, depending on whether x is nonnegative or negative. These blocks are then transformed back to the spatial domain to get the watermarked image. To find distortion for a challenged image, each embedded DCT coefficient $F_{i,j}^n$ and its corresponding masking value $M_{i,j}^n$ are calculated from the challenged image, and the distortion at the frequency bin is estimated as [183] $e_{i,j}^n = F_{i,j}^{n,T} - M_{i,j}^{n,T} \{ r_{i,j}^n \text{sign}(F_{i,j}^{n,T}) + \lfloor F_{i,j}^{n,T} / M_{i,j}^{n,T} \rfloor + (r_{i,j}^n - 0.5) \text{sign}(F_{i,j}^{n,T}) \}$. It is shown that the estimation is rather accurate if distortion to the frequency bin is up to half of $M_{i,j}^n$.

Local or global distortion can be found by a weighted sum of distortion to each frequency bin and then compared with a preset threshold to check authenticity and

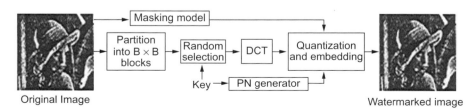

Figure 5.77 Quantization-based watermarking scheme for authentication [183]. (©2004 IEEE.)

find out tamper locations. The type of modification can also be estimated by examining the distortion pattern at different frequency bins and locations.

The above-estimated distortion for an undistorted watermarked image is small, but not the desired 0. This small error is caused by a small difference in masking values calculated from the original and watermarked images and by the integer-rounding error introduced when transforming back to the spatial domain. The first factor can be removed by replacing masking values with a signal-independent quantization vector such as a JEPG quantization table. The latter can be removed by embedding in the spatial domain.

Content-Based Authentication. This passes multimedia as authentic when the semantic meaning of the signal remains unchanged; that is, the content does not change. Media content is represented by a feature vector extracted from the media. It is this content, the extracted feature vector, instead of the multimedia signal itself, that is authenticated in content-based authentication. The general structures for multimedia authentication generation and verification for content-based authentication are shown in Figures 5.78 and 5.79, respectively.

In authenticator generation, a feature vector is extracted from media, followed by an optional data-reduction stage, and another optional lossless compression stage, to reduce the amount of data in the feature vector. The result is authenticated by a digital signature that is either tagged or embedded to the media with robust or semifragile watermarking. If watermarking is used, the media is typically partitioned depending on a secret key into disjoint signature and watermark subspaces for authentication data generation and embedding, respectively, without interference. If the two subspaces overlap, great care is necessary to avoid false alarm or reduced tolerance caused by watermarking distortion. Iteration is typically used in this case to reduce the adverse impact of watermarking on feature extraction, but it is difficult to prove that such iteration converges. In the verification stage, the embedded authentication data is extracted, decrypted, and decompressed if necessary, to get the original feature vector, which is compared to the feature vector calculated from the challenged signal. If their difference measured in some metric such as L_2 norm is smaller than a preset threshold, the content of the signal is deemed authentic. Tamper locations may be found by measuring the local distortion of the future vectors.

The major challenge in content-based authentication is to define a computable feature vector that can capture the major content characteristics from a human perspective. This remains a research challenge.

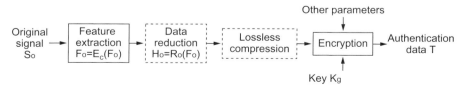

Figure 5.78 Generating multimedia authentication in content-based authentication [183]. (©2004 IEEE.)

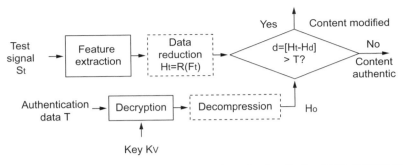

Figure 5.79 Authentication verification in content-based authentication [183]. (©2004 IEEE.)

Speech Authentication

Three types of features are proposed in speech authentication [187]: pitch information, the changing shape of the vocal tract, and the energy envelope. They are extracted with the help of a CELP codec. The first three LSP coefficients – except the silent portion – are used as the pitch information. One pitch coefficient, used as the changing shape of the vocal tract, is obtained from each frame as the average of the *lag of pitch predictors* of all subframes except the nontonal part. The starting and ending points of silent periods and also nontonal regions are included in the authentication data. At the verification phase, distortion is calculated independently for each type of feature except silence periods for the first feature and nontonal regions for the second feature. A low-pass filter is applied to the resulting difference sequences before being compared with a threshold to determine a signal's authenticity.

Future Directions

Hard authentication, usually based on fragile watermarks to detect modification to the underlying signal, has received the most coverage in the literature. Better tamper localization without sacrifice of security remains an issue to study. Soft authentication is broken into two categories: quality-based and content-based. Quality-based authentication, often based on watermarks to measure signal modification without perceptual tolerance, still needs work to improve robustness to incidental changes and yet remain sensitive to malicious modifications. Content-based authentication detects any manipulations that change a signal's semantic meaning. To be robust to content-preserving modifications and yet fragile to content-modifying manipulations, additional work is needed to define features that adequately describe the perceptual content of a multimedia signal. In soft authentication, lack of a clear-cut distinction between incidental and malicious modifications makes it difficult to accurately characterize incidental manipulations from malicious manipulations. Many proposed schemes reject manipulations that may preserve better perceptual quality or semantic meaning than acceptable manipulations. A typical example is geometrical manipulations such as image scaling and rotations, which preserve perceptual quality but are likely to cause misalignment in verification and thus are rejected by most proposed authentication schemes. Differentiating between malicious and incidental manipulations in soft authentication remains an open research issue. More effort should be directed to develop soft authentication schemes to accept quality or semantic meaning preserving manipulations, such as incidental geometrical manipulations, and also to authenticate a portion of a multimedia.

5.5.4 Watermarking Frameworks and Technologies

The watermark [240, 241] is a digital code unremovably, robustly, and imperceptibility embedded in the host data, and typically contains information about origin, status, and/or destination of the data. Although not directly used for copy protection, it can at least help in identifying the source and destination of multimedia data and, as a *last line of defense*, enable appropriate follow-up actions in case of suspected copyright violations.

While copyright protection is the most prominent application of watermarking techniques, others exist, including data authentication by means of fragile watermarks that are impaired or destroyed by manipulations, embedded transmission of value added services within multimedia data, and embedded data labeling for purposes other than copyright protection, such as data monitoring and tracking.

The development of watermarking methods involves several design trade-offs. Watermarks should be robust against standard data manipulations, including digital-to-analog conversion and digital format conversion. Security is a special concern, and watermarks should resist even attempted attacks by knowledgeable individuals. On the other hand, watermarks should be imperceptible and convey as much information as possible. In general, watermark embedding and retrieval should have low complexity because for various applications, real-time watermarking is desirable.

Watermarking is a technology that complements cryptography by embedding imperceptible signals in work. These signals remain in the work after decryption and even after conversion to the analog world, and their use has been proposed for a variety of digital rights management purposes. This feature topic focuses on this new rapidly maturing technology.

Watermarking imperceptibly alters a work to embed a message about the work. The content of the message can vary greatly depending on the application. The message might contain a unique identifier that can be used to identify the work. Or the message might contain information regarding ownership of the work, the recipient of the work, and/or a variety of additional annotations that pertain to the work's history. Or the message might contain information regarding copyright permissions, including information inhibiting or permitting copying. Applications of watermarking included content authentication, proof of ownership, usage control, broadcast monitoring, and annotation, including linking multimedia work to the Internet.

Clearly, watermarking is a form of communications, and a number of models has been proposed. A watermarking system consists of a watermarking embedder and a watermark detector, which are analogous to a transmitter and receiver. The most common model then views the original work as the communication channel and the watermarks as the modulated signal that conveys the message. However, unlike traditional communications, watermarking must also pay careful attention to the fidelity of the underlying cover work. This has sometimes been modeled as a power constraint at the transmitter.

This fidelity constraint, together with the conflicting requirements that a watermark be robust (i.e., survive common forms of signal processing that are applied to a work) and secure (i.e., survive intentional efforts to remove it), led many researchers to model watermarking as a form of spread spectrum communications. Spread spectrum communications systems transmit very little power with any single frequency, and have excellent interference and antijamming properties that make them very suitable for watermarking applications. The secret chip sequences of spread spectrum communications are considered key in the watermarking literature [188].

A more recent model considers watermarking as a form of communications with side information. In this model, the cover work is no longer treated as noise, but as side information available to either the transmitter and/or receiver. This communication perspective holds the promise of greatly improving the channel capacity of a watermarking system.

Watermark detection is increasingly utilizing common performance measures such as bit error rate (BER). Watermarking technology has many performance characteristics that rival one another in the system design space.

Realizing the full potential of this technology calls for innovation. Innovation requires two fundamental components: novel ideas and implementation. Novelty comes from new perspectives and concepts, along with the requisite research into their implications. Implementation converts these new perspectives, concepts, and research into applied technology, where engineering creates successful applications of a new technology. Novelty without implementation carries no economic benefits or rewards. Implementation without new ideas is simply incremental improvement, with no capacity to radically change or challenge existing technology paradigms.

As an exceptionally powerful technology, watermarking has many potential commercial applications, in many fields. And as digital content, such as images, music, and video, becomes more pervasive, the ability to imperceptibly embed digital information into the digital work will find growing numbers of imaginative commercial applications. For instance, the capability of some forms of digital watermarks to survive transition to the analog domain, and a subsequent redigitization, create additional attractive system capabilities.

The basic requirements in watermarking apply to all media and are very intuitive [189]:

- A watermark shall convey as much information as possible, which means the watermark data rate should be high.
- A watermark should in general be secret and should only be accessible by authorized parties. This requirement is referred to as security of the watermark and is usually achieved by the use of cryptographic keys.

- A watermark should stay in the host data regardless of whatever happens to the host data, including all possible signal processing operations that may occur, and including all hostile attacks that unauthorized parties may attempt. This requirement is referred to as robustness of the watermark. It is a key requirement for copyright protection or conditional access applications, but less important for applications where the watermarks are not required to be cryptographically secure, for example, for applications where watermarks convey public information.
- A watermark should, though being unremovable, be imperceptible.

Depending on the media to be watermarked and the application, this basic set of requirements may be supplemented by additional requirements.

- Watermark recovery may or may not be allowed to use the original, unwatermarked host data.
- Depending on the application, watermark embedding may be required in real time, for example, for video fingerprinting. Real-time embedding again may, for complexity reasons, require compressed-domain embedding methods.
- Depending on the application, the watermark may be required to be able to convey arbitrary information. For other applications, only a few predefined watermarks may have to be embedded, and for the decoder it may be sufficient to check for the presence of one of the predefined watermarks (hypothesis testing).

Basic Watermarking Principles

The basic idea in watermarking is to add a watermark signal to the host data to be watermarked such that the watermark signal is unobtrusive and secure in the signal mixture, but can partly or fully be recovered from the signal mixture later on if the correct cryptographically secure key needed for recovery is used.

To ensure imperceptibility of the modification caused by watermark embedding, a perceptibility criterion of some sort is used. This can be implicit or explicit, host data adaptive or fixed, but it is necessary. As a consequence of the required imperceptibility, the individual samples (e.g., pixels or transform coefficients) that are used for watermark embedding can only be modified by an amount relatively small compared to their average amplitude.

To ensure robustness despite the small allowed changes, the watermark information is usually redundantly distributed over many samples (e.g., pixels) of the host data, thus providing a *holographic* robustness, which means that the watermark can usually be recovered from a small fraction of the watermarked data, but the recovery is more robust if more of the watermarked data are available for recovery.

As described earlier, watermark systems do in general use one or more cryptographically secure keys to ensure security against manipulation and erasure of the watermark.

There are three main issues in the design of a watermarking system.

- Design of the watermark signal W to be added to the host signal. Typically, the watermark signal depends on a key K and watermark information I:

$$W = f_0(I, K)$$

Possibly, it may also depend on the host data X into which it is embedded

$$W = f_0(I, K, X)$$

- Design of the embedding method itself that incorporates the watermark signal W into the host data X yielding watermarked data Y

$$Y = f_1(X, W)$$

- Design of the corresponding extraction method that recovers the watermark information from the signal mixture using the key and with help of the original

$$I = g(X, Y, K)$$

or without the original

$$I = g(Y, K)$$

The first two issues, watermark signal design and watermark signal embedding, are often regarded as one, specifically for methods where the embedded watermark is host signal adaptive.

Figure 5.80 illustrates the generic digital watermarking scheme for the embedding process. The input to the scheme comprises the watermark, the host data, and an optional public or secret key. The host data may, depending on the application, be uncompressed or compressed; however, most proposed methods work on uncompressed data. The watermark can be of any nature, such as a number, text, or an image. The secret or public key is used to enforce security. If the watermark is not be read by unauthorized parties, a key can be used to protect the

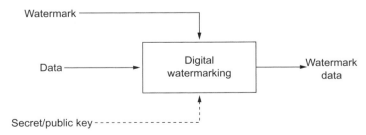

Figure 5.80 Generic digital watermarking scheme [189]. (©1999 IEEE.)

watermark. In combination with a secret or a public key, the watermarking techniques are usually referred to as secret and public watermarking techniques, respectively. The output of the watermarking scheme is the modified, that is, watermarked, data.

The generic watermark recovery process is depicted in Figure 5.81. Inputs to the scheme are the watermarked data, the secret or public key, and, depending on the method, the original data and the original watermark. The output of the watermark recovery process is either the recovered watermark or some kind of confidence measure indicating how likely it is for the given watermark at the input to be present in the data under inspection.

Many proposed watermarking schemes use ideas from spread-spectrum radio communications [190]. They embed a watermark by adding a pseudonoise (PN) signal with low amplitude of the host data. This specific PN signal can later on be detected using a correlation receiver or matched filter. If the parameters like amplitude and the number of samples of the added PN signal are chosen appropriately, the probabilities of false-positive or false-negative detection are very low. The PN signal has the function of a secret key. The scheme can be extended if the PN signal is either added or subtracted from the host signal. In this case, the correlation receiver will calculate either a high-positive or high-negative correlation in the detection. Thus, one bit of information can be conveyed. If several such watermarks are embedded consecutively, arbitrary information can be conveyed.

Text Documents Watermarking

For watermarking, we have to distinguish between methods that hide information in the semantics, which means in the meaning and ordering of the words, and methods that hide information in the format, which means in the layout and the appearance.

Text watermarking has applications wherever copyrighted electronic documents are distributed. Important examples are virtual digital libraries where users may download copies of documents, for example, books, but are not allowed to further distribute them or store them longer than for certain predefined periods. In this type of application, a requested document is watermarked with a requester-specific watermark before releasing it for download. If, later on, illegal copies are discovered, the embedded watermark can be used to determine the source.

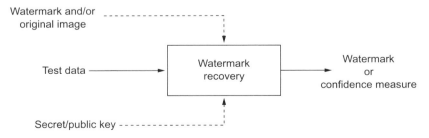

Figure 5.81 Generic watermark recovery scheme [189]. (©1999 IEEE.)

Three different methods for information embedding into text documents are proposed [191–193]: line-shift coding, word-shift coding, and feature coding. In line-shift coding, single lines of the document are shifted upwards or downwards by very small amounts. The information to be hidden is encoded in the way the lines are shifted. Similarly, words are shifted horizontally in order to modify the spaces between consecutive words in word-shift coding. Both methods are applicable to the format file or a document or to the bitmap of a page image. While line-shift coding can rely on the assumption that lines are uniformly spaced, and thus does not necessarily need the original for watermark extraction, the original is required for extraction in word-shift coding, since the spaces between words are usually variable. The third method, feature coding, slightly modifies features such as the length of the end lines in characters like b, d, h, and so on. Among the three presented methods, line-shift coding is most robust in the presence of noise, but also most easily defeated. The presented methods are robust enough to resist printing, consecutive photocopying up to ten generations, and rescanning.

Image Watermarking

Most watermarking research and publications are focused on images. The reason might be that there is a large demand for image watermarking products due to the fact that there are so many images available at no cost on the World Wide Web that need to be protected.

Meanwhile, the number of image watermarking publications is too large to give a complete survey of all proposed techniques. However, most techniques share common principles.

The watermark signal is typically a pseudorandom signal with low amplitude, compared to the image amplitude, and usually with spatial distribution of one information (i.e., watermark) bit over many pixels. A lot of watermarking methods are in fact very similar and differ only in parts or single aspects of the three topics: signal design, embedding, and recovery.

The information that is embedded is usually not important for the watermarking itself. However, there are methods that are designed to embed and extract one out of a codebook of codes, and thus cannot accommodate arbitrary information.

Other proposed schemes modulate the codes available in the codebook with arbitrary information bits and can accommodate arbitrary messages. The watermark signal is often designed as a white or colored pseudorandom signal with Gaussian, uniform, or bipolar probability density function (pdf). In order to avoid visibility of the embedded watermark, an explicit or implicit spatial shaping is often applied with the goal to attenuate the watermark in areas of the image where it would otherwise become visible [194–196]. The resulting watermark signal is sometimes sparse and leaves image pixels unchanged, but mostly it is dense and alters all pixels of the image to be watermarked. The watermark signal is often designed in the spatial domain, but sometimes also in transform domain like the full-image DCT domain or blockwise DCT domain.

The signal embedding is carried out by addition or single-adaptive (i.e., scaled) addition, mostly to the luminance channel alone, but sometimes also to color channels, or only to color channels [197, 198]. The addition can take place in the spatial domain or in transform domains such as the DFT domain, full-image DCT domain, the blockwise DCT domain, the wavelet domain, the fractal domain, the Hadamard domain, the Fourier–Mellin domain, or the Radon domain [199–204]. It is often claimed that embedding in the transform (mostly DCT or wavelet) domain is advantageous in terms of visibility and security. While some authors argue that the watermarks should be embedded into low frequencies [205], others argue that they should rather be embedded into the medium or high frequencies. In fact, for maximum robustness, watermarks should be embedded signal adaptively into the same spectral components that the host data already populated. For image and video, these are typically the low frequencies. Watermark signal generation and watermark embedding are often treated jointly. For some proposed methods they cannot be regarded separately, especially if the watermark is signal adaptive.

Watermark recovery is usually carried out by some sort of correlation method, like a correlation receiver or a matched filter. Since the watermark signal is often designed without knowledge of the host signal, crosstalk between the watermark signal and host data is a common problem in watermarking. In order to suppress the crosstalk, many proposed schemes require the original, unwatermarked data in order to subtract it before watermark extraction. Other proposed methods apply a prefilter instead of subtracting the original.

Video Watermarking

Video sequences consist of a series of consecutive and equally time-spaced still images. Thus, the general problem of watermarking seems very similar for images and video sequences, and the idea that image watermarking techniques are directly applicable to video sequences is obvious. This is partly true, and there are a lot of publications on image watermarking that conclude with the remark that the proposed approach is also applicable to video.

However, there are also some important differences between images and video that suggest specific approaches for video. One important difference is the available signal space. For images, the signal space is very limited. This motivates many researchers to employ implicit or explicit models of the HVS, in order to reach the threshold of visibility and to embed a watermark as robust as possible without sacrificing image quality.

For video, the available signal space, that is, the number of pixels, is much larger. On the other hand, video watermarking often imposes real-time or near-real-time constraints on the watermarking system. As a consequence, it is less important, and for many applications even prohibitively complex, to use watermarking methods based on explicit models of the HVS. Complexity, in general, is a much more important issue for video watermarking applications than it is for image watermarking applications.

For individual watermarking, that is, fingerprinting of video sequence, this problem is more severe, because video sequences are usually stored in compressed format. Uncompressed storage and on-the-fly compression, or decompression, watermarking, and recompression, are usually not feasible for this kind of application, unlike for images. Thus, such applications may require compressed-domain watermarking. Another point to consider is that the structure of video as a sequence of still images gives rise to particular attacks, for example, frame averaging, frame dropping, and frame swapping. At frame rates of 25–30 Hz, as are used in television, this would possiby not be perceived by the casual viewer. A good watermarking scheme, however, should be able to resist this kind of attack, for example, by distributing watermark information over several consecutive frames. On the other hand, it might be desirable to retrieve the full watermark information from a short part of the sequence depending on the application used for these two competing requirements (e.g., embedding a multiscale watermark with more than one temporal scale or progressive watermark transmission).

While a lot of research has been published on image watermarking, there are fewer publications that deal with video watermarking. However, the interest in such techniques is high, for example, in the case of the emerging digital versatile disk (DVD) standard, which will contain a copy protection system employing watermarking. The goal is to mark all copyright video material such that DVD standard compliant players or recorders will refuse to play back or record pirated material.

Audio Watermarking

Compared to images and video, audio signals are represented by much fewer samples per time interval. This alone indicates that the amount of information that can be embedded robustly and inaudibly is much lower than that for visual media. An additional problem in audio watermarking is that the human audible system (HAS) is much more sensitive than the HVS, and so inaudibility is much more difficult to achieve than invisibility for images.

For a spread-spectrum approach in audio watermarking, we can use a pseudo-noise (PN) sequence that is filtered [189] in several stages in order to exploit long-term and short-term masking effects of the HAS. In order to exploit long-term masking, a masking threshold for each overlapping block of 512 samples is calculated and approximated using a tenth-order all-pole filter, which is then applied on the PN sequence. Short-term masking is additionally exploited by weighting the filter's PN sequence with the relative time-varying energy of the signal in order to attenuate the watermarked signal where the audio signal energy is low. Additionally, the watermark is low-pass filtered in order to guarantee that it survives audio compression. A high-pass component of the watermark is also embedded, which improves watermark detection from uncompressed audio pieces, but is expected to be removed by compression. The authors denote the two spectral components of the watermark by *low-frequency watermark* and *coding error watermark*. The watermark can be extracted by hypothesis testing using the original and the PN sequence and by employing a correlation method. Experimental results show the robustness of the scheme to MPEG-1 layer III audio coding, to coarse PCM quantization using word length down to 6 bits/sample instead of 16 bit/samples as for the original, and additive noise.

A system for an interactive television has been developed in reference [206]. Information is embedded into the audio component of a television signal. The embedded information is detected from the acoustic signal emitted from the television receiver. The information to be embedded is partitioned in blocks of 35 bits. Each information bit is modulated using a sinusoidal carrier of a specific frequency and low amplitude and added to the audio signal. The simplified principle is that if the sinusoidal carrier for a specific bit is present in the signal, the bit is 1, otherwise it is 0. The frequencies of the sinusoidal carriers are above 2.4 kHz; thus at frequencies where the HAS is less sensitive, no explicit model of the HAS is employed. In order to reduce interference from the audio signal itself, the audio signal is attenuated at frequencies above 2.4 kHz. Thus, the principle involves a fidelity loss of the host signal that seems acceptable for the envisaged application. In order to increase the robustness, the information bits are protected by a cyclic redundancy code (CRC) and bit repetition. In order to compensate for the frequency shift of the whole signal, for example, after analog recording and playback with inaccurate speed, a frequency locking mechanism is applied using five special sinusoidal carriers of known frequency. Thus the scheme is robust against room noise and video tape recording.

Several techniques for watermarking that are applicable to audio are proposed in reference [207]. These techniques are called spread-spectrum coding, echo coding, and phase coding. Direct sequence spread-spectrum coding performs biphase shift keying on a carrier wave by using an encoder binary string and pseudorandom noise. The code introduces perceptible noise into the original sound signal, but by using adaptive coding and redundant coding, the perceptible noise can be reduced. Echo coding is a method that employs multiple decaying echos to place a peak in the cepstrum at known locations. The result is that moderate amounts of data can be hidden in a form that is fairly robust versus analog transmission. Phase coding is a method that employs the phase information as a data space. For the encoding, a Fourier transform is applied and the phase values of each frequency component are lined up as a matrix; binary information can be embedded into this matrix by modifying the phase components. Since the HAS is not very sensitive to the distortion to the phase of the sound, it can be used to encode data without introducing much audible distortion to the original sound.

Watermarking of Other Multimedia Data

Most watermarking research, publications, and products are dedicated to images. Less has been published on video, audio, and formatted text watermarking, and even less on watermarking of other media. However, the underlying basic ideas are certainly applicable to almost all kinds of digital data.

The video compression standard MPEG-4 features additional functionalities, besides common video compression, such as model-based animation of the 3D head model using so-called facial animation parameters (FAPs). These are parameters like *rotate head*, *open mouth*, or *raise right corner-lip*. The head model used at the receiver is either a predefined generic head and face model or a particular model that can be transmitted using so-called facial definition parameters (FDPs).

The tool for face animation, for example, in video telephony applications, has bit rates below 1000 bit/s.

A spread-spectrum method for watermarking of MPEG-4 FAPs was proposed in reference [208]. The watermarks are additively embedded into the animation parameters. Smoothing of the spread-spectrum watermark by low-pass filtering and an adaptive amplitude alternation prevents visible distortions.

5.5.5 Digital Rights Management Systems and Security Information for Content Distribution

The problem of digital content piracy is becoming more and more critical, and major content producers are risking seeing their business drastically reduced because of the ease by which digital contents can be copied and distributed. This is the reason why digital rights management (DRM) is currently paid much attention by industry and research. Among the various technologies that can contribute to set up a reliable DRM system, data hiding (watermarking) has found an important place, thanks to its potentiality of persistently attaching some additional information to the content itself. Many applications (ownership proofing, copy control, etc.) have been devised in this framework, exploiting data hiding techniques, and many problems have emerged. Some industrial initiatives have also been carried out that tried to exploit watermarking technology for particular DRM problems: for example, the Copy Protection Technical Working Group (CPTWG) for DVD copy protection [209] and the Secure Digital Music Initiative (SDMI) [210]. The time is right for accessing the effectiveness of data hiding technology with respect to the various applications it can serve, and to draw some conclusions from past experiences. DRM technology could benefit from data hiding in several ways, as is evident by the variety of watermarking-based systems addressing DRM problems.

Data Hiding System

The general model of a data hiding system is presented in Figure 5.82. The watermark code b is embedded into the host signal A, thus producing the watermarked asset A_w. Owing to possible attacks, A_w is transformed into A' [211]. Finally, the hidden information is recovered from A', either by extracting the hidden message b or by decoding whether A' contains a known code b or not. Watermark embedding and recovery require the knowledge of a secret key K. Watermark recovery may benefit from the knowledge of the original, nonmarked signal A.

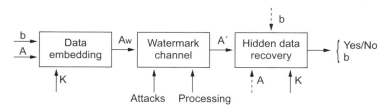

Figure 5.82 Data hiding scheme [211]. (©2004 IEEE.)

Let us discuss the data hiding system in some more detail. At the input of the system we find the information to be hidden and the original, nonmarked host signal A. The host signal, sometimes called the cover signal, may be an audio file, a still image, a piece of video, or a combination of the above. We assume that the information to be hidden takes the form of a binary string $b = (b_1, b_2, \ldots, b_k)$, with b_i taking values of $\{0,1\}$. We refer to b as the watermark code. The data embedding module, or simply the embedder, mixes the cover signal A and the watermark code b to produce a watermarked signal A_w. In order to increase the secrecy of the system, the embedding function e usually depends on the secret key K. Thus, in its more general form, e can be written as $A_w = e(A, b, K)$. Usually, the definition of e goes through the extraction from the host signal A of a set of features $f = (f_1, f_2, \ldots f_n)$, called host features, that are modified according to the watermark code. Possible choices of f include audio or image samples, discrete Fourier transform (DFT) or DCT coefficients, and wavelet coefficients. In many cases it is useful to describe the embedding function by introducing a watermark signal $w = (w_1, w_2, \ldots, w_n)$, which is added to the host feature set. In this case, by letting $f_w = (f_{w1}, f_{w2}, \ldots, f_{wn})$ indicate the watermarked host features, we have $f_{wi} = f_i + w_i$, $i = 1, \ldots, n$.

In the simplest case, w depends only on b and K. For instance according to the spread-spectrum approach, a pseudorandom sequence $s = (s_1, s_2, \ldots, s_n)$ is generated depending on K, then s is modulated by means of an antipodal version of b [212]. A data hiding system from which w does not depend on A is called a blind embedding system, since embedding is carried out blindly without taking into account the particular host signal in hand. Better results can be obtained by tailoring w to the host asset I [213–215]. More specifically, by assuming that the detector structure and the corresponding detection regions are known to the embedder, the embedding problem may be seen as the mapping of the host signal into a point within the correct detection region. If the role of the watermark signal has to be retained, we must now admit that W depends on the host signal A, since in this way it is possible to push the watermark more inside the detection region, given a desired level of distortion. The informed embedding principle may be pushed further, by letting detection regions to be composed by a set of nonconnected subregions spread over all the signal space, and by deciding to map the host signal inside the subregion that results in the lowest distortion. Data hiding systems obeying this strategy are collectively termed *informed watermarking* or *informed data hiding* systems.

The second element is the so-called watermark channel. This accounts for all the manipulations the host signal may undergo after information embedding. Note that both intentional and unintentional manipulations must be taken into account, with the former accounting for the possible presence of an enemy, usually called the attacker, acting with the explicit goal of damaging the hidden message, and the latter accounting for the manipulations the host signal may undergo during its normal lifecycle (e.g., lossy coding, resizing, filtering). The ability to survive intentional attacks is referred to as watermark security. For example, scenarios where pirates can freely access the detector are by far more complex than those for which the extraction device is not publicly available.

After the host signal has passed the watermark channel, it enters the detector, whose scope is to retrieve the hidden information. Extraction of the hidden information may follow two different approaches: the detector looks for the presence of a specific message, thus only answering yes or no, or the detector (which in this case is called a decoder) reads the information conveyed by the host signal without knowing it in advance. These two approaches lead to a distinction between algorithms embedding a message that can be read (readable or multibit watermarking) and those inserting a code that can only be detected (detectable or one-bit watermarking). An additional distinction may be made between systems that need to know the original, nonmarked signal A in order to retrieve the hidden information and those that do not retrieve it. In the latter case, we say that the detector is blind (the term oblivious detection may also be used) whereas in the former case the detector is said to be nonblind.

In all cases, the retrieval of b goes through the definition of a detection (decoding) function D. In oblivious detectable watermarking, D is a three-argument function accepting as input a digital asset A', a watermark code b, and a secret key K. D decides whether A' contains b or not, that is $D(A', b, K) =$ yes/no. In the nonoblivious case, the original asset A is a further argument of D. In blind, readable watermarking, the decoder function takes as inputs a digital asset A', a keyword K, and gives as output the string b it reads from A': $D(A', K) = b$. In the nonoblivious case, the original asset A is a further argument of D. Note that, in readable watermarking, the decoding process always results in a decoded bit stream; however, if the asset is not marked, decoded bits are meaningless.

So far we have implicitly assumed that the secret key K used in the decoding/detection process is the same used for embedding. We term these kinds of algorithms as symmetric watermarking schemes. A problem with symmetric watermarking is an intrinsic lack of security, especially if the decoder/detector is implemented in publicly available consumer devices. The knowledge of K, in fact, is likely to give attackers enough information to remove the watermark from the host signal. In order to overcome these problems, increasing attention has been given to the development of asymmetric schemes. In such schemes two keys are present, a private key K_s, used to embed the information within the host signal, and a public key K_p, used to detect/decode the watermark (often K_p is just a subset of K_s). Knowing the public key, it should be neither possible to deduce the private key nor to remove the watermark [216].

Copyright Protection

This has been one of the first industrial DRM applications of watermarking, surely the one that triggered the attention toward watermarking and data hiding. In the early days of watermarking research, it was thought, in fact, that simply embedding a flag within the work to be protected, for example, stating the work could not be copied, was enough to prevent fraudulent copying. It was soon realized, however, that a much deeper analysis is necessary before data hiding can be effectively used to enforce, or at least to help enforce, copyright laws. First of all, the embedded watermark must be robust against unintentional processing and secure against intentional attacks. This turned out to be an extremely challenging task that is far from being solved, especially with regard to security. Impressive progress has been made in terms of both achievable robustness and capacity; thus, even if security may still be out of reach, high-capacity, robust watermarking may soon become a reality. Yet the design of watermarking-based copy protection mechanisms is not a matter of robustness, since protocol issues must be considered as well.

In the following, we review the main watermarking-based copyright protection scenarios. According to one of them, a so-called copy difference mechanism is adopted to discourage unauthorized duplication and distribution. Copy difference is achieved by providing a mechanism to trace unauthorized copies to the original owner of the work or, more generally, to track the author of the infringement. In the most common case, distribution tracking is made possible by letting the seller insert a distinct watermark, which in this case is called a fingerprint, identifying the buyer, or any other addressee of the work, within any copy of data that is distributed. (It is worth noting here that the term *fingerprinting* has been recently used for another type of technology aimed at extracting from a digital document a distinctive set of unique characteristics – fingerprint – that can be later used for identifying it [217, 218].) If an unauthorized copy of the protected works is found, then its origin can be recovered by retrieving the unique watermark contained in it. Figure 5.83 shows a simplified scheme exemplifying a fingerprint protocol (the numbers indicate the sequence of the operations).

Before the multimedia document is legally acquired, the buyer has to provide his own identifier to the seller. The seller embeds this identifier into the document and gives it to the buyer. If the buyer illegally distributed the acquired content to another party (here the pirate), a control authority can detect his identifier into the document, and take actions against him. A similar scheme has been proposed in reference [219] for avoiding illegal copying and distribution of digital cinema.

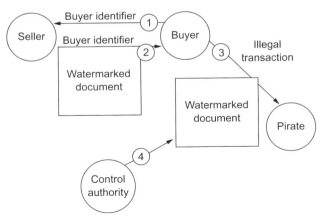

Figure 5.83 Block scheme for exemplifying a fingerprinting protocol [211]. (©2004 IEEE.)

The main requirement set by the fingerprinting application is security, since any attempt to remove the watermark or making it unreadable must be prevented. At the same time, a readable watermarking scheme is preferable, since in many cases it is not possible to guess in advance the watermark content.

When copy deterrence is not sufficient to effectively protect legitimate rights holders, a true copy protection mechanism must be envisaged. Having said that, a comprehensive solution of the copy protection problem goes well beyond watermarking technology. We describe a mechanism that has been considered for protection of DVD video. This scenario, in fact, represents a good example of how watermarking can be integrated in a complex copy protection system and effectively contribute to its efficacy.

A possible approach to make illegal duplication and distribution difficult enough to limit the losses caused by missed revenues relies on the distinction between copyright compliant devices (CC devices) and noncompliant devices (NC devices), where CC devices are designed to refuse to make copies when they are not explicitly allowed. Then, the copy control mechanism consists of keeping the CC and the NC worlds as separate as possible, for example, by allowing NC devices to play only illegal disks and CC devices to play only legal disks. In this way, users willing to play both legal and illegal disks must buy two series of devices (of course nonprotected disks would be playable on both kinds of devices).

Figure 5.84 depicts the separation between the worlds of copyright compliant and noncompliant devices. Only nonprotected disks can be played by both kinds of devices.

The possibility that a legal disk is played on an NC device is prevented by means of a proper content scrambling system (CSS). Descrambling requires a pair of keys, one of which is unique to the video file, while the other is unique to the DVD. Keys are stored on the lead-in area of the DVD, an area that is only read by CC devices.

The sensitivity attack coupled with the closest-point attack is a very general and effective attack that cannot be easily avoided. Two possible solutions have been proposed so far: (a) to design the watermarking system so that the boundary of the detection region is not a parametric one, and (b) to adopt an asymmetric watermarking scheme in which the key used to detect the watermark, if any, does not reveal the parameters used during the embedding phase. Although these are promising solutions, their effectiveness has not been proved yet; hence, further research is needed before the sensitivity plus the closest-point attack no longer hampers the practical implementation of copy control mechanisms relying on data hiding technology.

Item Identification

One of the main features of watermarking technology is that it provides a way to attach a code to a multimedia document in such a way that the code is persistent with respect to the possible changes of format the document may undergo. As an extreme case, the embedded code can be resistant even to digital-to-analog conversion. To fully exploit the potentiality of this characteristic, the concept of persistent association has been developed during the past few years. The basic idea is to associate a unique identifier (UI) to each multimedia creation. The UI is embedded inside

5.5 INFRASTRUCTURE FOR MULTIMEDIA CONTENT DISTRIBUTION

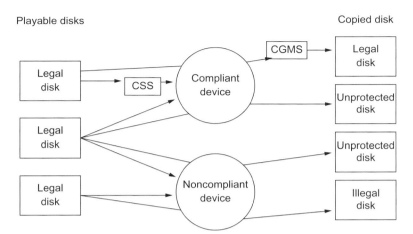

Figure 5.84 Separation between the worlds of copyright compliant and noncompliant devices [211]. (©2004 IEEE.)

the document itself by means of a watermarking primitive and is used for indexing a database where more detailed information (not only related to intellectual property rights, IPR) can be retrieved. The concept of persistent association is shown in Figure 5.85.

Thanks to the persistent association of a UI to each multimedia document, a player can request the appropriate licensing information from an IPR database and, as a consequence, apply the corresponding copyright policy to the document. The numbers indicate the sequence of the operations. The use of watermarking for tightly attaching a UI to a document is proposed in reference [220].

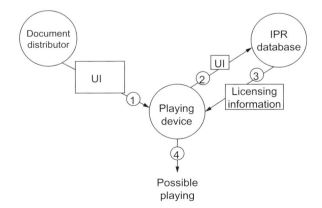

Figure 5.85 Concept of persistent association of a unique identifier (UI) to each multimedia creation [211]. (©2004 IEEE.)

5.5.6 Multimedia Advances in Media Commerce

Owing to the recent rapid technology advances in e-commerce, multimedia, and broadband Internet, media commerce has already begun enabling new applications and streamlining of core business processes in a number of industries. It has a strong potential to influence many aspects of our daily lives. Most of the better arranged e-commerce sites usually provide more than just the basic catalog browsing and transactional capability. Many of the e-commerce sites also provide some forms of portals including latest sale, personalized *stickiness* of the sites. Also, some of these promotions ads utilize enhanced multimedia capabilities (animated gifts, `MacromediaFlash`, or sometimes even streaming media). Media such as music, movies, and books have been traded and delivered directly through broadband networks. As a result, it has become apparent that content and media will play an increasingly essential role in e-commerce, both indirectly such as in the catalog campaign, promotion, and banner ads, or directly as the merchandise.

Since 1995, we have witnessed the emergence of the business-to-consumer (B2C) type of e-commerce. The pace of developing business-to-business e-commerce (B2B) applications, especially the establishment of an electronic marketplace, began to accelerate from 1998. Also, business-to-employee (B2E), peer-to-peer (P2P), and government-to-citizen (G2C) are picking up steam.

Content management provides the environment for authoring, editing, approval, and publishing of content for catalogs. In addition, many e-commerce sites also have to support pervasive/ubiquitous access of the sites from a wide variety of devices. The content for enabling pervasive access may be transcoded in advance to minimize the overhead of having to translate the content for each and every request from the pervasive devices. Portals provide a one-stop shop of all the relevant information. The multimedia content used for enhancing and enriching the user experience can also be used for search and retrieval [221].

When media becomes the goods that are being sold through a B2C storefront or traded through an electronic marketplace, many new issues arise. Similar to traditional goods, it is possible to define a supply chain for the media goods. Each stage of the media supply chain consumes input from the previous stage, and generates output for the next stage. All the trading mechanisms (fixed price, dynamic price, auction, reverse auction, request for quotes – RFQ) that have been used for real goods are applicable to media goods as well. On the other hand, media goods can be distributed through the same mechanism upon which the transactions are conducted. Consequently, trust and rights management become some of the most important issues. Furthermore, multiple media sources can be composed through overlay, mosaicing, or fusion. As a result, third-party service providers can participate in media commerce through providing value-added services such as processing of the media.

We will investigate two types of media commerce: *media in commerce* and *media for commerce*. Media in commerce is used to create richer user experience and to enable multimodel collaboration, and servers in a supplemental role for e-commerce transactions, such as in portals, catalogs, and advertisement. Media for commerce is traded as the merchandise in commerce transactions. Examples include music/audio, video/movies, electronic books, and electronic magazines. Furthermore, the fulfillment of the transactions, including personalization, can take place directly through the Internet. Media for commerce includes both B2C and B2B scenario. Roles of participants in a B2C media commerce environment are shown in Figure 5.86. In the B2C case, the entire supply chain of content publication and distribution starts with the content publisher, and includes aggregator, retailer, and end consumer. Additional mechanisms such as payment and rights clearance are also needed to enable the transaction to take place electronically.

In the B2B scenario shown in Figure 5.87, the media provider, media service provider, and media consumer will exchange the media goods through a media marketplace. The configuration of the B2B marketplace includes (1) one supplier to many consumers, the traditional sell-side marketplace; (2) one consumer to many suppliers, the buy-side marketplace, where the consumer *procures* the media goods through comparing of the bids or offerings from multiple suppliers, and chooses the best one according to a utility function; and (3) many suppliers to many consumers, the public exchange scenario, such as the stock exchange.

Based upon how media contributes to a business process, media commerce can be categorized as Maintenance, Repair, and Operation (MRO), in which media participates in peripheral aspects of the business such as education and training, and Direct, in which the media commerce is part of the core business process. Examples of the Direct category include media and entertainment.

In a manner similar to that for goods in traditional commerce, the pricing mechanisms for trading media merchandise include fixed pricing and dynamic pricing. In the former case, the pricing of the media goods is fixed and nonnegotiable. This is usually applicable to catalog-based transactions where the price of the media is

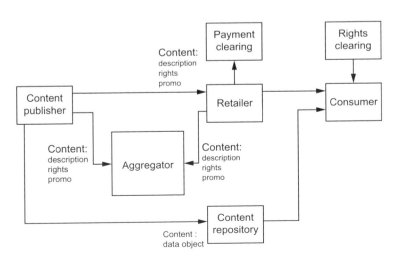

Figure 5.86 Participants in a B2C media commerce environment [221]. (©2002 IEEE.)

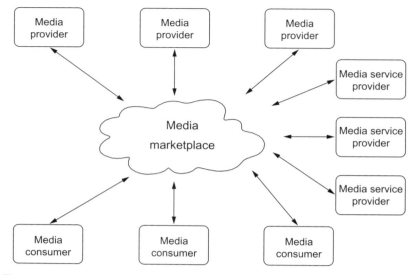

Figure 5.87 Participants in a B2B media commerce environment [221]. (©2002 IEEE.)

predetermined. In the latter case, the pricing of the media goods is determined through various negotiation mechanisms, such as request for quote (RFQ), reverse auction, and auction.

The most important differentiating aspects of media merchandise are that the media goods usually have large initial cost, and negligible marginal cost (the cost for producing additional copies of the same goods), and that media goods can be transported through broadband Internet, thus allowing seamless transactions and fulfillment to take place. Commonly used differential pricing mechanisms can be easily applied, in which different prices can be developed for different versions of the content such as progressive image representation.

The lifecycle of media goods, shown in Figure 5.88 includes the following steps.

- *Publish assets*. The content publishes the media and content assets that can be traded on a catalog.
- *Discover and search*. The customer of the contents searches for possible matches in the catalog.
- *Negotiate*. The supplier and consumer reach consensus, which is usually captured in a contract or service-level agreement on: terms and conditions (including the service-level agreement for long running contrast), and pricing (with an agreement in which the provider and consumer of the media engage in negotiation).
- *Provision/bind*. This step binds the service-level agreement established during contract negotiation and the run-time environment of the media and content delivery.

5.5 INFRASTRUCTURE FOR MULTIMEDIA CONTENT DISTRIBUTION

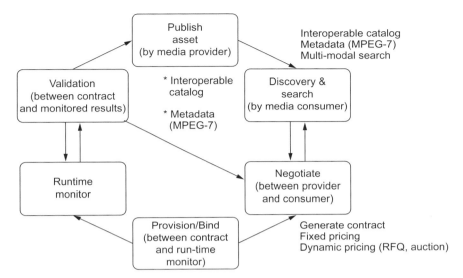

Figure 5.88 Lifecycle of media commerce [221]. (©2002 IEEE.)

- *Run time/monitor.* Monitor observes the quality of service of content delivery process during run time.
- *Validate.* Based on the statistics collected during the run time of content delivery, the quality service is compared with the original service-level agreement established during the contract negotiation.

To conclude, due to the recent rapid technology advances in e-commerce, multimedia and broadband networks, media commerce has already become prevalent during the past two years. Through introducing new applications and streaming existing core business processes in a number of industries, media commerce has the potential to have a profound impact on our daily lives.

Business Models
In addition to technical and legal means, owners of digital entertainment content can also make use of new, creative ways to bring their works to the market. A good understanding of the complexity and cost of protection is probably a prerequisite to be on the right track. With the wide availability of digital A/V devices and networks in the near future, it will become increasingly difficult to control individual behavior and detect infringements of copyrights.

In general, the selection of a business model depends on a number of factors, including:

- type of content
- duration of the economic value of the content

- fixation method
- distribution channel
- purchase mechanism
- technology available for protection
- extent of related legislation.

A good case study to explore the opportunities in a digital market is superdistribution [163]. E-commerce with superdistribution is shown in Figure 5.89. This figure shows the players in a DRM-supported e-commerce system: a content owner, a clearing house, several retailers, and many customers. The media files requested by the customers are hosted by the content owner or the retailers, and the licenses are downloaded from the clearing house.

In practice, every recording device (be it hardware or software) could be able to extract the hidden identification information from the document to be recorded and to send it to a transport organization monitoring and relating on the recording activities; this exchange could be exploited to increase the diffusion of the documents themselves. In this way, no user would be motivated to remove the watermark from the document as this does not give any advantage, and watermark robustness would be enough. The problem of security is translated from the users' to the producers' side, as it should be granted that they are not allowed to deceive the monitoring service.

The main advantage of the above approach resides in the fact that the problems raised by peer-to-peer networks allowing an extremely rapid diffusion of copyrighted material all around the world can be transformed into a great opportunity for enhancing the market of the products, and, in parallel, reducing the costs of distribution. Of course, it remains to be evaluated if these increased business opportunities are enough to compensate the reduction of income that a levies-based model causes on a single recording.

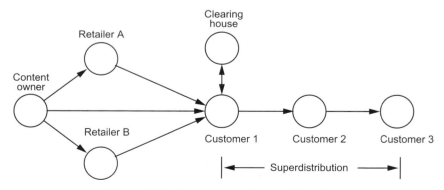

Figure 5.89 E-commerce with superdistribution.

A further requirement of the proposed model is that the recording devices must always be able to communicate the recording information to the monitoring organizations. However, with the diffusion of home DSL, wireless communications, and (in general) *always on* systems, this does not seem to be a major limitation. It is also worth highlighting that some privacy issues could emerge, even if, at least in principle, the system would require that only the information related to the multimedia document is transmitted to the monitoring agencies, and no private information regarding the document user is required.

5.6 MIDDLEWARE TECHNOLOGIES FOR MULTIMEDIA NETWORKS

Middleware represents the convergence of two key areas of information technology (IT): distributed systems and component-based design and programming. Techniques for developing distributed systems focus on integrating many computing devices to act as a coordinated computational resource. Likewise, techniques for developing component-based systems focus on reducing software complexity by capturing successful patterns and creating reusable component frameworks. Middleware is therefore the field dealing with component-based systems that can be distributed effectively over a myriad of computer devices and communication networks to provide developers of networked applications with the necessary platforms and tools to:

- Formalize and coordinate how parts of applications are composed and how they interoperate.
- Monitor, enable, and validate the (re)configuration of resources to ensure appropriate applications end-to-end quality of service (QoS), even in the face of failures or attacks.

During the past decade, IT developers and end users have benefited from the commoditization of commercial off-the-shelf (COTS) hardware (e.g., CPUs and storage devices) and networking elements (e.g., IP routers). More recently, the maturation of programming languages (e.g., Java, C++), operating environments (e.g., POSIX and Java Virtual Machines), and middleware (e.g., CORBA, Java 2 Enterprise Edition, and SOAP/Web services) are helping to commoditize many COTS software components and architectural layers. Although the quality of COTS commodity software has often lagged behind hardware, recent improvements in software frameworks, component models, patterns and development processes have encapsulated the knowledge that enables COTS middleware to be developed, integrated and used successfully with the fixed networks in an increasing number of real-worlds networked application domains, including telecom/datacom, enterprise e-commerce systems, desktop business applications, aerospace and defense systems, industrial process control, and financial services.

COTS middleware platforms have generally expected static connectivity, reliable communication channels, and relatively high bandwidth. Significant challenges remain, however, to design, optimize, and apply middleware for more flexible network environments, such as self-organizing peer-to-peer (P2P) networks, mobile settings, and highly resource-constrained sensor networks. For example, hiding network topologies and other deployment details from networked applications becomes harder (and often undesirable) in wireless sensor networks, since applications and middleware often need to adapt according to changes in location, connectivity, bandwidth, and battery power.

Middleware generally supports two models for component interaction:

- a request-response communication model, in which a component invokes a point-to-point operation on another component
- an event-based communication model, in which a component transmits arbitrarily defined messages, called *events*, to other components.

Event-based middleware is becoming popular for large-scale heterogeneous distributed systems because it helps reduce software dependencies, and enhance system composability and evolution [222].

The next generation of large-scale networked applications (e.g., streaming video, Internet telephony, and interactive simulation systems) has increasingly stringent QoS requirements. In today's horizontally integrated and commoditized IT environment, these types of applications are being developed using multiple layers of hardware, operating systems, and middleware components. Historically, it has been hard to simultaneously satisfy multiple QoS properties, such as security and reliability. Moreover, in the face of changing requirements and environments, there is a need for a more flexible software infrastructure that can be (re)configured dynamically to react to changing context [223].

5.6.1 Middleware to Support Sensor Network Applications

Sensor network applications represent a new class of applications that are

- *Data-driven*, meaning that the applications collect and analyze data from the environment, and, depending on redundancy, noise, and properties of the sensor themselves, can assign a quality level to the data.
- *State-based*, meaning that an application's needs with respect to sensor data can change over time, based on previously received data.

Typically, sensors are battery-operated, meaning they have a limited lifetime during which they provide data to the application. A challenge of the design of sensor networks is how to maximize network lifetime while meeting application QoS requirements. For these types of applications, the needs of the application should dictate which sensors are active and the role they play in the network topology.

The sensor network may consists of sensors with overlapping coverage areas providing redundant information. If the application does not require all this redundant information, it would be desirable to conserve energy in some sensors by allocating them to sleep, thereby lengthening the lifetime of the network.

One of the distinguishing characteristics of sensors networks is their reliance on nonrenewable batteries, despite their simultaneous need to remain active for as long as possible. Therefore, initial work has been done to create network protocols tailored to sensor networks that extend network lifetime considering the energy constraints of the individual sensors.

Middleware has often been useful in traditional systems for bridging the gap between the operating system (a low-level component) and the application, easing the development of distributed applications. Wireless sensor networks share many properties with traditional distributed computing middleware for use in sensor networks.

5.6.2 Middleware for Wireless Sensor Networks

Wireless sensor networks (WSNs) are a significant technology attracting considerable research attention in recent years. They are being developed for a wide range of civil and military applications, such as object tracking, infrastructure monitoring, habitat sensing, and battlefield surveillance. Typically, a WSN consists of hundreds to thousands of tiny sensor nodes that communicate over wireless channels and perform distributed sensing and collaborative data processing.

State-of-the-art techniques for WSNs focus on simple data-gathering-style applications, and in most cases support one application per network. Therefore, design of the network protocols and applications are usually closely coupled, or even combined as a monolithic procedure. However, such procedures are sometimes ad hoc and impose direct interaction with the underlying embedded operating system, or even the hardware components, of sensor nodes. We envision that the development of WSNs will finally demand systematic application of the system. In addition, multiple applications will be required to be concurrently executed over a single WSN. For instance, a building monitoring system may need to simultaneously monitor the temperature and luminance, check cracks on the wall, track traversing persons, and even communicate with systems in nearby buildings. Thus, middleware sitting between the network hardware, operating systems, and network stacks and the application is required to provide:

- standardized system services to diverse applications
- a run-time environment that can support and coordinate multiple applications
- mechanisms to achieve adaptive and efficient utilization of system resources.

The middleware design needs to address the unique operating modes of WSNs that are significantly different from traditional networks, including ad hoc deployment and dynamic operating environments. A cluster-based middleware architecture is shown in Figure 5.90 [224].

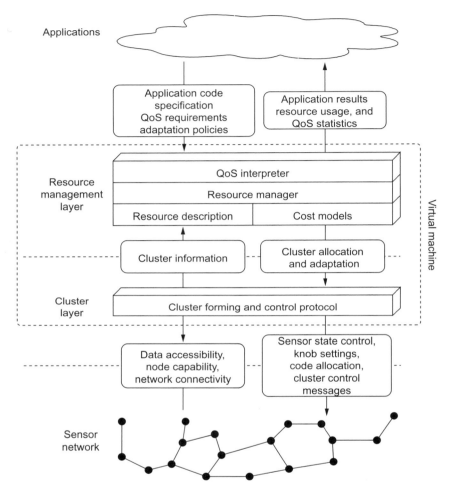

Figure 5.90 Cluster-based middleware architecture for wireless sensor networks [224]. (©2004 IEEE.)

The middleware infrastructure is divided into two layers. We call the abstraction provided by the middleware a virtual machine, because of its similarity to the virtual machine concept in traditional distributed systems in terms of providing application semantic transparency from the physical infrastructure. While the cluster forming and control protocol is distributed among all sensor nodes, the code for the resource management layer resides at the cluster head. A multicluster hierarchical architecture can be analogously developed.

The cluster layer is responsible for forming a cluster from a pool of sensor nodes that are around the target phenomena. Typically, the data accessibility, node capability (including remaining energy and CCS capabilities), and network connectivity are the criteria for determining the membership of a sensor node. Obviously, the gathering and exchange of such information should be performed in a distributed way. Application-related knowledge is embedded in the application specification and passed down to the cluster layer after being interpreted by the resource management layer. In addition, the cluster layer distributes the commands issued from the cluster head for resource management and cluster control purposes.

As the key component of the middleware, the resource management layer commands the allocation and adaptation of resources, such that the QoS requirements specified by the applications can be met. Resource allocation focuses on generating an initial solution when the cluster is formed, while resource adaptation controls the run-time behavior of the cluster. Both of these steps need to solve the problem of determining the scheduling of applications onto corresponding resources and the adjustment of systems controls. In general, such problems turn out to be computationally hard. Thus, heuristics providing near-optimal solutions are needed.

It is crucial to ensure the robustness of the system. System and environmental variations may cause deterioration of node capabilities and the quality of communication channels, or even node failure. Therefore, this information needs to be periodically gathered through the cluster control protocol and updated at the cluster head. The cluster head is then responsible for taking adaptation actions at the sight of possible failure of the functionality or QoS requirements of applications. Directions from applications will be necessary in selecting the right adaptation actions, especially when the resource availability is tight. On one hand, adaptive fidelity algorithms are extremely useful in trading the quality of a single application for its resource usage. On the other hand, interapplication coordination is necessary to perform systemwide trade-offs, and achieve optimized and balanced resource utilization among multiple applications. For instance, different priorities associated with applications will affect the fidelity adaptation of individual applications.

To maintain the robustness of the system under stringent resource constraints and high dynamics, the key is to provide flexible and adaptive resource management mechanisms to efficient trade-offs among multidimensional performances of multiple applications. Intercluster coordination mechanisms are also needed to avoid unfairness, starvation, and deadlock when two clusters compete for the same resource pool in terms of sensor node.

To implement successful middleware still requires significant work to resolve a set of technical issues. The middleware architecture also needs to adapt to newly emerging applications, such as mobile robot-based networks. The mobility of sensor nodes would impose additional challenges in both cluster control and resource management. The performance of the middleware can be enhanced by identifying proper system parameters for a wide range of environments and application scenarios.

5.6.3 Peer-to-Peer Middleware

Peer-to-peer systems that dynamically organize, interact, and share resources are increasingly being deployed in large-scale environments. They present the evolution of the client–server model that was primarily used in small-scale distributed environments to accommodate a vast number of users. The most distinct characteristic of the peer-to-peer model is that there is symmetric communication between the peers. Each peer has both client and server role [225]. The advantages of peer-to-peer systems are multidimensional:

- They enable knowledge sharing by aggregating information and resource nodes located at geographically distributed and potentially heterogeneous platforms.
- They provide high availability by eliminating the need for a single centralized manager.

Peer-to-peer systems emphasize heterogeneity in computing capabilities, communication bandwidths, and connection times among peers. Thus, P2P systems have the following properties:

- They can be deployed over heterogeneous operating systems, networks, and platforms. Peer-to-peer applications can run on multiple devices ranging from powerful workstations to laptops and wireless personal digital assistants (PDAs).
- Peers are autonomous and highly dynamic and have various joint and disconnect cycles. Each peer has only a limited view of the state and topology of the network.
- Resources and applications are highly replicated and available at multiple peers in the system.

A P2P system is typically built on existing network infrastructure, providing end-to-end connectivity. A systems view and layering for P2P networks are shown in Figure 5.91 [226]. P2P applications like file sharing, grid computing, or instant messaging require search middleware building on overlay networks to overcome the hurdles of decentralization and search the network (e.g., for files, computing resources, or users). Other P2P middleware functionality may be necessary, for

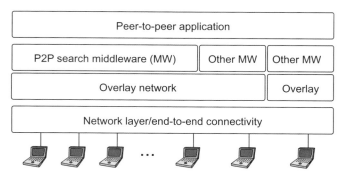

Figure 5.91 Systems view and layering in P2P networks [226]. (©2004 IEEE.)

instance, for peer and content reputation information handling, and can build on the same or separated overlay networks; the scope of this work, however, is restricted to search middleware. Innumerable efforts have been started to build such middleware, construct suitable overlay networks, and hence design P2P search, lookup, and routing systems.

Peer-to-peer networks are unpredictable and may impact the message delivery and thus disrupt applications. Quality of service (QoS) parameters can compensate some issues resulting from the dynamic peer-to-peer network, because the application can specify its primary requirements for the message delivery.

The P2P messaging system supports QoS parameters for messages: priority, preservation priority, and expiration time. A key parameter is the priority, which significantly influences the delivery process [227]. The application can further determine messages that should be kept when message queues get full using the preservation priority. Expiration time specified when a message loses its relevance can thus be discarded by the system to reduce network traffic.

REFERENCES

1. B. W. Moore, The ITU's role in standardization of the GII, *IEEE Comm. Magazine*, 36, 98–104 (1998).
2. ITU-T IP Project, *Project Description – Version 5*, ITU, Geneva, May 2001.
3. MEDIACOM2004 Framework for multimedia standardization, *Framework Description, Version 2.0*, June 2001.
4. ITU-T SG13, *IP Project*, available at http://www.itu.int/ITU-T/com13/IP/index.html.
5. ITU-T MEDIACOM2004, *Project description, Version 3.0*, ITU-T, March 2002.
6. I. M. Pao and M. T. Sun, Encoding stored video for streaming applications, *IEEE Trans CSVT*, 11, 199–209 (2001).
7. S. F. Chang et al., Multimedia search and retrieval, in A. Puri and T. Chen, Eds., *Advances in Multimedia: Systems, Standards and Networks*, Marcel-Dekker, New York, 1999.
8. H. Kalva, *Delivering MPEG-4 Based Audio-Visual Services*, Kluwer, Boston, 2001.
9. A. Busso et al., *MPEG-4 Integrated Intermedia Format (IIF): Basic Specification*, ISO/IEC/SC29/WG11 MPEG98/2978, February 2001.
10. S. Jacobs and A. Eleftheriadis, Streaming video using TCP flow control and dynamic rate shaping, *J. Visual Communication and Image Representation, Special Issue on Image Technology for WWW Applications*, 9, 211–222 (1998).
11. A. T. Campbell, G. Coulson, and D. Hutchinson, Transporting QoS adaptive flows, *ACM/Springer Multimedia Systems Journal, Special Issue on QoS Architecture*, 6, 167–178 (1999).
12. ISO/IEC/SC29/WG11, *Delivery Multimedia Integration Framework*, DMIF (ISO/IEC14496-6), April 1999.
13. B. Furth and H. Kalva, Multimedia networks, in B. Furth, Ed., *Multimedia Systems and Techniques*, Kluwer, Norwell, MA, 1996.

14. Z. Liang and R. Talluri, Tools for robust image and video coding in JPEG2000 and MPEG-4 standards, *Proc. IS&T-SPIE Visual Comm., Image Processing*, 3653, 40–51 (1999).
15. Available at http://www.itu.int/ITU-T/news/jvtpro.html
16. J. Moccagatta et al., Error-resilient coding in JPEG2000 and MPEG-4, *IEEE J. Selected Areas in Comm*, 18, 899–914 (2000).
17. I. E. G. Richardson, *H.264 and MPEG-4 Video Compression*, Wiley, Hoboken, NJ, 2003, pp. 228–230.
18. Available at http://article.gmane.org/gmane.comp.video.uci.devel/20
19. T. Stockhammer, M. Hannuksela and S. Wenger, H.26L/JVT coding network abstraction layer and IP-based transport, Special Session: The Emerging JVT/H.26L Video Coding Standard, in *Proc. IEEE ICIP 2002*, vol. 2, pp. 485–488, Rochester, New York, September 2002.
20. K. R. Rao and J. J. Hwang, *Techniques and Standards for Image, Video and Audio Coding*, Prentice Hall, Upper Saddle River, NJ, 1996.
21. J. Ribas-Corbera, P. A. Chou and S. L. Regunathan, A flexible decoder buffer model for JVT video coding, Special Session: The Entropy JVT/H.26L Video Coding Standard, in *Proc. IEEE ICIP*, vol. 2, pp. 493–496, Rochester, New York, September 2002.
22. ITU-T Rec.H.264/ISOIEC 14496-10 AVC, *Editors proposed modifications to Joint Committee Draft (CD) or Joint Video Specification*, 4th JVT Meeting, Klagenfurt, Austria, July 2002.
23. B. Girod and M. Flier, Multiframe motion compensated video compression for the digital set-top-box, in *Proc. IEEE ICIP*, vol. 3, pp. 1–4, Rochester, New York, September 2002.
24. D. Tian et al., Coding faded scene transitions, Special Session: The Emerging JVT/H.26L Video Coding Standard, in *Proc. IEEE ICIP*, pp. 505–508, Rochester, New York, September 2002.
25. Available at http://standards.pictel.com/fip/video-site/0201_Gen/JVT-B067.doc
26. Joint Video Team, *JVT IPR Status Report* (ISO/IEC JTC1/SC29/WG11 and ITU-T SG16 Q.6), 2nd Meeting in Geneva, February 2002.
27. G. J. Sullivan, P. Topiwalla, and A. Luthra, The H.264/AVC advanced video coding standard: overview and introduction to fidelity range extensions, *SPIE Special Session on the Advances in the New Emerging Standard: H.264/AVC*, Denver, CO, August 2004, 53–74.
28. Joint Video Team, *Joint Model Number 1, revision (JM)*, JVT-A003r1, Pattaya, Thailand, December 2001.
29. *Envivo* Web site, available at http://www.envivo.com.
30. G. J. Sullivan (plenary speaker), Advances in video compression and emerging JVT/H.26L/AVC standard, in *Proc. IEEE ICIP*, Rochester, New York, September 2002.
31. ISO/IEC MPEG and ITU-T VCEG, JVT-B068, *New Interlace Coding Tools*, Geneva, February 2002.
32. JVT_F100d2.doc, *Study of Final Committee Draft of JVT* (ITU-T Rec. H.264 and ISO/IEC 14496-10 AVC) *Draft 2*, 6th Meeting, Awaji Islands, December 2002.
33. ITU-T Q.15/SG16, Doc.Q15-J-14, T. Wedi, *Interpolation Filters for Motion Compensated Prediction with 1/4 and 1/8 Pel Accuracy*, Osaka, Japan, May 2000.
34. A. Joch et al., Performance comparison of video coding standards using Lagrangian coder control, Special Session: The Emerging JVT/H.26L Video Coding Standard, in *Proc. IEEE ICIP*, vol. 2, pp. 501–504, Rochester, New York, September 2002.

35. T. Wiegand et al., Overview of the H.264/AVC video coding standard, *IEEE Trans CSVT*, 13, 560–576 (2003).
36. J. Osterman et al., Video coding with H.264/AVC: tools, performances and complexity, *IEEE Circuits and Systems Magazine*, 4, 7–28 (2004).
37. B. G. Haskell, A. Puri, and A. N. Netravali, *Digital Video: An Introduction to MPEG-2*, Chapman & Hall, New York, 1996.
38. A. Puri and T. Chen (Eds.), *Multimedia Systems, Standards and Networks*, Marcel Dekker, New York, 2000.
39. M. Domanski, A. Luczak, and S. Mackowiak, Spatial-temporal scalability for MPEG video coding, *IEEE Trans CSVT*, 10, 1088–1093 (2000).
40. G. J. Connlin and S. S. Hemami, A comparison of temporal scalability techniques, *IEEE Trans CSVT*, 9, 909–919 (1999).
41. D. Wilson and M. Ghanbari, Exploiting interlayer correlation of SNR scalable video, *IEEE Trans CSVT*, 9, 783–797 (1999).
42. U. Benzler, Spatial scalable video coding using a combined subband-DCT approach, *IEEE Trans CSVT*, 10, 1080–1087 (2000).
43. K. Shen and E. J. Delp, Wavelet based rate scalable video compression, *IEEE Trans CSVT*, 9, 109–122, (1999).
44. Q. Wang and M. Ghanbary, Scalable coding of very high resolution video using the virtual zerotree, *IEEE Trans CSVT*, 7, 719–727 (1997).
45. P. J. Cheng, J. Li, and C. C. J. Kuo, Rate control for an embedded wavelet video coder, *IEEE Trans CSVT*, 7, 692–702 (1997).
46. S. A. Martucci et al., A zerotree wavelet video coder, *IEEE Trans CSVT*, 7, 109–118 (1997).
47. D. Taubman and A. Zakhor, A common framework for rate and distortion based scaling of highly scalable compressed video, *IEEE Trans CSVT*, 6, 329–354 (1996).
48. W. Li, *Bit-plane coding of DCT coefficients for fine granularity scalability*, ISO/IEC JTC1/SC29/WG11 MPEG98/M3989, October 1998.
49. B. Schuster, *Fine granular scalability with wavelet coding*, ISO/IEC JTC1/SC29/WG11 MPEG98/M4021, October 1998.
50. S. C. Scheng and A. Zakhor, *Matching pursuit coding for fine granularity video scalability*, ISO/IEC JTC1/SC29/WG11 MPEG98/M3991, October 1998.
51. F. Ling and X. Chen, *Report of fine granularity scalability using bit plane coding*, ISO/IEC JTC1/SC29/WG11 M4311, November 1998.
52. W. Li, Overview of fine granularity scalability in MPEG-4 video standard, *IEEE Trans CSVT*, 11, 301–317 (2001).
53. S. Wenger, H.264/AVC over IP, *IEEE Trans CSVT*, 13, 645–656 (2003).
54. R. Fielding et al., *Hypertext Transport Protocol – HTTP*, RFC 2616, 1999.
55. T. Wiegand et al., Rate constrained coder control and comparison of video coding standards, *IEEE Trans CSVT*, 13, 688–703 (2003).
56. 3GPP Technical Specification, *Multimedia Messaging Service (MMS): Media Formats and Codecs*, TR26.140.
57. T. Stockhammer, M. Hannuksela, and T. Wiegand, H.264/AVC in wireless environments, *IEEE Trans CSVT*, 13, 657–673 (2003).

58. H. Holam and A. Taskala (Eds.), *WCDMA for UMTS: Radio Access for Third Generation Mobile Communications*, Wiley, New York, 2000.
59. J. Ribas-Corbera, P. A. Chou, and S. L. Regunathan, A generalized hypothetical reference decoder for H.264/AVC, *IEEE Trans CSVT*, 13, 674–678 (2003).
60. E. G. Steinbach, N. Farber, and B. Girod, Adaptive play-out for low latency video streaming, presented at the IEEE ICIP 2001, Thessaloniki, Greece.
61. Y. J. Laing and B. Girod, Rate-distortion optimized low latency video streaming using channel-adaptive bitstream assembly, presented at the ICME 2002, Lausanne, Switzerland.
62. M. Marczewicz and R. Kurceren, The SP and SI frames design for H.264/AVC, *IEEE Trans CSVT*, 13, 637–644 (2003).
63. H. Radha, M. vanderSchaar, and Y. Chen, The MPEG-4 fine-grained scalable video coding for multimedia streaming over IP, *IEEE Trans Multimedia*, 3, 53–68 (2001).
64. A. Vetro, Ch. Christopoulos, and H. Sun, Video transcoding architecture and technologies: an overview, *IEEE Signal Processing Magazine*, 20, 18–29 (2003).
65. T. D. Wu, Y. N. Hwang and M. T. Sun, Video transcoding, in L. Guan, S. Y. Kung and J. Larsen, Eds., *Multimedia Image and Video Processing*. CRC Press, Boca Raton, FL, 2001, pp. 467–486.
66. P. Assuncao and M. Ghanbari, Post-processing of MPEG-2 coded video for transmission in lower bit rates, *Proc. IEEE ICASSP*, 4, 1998–2001 (1996).
67. A. Lan and J. N. Hwang, Context dependent reference frame placement for MPEG video cdoing, *Proc. IEEE ICASSP*, 4, 2997–3000 (1997).
68. M. T. Sun, T. D. Wu, and J. N. Hwang, Dynamic bit-allocation in video combining for multipoint conferencing, *Proc. Int. Symp. on Multimedia Information Processing ISM*IP, pp. 350–355, Taipei, Taiwan, December 1997.
69. H. Sun, W. Kwok, and J. Zdepsei, Architectures for MPEG compressed bitstream scaling, *IEEE Trans CSVT*, 6, 191–199 (1996).
70. G. Kessman et al., Transcoding of MPEG bitstreams, *Signal Processing: Image Communication*, 8, 481–500 (1996).
71. N. Bjork and C. Christopoulos, Transcoder architectures for video coding, *IEEE Trans Consumer Electronics*, 44, 88–98 (1998).
72. P. A. A. Assuncao and M. Ghanbari, Post-processing of MPEG-3 coded video for transmission at lower bit-rates, *Proc. IEEE ICASSP*, Atlanta, GA, pp. 1998–2001, May 1996.
73. S. F. Chang and D. G. Messerschmit, Manipulation and compositing of MC-DCT compressed video, *IEEE J. Selected Areas in Comm.*, 13, 1–11 (1995).
74. P. A. A. Assuncao and M. Ghanbari, A frequency-domain video transcoder for dynamic bit-rate reduction of MPEG-2 bitstreams, *IEEE Trans CSVT*, 8, 953–967 (1998).
75. H. Sun et al., A new approach for memory efficient ATV decoding, *IEEE Trans Consumer Electronics*, 43, 517–525 (1997).
76. N. Merhav, Multiplication-free approximate algorithms for compressed-domain linear operations on images, *IEEE Trans. Image Processing*, 8, 247–254 (1999).
77. C. W. Lin and Y. R. Lee, Fast algorithms for DCT-domain video transcoding, *Proc. IEEE ICIP*, Thessaloniki, Greece, vol. 1, pp. 421–424, September 2001.
78. K. Stuhlmuller et al., Analysis of video transmission over lossy channels, *IEEE J. Selected Areas in Comm.*, 18, 1012–1032 (2000).

79. P. Yin et al., Drift compensation for reduced spatial resolution transcoding, *IEEE Trans CSVT*, 12, 1009–1020 (2002).
80. G. deLosreyes et al., Error-resilience transcoding for video over wireless channels, *IEEE J. Selected Areas in Comm.* 18, 1063–1074 (2000).
81. S. Dogan et al., Error-resilient video transcoding for robust inter-network communications using GPRS, *IEEE Trans CSVT*, 12, 453–464 (2002).
82. S. We and J. Aposotlopoulos, Secure scalable streaming enabling transcoding without decryption, *Proc. IEEE ICIP*, Thessaloniki, Greece, vol. 1, pp. 437–440, September 2001.
83. M. S. Lei, T. C. Chen, and M. T. Sun, Video bridging based on H.261 standard, *IEEE Trans CSVT*, 4, 425–437 (1994).
84. M. E. Lukacs and D. G. Boyer, A universal broadband multiport teleconferencing service for the 21st century, *IEEE Comm. Magazine*, 33, 36–43 (1995).
85. R. Caglianello and G. Cash, Montage: continuous presence teleconferencing utilizing compressed domain video bridging, *IEEE International Conference on Communications*, Seattle, WA, vol. 1, pp. 578–581, June 1995.
86. M. K. Willebeek-LeMar and Z. Y. Shal, Videoconferencing over packet based networks, *IEEE J. Selected Areas in Comm.*, 15, 1101–1114 (1997).
87. M. Ghanbari, *Standard Codecs: Image Compression to Advanced Video Coding*, IEE, UK, 2003.
88. Y. H. Luo, Ch. N. Wang, and T. Chiang, A novel all-binary motion estimation (ABME) with optimized hardware architecture, *IEEE Trans CSVT*, 12, 700–712 (2002).
89. J. K. Jain and A. K. Jain, Displacement measurement and its application in interframe image coding, *IEEE Trans Commun.*, COM-29, 1799–1808 (1981).
90. R. Srinirasan and K. R. Rao, Predictive coding based on efficient motion estimation, *IEEE Trans Commun.*, COM-33, 1011–1014 (1985).
91. K. Chow and M. L. Liou, Genetic median search algorithm for video compression, *IEEE Trans CSVT*, 3, 440–445 (1993).
92. C. H. Lin and J. L. Wu, A lightweight genetic block-matching algorithm for video encoding, *IEEE Trans CSVT*, 8, 386–392 (1998).
93. J. Y. Than, S. Ranganath, and A. Kassim, A novel unrestricted center-based diamond search algorithm for block motion estimation, *IEEE Trans CSVT*, 8, 369–377 (1998).
94. A. M. Tourapis et al., Optimizing the MPEG-4 encoder advanced diamond zonal search, *Proc. Int. Symp. Circuits and Systems*, vol. 3, pp. 674–677, Geneva, Switzerland, May 2000.
95. S. Zhu and K. K. Ma, A new diamond search algorithm for fast block matching motion estimation, *IEEE Trans Image Processing*, 9, 287–290 (2000).
96. J. S. Kim and R. H. Park, A fast feature-based block matching algorithm using integral projections, *IEEE J. Selected Areas in Comm.*, 10, 968–971 (1992).
97. B. Lin and A. Zaccarin, New fast algorithms for the estimation of block motion vector, *IEEE Trans CSVT*, 3, 148–157 (1993).
98. V. G. Moshnyage, Reducing energy dissipation of frame memory by adaptive bit-width compression, *IEEE Trans CSVT*, 12, 713–719 (2002).
99. Y. S. Tung et al., An efficient streaming and decoding architecture for stored FGS video, *IEEE Trans CSVT*, 12, 730–735 (2002).

100. R. Garg et al., Boundary macroblock padding in MPEG-4 video decoding using a graphics coprocessor, *IEEE Trans CSVT*, 12, 719–723 (2002).
101. C. Herpel, G. Franceschini, and D. Suger, Transporting and storing MPEG-4 content, in F. Pereira and T. Ebrahimi, Eds. *The MPEG-4 Book*, Prentice Hall PTR, Upper Saddle River, NJ, 2002, pp. 227–292.
102. J. Postel, RFC768 *Datagram Protocol*, August 1980.
103. H. Schulrzinne et al., RFC1889 *RTP: A Transport Protocol for Real Time Applications*, ETF, January, 1996.
104. ISO/IEC 14496-6, Information technology, *Coding of audio-visual objects, Part 6: Delivery Multimedia Integration Framework (DMIF)*, 2000.
105. D. Wu et al., Streaming video over the Internet: approaches and directions, *IEEE Trans CSVT*, 11, 282–300 (2001).
106. D. Wu, Y. T. Huan, and Y. Q. Zhang, Transporting real-time video over the Internet: challenges and approaches, *Proc. IEEE*, 88, 1855–1875 (2000).
107. S. McCanne, V. Jacobson, and M. Vetterly, Receiver-driven layered multicast, *Proc. ACM SIGCOM*, pp. 117–130, August 1996.
108. B. Girod, V. Horn, and B. Belzer, Scalable video coding with multiscale motion compensation and unequal error protection, *Proc. Symp. Multimedia Communications and Video Coding*, New York, pp. 475–482, October 1995.
109. S. Li, F. Wu, and Y. Q. Zhang, *Study of a new approach to improve FGS video coding efficiency*, ISO/IEC JTC1/SC29/WG11, Doc. M5583, December 1999.
110. W. Li, *Bit-plane coding of DCT coefficients for fine granularity scalability*, ISO/IEC JTC1/SC29/WG11, Doc. M3989 October 1998.
111. F. Wu, S. Li, and Y. Q. Zhang, A framework for efficient progressive fine granularity video coding, *IEEE Trans CSVT*, 11, 172–194 (2001).
112. G. J. Conklin et al., Video coding for streaming media delivery on the Internet, *IEEE Trans CSVT*, 11, 269–281 (2001).
113. Y. Wang, M. T. Orchard, and A. R. Reibman, Multiple description image coding for noisy channels by pairing transform coefficients, *Proc. IEEE Workshop on Multimedia Signal Processing*, pp. 419–424, June 1997.
114. A. Eleftheriadis and D. Anastassiou, Meeting arbitrary QoS constraints using dynamic rate shaping of coded video, *Proc. 5th Int. Workshop Network and Operating System Support for Digital Audio and Video NOSSDAV*, pp. 95–106, April 1995.
115. X. Wang and H. Schulrzinne, Comparison of adaptive Internet multimedia applications, *IEICE Trans Comm.*, E82-B, 806–818 (1999).
116. D. Wu et al., On end-to-end architecture for transporting MPEG-4 video over the Internet, *IEEE Trans CSVT*, 10, 923–941 (2000).
117. J. C. Bolot, T. Turletti, and I. Wakeman, Scalable feedback control for multicast video distribution in the Internet, *Proc. ACM SIGCOM*, London, UK, pp. 58–67, September 1994.
118. T. Turletti and C. Huitema, Videoconferencing on the Internet, *IEEE Trans Networking*, 4, 340–351 (1996).
119. S. Floyd and K. Fall, Promoting the use of end-to-end congestion control in the Internet, *IEEE Trans Networking*, 7, 458–472 (1999).

120. S. Y. Cheung, Ma. Annuar, and X. Li, On the use of destination set grouping to improve fairness in multicast video distribution, *Proc. IEEE INFOCOM*, pp. 553–560, March 1996.
121. N. Yeadon et al., Filters: QoS support mechanisms for multipeer communications, *IEEE J. Selected Areas in Comm.*, 14, 1245–1262 (1996).
122. A. Albanese et al., Priority encoding transmission, *IEEE Trans Information Theory*, 42, 1737–1744 (1996).
123. J. C. Balot and T. Turletti, Adaptive error control for packet video in the Internet, *Proc. IEEE ICIP*, pp. 25–28, September 1996.
124. ISO/IEC JTC1/SC29/WG11 FCD 14496, *Part 1, 2, 3*, December 1998.
125. A. Puri et al., Application of FEC based multiple description coding for Internet video streaming and multicast, *Proc. Packet Video Workshop*, Cagliari, Sardinia, Italy, May 2000.
126. Y. Wang and Q. F. Zhu, Error control and concealment for video communication: a review, *Proc. IEEE*, 86, 974–997 (1998).
127. S. Deering, Multicast routing in internetworks and extended LANs, *Proc. ACM SIGCOM*, Stanford, CA, pp. 55–64, August 1988.
128. L. Fan et al., Summary cache: a scalable wide-area web cache sharing protocol, *IEEE Trans Networking*, 8, 281–293 (2000).
129. R. Steinmetz and K. Nahrstedt, *Multimedia Computing, Communications and Applications*, Prentice Hall, Englewood Cliff, NJ, 1995.
130. G. Blakonski and R. Steinmetz, A media synchronization survey: Reference model, specification and case studies, *IEEE J. Selected Areas in Comm.*, 14, 5–35 (1996).
131. J. P. Jarmasz and N. D. Georganas, Designing a distributed multimedia synchronization scheduler, in *Proc. IEEE Int. Conf. Multimedia Computing and Systems*, pp. 451–457, June 1997.
132. H. Schulrzinne et al., RFC1889 *RTP: A Transport Protocol for Real Time Applications*, IETF, January 1996.
133. H. Schulrzinne, A. Rao, and R. Lanphier, RFC2326 *RTSP: Real Time Streaming Protocol*, IETF, April 1998.
134. M. Handley et al., RFC2543 *SIP: Session Initiation Protocol*, IETF, March 1999.
135. R. Osso (Ed.), *Handbook of Communication Technologies: The Next Decade – Multimedia over IP: RSVP, RTP, RTCP, RTSP*, CRC Press, Boca Raton, FL, 1999, pp. 29–46.
136. Z. L. Zhang et al., Efficient server selective frame discard algorithm for stored video delivering over resource constrained networks, *Proc. IEEE INFOCOM*, pp. 472–479, March 1999.
137. D. Wu et al., An end-to-end approach to optimal mode selection in Internet video communication: theory and application, *IEEE. J. Selected Areas in Comm.*, 18, 977–995 (2000).
138. A. Eleftheriadis and D. Anastassiou, Meeting arbitrary QoS constraints using dynamic rate shaping of coded digital video, *Proc. Int. Workshop on Network and Operating Support for Digital Audio and Video NOSSDAV*, pp. 95–106, April 1995.
139. J. D. Angelopoulos et al., Efficient transport of packets with QoS in an FSAN aligned GPON, *IEEE Comm. Magazine*, 42, 92–98 (2004).

140. ITU Recomm. G.984.3, Study Group 15, *Gigabit-Capable Passive Optical Networks (G-PON): Transmission Convergence Layer Specification*, ITU, Geneva, October 2003.
141. I. Van de Vorde et al., The Super PON demonstration: an exploration of possible evolution paths for optical access networks, *IEEE Comm. Magazine*, 38, 74–82 (2000).
142. D. Angelopoulos, I. S. Venieris, and G. I. Stassinopoulos, A TDMA based access control scheme for GPON's, *IEEE/OSA J. Lightwave Tech.*, 11, 1095–1103 (1993).
143. W. Webb, Broadband fixed wireless access as a key component in the future integrated communications environment, *IEEE Comm. Magazine*, 39, 115–121 (2001).
144. D. Gesbert et al., Technologies and performance for non-line-of-sight broadband wireless access networks, *IEEE Comm. Magazine*, 42, 86–95 (2002).
145. M. Zhang and R. S. Wolf, Grossing the digital divide: cost-effective broadband wireless access for rural and remote areas, *IEEE Comm. Magazine*, 42, 99–105 (2004).
146. D. Jones, Fixed wireless access: a cost effective solution for local loop service in underserved areas, *Proc. IEEE Int. Conf Selected Topics in Wireless Communications*, Vancouver, pp. 240–244, June 1992.
147. P. Heury and H. Luo, WiFi – What's next?, *IEEE Comm. Magazine*, 40, 66–72 (2002).
148. D. O'Here et al., *Voice Over DSL Requirements: MSDN*, ADSL Forum 1999–217.
149. A. Habib and H. Saiedian, Channelized voice over digital subscriber line, *IEEE Comm. Magazine*, 40, 94–100 (2002).
150. W. Y. Chen, The development and standardization of asymmetrical digital subscriber line, *IEEE Comm. Magazine*, 37, 68–72 (1999).
151. J. H. Park, Wireless Internet access for mobile subscribers on the GPRS/UMTS network, *IEEE Comm. Magazine*, 40, 38–49 (2002).
152. 3GPP, *Combined GSM and MobileIP Mobility Handling in UMTS IP CN*, 3G TR23.923, V3.0.0, May 2000.
153. 3GPP, *GPRS Service Description – Stage 2*, 3G TS22.060, V3.3.0, March 2000.
154. 3GPP, *GPRS Service Description – Stage 1*, 3G TS22.064, V3.3.0, March 2000.
155. D. Gesbert et al., Technologies and performance for non-line-of-sight broadband wireless access networks, *IEEE Comm. Magazine*, 40, 86–95 (2002).
156. J. Hillebrand et al., Quality-of-service signaling for next-generation IP-based mobile networks, *IEEE Comm. Magazine*, 42, 72–79 (2004).
157. S. Blake et al., IETF RFC1633, *An Architecture for Differentiated Services*, IETF, December 1998.
158. R. Barden, IETF RFC1633, *Integrated Services in the Internet Architecture: An Overview*, IETF, June 1994.
159. S. I. Maniatis, E. G. Nikolouzou, and J. S. Venieris, End-to-end QoS specification issues in the converged all-IP wired and wireless environment, *IEEE Comm. Magazine*, 42, 80–86 (2004).
160. A. Kajanes et al., Individual QoS rating for voice services in cellular networks, *IEEE Comm. Magazine*, 42, 88–93 (2004).
161. 3GPP TS23.107, V5.9.0, *Quality of Service (QoS) Concept and Architecture*, June 2003.
162. ITU-T Rec. E.860, *Framework of a Service Level Agreement*, June 2002.
163. A. M. Eskicioglu, J. Town, and E. J. Delp, Security of digital entertainment content from creation to consumption, *Signal Processing: Image Communication*, 18, 237–262 (2003).

164. *Copyright law of the United States of America*, available at http://www.loc.gov/copyright/title17/92chapt.html.
165. National Research Council, *The Digital Dilemma: Intellectual Property in the Information Age*, National Research Council, The National Academy Press, 2000.
166. R. deBruin and J. Smits, *Digital Video Broadcasting: Technology, Standards and Regulation*, Artech House, Norwood, MA, 1999.
167. L. C. Guillon and J. L. Giachetti, Enrichment and conditional access, *SMPTE Journal*, 103, 398–406 (1994).
168. G. Rossi, Conditional access to television broadcast programs: technical solutions, *ABU Technical Review*, 166, 3012–3019 (1996).
169. H. Benoit, *Digital Television: MPEG-1, MPEG-2 and Principles of DVD Systems*, Arnold, 1997.
170. ISO/IEC 13818-1, *Generic Coding of Moving Pictures and Associated Audio Information: Systems*, First Edition, Geneva, 1996.
171. *Digital Video Broadcasting Project (DVB)*, available at http://www.dvb.org.
172. *The Advanced Television Systems Committee (ATSC)*, available at http://www.atsc.org.
173. A. M. Eskicioglu and E. J. Delp, Overview of multimedia content protection in consumer electronics devices, *Signal Processing: Image Communication*, 16, 681–699 (2001).
174. *Video Electronics Standards Association (VESA)*, available at http://www.vesa.org.
175. *Universal Plug and Play Forum*, available at http://www.upup.org.
176. W. Li, Scalable video coding with fine granularity scalability, Digest of Technical Papers, in *Proc. IEEE Int. Conf. Consumer Electronics*, Los Angelos, CA, pp. 306–307, 1999.
177. A. B. Benitex et al., Description of a single multimedia document, in B. S. Manjunath, P. Salembier and T. Sikora, Eds., *Introduction to MPEG-7 Multimedia Content Description Interface*, John Wiley, Chichester, 2002, pp. 111–138.
178. ISO/IEC JTC1/SC29/WG11 Doc.N3933, *MPEG-7 Requirements V.13*, January 2001.
179. Ph. Salembier and J. R. Smith, MPEG-7 multimedia description schemes, *IEEE Trans CSVT*, 11, 748–759 (2001).
180. F. Pereira and R. Koenen, Content, goals and procedures, in B. S. Manjunath, P. Salembier and T. Sikora, Eds., *Introduction to MPEG-7 Multimedia Content Description Interface*, John Wiley, Chichester, 2002, pp. 7–29.
181. D. R. Stinson, *Cryptography: Theory and Practice*, CRC Press, Boca Raton, FL, 1995.
182. B. B. Zhu and M. Swanson, Multimedia authentication and watermarking, in D. Feng, W. C. Sin and H. Zhang, Eds., *Multimedia Information Retrieval and Management*, Springer-Verlag, 2003, ch.7, pp. 148–177.
183. B. B. Zhu, A. H. Tewfik, and M. D. Swanson, When seeing isn't believing, *IEEE Signal Processing Magazine*, 21, 40–49 (2004).
184. M. Holliman and N. Memon, Counterfeiting attacks on oblivious blockwise independent invisible watermarking schemes, *IEEE Trans Image Processing*, 9, 432–441 (2000).
185. J. Fridrich, M. Goljan, and R. Du, Lossless data embedding – new paradigm in digital watermarking, *EURASIP J. Applied Signal Processing*, Special issue on Emerging Applications of Multimedia Data Hiding, 2002 (2), 185–196 (2002).

186. B. Zhu, M. D. Swanson, and A. H. Tewfik, A transparent authentication and distortion measurement technique for images, *Proc. IEEE Digital Signal Processing Workshop*, Loen, Norway, pp. 45–48, September 1996.
187. C. P. Wu and C. C. J. Kuo, Speech content authentication integrated with CELP speech coders, *Proc. IEEE Int. Conf. Multimedia Expo ICME*, pp. 1009–1012, August 2001.
188. M. Barni et al., Digital watermarking for copyright protection: a communications perspective, *IEEE Comm. Magazine*, 39, 90–91 (2001).
189. F. Hartung and M. Kutter, Multimedia watermarking techniques, *Proc. IEEE*, 87, 1079–1107 (1999).
190. P. G. Flikkema, Spread-spectrum techniques for wireless communications, *IEEE Signal Processing Magazine*, 14, 23–26 (1997).
191. J. Brassil et al., Electronic marking and identification techniques to discourage document copying, *IEEE J. Selected Areas in Comm.*, 13, 1495–1504 (1995).
192. S. Low and N. F. Maxemchuk, Performance comparison of two text marking methods, *IEEE J. Selected Areas in Comm.*, 16, 561–572 (1998).
193. N. F. Maxemchuk and S. Low, Marking text documents, *Proc. IEEE ICIP*, 3, 13–16, Santa Barbara, CA, October 1997.
194. D. Benham et al., Fast watermarking of DCT-based compressed images, *Proc. CISST*, 243–253, Las Vegas, NV, June 1997.
195. M. Swanson, B. Zhu and A. Tewfik, Multiresolution video watermarking using perceptual models and scene segmentations, *Proc. ICIP*, 2, 558–561, Santa Barbara, CA, October 1997.
196. K. R. Rao, Z. S. Bojkoric, and D. A. Milovanovic, *Multimedia Communication Systems*, Prentice Hall, Upper Saddle River, NJ, 2002.
197. C. Langelaar, J. C. A. Van Der Lubble, and R. L. Ladendijk, Robust labeling for copy protection of images, *Proc. SPIE Electronic Imaging*, 3022, 298–309, San Jose, CA, February 1997.
198. M. Barni et al., A DCT-domain system for robust image watermarking, *Signal Processing*, 66, 357–372 (1998).
199. F. Hartung and B. Girod, Digital watermarking of raw and compressed video, in *Proc. SPIE Digital Compression Technologies and Systems for Video Communications*, 2952, 205–213 (1997).
200. X. Xia, C. Boncelet, and G. Arce, A multiresolution watermark for digital images, *Proc. IEEE ICIP*, 1, 548–551, Santa Barbara, CA, October 1997.
201. D. Kundor and D. Hatzinekos, A robust digital image watermarking method using wavelet-based fusion, in *Proc. ICIP*, 1, 544–547, Santa Barbara, CA, October 1997.
202. H. Wang and C. C. J. Kuo, An integrated progressive image coding and watermark system, *Proc. IEEE ICIP*, 6, 3721–3723, Seattle, WA, May 1998.
203. P. Daver and M. Scott, Fractal based image steganography, *Lecture Notes in Computer Science: Information Hiding*, 1174, 279–294, Springer, Berlin, Germany, 1996.
204. A. Johnston and M. Bigger, *Digital Watermarking of Video/Image Content for Copyright Protection and Monitoring*, ISO/IEC JTC1/SC29/WG11 Doc.M2228, July 1997.
205. J. J. K. O. Ruanaidh and T. Pun, Rotation scale and translation invariant digital image watermarking, *Proc. IEEE ICIP*, 1, 536–539, Santa Barbara, CA, October 1997.

206. J. F. Tilki and A. A. Beex, Encoding a hidden digital signature onto an audio signal using psychoacoustic masking, *Proc. Int. Conf. Digital Signal Processing Applications and Technology*, pp. 476–480, Boston, MA, October 1996.
207. W. Beduer, D. Guchl, and N. Morimoto, Technologies for data hiding, *Proc. SPIE*, 2420, 40–44, San Jose, CA, February 1995.
208. F. Harting, P. Eisert, and B. Girod, Digital watermarking of MPEG-4 facial animation parameters, *Comp. Graphics*, 22(3), 425–435 (1998).
209. J. A. Bloom et al., Copy protection for DVD video, *Proc. IEEE*, 87, 1267–1276 (1999).
210. S. Grever and J. P. Stern, Lessons learned from SDMI, *Proc. IEEE Workshop Multimedia Signal Processing MMSP*, pp. 213–218, October 2001.
211. M. Barni and F. Bartdini, Data hiding for fighting piracy, *IEEE Signal Processing Magazine*, 21, 28–39 (2004).
212. I. J. Cox, M. L. Miller, and J. A. Bloom, *Digital Watermarking*, Morgan Kaufmann, San Mateo, CA, 2001.
213. B. Chen and G. Wornell, Optimization index modulation: a class of provably good methods for digital watermarking and information embedding, *IEEE Trans Information Theory*, 47, 1423–1443 (2001).
214. A. S. Cohen and A. Lapidoth, The Gaussian watermarking game, *IEEE Trans Information Theory*, 48, 1639–1667 (2002).
215. J. J. Eggers and B. Girod, *Informed Watermarking*, Kluwer, Norwell, MA, 2002.
216. T. Furon, J. Venturini, and R. Duhamel, An unified approach of asymmetric watermarking schemes, *Proc. SPIE Security and Watermarking of Multimedia Contents*, 4314, 269–279 (2001).
217. J. Haitsma, T. Kaluer, and J. Oostreen, Robust audio hashing for content identification, *2nd Int. Workshop Content Based Multimedia Indexing CBMI*, Brescie, Italy, September 2001.
218. H. Neuschmied, H. Mayer, and E. Balle, Content-based identification of audio titles on the internet, *Proc. Int. Conf. WEB Delivering of Music*, pp. 96–100, Firenze, Italy, November 2001.
219. J. Hatisms and T. Kaler, A watermarking scheme for digital cinema, *Proc. IEEE ICIP*, 2, 487–489, Thessalonike, Greece, October 2001.
220. The Content ID Forum, *CIDF Specification v.1.1*, Tokyo, Japan, September 2002.
221. C.-S. Li, Media commerce, *IEEE Circuits and Systems Magazine*, 2, 4–22 (2002).
222. P. R. Pietzuch, B. Shand, and J. Bacon, Composite event detection as a generic middleware extension, *IEEE Network*, 18, 44–55 (2004).
223. Q. Han, S. G. Nobesco, and N. Venkatasubramanian, Reflective middleware for integrating network monitoring with adaptive object messaging, *IEEE Network*, 18, 56–65 (2004).
224. Y. Yu, B. Krishnamachari, and V. K. Prasanna, Issues in designing middleware for wireless sensor networks, *IEEE Network*, 18, 15–21 (2004).
225. V. Kologeraki and F. Chen, Managing distributed objects in peer-to-peer systems, *IEEE Network*, 18, 22–29 (2004).
226. J. Mischke and B. Stiller, A methodology for the design of distributed search in P2P middleware, *IEEE Network*, 18, 30–37 (2004).

227. M. Oliver, J. Juginger, and Y. Lee, A self-organizing publish/subscribe middleware for dynamic peer-to-peer networks, *IEEE Network*, 18, 38–43 (2004).
228. A. Luthra, *MPEG-4 VC/ITU-T H.264 An overview*, Motorola, February 2002.
229. Available at http://www.stanford.edu/class/ee389b/handouts/09_VideoCodingStandrs.pdf
230. H. Malvar, et al., Low-complexity transform and quantization with 16-bit arithmetic for H.26L, *Special session: The emerging JVT/H.26L: video coding standard*, in *Proc. IEEE ICIP*, Rochester, New York, vol. 2, pp. 489–492, September 2002.
231. Available at http://www.eurasip.org/phd_abstracts/frossard_pascal.htm
232. I. E. G. Richardson, *Video Codec Design*, Wiley, Chichester, 2002, pp. 64–75.
233. S. F. Chang and D. G. Messerschmitt, A new approach to decoding and compositing motion compensated DCT-based images, *Proc. IEEE ICASSP*, 5, 421–424 (1993).
234. J. Youn, M. T. Sun, and C. W. Lin, Motion vector refinement for high performance transcoding, *IEEE Trans Multimedia*, 1, 30–40 (1999).
235. T. D. Wu, Y. N. Hwang, and M. T. Sun, Video transcoding, in L. Guan, S. Y. Kung and J. Larsen, Eds., *Multimedia Image and Video Processing*, CRC Press, BocaRaton, FL, 2001, pp. 467–486.
236. H. Schulrzinne, RFC1890 *RTP Profile for Audio and Video Conference With Minimal Control*, IETF, January 1996.
237. J. DeVriend et al., Mobile network evolution: a revolution on the move, *IEEE Comm. Magazine*, 40, 104–111 (2002).
238. R. Braden et al., IETF RFC2205, *Resource Reservation Protocol (RSVP) – Version 1*, IETF September 1997.
239. EIA-679B, *National Renewable Security Standard*, September 1998.
240. P. W. Wong and N. Memon, Secret and public key image watermarking schemes for image authentication and ownership verification, *IEEE Trans Image Processing*, 10, 1593–1601 (2001).
241. J. Fridrich, M. Golijan, and R. Du, Invertible authentication, *Proc. SPIE, Security Watermarking of Multimedia Contents*, San Jose, CA, vol. 3971, pp. 197–208, 2001.

6

NETWORK LAYER

This chapter concentrates on the multimedia communication network layer. Many issues relating to the networking are presented and discussed. The development of the network layer on a large scale presents numerous technical challenges in a variety of multimedia areas. As progress continues to be made in multimedia communications, it is necessary to ascertain consumer interest in this new form of communications. After an introductory discussion including network aspects of standardization projects, network functions such as control, signaling, management, transport, routing, and security are described. We then speak about network traffic analysis. The emphasis is on connection admission control, resource and bandwidth allocation, congestion control for multicast communications, as well as traffic modeling. After that, we outline the issues concerning quality of service (QoS) in networked multimedia systems, especially IP oriented. After presenting QoS parameters, we discuss QoS requirements, while maintaining QoS guarantees. The concept of generic networks is presented and illustrated. Also, access broadband networks including multimedia digital subscriber lines, and cable and wireless access networks are invoked, together with the technology and details of current solutions. The technologies that make up the core broadband networks are presented, too. At the heart of these networks are technologies that function as high-speed, high-performance bit pumps, pumping data through high-capacity pipes among different points in the network. Finally, we examine content delivery networks before giving some concluding remarks.

Introduction to Multimedia Communications, By K. R. Rao, Zoran S. Bojkovic, and Dragorad A. Milovanovic
Copyright © 2006 John Wiley & Sons, Inc.

6.1 INTRODUCTION

Networked multimedia systems offer many advantages over stand-alone systems, whereas multimedia networks pose many design challenges not faced in other sorts of telecommunications systems. In a networked multimedia system, the user's equipment is connected to remote sources of information. Some or all of the content originates with storage devices or real-time input transducers that reside elsewhere. A typical user expects to see no significant quality difference between locally and remotely stored contents. Because of the immediacy of multimedia data, and the audiovisual ways in which network problems manifest themselves to the user, the task of designing network hardware and software architectures appropriate for these applications is critical challenging. Concerns that must be addressed include providing the necessary bandwidth, keeping delay within known and manageable bounds, and either minimizing data losses or designing applications and data representations with some insensitivity to lost data [1].

Since the advantage of the advent of computer communication networks, there have been considerable advances in the provision of standards and facilities to allow computers, at an application level, to freely communicate in an open way. This is often referred to as open system interconnection (OSI) and is intended to allow computers to communicate and change data in an unconstrained way, independent of the manufacturer or of the operating system used within the computers. This latter issue dominated the early computer communication subsystems produced by computer manufacturers [2].

The various network layer protocols are concerned with the establishment of connections over or through the underlying network(s) and hence embrace addressing and routing functions. Also, if intermediate networks are used with gateways (routers) between the different networks, providing that both comply with the OSI model and are based on ISO protocols, the network layer provides its services independently of the number of networks separating the two communicating computer nodes. The network layer also enables two systems to interconnect across one or more subnetworks, thus providing a uniform end-to-end service to the transport layer. Two types of services are defined: connection oriented (CO), in which a permanent connection, either physically or logically, is set up for the duration of the communication, and connectionless (CL), in which such a connection is not required and a best try approach is adopted.

6.2 NETWORK ASPECTS OF STANDARDIZATION PROJECTS

The ITU has made itself responsive to industry needs by rapidly developing a number of standards that cover new technologies that are likely to act as fundamentals in building up the network fabric of the global information infrastructure (GII). For example, the emergence of new technologies like asynchronous transfer mode (ATM) and internet protocol (IP), which lie at the heart of many of today's high-speed computer networks, has already prompted the development of new ITU Recommendations and the modification of some existing ones.

Broadly speaking, the technological building blocks that make up the inner working of the GII can be divided into a number of distinct areas:

- transport systems
- network technologies
- user interfaces
- multimedia
- radio-communications and satellite systems.

Work is under way in each of these areas to ensure that progress in implementing GII systems takes place as quickly as possible.

6.2.1 Technological Building Blocks of the GII

Some of the technological building blocks of the global information infrastructure include transport, network technologies, as well as radio and satellite systems.

Transport

The transport technologies that will underpin information exchange across the GII include X.25, frame relay, ISDN, and ATM.

One of the oldest transport systems still in use, X.25 will remain an important element in systems where accuracy is vital, such as financial transactions. First standardized by the ITU in 1976 and reviewed and revised several times since, X.25's slow speed – just 64 kbps – is offset by its robustness, reliability, and cost effectiveness.

Frame relay, as defined by ITU-T Recommendation I.233, takes over where X.25's laggardly pace becomes an impediment. It makes use of existing technologies to send data at higher speeds by reducing the protective overhead imposed by X.25 needed to guarantee information accuracy. As frame relay finds growing popularity as a bridge between older, slower systems and very fast transmission technologies like ATM, ITU study groups continue to fine-tune the technology and to work on ways of integrating it into the vision of future global networks.

ISDN is an ITU-developed standard, which, while initially being slow in taking off in some regions, is now finding wider appeal because of its support for higher bandwidth data transfer and improved call quality. Because it excels in transferring large amounts of data very quickly from point to point, it is highly suitable for use in data-intensive applications, videoconferencing, and multimedia applications. Its added ability to support sophisticated supplementary networks services such as caller line identification also means that it can have important future applications in sales and customer support – for example, by automatically linking a business' incoming calls to a screen displaying the customer's history.

B-ISDN, or Broadband ISDN, is a faster version of the original flavor, supporting standardized data rates of up to 155 Mbps over fiber optic networks. Both standards continue to be upgraded within the study groups of the ITU-T. Furthermore, transport capacity has dramatically increased, reaching several gigabits-per-second.

ATM is perhaps the most famous transport technology of all, having enjoyed much ballyhoo when it was first launched early in the 1990s. The culmination of some 20 years research into high-speed transport technologies, this cell-based fast packet-switching system is certain to form one of the key elements of the GII. Its unique ability to move a range of different traffic types along multiple virtual traffic paths at up to several gigabits-per-second transmission speeds makes it the ideal medium for delivering true multimedia applications – real-time video, very large data files, and new audio and voice-based applications [3]. Standardized in 1991, development of ATM continues to be accorded a very high priority within the ITU and is the object of work to provide enhanced capabilities.

Network Technologies

The network transmission system known as synchronous digital hierarchy (SDH) will be one of the most important elements of the emerging GII. This new-generation technology increases the usable bandwidth in telecommunications networks and allows them to be flexibly reconfigured to adapt to changing traffic loads and user demands. Under SDH, network intelligence will be decentralized across the network, which will also be better able to handle different types of traffic traveling at different speeds. As well as supporting network reconfiguration on-the-fly, SDH can also greatly improve network capacity by more efficiently packing traffic into the available bandwidth. The ITU has already developed a wide range of standards for SDH, covering system operation and architectures, the performance and management capabilities of SDH networks and their interfaces, SDH multiplexing equipment, and international interconnection between SDH and the older networks.

Improved network management will also be an important future consideration because of the increasing complexity of equipment and the need to integrate an ever-widening range of maintenance routines as new types of equipment are added to the network. To address this issue, the ITU standardized the Telecommunications Management Network (TMN) model back in 1988, and continues to upgrade it to meet the needs of the world's evolving network technologies. The Union is also working on human computer interfaces to deal with the complex task of managing the large global networks that will be intrinsic to proper functioning of the GII. Development of these new interfaces will address the need for large amounts of real-time user interaction with the network, and the possible threats to security and network performance that it can pose.

Radio and Satellite Systems

The ITU plays a vital role in management of the world's radio frequency spectrum, an increasingly scare resource due to the rapid development and popularity of a wide range of mobile communication systems. Over the last years, the Union has also worked to facilitate the development of new kinds of satellite systems, which will almost certainly represent an important component of the future GII.

Network infrastructure for GII can be specified as:

- wideband/broadband access techniques for GII
- network interworking for the GII
- internet access/interworking
- advanced network platforms
- intelligent mobility for GII
- B-QSIG/DSS2 harmonization
- enhanced network intelligence for the GII.

N Projects (network aspects) include

- N.1 – architecture and layer 1 aspects of wideband/broadband access infrastructures for GII
- N.2.1 – signaling and control aspects of wideband/broadband access interfaces for GII
- N.2.2 – signaling and control aspects of wideband/broadband network element to network element interface for GII
- N.3 – network interworking for the GII
- N.4 – access to and interworking with IP-based networks
- N.5.1 – intelligent mobility for the GII, IMT2000 (former FPLMTS)
- N.5.2 – intelligent mobility for the GII, global mobility
- N.6 – harmonization of B-ISDN signaling protocols and their interfaces to public broadband networks
- N.7 – enhanced network intelligence for the GII
- N.8 – quality of service and network performance
- N.9 – addressing for the GII.

6.2.2 ETSI EII Network-Related Projects

The following ETSI EII standardization and specification highways have been identified, using the Enterprise Models, as:

- *Network* oriented standardization/specification highway
- *Middleware* oriented standardization/specification highway
- *Applications* oriented standardization/specification highway
- *Architecture* oriented standardization/specification highway.

Network-related projects concern the development of standards for provision of basic and enhanced telecommunication services required by EII. This area is covered by the existing expertise of ETSI members and experts. It covers aspects of network technology and interconnection, including naming and addressing, telecommunications management network (TMN). These projects have been assigned to the appropriate ETSI Technical Committee. Given the worldwide importance of the Internet, a key collaborative body is the IETF. The IETF is a technical working body, which develops the specifications used in the Internet itself and for applications that run over it.

Network (Highway 1) includes:

- access networks for residential environment
- telecommunication network interfaces for residential environment
- internetworking
- European backbone telecommunications network
- naming, addressing, numbering, and routing
- IN/TMN support for EII
- APIs for native ATM.

6.2.3 Network Design in the MEDIACOM2004 Project

The MEDIACOM2004 project is divided into a number of main framework study areas (FSAs), each of which has an associated study question in SG16. They are [4]:

- multimedia architecture (Annex B)
- multimedia applications and services (Annex C)
- interoperability of multimedia systems and services (Annex D)
- media coding (Annex E)
- quality of service and end-to-end performance in multimedia systems (Annex F)
- security of multimedia systems and services (Annex G)
- accessibility to multimedia systems and services (Annex H).

Interoperability can be considered in terms of reliable end-to-end multimedia operations across a number of different networks. This view of interoperability is discussed at length in Annex A. However, there is an alternative view in terms of different applications and services (either network or end system based) interoperating efficiently and reliably in a given multimedia environment. Support of such interoperability requires agreement on a framework within which common tasks can locate and establish communications with their peers, and similar tasks can exchange media streams of mutual interest.

SG16 will undertake the task of managing the process of harmonization of new multimedia systems and services and ensuring their end-to-end interoperability. Interoperability can be considered in terms of reliable end-to-end multimedia operation across a number of different networks. Interoperability between information appliances depends on two terminals being able to operate in a compatible manner and to use compatible interfaces, coding schemes, and information media after connection is established.

In a heterogeneous environment, networks and end systems have to interoperate efficiently for different applications and services to deliver good performance. Gateways are vital elements to ensure interoperability or interworking of legacy systems with the new IP-based networks.

SG16, in managing the process of harmonization of new multimedia systems and services and ensuring their end-to-end interoperability, will create and maintain the necessary service definition databases.

In obtaining the desirable end-to-end interoperability, common interface and media components will ideally be used. However, for historical reasons, this is rarely possible.

In general, the communication between two terminals can be realized by the following procedure:

- connection of the terminals through the network
- identification and selection of the partner's terminal
- connection establishment and selection of the known parameters about the performance of the network
- selection of the QoS (network QoS and terminal QoS)
- transfer of the information data, coded data, and so on
- disconnecting the network.

In the case of communication between two terminals, it is necessary that both terminals can identify the capabilities of the terminals at the other end and know the appropriate terminals. In order to know the partner's capability, definition of the service/applications of the multimedia terminals is important. The defined capabilities of the service/applications should be registered through an information database for the terminals.

The aims of the framework study area (FSA) for QoS and performance (QF/16) are to

- Ensure that required QoS levels for the various media types are established and defined.
- Ensure that the necessary mechanisms and protocols for providing these multimedia QoS levels are provided.

This FSA will focus on the end-to-end performance as perceived from an end user of a multimedia service or system and will identify the appropriate methods and guidelines suitable for

- measurement of the quality of media coding
- measurement of the end-to-end quality of multimedia services.

This FSA will identify suitable end-to-end performance guidelines to assist the implementation of new multimedia systems and services.

It should be mentioned that there are a number of security considerations that need to be addressed when developing an architecture for the multimedia information infrastructure. Such considerations include end-to-end privacy of data, authentication (user identification), anonymous access, access control intrusion detection, electronic signature, encryption, nonrepudiation, and lawful intercept. Within a telecommunications context, security issues can be grouped in terms of four roles: user, network operator, third party, and government.

6.3 NETWORK FUNCTIONS

Efficient utilization of network resources is essential in the provision of cost-effective multimedia services where quality-of-service (QoS) requirements for the different network applications are met and are appropriately differentiated. Efficient traffic management and efficient network design are strongly related. During the network planning phase, it is important to consider the benefit provided by traffic management (e.g, routing and congestion control) in the efficient use of network resources. When a decision is made regarding which traffic management functions are required in the network, the network topology and design need to be considered. For example, if the network is overdimensioned, the network provider may not need complex and sophisticated traffic management features. In some cases, when the new sophisticated features are not available, or are available at a high cost, a decision may be made to use only simple traffic management schemes and to invest more in transmission and switching.

To deliver multimedia services efficiently, cost-effectively, and within acceptable performance levels, a robust transport system is required. This transport system ensures the connection between a customer and a network provider, among customers, and among network providers. It includes both the access systems and trunk transport system.

As multimedia communication networks (MCN) evolve, they will encompass most aspects of wireline and wireless networks. They will also be predominantly digital networks. The implementation of new technologies (e.g., ATM), increased use of existing technologies (e.g., IP), planning for more distributed network architectures, and the introduction of new services (e.g., desktop multimedia conferencing) are all factors driving multimedia network evolution.

Increased interconnection of various networks, such as telcos, interexchange carriers, Internet providers, information service providers, and video service providers are a key aspect of multimedia network evolution. Another important factor is that multimedia networks will have an increased dependency on software to control critical network elements and some will increasingly rely on the use of external service logic (separate from switch fabric) and external databases.

Additionally there will be increased access by customers to management information (e.g., customer network management services). The integration of old networks and new networks, the coexistence of older technology and newer technology, and the addition of new network designs all provide significant opportunity for security problems. There will be increased risks in areas such as disclosure of information, fraud, unauthorized usage, loss of service, loss of data, and misdirected information. There is an increased need for protection of Internet work interfaces, protection and proper use of network elements, and protection of network and customer data. Network planners need to implement the necessary security capabilities and features to protect networks, and to provide security to customers using those networks.

6.3.1 ISO Reference Model

The seven layers that make up the reference model are shown in Figure 6.1. The reference model is intended to act as a framework or template against which the actual protocol standards relating to each layer can be based [5, 6]. It identifies the standards to be produced, relating them together and providing common methods of description and terminology. Also, the ISO reference model provides a framework for the development of the OSI standard. It subdivides the functions of computer communications and networking into two functional groups – the

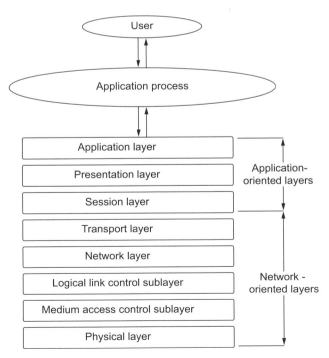

Figure 6.1 ISO reference model.

network-oriented layers and the application-oriented layers. The two functional groups are each comprised of a number of protocol layers each providing a complementary function in the context of the overall communication task. The function of the network-oriented layers – layers 1 through 5 – is first to establish a connection between the two systems that wish to communicate through the underlying network(s) being used and secondly, to control the transfer of messages across this connection without errors, losses, or replication. Collectively, they provide the higher, application-oriented layers with a network-independent, reliable message interchange service.

The application-oriented layers – layers 6 through 8 – are concerned with the representation of the data being exchanged and the support services required to perform the desired distributed processing functions; that is, they provide user application processes with the necessary support functions to perform a range of distributed processing tasks.

The layering principle is an important concept since it allows the complex functions associated with a complete communication subsystem to be broken down into a number of simpler functional units. Ideally, each layer function is made transparent, enabling layers to be modified without disturbing the other layers.

Each layer plays its part in data transfer by adhering to the rules of its own protocol, communicating only with its peer layer in the corresponding system. The only direct link between two communicating systems is the actual physical network connection. Within each system, however, there is a well-defined interface associated with each layer that allows the layer to accept, process, and pass on data during a transfer. In this way, a layer communicates with its peer layer using the services provided by the layer immediately below. For example, the application layer communicates with a peer application layer using the services provided by the presentation layer. Similarly, the presentation layer communicates with a peer presentation layer using the services provided by the session layer, and so on.

The function of each layer of the ISO reference model is as follows:

- *Application Layer* – forms the user interface to the protocol stack and the various protocol entities. The application layer provides the support for application programs (processes) to perform a range of distributed information processing functions in an open way.
- *Presentation Layer* – is concerned with the syntax of the data being exchanged between two user application processes. It provides the means by which a mutually agreed and understood syntax for the transfer of the user data can be negotiated. The agreed syntax is referred to as the transfer syntax and examples include character sets and document structure formats.
- *Session Layer* – provides the means by which two cooperating application processes running on different network nodes (computers) can organize, synchronize, and regulate the orderly exchange of data. The session layer thus provides facilities for managing and coordinating the dialog between two application processes.

- *Transport Layer* – provides the session layer above it with a reliable message transfer facility (that is, error-free, no losses, or replication), that is independent of the underlying network(s) being used.
- *Network Layer* – provides for the establishment and clearing of network connections and message routing (addressing) between two transport layer protocol entities.
- *Data Link Layer* – provides the means of transmitting data – network layer message units in practice – over the underlying physical connection. It thus includes bit (clock), byte, frames synchronization, and error detection.
- *Physical Layer* – establishes the physical connection between the computer and network termination equipment and is specifically concerned with the electrical and mechanical properties of this connection.

The aim of various OSI standards developed is to define protocols for interconnecting computer systems and the supporting interworking between application programs in those systems that are independent of any single manufacturer. The freedom offered by widespread implementation of OSI standards means that one can choose a computer on the basis of the application it supports, rather than its communications facility. Furthermore, new systems can be more readily designed to interwork with existing systems, making efficient use of existing resources.

Most of the ISO lower layers enjoy a limited number of properly defined standards, effectively stable and universally accepted by most manufacturers and many users. ISO has also developed standards for most of the application-oriented layer entities. Progress at the application layer has branched into many different application domains.

6.3.2 ISO Transport Layer

The transport layer provides the application-oriented layers with a reliable message transfer facility that is independent of the underlying network being used. Thus, it performs such functions as flow control, error control, and message fragmentation. Two alternative models are supported: connection oriented (CO) and connectionless (CL) [7]. In the CO mode of operation, the transport protocol provides reliable full duplex transfer of transparent byte streams between two systems and enhances the quality of the underlying network service to meet the application needs. In the CL mode of operation, the transport protocol offers a best try approach to transfer the offered data and is, therefore, intended for use with high-quality networks. In order to interwork with underlying networks of differing quality, the CO standard defines five classes of protocols, each intended for a different underlying network type. The protocol class to be used in a particular open system interconnection environment (OSIE) is fixed by the management authority and chosen with due regard to the network(s) being used. There are five classes (and hence protocols):

- Class 0 – provides no error recovery and is, therefore, used where the quality of the supporting network is adequate to support the application and multiplexing is not required.

- Class 1 – provides recovery from network errors or disconnections and is used where they are too frequent for the application.
- Class 2 – provides the basic facilities of class 0 but with multiplexing.
- Class 3 – combines the error recovery facilities of class 1 and the multiplexing of class 2.
- Class 4 – provides the facilities of class 3 but also detects unsignaled errors and out-of-sequence data. It should be noted that the kinds of errors detected include transport protocol data unit (TPDU) loss, TPDU delivery out-of-sequence, TPDU duplication, and TPDU corruption. These errors may affect control TPDUs as well as data TPDUs. This class also provides for increased throughput capability and additional resilience against network failure.

6.3.3 ISO Network Management Framework

The ISO network management framework on the ISO basic reference mode has been defined to provide the means for the communication subsystems in all the nodes connected to a network to be remotely monitored and controlled. A user interacts with the network management application processes (NMAPs) to monitor and control the communication subsystems in all the network nodes, as shown in Figure 6.2.

The bulk of the intelligence (and hence processing) associated with the management of such a network is in the NMAP. In practice, to allow for large, perhaps multisite, systems to be managed, the management framework is divided into a number of domains, each with its own domain manager. Management domains can overlap and a network manager can potentially cross many domains to manage a managed object (MO).

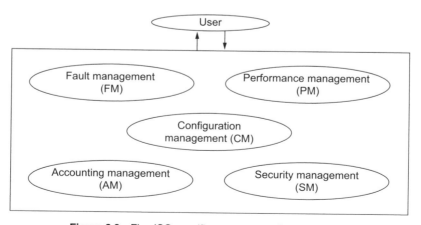

Figure 6.2 Five ISO specific management function areas.

ISO defines five specific management functional areas (SMFAs). These SMFAs resides in the NMAP that may be invoked by the network manager. These are:

- *Fault Management (FM)* – FM facilities alert a network manager when a fault is detected. It also provides isolation, examines error logs, accepts and acts upon error detection notification, traces faults, and corrects faults arising from abnormal operation.
- *Accounting Management (AM)* – AM facilities calculate the amount of network time used by each segment of the network and facilitates a billing system for the usage of resources.
- *Configuration Management (CM)* – CM facilities enable a network manager to exercise control over the configuration of a communication subsystem. It allows a manager to close down nodes at will should a fault occur or workloads change.
- *Performance Management (PM)* – PM facilities evaluate the behavior of network and layer entity resources and the effectiveness of communication activities. It can also adjust operating characteristics and generate network utilization reports by monitoring a station's performance.
- *Security Management (SM)* – SM facilities provide for the protection of the network resources. It includes authorization facilities, access controls, encryption, authentication, maintenance, and examination of security logs.

The network management framework defined by ISO refines the concept of network management (that is, the SMFAs) and defines a model for the management of the communications subsystems in each network node. It establishes activities and defines a set of management information services.

6.3.4 Advanced Control

The future vision of *information and communication anytime, anywhere and in any form* is starting to come into focus as major players begin to position themselves for radical transformation of their network and service infrastructures. It has become increasingly clear that a prerequisite for realization of such a vision is the convergence of the current multiple networks into a unified, multiservice, data-centric network offering services at different qualities and costs on open service platforms. The evolution in such a vision is influenced by the growing trends of deregulation in the telecommunications industry, on one hand, and the rapidly evolving technological convergence between distributed computing and communications on the other. In a real sense, the necessary technology and environment exist today for next-generation service providers to begin the process of transforming their infrastructures in ways that will enable provision of many new innovative services rapidly and economically, on a mass scale, and with the required quality of service. System, hardware, and especially software vendors will need to return their production processes to meet the challenges of convergence in the next-generation networks (NGN) [8].

Current networking infrastructures employ several systems in a nonharmonized fashion. Telecommunications operators invested substantially in systems using traditional signaling and control protocols. The most common is Signaling System No.7 (SS7) for analog and integrated services digital network (ISDN), the extensions for mobile and wireless environments and the protocols for broadband networks. Important investments have been made in supplementary intelligence systems such as the intelligent network (IN).

Network protocols define the set of formats needed to control the use of networks. Both telephony and IP networks conform with the ultimate goal of enabling communication between end users, and must consequently handle naming, addressing, and routing issues. Unfortunately, the telephony and IP worlds provided different solutions and addressing schemes, so universal naming and addressing has become a real issue.

The routing solutions differ quite fundamentally as well. The telephony network is well designed for audio and video types of traffic, with an excellent quality of service (QoS) guarantee. On the other hand, the Internet is much better designed for variable and dynamic data traffic, but offers best-effort quality only. The Internet appears to be more of a treat to traditional telephone systems because the telephone network is less adaptable to data traffic than the Internet seems to be to voice traffic [9].

The true objective of a communication network is the same as it has been for the past ten years, continuously expanding the service offered (deploying new services) while capitalizing on the installed base (current services and network platforms). The services that have been defined in the public switched telephone network (PSTN), such as closed user group (CUG) and calling line identification presentation (CLIP) as well as calling line identification restriction (CLIR), are quite difficult to define, deploy in the network, and manage. In order to deploy new services more easily and rapidly, the service intelligence was placed outside the network with the definition of the intelligent network (IN). This solution constitutes an important source, but brought its own complexity with a new series of protocols defined in the IN application protocol (INAP).

Similarly, the Internet brings services together with its related applications (e-mail, the Web, chat service). The Internet is defined in its own connectionless world and exploits a different network infrastructure (servers and routers). While data are ideally transported on IP networks, the provision of voice over IP (VoIP) added a whole new level of complexity. Since the majority of end users access the network with PSTN standards, the use of VoIP implies the development of the required gateways, gatekeepers, multipoint controllers, and so on.

In order to ultimately reach convergence, fundamental objectives must be defined, such as synergy between IN and IP services, service components reuse, genericity, flexibility, and ease of deploying new services onto any network resource. These objectives imply a list of basic requirements any advanced communication network architecture should comply with in order to fulfill the objectives.

Rate Control Function

The purpose of rate control is to enforce the specification of the bit stream. The general system is shown in Figure 6.3. The bit stream from the coder is fed into a buffer at a rate $R'(t)$ and it is served at some rate $\mu(t)$ so that the output rate $R(t)$ meets the specified behavior [10]. The bit stream is smoothed by the buffer whenever the service rate is below the input rate. The size of the buffer is determined by delay and implementation constraints. It is therefore common to add a back pressure signal from the buffer to the encoder, so that the compression rate is increased when buffer overflow is at risk. The issue is to reduce the variability of the rate function $R(t)$ while minimizing the effects of the consistency of the perceptual quality [11, 12]. The joint problem of traffic characterization and rate is to find a suitable description of the bit stream that is sufficiently useful to the network and that can be enforced without overly throttling the compression rate. Provided that a model has been chosen, the user should not estimate the parameters for it and find a way to regulate the service rate $\mu(t)$ so that $R(t)$ strictly obeys the specification. The network would then have the possibility to verify that the traffic is in accordance with its specification.

Rate Control Techniques. In developing a rate control technique, there are two widely used approaches:

- analytical model based approach
- operational rate distortion $R(D)$ based approach.

In the model-based approach, various distributions and characteristics of signal source models with associated quantities are considered. Based on the selected model, a closed-form solution is derived using optimization theory. Such a theoretical optimization solution cannot be implemented easily because there is only a finite discrete set of quantizers and the source signal model varies spatially. Alternatively, an operational $R(D)$ based approach may be used in the practical coding environment. For example, to minimize the overall coding distortion to a total bit budget constraint, lots of techniques based on dynamic programming or Lagrangian multiplier for optimization solutions have been developed [13–16]. These methods share the similar concept of data preanalysis. By analyzing the $R(D)$ characteristics of future frames, the bit allocation strategy is determined afterwards. The Lagrangian multiplier is a well-known technique for optimal bit allocation in image and video coding, but with an assumption that the source consists of statistically independent

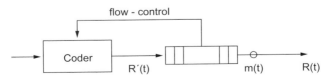

Figure 6.3 Bit rate control.

components. Thus, an interframe-based coding may not find the Lagrangian multiplier approach applicable because of the bit rate control.

Frame dependencies are taken into account in bit rate control. However, potentially high complexity with increasing operating $R(D)$ points make this method unsuitable for applications requiring interactivity or low encoding delay. In reference [17], a joint encoder and channel rate control scheme are investigated for VBR video over ATM networks. Also, it is claimed that the rate control in the encoder side and bit stream smoothness for statistical multiplexing gain in the network side. A parametric $R(D)$ model for MPEG encoders, especially for picture-level rate control, is proposed in reference [18]. Based on the bit rate *mquant* model, the desired *mquant* is calculated and used for encoding every macroblock (MB) by combining with the appropriate quantization matrix entry in a picture. A normalized $R(D)$ model based approach has also been developed for H.263 compatible video codecs. By providing a good approximation of all 31 rate distortion relations, the authors claim that the proposed model offers an efficient and less memory requirement approach to approximate the rate and distortion characteristics for all quantization parameters (QPs) [19]. Rate control techniques for MPEG-4 object level and macroblock level video coding were proposed in references [20, 21], respectively. However, most of the aforementioned techniques only focus on a single coding environment, either frame-based level, object level, or macro level. None of these techniques demonstrates its applicability to MPEG-4 video coding including the above three coding granularities simultaneously.

In reference [22], based on a revised quadratic $R(D)$ model, scalable rate control (SRC) proposes a single framework that is designed to meet both VBR without delay constraints and CBR with low latency and buffer constraints. With this scalable framework based on a new $R(D)$ model and several new concepts, not only more accurate bit rate control with buffer regulation is achieved, but scalability is also preserved for all test video sequences in various applications [23]. By considering video contents and coding complexity in the quadratic $R(D)$ model, the rate control scheme with joint buffer control can dynamically and appropriately allocate the bits among VOs to meet the overall bit rate requirement with uniform video quality.

Because of the precision of the $R(D)$ model and ease of implementation, the rate control scheme with the following new concepts and techniques has been adopted as part of the rate control scheme in the MPEG-4 standard [20]. These new concepts and techniques are:

- a more accurate second order $R(D)$ model for target bit rate estimation
- a sliding-window method for smoothing the impact of scene change
- an adaptive selection criterion of data points for a better model updating process
- an adaptive threshold shape control better use of bit budget
- a dynamic bit rate allocation among VOs with different coding complexities.

This rate-control scheme provides a scalable solution, meaning that the rate-control techniques offer a general framework for multiple layers of control for objects, frames, and MBs in various coding contexts.

Rate Control in MPEG-4 Visual and H.264 Standards

The MPEG-4 Visual and H.264 standards require each video frame or object to be processed in units of macroblocks. If the control parameters of a video encoder are kept constant (e.g., motion estimation, search area, quantization step size, etc.), then the number of coded bits produced for each macroblock will change depending on the content of the video frame, causing the bit rate of the encoder output (measured in bits per coded frame or bits per second of video) to vary. Typically, an encoder with constant parameters will produce more bits when there is high motion and/or detail in the input sequence and fewer bits when there is low motion and/or detail [24].

This variation in bit rates can be a problem for many practical delivery and storage mechanisms. For example, a constant bit rate channel (such as a circuit-switched channel) cannot transport a variable bit rate data stream. A packet-switched network can support varying throughput rates, but the mean throughput at any point in time is limited by factors such as link rates and congestion. In these cases it is necessary to adapt or control the bit rate produced by a video encoder to match the available bit rate of the transmission mechanism. CD-ROM and DVD media have fixed storage capacities and it is necessary to control the rate of an encoded video sequence (for example, a movie stored on DVD-Video) to the capacity of the medium.

The variable data rate produced by an encoder can be *smoothed* by buffering the encoded data prior to transmission [24]. Figure 6.4 shows a typical arrangement in which the variable bit rate output of the encoder is passed to a first-in/first-out (FIFO) buffer. This buffer is emptied at a constant bit rate that is matched to the channel capacity. Another FIFO is placed at the input to the decoder and is filled at the channel bit rate and emptied by the decoder at a variable bit rate.

It can be seen that a variable coded bit rate can be adapted to a constant bit rate delivery medium using encoder and decoder buffers. However, this adaptation comes at a cost of buffer storage space and delay and the wider the bit rate variation, the larger are the buffer size and decoding delay. Furthermore, it is not possible to cope with an arbitrary variation in bit rate using this method, unless the buffer sizes and decoding delay are set at impractically high levels. It is usually necessary to

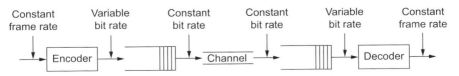

Figure 6.4 Encoder output and decoder input buffers.

implement feedback mechanisms to control the encoder output bit rate in order to prevent the buffers from over- or underflowing.

Rate control involves modifying the encoding parameters in order to maintain a target output bit rate. The most obvious parameters to vary are the quantizer parameter (QP) or step size since increasing QP reduces coded bit rate (at the expense of lower decoded quality) and vice versa. A common approach to rate control is to modify QP during encoding in order to

- maintain a target bit rate (or mean bit rate)
- minimize distortion in the decoded sequence.

Optimizing the trade-off between bit rate and quality is a challenging task and many different approaches and algorithms have been proposed and implemented. The choice of rate control algorithm depends on the nature of the video application [24].

Example 6.1. Consider offline encoding of stored video for storage in a DVD. Processing time is not a particular constraint and so a complex algorithm can be employed. The goal is to *fit* a compressed video sequence into the available storage capacity while maximizing image quality and ensuring that the decoder buffer of a DVD player does not overflow or underflow during decoding. Two-pass encoding (in which the encoder collects statistics about the video sequence in a first pass and then carries out encoding in a second pass) is a good option in this case.

Example 6.2. Have a look at encoding of live video for broadcast. A broadcast program has one encoder and multiple decoders. Decoder processing and buffering is limited, whereas encoding may be carried out in expensive, fast hardware. A delay of a second is usually acceptable and so there is scope for a medium-complexity rate control algorithm, perhaps incorporating two-pass encoding of each frame.

Example 6.3. Let us see about rate control in the case of encoding for two-way videoconferencing. Each terminal has to carry out both encoding and decoding. Thus, processing power may be limited. Delay must be kept to a minimum (ideally less than around 0.5 seconds from frame capture at the encoder to display at the decoder). In this scenario, a low-complexity rate control algorithm is appropriate. Encoder and decoder buffering should be minimized in order to keep the delay small. Hence the encoder must tightly control output rate. This, in turn, may cause decoded video quality to vary significantly. For example, it may drop significantly when there is an increase in movement or detail in the video scene.

Recommendation H.264 does not specify at present a rate control algorithm. However, a proposal for H.264 rate control is described in reference [25]. MPEG-4 Visual describes a possible rate control algorithm in an Informative Annex [26]. This algorithm, known as the scalable rate control (SRC) scheme, is appropriate for a single video object and a range of bit rates and spatial/temporal resolutions. The SRC attempts to achieve a target bit rate over a certain number of frames (a *segment* of frames, usually starting with an I-VOP) and assumes the following model for the encoder rate R:

$$R = \frac{X_1 S}{Q} + \frac{X_2 S}{Q^2} \qquad (6.1)$$

where Q is the quantizer step size, S is the mean absolute difference of the residual frame after motion compensation (a measure of frame complexity), and X is a model parameter. Rate control consists of the following steps, which are carried out after motion compensation and before encoding of each frame i:

1. Calculate a target bit rate R_i, based on the number of frames in the segment, the number of bits that are available for the remainder of the segment, the maximum acceptable buffer contents, and the estimated complexity of frame i.
2. Compute the quantizer step size Q_i, to be applied to the whole frame. Calculate S for the complete residual frame and solve equation (6.1) to find Q.
3. Encode the frame.
4. Update the model parameters X based on the actual number of bits generated for frame i.

The SRC algorithm aims to achieve a target bit rate across a segment of frames, rather than a sequence of arbitrary lengths, and modulate the quantizer step size within a coded frame, giving a uniform visual appearance within each frame but making it difficult to maintain a small buffer size and hence a low delay. An extension to the SRC supports modulation of the quantizer step size at the macroblock level and is suitable for low-delay applications that require *tight* rate control. The macroblock-level algorithm is based on a model for the number of bits B_i required to encode macroblock i, as follows:

$$B_i = A\left(K \frac{\sigma_i^2}{Q_i^2} + C\right) \qquad (6.2)$$

where A is the number of pixels in a macroblock, σ_i is the standard deviation of luminance and chrominance in the residual macroblock (i.e., a measure of variation within the macroblock), Q_i is the quantization step size, and K and C are constant model parameters. The following steps are carried out for each macroblock i:

1. Measure σ_i
2. Calculate Q_i based on B_i, K, C, σ_i and a macroblock weight α_i.
3. Encode the macroblock.
4. Update the model parameters K and C based on the actual number of coded bits produced for the macroblock.

The weight α_i controls the *importance* of macroblock i to the subjective appearance of the image, and a low value of α_i means that the current macroblock is likely to be highly quantized [24]. These weights may be selected to minimize changes in Q_i at lower bit rates since each change involves sending a modified quantization parameter DQUANT, which means encoding an extra five bits per macroblock. It is important to minimize the number of changes to Q_i during encoding of a frame at low bit rates because the extra five bits in a macroblock may become significant. At high bit rates, this DQUANT overhead is less important and Q may change more frequently without significant penalty. This rate control method is effective at maintaining good visual quality with a small encoder output buffer, keeping coding delay to a minimum. This is important for low delay applications such as the scenario concerning encoding for two-way videoconferencing described in Example 6.3.

Further information on some of the many alternative strategies for rate control can be found in reference [27].

6.3.5 Signaling in Communications Networks

Signaling is one of the most important functions in the telecommunications infrastructure because it enables various network components to communicate with one another to set up and tear down calls. Significant efforts were undertaken in past decades to develop the signaling protocols in use in today's telephone network, also known as the public switched telephone network (PSTN). These protocols, such as Signaling System No. 7 (SS7) and Q.931, are defined in detailed specifications developed by various standardization organizations.

H.323 was the first VoIP standard that helped move the VoIP industry away from proprietary solutions and towards interoperable products. The H.323 architecture is still evolving in several areas such as the gateway decomposition architecture and integration of H.323 with IN. This evolution is addressing some of the original limitations of H.323. Other protocols such as Session Initiation Protocol (SIP) have also been introduced as alternatives to H.323. The industry debate on the VoIP signaling architecture will continue to attract a lot of attention, and the evolution will be determined by the VoIP market forces. Since the very beginning of the VoIP industry, issues around signaling protocols for VoIP have been the focal point of industry debates. So far, the VoIP industry has gone through three stages in terms of signaling protocol evolution: precommercial, PC-centric, and carrier grade [28].

The precommercial stage was characterized by research activities in various universities and research organizations of the Internet community. Much of the work was coordinated by two working groups in the IETF. The audio/video transport (AVT) working group produced Real-Time Transport Protocol (RTP) [29]. The Multiparty Multimedia Session Control (MMUSIC) working group designed a family of protocols for multimedia conferencing over the Internet, including the Session Initiation Protocol (SIP), for session setup and teardown [30]. The primary focus in this stage was on audio and video conferencing over the Internet.

Interworking with the PSTN was only a small part of the overall effort. Until 1996 SIP was the only signaling protocol for multimedia conferencing over the Internet.

The PC-centric stage started in early 1995 when commercial VoIP software first appeared in the market. Initially, these products allowed a user to place a call over the Internet from a multimedia PC to another multimedia PC. All the signaling and control functions resided in the PCs. Each product relied on a proprietary signaling protocol for call setup and teardown, which made it virtually impossible for the ITU, which started work on standardizing VoIP signaling. In June 1996, Study Group 16 of ITU-T decided on H.323 v.1, referred to as a standard for real-time videoconferencing over nonguaranteed quality of service (QoS) LANs.

The carrier-grade stage started around early 1998. As IP telephone service providers began to deploy networks of H.323 gateways to offer VoIP services, they soon realized that H.323 has some limitations. H.323 assumed that a gateway handles signaling conversion, call control, and media transcoding in one box, which poses scalability problems for large-scale deployment. In order to provide carrier-grade VoIP services, in May 1998, the concept of a decomposed gateway was introduced where call control resides in one box, called the *media gateway controller*, while media transformation resides in another box called the *media gateway*. The Media Gateway Control Protocol (MGCP) was introduced in 1998. After about two years of extensive work, ITU-T SG16 and IETF defined the media gateway control standard called H.248.

In order to ultimately reach convergence, fundamental objectives must be defined such as synergy between IN and IP services, service component reuse, genericity, flexibility, and ease of deploying new services on top of any network resource [31]. These objectives imply a list of basic requirements any advanced communication network architecture should comply with in order to fulfill the objectives.

6.3.6 Network Management

During the last decade, innovations in optical networking technology as well as advances in digital compression and transmission over copper and cable have increased the capability limits and capacity per unit cost in almost all parts and all levels of network infrastructure: access, metro, and coder; transport and switching; wireline and wireless access. Deployment of new technologies in core optical transport and massive deployment of second-generation cellular systems stand out as major breakthroughs in capabilities already available in the field. Broadband wireline and metro networks are now getting ready for major increases in deployed capacity. The technology front is not standing still. Third-generation wireless systems, targeting the transmission of voice, video, and high-speed data, are already under preliminary deployment. Furthermore, researchers and vendors are expressing a growing interest in fourth-generation wireless systems that will support even higher rates and cater to global roaming across multiple wireless networks. Metro optical network equipment vendors are driving towards low-cost, high-capacity networking, exploiting short distances, passive components, and fiber abundance.

Researchers are continuously increasing the ability to extract transmission capacity out of copper and cable.

Meanwhile, based on the TCP/IP protocol suite, the Internet has evolved from a research network, targeting a limited audience of academic and military users, to a huge commercial network of loosely connected subnetworks. The TCP/IP protocol suite has also been deployed in a large number of private intranets and public data networks, built by established carriers to offer public IP services. This has blurred the difference between the global Internet and thousands of private IP networks. Only access controls and firewalls separate the two. In recent years, IP networks are increasingly used for mission-critical data applications and are getting ready for real-time multimedia applications.

Despite the fact that network operators and manufacturers face a major financial crisis, the size and importance of the Internet continue to grow at amazing speeds. At the core of the Internet, optical switches are replacing traditional switches, allowing data flow at unprecedented speeds. At the edges, digital subscriber lines (DSL) are becoming the technology of choice; within every home it is now possible to permanently connect multiple devices to the Internet. Wireless LAN (WLAN) hotspots are becoming commonplace, allowing people to connect to the Internet independent of whether they are flying in a plane, traveling on a train, or sitting on a terrace.

Considering the importance of the Internet in our economic and daily life, it is remarkable that technologies to manage the Internet get limited attention. Since the late 1980s, the Internet standard management framework has been based on the Simple Network Management Protocol (SNMP). In the 1990s, the IETF defined several new versions; the latest one, SNMPv3, has already been a full Internet standard for many years. However, despite its success, there is common agreement that SNMP technology will be unable to solve all future Internet management problems. Although it is likely that SNMP will remain the protocol choice for many monitoring tasks, it is not the ideal technology for tasks such as configuration management. In fact, further progress in SNMP technology is unlikely to occur, and a new version of SNMP should no longer be expected.

Of such approaches, the most promising ones are based on extensible markup language (XML) and Web services technologies. These directions have created two major needs that recent and ongoing research is trying to fill [32]. One of these needs is for the support of multiple grades of quality of service (QoS) in terms of both sunny day measures like packet delay, jitter, and loss, and rainy day measures such as service availability and restoration speed. Much work has been done on providing bearer plane and control plane capabilities to allow multiple grades of QoS in IP and IP/multiprotocol label switching (MPLS) networks. Many of these have been standardized. Examples are differentiated service (`Diff-Serv`) and integrated services (`IntServ`), Resource Reservation Protocol (RSVP) and Constrained-Based Routing Label Distribution Protocol (CR-LDP), and traffic engineering extensions [33].

Equally important is the need to provide management plane functionality to allow multiple grades of QoS to be managed effectively. The needed management plane capabilities include network and service configuration; performance and fault management over heterogeneous underlying technologies and multilayered networks; end-to-end QoS and traffic management; service control platforms; and multiplayer management. Given the free-for-all heritage of the Internet, these management plane functions are challenging areas of research and deployment.

Nevertheless, providing management functionality at the service level implies that there must be appropriate support at the network management level, and the two must be conveniently integrated. This is not yet solved because, among other problems, integrated and seamless management among different network technological layers intervening in the network infrastructure is still an open issue. Furthermore, provisioning of multimedia may make this integration process even more complex.

Different management paradigms and management technologies need to be investigated because the traditional approaches are no longer valid. Policy-based management has been proposed as a way to facilitate the automation of management processes. XML-based management appears to be very promising in helping in the process of integrating management functions.

Evolution of Network Management

Network management is the sum of all activities related to configuring, controlling, and monitoring networks and systems with the aim of ensuring their effective operation. The complexity of this task lies in the fact that the managed components have evolved from an isolated, homogenous, controllable set of systems within one domain to a large, heterogeneous, distributed communication environment spanning multiple domains. We are thus dealing with a variety of network components, multiple operating domains, integrated service environments, and highly heterogeneous systems. Being faced with such challenges, different network management solutions have arisen over the years to cater for the heterogeneous and ubiquitous communication networks of the present.

In conforming to this need, standard specifications have been decided on with the goal of enabling a cross-system and multivendor orientation in the management field. Management communication frameworks like Simple Network Management Protocol (SNMP) and Common Management Information Protocol (CMIP) [34, 35] have dominated from the early days of management by giving a promising solution to networks that had not met at that time the complexity and heterogeneity of the present communication infrastructures and service environments. In the meantime, the emergence of advanced technologies, like the Common Object Request Broker Architecture (CORBA) and Remote Method Invocation (RMI), destined to address the distribution of software environments and the interoperability of systems, has undoubtedly influenced the management area. Owing to these technologies, the development of more open, flexible, and scalable management architectures in conjuction with their ease of interoperability has become a reality.

However, there are still steps to be taken when it comes to the management of networks that are continuously changing their functionality and consequently their expectations from a management platform. The aforementioned technologies provide a convenient underlying infrastructure for the communication of a management information model among others and are particularly important in networks that support advanced quality of service (QoS) [36]. Therefore they introduce a high degree of complexity, dynamic behavior, and informational power. It is very difficult to build management systems that can easily program managed resources to adapt to emerging requirements and are able to dynamically change the behavior of the whole system to support modified or additional functionality.

QoS Management Tool (QMTools) is a platform independent of the underlying technologies or vendor-specific equipment and endows the network administration with a set of basic functionalities for (re-)configuration, fault management, and monitoring of the managed components. The QMTool can be viewed as a network management application that adopts a number of novel technologies aspiring to provide dynamic QoS management. The introduction of XML offers an abstract definition model for configuration and monitoring network elements [37].

Management research and standardization started in the mid-1980s, but after almost 20 years there is no widely adopted framework and technology that satisfy the needs of network, system, service, and application management. Initial technologies such as open systems interconnection (OSI) systems management and Simple Network Management Protocol (SNMP) were based on the *manager-agent* model. In the early to mid-1990s, general distributed object technologies, most notably the CORBA, spurred a lot of research and standardization activity regarding their use for management. These technologies are based on the principles of the open distributed processing model.

All these management technologies found their niche markets: OSI systems management for network management in telecommunication environments; CORBA for service management in telecommunication as well as some IP environments; and SNMP for network and system management in IP environments. Recently, Web services have emerged as a promising XML-based technology for standardizing e-service interfaces. However, careful examination shows Web services to be very similar to distributed object technologies, so potentially it can be used for management.

Two management schemes have dominated the Internet and telecommunications fields from the very early needs for standardization in both areas. The principles and models of Internet SNMP and OSI systems management are primarily influenced by the capabilities and supported services of the networks to be managed. The former aims to manage an unreliable network with minimal expectations from its customers. The latter operates in a connection-oriented environment where QoS provisioning is of prime importance.

Simple Network Management Protocol (SNMP). The SNMP of the IETF [38] is based on three major components: *manager*, *agent*, and *management information base* (MIB). The agents export an abstract view of the managed resources; based on this view, the manager can perform management actions. The structuring of the objects representing the network resources constitutes the MIB. The information model of SNMP exhibits simplicity and, albeit object-based, does not adopt the classes and inheritance capabilities of the object-oriented approaches. Accordingly, the protocol (SNMP) used to access and control the management information has been influenced by the simple and linear form of the information model. In addition, the adoption of User Datagram Protocol (UDP) for the underlying transportation means fulfilling the design goals of the SNMP framework of simplicity at the agent and communication part.

The SNMP information model is very simple, with scalar variables (of integer or string type) used to model managed entities. SNMP objects are formally specified in a language known as structure of management information (SMI). Although we talk of objects in SNMP, they are effectively equivalent to object attributes in other frameworks. There is no inheritance and the only operations allowed on them are read and write. These objects can be either single- or multiple-instances, the latter for tables consisting of a series of rows. Table rows are the only composite objects that can be created and deleted. This very simple variable-based rather that object-oriented approach may result in significant awkwardness and complexity when trying to model complex real entities. There is no information reuse through inherence, while complex data types and imperative methods can only be modeled indirectly.

A transport protocol is modeled in SNMP as a group of scalar value objects, with each such *object* having an individual name, while connections are modeled through a table row of such objects. The TCP protocol and connection modeling can be found in IETF RFCT 2012.

On the other hand, the OSI management model of the International Organization for Standardization (ISO) / International Telecommunications Union–Telecommunication Standardization Sector (ITU-T), although adopting the same manager-agent paradigm, uses a more powerful information model for encapsulation of the underlying resources. Here, an object-oriented approach has been adopted that fully exploits the concepts of inheritance and polymorphism. The use of a management information tree (MIT) for the organization of managed objects as well as their object-oriented functionality necessitates a more powerful management protocol for supporting remote method execution. OSI management places a greater burden on the agent and communication parts than does Internet management.

OSI System Management. OSI systems management (OSI-SM) was the first management technology to be standardized [39]. It is a sophisticated and powerful technology, but complicated and expensive to deploy, so it has only found use in telecommunication environments. The OSI management model was adopted for the information architecture of the telecommunication management network (TMN). The latter was developed by the ITU to provide methods for end-to-end management of networks and services supported by different service providers [40].

In particular, its functions cover five functional areas: fault, configuration, accounting, performance, and security management (FCAPS). The TMN is a framework defined by three basic architectures. The functional architecture specifies the management functionality; the physical architecture specifies the physical components and their interfaces; and the information architecture, described in ITU-T M.3010 [41] highlights OSI system management as the best candidate to be applied.

One of the key aspects of the TMN as a management framework is that it transforms the old flat centralized management paradigm to a hierarchical distributed one. This is accomplished by the logical layering of management functionality, resulting in increased abstraction of management information as we move to the higher layers. More specifically, the management functions are classified into business, service, network, and network element ones. The need for a layered/distributed architecture, which would also enable the integration of the different network management schemes, has led several research groups to seek a technology for realizing the TMN framework.

Manager-Agent Model. This model defines the principles of operation for protocol-based management frameworks. Managed resources are modeled through *managed objects* (MO), which encapsulate the underlying resource and offer an abstract access interface. Any managed network/service element or distributed application should expose a *cluster* of managed objects modeling its resources across a management interface provided by an *agent*. The interface is formally defined through specification of the available managed object types or classes and the management access service/protocol/*manager* applications access managed objects across interfaces for implementing management policies. A management application may act in both agent and manager roles; this is the case for peer-to-peer management interactions or hierarchical management environments [39].

An agent is a software entity that administers managed objects, responds to management requests, and disseminates spontaneous events through the management protocol. It uses an implementation-specific mechanism to access the managed objects and, given its collective view of the relevant cluster, can provide sophisticated object access facilities for bulk data transfer or selective information retrieval. In addition, it can evaluate event notifications at the source and only send them to interested managers according to predefined, potentially sophisticated, criteria. The manager-agent model is shown in Figure 6.5.

The management protocol supports access or multiple attributes from multiple objects through one request. The manager-agent model projects a communication framework in which standardization affects only the way in which management information is modeled and carried across systems, deliberately leaving aspects of their internal structure unspecified. This may result in highly optimized implementations given that there are no internal application programming interfaces (APIs) to be adhered to.

Network Management Technologies. OSI system management (OSI-SM) was the first management technology to be standardized; in fact, the manager-agent model was devised in its context. It is a sophisticated and powerful technology, but also complicated and expensive to deploy, so it has only found use in telecommunication environments. It was the technology behind the telecommunications management network (TMN) Q/X interfaces, but even in these environments it is gradually being phased out in favor of CORBA. Still, it is accepted as the most

6.3 NETWORK FUNCTIONS

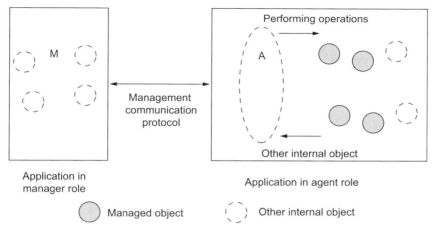

Figure 6.5 The manager-agent model [33]. (©2004 IEEE.)

powerful technology thus far to support features that should be essential in any management framework, so it is worth examining.

OSI-SM uses a fully object-oriented information model supporting inheritance. Managed object classes are specified in the Guidelines for the Definition of Managed Objects (GDMO) abstract language. Key deviations from object-oriented concepts are the run-time specialization of an MO instance through conditional packages and the fact that imperative commands are modeled through a generic method called an *action* with a single parameter and result, which may result in awkward parameter modeling. Telecommunication information models were initially specified in GDMO, although there is a push today towards technology-neutral information specification in the unified modeling language (UML), with reverse engineering of existing models.

ISO/ITU-T open distributed processing (ODP) is a general framework for specifying and building distributed systems [42], while CORBA can be seen as its pragmatic counterpart. ODP came about in response to the recognition that although ITU-T and IETF protocol-based solutions addressed the problem of heterogeneous system interconnection, the proliferation of application-layer standards and distributed applications meant that application intercommunication needed to be addressed as well. This was further fueled by the convergence of the information technology and communication sector, and the resulting demand for standardized APIs between distributed application components and underlying platforms. Hence, the target for ODP is not only to facilitate distribution and interoperability, but also to achieve software portability.

ODP projects a client-server model, with distributed applications composed of objects interacting solely by accessing each other's *interfaces*. The underlying ODP platform provides a number of transparencies, such as access, location, replication, failure, and resource. Clients access server objects through *interface references*, obtained through access to well-known special servers such as name server and traders. A name server keeps a name space with interface references *advertized* by server objects, while clients could resolve names to interface references. Traders, on the other hand, keep interface references together with object *properties* (i.e., attributed with static values), and may also search the administered object space to evaluate assertions on dynamic attribute values. Clients may search for an interface reference through a predicate on properties and attribute values of the sought object. Finally, there is an underlying protocol for interoperability hidden inside the platform, with objects *sitting* on the platform through a standardized API. The ODP model is depicted in Figure 6.6.

Web services is an emerging Internet-oriented technology that has strong analogies to CORBA. It is developed and standardized by the World Wide Web Consortium (W3C) so that Web-based e-services expose standard interfaces and are accessed in an open interoperable manner. Web services aim to put structure in Web content and associated services so that the latter are accessible by distributed applications [43].

Service interfaces are specified in the Web services description language (WSDL), which constitutes a general XML-based framework for the description of services as communication endpoints capable of exchanging messages. It describes the service location through a uniform resource identifier (URI), supported operations, and messages to be exchanged. Service inheritance is also supported. WSDL does not mandate a specific communication protocol, but can support different protocol bindings; despite this, the default binding is usually Simple Object Access Protocol (SOAP). WSDL can be considered as broadly equivalent to CORBA interface definition language (IDL). In this context, URIs are broadly equivalent to CORBA interoperable references (IORs). SOAP is a stateless protocol with XML-based encoding. It can support request/response/error remote call interactions and is broadly equivalent to CORBA General Inter-Operability Protocol (GIOP) when used that way. The default SOAP mapping on HTTP/TCP/IP can be considered equivalent to CORBA Internet Inter-Operability Protocol (IIOP).

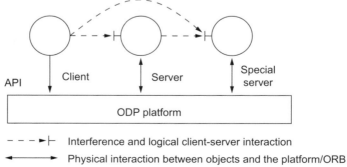

Figure 6.6 The open distributed processing (ODP) model [33]. (©2004 IEEE.)

Service specification and interface discovery are supported through Universal Description Discovery and Integration (UDDI). This provides a mechanism for service providers to advertize (i.e., register) services with it in a standard form so that service consumers query services of interest and discover their locations. UDDI is itself implemented as a well-known Web service in terms of interface and location. When used for service specification discovery, it is broadly equivalent to the CORBA Interface Repository; when used for interface location discovery, it is broadly equivalent to the CORBA Naming and Trading services.

It should be obvious from the above comparison that Web services can be used as a distributed object technology. Some key differences from CORBA are the following. In CORBA, the default client-server coupling is tight, with the client having precompiled knowledge of the server's interface, which supports compile time type-checking.

The use of distributed object technologies for network, service, and application management was a subject of intense research in the mid- to late 1990s. It is now widely accepted that distributed objects are naturally suited to service and application management: service management involves mostly business process reengineering and automation, for which technologies like CORBA or J2EE are well suited; in addition, distributed applications are typically realized using distributed object technologies, so it makes sense to use the same technology to manage them. On the other hand, network and system management have relatively different requirements: large amounts of information need to be accessed, some of it real time in nature, while concerted configuration changes must be supported across devices. So fundamental requirements of network management are support for flexible information retrieval, both in bulk but also selectively; support for fine-grained event notifications through selective criteria; and support for transactions that involve many operations to one or more devices.

The popularity of CORBA led the Tele-Management Forum (TMF) and OMG to set up the Joint Inter Domain Management (JIDM) task force, which produced a static mapping (i.e., specification translation) between SNMP SMI/OSI-SM GDMO and CORBA IDL, and a dynamic mapping (i.e., interaction translation) to support generic gateways.

A number of management technologies have emerged over the last 15 years, but none of them satisfies all the requirements of management environments. At one end of the spectrum, OSI SM has been the most powerful technology, but is complicated and expensive, and relies on OSI protocols that have gone out of fashion. It was used in telecommunication environments, but is gradually being phased out in favor of CORBA. On the other end of the spectrum, SNMP has been a very simple framework that became widely deployed in IP environments but fell victim to its own simplicity: its information modeling capabilities are rudimentary, it does not support bulk data retrieval and event-based management well, and, most importantly, it does not support configuration management well due to its lack of transaction capabilities.

Distributed object technologies, and CORBA in particular, have a number of advantages, but were designed with distributed systems in mind, lacking support for bulk data retrieval; they also suffer from potential scalability problems for large managed object populations. They can be used for management, but this requires special modeling to support predefined bulk transfer through special methods and avoid modeling large populations of dynamic objects through separate object/interfaces. These problems of distributed object technologies and the ways around them through special modeling became evident from the performance evaluation in the previous section.

Web services can be seen as a distributed object technology; in fact, platform providers have been taking a CORBA-like approach with stub objects, which reinforces this view. Its use for network and systems management presents the same problems as CORBA, so exactly the same solutions can be adopted. Its usability is similar to CORBA due to the stub-based APIs and arguably better than SNMP. On the other hand, there is no security and notification support at present, which means this technology is not yet ready to be used for network management. The initial performance evaluation is encouraging, but also highlights some expected problems. Information retrieval times are approximately twice those of CORBA, but the key problem is the amount of management traffic incurred due to the XML-based encodings, which can be up to eight times that of CORBA. This can be reduced through compression at the expense of slower retrieval times. The footprint of managed systems is also relatively large, but smaller than CORBA.

In summary, Web services is a promising technology but, being XML-based, has more overheads than SNMP and CORBA. On the other hand, being XML-based is also its biggest attraction, due to potential easy integration with other applications.

Digital Subscriber Lines (DSL) Management

Digital subscriber lines (DSL) transmit over ordinary copper twisted-pair telephone loops at high frequencies (up to several MHz) to provide broadband digital services. Advanced DSL management combines gathering and storing data about DSL frequencies with analysis of this data to deploy and maintain DSL. It has the potential for dramatic increases in DSL performance, and is a compelling way to manage the telephone plant as it transitions to digital services. Rather than throwing extra crosstalk margin at a DSL line to handle most problems, isolating the particular difficulties and handling them properly results in better service and less effort wasted on dealing with provisional errors. Carrier-grade service may be assured [44].

Crosstalk is created by DSL lines coupling into each other, and dynamic spectrum management (DSM) can balance multiple DSL signals, minimize crosstalk, and jointly optimize the loop plant [45]. DSL is a relatively new service from the local exchange carriers (LECs). Current practice assumes that there is little knowledge about a particular loop's transmission parameters except a rough estimate of loop length. All DSL services must withstand a statistical worst case environment assuming 99 percent worst case crosstalk coupling that is only exceeded on 1 percent of cables, and binders filled with the worst case types of crosstalkers. This conservative practice denies some customers DSL service who could have otherwise have received it (false negatives), in order to achieve a low number of expensive unexpected failures (false positives). However, it fails to completely eliminate false positives, since it does not account for the many different factors that can cause failures such as high levels of radio ingress or impulse noise. Worse, many DSLs are set to transmit higher power than necessary, creating unnecessarily high levels of crosstalk, instead of responding properly to the actual impairments on each particular loop.

Advanced DSL management measures the loop, crosstalk couplings, and received noise on an individual basis. The measurements can identify pairs with crosstalk and noise well below the worst case, and systems on these pairs may transmit at higher bit rates or even longer distances than current practice. Failures may be predicted and stopped before they ever occur.

Properties of most copper loops are generally time invariant, so they can be measured at DSL frequencies and stored in a database. Such a database allows DSL service provisioning with high accuracy, relatively easy diagnosis of failures, and opens the door for future joint optimization of multiple DSLs and controlling crosstalk. Advanced DSL management can lower the cost of DSL provisioning and maintenance, while also providing a platform for future services.

Some infrastructure is needed for advanced DSL management, as shown in Figure 6.7. Data may be collected by installing automated test equipment in the central office (CO), or extracted from DSL modems and DSL access multiplexers (DSLAMs), which may need upgrading. There should be a communications path from the DSLAMs to a station that can access the database and analysis engine, as well as communications with existing software systems and databases. The DSL database will need to be populated and maintained. There is a cost for this. However, it can be shared over the many lines in a CO, and it should be considerably less costly than a brute force manual upgrade of the outside plant. Adding communications and knitting them together with intelligent algorithms and control creates a management system that is a *force multiplier*, leveraging the existing copper plant

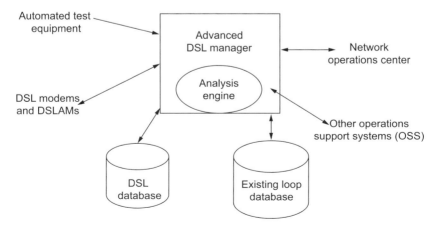

Figure 6.7 Infrastructure components of advanced DSL management [46]. (©2003 IEEE.)

and outside plant maintenace craft forces to obtain precise control over facilites and services.

A coordinated implementation of advanced measurement, database storage, analysis, and control of DSL loop and transmission parameters enables a *force multiplier* effect, leveraging existing copper by adding intelligence, control, and communications. This is less costly than physical plant upgrades, and is complementary as the management system can grow along with plant upgrades. Investment is needed for enabling more data exchange and control with DSL modems, populating and maintaining a DSL database, maybe new test equipment, and implementing an analysis system; but this should be recovered by virtue of savings on DSL provisioning and maintenance costs alone. Additional revenue from new service offerings, such as a 10 Mb/s symmetric enterprise service and video services, increases motivation.

Management of Location Information

Next-generation wireless cellular networks and services will demand much greater service differentiation and customization, which requires new intelligent location-based faster service provisioning techniques. Managing the location information itself will need new approaches, since many new differentiating services will rely on the location information.

The emergence of broadband and third-generation (3G) mobile communications, as well as the imminence of 4G systems, introduces an increasingly strong trend to renounce proprietary monolithic approaches and endorse more flexible modular architectures in communication network elements. Within that scope the concepts of reconfigurability and location information management are nominated as critical enablers for the introduction of ubiquitous services and applications to mobile users. Although reconfigurability research at its first steps focused primarily on the radio domain (RF processing, analog-to-digital conversion, etc.), currently a more innovative and forward-looking view is increasingly drawing interest. According to that, reconfigurability encompasses the entire service provision domain, extending from the mobile terminal through the network infrastructure to application and services. The most significant near-term impact of reconfigurability is likely to be in the field of service and application innovation, as a tool to allow rapid and flexible service customization and new degrees of operator differentiation. Furthermore, the importance of location information as a key enabler of service customization, network reconfigurability, and operator differentiation has motivated many telecommunication companies and institutions to develop or integrate positioning systems into their networks. The exploitation of location information presents a powerful new dimension to the range of information services that can be offered. By combining positional mechanisms with location- and mobility-specific user information, it is possible to offer truly customized personal communication services through various mobile terminals.

In that context, the convergence of the IP and telecom worlds enjoying the business models to be adopted in the new era to encompass the active participation of third-party value-added service providers (VASPs). The VASPs should be able to offer their value-added services (VAS) through the operator network under the respective business agreements. Incipient architectures enabling advanced business model support and flexible service provision comprise the specification of standardized open application programming interfaces (APIs) (e.g., the Open Service Access, OSA, and Parlay APIs) and frameworks. Such APIs hide the network heterogeneity that is likely to dominate the forthcoming mobile communications era by providing independence from the underlying network infrastructure to trusted third parties [47, 48].

Therefore, authorized third parties and VASPs can access network services, as well as develop and deploy their services and applications seamlessly, simply by using the standardized execution environment and the respective methods inherent to these APIs. The challenge for network operators is to attract and engage third-party application providers while protecting their networks from harmful misuse. The support of standardized APIs combined with the employment of intelligent service mediations is envisioned to enable various business entities by actively participating in flexible service provision, as well as to reduce the complexity involved in delivering applications developed by third parties over public switched and mobile networks.

Hence, the provision of an integrated framework for the support and management of flexible service provision, reconfigurability control aspects, and location information management in a coherent manner is a very important issue for the emergence of advanced service provision in 3G systems and beyond.

The highly demanding service provision features for future mobile communications impose the introduction of advanced mechanisms and components for the support of reconfigurability management, context awareness, and flexible service provision. Figure 6.8 depicts the general architecture and an example of physical placement of the reconfiguration control and service provision platform (RCSPP) to meet such requirements. It is assumed that the independent service provided by the underlying Universal Mobile Telecommunication Systems (UMTS) network.

The RCSPP architecture provides a framework for flexible service deployment, provision, charging, and management in a reconfigurable mobile environment. It can be viewed as an intelligent service middleware that mediates between VASPs and network operators in order to provide VASPs and end users with advanced context-aware services. The RCSPP takes into account location and mobility information for subscribers, along with the current terminal and network capabilities to perform reconfiguration actions on the network nodes and end user equipment. Hence, service provisioning offered by the RCSPP is adapted and tailored to context attributes, providing intelligent personalization and customization in service offering.

The functional architecture of RCSPP incorporates the reconfiguration control and service provision manager (RCSPM), charging, accounting, and billing system (CAB) and metering devices (MDs). These components have high involvement in the overall service provision chain. Although business roles may be integrated into one entity (e.g., the RCSPP provider and network access provider business roles may be undertaken by the mobile operator), we see that the business domains

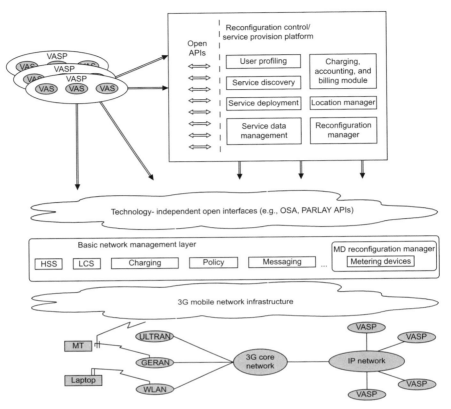

Figure 6.8 General architecture of the reconfiguration control and service provision platform (RCSPP) framework [49]. (©2003 IEEE.)

can also be physically separate. More specifically, the RCSPM is the main entity responsible for managing context-aware service provision and controlling the respective reconfiguration actions. It may reside in a third-party domain and may coordinate the required procedures for dynamic deployment and personalized, consistent, and reconfigurable discovery, provision, and execution of VAS to mobile users. For that reason, it maintains a database with information about the services offered by the platform (the service database), as well as user profile data. It hosts the reconfiguration and location manager modules. The former is responsible for interacting with other components of the proposed architecture (e.g., MDs) in order to configure them properly as well as to perform reconfiguration actions on them. Reconfiguration actions are based on policies, tailored to user location, terminal capabilities, user preferences, and usage data. On the other hand, the location manager interacts with the location information's sources of the underlying network infrastructure, such as the location service (LCS) server or the presence server, to track the location and mobility subscribers. Hence, it enhances the service provisioning approach adopted by the RCSPP with location information features, enabling better customization and personalization of the whole service offering [50].

Administratively, the RCSPP may be situated on and managed by entities independent of the operator network. Therefore, interactions among the modules of the platform and basic network elements should take place through open network interfaces. To this end, various industry initiatives, such as OSA and PARLAY, address the introduction and standardization of open network interfaces to third-party providers that enable, for example,

- retrieval of location and presence information, as well as registration for specific location- or presence-sensitive event notifications by the LCS and presence servers of the network operator
- retrieval of user-specific information kept in the home subscriber system (HSS) of the network operator
- the transmission of simple or multimedia messages to mobile users through the respective messaging server
- access to the charging service feature of the network for addition of charges to subscriber's bills related to content or products consumption by respective subscribers
- access to the policy management infrastructure of the mobile operator for the application of context-aware policies to the underlying network infrastructure.

Common to all these architectures is the provision of a basic network management layer by the mobile operator, which acts as a gateway (mediator) among third parties and provides basic network services. Hence, in order for the RCSPP to access the network services offered by the underlying network infrastructure, interaction with the basic network management layer provided by the mobile operator is required. By accessing the standard open network interface provided by the basic management layer, the RCSPP is then able to build its services based on network features and functionality offered by the network operator. In that context, policies toward various reconfigurable components can be applied (e.g., MDs).

The coordination of MDs is under the supervision of the MD reconfiguration manager, which is responsible for the policy-based configuration and reconfiguration of MDs. Since MDs are likely to be included in the standard network infrastructure, access to reconfiguration of MDs should be offered to the RCSPP and other authorized entities (e.g., third-party VASPs) through a standardized open API provided by the MD reconfiguration manager. The dynamic (re-)configuration of MDs can then be performed by the RCSPP-based policies for processing and monitoring the IP traffic as well as metering data about resource consumption in the network (e.g., the transmitted volume). In the presented architecture, the MDs are located at the edge of the network access provider so that they process all traffic between VASPs and end users. The respective metering data collected by the MDs are formatted into appropriate records, VAS data records (VASDRs) and sent to the CAB for further processing. The VASDRs, apart from calculating charges, are also used by the MDs and are formatted into appropriate records, VAS data records (VASDRs), and sent to the CAB for further processing. The VASDRs, apart from calculating charges, are also used by the CAB for retrieval of statistical information, such as the VAS a specific user currently executes.

A very important feature of the platform is the management of location information and support of location-based reconfiguration. The MD reconfiguration manager mediates between authorized entities and MDs to securely grant MD services to the respective parties. Since reconfiguration of MDs is policy-based, communication with MDs is built on the common protocols for policy provisioning such as Common Open Policy Service (COPS) or COPS-PR (COPS Usage for Policy Provisioning). The MD reconfiguration manager maps all incoming calls on the aforementioned open interface to the appropriate COPS/COPS-PR messages for interacting with MDs.

A very important feature of the platform is the management of location information and support of location-based reconfiguration policy provision. The location manager can be considered as an internal functional entity of the RCSPM responsible for retrieving, managing, and exploiting information related to the location and mobility of subscribers. This is performed in a generic way, enabling the location manager to interact with the location information's sources in the underlying network infrastructure (e.g., the LCS or presence server), independent of the type of network, in order to track the location and mobility of subscribers. Then, location and mobility data, along with the preferences of the corresponding subscriber taken from his/her user profile, are processed properly to form new advanced location-aware services and policies [51]. By spreading the location information to any authorized entity internal and external to the RCSPM, the location manager enriches the service provisioning approach adopted by the RCSPP with location information features, enabling better customization and personalization of the whole service offering. Location tracking takes place with respect to the user privacy settings included in the user profile [52]. Hence, the location manager prompts the involved user for authorization prior to any location retrieval action, if his/her profile requests this.

Resource Management for Quality of Presentation (QoP)-Based Multimedia Applications in Mobile Networks

Next-generation wireless networks and services will be drastically more complex than today's so-called second-generation (2G) systems. Global roaming and multimedia will be the key driving factors that will have a profound impact on how these networks will be managed. New technologies, new networking schemes, and new services will be part of this changing environment. In offering new services, operators will need to employ more powerful and efficient but more user-friendly ways of managing the underlying networks. Evolution, not revolution, will be the dictating factor. Not only will network elements and communication devices evolve, but also the management system and the way of managing.

Groundwork has been completed over the past several years for third-generation (3G) cellular networks and services that will achieve global roaming and data everywhere at higher speeds (344 kbps and higher). Owing to economic realities, there is an industry-wide consensus that deployment of 3G networks will be delayed.

Cellular networks are not the only game in town anymore. Long-awaited wireless LANs (WLANs) are finally here, and their success is overshadowing 2.5G and 3G-based efforts. Technical magazines and conferences are overcrowded with papers and talks comparing WLANs and 3G. This is based on the phenomenon that WLANs are being deployed not only in residual and small offices, but also on a large scale in enterprises and public areas such as airports, malls, and neighborhoods. Carriers are struggling to take advantage of this by offering public WLAN services and linking them and integrating them with their own infrastructure. In an integrated environment where a large number of wireless local area networks (WLANs) and cellular networks cooperate, Mobile IP, which has become the protocol of choice to support mobility, needs to be enhanced to support high micromobility with fast handoff support.

The existing wireless network infrastructure and communication protocols provide mobile users limited Internet-based services including Web browsing, short message service (SMS), and e-mails. This is mainly because of the low bandwidth capacity and high error rates associated with wireless networks. However, with rapid advances in wireless technologies and the emergence of third-generation (3G) systems that can support data rates up to 2 Mbps, the service spectrum for mobile users is being widened to include novel multimedia applications such as e-commerce, distance learning, videoconferencing, digital libraries, online TV/radio, and many others.

The future network system can be envisioned as a global network with a high degree of heterogeneity, providing advanced Internet services to users, irrespective of their points of connectivity to the network. To accommodate mobile users, a wireless network must provide continuous and mobility-independent connectivity [53]. Mobility is a generic term and does not fully characterize users' roaming behavior. For instance, a stockbroker can have high indoor mobility in the stock exchange building. On the other hand, commuters driving on interstate highways have long-range mobility, and passengers on a plane have global mobility. The future network system is expected to integrate all these levels of mobility and will provide ubiquitous access to users as depicted in Figure 6.9.

Conceptually, a multimedia mobile network is similar to a cellular communication system in which a given geographical area is divided into multiple cells. Each cell has its own base station that communicates with the mobile users in that cell on an RF channel. The base stations can be connected to a backbone land-based network consisting of high-speed routers/switches and multimedia database servers [54]. Like cellular phone systems, a user in this network can roam in any direction and migrate to any of the neighboring cells. Typically, the lifetime of a multimedia session involving browsing through a multimedia Web environment can be greater than the residence time of a mobile user in a single cell. In other words, a mobile user subscribing to a multimedia service in some cell is likely to migrate to a different cell during a session. In that case, the land-based route needs to be established to the new cell so that the multimedia session can continue.

572 NETWORK LAYER

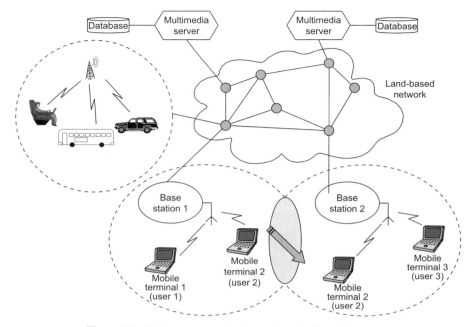

Figure 6.9 Heterogeneous wireless network with mobile users.

However, if sufficient RF bandwidth resources are not available in the new cell, the session may be terminated. To avoid such handoff failures, admission control must be performed based on the long-term availability of resources.

Another major challenge is the characterization of the bandwidth requirements of multimedia documents. Different objects such as video clips, text, and images in a multimedia document can have different bandwidth requirements. Depending on the concurrency level of objects, the quality parameters associated with an individual's multimedia objects, the overall bandwidth profile of a multimedia document may change considerably over a period of time. Accordingly, the resource requirements also vary. In order to provide quality-based multimedia services, the underlying network must accommodate such changes by allocating bandwidth resources in an efficient and timely manner.

The support of multimedia applications over a heterogeneous network requires communication protocols and networking infrastructure capable of providing high-quality services. Prior to connection establishment, such services need to specify connection requirements in terms of network resources, and the duration for which these resources are needed. Multimedia applications in general are not monolithic in nature, and can consist of several objects integrated together such as video clips, images, text, and audio segments. A multimedia application consisting of multiple objects can be represented as a multimedia document [55]. An example of a multimedia document is an MPEG-4 based application. MPEG-4 encompasses all types of media, and uses scene and object descriptors (SDs and ODs) to define the spatiotemporal features of the component media objects. To support multimedia document applications with guaranteed QoP, resource must be reserved at each node along the land-based route and the base stations that serve as an interface between the land-based network and mobile users. For simplicity of discussion, we assume that the land-based network is a resource-sufficient system in the sense that it has enough resources to guarantee QoP required by the multimedia application. The resource management problem is then primarily confined to the base stations that transmit multimedia data to mobile users. We assume that the reverse channel from mobile user to base station is primarily used for communicating control signals for browsing and requires negligible bandwidth.

To synchronize multimedia objects within a document, their data streams can be divided into fine-grained data units. The smallest unit is referred to as a synchronization interval unit (SIU). For example, the synchronization interval for a video object can be taken as $1/30$ of a second, which corresponds to the playout duration of a single video frame. For audio data, the smallest unit can be an audio sample.

For synchronization, one option is to cache enough multimedia data to compensate for a slower transfer rate. However, for mobile devices, providing a large buffer for multimedia data caching may not be feasible. Owing to limited buffering capability and restricted bandwidth capacity at the base station, objects can be delivered partially by dropping some SIUs at the base station. For concurrent object data streams, determining the number of SIUs or each object to be dropped by the base station is equivalent to distributing some data reliability penalty among these objects. Criteria for such a decision can be based on the user's specified reliability parameter for multimedia objects. This parameter indicates the maximum acceptable loss of SIUs for different multimedia objects so that compensation for slow data streams can be provided.

The traffic load at each base station can change dynamically due to various factors, such as the number of connections concurrently served by the base station, the changing level of concurrency of objects in a document, requests for new connections, and the migration of connection from/to other cells.

Newly originated connections are less sensitive to initial setup delay. Therefore, they can be handled in such a way that other ongoing sessions are not disturbed by them. Depending on the availability of RF channel bandwidth, the possibility of denying a new connection request cannot be ruled out. When a base station receives a request for a new connection, the request can be defined if sufficient bandwidth is not available to accommodate it without degrading the presentation quality of the already established connections beyond an acceptable level.

Another factor that determines the channel requirements at the base station is the migration of users among cells. In order to avoid handoff drops, the new cell must accommodate the migrating connection. One possible way to handle a handoff connection is to treat it as a new connection request and invoke a bandwidth assignment procedure in the new cell. However, a migratory connection may be terminated if sufficient resources are not available. A viable solution to this problem is to reserve enough resources for mobile users in advance in all the cells to be visited by a mobile user during his/her connection lifetime. Advance reservation of resources in multiple cells for the entire duration of a multimedia session can be achieved using *a priori* information of the mobile user's arrival and departure times in each cell. Such a mobility profile of a user can contain estimated time of arrival and departure within a cell, using already known information such as the connection initiation point, size, structure, and geographical location of cells, and the maximum speed of mobile users.

Mobility Management in a 4G System

While 2G was mainly designed for voice and 2.5G for packet-switched data, 3G needs to support multimedia application in addition to all the services of previous generations. The convergence of services and delivery platforms in the ongoing enhancement of 3G wireless will lead to more intelligent use of the communications media, where 3G will be able to offer users what they need in any specific mobile environment. The range of applicability of 3G wireless is very much wider than earlier mobile systems and is expected to include future enhancement, which will offer increasing capability and performance in low-mobility environments.

Systems beyond 3G (or 4G mobile) will support a wide range of data rates according to economic and service demands in multiuser and multicell environments, with terminals moving at vehicular speeds, and support 50–100 Mbps maximum. 4G mobile will support a wide range of symmetrical and asymmetrical services, and also provide QoS for real-time services and efficient transport of packet-oriented services, as well as efficient support of broadcast and distributed services. Future mobile networks will mainly be characterized by a horizontal communication model, where different access technologies such as cellular mobile, broadband wireless access, wireless LANs, short-range connectivity, and wired systems will be combined on a common platform to complement one another in an optimum way for different service requirements and radio environments.

Mobility management is very important in 4G wireless network systems. Mobility usually involves mobile users roaming among different network segments. Link-layer mobility support is usually restricted to homogeneous networks, while network-layer mobility support is provided for any kind of network without regard to the link-layer techniques employed. Moreover, Internet mobility support is one of the main trends. Hence, it seems very natural to think that network-layer mobility will be a key issue in mobility management in 4G systems. An example for 4G core network management is shown in Figure 6.10.

Mobile IP represents a simple and scalable global mobility solution. However, it lacks support for real-time location management and a fast and seamless handoff mechanism. Additionally, QoS in the mobile IP environment using `DiffServ` and/or `IntServ`/RSVP needs to be addressed.

When a mobile node moves among networks, networks must update the mobile node's location and its routes. Mobile IP provides a simple and elegant location registration and update scheme. However, it lacks support for real-time location tracking and optimal route selection.

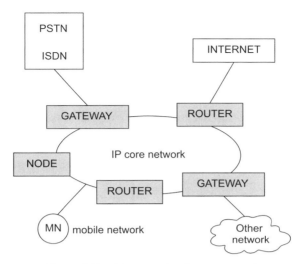

Figure 6.10 4G core network management.

Mobile IP does not provide a fast and seamless handoff scheme. Therefore, a mobile node experiences data losses and delays during the handoff process. For real-time multimedia applications and reliable transport protocols, those data losses should be avoided and delays should be reduced.

6.3.7 Transport Network Layered Architecture

The capacity of transmission systems has been increasing rapidly with the advancement of recent technology. Transport node throughput has also been enhanced with the development of SDH/SONET and ATM technologies. New services need to be developed and provided through the network in pace with the increasing demands. A telecommunications network and the services provided by the network should gracefully evolve to incorporate the advances in technologies and to respond to diverse demands.

The increased transmission capacity, on the other hand, causes severe damage on single transmission link/system failure. The development of effective network restoration techniques against such network failures is, therefore, becoming more and more important. Moreover, the increasing volume of data transmission between computer systems urges rapid network restoration to minimize social economic loss. The cost of the network operations needed to satisfy the requirement, however, must be minimized. Flexibility of the network is another key issue, since it permits a variety of cost-effective end-customer controls and provides the network with adaptability in the face of unknown service requirements. A transport network architecture supports rather diverse demands including those already mentioned by layering the network on the basis of network functionality. This layered transport network architecture simplifies the design, development, and operation of the network, and allows smooth network evolution. The layered concept also makes it easy for each network layer to evolve independently of the other layers by capitalizing on the introduction of new technology specific to each layer. The layering concept has been extensively discussed within ITU-T for the SDH transport network [56].

Figure 6.11 illustrates a layered network structure for a public telecommunication network. It comprises a service network layer and a transport network layer. The service network layer consists of different service networks, each of which is dedicated to a specific service. This layer provides circuits or channels. A transport network is realized with paths and physical transmission media, and is less service dependent. Thus, the telecommunication network can be divided into three layers from the viewpoint of functions: a physical media layer, a transmission path layer, and a communication circuit layer.

In connection-oriented communication, an end-to-end connection is established/released dynamically on the basis of short-term provisioning among users via service nodes. In connectionless communication, a block of data (packet, cell, frame, etc.) is transferred to the neighboring service node without establishing a connection. The circuit layers are dedicated to a specific service such as the public switched telephone service, packetized data communication service, frame relay service, and ATM cell relay service.

The transmission media, which interconnects nodes and/or subscribers, is constructed based on long-term provisioning; geographical conditions are taken into consideration. The point-to-point transmission capability has increased significantly during this decade through the introduction of optical fiber transmission; 10 Gb transmission systems are now being introduced. However, traffic demands between two service nodes are not always large enough to fill up the bandwidth of a high-speed transmission system that connects the service nodes directly.

The path layer bridges the circuit and transmission media layers and provides a common signal transport platform supporting different service-dependent circuit

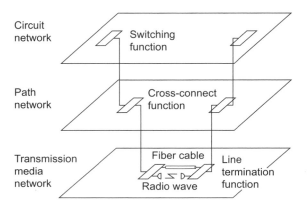

Figure 6.11 Layered public telecommunication network structure [57]. (©1994 IEEE.)

networks. A path can be a grouped circuit serving as the unit of network operation, design, and provisioning, and can be an object to be manipulated for restoring node and transmission line/system failures. Here, a path used as the unit of a group of connections between service nodes (or node systems) is called a service access path.

Path layer functions can be performed with intelligent ADM (add/drop multiplexer) systems and/or digital cross-connect systems and control systems. Network flexibility can be enhanced with path layer control.

Transport Mechanisms

There are a number of possible transport solutions depending on the method of transmission, including the following.

MPEG-2 Systems Part 1 of the MPEG-2 standard [58] defines two methods of multiplexing audio, video, and associated information into streams suitable for transmission (program streams or transport streams). Each data service or elementary stream (e.g., coded video or audio sequence) is packetized into packetized elementary stream (PES) packets.

PES packets from the different elementary streams are multiplexed together to form a program stream (typically carrying a single set of audio/visual data such as a single TV channel or a transport stream), which may contain multiple channels is shown in Figure 6.12.

The transport stream adds both Reed–Solomon and convolutional error control coding and so provides protection from transmission errors. Timing and synchronization is supported by a system of clock reference and time-stamps in the sequence of packets. An MPEG-4 visual stream may be carried as an elementary stream within an MPEG-2 program of transport stream. Carriage of an H.264/AVC (MPEG-4 Part10) stream over MPEG-2 systems is covered by Amendment 3 to MPEG-2 systems.

Real-Time Protocol (RTP) is a packetization protocol that may be used in conjuction with the User Datagram Protocol (UDP) to transport real-time multimedia data across networks that use the Internet Protocol (IP) [59].

UDP is preferable to the Transmission Control Protocol (TCP) for real-time applications because it offers a low-latency transport across IP networks. However,

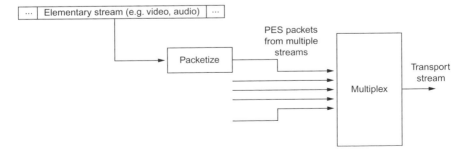

Figure 6.12 MPEG-2 transport stream.

| Payload type | Sequence number | Time stamp | Unique identifier | Payload (e.g. Video Packet) | ... |

Figure 6.13 Simplified RTP packet structure.

it has no mechanisms for packet loss recovery or synchronization. RTP defines a packet structure for real-time data as shown in Figure 6.13, which includes a type identifier (to signal the type of CODEC used to generate the data), a sequence number (essential for reordering packets that are received out of order), and a timestamp, which is necessary to determine the correct presentation for the decoded data. Transporting a coded audiovisual stream via RTP involves packetizing each elementary stream into a series of RTP packets, interleaving these, and transmitting them across an IP network (using UDP as the basic transport protocol). RTP payload formats are derived for various standard video and audio CODECs, including MPEG-4 Visual and H.264. The network abstraction layer (NAL) structure of H.264 has been designed with efficient packetization, since each NAL unit can be placed in its own RTP packet.

MPEG-4 Part 6 defines an optional session protocol, the Delivery Multimedia Integration Framework, which supports session management of MPEG-4 data streams (e.g., visual and audio) across a variety of network transport protocols. The FlexMux tool (part of MPEG-4 Systems) provides a flexible, low-overhead mechanism for multiplexing together separate elementary streams into a single, interleaved stream. For example, this may be useful for multiplexing separate audiovisual objects prior to packetizing into MPEG-2 PES packets.

6.3.8 Multicast Protocols Classification

For point-to-point (unicast) communication, the Transmission Control Protocol (TCP) has dominated the Internet landscape for many years [60]. TCP provides absolutely reliable service unless the underlying network fails. It assumes the user requires that all data be delivered, in sequence.

To provide multicast services (point-to-multipoint and multipoint-to-multipoint), a variety of protocols is available. However, when considering reliable multicast equivalents to TCP, the situation is vastly different from the unicast situation, as the number of possible failure modes is larger, and the definition of *reliability* can take on many shades of meaning. Each protocol proposed in the literature is intended to meet the needs of a particular set of applications. Each has a slightly different definition of reliability, and each operates in a slightly or significantly different environment. Given the wide variety of requirements and environments, there will never be a *one size fits all* multicast protocol design. However, this does not mean it is impossible to develop a family of protocols (or a single protocol with a variety of selectable features) that satisfies a wide range of requirements [61]. Traditional protocol design has started with a statement of the requirements, and followed with the exploration of possible designs that can meet these requirements.

Multicast applications vary in their requirements along a number of dimensions [62]:

- number of senders
- group organization and receiver scalability
- data reliability
- congestion control
- group management
- ordering.

These parameters all interact with one another: for example, the data reliability requirements that can be met are dependent on the receiver scalability required.

The first dimension of our classification is based on the number of senders: single-sender (point-to-multipoint) and multiple-sender (multipoint-to-multipoint) applications. Point-to-multipoint or 1-to-N multicast applications require data delivery from a single source to multiple receivers, and usually run without human interaction. A few example applications for this set include software distribution, data distribution and replication, and mailing list delivery. Multipoint-to-multipoint or M-to-N applications require data delivery from multiple sources to multiple receivers. They may also have a requirement to order the delivery of the data from the multiple senders. In most cases, an M-to-N protocol operates in M-to-M mode (all senders are also receivers). Example applications include interactive distributed simulations, networked gaming, and other collaborative applications such as teleconferencing and videoconferencing.

Multicast protocols have very different characteristics depending on the number of senders and the number of receivers. We will use the categories *one*, *few*, and *many* to describe these two dimensions. Few is less than 20. Many can be several hundred or several thousand. One-to-one is unicast transmission. Most collaboration applications fall in the few-to-few category. Given the nature of human interaction, there is a practical limit on the number of users who can simultaneously carry on a useful session. Hence, it is unlikely that such an application would ever require the protocol to be scalable to even 20 users. One-to-few and one-to-many represent data distribution applications. One-to-few applications are relatively simple to design, but one-to-many applications may require communication support for groups of hundreds or thousands of members, which may undergo a very high number of group changes per second. Few-to-many could be a collaboration application with a small number of senders and a large number of receivers [62]. We differentiate four levels of scalability: small groups with few members; medium groups, where the group is on a single LAN or within a single administration, and the cost of multicasting from a receiver is small; large groups, where there is geographical distribution, or the cost of multicasting from a receiver is high; and enormous groups, where the losses on separate branches of the distribution tree are uncorrelated, or there is *no* reverse path to provide feedback for error control.

There are many *reliable multicast* protocols, each of which has its own view of what data reliability is (or how important certain data reliability features are). The mechanisms used in a protocol will be determined to a great extent by the actual data reliability requirements. For unicast transmission, the notion of data reliability is simple: there is only one recipient, and that recipient either does or does not receive all the data sent by the sender. For multicast transmission, the situation is much more fluid, given that there are many more ways in which the communication can fail. For small groups with a single sender, a very strong meaning of reliability can be assigned, similar to the meaning assigned for unicast transmission using TCP: communication is reliable if all data provided by the sender are received, in order and without duplication, by the receivers. For multiple senders, this definition must be applied to each individual sender. The test of reliable reception can be made by the sender or receiver. The term *strong* reliability is used for this case. For large groups, a relaxed definition of data reliability must be accepted, because the test can be applied only by the receivers, to avoid the possibility of overwhelming the sender(s) with status-response packets. This also implies weak or nonexistent knowledge at the sender(s) about the membership of the group and the progress of data reception by these members. In reference [63] this is called semireliable delivery, but weak reliability is the term often used.

Congestion control is one of the most difficult requirements to satisfy in multicast due to its multipoint nature. Uncontrolled propagation of multicast data by even a few sources can potentially cause havoc on large-scale Internet works. Thus, for reliable multicast to be accepted and embraced, it must address congestion control requirements in an efficient manner. Multicast protocols must also coexist with more common unicast protocols such as TCP. The need for multicast congestion control can be expressed in two forms. The first is to require maintenance of the same speed for all receivers, in the presence of congestion. The second is to permit certain receivers to lower their rate requirements on experiencing congestion. Congestion control uses evidence of packet loss as its signal to reduce traffic flow. Since packet loss is also indicative of the need to repair, congestion control will often be integrated with error recovery, using the same mechanisms.

One way to express the requirement for group management is to define the set of conditions that must be true for a transport connection to enter or remain in the data transfer phase. The term *active group integrity* (AGI) identifies the necessary conditions in accordance with various levels of group reliability. Multicast group management deals with managing the group in accordance with the AGI. The level of reliability provided by a multicast protocol depends on the control algorithms and group policies used to manage the group. In keeping with our use of data reliability, we use the terms *weak* and *strong* for different protocols that make some effort to manage the group membership from those that ensure complete knowledge, supplemented with the use of one for those that provide no support at all.

Few-to-few applications often carry with them a requirement for global ordering. This applies only to applications with multiple senders, since a single-sender application only has one possible ordering. A protocol serving this class of applications will typically provide at least mechanisms for ensuring complete ordering among the packets issued by the session members.

Packets may arrive out of sequence at their destination due to packet losses or changes in the datagram routes. Many distributed applications require ordered reception of packets, because misordering may give a different view of the state of the group. Ordered reliable multicast communication among a set of users is defined based on how the objects of a sending user are presented to the receiving user and how the receiving user gets objects from the sender.

In the case of single-sender ordering, the objects generated by the sending user must be delivered to each receiving user in the active group in the same order in which they were sent. In the case of multiple sending users, ordering is determined in terms of the relative sequencing of objects received from multiple sending users. The ordering relationship defines the arrangement of interleaving of objects from multiple senders, and can be classified as no ordering, local ordering, causal ordering, partial ordering, or total ordering.

A protocol that ensures, in addition to reliability, a total ordering of the delivered messages is called an *automatic* protocol. Such a protocol may be used for reliable validation, atomic operations, group memberships, and so on, while a causal protocol (ensuring causal and not total ordering) may be used for ensuring consistency in updates to replicated data.

UDP provides no ordering guarantees. (In fact, it provides no guarantee that the packets will be delivered at all.)

All *reliable* multicast protocols must, of necessity, provide local ordering; guaranteeing that a sequence of packets will be delivered requires that the packets be sequenced (ordered). Partial, causal, and total ordering are meaningful only when multiple senders are allowed, and reasonable only when there are few senders.

Examples of Multicast Protocols

We will present some examples of multicast protocols that will illustrate the range of requirements and solutions that have been adopted.

User Datagram Protocol (UDP). This protocol provides an end-to-end variant of the network-level best-effort service, with no error recovery mechanisms, group control, or ordering. There can be an arbitrary number of senders and receivers, as long as the group address is known.

Xpress Transport Protocol (XTP). This was originally introduced to be a network and transport layer protocol, with unified multicast and multicast features [64]. The design was pure mechanism, and was based on a set of independent functions (error control, flow control, rate control, priorities, etc.) that could be individually selected. With the introduction of Revision 4.0, the name was changed to Xpress Transport Protocol, and the multicasting features were considerably strengthened. Revision 4.0b further strengthened the multicast features. All revisions allow only a single multicast sender. We use the short forms XTP-3 and XTP-4 for XTP up to revision 3.6 and XTP since Revision 4.0, respectively [65].

XTP is a single-sender (point-to-multipoint) protocol, intended to serve small to medium groups. XTP-3 provides weak reliability (because it uses slotting and damping to suppress redundant error reports), while XTP-4 provides strong reliability. XTP-3 has relatively weak group management capabilities, while XTP-4 provides strong *mechanisms* for group management.

The error reports are multicast in XTP-3 and unicast to the sender in XTP-4. Status is only reported when the sender explicitly asks for it. Error recovery is achieved by resending the missing packet(s); the policy is sender-reliable.

There are no explicit congestion control policies in the specification, although all implementations provide this capacity.

An XTP multicast group is normally formed when an invitation to join is issued by the sender. This invitation is a FIRST packet with the multicast bit set in the header. The receiver responds to the invitation with a JCNTL (join control) packet (XTP 4.0b and later), and the sender acknowledges the member with another JCNTL. This JCNTL has specified fields to negotiate a service. It is possible to join late by multicasting a JCNTL to the group.

Receivers depart by exchanging the CLOSE bits in the header, or can be forced to leave with the END bit in the header.

There are no *policies* for managing the group; this is left to the application.

Reliable Multicast Protocol (RMP). This protocol provides a totally ordered, reliable atomic multicast service on top of an unreliable multicast service such as IP multicasting. RMP provides a wide range of guarantees, from unreliable delivery to totally ordered delivery, to K-resilient, majority resilient, and totally resilient atomic delivery.

Each packet is marked with a unique identification, and multicast to the entire group. Negative acknowledgements (NACKS) are used for faster error recovery, and a limited number of ACKs is used to guarantee delivery. These ACKs are sent by a *token site* that constantly rotates among the members of the multicast group, thus making the protocol fully and symmetrically distributed so that no site bears an undue portion of the communication load.

Responsibility for managing join and leave rests with the current token site, and a new member becomes the next token site to ensure that it has really joined the token ring.

RMP provides a windowed flow and congestion control algorithm that allows RMP to provide high performance over both LANs and WANs, even in the face of congestion.

Scalable Reliable Multicast Protocol (SRM). This protocol is a reliable M-to-N multicast framework that has been designed for lightweight sessions and application-level framing. The algorithms of this framework are efficient and robust, and scale well to both very large networks and very large sessions. SRM has been prototyped in the distributed whiteboard application (WB). The design of SRAM is based on the application-level framing (ALF) concept, which says that the best way to meet diverse application requirements is to leave as much functionality and flexibility as possible to the application. Therefore, SRM is designed to meet only the minimum definition of reliable multicast; that is, eventually, delivery of all data to all the group members, without enforcing any delivery order [66].

SMR is a multiple-sender (multipoint-to-multipoint) protocol serving large groups. It provides weak data reliability and minimal group management. When a receiver detects a hole in the data stream, it multicasts a repair request to the entire group. However, to avoid implosion as well as identical requests for the same data, SMR uses the slotting and damping mechanisms used by XTP-3. Receivers delay themselves randomly before sending a repair request, and refrain from sending if a similar request is sent within the delay time. As with the original data, repair requests and retransmissions are always multicast to the entire group. To avoid transmitting duplicate repairs, hosts that wish to transmit a repair delay themselves randomly before transmitting the repair. The delay time is a function of the distance in seconds between the member that wishes to send the repair and the one that triggered the request or repair. Thus, it is more likely that a host closer to the point of failure will time out first and multicast the request [66].

Reliable data delivery in SRM is ensured as long as each datum is available from at least one member. This has the advantage of reducing the buffering requirements on all members within the communication session.

Pragmatic General Multicast (PGM). This is a reliable multicast transport protocol for applications that *require ordered or unordered, duplicate-free, multicast data delivery from multiple sources to multiple receivers.* (Note that the ordering referred to in this quotation is single-source ordering; multiple sources are considered by PGM to be independent.) PGM guarantees that a receiver in the group either receives all data packets from transmissions and repairs, or is able to detect unrecoverable data packet loss. PGM is specifically intended as a workable solution for multicast applications with basic reliability requirements. Its central design goal is simplicity of operation with due regard for scalability and network efficiency [67].

PGM uses hierarchical receiver-oriented error recovery. Under the premise that NACKs are for repair and ACKs for buffer management, the choice is made to use NACKs exclusively, and use something else to decide when the buffer expires. (The transmit window is advanced by a source according to a purely local strategy.) The NACKs are unicast towards the network element that the multicast data are coming from, aggregated and forwarded up towards the source, unless the repair data are provided by a *designated local repairer.* NACKs from other receivers on the same subnet are suppressed through receipt of NACK confirmations, which are multicast to the subnet. Since there is no concept of group management in PGM, there are no group management mechanisms or policies. Similarly, there are no policies for multisender ordering.

Local Group Based Multicast Protocol (LGMP). This supports reliable and semi-reliable transfer of both continuous media and data files. The protocol improves scalability and performance by using subgroups (also called *local groups*) for local ACK processing and error recovery, as defined by the local group concept.

Local groups are formed by dynamic organization of receiver sets and managed in turn by a special receiver called the *group controller* (GC), which handles status responses and coordinates local retransmission. The selection of appropriate receivers as GCs is based on the current state of the network and the receivers themselves. This process is not part of the LGMP itself, and is implemented by a separate configuration protocol called Dynamic Configuration Protocol (DCP).

LGMP is a single-sender (point-to-multipoint) protocol, intended to serve large groups. It provides reliable and semireliable transfer.

If a member of a local group does not receive a data unit, local error recovery is first attempted. In this case the GC will send a SNACK to its parent, indicating that the packet has been received in the group, but not by all members. The local recovery is managed by the GC, which will unicast or multicast the repair, depending on how many members of the subgroup have indicated data loss.

LGMP defines two different modes of performing local retransmissions: *local-sensitive* and *delay-sensitive*. It is up to the application to choose the appropriate mode according to its requirements. It is also possible for different CGs to operate in different modes.

LGMP separates the signal for indicating congestion from the algorithm for congestion control. It provides mechanisms to detect network congestion based on the status reports of each receiver, but also it is up to the application to choose the algorithm to deal with congestion.

Instead of integrating mechanisms into LGMP to define local groups, and establish and maintain these logically structured group hierarchies, a new protocol named the Dynamic Configuration Service (DCS) that performs these tasks has been defined [68].

Reliable Multicast Transport Protocol (RMTP). This protocol permits a single sender to deliver bulk information to a large community of users [69]. A new version of this protocol called RMTP-II improves RMTP by providing for multiple senders and introduces a number of new features to meet the needs of emerging applications [70].

The primary requirement for RMTP and RMTP-II is receiver scalability. RMTP provides indefinite availability of data through the use of a two-level cache. RMTP-II provides only bounded time reliability. Within the bounded time, RMTP-II provides a guarantee of delivery. The RMTP sender has no knowledge of the set of receivers; RMTP-II provides mechanisms to ensure reporting of the set of receivers.

An RMTP sender divides the data to be sent into fixed-size packets and transmits them using the global multicast tree. Every packet is assigned a unique sequence number that defines the overall order before they are multicast to the group.

Designated receivers learn of missing packets from the ACKs unicast to them by the receivers in their local region. The ACK packets contain the sequence number of the first in-sequence packet not received by the receiver, plus an ACK bitmap indicating which out-of-sequence packets have been successfully received.

RMTP uses a rate-based windowed flow control mechanism to avoid overloading slow receivers and links with low bandwidth. RMTP also implements the slow-start algorithm to prevent a sender from flooding an already congested network.

6.3.9 Routing Procedure for Multimedia Communications

Path selection has long been one of the most important issues in network design and management, which is also referred to as routing. It directly impacts network throughput. In addition to finding a physical route containing sufficient bandwidth (or available wavelength channels), the design objective of the path selection algorithms is to spread traffic flows homogeneously onto the whole network while avoiding resource fragmentation. Most reported schemes take routing as a task that is sequentially performed to simplify the path selection process [71].

In general, routing is to select a physical route through which the lightpath traverses in the *horizontal* direction, while wavelength assignment is an effort to decide which wavelength plane the lightpath is to take in the *vertical* direction. When the network infrastructure contains multiple fibers in a directional link (a multifiber network), a single wavelength plane for each directional link contains multiple interchangeable wavelength channels in the path selection process. Each wavelength plane is not interchangeable due to the wavelength continuity constraint. In such case, a dilemma in traffic engineering emerges: the effort to *spread* working capacity can horizontally balance the load distribution on a wavelength plane; however, resource fragmentation may be induced among different wavelength planes. On the other hand, the effort to vertically pack working capacity can reduce resource fragmentation among different wavelength planes; however, the horizontal load balancing characteristic may be lost [72].

The path selection process taking the changing link state into consideration is called *adaptive routing*. According to whether or not the global network link state is considered in the decision-making process, a routing scheme can be either fully or partially adaptive.

Fully adaptive routing can greedily find the networkwide least cost based on the dynamic link state and a custom-designed cost function. The most commonly used adaptive routing scheme takes Dijkstra's shortest path first algorithm, which has a reasonable computational complexity, $O(N^2)$, where N is the number of network nodes. To perform fully adaptive routing in a fully wavelength-convertible network, the problem is equivalent to the routing process in conventional communication networks without any buffer, where wavelength channels along a directional link are interchangeable and can be treated as a bulk of bandwidth. However, in the event that networks have no wavelength conversion capability or are only partially wavelength-convertible, the use of the layered wavelength graph approach is necessary before Dijkstra's algorithm can be applied. With this, the tasks of routing and wavelength assignment are completed in a single step.

Fully adaptive routing suffers mainly from the following four problems. First, a stale link state may result in a conflict of network resources during resource reservation between two lightpaths. In the optical layer, this problem can be solved by the holding/preemption parameters in the path setup message defined in the Resource Reservation Protocol (RSVP). However, provisioning dynamicity may be impaired due to the delay, which degrades the quality of service. The second problem is scalability in link state dissemination. The flooding of the link state from every node to the whole network may congest the control plane and disturb the other signaling mechanisms, especially in the event that flooding is performed at the same time by each node. The third problem is that the fully adaptive scheme can hardly take the path-based metrics into consideration, such as the potential bottleneck between a node pair and the correlation of the shared risk link group (SRLG) constraint [73]. The fourth problem is the computational complexity induced by fully adaptive routing. When the wavelength graph technique is applied, the network is expanded at least W times, where W is the number of wavelength planes in the network.

Partially adaptive routing includes fixed routing (FR) and fixed alternate routing (FAR). FR and FAR are distinguished by the number of prescheduled paths for each source–destination (S–D) pair. The prescheduled paths are also termed *alternate paths*. In general, FR equips each S–D pair with a single physical path for path selection. With FAR, nodes are equipped with multiple alternate paths. For both schemes, each source node is provided with a routing table, in which the alternate paths to all its destinations are defined offline. As a connection request arrives at a node pair, the best lightpath is selected from all the available ones along the alternate paths defined in the routing table.

To operate a network under the architecture of FAR, path selection is based on a predefined cost function and dynamic link state. The dynamic link state can be gathered in three ways. First, the source sends a probing packet along each alternate path to the destination after a connection request arrives. The destination coordinates the probing packets, makes a routing decision, and initiates a resource reservation process along the selected path. Secondly, periodic probing is performed by each source to all its potential destinations to gather the dynamic link state along the alternate paths defined in the routing table. Thirdly, every node periodically disseminates the link state of its corresponding network resources to the whole network. The three methods are subject to different problems that should be further elaborated on. The first approach incurs extra signaling latency before the routing decision can be made. On the other hand, periodic probing and dissemination may suffer from the stale link state problem, which impairs performance and dynamics if a conflict occurs during resource reservation.

Intrinsically, the FAR architecture has the following two design criteria that need to be further engineered. First, the performance behavior is strongly determined by the alternate paths arranged for each node pair, which may yield a load balancing problem if the alternate paths are not properly designed. Secondly, since the alternate paths are fixed and cannot adapt to any traffic variation, the performance is impaired if the alternate paths cannot be adaptive to the changing traffic pattern. It is clear that the planning of alternate paths are fixed and cannot be adaptive to the changing traffic pattern. It is clear that the planning of alternate paths for each S–D pair is important, which, however, has rarely been investigated before. The study in reference [74] provided an approach called Capacity-Balanced Alternate Routing (C-BAR) for the deployment of alternate paths according to the physical topology, potential traffic load, and location of each S–D pair in the network. Performance was reported to be greatly improved compared to the other strategies for planning alternate paths.

Routing and Quality of Service (QoS) Requirements

QoS-driven routing algorithms are needed for efficient establishment reservation. These algorithms suggest one or multiple suitable paths towards a given target considering a given set of QoS requirements. Then one attempts to make a reservation on such a path. Without appropriate routing mechanisms that take QoS requirements into account, the setup reservation becomes a trial-and-error approach [75].

A QoS-driven routing algorithm has to consider the currently available capacity of a resource to avoid an immediate rejection of the reservation attempt and the QoS requirements of the reservation to find a route best suited for this QoS. It should also consider the resource load after the routing decision to avoid using up the majority of resources on this route. Some of the problems to be solved with QoS routing are:

- How much state information should be exchanged among routers?
- How often should this system information be updated?
- Must there be a distinction between exterior and interior systems and, if yes, how can it be made?
- Is it possible to hide internal details of an autonomous system?
- Can the complexity of path computation be managed?

Routing is a collection of algorithms that determines the routes that data packets will traverse until reaching their destination nodes. Virtual circuit networks, like ATM, use connection-oriented routing. This means that a route is determined once, upon VC setup, and is then used by all the cells of the VC. By contrast, datagram networks, like the Internet, use connectionless routing. This means that a routing decision is made for each individual packet, independently of the previous packets of the same session.

Most of the routing algorithms used in traditional networks, which are not supposed to support QoS, try to find the shortest path from the source to the destination. These algorithms assign to each link a positive length, and associate with each path a length equal to the sum of its links. If each link is assigned the same length, the shortest path is the one with the minimum number of hops, and therefore the one that consumes minimum bandwidth from the network. However, many other metrics are possible, like distance, bandwidth, load, or cost. When the length represents loads, the shortest path is the one over which packets will suffer lower average queuing delay. In such a case the length assigned to each link will change over time, depending on the prevailing congestion level of the link. Some algorithms use a composite metric of several parameters like propagation delay, bandwidth, reliability, and so on.

To support a wide range of QoS requirements, a routing protocol needs to consider multiple metrics, such as bandwidth, average delay, maximum delay, loss probability, and so on. A typical route for a multimedia session should have sufficient bandwidth, a bounded delay, a bounded delay jitter, and a bounded loss rate. Moreover, from all the routes that fulfill each of these, requirements, the one with the minimum number of hops is preferred, because it wastes resources on a minimum number of links.

To make routing decisions, the network nodes should get topology information and maintain routing tables. Two approaches are used to perform this task distributedly. In the first approach, called *distance-vector routing*, each node sends to its neighboring nodes only its entire routing table. The receiving nodes use the received information in order to update their own routing tables, which they then send to their own neighbors. In the second approach, called *link-state routing*, each node broadcasts to all the network nodes information regarding the status of its local links only. The networked nodes use the received information in order to create and maintain an up-to-date network map, from which they deduce their routing tables.

Internet Routing Protocols

At a fundamental level, all Internet-based applications rely on a dependable packet delivery service provided by the Internet routing infrastructure. However, the Internet is a large-scale, complex, loosely coupled distributed system made of many imperfect components. Faults of varying scale and severity occur from time to time.

At the routing protocol level, the Internet is composed of thousands of autonomous systems (ASs) loosely defined as networks and routers under the same administrative control. Border Gateway Protocol (BGP) is the inter-AS routing protocol.

The routing protocol running within an AS is called Interior Gateway Protocol (IGP), typically Open Shortest Path First (OSFP), Intermediate System to Intermediate System (IS-IS), Routing Information Protocol, or Interior Gateway Routing Protocol. These routing protocols can be divided into three general classes: distance vector protocols, link state protocols, and path vector protocols.

In a link state protocol (e.g., OSFP and IS-IS), each router floods its local connectivity information (i.e., link state) to every other router in the same network. Each router collects the updates, builds a complete network topology, and uses this topology to compute paths to all destinations. Because each node has knowledge of the full topology, there is minimal dependence among nodes in the routing computation; thus, link state routing protocols are generally considered most promising for detecting faults [76].

In a distance vector protocol (e.g., BGP), a router announces the full path to each destination. Path information provides each router with partial information regarding topological connectivity; this partial information makes a fundamental difference between path vector and distance vector protocols, although path information is not sufficient to construct complete topological connectivity. It can be used effectively for fault detection. Owing to BGP's critical role in routing packets across loosely coupled ASS in the global Internet, the majority of the research efforts cited in this survey are related to BGP resilience.

6.3.10 Security Issues

Regardless of how we describe multimedia, security is a very critical part of networking today. Security is a very complex and broad topic. It is important to recognize the significance of security for future and current users and implementers of multimedia networks.

In the last decade we have seen dramatic advances in technologies that enable and support multimedia services. These advances provide both benefits and challenges when it comes to security. For example, increases in computational power at the desktop will aid in faster processing of complex mathematical calculations that permit cryptography in a more practical real-time environment. This same advance also provides some additional capability for those intent on cracking (for better or worse) cryptographic algorithms and compromise the confidentiality of the data stream.

Multimedia itself may not bring to mind any new considerations for security, if you simply regard it as the compilation of the voice, data, image, and video networks, areas that have at least some security mechanisms at present. With the advent of this consolidation/integration we do see a greater amount of Internet working taking place. So what is of particular interest is the networking of multimedia – simply stated, when multiple networks are connected (at least two) with multiple users (at least one), a level of Internet working has been achieved. Also of particular interest is the level of interactivity that you experience with multimedia services. So two of the key multimedia network characteristics that we focus on are Internet working and interactivity and their relation to security of multimedia networks.

Protection of the network is an important security issue for multimedia service providers, as is the ability to offer customers value added security services as needed or desired. Service or network providers planning for implementation of security mechanisms need to consider

- the nature of the security threat
- the strength of security needed
- the location of security solution(s)
- the cost of available mechanisms
- the speed and practicality of mechanisms
- interoperability.

Security is a critical technology in conventional and emerging networks and their applications and services today.

Networks are being used in a wide range of computing environments, ranging from distributed embedded systems and system area networks to private enterprise networks and the Internet. Their widespread use has been triggered by the significant development in high-speed transmission systems and system area networks to private enterprise networks and the Internet. Their widespread use has been triggered by the significant development in high-speed transmission systems with low bit rates at a low cost. However, their increasing use in application areas requires the provision of several properties at the network or application layers, which are typically considered security properties: privacy of communication, nonrepudiation of actions of communication members, authentication of parties in an application, high availability of network links and services, among several others.

Security technologies enable the provision of several of these properties at the required level. The criticality of security becomes clear when considering that several applications and services, for example, military, e-commerce, and so on, require the existence of such properties. This indicates that security technology constitutes an important enabling technology for next-generation applications and services. As an enabler, security has received the attention of a body of scientists and engineers who are working to address a large number of problems in the technology. The problems range from efficient and effective cryptographic algorithms to e-commerce protocols, and from secure thin devices such as smartcards to intellectual property protection. Clearly, the problems are significant and numerous and, importantly, each problem may require a different solution depending on the environment to which is targeted: a different protocol and/or algorithm will be used in a video (or TV) distribution system with subscribing customers from the ones used in a secure telephone system using smartcards for authentication and billing, as used in many countries worldwide.

Considering the advent of the Internet, its wide acceptance and penetration to a continuously increasing population, an interesting issue is the provision of security in the Internet. The recent explosive growth of the Internet as well as the identification and exploitation of its lack of security, has triggered a large number of attacks, leading to a significant number of problems such as denial-of-service, eavesdropping of communicated data, and so on. Security in (or over) the Internet will enable a wide range of services and applications that have not been realized in the classical, nonsecure Internet: distribution of sensitive material (protected intellectual property), reliable subscription services, highly available e-commerce services, and so on. Importantly, a secure Internet will provide a reliable, highly available infrastructure for all applications.

The following list of threats and countermeasures is conveyed to provide a high-level view on the types of issues that are of concern, and the mechanisms or tools that can be used to prevent, or at least detect, these security events [77].

- *Threats.* Some of the industry segregates threats into passive and active. Passive attacks can involve acts of wiretapping or analysis of user traffic (to determine meaningful patterns in type, volume, timing, source, and destination), where the data in question are not necessarily modified. Active attacks can take the form of masquerades, replays of old messages, denial of service, or alteration of a message, where the attacker is taking a proactive approach. The following series of threats are examples and are not limited to be an exhaustive list.

- *Disclosure of information/interception.* This is an attack on data confidentiality, where the unauthorized party gains access to information or resources. This can take two forms: an attack on the content, or an attack where communication patterns are analyzed. Wanting to keep transmission of information secret is of great concern – whether this is for business reasons, personal privacy reasons, concern for fraud, or for reasons of national defense. And not having communication patterns analyzed is also of concern.
- *Unauthorized modification.* Any transfer of data can be subject to modification en route. This can be done in any number of ways, such as accessing a copy of the data at an interim node, or by tapping a facility; either way an unauthorized party has gained access to, and may then modify, the data in question.
- *Unauthorized usage.* Gaining access to network resources and services without proper authorization is the focus of this threat. This can range from logging into remote network elements to tapping into a shared medium, to stealing services (e.g., cable). Access control and authentication countermeasures help mitigate these issues.
- *Denial, loss of service, and interruption.* This is an attack on availability; the service or system in question is not performing as designed. In a telephony network this may equate to not being able to get a dial tone when trying to make a phone call; or for cable services this may mean that the cable system is down; and for corporate data services an example can be having the LAN go down. The basic concept here is that some security element has contributed to the interruption (either short- or long-term) of services. Implementing access control and authentication can be helpful in this area.
- *Repudiation and nonrepudiation.* In many instances it is important to know that the receiving party has indeed received the intended data. A common example of this is with e-mail. Viewed another way, this can be the verification that the originating party is indeed who he/she they claim to be – for example, when ordering products or services via the WWW. Proper authentication techniques support this area.

Network Security Services
A number of security services are available that can help prevent attacks like those just noted. And there are additional services that can help detect when either malicious or nonmalicious security problems occur.

Data Confidentiality. With networked systems, it is possible that there is not complete confidence in and control over the path that your data will be taking. This is especially true for users sending data across a public network (the Internet, for example) or if you are a network provider interconnecting with another network.

Encryption is a primary means of providing confidentiality. Encryption typically involves the use of complex mathematical algorithms and unique values to transform material into an unusable form – hopefully unusable for those without the algorithm and unique value. This value is the key – figuratively and literally. The key can be of any length, but the longer it is, the harder it is to figure out; this greater key length is also more cumbersome for the user to use and manage. So a good balance should be struck – something that is practical to use and manage but also hard to crack.

The basic concept here is that the original material (plaintext) is transformed by the algorithm and key to a protected state (ciphertext). Here are some key concepts and tools that are important with cryptography:

- *Secret key cryptography*. Parties involved both know the single secret key, and use this one key to both encrypt and decrypt the material to be kept in confidence. Since only one key is used, and both parties use it, this method is sometimes called *symmetric cryptography*. It is imperative that secret keys be kept secret, and that users consider changing this key with some frequency.
- *Public key cryptography*. This involves the use of two keys: the encrypting party can use his/her private key (known only to the sender) to encrypt the material such that the decrypting party using the sender's public key to decrypt is assured that the message is confidential. Additionally, the sender can use the public key of the intended recipient so that he/she is the only party that can decrypt, since the recipient's private key is the only key able to decrypt the message. Public key cryptography is also called *asymmetrical cryptography*.
- *Session key*. Used with public key encryption systems, this scheme involves a third key, randomly generated for every message, or session, encrypted. What is of special note in this process is that the most processor intensive algorithm (public key) is only encrypting with a one-time only session key (that may not even have been disclosed to the user).
- *Hash function*. Hashing is a simple form of security transformation, used primarily for message authentication and integrity assurance. Hashing takes a message (plaintext) and transforms it into a fixed-length value. One of the criteria for a hashing function to be secure is that even if the hashing function, the message, and its digest are known, it is very difficult to construct another message that produces the same digest. By using a password in the process, and computing the hash over the message plus a secret password, the receiving end (that knows the secret password) can be assured that the sender is the claimed one and that the message has not been modified when they compute the same message digest.
- *Key management*. Managing keys used with encryption is very important for obvious reasons. But with public key encryption it is important to have a well-known process for sharing public keys. For this process a key certificate is used, which contains the public key, user identifier for the key creator, and digital signature(s) of those who can attest to the key being genuine. A key distribution center (KDC) can be used with this process. A KDC is a trusted node that knows all the secret keys of the parties and can easily manage changes and additions to the keys. This central design also sets the KDC up as a

potential single point of attack/failure, so it is wise to consider implementing a backup/paired KDC although this increases complexity and points of attack.
- *Authentication.* This is a service that assists in assuring the users that the other parties involved in the communication are who they claim to be. This can either be one-way or mutual (two-way) authentication. Having this assurance helps to undermine attempts at impersonation and spoofing that can occur. Authentication is important not only as a service on its own, but also in support of other security services, such as key exchange. Cryptography can provide a means for supporting strong authentication and either public key or secret key algorithms can be used. A simple exchange might proceed as follows. User A wants to establish a secure connection with User B. User A sends an encrypted message (chipertext) to B using the public key B. User B in turn decrypts the message with his/her private key, and proceeds to encrypt a return message with A's public key. Now that A and B are confident of each other's identity, they may even want to exchange a secret key this way, and proceed to have an encrypted session based on that secret key.
- *Access control.* This is a means of controlling who has access to particular services or elements in a network. Access can depend on the identity of the calling party, the services requested, and/or the type of data being accessed. Assuming that information is given different levels of sensitivity, access can be granted at a given level. A traditional access policy prohibits read-up and write-down. That is, if you only have permission to read a certain level of material, or gain access to it, you cannot read or access higher, more secure levels. Also, if you have material that is sensitive at a certain level, you can not provide it down to those with lesser levels of access, hence write-down is prohibited. Access lists, closed user groups (CUGs), and even virtual private networks (VPN) are examples of access mechanisms of a different sort. Generally speaking, they all restrict access to/from users and applications and pieces of the network. Access lists can be set up for permitting or denying access to specified interfaces, traffic types, and applications/ports. CUGs, which have been defined in technologies such as ISDN as well as B-ISDN/ATM, determine access capabilities of a group of users, among themselves and with those outside the group. VPNs can help to extend private network configurations over public network facilities in a manner that is not evident to the users on the VPN.
- *Integrity.* Digital seals and digital signatures can provide assurances for integrity and message authentication. A digital seal refers to the process in which a message digest, or hash, is computed for a message, and then the digest is encrypted using a secret key encryption process. A digital signature is similar, however, in that a public key encryption process is used, encrypting the digest with the sender's private key. The receiver then uses the sender's private key. The receiver then uses the sender's public key to decrypt the digest, compute the hash, and verify that the message was from the intended party and was intact. Just because multimedia communication networks may be high speed, and may be running over optical facilities, it is not safe to assume that there

are no security risks. While the higher data rates and optical media may pose more challenges for would-be attackers, there are still ways in which these areas can be compromised. Optical taps are possible, although challenging. High-speed analyzers, encryptors, and decryptors provide an indication that the high-speed rates will not be much of an issue. Network security requirements include accountability, alarming, archiving, assurance, audit and data analysis, authentication, authorization, availability, confidentiality, damage limitation, error recovery, integrity, logging, nonrepudiation, privacy, reporting mechanisms, robustness, short-term data backup, and survivability [78]. New technologies are prone to attacks since many are implemented early, with a focus on first to market, and they have not been put to the test of time and exposure that older technologies have.

General Security Analysis

The goal of security analysis is to determine what requirements a security system must meet in a given context to provide adequate protection for a valued set of resources. Depending on the context, the resources can be real objects with intrinsic value, or objects whose values are basically determined by the information that they represent. The dominant focus in analyzing network security requirements is on objects that represent intellectual property or assets, and not on other types. The scope of the following discussion is limited to such information objects.

The fundamental mechanism used in secure systems to protect information is *access control*. With adequate access controls in place, both the confidentiality and the integrity of an information object can be assured to some defined confidence level. Confidentiality is at risk whenever information flows out of an object. Integrity, on the other hand, is at risk whenever information flows into an object. Information flows in and out of an object can be characterized as basic read and write operations on the object, regardless of the type of actual physical means used to access the object.

When adequate physical access controls exist, information can be stored, and moved from one place to another, without taking additional steps to hide it or to disguise it. If adequate physical path protection cannot be assured, then information needs to be transformed in some way to make it unintelligible to unauthorized parties, and able to indicate to an authorized recipient that it has not been altered in transit. To ensure the confidentiality of information flows sent via insecure paths, it is necessary to encrypt the information at the source and to decrypt it at the destination. To ensure integrity, validation methods must be used to verify that stored or transferred information is not contaminated.

A security policy establishes the scope, terms of reference, and requirements for the protection of valued resources in a given domain. The scope of a security policy can apply to a complete system, or selectively to parts of a system.

A security policy can stipulate the use of particular security methods or mechanisms, or it can leave such matters up to the manufacturer of a security product or system. It can also explicitly identify resources to be protected, and the level of protection each requires, or just provide a set of rules that need to be satisfied. In any case, a candidate product or system must comply with all the requirements specified in an applicable security policy to qualify. Security criteria establish discrete reference levels for evaluating how well a candidate product or system element meets the requirements of a security policy. The criteria do not establish what protection needs to be provided in a given context: the information is provided by an applicable security policy.

Various standards for security criteria have been published. The standards reflect a general trend toward harmonization, but no universal set of security criteria has emerged to date. Newer standards tend to refine and enhance earlier work. The following list is an attempt to merge various current points of view:

- *General criteria.* Define general compliance requirements.
- *Confidentiality criteria.* Define levels of protection against unauthorized disclosure.
- *Integrity criteria.* Define levels of protection against unauthorized modification.
- *Accountability criteria.* Define levels of ability to correlate events and users.
- *Reliability criteria.* Establish levels to measure how well a product performs.
- *Assurance criteria.* Define overall trust levels for products or systems.

To qualify as compliant with respect to a particular criterion, a candidate product or a system must satisfy all of the requirements that apply to the criterion.

Security Criteria

Security criteria establish a finite, discrete set of terms and conditions for evaluating the ability of product, system block, or a complete system to protect designated types of resources. The complexity of a system and the tasks it needs to perform determine how granular an appropriate set of security criteria needs to be.

A security policy specifies which security criteria must be satisfied by a qualifying product to ensure that an adequate level of protection is provided to resources. A compliant product meets or exceeds all of the requirements specified in the policy.

A product is evaluated for compliance with each criterion specified in an applicable security policy. For total compliance, a product must meet or exceed all the requirements of the criterion.

The overall rating of a product is based on the set of criteria that the product can satisfy. If the capabilities of a product match the requirements of a security policy, then the product can be certified as compliant at a particular level.

To evaluate systems that provide more than one level of protection, different subcategories of security criteria requirements need to be defined. Security policies define which security criteria requirements apply to each defined protected information object and user class. A product is evaluated against each required security criterion subcategory for compliance one by one. If the product satisfies all applicable security criteria requirements, then it is rated as compliant with the security policy.

Security Aspects of ATM Networks

The challenge with security in ATM networks is to identify the type of security services and how best to implement them; considering the weakest link guide, if the ATM network is not end-to-end, then it is not of much value to only implement security mechanisms at the ATM protocol layer (although support to higher layer services can still be valuable). It is likely that a combination of mechanisms, at various layers, may be most useful. Scalable solutions are needed due to the expected number of ATM users and the size of ATM networks. And finally, security solutions must be implemented in a fashion that does not prohibit compatibility with nonsecurity aware/capable devices.

In general, there is a security advantage in that ATM is connection oriented in nature as opposed to being connectionless. The greatest advantage with this condition is when the circuit is preprovisioned and this is because the route that the cell takes is preestablished and well known. This does not prevent attacks in itself, but it is more assured than connectionless packets traversing any route that happens to have good metrics at the time. However, the signaling associated with establishment of on-demand connections is itself subject to attack, and should undergo security analysis to identify risks and potential threats.

One of the most valuable assets of ATM is the quality of service (QoS) component it brings to the transport protocol world. Because QoS is one of the critical aspects to ATM, it may be the focus of future threats. The network needs to assure that the QoS agreed to for the customer/service can be maintained – either unintentionality or intentionally no other user can consume so many resources as to degrade the QoS for others. Quite simply the customers should get what they paid for. If unauthorized changes to the QoS parameters can be made, then can some parameters can either be *lowered* to damage a user's service, or they can be improved, which is the equivalent of theft service, and can contribute toward overloading the network and degrading service for all customers.

How ATM networks manage congestion is also important. If those mechanisms can be tampered with then the network may not operate properly, buffers can overflow, and service can deteriorate.

For ATM protocol the most practical layer on which to provide data confidentiality is the ATM layer. The ATM cell payload can be encrypted easily because it is always a fixed length (48 bytes). More importantly, unless one plans to do encryption on a hop-by-hop basis with each node in the network (and establish trust with each of those nodes), the ATM cell header will have to be in the clear, or be readable by each node. This confidentiality service is likely to be useful in untrusted networks spanning among trusted networks, typically at the points where private networks attach to the public network.

Authentication in ATM networks is important among communicating nodes and is a foundation or supporting service to other security services such as confidentiality and integrity. It is likely the nodes will be security agents (SAs) associated with end user virtual path connections (VPCs) and/or virtual channel connections (VCCs). Authentication should be considered for both the user plane (user data) and the control plane (signaling), and is typically performed once at the start of the connection.

Signaling needs to be authenticated, either bilaterally or unilaterally. Signaling must also support higher layer security exchanges, when the customer negotiates security services on an end-to-end basis.

This service is most useful at the ATM adaptation layer (AAL). The ATM service data unit (SDU) offered to the ATM layer can be taken as one data block for this service. Currently both the AAL5 SDU and AAL3/4 SDUs are supported in the ATM Forum security specification. And for these, the option of replay ordering can either be used or not. If it is used, then the signature of the SDU is computed after adding the sequence number. In this way the signature will both protect and preserve the sequence number. If reply/reordering is not desired, then the signature can just be computed over the SDU by itself.

This would generally need to be done on an endpoint-to-endpoint basis because of being supported at the AAL, which is typically transparent to the intermediate switching nodes. However, it is likely that the SA will be considered the endpoint in this scenario, and due to a number of considerations the SA will not always be coincident with the end user.

Security Aspects of IP Networks

IP-based networks generally have the same security concerns that other networks have. One point that should be highlighted is that IP networks are connectionless; that is, they do not require a preestablished route for the data to traverse. With connectionless services integrity, the focus is on the transmission from source to destination without undetected alternation; in connection-oriented networks (such as ATM-based networks) this same focus exists, but there is also concern with the data sequencing – making sure the data arrive in the proper order.

The Internet Protocol Security (IPSEC) group develops mechanisms that protect client protocols of IP. A network-layer security protocol will be developed to provide cryptographic security services that support combinations of authentication, integrity, access control, and confidentiality. The IPSEC has two parts: IP authentication header (AH) and the IP encapsulating security payload (ESP). These will be independent of any encryption algorithm used by the client. IPSEC's preliminary goal is to pursue host–host security, followed by subnet–subnet and host–subnet topologies.

Secure sockets layer (SSL) is a security that assists in providing secure communications over the Internet. The main focus is on privacy, to prevent eavesdropping, tampering, and message forgery. The two-layer protocol (SSL record protocol is at the lowest layer, layered on top of reliable transport such as TCP) does encapsulation of higher layer protocols such as the SSL handshake protocol, allowing a client and server to mutually authenticate and negotiate use of an encryption algorithm and keys. This authentication and negotiation takes place before any user data are sent by the application. Another key focus of SSL is on keeping the connection reliable, that is, to do a message integrity check with message transport.

Domain name systems (DNS) security will ensure enhancement to the secure DNS protocol to protect the dynamic update operation of the DNS. This is important to protect the validity of the address mapping on the DNS server. The IETF working group is focusing on the need to be able to detect the reply of update transactions and the ability to order update transactions. They also intend to address clock synchronization as well as all of the dynamic update specifications. Note that this IETF group considers data in the DNS to be public – as a result they consider data confidentiality and access control proposals outside the scope of their work.

The combination of IP and ATM tends to follow the same issues mentioned for each technology separately. However, there are also several topics of particular interest with supporting IP-based applications in an ATM network. These include the areas of addressing, routing, and member groups.

With addressing, there are traditional addressing concerns related to spoofing and masquerading, but address resolution between IP and ATM addresses adds an additional area of analysis. Two items in particular to consider relate to the trust relationship with users and other address servers. An address server should deny requests for changes to address mapping if the user requesting the change is not considered a trusted source. Likewise, address mapping updates from other address servers should not be accepted if those servers have not established their trustworthiness. An attack on addressing can redirect data from the proper destination to a fraudulent destination, or could cause service performance delay or even service loss for affected customers. While address resolution is one if the most critical functions with ATM networks, it can also be one of the most vulnerable and risky. A likely mechanism to assist with these concerns is authentication.

For the routing area, again the issues mentioned for IP and ATM independently are still of concern. But for IP/ATM implementations, integrated routing techniques must be examined as well. Attacks on the routing tables or algorithms can cause the packets and cells to be misdirected or diverted to improper destinations. A key area for protection is the control plane because signaling to establish *short-cut* connections, control messages that provide updates to address resolution, and routing tables need to be analyzed for authentication services. The grouping concept (e.g., subnets, logical IP subnets, emulated LANs) needs to be managed carefully. For group memberships/associations, any member joining the group who is untrusted and unauthenticated can gain access to all the information privileges to which the group has access. This is similar to the issues with closed user groups and virtual private networks. Strict control of group access and privileges needs to be undertaken.

Technologies such as ATM and IP are expected to be used heavily in multimedia networks, and both have some security tools that can be used or are being developed. These tools will support security functions directly in the lower layers of multimedia networks, as well as support higher layer security applications implemented by end users. Industry standards assist in the development of needed security mechanisms and in promoting the interoperability of various mechanisms and platforms.

Internet Infrastructure Security

Securing the Internet, as in any other field of computing, is based on the principle of *confidentiality* and *integrity*. Confidentiality indicates that all data sent by users is accessible to only legitimate receivers, and integrity indicates that all data received is sent/modified by *legitimate* senders. These principles exist in every field, but the presence of *packet sniffers*, *malicious routers*, *covert channels*, and *eavesdroppers* in the Internet makes this extremely important problem quite challenging [79].

The past several years have seen a urge of Internet security research in the field of *information assurance*, which primarily focuses on protecting data using techniques such as *authentication* and *encryption*. However, information assurance assumes that the devices responsible for encrypting, forwarding, and sending packets are trustworthy. Scientists are now questioning these assumptions, since there have been instances where the network infrastructure (e.g., routers, servers) has been compromised by malicious adversaries. Thus, network infrastructure security is clearly a pressing need, especially in light of recent national attacks, since the attacks have the potential for affecting the entire Internet infrastructure, which may have serious consequences for the security and economic vitality of society.

Attacks on the Internet infrastructure can lead to enormous destruction, since different infrastructure components of the Internet have an implicit trust relationship with one another. As an example, consider the scenario shown in Figure 6.14. In this scenario, an intruder wishes to attack domain Z, which contains a high-profile server. Most of the links are fairly heavily loaded but have under capacity (70–80 percent usage). The attacker compromises router A, so the router increases the cost of link B to an artificially high value, say 10,000. Internet traffic is generally routed along the shortest path. Since link B has a high cost, packets will be routed around B. Thus, packets will be routed through the border of domain Z. Artificial congestion thus created will slow down the services to clients of domain Z (also W, X, and Y), and many clients will be denied access to the server.

As shown by this fairly simple example, it is possible for an attacker to create a large amount of service disruption. Such service disruption has already been noticed

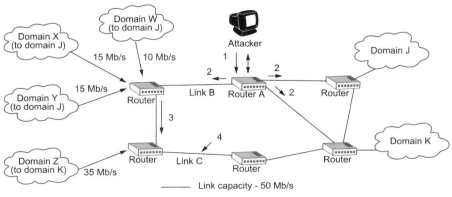

Figure 6.14 Router attack and its consequences [79]. (©2002 IEEE.)

with untargeted breaches in the infrastructure such as fiber cuts, as well as Border Gateway Protocol (BGP) routing flaps due to Nimbda/Code Red. Also, these types of attacks are very difficult to detect because the attacker is hidden during the actual transmission of packets. The attacker can achieve the same results as above by sending more packets than what the border router of domain Z can handle. In addition, router A can also deliberately misroute packets, causing congestion at the border router of domain Z. Thus, compromising the infrastructure can lead to potentially dangerous attacks on the Internet.

The effect of the types of attacks mentioned above is dangerous because the attacker knows the network topology and intelligently takes advantage of the basic flows of the networking protocols. Although security research in this area is considered absolutely necessary, there has been no framework that would encompass all the possible attack scenarios in the Internet. Because of the absence of a guiding framework, research efforts in Internet security have lacked direction.

Internet infrastructure attacks can be broadly classified into the following four categories: DNS hacking, routing table poisoning, packet mistreatment, and denial of service.

DNS Hacking Attacks. Domain Name System (DNS) is a distributed hierarchical global directory that translates machine/domain names to numeric IP addresses. The DNS infrastructure consists of the 13 root servers at the top layer, top-level domain (TLD) servers at the top layer, top-level domain (TLD) servers (.com and.net), as well as country code top-level domains (.us, .uk, etc.) at the lower layers. Owing to its ability to map human memorable names to numerical addresses, its distributed nature, and its robustness, DNS has evolved into a critical component of the Internet. Therefore, an attack on the DNS infrastructure has the potential to affect a large portion of the Internet [80].

Routing Table Poisoning Attacks. Routing tables are used to route packets over the Internet. They are generated by exchange of routing information or updates between routers. Poisoning attacks refer to the malicious modification or *poisoning* of routing tables. This can be achieved by maliciously modifying the routing information update packets sent by the routing protocols. This can result in wrong entries in the routing table and can lead to a breakdown of one or more domains of the Internet.

Packet Mistreatment Attacks. In this type of attack the malicious router mishandles packets, thus resulting in congestion, denial of service, and so on. The problem becomes intractable if the router selectively interrupts or misroutes packets resulting in triangle routing, that is, loop information. An example of triangle routing is shown in Figure 6.15. The shortest path from 1 to 4 is 1−2−4 and the shortest path from 3 to 4 is 3−1−2−4. Let 2 be the malicious router. Whenever 2 gets a packet from 1 destined for 4, it routes it to 3. Since the shortest path from 3 to 4 is through 1, a loop is created. This type of attack is very difficult to detect.

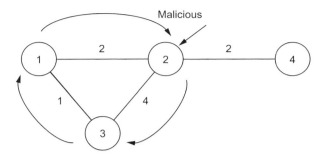

Figure 6.15 An example of triangle routing.

Denial of Service Attacks. These attacks are directed at specific hosts with an intention of breaking into the system or causing denial of service (DoS). These attacks may be carried out by individuals or groups who may use such attacks for personal gain or notoriety. These attacks become extremely dangerous and hard to prevent if a group of attackers coordinate in DoS. These types of attacks are called distributed DoS (DDoS) attacks. It is to be noted that DoS can also result from routing table poisoning and packet mistreating. We categorize DoS attacks as those attacks that are directed towards the end system rather than towards the transmission infrastructure like the routers/links.

Among these attacks, the last type (DoS attacks) is related to end systems, while the other three are related to the network infrastructure (DNS, backbone routers, and communication links). In fact, most of the traditional security research has not focused on transmission-system-related attacks, but rather focused on the security of the end system. Although DoS attacks against specific machines are an important threat, the potential for attacks against the transmission infrastructure can result in a massive DoS attack against entire groups or whole portions of the Internet. Thus, the routers and the other networking infrastructure components represent ideal targets for disrupting the national infrastructure.

General Packet Radio Service (GPRS) Security

The most widely deployed public mobile data network, which enables the integration of IP with mobile networks and constitutes a migration step toward third-generation communication systems, is the General Packet Radio Service (GPRS) [81]. GPRS technology supports a set of security mechanisms to protect network operation and data transfer through it. Such mechanisms mainly protect the air interface between the mobile devices and the base stations; however, the IP-based core network and the interconnection points are potentially vulnerable, and are likely points for attacks [82].

Mobile Internet is becoming available with the deployment of the enhanced version of second-generation mobile communication systems, such as GPRS. GPRS attempts to reuse the existing Global System for Mobile Communication (GSM) network elements as much as possible. But in order to effectively build a packet-based mobile cellular network, some new network elements, interfaces, and protocols are required. The new network nodes are called GPRS support nodes (GSNs). The GSNS are responsible for the delivery of data packets from and to the mobile station (MS) within its service area. The gateway GSN (GGSN) acts as an interface between the GPRS backbone network and the external packet data network. Communication between GSNs is based on IP tunnels through the use of the GPRS Tunneling Protocol (GTP). The GPRS network architecture is illustrated in Figure 6.16.

Given that the GPRS is built on the GSM infrastructure, it employs the same security functions used in GSM [7], slightly modified to adapt to packet-oriented traffic and to the GPRS network components. These function aim at two goals:

- to protect user privacy
- to protect the network against unauthorized access.

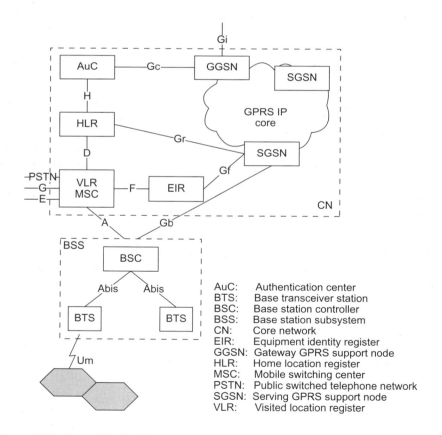

Figure 6.16 General Packet Radio Service (GPRS) system architecture [83]. (©2002 IEEE.)

The GPRS backbone network utilizes private IP addressing and network address translation (NAT) to restrict unauthorized access to it. Furthermore, firewalls guard the traffic to and from other networks, protecting the GPRS backbone from IP spoofing.

Since the GPRS core network is based on IP and is connected to the public Internet, data in transit are subject to various security threats, such as interception, unauthorized disclosure, and malicious alternation. The GPRS encryption/decryption mechanisms do not extend far enough toward the core network, resulting in the clear-text transmission of user and signaling data. Border firewalls and private IP addressing attempt to protect the clear-text transmitted data within the GPRS core from external attacks. However, these measures are inadequate against attacks that originate from malicious mobile network subscribers, as well as third parties that get access to the GPRS core network.

Security Mechanisms Provided with Session Initiation Protocol (SIP)

Session Initiation Protocol (SIP) is the most promising candidate for call setup signaling for IP-based telephony services and it has been chosen by the Third-Generation Partnership Project (3GPP) as the protocol for multimedia application in 3G mobile networks [84]. Within the traditional public switched telephone network (PSTN) a good level of quality of service (QoS) and security has been established over the years, and it is now widely guaranteed. If the new IP telephony architecture and SIP want to replace the PSTN, proposing new service scenarios, they should provide the same basic telephony service with a comparable level of QoS and network security. While the problem of QoS support mainly concerns the IP network layer, the problem of security involves the network layer, the service and control architecture, and its signaling protocols. The following security characteristics should be guaranteed: high service availability, stable and error-free operation, and protection of the user-to-network and user-to-user traffic (for both control and user data).

Although security and privacy should be mandatory for an IP telephony architecture, most of the attention during the initial design of the IETF IP telephony architecture and its signaling protocol, SIP, has been focused on the possibility of providing new dynamic and powerful services, and simplicity. A very hot topic in the SIP and IP telephony standardization track is how security support can be enhanced to an acceptable level for any type of service that must be provided. Two main security mechanisms are used with SIP: authentication and data encryption.

Data authentication is used to authenticate the sender of the message, and to ensure that some critical message information was unmodified in transit. This is to prevent an attacker from modifying and/or replaying SIP requests and responses. SIP makes use of Proxy-Authenticate, Proxy-Authorization, Authorization, and WWW-Authenticate header fields, similar to those of HTTP, for authentication of the end system by means of a digital signature. Instead, hop-by-hop authentication can be performed using transport- or network-layer authentication protocols.

Data encryption is used to ensure confidentiality of SIP communications, letting only the intended recipient decrypt and read the data. This is usually done using encryption algorithms such as Data Encryption Standard (DES) and Advanced Encryption Standard (AES). SIP supports two forms of encryption: end-to-end and hop-by-hop. End-to-end encryption provides confidentiality for all information (some SIP headers and the message body) that does not need to be read by intermediate proxy servers.

On the other hand, hop-by-hop encryption of entire SIP messages can be used in order to protect the information that should be accessed by intermediate entities, such as *From*, *To*, and *Via* headers. Encryption of such information can prevent malicious users from determining who calls who, or accessing route information. Hop-by-hop encryption can be performed by security mechanisms external to SIP (transport-level security, TLS).

The security mechanisms must be combined properly to obtain a trusted network scenario. Just to give an example of this combination, we consider the scenario shown in Figure 6.17. User agents authenticate themselves to local outband proxies using SIP authentication; servers authenticate themselves to other servers one hop away or to user agents with a site certificate delivered by TLS. In such a way, a trusted network architecture can be built, also covering end-to-end SIP paths. Of course, this can be complicated if different servers belong to different administrative domains, crossing the so-called trust boundaries. On a peer-to-peer level, user agents ordinarily trust the network to authenticate remote user agents.

6.4 NETWORK TRAFFIC ANALYSIS

By the term telecommunications traffic or teletraffic we mean the flow of information, or data, in telecommunications networks of all kinds [186]. From its origins as an analog signal carrying encoded voice over a dedicated wire or circuit, traffic now covers information of all kinds, including voice, video, text, telemetry, and real-time versions of each, including distributed gaming. Instead of the dedicated circuits of traditional telephone networks, packet-switching technology is now

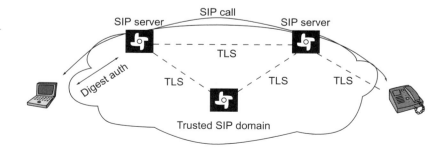

Figure 6.17 Trusted network scenario [85]. (©2002 IEEE.)

6.4 NETWORK TRAFFIC ANALYSIS

used to carry traffic of all types in a uniform format (to a first approximation): as a stream of packets, each containing a header with networking information and a payload of bytes of data.

Although created by man and machine, the complexity of teletraffic is such that in many ways it required treatment as a natural phenomenon. It can be likened to a turbulent, pulsating river flowing along a highly convoluted landscape, but where streams may flow in all directions in defiance of gravity. It consists of a deep hierarchy of systems with complexity at many levels.

The teletraffic has more levels of complexity than the underlying network. Three general categories can be distinguished [86].

- *Geographic* complexity plays a major role. Although one can think of the Internet as consisting of a *core* of very high bandwidth links and very fast switches, with traffic sources at the network *edge*, the distance from the edge to the core varies greatly, and the topology is highly convoluted. Access bandwidths vary widely, from slow modems to gigabit Ethernet local area networks, and mobile access creates traffic that changes its spatial characteristics. Sources are inhomogeneously distributed; for example, concentrations are found in locations such as universities and major corporations. Furthermore, traffic streams are split and recombined in switches in possibly very heterogeneous ways, and what is at one level a superposition of sources can be seen at another level, closed to the core, as a single, more complex kind of *source*.
- *Temporal* complexity relates to the multilayered nature of traffic demands. Users, generating Web browsing sessions for example, come and go in random patterns and remain for widely varying periods of time, during which their activity levels (number of pages downloaded) may vary both qualitatively and quantitatively. The user's applications will themselves employ a variety of protocols that generate different traffic patterns, and finally, the underlying objects themselves, such as text, audio, images, video, have widely differing properties.
- *Temporal* complexity is omnipresent. All the above aspects of traffic are time varying and take place over a very wide range of time-scales, from microseconds for protocol acting packets at the local area network level, through daily and weekly cycles, up to the evolution of the phenomena themselves over months and years. The huge range of time-scales in traffic and the equally impressive range of bandwidths, from kilobytes up to terabytes per second over large optical backbone links, offers enormous scope for scale-dependent behavior in traffic.

The scale-invariant features of traffic can be thought of as giving precise meaning to the important but potentially vague notion of traffic burstiness, which means, roughly, a lack of smoothness. In very general terms, burstiness is important because from the field of performance analysis for networks, and in particular that of switches via queuing theory, we know that increased burstiness results in lower levels of resource utilization, offers a fixed quality of service, and therefore higher costs. At the engineering level, service quality refers to metrics such as available bandwidth, data transfer delay, and packet loss. The impact of scale invariance extends to network management issues such as call admission control, congestion control, as well as policies for fairness and pricing.

The Internet new services generate a traffic explosion. As the telecommunication infrastructure is currently undergoing radical and rapid changes, new services will be added. Consequently, there will be much higher demands for bandwidth. The new services will span all areas of telecommunications from telephone networks to cable TV, local area networks, and cellular and satellite communications. Users in rural areas all over the globe will have access to full Internet services, and mobile customers should be able to make multimedia calls and download video files at speeds in the order of 20 Mbps. Network traffic is more dynamic and complex as new services are added. Accordingly, efficient allocation of network resources (bandwidth buffers) will be an increasingly difficult task. Novel adaptive techniques that can predict traffic variability and dynamically allocate resources will be very useful.

The analysis and modeling of high-speed network traffic is an important problem for a number of networking and communication tasks, such as designing, controlling, planning, and maintaining a network, determining pricing schemes and quality of service (QoS), as well as providing network security. When circuit-switched, voice-carrying networks are utilized, then reasonable models for network behavior exist. However, the traffic found in packet-switched, high-speed networks is considerably more complicated and difficult to describe analytically.

In modeling various real-world phenomena, the fact that observations made on different scales carry essential information must be utilized. Networks traffic is complex as it exhibits strong dependencies and self-similarity, and therefore classical models of time series such as Poisson and Markov processes (relying heavily on independence or weak independence assumptions) are not appropriate for its modeling. Instead multiresolution techniques provide a natural framework for modeling and analyzing the network traffic. For example, fractal and multifractal models have been utilized to model the long-range dependencies while the multifractal spectrum has been utilized to model the *burstiness* of the data. Actual network traffic data drive the development of appropriate traffic models. Similarly, any analytical model should be validated with actual data. A twist comes from the fact that current data networks such as the Internet are complex and heterogeneous, and, therefore, a single analytical model is not sufficient.

6.4.1 Traffic Engineering

Traffic engineering (TE) is considered one of the hottest topics in the framework of new-generation networks. The goal of TE is to improve the efficiency and reliability of network operations while optimizing network resource utilization and traffic performance as stated in IETF Requests for Comments (RFC) 2702. Before introducing the different subjects of TE, it is meaningful to review the main motivations for TE and the main aspects related to its application in communication networks.

In recent years, the amount of traffic due to Internet-based services has become more and more evident. Besides the traditional services carried by the Internet, essentially with no guaranteed quality of service (QoS), a migration to a number of real-time services over Internet Protocol (IP) is foreseen in the near future. Therefore, future network infrastructures will have to handle a huge amount of IP data traffic, including a significant amount of real-time traffic with demands for assured QoS. Moreover, given the fact that Internet traffic is much more variable with time than traditional voice traffic, and thus not easily predictable, such networks have to be flexible enough to react adequately to traffic changes. Future infrastructure must also have multiservice capability, in order to support different types of services with different requirements in terms of QoS. Besides the requirements of flexibility and multiservice capabilities that lead to different levels of QoS requirements, there is another key aspect that needs to be taken into account: cost-effectiveness. In fact, a dilemma emerges for carriers and network operators: the cost to mobile telephone networks is too high to be supported by revenues coming from Internet services. Actually, revenue coming from voice-based services is usually much higher than that derived by current Internet services. Therefore, to obtain cost-effectiveness, it is necessary to design networks that make effective use of bandwidth or, in a broader sense, of network resources.

TE is a solution that enables the fulfillment of all these requirements, since it allows network resources to be used when necessary, where necessary, and for the desired amount of time. A network with TE capability can dynamically control traffic flows in order to prevent congestion and optimize the availability of resources. More specifically, TE allows a network to choose routers for traffic flows while taking into account traffic loads and network state, to move traffic flows toward less congested paths, and to react to traffic changes or failures in a timely way. In order to adopt TE solutions, it is necessary to create an intelligent control plane that is able to adequately handle network resources. Such an intelligent control plane will require a paradigm shift in the design of network architecture: some features of traditional synchronous digital hierarchy/synchronous optical network (SDH/SONET) or asynchronous transfer mode (ATM) networks have to be imported into the IP world and suitably adapted. A relevant network paradigm considered worldwide is based on the multiprotocol label switching (MPLS) technique and its generalization, GMPLS. In practice, this paradigm consents to the reintroduction of the virtual connection into IP-based networks that are intrinsically connectionless. The GMPLS control plane allows harmonization among the Internet world, based on packet switching, and with the optical world, which is intrinsically circuit-switched. Several network architectures and deployment scenarios have been proposed in the literature. Constraint-based routing algorithms are also a key component for realizing TE strategies.

As far as transport technology is concerned, it is widely recognized that wavelength-division multiplexing (WDM) optical networks will play a significant role in the realization of the next-generation transport infrastructure, which will have to support both traditional and Internet-based services. Besides the general components, TE in optical networks also includes how routing of lightpaths is achieved, how data flows coming from either a circuit or packet bearer network layer are groomed and routed onto the lightpaths, and how lightpath recovery is performed.

6.4.2 Multimedia Traffic Management

Multimedia traffic can be divided into three categories according to the timeliness requirements: (1) interactive audio or video communications; (2) transmission of stored information to another temporary storage; and (3) transmission of stored information for immediate use.

The interactive video or audio communication category includes bidirectional, multiparty, and conference traffic. Owing to the requirement for real-time responsiveness, these services impose the most stringent delay and jitter requirements. Speech traffic alternates between spurts and periods of silence and hence may be modeled by an on/off process. Because the *on* (or *off*) periods do not have a heavy-tailed distribution, a nonfractal version of the on/off model is suitable. On the other hand for interactive VBR video, a self-similar model may be required.

The second category includes transfer of text (e-mail messages, files, documents), movies (offline), and images. It also includes transmissions of data to update databases. By comparison to the first category, services of the second category do not have strict delay and jitter requirements. Hence they may be transferred at a lower delay priority than the real-time services of the first and third categories. The size of files to be transferred varies significantly from a few bytes in an e-mail message to many gigabytes in a movie to be transferred offline. Images may require storage of the order of megabytes (for example, an x-ray may require 50 Mbyte storage). This huge variety in size in the different applications indicates that the distribution of a *chunk of work* introduced by a user may have tail, which in turn leads to long-range dependent traffic. This conjecture is consistent with statistical studies performed on WWW traffic [87].

The third category includes audio and video broadcasting, video on demand, and electronic mail (which may belong either to the second or third category). Because they do not have to cater to interactive communication needs, these services have less stringent delay requirements than those of the first two categories. A large enough buffer can be used to absorb the jitter and to present the traffic stream to the end user without delay fluctuation. The size of such a buffer will dictate the delay requirements of these services, which usually have more stringent delay requirements than those of the second category. The services under this category, which involve VBR video, are both bandwidth hungry and bursty.

The traffic models used for VBR video sequences can be divided into three categories: Markov chains, autoregressive processes (not necessarily Gaussian) and self-similar models.

The challenges of traffic management in telephone networks, data networks, and high-speed multimedia networks that have evolved during the 20th century have led to a vast amount of academic and industrial research in this area. Notice that in telephony, all connections require the same capacity, while in ATM and multimedia networks, different connections require different bandwidths, which may also vary. The fact that ATM and multimedia networks support a multiservice environment, bursty and unpredictable traffic streams, and nonhomogeneous demands for bandwidth has introduced significant additional complexity in traffic management, which in turn has resulted in vast amounts of research being done. In this section, we report on many recent important developments.

The set traffic management functions can be divided into two groups: (1) those that take place during connection setup and (2) those that take place during the connection. Examples of those belonging to the first group are connection admission control and routing. Functions belonging to the second group are aimed at protecting the network from congestion and increasing efficiency; for example, policing to confirm that there are no violations of negotiated traffic parameters by the user, traffic shaping to achieve desired modification of traffic characteristics, selective discard, ABR flow control, and cell scheduling.

6.4.3 Connection Admission Control

Connection admission control (CAC) is set of actions taken by the network during call setup to determine whether a connection request is to be admitted or rejected. If a call requires more than one connection, CAC should be performed for each connection. Based on the CAC scheme, a connection request is admitted only when sufficient capacity is available end to end so that required QoS can be guaranteed for that connection as well as for all other existing connections.

At connection setup, the user and the network negotiate the traffic parameters such as peak cell rate (PCR), sustainable cell rate (SCR), minimum cell rate (MCR), intrinsic burst tolerance (IBT), the cell delay variation tolerance (CDVT), the requests for QoS requirements, and the transfer capability. The result of this negotiation is a traffic contract. The role of the traffic contract is to enable an efficient network operation where QoS requirements of each connection are met. Different transfer capabilities require different traffic parameters. Hence, the set of connection descriptors may change from service to service. For example, a service requiring a CBR transfer capability does not need to specify the SCR like a VBR service. Also, only an ABR connection can specify minimum cell rate. The CAC can be regarded as a system whose input is the traffic parameters, the CDVT, the QoS, and the ATC of the proposed connection. The CAC input parameters are henceforth called *connection descriptors*.

Although many proposals for CAC have been studied and analyzed in the literature, there is no clear agreement on which is the best one for a given set of services. In this section we discuss three fundamentally different approaches for implementing CAC. The first is the simplest and the most conservative. It is based on peak rate allocation. The second is based on the zero buffer approximation or rate envelope multiplexing (REM). The third is based on traffic and queuing models and takes into consideration the buffering capacity on the network to share the excess traffic load during periods of heavy traffic (rate sharing).

CAC Based on Peak Rate Allocation

The description of this approach for implementing CAC is as follows. We consider the peak rate of every new connection. If for every network link, on the route of the connection, the sum of the peak rates of all connections already in progress, plus the peak rate of the new connection, is lower than the total link capacity, admit the new connection, otherwise reject it. This simplistic description does not take into consideration the effect of cell delay variation (CDV). Due to CDV, cell loss may occur even if the sum of the peak rates is lower than the link capacity.

CAC Based on Rate Envelope Multiplexing (REM)

To obtain higher efficiency, the CAC should admit connections such that the total peak rate may exceed the available capacity, and to rely on the fact that, most of the time, there is enough bandwidth to satisfy connections transmitting at their peak because others transmit at a lower rate than their peak. In particular, REM is based on the conservative assumption that there is no buffering, or that the amount of buffering is very small. Accordingly, this assumption is also called the *zero buffer approximation*. This assumption is appealing for the following reasons. First, in many real-time applications the buffers are too small to absorb burst scale fluctuations. It is therefore convenient to assume that the buffers will avoid all cell losses that may be due to cell scale fluctuations. In this case we can use burst scale traffic models such as on/off models together with the zero buffer approximation to obtain an accurate evaluation of the cell loss. Secondly, the zero buffer approximation makes the derivation of CLR much easier because there is no need for queuing analysis. Thirdly, under the zero buffer approximation there is no need to consider the correlation in the arrival process; hence, processes with independent arrivals can be considered. This makes the process of traffic modeling and characterization much simpler. Fourthly, the zero buffer approximation is a conservative assumption, but not too conservative in certain important cases of real-time traffic as discussed earlier.

Let S be a random variable representing the total amount of work arriving in a small time interval (which includes the amount of work arriving by existing connections plus that of the new connection). Let C be the available link capacity. The CLR is given by

$$P_{\text{loss}} = \frac{E\{(S-C)^2\}}{E\{(S)\}} \qquad (6.3)$$

In this equation, cell loss ratio (CLR) is simply the ratio between the amount of work lost and the amount of work arrived. This is the way CLR is defined by the different standards bodies. It is important to recognize the difference between loss probability in terms of amount of work, or number of cells, and the proportion of time that the system is losing cells. The latter is defined by $P(S > C)$. Many authors use the latter to approximate P_{loss} as defined by equation (6.3). It is important to realize that the difference between the two may be very significant.

Example 6.4. Consider the very bursty process in which a very huge burst arrives during the first cell-time, but there are no arrivals afterwards. In other words, the probability $P(S > 0)$ is almost zero. Hence, $P(S > C)$ is almost zero. However, most the cells that arrive are lost, that is, P_{loss}, as given in equation (6.3), is very close to one.

Example 6.5. To implement a CAC based on REM, we need to have the distribution of S at every point in time when a new connection request arrives. This can be estimated either by traffic measurements, or we can rely on traffic models. For example, if the traffic of each of the sources is modeled as a discrete time on/off source with geometric *on* and *off* times, then the distribution of S is binomial. If the total capacity is much larger than the peak rate of each of the active sources, S may be assumed to have a Gaussian distribution. Having the distribution of S, for a newly arriving connection request, we compute the CLR using equation (6.3). If the predicted CLR is lower than the required CLR for all existing connections as well as for the new connections (as specified in their traffic contract), the connection is accepted, otherwise it is rejected.

CAC Based on Rate Sharing

There is a need for a CAC that takes into consideration the buffering capabilities of the switches. One of the important functions of such a CAC is a cell loss ratio (CLR) predictor. Let $S(t)$ be a random variable that represents the amount of work arriving in an interval time t. We know that if during time t, the amount of work arriving is more than what can be served plus what can be buffered, some work must be lost. We also know that the loss must be at least the excess of work over what can be served and buffered. Let $CLR(t)$ be lower bound for the cell loss ratio based on the random variable $S(t)$, given by

$$CLR(t) \geq \frac{E\{(S(t) - C(t) - b)^+\}}{E\{(S(t))\}} \quad (6.4)$$

where $C(t)$ is the amount of work that can be served during time t and b is the buffer size. Because $CLR(t)$ for all t is a lower bound of CLR, we have that

$$CLR \geq \max_t CLR(t) \quad (6.5)$$

If we consider the inequality in (6.5) as an equality, we obtain an estimate for cell loss ratio (CLR). The right-hand side of (6.5) is an accurate estimator for the CLR.

The key problem here is to have accurate estimation of the distributions of $S(t)$ for different t values. These can be obtained using traffic measurements by which moments or histograms are produced and the required distributions can be estimated. Another important factor in estimating the distribution of $S(t)$ for different t is the traffic contract.

If the peak cell rate (PCR) and its cell delay variation tolerance (CDVT) are specified, we have the joint where required distributions should be truncated for small t while the sustainable cell rate (SCR) and the intrinsic burst tolerance (IBT) tell us where the distributions are truncated for large t values.

6.4.4 Resource Allocation

Resource allocation is the key to achieving a certain QoS level objective for a connection requesting a certain amount of bandwidth. There are various resource allocation strategies. We can categorize them as either dynamic or static [88]. Consider an ATM network that contains both physical resources (buffers, bandwidths, processors, etc.) and logical resources (ATM addresses, virtual path and channel identifiers VPCIs, virtual path identifiers VPIs, virtual channel identifiers VCIs, etc.).

QoS and bandwidth are mainly related to physical resources. Objects using allocated physical resources in an ATM network include cells, bursts, virtual channels (VCs), virtual paths (VPs), and signaling messages. A cell occupies a buffer, and cell transmission requires bandwidth. Thus, buffer and bandwidth allocation to cells is required. A burst (i.e., a group of cells) occupies buffers, and its transmission also requires bandwidth. Similarly, buffer and bandwidth allocation to bursts is needed. A whole stream of cells in a VC (or VP) uses buffers, and their transmission also requires bandwidth. In other words, this stream must be transmitted within a QoS objective using buffers and bandwidth. In this sense, cells, bursts, VCs, and VPs use buffers and bandwidth.

Similarly, signaling messages carried through a VC for a signal use processor resources in switching nodes, where normally they are handled with a lower priority than emergency signals or emergency alarms and with a higher priority than background management messages. Among SETUPs, the first-in first-out (FIFO) or the last-in first-out (LIFO) discipline is used. Therefore, processor resources in a switching node are allocated to each signaling message through this prioritized-FIFO (that is, head-of-the-line) or -LIFO discipline (often with periodical interruption). In the future, the technology initiated by load sharing among multiple switching nodes will be possible, for example, in advanced intelligent networks [89].

Normally, buffers and bandwidth are allocated to each cell and each burst allocates buffers and bandwidth to the VC containing the cell or burst. Allocation of buffers and bandwidth to VC connections is handled by connection admission control (CAC). Using the traffic characteristics received from the user through a source traffic descriptor, the QoS objectives, the cell delay variation tolerance (CDVT), conformance definition, and the network conditions (bandwidth, buffers, number of established VC connections, etc.), CAC judges whether a VC connection can be accepted while still meeting the QoS objective of each established VC connection. This means that the cells in each VC connection share and compete for the bandwidth of a VP and for the buffers. (Some switches use per-VC queuing, isolating each VC from the other VC.) Combined with CAC, usage parameter control (UPC) forces each VC connection to conform to the source traffic descriptor; that is, to keep the number of cells and bursts below the level established at setup. As a result, each burst or cell can use bandwidth and buffers because the total number of bursts or cells is less than that which would exhaust the bandwidth and buffers. Therefore, CAC implicitly allocates bandwidth and buffers to each burst or cell [90].

6.4.5 Bandwidth Allocation

For effective bandwidth to be useful it should have the following characteristics:

- *Independence*. Effective bandwidth of a given traffic stream is only related to the statistical characteristics of that traffic stream and the network equipment (e.g., buffering capacity). It should not be dependent on the traffic characteristics of any other stream.
- *Additivity*. The effective bandwidth of the superposition of k streams is equal to the sum of the effective bandwidths of each of the streams.
- *Efficiency*. It should lead to an efficient operation. For example, if we mix k VBR streams where, for each of the streams the CDVT of the PCR is equal to zero, we could set the effective bandwidth to be equal to the PCR for each of the streams. This will be consistent with the definition of effective bandwidth and the first two requirements listed here, but may not necessarily lead to efficient network operation.

Effective bandwidth is a number w_i with connection i such that the grade of service requirement is met if

$$\sum_i w_i \leq C \qquad (6.6)$$

where C is the total available capacity.

If we have an effective bandwidth value for each connection, the CAC function will be simply to compare the sum of these effective bandwidth values (including the one for the connection yet to be admitted), in all networks links on the route of the new connection, with the total available capacity. If they exceed the total capacity on at least one of these links, the connection is rejected, otherwise it is admitted.

For the case of REM, we can obtain effective bandwidth by using (6.6), simply, for given distribution of S to find the smallest value for C such that P_{loss} is not higher than the required CLR.

For the case of rate sharing CAC, the derivation of effective bandwidth requires a formula for the loss probability that takes into consideration the buffer as well as realistic models of the traffic. This is a very difficult problem. A practical approach would be to use traffic traces generated by the different services and to keep a record of CLR versus buffer size for a large range of traffic streams.

Bandwidth allocation delays determine the amount of bandwidth required by a connection for the network to provide the required QoS. There are two alternative approaches for bandwidth allocation: deterministic and statistical multiplexing.

In deterministic multiplexing, each connection is allocated its peak bandwidth. Doing so causes large amounts of bandwidth to be wasted for bursty connections, particularly for those with large peak to average bit rate ratios.

An alternative methods is statistical multiplexing. In this scheme, the amount of bandwidth allocated in the network to a VBR source (hereafter referred to as the *statistical bandwidth* of a connection) is less than its peak but necessarily greater that its average bit rate. Then the sum of peak rates of connections multiplexed onto a link can be greater than the link bandwidth as long as the sum of their statistical bandwidths is less than or equal to the provisioned link bandwidth.

Estimating the value of the statistical bandwidth for an incoming call request must address the following issues:

- QoS requirements of the new connection must be guaranteed.
- QoS provided to preexisting connections must not be degraded to unacceptable levels when they are multiplexed with the new connection.

Accordingly, the statistical bandwidth of a connection depends not only on its own stochastic characteristics, but also strongly on the characteristics of existing connections in the network. The bandwidth allocation mechanisms are summarized in Table 6.1. The bandwidth and buffer allocation mechanisms determine the amount of bandwidth and the number of buffers. Routing determines the location of the bandwidth and buffers to be allocated.

Every bandwidth allocation procedure uses either a static or dynamic allocation strategy. In static allocation a reference model is used to determine the bandwidth allocation. The reference model, such as the source traffic descriptor used for CAC, is given *a priori*. Therefore, static allocation does not modify the allocated bandwidth when traffic conditions change. In dynamic allocation, actual traffic conditions are monitored, and bandwidth is reallocated based on the changing conditions. Thus, with dynamic allocation the allocated bandwidth may be modified and/or reallocated if current conditions are different from the *a priori* ones. ABR source-cell rate control is a typical example of a dynamic allocation strategy.

The CAC used for CBR and VBR ordinarily uses a static allocation strategy. It uses a source traffic descriptor for every connection, but does not use the information for how many cells are actually offered. It evaluates the QoS performance under the assumption that cells from each connection arrive in a process defined by the source traffic descriptor and that such arrival processes are multiplexed. The cell arrival process defined by the source traffic descriptor is the reference model used by CAC. In practice, the source traffic descriptor does not uniquely specify the arrival process, and the transform from the source traffic descriptor to the arrival process is part of the CAC algorithm. There are many examples of CAC methods of this type.

Table 6.1 Bandwidth and buffer allocation

Object Allocation	Example of Allocation Mechanisms
Cell	Source cell rate control (ABR)
Burst	Bandwidth reservation (ABT)
VC	CAC + UPC
VP	Provisioning
ATM layer service	Scheduler/queuing discipline

For example, a CAC evaluates the mean cell rate and cell rate variance using the peak and sustainable cell rates, where the maximum variance is normally used because the variance is not uniquely determined. The CAC assumes that the cell arrival process is independent and the multiplexed cell arrival processes are Gaussian. Then the CAC evaluates the quantile of this Gaussian distribution, where the quantile corresponds to the VP bandwidth. If this quantile, which is an approximated value of the cell loss ratio without buffers, is larger than the cell loss ratio objective, the VC is rejected [91].

Since in static allocation the bandwidth is determined based on a fixed reference model, the determination is usually simple and can be performed offline. If the reference model is valid, the resulting bandwidth allocation should be accurate and effective. On the other hand, the static allocation strategy is inaccurate and ineffective when the reference model is invalid because it does not have functions for monitoring the traffic and modifying the allocated bandwidth as needed. Hence, this strategy does not enable the system to adapt to changing situations.

Dynamic allocation was proposed to overcome the deficiencies of static allocation. Dynamic allocation adapts to the actual situation and dynamically modifies the allocated bandwidth accordingly. It thus needs more advanced functions to monitor the current situation, determine the required bandwidth, allocate the bandwidth, and modify the bandwidth. Dynamic allocation is particularly effective when valid accurate reference models are unavailable. For example, if the users are unable to provide the network with accurate traffic characteristics at connection setup, CAC based on dynamic allocation or ABR source-cell rate control is more effective. Also, when it is difficult to forecast demand for each VP bandwidth, dynamic allocation is better than static provisioning. When the reference model is valid, a static allocation strategy is effective enough.

6.4.6 Congestion Control for Multicast Communications

The demand for services that simultaneously involve groups of users or apply to entire subnets is playing a fundamental role in the current evolution of the Internet. Driven by commercial interests, many new applications are no longer based simply on computer-to-computer interaction, but require interconnection between several computers or distribution of data over entire subnets. It is widely recognized that specific networking services are required both to enable emerging multimedia applications and to accommodate nomadic users and teleworkers.

New challenges arise essentially from the specific requirements of group communication and subnet interconnection that differ from those of standard unicast communication. Multicast service – delivering the same data to a group of receivers – constitutes a major building block in this area. In the Internet, the adoption of IP Multicast and its coexistence with IP Unicast present a new set of challenges that require new approaches. IP Multicast is already a reality, but in order to be widely adopted, it remains to define a completely satisfactory approach to congestion control and provide a range of quality differentiated services.

Widespread deployment of multicast communication in the Internet depends critically on the existence of practical congestion control mechanisms that allow multicast and unicast traffic to share network resources fairly. Most service providers recognize multicast as an essential service to support a range of emerging network applications including audio and video broadcasting, bulk data delivery, and teleconferencing. Nevertheless, network operators have been reluctant to enable multicast delivery in their networks, often citing concerns about the congestion such traffic may introduce. The basic conflict is this: it is desirable to encourage use of multicast where appropriate, to reduce the overall bandwidth demand of applications that transmit high-bandwidth data to many receivers, but the introduction of multicast sessions into the network must not deteriorate the performance of existing unicast traffic. The specific worry of operators is that multicast congestion control protocols may interact too aggressively with the standard unicast congestion control mechanism of Transmission Control Protocol (TCP). This Internet standard unicast transport protocol includes mechanisms for both reliable data delivery and congestion control. There is a clear need for multicast congestion control algorithms that are probably fair to unicast traffic.

Multicast is a service that delivers packets from a sender to a group of receivers. IP Multicast (the network-level multicast standard in the Internet) provides this service by transmitting a copy of each offered packet once on each link of a tree with its root at the sender and a leaf at each receiver. The links in this tree are shared by other traffic, and some of these links may be congested. Multicast congestion control protocols are designed to detect this congestion and adapt the session's flow rate on each link to the available bandwidth. Well-behaved protocols must also ensure that bandwidth on bottleneck links is shared fairly with other competing flows.

So-called *single-rate* multicast protocols require that a given session use the same transmission rate on every link. Thus, the rate for the entire session is limited by the most bandwidth-constrained receiver. In single-rate protocols, rate adaptation is performed by the sender in response to receiver feedback. Single-rate multicast is widely used in real-world multicast applications. In contrast, so-called *multirate* protocols allow the session's flow rate on each link to vary depending on the bandwidth available to downstream receivers. This flexibility is achieved at the expense of degrading the data stream along bandwidth-constrained paths and is thus only appropriate for applications that use highly loss-tolerant data encoding techniques. Rate adaptation in multirate protocols is performed with support from the network. In this article we focus on single-rate multicast.

Optimization-based congestion control employs a simple network model augmented with economic features. Consider a network modeled as a set of directed links with fixed but not, generally, equal capacities. The workload for the network is generated by a set of sessions. A session is described by the subset of network links over which it transmits data and by a variable transmission rate, denoted x. The aggregate data rate on any link, then, is the sum of rates for all sessions using that link. Each session is characterized by a *utility function* $u(x)$, which represents the value of bandwidth to the session. Utility functions have two important properties that capture natural intuitions about session behavior. First, utility is an increasing function of transmission rate because we assume that each session would prefer as high a rate as possible. Secondly, the utility function has decreasing marginal returns, which models the idea that the value of a small increase in transmission rate is high for a session currently transmitting at a low rate, but decreases as the session's rate increases. Two widely used utility functions in congestion control models are *logarithmic utility*, $u(x) = \log x$ and a utility function that is a good model for the behavior of TCP at lower loss rate, which we will call *TCP utility*, $u(x) = -1/x$ [92].

Associated with each link is a real value, which for reasons that will become clear shortly, we informally interpret as a *price* per unit bandwidth and refer to it as the *link price*. For each session, then, we may define the *session price* as the sum of prices over all of the links it used.

In this model, the problem of congestion control can be cast as one of utility maximization. From a networkwide perspective, it is desirable to maximize the *aggregate utility* of all sessions. To make this notion precise, we define aggregate utility as the sum of session utilities, allowing a precise statement of the congestion control problem.

Congestion Optimization Problem

Find a set of rates for all sessions to maximize *aggregate utility* $= \sum_{\text{all sessions}} u(x_i)$ subject to set of rates for all sessions to maximize subject to the constraint that the traffic on each link may not exceed its capacity.

In designing congestion control protocols, we are essentially looking for a distributed algorithm whose rate setting behavior can be naturally interpreted as a decentralized iterative solution to this problem. For such algorithms to be practical, it is essential that each individual session need only rely on locally available information to set its own rate. We therefore assume that each session behaves greedily by setting its rate to maximize its own utility minus the total price assessed by the network. (Recall the session price multiplied by the session's transmission rate.) It is worth emphasizing that despite suggestive names for quantities such as price in the model, one need not think of the link prices as representing actual monetary charges. Rather, these quantities can be thought of as congestion signals, which play the role of aligning supply and demand, similar to the way prices behave in economic models of markets. The rate chosen by a session depends on the session price in an intuitive way: the lower the price, the higher the rate. The key point of optimization-based congestion control is that if the network sets the link prices to accurately reflect the current level of congestion, the sessions will converge to rates that jointly maximize the sum of their utilities, thereby optimizing the networkwide objective subject to the capacity constraints.

6.4.7 Traffic Modeling

The goal in service traffic modeling is to find a traffic model, defined by a small set of statistics that can still capture the significant statistical information of the real traffic, so that an accurate queuing performance evaluation (e.g., evaluation of loss probability of delay) can be obtained if the model is used as input into a queuing system.

The art of modeling, of any system or object, usually involves an understanding of the different components of that system or object. When we analyze a traffic stream generated by a source, we observe that three different components are related to the hierarchy of time-scales: call scale, burst scale, and cell scale. In addition to these three lower time-scales, there are higher layer time-scales related, for example, to work habits, holiday periods, and so on.

An interesting question is how self-similar traffic occurs. The answer is probably related to the hierarchy of time-scales and to the different levels of activities that characterize multimedia traffic. For example, the variance of the amount of work arriving in a time-scale of seconds can be high in VBR video connection due to the fact that some video frames carry much more information than others (e.g., due to scene change). Some Web browsers generate more traffic than others transmitting complex images. Some FTPs generate more traffic than others simply because some data files are much larger than others. On a higher time-scale, burstiness is created by busy hours, time variability in work pressures and requirements, banks transmitting their files and updating their records at night, disasters, and special occasions that generate unusual amounts of traffic.

A large number of traffic models have been proposed over the years that attempt to capture the different statistical characteristics of real traffic for the purpose of performance evaluation and dimensioning.

On/Off Model

According to the on/off model, traffic is altering between two states: on and off. During the on period the traffic is generated (transmitted) at a constant rate r, and during the off period no traffic is generated. The lengths of the *on* and *off* periods are independent. The *on* periods have a common distribution, and the *off* periods have another common distribution.

Let $E(on)$ and $E(off)$ be the means of the *on* and *off* periods, respectively. Then the probability that the process is in the *on* state is given by

$$P_{on} = \frac{E(on)}{E(on) + E(off)} \tag{6.7}$$

Let X_t be a process representing the state of the on/off process at time t with $X_t = 1$ and $X_t = 0$ denoting the process being *on* and *off* at time t, respectively. Let R_t be the process representing the rate of the on/off process at time t. Clearly, $R_t = rX_t$. Because the random variable X_t is a *Bernoulli* random variable with parameter P_{on} the mean of R_t is given by $E(R_t) = P_{on}$ and its variance is given by $var(R_t) = r^2(P_{on})(1 - P_{on})$.

The latter two simple equations for the mean and the variance of an on/off process are very useful in traffic modeling. In many cases, we are given the peak cell rate (PCR) (which is r in our case) and the mean cell rate. Using these two equations, we can estimate the variance, which can be further used when we consider many independent sources multiplexed together. In this case, we can benefit from the fact that the variances of independent random variables are additive and can calculate the variance of the total multiplexed traffic. Such a variance can be an important parameter for evaluation of CLR. Notice that, so far, we have not made any assumption concerning the distribution of the *on* and the *off* periods, so that the result may also be applicable to a wide range of processes including those with long-range dependence.

If we further assume that the *on* and *off* periods are exponentially distributed, then results for the densities of the amount of work arriving in any time interval t can also be obtained.

Markov Modulated Model

A Markov modulated process is a process in which the arrival rate is modulated by an embedded *Markov chain*. The advantage of Markov modulated processes is that they are flexible and allow for the incorporation of different correlation patterns. Markov modulated processes are defined at both continuous and discrete time intervals. They can also be classified based on the possibility of group arrivals. The most used Markov modulated processes are:

- Batch Markovian arrival process (BMAP) – continuous time with batch arrivals.
- Markovian modulated *Poisson* process (MMPP) – continuous time with single arrivals.
- Discrete-time batch *Markovian* arrival process (D-BMAP) – discrete time with batch arrivals.
- Markov modulated *Bernoulli* process (MMBP) – discrete time with single arrivals.

Gaussian Autoregressive Model

A traffic model based on a *Gaussian* process can be described as a traffic process where the amount of traffic generated within any time interval has a Gaussian distribution. There are several ways to represent a Gaussian process. In this section, we define the process as a discrete time one. A discrete time representation may be justified by the nature of ATM where time is usually considered discrete with the smallest time unit being the cell-time.

Let time be divided into fixed-length intervals. Let A_n be a continuous random variable representing the amount of work entering the system during the nth interval. The variable A_n may represent the number of bits or ATM cells entering the system during the nth interval.

According to the Gaussian autoregressive model we assume that A_n, $n = 1, 2, 3, \ldots$ is a kth-order autoregressive process, that is,

$$A_n = a_1 A_{n-1} + a_2 A_{n-2} + \cdots + a_k A_{n-k} b \tilde{U}_n \tag{6.8}$$

where $\tilde{U}_n$ is Gaussian with mean η and variance $\tilde{\sigma}^2$, and a_i and b are real numbers with $|a_i| < 1$.

To characterize real traffic, we need to find the best fit for the parameters $a_1, \ldots, a_k, b$, as well as η and $\tilde{\sigma}^2$. On the other hand, it has been shown in references [93, 94] that in any Gaussian process only three parameters are sufficient to estimate queuing performance. It is therefore sufficient to reduce the complexity involved in fitting all the parameters required in a case of the kth-order autoregressive model and use only the first-order autoregressive process. In this case we assume that the A_n process is given by

$$A_n = a_n A_{n-1} + b \tilde{U}_n \tag{6.9}$$

where $\tilde{U}_n$ is Gaussian with mean η and variance $\tilde{\sigma}^2$, and a_n and b are real numbers with $|a_n| < 1$. This model is proposed in reference [95] for a VBR traffic stream generated by a single source video telephony, ($\tilde{\sigma}^2 = 1$). Let $\lambda = E\{A_n\}$ and $\sigma^2 = \text{VAR}\{A_n\}$.

The A_n values can be negative with positive probability. This may seem to hinder the application of this model to real traffic processes. However, in modeling traffic, we are not necessarily interested in a process that is similar in every detail to the real traffic. What we are interested in is a process that has the property that when it is fed into a queue, the queuing performance is sufficiently close to that of the queue fed by the real traffic.

Self-Similar Models

There are many ways to generate self-similar traffic. We present three models, each of which has its own justification to be a *realistic* traffic model. The first will be based on the on/off model, the second on the M/Pareto/∞ model, and the third is the *fractional Brownian motion* (FBM) process, which is in fact a Gaussian process with long-range dependence characteristics.

An on/off model with the *off* (silence period) and/or the *on* (activity) period having a distribution with a heavy tail is self-similar. An example of such a distribution is the Pareto distribution. A random number X has a Pareto distribution with parameters $\theta > 0$ and $1 < \gamma < 2$ if

$$P(x > t) = [\theta/(t+\theta)^\gamma] \tag{6.10}$$

A heavily tailed distribution of the *on* period can be justified by transmission of extremely long files (e.g., complex images) occasionally by a source. A heavy-tailed *off* period is related to human/business behavior such as people going on holiday, conferences, and so on. To reduce the number of parameters it is convenient to use the Pareto distribution with $\theta = 1$, and we talk about Pareto distribution with parameter $1 < \gamma < 2$. In this case the complementary distribution function is given by

$$P(x > t) = [(t+\theta)]^{-\gamma} \quad \text{for } t > 0 \tag{6.11}$$

The mean of the Pareto distribution is given by

$$E\{X\} = \int_0^\infty x(x+1)^{-\gamma} dx = 1/(\gamma - 1) \tag{6.12}$$

The M/Pareto/∞ model has been used to model traffic generated by a large number of sources accessing a certain buffer in a switch. According to this model, chunks of work (e.g., files) are generated at points in time in accordance with a *Poisson* process with parameter λ. The size of these chunks of work has a Pareto distribution, and each of them is transmitted at a fixed rate *r*. At any point in time, we may have any number of sources transmitting at rate *r* simultaneously because according to the model, new sources may start transmission while others are active. If *m* sources are simultaneously active, the total rate equals *mr*. The symbol ∞ in the notation of the model is a representation of the unlimited number of active resources transmitting simultaneously.

Again, let time be divided into fixed-length intervals, and let A_n be a continuous random variable representing the amount of work entering the system during the *n*th interval. For convenience, we assume that the rate *r* is the amount transmitted by a single source within one time interval if the source was active during the entire interval. We also assume that the Poisson rate λ is per time interval. That is, the total number of transmissions to start in one time interval is λ.

To find the mean of A_n for the M/Pareto/∞ process, we consider the total amount of work generated in one time interval. The reader may notice that the mean of the total amount of work generated in one time interval is equal to the mean of the amount of work transmitted in one time interval. Hence,

$$E\{A_n\} = \lambda r/(\gamma - 1) \tag{6.13}$$

Also, another important relationship for this model, which is provided here without proof, is

$$\gamma = 3 - 2H \tag{6.14}$$

where *H* is the *Hurst* parameter.

Using the last two equations, we are able to fit the overall mean of the process $E\{A_n\}$ and the Hurst parameter of the process with those measured in a real-life process, and generate traffic based on the M/Pareto/∞ model.

Pseudo-Self-Similar Models

Models that are not self-similar, such as those based on Markov chains and autoregressive Gaussian, are more amenable to queuing analysis. On the other hand, self-similar models better capture statistical characteristics of real traffic. It is therefore useful to use models that experience short-range dependence in the mathematical sense, with an exponential decaying autocorrelation function, and are amenable to queuing analysis but do exhibit enough long-range dependence to capture the heavy autocorrelation for the range of interest. Notice that it is not very important to find a traffic model that has the same autocorrelation function as that of the real traffic for its entire range. For example, if we consider the case of the zero buffer approximation, correlation does not have any effect on queuing performance. The larger the buffer, the longer the range of correlation that should be matched. Now, since no buffer is of infinite size, the significant range of the autocorrelation function is also limited.

Modeling the Aggregated Traffic

Several video traffic models are proposed to model a single source at frame level and for subdivisions of the frame. However, the models for aggregate traffic should reproduce the distribution of cells in time, caused by the aggregation of the traffic of several sources with a determined packetization process. Previously, the aggregate video traffic was modeled with the sum of bits generated by frame. These models do not consider the possibility of services with random offset in time [96].

To model the aggregated traffic more accurately, we assume cells generated in a scale interval. A model should capture the essential properties of the source in order to generate an equivalent traffic artificially. In this way, the model can be used simultaneously at the place of real traffic, with the computational load saving that this implies. The autocorrelation function is the most important statistical function related to the time. The marginal distribution also has an important impact in the behavior of the model. The series number of cells generated in a slice interval can be seen as a time series that can be described with an autoregressive moving average (ARMA) model [97]. The ARMA model, unlike Markovian and autoregressive models, allows us to generate sequences with a more accurate autocorrelation function. A transformation of the sequence is required if another type of distribution is desired.

6.5 QUALITY OF SERVICE (QoS) IN NETWORK MULTIMEDIA SYSTEMS

Meeting quality of service (QoS) guarantees in multimedia systems is an end-to-end issue, that is, from application to application. A key observation is that for applications relying on the transfer of multimedia and, in particular, continuous media flows, it is essential that quality of service be configurable, predictable, and maintainable systemwide, including the end-system devices, communication subsystems, and networks. Furthermore, it is also important that all end-to-end elements of distributed systems architecture work in unison to achieve the designed application-level behavior.

To date, most of the developments in the area of QoS support have occurred in the context of individual architectural components [98]. Much less progress has been made in addressing the issue of an overall QoS architecture for multimedia communications. However, considerable progress has been made in the separate

6.5 QUALITY OF SERVICE (QoS) IN NETWORK MULTIMEDIA SYSTEMS

areas of distributed systems platforms [99], operating systems [100], transport systems [101], and multimedia networking [102] support for quality of service. In end systems, most of the progress has been made in the areas of scheduling [103], flow synchronization [104], and transport support. In networks, research has focused on providing suitable traffic models and service disciplines, as well as appropriate admission control and resource reservation protocols [105]. Many current network architectures, however, address quality of service from a provider's point of view and analyze network performance, failing to address comprehensively the quality needs of applications.

The current state of QoS support on architectural frameworks can be summarized as follows:

- *Incompleteness.* Current interface (e.g., application programming interfaces) are generally not QoS configurable and provide only a small subset of the facilities needed for control and management of multimedia flows.
- *Lack of mechanisms to support QoS guarantees.* Research is needed in distributed control, monitoring, and maintenance of QoS mechanisms so that contracted levels of service can be predictable and assured.
- *Lack of overall framework.* It is necessary to develop an overall architectural framework to build on and reconcile the existing notion of quality of service at different system levels and among different network architectures.

In recognition of these limitations, a number of research teams have proposed systems architectural approaches to QoS support. In this chapter these are referred to as QoS architectures [106]. The intention of QoS architecture research is to define a set of QoS configurable interfaces that formalize quality of service in the end system and network, providing a framework for the integration of QoS control and management mechanisms.

A number of QoS principles motivate the design of a generalized QoS framework:

- The *transparent principle* states that applications should be shielded from the complexity of underlying QoS specification and QoS management. The benefits of transparency are that it reduces the need to embed functionality in the application, hides the detail of underlying service specification from the application, and delegates the complexity of handling QoS management activities to the underlying framework.
- The *integration principle* states that QoS must be configurable, predictable, and maintainable over all architectural layers to meet end-to-end QoS [107]. Flows traverse resource modules (e.g., CPU, memory, multimedia devices, network, etc.) at each layer from source media devices, down through the source protocol stack to the playout devices. Each resource module traversed must provide QoS configurability (based on QoS specification), resource guarantees (provided by QoS control mechanisms), and maintenance of ongoing flows.

- The *separation principle* states that media transfer, control, and management are functionally distinct architectural activities. The principle states that these tasks should be separated in architectural QoS frameworks. One aspect of this separation is the distinction between signaling and media transfer. Flows (which are isochronous in nature) generally require a wide variety of high-bandwidth, low-latency, nonassured services with some form of jitter correction. On the other hand, signaling (which is full duplex and asynchronous in nature) generally requires low-bandwidth, assured-type services [108].
- The *multiple timescale principle* guides the division of functionality between architectural modules and pertains to the modeling of control and management mechanisms. It is necessitated by, and is a direct consequence of, fundamental time constraints that operate in parallel between resource management activities (e.g., scheduling, flow control, routing, and QoS management) in distributed communications environments.
- The *performance principle* subsumes a number of widely agreed rules for the implementation of QoS-driven communication subsystems, which guide the division of functionality in structuring communication protocols for high performance in accordance with system design principles, avoidance of multiplexing, recommendations for structuring communications protocols, and the use of hardware assists for efficient protocol processing [109].

QoS specification encompasses, but is not limited to the following:

- *Flow performance specification* characterizes the user's flow performance requirements. The ability to guarantee traffic throughput rates, delay, jitter, and loss rates is particularly important for multimedia communications. These performance-based metrics are likely to vary from one application to another. To be able to commit necessary end system and network resources, QoS frameworks must have prior knowledge of the expected traffic characteristics associated with each flow before resource guarantees can be met [110].
- *Level of service* specifies the degree of end-to-end resource commitment required (e.g., deterministic, predictive, and best effort). While the flow performance specification permits the user to express the required performance metrics in a quantitative manner, level of service allows these requirements to be refined in a qualitative way to allow a distinction to be made between hard and soft performance guarantees. Level of service expresses a degree of certainty that the QoS levels requested at the time of flow establishment or renegotiation will be honored [111].
- *QoS management policy* captures the degree of QoS adaptation that the flow can tolerate and the scaling actions to be taken in the event of violations in the contracted QoS. By trading off temporal and spatial qualities to available bandwidth, or manipulating the playout time of continuous media in response to variation in delay, audio and video flows can be presented at the playout device with minimal perceptual distortion. The QoS management policy also

includes application-level selection for QoS indications in the case of violations in the requested quality of service and periodic QoS availability notifications for bandwidth, delay, jitter, and loss.
- *Cost of service* specifies the price the user is willing to incur for the level of service. Cost of service is a very important factor when considering QoS specification. If there is no notion of cost of service involved in QoS specification, there is no reason for the user to select anything other than the maximum level of service, for example, guaranteed service.
- *Flow synchronization specification* characterizes the degree of synchronization (i.e., tightness) among multiple related flows. For example, simultaneously recorded video perspectives must be played in precise frame-by-frame synchronism so that relevant features may be simultaneously observed. On the other hand, lip synchronization in multimedia flows does not need to be absolutely precise when the main information channel is auditory and video is only used to enhance the sense of presence.

6.5.1 Selection and Configuration of QoS Mechanisms

QoS mechanisms are selected and configured according to user-supplied QoS specifications, resource availability, and resource management policy. In resource management, QoS mechanisms can be categorized as either static or dynamic in nature. Static resource management deals with flow establishment and end-to-end QoS renegotiation phases (which we describe as QoS provision), and dynamic resource management deals with the media-transfer phase (which we describe as QoS control and management). The distinction between QoS control and QoS management is characterized by different time-scales over which they operate. QoS control operates on a faster time-scale than QoS management.

QoS Provision

This provision is comprised of the following components.

- *QoS mapping* performs the function of automatic translation among representations of QoS at different levels (i.e., operating system, transport layer, and network) and thus relieves the user of the necessity of thinking in terms of lower level specification. For example, the transport-level QoS specification may express flow requirements in terms of level service, average and peak bandwidths, jitter, loss, and delay constraints. For admission testing and resource allocation purposes, this representation must be translated to something more meaningful to the end system.
- *Admission testing* is responsible for comparing the resource requirement arising from the requested QoS against the available resources in the system. The decision about whether a new request can be accommodated generally depends on system-wide resource management policies and resource availability. Once admission testing has been successfully completed on a particular resource

model, local resources are reserved immediately and then committed later if the end-to-end admission control is successful.

- *Resource reservation protocols* arrange for the allocation of suitable end-system and network resource s according to the user's QoS specification. In doing so, the resource reservation protocol interacts with QoS-based routing to establish a path through the network in the first instance; then, based on QoS mapping and admission control at each local resource module traversed (e.g., CPU, memory, I/O devices, switches, and routers), end-to-end resource are allocated. The end result is that QoS control mechanisms such as network-level cell/packet schedulers and end-system thread schedulers are configured accordingly.

QoS Control

QoS control mechanisms operate on time-scales at or close to media transfer speeds. They provide real-time traffic control of flows based on requested levels of QoS established during the QoS provision phase. The fundamental QoS control mechanisms include the following.

- *Flow scheduling* manages the forwarding of flows (chunks of media based on application-layer framing) in the end system [112] and network (packet/or cells) in an integrated manner. Flows are generally scheduled independently in the end systems but may be aggregated and scheduled in unison in the network. This is dependent on the level of service and the scheduling discipline adopted.
- *Flow shaping* regulates flows based on user-supplied flow performance specifications. Flow shaping can be based on a fixed-rate throughput (i.e., peak rate) or some form of statistical representation (i.e., sustainable rate burstiness) of the required bandwidth. The benefit of shaping traffic is that it allows the QoS frameworks to commit sufficient end-to-end resources and to configure flow schedulers to regulate through the end systems and network. It has been mathematically proven that the combination of traffic shaping at the edge of the network and scheduling in the network can provide hard performance guarantees.
- *Flow policing* can be viewed as the dual monitoring. Monitoring, which is usually associated with QoS management, observes whether the QoS contracted by a provider is being maintained, whereas policing observes whether the QoS contracted by a user is being adhered to. Policing is often only appropriate where administrative and charging boundaries are being crossed; for example, at a user-to-network interface flow shaping schemes at the source allow the policing mechanisms to detect misbehaving flows.
- *Flow control* includes both open-loop and closed-loop schemes. Open-loop flow control is used widely in telephony and allows the sender to inject data into the network at the agreed levels given that resources have been allocated in advance. Closed-loop flow control requires the sender to adjust its rate

based on feedback from the receiver or network. Applications using closed-loop flow control based protocols must be able to adapt to fluctuations in the available resources. On the other hand, applications that cannot adjust to changes in the delivered QoS are more suited to open-loop schemes where bandwidth, delay, and loss can be deterministically guaranteed for the duration of the session.
- *Flow synchronization* is required to control the event ordering and precise timings of multimedia interactions. Lip-sync is the most commonly cited form of multimedia synchronization (i.e., synchronization of video and audio flows at a playout device). Other synchronization scenarios reported include event synchronization with or without user interaction, continuous synchronization other than lip-sync, and continuous synchronization for disparate sources and sinks. All scenarios place fundamental QoS requirements on flow synchronization protocols.

QoS Management

In order to maintain agreed levels of QoS, it is often insufficient to adjust committed resources. Rather, QoS management is frequently required to ensure that the contracted QoS is sustained. QoS management of flows is functionally similar to QoS control. However, it operates on a slower time scale, that is, over longer monitoring and control intervals. The fundamental QoS management mechanisms include the following.

- *QoS monitoring* allows each level of the system to track the ongoing QoS levels achieved by the lower layer. QoS monitoring often plays an integral part in a QoS maintenance feedback loop, which maintains the QoS achieved by resource modules. Monitoring algorithms operate over different time-scales. For example, they can run as part of a scheduler (as a QoS control mechanism) to measure individual performance of ongoing flows. In this case, measured statistics can be used to control packet scheduling and admission control. Alternatively, QoS monitoring can operate on an end-to-end basis as part of transport-level feedback mechanisms or as part of the application itself [113].
- *QoS availability* allows the application to specify the interval over which one or more QoS parameters (e.g., delay, jitter, bandwidth, loss, synchronization) can be monitored and the application informed of the delivered performance via a QoS signal. Both single and multiple QoS signals can be selected based on the user-supplied QoS management policy.
- *QoS degradation* issues a QoS indication of the user when it determines that the lower layers have failed to maintain the QoS of the flow and nothing further can be done by the QoS maintenance mechanism. In response to such an indication, the user can choose either to adapt to the available level of QoS or scale back to a reduced level of service (i.e., end-to-end renegotiation).
- *QoS maintenance* compares the monitored QoS against the expected performance and then exerts tuning operations (i.e., fine- or coarse-grain resource

adjustments) on resource modules to sustain the delivered QoS. Fine-grain resource adjustment counters QoS degradation by adjusting local resource modules (e.g., loss via the buffer management, throughput via the flow regulation, and queuing delays and continuous media playout calculation via the flow scheduling).

- *QoS scalability* comprises QoS filtering (which manipulates flows as they progress through the communications system) and QoS adaptation (which scales flows at the end systems only) mechanisms. Many continuous media applications exhibit robustness in adapting to fluctuations in end-to-end QoS. Based on the user-supplied QoS management policy, QoS adaptation in the end systems can take remedial actions to scale flows appropriately. Resolving heterogeneous QoS issues is a particularly acute problem in the case of multicast flows. Here, individual receivers may have differing QoS capabilities to consume audiovisual flows; QoS filtering helps to bridge this heterogeneity while simultaneously meeting individual receiver's QoS requirements.

6.5.2 QoS Architecture

Until recently, research in providing QoS guarantees has focused primarily on network-oriented traffic models and service scheduling disciplines. These guarantees are not, however, end-to-end in nature. Rather they preserve QoS guarantees only among network access points to which end systems are attached. Work on QoS-driven end-system architecture needs to be integrated with network configurable QoS services and protocols to meet application-to-application QoS requirements. In recognition of this, researchers have recently proposed new communication architectures that are broader in scope and cover both network and end-system domains.

The quality of service architecture (QoS-A) is a layered architecture of services for quality-of-service management and control of continuous media flows in multiservice networks. The architecture incorporates the following key notions: *flows*, which characterize the production, transmission and eventual consumption of single media streams (both unicast and multicast) with associated QoS; *service contracts*, which are binding agreements of QoS levels between users and providers; and *flow management*, which provides for the monitoring and maintenance of the contracted QoS levels. The realization of the flow concept demands active QoS management and tight integration among device management, end-system thread scheduling, communications protocols, and networks.

In functional terms, the QoS-A (as illustrated in Figure 6.18) is composed of a number of layers and planes. The upper layer consists of a distributed applications platforms augmented with services to provide multimedia communications and QoS specification in an object-based environment. Below the platform level is an orchestration layer that provides jitter correction and multimedia synchronization services across multiple related application flows. Supporting this is a transport layer that contains a range of QoS configurable services and mechanisms. Below this, an Internet working layer and lower layers form the basis for end-to-end QoS support [114]. QoS management is realized in three vertical planes in the QoS-A. The protocol

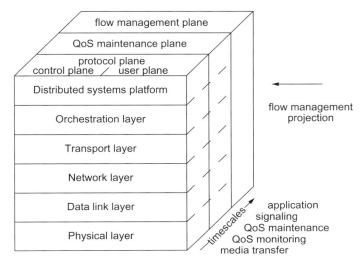

Figure 6.18 Quality of service architecture [114]. (©1997 IEEE.)

plane, which consists of distinct user and control subplanes, is motivated by the principle of separation. QoS-A uses separate protocol profiles for the control and media components of flows because of the different QoS requirements of control and data. The QoS maintenance plane contains a number of layer-specific QoS managers. These are each responsible for the fine-grained monitoring and maintenance of their associated protocol entities.

For example, at the orchestration layer, the QoS manager is illustrated in the tightness of synchronization between multiple related flows. In contrast, the transport QoS manager is concerned with intraflow QoS such as bandwidth, loss, jitter, and delay. Based on flow monitoring information and a user-supplied service contract, QoS managers maintain the level of QoS in the managed flow by means of fine-grained resource running strategies. The final QoS-A plane pertains to flow management, which is responsible for flow establishment (including end-to-end admission control, QoS-based routing, and resource reservation), QoS mapping (which translates QoS representations among layers), and QoS scaling (which constitutes QoS filtering and QoS adaptation for coarse-grained QoS maintenance control).

6.5.3 OSI QoS Framework

One early contribution to the field of QoS-driven architecture is the OSI QoS framework, which concentrates primarily on quality-of-service support for OSI communications. The OSI framework broadly defines terminology and concepts for QoS and provides a model that identifies objects of interest to QoS in open system standards. The QoS associated with objects and their interactions is described through the

630 NETWORK LAYER

definition of a set of QoS characteristics. The key OSI QoS framework concepts include the following:

- *QoS requirements* which are realized through QoS management and maintenance entities.
- *QoS characteristics*, which are a description of the fundamental measures of QoS that have to be managed.
- *QoS categories*, which represent a policy governing a group of QoS requirements specific to a particular environment such as time-critical communications.
- *QoS management functions*, which can be combined in various ways and applied to various QoS characteristics in order to meet QoS requirements.

A block scheme of the OSI QoS framework is illustrated in Figure 6.19. This framework is made up of two types of management entities, *layer-specific* and *systemwide* entities, that attempt to meet the QoS monitoring requirements by monitoring, maintaining, and controlling end-to-end QoS. The task of the policy control function is to determine the policy that applies at a specific layer of an open system. The policy control function models any priority actions that must be performed to control the operation of a layer. The definition of a particular policy is layer specific and therefore cannot be generalized. Policy may, however, include aspects of security, time-critical communications, and resource control. The role of the QoS control function is to determine, select, and configure the appropriate protocol entities to meet layer-specific QoS goals. The system management agent is used in conjunction with OSI systems management protocols to enable system resources to be remotely managed. The local resource manager represents end-system control of resources. The system QoS control function combines two systemwide capabilities: to tune

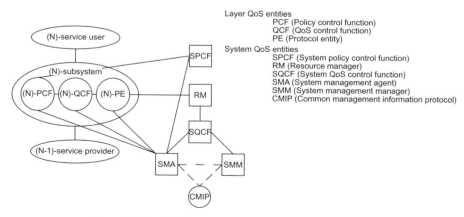

Figure 6.19 OSI QoS framework [115]. (©1995 ISO/IEC.)

performance of protocol entities and to modify the capability of remote systems via OSI systems management. The OSI systems management interface is supported by the systems management manager, which provides a standard interface to monitor, control, and manage end systems. The system policy control function interacts with each layer-specific policy control function to provide an overall selection of QoS functions and facilities.

6.5.4 QoS from Providers' and Customers' Viewpoints

The existing definition of QoS lacks the clarity required to express separately the service providers' and customers' viewpoints, as illustrated in Figure 6.20. QoS required by the customer is a statement of the level of quality of a particular service required or preferred by the customer. The level of quality may be expressed by the customer in technical or nontechnical language. A typical customer is not concerned with how a particular service is provided or with any of the aspects of the network's internal design, but only with the resulting end-to-end service quality. It must be recognized that the customer's QoS requirements can be sometimes subjective. These requirements are useful, although subjective. It is up to the service provider to translate them into something of objective use [116].

QoS offered by the service provider is a statement of the level of quality that will be offered to the customer. This is the level of service that the service provider can achieve with the design of the network. The level of quality will be expressed by values assigned to network performance parameters, which not only cover the network, but also network support [117].

QoS achieved by the service provider is a statement of the level of quality achieved by the service provider. It is a record of the levels of quality that have

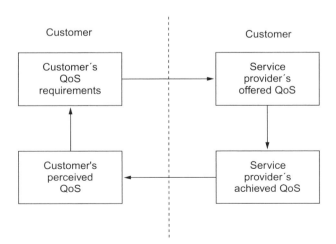

Figure 6.20 The four QoS viewpoints.

actually been achieved. These are expressed by values assigned to the parameters specified for the offered QoS. These performance values are summarized for specified periods of time, for example, for the previous three months and/or on an annual basis.

QoS perceived by the customer is a statement expressing the level of quality experienced by the customer. The perceived QoS is expressed usually in terms of degrees of satisfaction and not in technical terms. The perceived QoS is accessed by various methods including customer surveys, customer comments, and customer complaints. Figure 6.21 shows how the various QoS view points interrelate with one another. The service provider and the network provider have been separated in the diagram to illustrate the fact that the service provider need not always be the network provider.

However, the service provider must always take full responsibility for the QoS offered to the customer. From the intrarelationships among the QoS viewpoints, it can be concluded that both the customer's and service provider's quality interest must be in a state of equilibrium in order for successful business relationships to be achieved. Therefore, it is necessary to manage the activities and relationships associated with QoS viewpoints to obtain the optimum quality levels in accordance with the price the customer is willing to pay.

The effect on end-to-end image quality of packet loss is not yet well defined. In early MPEG reference models, cell loss rates lower than 10^{-9} were proposed, but rates of 10^{-4} are currently being considered as acceptable. The effect of cell loss is not only dependent on the average cell loss rate, but also on the distribution of cell loss over time. Periods of high cell loss due to network congestion can have a seriously detrimental impact on image quality [118].

Delay requirements vary depending on the application. For interactive video applications, a maximum end-to-end delay of some 100 ms is appropriate, while a much longer delay would be tolerable for a user simply watching a recorded clip

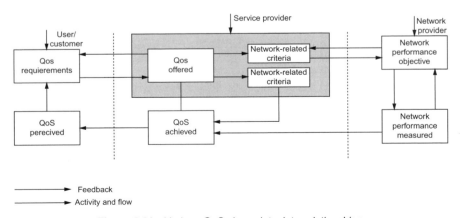

Figure 6.21 Various QoS viewpoints: intrarelationships.

or movie in a video playback application. Delay requirements have a strong impact on the type of network service to be provided [3, 119]. For example, in the case of video playback, considerable variation in network delays of successive cells can be absorbed by a large buffer in a set-top box. On the other hand, the tight delay constraints for real-time communication limit the possibility of dealing with the congestion on network lines by cell buffering. Coding delays must be included in the overall delay budget, thus limiting the scope for rate smoothing in a closed-loop coder producing CBR output.

6.5.5 Quality of Service Parameters

QoS is defined by specific parameters for cells that are conforming to the traffic contract. It is defined on an end-to-end basis. This perspective is actually meaningful to an end user. The definition of end can be the end workstation, a customer premises network, a private ATM UNI, or a public ATM UNI. QoS is defined in terms of one of the following measurement outcomes, where the measurement is done with respect to cells sent from an originating user to a destination user:

- a *transmitted cell* from the originating user
- a *successfully transferred cell* to the destination user
- a *lost cell*, which does not reach the destination user
- an *error cell*, which arrives at the destination but has errors in the payload
- a *misinserted cell*, which arrives at the destination but was not sent by the originator. This can occur due to an undetected cell header error or a configuration error.

The QoS parameters are defined in terms of the above outcomes by the following definitions [3]:

$$\text{Cell loss ratio} = \frac{\text{Lost cells}}{\text{Transmitted cells}} \tag{6.15}$$

$$\text{Cell error ratio} = \frac{\text{Errored cells}}{\text{Successfully transmitted cells} + \text{Errored cells}} \tag{6.16}$$

$$\text{Severely errored cell block ratio} = \frac{\text{Severely errored cell blocks}}{\text{Total transmitted cell blocks}} \tag{6.17}$$

A severely errored cell block is defined as the case where more than M out of N cells are in error, lost, or misinserted.

$$\text{Cell misinsertion rate} = \frac{\text{Misinserted cells}}{\text{Time interval}} \tag{6.18}$$

Delay can occur on the sending and receiving sides of the end terminal, in intermediate ATM nodes, and in the transmission links connecting ATM nodes.

Cell delay variation (CDV) is currently defined as a measure of cell clamping, which is heuristically how much more closely the cells are spaced than the nominal interval. CDV can be computed at a single point against the nominal intercell spacing, or from an entry to exit point. Cell clamping is of concern because if too many cells arrive too closely together, then cell buffers may overflow.

The error rate is principally determined by fiber optic error transmission characteristics and is common to all QoS classes. Average delay is largely impacted by the propagation delay in the WAN, and average queuing behavior. A lower bound on loss is determined by fiber optic error characteristics, with higher values of loss dominated by the effects of queuing strategy and buffer sizes. Delay, delay variations, and loss are impacted by buffer size and buffering strategy.

For those connections that do not (or cannot) specify traffic parameters and QoS class, there is a capability defined by the ATM Forum as best effort where no QoS guarantees are made and no specific traffic parameters need be stated. This traffic can also be viewed as *at risk*, since there are no performance guarantees. In this case, the network admits this traffic and allows it to utilize capacity unused by connections which have specified traffic parameters and have requested a QoS class. It is assumed that connections utilizing the best effort capability can determine the available capacity on the route allocated by the network.

The main problem of QoS parameters estimation is essentially that ATM traffic has almost no statistical structure. The underlying processes we are interested in are the occurrences of losses, the cell delay, and the cell delay variation processes in an ATM connection. If we assume wide sense stationarity (WSS) in these processes, even fulfilled for self-similar traffic, the estimation of the delays' distribution is biased since at least the variance and all ordinary moments of degree two and higher are biased. The bias of the variance estimator of any random variable X is given by

$$E[\text{Var}X] = \text{Var}X \left(1 - 2 \sum_{j=1}^{N-1} \left(\frac{(1 - \frac{j}{N})\rho_j}{N - 1}\right)\right) \tag{6.19}$$

where N is the number of samples and ρ_j is the normalized autocovariance function (ACF) of the process X at log j. As the number of samples N (the measurement window) are increased, this bias tends towards zero, even if the ACF does not decrease quickly, as is the case in self-similar processes. The online measurement of the losses shows, due to correlation effects, similar behaviors as the delays' process. In other words, the measured QoS parameters only converge towards a constant value as long as the traffic is at least ergodic and WSS. These conditions are generally fulfilled in simulations but will certainly not be in reality. Thus, since nonstationarity means that the distributions are time-dependent, all moments of the measured process are also time-dependent.

The main consequence is that any window-based measurement will give wrong results. The longer the measurement is, that is, the wider the window is, the more the effects of nonstationarity will be taken into account.

Example 6.5. Let us define the occurrences of losses of stochastic process $X_k = n$ $(n \geq 0)$, the number of losses in one connection at slot time k. Since from the beginning until the end of the connection, the random variable X has an assigned value for each time slot, the number of samples N is the number of ATM slots in the connection. The mean of this process is an estimator of the CLR. This estimator converges with increasing N since the bias of the variance becomes negligible.

Example 6.6. Consider the time needed to measure a CLR of 10^{-5} in a 1.5 Mbps connection. For about a hundred cells lost, this time is about 45 minutes. For the same connection, the 10^{-6} delay quantile is in the range of 10 seconds. The losses and delays in such a connection are correlated. The same estimator used in different time periods over a connection will provide different estimations in the same QoS parameter.

6.5.6 Quality of Service Classes

In order to make things simpler on users, small numbers of predefined QoS classes are defined, with particular values of parameters prespecified by a network in each of a few QoS classes. The ATM Forum *UNI Specification Version 3.0* defines the five numbered QoS classes and example applications summarized in Table 6.2.

In the unspecified QoS class, no objective is specified by the network operator for the performance parameters. Services using the unspecified QoS class may have explicitly specified traffic parameters. An example application of the unspecified QoS class is the support of best effort service, where effectively no traffic parameters are specified. For this type of *best effort* service, the user does not effectively specify any traffic parameters and does not expect a performance commitment from the network. One component of the best effort service is that the user application is expected to adapt to the time-variable, available network resources. The interpretation and clearer definition of the best effort service is an ongoing activity in the ATM Forum. The current name for this type of service is the unspecified bit rate (UBR).

For each specified QoS class, there is one specified objective value for each performance parameter, where a particular parameter may be essentially unspecified – for example, a loss probability of 1. Initially, each network provider should define the ATM performance parameters for at least the following service classes from ITU-T Recommendation I.362 in a reference configuration:

- *Service Class A*: circuit emulation, constant bit rate video
- *Service Class B*: variable bit rate audio and video
- *Service Class C*: connection-oriented data transfer
- *Service Class D*: connectionless data transfer.

Table 6.2 ATM Forum QoS classes [120]

QoS Class	QoS Parameters	Application
0	Unspecified	*Best Effort At Risk*
1	Specified	Circuit emulation, CBR
2	Specified	VBR Video/Audio
3	Specified	Connection-oriented data
4	Specified	Connectionless data

In the future, more QoS classes may be defined for a given service class [3]. The following specified QoS classes are defined by the ATM Forum:

- Specified QoS Class 1 supports a QoS that meets service class A performance requirements. This class should yield a performance comparable to current digital private line performance.
- Specified QoS Class 2 supports a QoS that meets service class B performance requirements. This class is intended for packetized video and audio in teleconferencing and multimedia applications.
- Specified QoS Class 3 supports a QoS that meets service class C performance requirements. This class is intended for interoperation of connection-oriented protocols, such as Frame Relay.
- Specified QoS Class 4 supports a QoS that meets service class D performance requirements. This class is intended for interoperation of connectionless protocols.

Figure 6.22 gives a concrete example of how the QoS parameters for cell loss ratio for the CLP = 0 flow and cell delay variation may be assigned for the four specified QoS classes. A network operator may provide the same performance for all or a subset of specified QoS classes, subject to the constraint that the requirements of the most stringent service class are met.

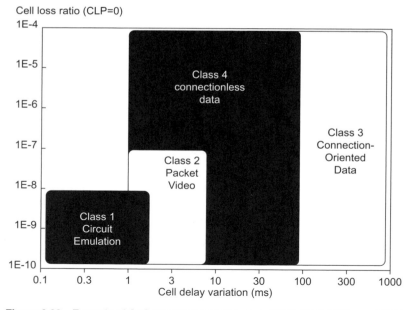

Figure 6.22 Example of QoS class value assignments [120]. (©ATM Forum 1996.)

6.5 QUALITY OF SERVICE (QoS) IN NETWORK MULTIMEDIA SYSTEMS

Various applications have different QoS requirements. For example, some well-known requirements exist for voice after 30 years of experience in telephony. If voice has greater than about 15 ms of delay, then echo cancellation is usually required. Packetized voice can accept almost a 1 percent cell loss rate without being objectionable to most listeners. Newer applications do not have such a basis, or well-defined requirements; however, there are some general requirements.

Video application requirements depend upon several factors, including the video coding algorithm, the degree of motion required in the image sequence, and the resolution required in the image. Loss generally causes some image degradation, ranging from distorted portions of an image to loss of an entire frame, depending upon the extent of the loss and the sensitivity of the video coding algorithm. Also, variations in delay of greater than 20 to 40 ms can cause perceivable jerkiness in the video playback.

Users of interactive applications are also sensitive to loss and variations in delay due to retransmissions, and inconsistent response time, which can decrease productivity. Consistent response time (or the lack thereof) can affect how users perceive data service quality.

Distributed computing and database applications can be very sensitive to absolute delay, loss, and variations in delay. The ideal for these types of applications is infinite bandwidth with latency close to that of the speed of light in fiber. A practical model is that of performance comparable to a locally attached disk drive or CD ROM, which ranges from 10 to 100 ms.

In ATM networks, QoS will be selectable and controllable from both terminal capability and network performance. Thanks to VBR communication realized by ATM, the terminal can generate its coded data at any bit rate regardless of the constants or variables during communication. This means that terminals will be able to demand any QoS they wish.

However, it is not possible to use the maximum QoS for all communications without any limitations. The important constraint may be the cost or charging of ATM network services. The better QoS will require higher costs when the amount of data transmitted during communication is the same. Users will have to consider a trade-off between QoS and cost of communications. This implies that, in principle, QoS of communication will be left to the user [121].

Since ATM networking and video coding techniques are closely related in the context of end-to-end QoS, interactions and relationships of terminal capability and the network performance need to be examined to assure a flexible and reasonable QoS for video communications. One of the goals of ATM networks is to support different traffic superpositions with different bandwidths and QoS requirements. At the same time, network resources have to be optimized.

The BISDN should be able to meet different requirements of the ATM layer [122, 123]. These QoS requirements are specified in terms of objective values of some of the network performance parameters specified in reference [124]. These parameters are cell loss ratio (CLR), cell transfer delay (CTD), and cell delay variation (CDV). The negotiation of the QoS class takes place at connection establishment and is part of the traffic contract. From the legal point of view, it is a commitment for the network to respect this contract and offer the required QoS to the user [125]. This means that the connection acceptance control (CAC) function must know whether the network can respect a contract under negotiation [126]. This function must access some measured parameters that will allow to decide whether required resources may be allocated or not, while respecting all other contracts of the already established connections.

6.5.7 QoS Maintenance and Monitoring

A systemwide resource manager is required to manage the various system components and to perform admission control to guarantee the QoS of the existing connections. The packet loss model accounts for packet losses of both the server and the network and can be used as the basis for systemwide admission control. QoS maintenance for each component may include congestion control and flow control in the network as well as real-time scheduling at the host systems and media servers.

QoS monitoring performs traffic policing to prevent the user from violating the negotiated traffic characteristics. For example, traffic-shaping mechanisms such as the leaky bucket scheme may be applied. If the user violates the service contract, penalties may be imposed, such as lowering the service priority or charging more [127].

Since the traffic is dynamic, sometimes the system may be overloaded. If the negotiated QoS cannot be guaranteed or the user would like to change the negotiated parameter values during the connection, renegotiation between the user and the system will be required.

6.5.8 Framework for QoS-Based Routing

There has been a tremendous amount of work in the QoS routing area [128]. In recent years, the concept of shortest-widest path has been proposed and discussed for QoS routing [129]. Constraint-based path selection aims at identifying a path that satisfies a set of quality of service constraints.

In response to demand for Internet-based multimedia applications, the research community has been extensively investigating several QoS-based networking frameworks, such as integrated services (`IntServ`), differentiated services (`DiffServ`), and multiprotocol label switching (MPLS). One of the key issues in all of these frameworks is how to identify efficient paths that can satisfy the given QoS constraints, commonly known as the QoS-based routing problem.

In general, routing (QoS-based or not) involves two entities: routing protocols and routing algorithms. Routing protocols capture the network state information (e.g., available resources) and disseminate it throughout the network, while routing algorithms use this information to compute appropriate paths. While current best-effort routing simply performs these tasks based on a single, relatively static measure, QoS routing takes into account both the application's requirements and the availability of network resources. As a result, QoS routing has to deal with some challenging issues that are not present in the best-effort routing, including scalable dissemination of dynamic (state-dependent) information, state aggregation, and computation of constrained paths.

QoS routing has received significant attention in the past few years with the emergence of service offering that requires QoS guarantees on the Internet. The QoS guarantee refers to meeting service requirements such as bandwidth guarantee, packet loss rate, and/or jitter delay for a request. On the other hand, typically these requirements are not required to be satisfied by best-effort services.

Currently, the primary routing decision for best-effort services on the Internet is at the packet level, and for this, shortest-path-based routing is commonly deployed. In most cases, best-effort services operate in a destination-oriented routing mode. What this means is that the source of the packet is not taken into account when a routing decision is made at any router; instead, the destination is taken into account on a hop-by-hop basis. Further, currently most routing protocols (e.g., Open Shortest Path First) are deployed with a static link metric (e.g., the interface cost being related to the speed of the link) to determine the shortest path [130].

While the destination-oriented shortest-path-based routing paradigm has worked well for current best-effort services, a new or different mode of operation is needed for services that require QoS guarantee. First, a source-destination-based routing mode is more desirable that purely destination-based routing. With this, links that are congested can be avoided with routing decisions made at the source (based on network information available to routers). Second, the use of link metrics such as the static link-speed-based interface cost is also problematic since services that require QoS guarantees cannot effectively use this information to select a good path from the source-destination-based routing paradigm. The ability to provide a set of possible paths that meet QoS guarantees rather than just the shortest path is also desirable [131].

Now we will briefly review the QoS requirement issue. Services that need QoS guarantees operate at the flow level and require QoS guarantees for the duration of the flow. A flow is defined as an IP packet stream from a source to a destination with an associated QoS and a finite duration. While it is not necessary that the same path be used for the entire duration of the flow, we will assume that the path stays the same for the entire duration; this is often referred to as route pinning. When a flow request arrives, there is a flow setup phase that can typically perform functions such as route determination, signaling along the path to the destination, and QoS checking before the flow is set up; in essence, a flow request faces a flow setup time before it can be connected. For services that require QoS guarantees, the flow setup phase must ensure that the QoS guarantees can be provided for the entire duration of the flow; otherwise, the flow request is denied. It is possible that at the user level, there is a retry attempt for a denied flow request. An important point to observe is that the route selection decision is at the flow level rather than at the packet level. The packet forwarding function is only activated at the packet level along the path, already established during the flow setup phase, and for the entire duration of the flow due to route pinning.

Our framework for QoS routing computation is motivated by the following goals.

- We want to reduce the impact on flow setup time, at least from the routing computation perspective. That is, an important goal is to minimize flow setup time.
- We want to avoid user-level retry attempts; it is preferable to do retry internal to the network as long as the flow setup time is not dramatically affected. It is important to note that user-level retry attempts cannot be completely avoided, at least in a heavily loaded network (i.e., a network where the ratio of traffic to network bandwidth is at a level beyond the normally acceptable tolerance for service guarantees).
- To be able to do internal-to-the-network retry, the network must have the capability to select a route from a number of possible paths very quickly, and also possibly have the *crankback* capability. Crankback refers to the network functionality that allows a flow that is in the process of being set up along another path [131].
- Sometimes a flow request attempting the very first outgoing link (for a path identified by the routing process) may not be able to access the link; this is referred to as *source-link overflow*. If this is the case, it should be able to try a second path as dictated by the routing process for overflow request.
- The framework should have the ability to incorporate several routing schemes so that network providers can choose the appropriate one depending on their performance goals.

The source-link-overflow idea along with the desire to avoid user-level retry and allow crankback suggests that having multiple path choices can be beneficial in a QoS routing environment; this is often referred to as the *alternate path routing* concept.

The basic idea behind the routing computation framework addresses the following: how is the selection of paths done, when are they selected, and how are they used by newly arrived flows. For flows with bandwidth guarantees, another important component that can complicate matters is the definition of the cost of a path based on possibly both additive and nonadditive properties. We will describe the three phases of the observed framework [131]:

- preliminary path caching (PPC) phase
- updated path reordering (UPO) phase
- actual route selection (ARS).

Each of these phases operates at different time-scales.

The first phase. PPC does preliminary determination of a set of possible paths from a source to a destination node, and their storage (caching). A simple case for this phase is to determine this set at the time of major topological changes. PPC, in the simplest form, can be thought of as topology-dependent: if there is a change in major topological connectivity, the PPC phase may be invoked. On the other hand, it can be somewhat intelligent: if link availability is expected to be less than a certain threshold for a prolonged duration or the link is scheduled for some maintenance work, PPC can also be used for pruning the link and determining a new set of cached paths. Essentially, PPC uses a coarse-grained view of the network and determines a set of candidate paths to be cached. A simple mechanism to determine the set of paths for each source node to each destination node may be based on hop count or interface cost or some administrative weight such as the cost metric using the k-shortest paths algorithm. Thus, for this phase we assume the cost metrics are additive.

The second phase. UPO narrows the number of QoS acceptable paths; this module uses the most recent status of all links available to each source node. Since the PPC phase has already cached a set of possible paths, this operation is more of a comparison or provides a set of QoS acceptable paths. Furthermore, for a specific service type or class, this phase may also order the routes from most to least acceptable (e.g., based on path residual bandwidth), and in general have a subset of the routes *active* from the list obtained from the PPC phase. In this phase, the cost metric can be both additive (e.g., delay requirement) and nonadditive (bandwidth requirement). Another important factor to note about the UPO phase is that the value of the update interval may vary, even for the same routing scheme; for simplicity, we refer to this as the *routing update interval* (RUI).

The third phase. From the UPO phase, we already have a reasonably good set of paths. The ARS phase selects a specific route on which to attempt a newly arrived flow. The exact rule for selecting the route is dependent on a specific route selection procedure. The main goal in this phase is to select the actual route as quickly as possible based on the pruned available paths from the UPO phase.

There are several advantages of the three-phase framework:

- Several different routing schemes can be cast in this framework.
- It avoids on-demand routing computation (from scratch); this reduces impact on the flow setup time significantly since paths are readily available; there is no cost incurred by needing to compute routes from scratch after a new flow arrives.
- The framework can be implemented using a link state routing protocol (with some extension). For the PPC phase, some topology information is needed to be exchanged in coarse-grained time windows. During the UPO phase, periodic update on status of link usage is needed in a finer-grained time window. Since different informations about links is needed at different time granularities for use by the PPC and UPO phases, we refer to this as the extended link state protocol concept.
- Utilizing the extended link state protocol, the entire environment can work in a distributed manner (without requiring centralized computation).

- Furthermore, each of the three phases can be independently modified without affecting the others. For example, in the PPC phase the k-shortest paths can be computed based on either pure hop count or other costs such as link-speed-based interface cost. In some schemes the UPO phase may not be necessary.

A possible drawback of the framework is that path caching will typically require more memory at the routers to store multiple paths; this will certainly also depend on how many paths are stored. On the other hand, with the drop in memory prices, a path caching concept is more viable than ever before. Additionally, there is some computational overhead due to k-shortest path computation in a coarse-grained time window. If needed, a router architecture can be designed to include a separate processor to periodically do this type of computation.

6.5.9 IP Oriented Quality of Service

The diversity of network assumptions indicates that QoS remains an increasingly important issue in modern networking. There exists the entire spectrum of QoS concepts, algorithms, targeting the core as well as access network segments. Also, new technologies and new standards are necessary to offer QoS for new multimedia applications. Therefore, new communication architectures integrate mechanisms allowing guarantees of specific quality to services as well as high data rates for communication systems. Quality of service must be seen as an end-to-end process. Assume, for example, that you want to establish a video conferencing over the Internet using the UMTS (the third-generation Universal Mobile Telecommunication Systems) as the access network for your communications. At the two very endpoints of this communication process we have the end user terminals (e.g., a cellular phone at one end and a desktop computer connected to the wired Internet at the other). The access technology here thus comprises several systems: the local bearer service providing the service to the cellular phone user, the UMTS bearer service, and the external bearer service providing service to the desktop user.

Without the support of the required QoS indicators (e.g., delay and bandwidth in our current example) by all segments of the network from end to end, we cannot claim we have QoS support. UMTS has its own share providing QoS, but the endpoint bearer services also need to support similar QoS indicators in order to complete the end-to-end process. Although it is possible to provide QoS with different ways of support by the individual segments in the network, it will be much more efficient and reliable to provide QoS with close interrelation among the individual segments.

The IETF started working on providing the QoS in IP networks in the mid-1990s. Two different approaches have been introduced: integrated services (`IntServ`) in 1994 and differentiated services (`DiffServ`) in 1998. `IntServ` was introduced in IP networks in order to provide guaranteed and controlled services in addition to the already available best-effort service. It is an extension to the Internet architecture to support both nonreal-time and real-time applications over IP. Each traffic flow in this service can be classified under one of three service classes: guaranteed, controlled load, and best effort.

Guaranteed service provides delay-bounded service agreements for voice and other real-time applications requiring severe delay constraints. Controlled load service provides a form of statistical delay service agreement (e.g., with a nominal mean delay). Finally, best-effort services are included to match current IP service, mainly for interactive burst traffic (e.g., Web), and background or asynchronous traffic (e.g., e-mail).

`IntServ` and reservation protocols such as RSVP have failed to become an actual end-to-end QoS solution, mostly because of the scaling problems in large networks and because of the need to implement RSVP in all network elements from the source all the way to the destination.

`DiffServ` came to remedy the disadvantages of `IntServ` in providing QoS in IP networks. `DiffServ` aims to provide simple, scalable, and flexible service differentiation using a hierarchical model. That is, resource management is now divided into two areas: interdomain and intradomain.

In `DiffServ`, all of the customer's local network requirements for QoS are aggregated, and then a service level agreement (SLA) is made with the network service provider. The SLA may be static (negotiated and agreed on a long-term basis, e.g., monthly) or dynamic, changing more frequently. The local network is then responsible for providing different services to end users within the network. This is usually done through marking packets with specific flags used in the type of service field of IP4 or the traffic class field of IPv6.

The cellular wireless system approaches to providing QoS are completely different from their Internet counterparts. General Packet Radio Service (GPRS) and UMTS, for example, use subscriber QoS profiles and traffic classifications, respectively, to manage QoS in their wireless systems.

At this time, we see little relation between the Internet and cellular approaches to providing QoS. They have aimed at QoS by their own approaches without much attention to each other. Providing end-to-end QoS with such a configuration would be a very difficult task, if not impossible, and therefore it will be a long time before we can see end-to-end QoS support for wireless Internet. Harmonization between the two approaches so that the service of one technology can cooperate and complement the service of the other remains the main issue toward QoS establishment for future networks.

QoS in the Internet Backbone

Given the background information, we are ready to analyze causes of QoS problems. They can generally be divided into two categories: nonnetwork-related and network-related causes [132].

Nonnetwork-related causes include the following.

Overloaded Servers. These overloaded servers include, for example, Web or e-mail that users are trying to access. In this case, common ways to improve QoS are to upgrade the servers, or to add servers and use a better load-balancing scheme among them.

Network Operation Errors. Configuring routers/switches is a complex and error-prone process. For example, duplicate IP addresses can be mistakenly configured and cause routing problems.

Lack of Access Capability. For economic reasons, there are always customers with slow access links (e.g., dialup modems) or oversubscribed uplinks. The technical solution for this kind of problem is clear:

- adding capacity
- classifying traffic and marking it differently for subsequent treatment using policing, shaping, and so on.

However, it should be pointed out that providing QoS may not make economic sense here if users are not willing to pay for it.

The most common cause of network-related QoS problems in the backbone is uneven traffic distribution that causes some links to be congested. Even though the average link utilization of a network can be low, say 30 percent during the peak hour, a small number of links can still have very high link utilization (close to 100 percent). Such congested links will cause long packet delay and jitter or packet loss. The causes of such hot spots in the network can be

- Unexpected events such as fiber cut or equipment failure:
- Traffic pattern shift while the network topology and capacity cannot be changed as rapidly. In the backbone, new capacity may not always be available at the right place and the right time. For example, the sudden success of a Web site or an unplanned broadcast of multimedia traffic can cause some links to become congested.

A practical approach for providing QoS can be discussed step by step. We will describe the steps in order of decreasing importance. As for the network service providers (NSP), they should start from the first step and add additional steps if needed.

Step 1 — Clean up the Network. Networks are generally well designed and provisioned in the beginning. But, over time, problems caused by quick-and-dirty fixes will accumulate. Therefore, regular cleanups should be performed. Single points of failure and bottlenecks should be removed. Capacity should be added at appropriate places so that even the most critical router/link failure will not cause traffic congestion. Interior Gateway Protocol (IGP) metrics, Border Gateway Protocol (BGP), and peering policies should be evaluated and readjusted. Logs should be examined and security measures checked. Audits should be performed to correct configuration mistakes. NSPs should also educate their customers to tighten up security and upgrade congested last mile circuits. These are by far the most important and useful things to do for providing QoS in the Internet. This step serves to prevent QoS-related problems from happening.

Step 2 — Divide Traffic into Multiple Classes. Three classes of services are proposed: premium, assured, and best effort. Premium service provides reliable, low-delay, low-jitter service. Real-time traffic (e.g., videoconferencing) and mission-critical traffic (e.g., financial or network control traffic) can benefit from such a service. Assured service provides reliable and predictable service. Nonreal-time virtual private network (VPN) traffic can benefit from it. Best effort service is used for the traditional Internet service.

Step 3 — Protection for Premium Traffic and Traffic Engineering. Multiprotocol label switching (MPLS) is used for traffic protection and traffic engineering. MPLS extends IP routing with respect to signaling and path controlling [133]. MPLS label switch paths (LSPs) are configured in the network. Each ingress label switch router (LSR) will have two LSPs for the egress. This process is illustrated in Figure 6.23, which represents fast reroute. One LSP is for premium traffic, and the other is shared by assumed and best-effort traffic. The premium LSP will have first reroute enabled. The basic idea of fast reroute is to have a patch LSP preconfigured for a link, a router, or a segment of the path consisting of multiple links and routers. This link or router or path segment is called a protected segment. When there is a failure in a protected segment, the router immediately upstream of the protected segment (called a protecting router) will detect the failure from layer 2 notification. The patch LSP will then be used to get around the failure. This protection can take effect within 50–100 ms. During fast reroute, the path taken by the LSP can be sub-optimal. To correct that, the protecting router will send a message to the ingress router of the LSP, which will then compute a new path for the LSP and switch traffic to the new LSP.

Fast reroute is essential for applications that cannot tolerate packet loss. However, fast reroute adds considerable complexity to the network. In the future, if IGP can converge faster (e.g., subsecond), the need for fast reroute will be reduced.

In our approach, traffic protection serves to provide high availability for premium traffic. Because network topology and capacity cannot and should not be changed rapidly, uneven traffic distribution can cause congestion in some parts of a network, even when total capacity of the network is greater than total demand. In our approach, each ingress router will have two LSPs to an egress, and the other is shared by assured and best effort traffic. Traffic from customers (including other NSPs) is classified at the ingress router based on the incoming interface and put into the appropriate LSPs. Network operators can also provide multifield (source and destination IP addresses, port numbers, protocol ID, etc.) classification as a value-added service. In addition, the experimental (EXP) fields of packets are served accordingly.

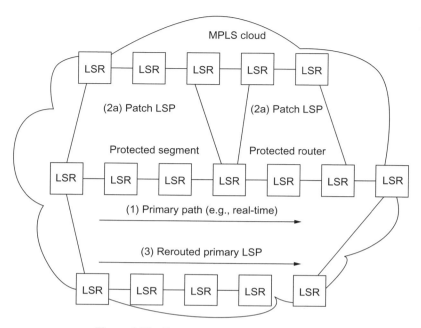

Figure 6.23 Fast reroute [132]. (©2002 IEEE.)

In order to avoid concentration of premium traffic at any link, an upper limit is set for each link regarding how much bandwidth can be reserved by premium traffic. When that portion of bandwidth is not used, it can be used by other traffic classes if that is desirable. The percentage of the premium traffic should be determined by NSP policy and premium service demand. `DiffServ`-aware traffic engineering is done for these two sets of LSPs to avoid congestion at each link [134].

In our approach, traffic engineering serves two purposes:

- To prevent (as much as possible) congestion caused by uneven traffic distribution from happening:
- If congestion does happen, to relieve it quickly.

By doing traffic engineering in a `DiffServ`-aware manner, a third purpose is also served:

- To make the percentage of premium traffic reasonably low in each link so that delay and jitter in each traffic are low; if necessary, premium traffic can preempt resources of low-priority traffic (which is not possible if all traffic is of high priority).

Compared to the traffic management schemes such as policing, shaping, and buffer management, traffic engineering can control traffic and network performance on a much larger scale. It can be considered as macro control mechanisms.

Step 4 — Class-Base Queuing and Scheduling. Queuing and scheduling are the actual mechanisms to ensure that high-priority traffic is treated preferably. It is important to prevent congestion of low-priority traffic, if any, from affecting performance of high-priority traffic. This is useful when network capacity becomes insufficient in meeting demand because of fiber cut or other equipment failure.

Another alternative is to use a priority queue for premium traffic. That is, premium traffic will always be sent before other traffic. In fact, this can be more effective in distinguishing premium service from other services. However, care must be taken to ensure that premium traffic will not starve other traffic.

Step 5 — Deploy Other Traffic Management Schemes. In this step we will discuss applicability of policing, shaping, and random early detection (RED). We will start with policing and shaping. When a customer signs up for network service, it will have a service level agreement (SLA) with its NSP. The SLA specifies the amount of traffic (in each class if applicable) the customer can send/receive. Many people think this means that NSPs will always do policing or shaping. But whether this is true or not actually depends on how a network is provisioned, which is in turn determined by the billing model.

In the access network where the flat-rate billing model is applied, the network is generally oversubscribed. Therefore, policing and shaping are useful to ensure that no customer can consume more bandwidth than he/she has signed up for. The policing/shaping parameters are usually fairly static in such a case. However, policing and shaping may affect performance of the access device. In that case, an alternative is to aggregate traffic from many customers and police/shape it collectively. The individual customer's traffic is only policed/shaped when that customer is causing trouble for others.

With bandwidth-based or data-based billing, because the NSP can increase revenue when the customer sends/receives more traffic than the SLA specifies, there is no need for the NSP to police or shape a customer's traffic (bad accounting will always be done), unless the customer is causing problems for others.

In some cases, a customer may request an upper bound on the amount of money he/she pays for bandwidth usage. In these cases, the NSP may need to do policing/shaping.

RED is a buffer management scheme to prevent tail drop caused by traffic burst. Backbone routers can generally buffer traffic for up to 100 ms per port. If traffic bursts over output line rate for longer than that, the buffer will become full and subsequent packets will be dropped. Tail drop causes many TCP flows to decrease, and later increase, their rate simultaneously, and can cause oscillating network utilization. By preventing tail drop, RED is widely believed to be useful for exchanging network performance. Weighted RED (WRED) is a more advanced RED scheme. It depends on other mechanism(s) such as classification/marking/policing to mark packets with different drop priorities. WRED will then drop them with different probabilities (which also depends on the average queue length).

RED/WRED is useful to prevent transient burst from causing tail drop. However, it should be noted that it is quite difficult to set the WRED parameters (e.g., different drop probabilities at different queue lengths) in a scientific way. Guidelines are yet to be developed on how to set such parameters. If a backbone's link utilization (time-averaged over a period of 1–5 min) can be maintained at 50 percent or lower, there should be sufficient capacity to accommodate transient burst to avoid drop. The need for WRED can be reduced.

In general, traffic engineering is more effective in controlling traffic distribution in a network and has a bigger impact on network performance than traffic management schemes such as policing, shaping, and WRED. It should be invoked before traffic management schemes. In our approach, policing, shaping and WRED are the enforcing mechanisms. They are used only when all other techniques such as traffic engineering fail to prevent congestion from happening. In that case, these traffic management schemes make sure that high-priority traffic is treated preferably compared to low-priority traffic.

QoS in DiffServ IP Networks

The enlargement of the Internet user community has generated the need for IP-based applications requiring guaranteed quality of service (QoS) characteristics. The integrated services (IntServ) and differentiated services (DiffServ) frameworks have been proposed to address QoS. While IntServ operates on a per-flow basis and hence provides a strong service model that enables strong per-flow QoS guarantees, it suffers from scalability problems. On the other hand, DiffServ keeps per-flow information only at the edge of a domain and aggregates flows into a limited set of traffic classes within the network, resolving the scalability problem at the expense of looser QoS guarantees [135].

Beyond the standardized functionality at the IP layer, a large body of work has been devoted to architectures and functions necessary to deliver end-to-end QoS. These functions can be categorized into traffic engineering (TE) functions. TE functions are mainly concerned with the management of network resources with the purpose of accommodating offered traffic in an optimal fashion. SrvMgt functions deal with the handling of customer service requests, trying to maximize incoming traffic, in terms of number of contracts and throughput, while respecting the service provider's (SP's) commitments on the agreed QoS guarantees. SrvMgt mechanisms for service offering, agreement, and activation need to be in place. In addition, in order to guarantee the agreed QoS requirements, SrvMgt needs to avoid overloading the network beyond loads it can gracefully sustain. SrvMgt functions that deal with the latter task are referred to as admission control [136].

A bandwidth broker (BB) architecture implies that admission control decisions are made at a central location for each administrative domain. Although the cost of handling service requests is significantly reduced, it is unlikely that the approach scales for large networks. In order to cope with scalability, most relevant studies adopt distributed admission control schemes, which are further distinguished into

6.5 QUALITY OF SERVICE (QoS) IN NETWORK MULTIMEDIA SYSTEMS

model-based and measurement-based approaches. Both approaches assess QoS deterioration probability upon service request arrivals; model-based approaches maintain state information for active services and employ mathematical models, whereas measurement-based approaches rely on either passive or active aggregate measurements [137].

SrvMgt and TE functions do not act in isolation, as shown in Figure 6.24. TE functions provide the grounds on which the SrvMtg functions operate, while SrvMtg functions set the traffic-related objectives for the TE functions to fulfill. Specifically, SrvMtg establishes subscription based on which a Traffic Forecast function produces the *traffic matrix* (TM), which specifies anticipated QoS traffic demand among network edges. Traffic demand is forecast from historical data and/or SP expectations (e.g., sales targets). Based on the forecasted traffic demand, the network is appropriately dimensioned by the TE functions, in terms of PHB configuration parameters and QoS route constraints. In turn, TE functions produce the *resource availability matrix* (RAM), which specifies estimates of the availability of the engineered network to accommodate QoS traffic among network edges. Based on the availability estimates, SrvMgt functions, also utilizing network state information, handle the admission of service requests so as not to overload the network.

The above interactions occur in the resource provisioning cycle (RPC) epoch. Should anticipated traffic demand significantly change, a new TE is produced, the network is appropriately redimensioned, the RAM is in turn produced, and a new RPC starts.

Admission control is an integral part of an IP QoS delivery solution. Hence, the following apply.

- It is dependent on the informational model used to describe QoS-based services. From the perspective of admission control, service models should include specification of the QoS parameters as well as user traffic conformance for receiving the specified QoS.
- Its operation is bound to the QoS capabilities available in the network.

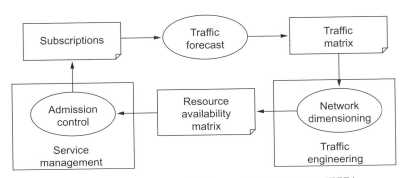

Figure 6.24 Resource provisioning cycle [136]. (©2003 IEEE.)

Furthermore, admission control should be policy-driven in order to adjust to the specific service provisioning strategies of the particular SP. Considering that providing hard QoS guarantees is prohibitively costly and that probabilistic QoS guarantees cannot be safely given at service request times, the proposed admission control scheme adopts a feedback-based model for asserting the risk of QoS deterioration as a function of admitted traffic. To increase efficiency, admission control logic is applied at both service subscription and invocation epochs. Feedback information is used at different levels of abstraction: input for offline TE functions on the ability of the engineered network to deliver QoS given subscription admission control decisions, and input from measurements on the actual status of the network given invocation admission control decision. Furthermore, our approach allows for policy-driven operation based on a best-practice paradigm.

QoS in Best-Effort Networks

The best-effort network [188] does not need information about each connection to be stored in the routers and switches on the data path, because there is no resource reservation for a connection. The sources themselves decide how much they should send. All requested connections are admitted, and the available capacity is shared between the connections. As a result, no explicit guarantees can be given about the bandwidth available to each connection. However, if the sources are designed properly, the bandwidth each source receives can be in accordance with the user's QoS need relative to other users' needs. Best-effort networks are capable of delivering elastic QoS with high performance in terms of packet delay and delay jitter while maintaining high utilization. Today, however, they do not deliver on these promises, but new research points out that with just a few changes in the links and sources, a highly elegant and simple approach to QoS is possible.

There are two objectives for a best-effort network as far as bandwidth allocation is concerned. The first objective of the best-effort network is to ensure that the available capacity of the network is utilized, and the second objective is to maximize the QoS perceived by users.

To achieve these objectives, the machinery behind a best-effort network needs just two fundamental elements, a *source flow control* algorithm and a link *active queue management* (AQM) algorithm. The flow control algorithm resides in the operating system of the host transmitting information (e.g., a Web server or PC), and decides how much information to transmit based on the congestion level of the network. The link AQM algorithm lives inside a router/switch, and monitors the queues that buffer packets awaiting transmission on the outgoing links connecting the router/switch to the next router/switch in the network. The AQM algorithm estimates the level of congestion the link is experiencing and generates a congestion signal, which communicates the congestion level to the source. Each AQM on the source to destination path to the total congestion signal is received by the source. The source algorithm can be thought of as the buyer of capacity, and the link AQM algorithm the seller of capacity. Today, the source flow control algorithm that accounts for the majority of the traffic volume on the Internet is TCP, and the dominant AQM algorithm is the tail-drop [133].

6.5 QUALITY OF SERVICE (QoS) IN NETWORK MULTIMEDIA SYSTEMS

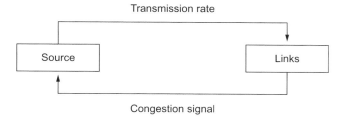

Figure 6.25 Congestion control feedback loop.

The congestion control feedback loop is shown in Figure 6.25. The two elements' source and links are connected in a closed loop system, which allows the buyer and seller to negotiate.

The most common way the signal is communicated on the Internet is by dropping packets. Depending on the level of congestion at the link, the AQM drops packets at a rate that reflects the severity of the congestion. When a packet is dropped from a stream of packets in the TCP connection, the destination host can detect this, and the destination communicates the loss of the packets back to the source host using

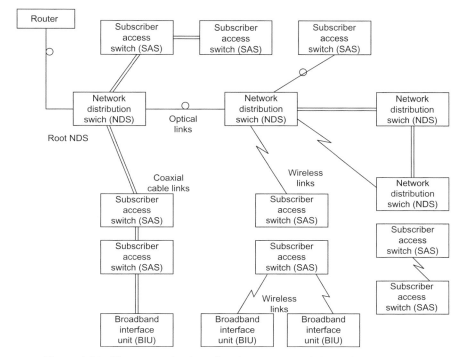

Figure 6.26 Next-generation broadband access network [139]. (©2002 IEEE.)

Acknowledgement packets. Therefore, the source can detect the number of lost packets, and the strength of the congestion control feedback signal, or price, is simply the rate of lost packets. So although there is no physical communication link between the link and the source to communicate the congestion signal, there is a logical communication link by courtesy of the destination host. Packet dropping is a natural mechanism of signaling congestion, because congestion occurs when more packets arrive than can be transmitted on the link, which leads to buffer overflow and, inevitably, packet loss. However, other modes of communicating the feedback signal have been developed and one is explicit congestion notification (ECN) [138]. With ECN, to avoid packet loss, packets are marked by setting a single bit in the packet header, instead of being dropped, to communicate the presence of congestion.

QoS in Ethernet Networks

High-bandwidth switched networks are an efficient transport mechanism for data services. Compared with shared, contention-based, or time-division multiple access media, switched networks provide strong privacy for user data and simplify scalability. An ideal access solution provides symmetric high-data-rate service at low cost. Ethernet satisfies all these requirements. Low-cost, high-speed core and edge routers with 100 Mbps/1 Gbps/10 Gbps Ethernet interfaces are available from multiple vendors. Ethernet offers other benefits too, such as easy migration to higher performance levels, low acquisition and operating cost, adaptability to new applications and data types, flexibility in network design, and the availability of a large trained technical work force. Ethernet, long a technology for LANs, is ubiquitous in enterprise networking and has already expanded its reach to the WAN.

Design of low-cost Gigabit Ethernet transport over short-haul microwave links is currently under way. Microwave frequencies permit the use of small, unobtrusive antennas; moreover, the large amount of spectrum available at high frequencies allows the use of simple modulations and low-complexity receiver structures. The addition of wireless links to the broadband access network leads to the next-generation broadband access network (NGBAN), shown in Figure 6.26. It comprises Ethernet switching nodes connected by point-to-point links operating at 100Mbps, 1 Gbps or 10 Gbps. Any technology – fiber, coaxial cable, or wireless – may be used to provide the physical medium underlying these links, the choice being based on economic or regulatory considerations.

Packet transport within the NGBAN takes place as follows. In the upstream direction, switched Ethernet transmission from a broadband interface unit (BIU) installed at the customer premises terminates at a neighborhood subscriber access switch (SAS). At the SAS, data packets received from one or more BIUs, as well as packets received from the downstream trunk/feeder port, are switched from the ingress ports to the upstream egress port of the next device, which may be another SAS or a network distribution switch (NDS). The NDS aggregates packets from several ingress trunk/feeder ports and switches them to the upstream egress port. The NDS at the *root* of the NGBAN, which is referred to as the root NDS, connects to the WAN router. The link between the root NDS and the WAN router is special, since it sees the

6.5 QUALITY OF SERVICE (QoS) IN NETWORK MULTIMEDIA SYSTEMS

maximum amount of aggregation and thus plays a role in admission control to the NGBAN. We refer to this link as the main trunk. In the downstream direction, each SAS or NDS switches data packets to appropriate ports on their way to the destination BIU, which forwards them to the appropriate subscriber device (set-top box, computer, telephone, etc.) [139].

Each trunk/feeder or drop link is selected to be coax, optical, or wireless link depending solely on economic considerations. The QoS management scheme for NGBAN combines the `IntServ` and `DiffServ` approaches to provide a flexible and scalable platform that can be used to deliver the desired QoS to a wide variety of applications. We define four QoS or priority classes.

- *Constant bit-rate (CBR)*. This is the highest priority designed for real-time applications such as voice over IP (VoIP) or T1 (circuit) emulation. Even though it is named CBR, there is no time-division multiplexed (TDM) bandwidth dedicated to the flow in the switched architecture. Delay objective: 5 ms.
- *Variable bit rate – real-time (VBR-rt)*. This is the second highest priority class designed for real-time VBR services such as streaming video. Delay objective: 15 ms.
- *Variable bit rate – nonreal-time (VBR-nrt)*. This service class is intended for nonreal-time applications such as connections between office sites for large data transfers and video downloads. Throughput is guaranteed when measured over a suitably long time period.
- *Unspecified bit rate (UBR)*. This service is similar to the best-effort service available to typical Internet users today. There are no delay or throughput guarantees.

QoS management in NGBAN is based on the following guiding principles:

- explicit admission control for connections requiring guaranteed QoS
- elimination of packet losses due to buffer overflows
- elimination of forced idleness of network resources
- fairness in the allocation of the leftover bandwidth to best-effort traffic
- stateless packet handling at intermediate nodes.

The first of these principles is related to `IntServ`, the last is `DiffServ`.

The following features enable NGBAN to provide explicit packet delay, jitter, and loss guarantees while achieving full statistical multiplexing gain:

- Traffic classification and policing at the customer premises to ensure that only compliant traffic enters the NGBAN.
- Priority scheduling at each aggregation switch to achieve guaranteed low delays.

- Link-by-link flow control to ensure that admitted traffic is not dropped within the NGBAN.
- Centralized admission control to ensure that only a manageable percentage of the bandwidth is allocated to low-delay traffic.

The broadband interface unit (BIU) represents the point of entry into NGBAN for all upstream traffic. As mentioned earlier, the `DiffServ`-based data transport within NGBAN cannot deliver absolute QoS guarantees unless it is accompanied by tight control of the traffic that is allowed to enter the network. Admission control ensures that the traffic *contracted* to be carried at each QoS level is within the capacity of the network. It is the responsibility of the BIU to ensure that the packets permitted to enter the network are within the limits laid out in the corresponding service-level agreements (SLAs).

A BIU handles upstream traffic originating from a single subscriber. This permits easy scalability for per-flow policing.

The QoS management tasks performed by the BIU on upstream traffic are as follows:

- *Packet classification* to identify flows is done on the basis of the medium access control (MAC) address, source or destination IP addresses, port numbers, and combinations thereof. Each packet is marked with the QoS class associated with the flow.
- *Token-bucket-based policing* for every flow. Each token bucket is characterized by two parameters: its sustained rate and maximum burst size. Packets violating the flow constraints are dropped, ensuring that the traffic entering NGBAN is consistent with the corresponding SLAs.
- *Egress buffers.* The egress buffer size for each QoS class is chosen according to the delay budget allocated at the BIU for the QoS class. If the egress buffer does not have sufficient space to accommodate an admitted packet, it is dropped.
- *Transmission scheduling.* Packet scheduling at the egress buffers is on the basis of QoS class and does not account for flows. The scheduling discipline is based on strict nonpreemptive priorities and obeys per-QoS class flow control from the upstream device.

QoS in IP-Over-WDM Networks

The proliferation of Internet Protocol (IP) technology coupled with the vast bandwidth offered by optical wavelength division multiplexing (WDM) technology are paving the way for IP over WDM to become the primary means of transporting data across large distances in the next-generation Internet. WDM is an optical multiplexing technique that allows better exploitation of the fiber capacity by simultaneously transmitting data packets over multiple frequencies or wavelengths. The tremendous bandwidth offered by WDM is promising to reduce the cost of core network equipment and simplify bandwidth management. However, the problem of providing QoS guarantees for several advance services, such as transport of real-time packet voice and video, remains largely unsolved for optical backbones. The QoS problem in optical WDM networks has several fundamental differences from QoS methods in electronic routers and switches. One major difference is the absence of the concept of packet queues in WDM devices, beyond the number of packets that can be buffered (while in flight) in fiber delay lines (FDLs). FDLs are long fiber lines used to delay the optical signal for a particular period of time. As an alternative to queuing, optical networks use additional signaling to reserve bandwidth on a path ahead of the arrival of optically switched data.

Since Internet traffic will eventually be aggregated and carried over the core networks, it is imperative to address end-to-end QoS issues in WDM networks. However, previous QoS methods proposed for IP networks are difficult to apply in WDM networks mainly due to the fact that these approaches are based on the store-forward model and mandate the use of buffers for contention resolution. Currently there is no optical memory, and the use of electronic memory in an optical switch necessitates optical-to-electrical (O/E) and electrical-to-optical (E/O) conversions within the switch. Using O/E and E/O converters limits the speed to the optical switch. In addition, switches that utilize O/E and E/O converters lose the advantage of being bit-rate transparent. Furthermore, these converters increase the cost of the optical switch significantly. The only means currently of providing limited buffering capability in optical switches is the use of FDLs. However, FDLs cannot provide the full buffering capability required by the classical QoS approaches. In addition to FDLs, the wavelength domain provides a further opportunity for contention resolution based on the number of wavelengths available and the wavelength assignment method.

Three major switching techniques have been proposed in the literature for transporting IP traffic over WDM-based optical networks. Accordingly, IP-over-*WDM* networks can be classified as *wavelength routing* (WR) networks, *optical packet switching* (OPS) networks, and *optical burst switching* (OBS) networks.

As a result, data transmitted among lightpath endpoints require no processing, no E/O conversion, and no buffering at intermediate nodes. However, as a form of circuit-switching network, WR networks do not use statistical sharing of resources and therefore provide lower bandwidth utilization.

In packet-switching networks, IP traffic is processed and switched at every IP router on a packet-by-packet basis. An IP packet contains a payload and header. The packet header contains the information required for routing the packet, while the payload carries the actual data. The future and ultimate goal of OPS networks is to process the packets header *entirely* in the optical domain. With the current technology this is not possible. A solution to this problem is to process the header in the electronic domain and keep the payload in the optical domain. Nevertheless, many technical challenges remain to be addressed for this solution to become viable. The main advantage of OPS is that it can increase the network's bandwidth utilization of statistical multiplexing for bandwidth sharing. OBS networks combine the advantage of both WR networks and OPS networks. As in WR networks, there is no need for buffering and electronic processing for data at intermediate nodes. At the same time, OBS increases the network utilization by reserving the channel for limited time periods. The basic switching entity in OBS is a burst. A burst is a train of packets moving together from one ingress node to one egress node. A number of approaches exist for burst forming, such as the containerization with aggregation-timeout (CAT) technique. A burst consists of two parts, header and data. The header is called the control burst (CB) and is transmitted separately from the data, which is called the data burst (DB). The CB is transmitted first to reserve the bandwidth along the path for the corresponding DB. Then it is followed by the DB, which travels over the path reserved by the CB [140].

A general framework for providing differentiated services in wavelength routing (WR) networks extends the differentiated optical services (DoS) model presented in reference [141]. The DoS model considers the unique optical characteristics of lightpaths. A lightpath is uniquely identified by a set of optical parameters such as bit error rate (BER), delay, and jitter, and behaviors including protection, monitoring, and security capabilities. These optical parameters and behaviors provide the basis for measuring the quality of optical service available over a given path.

The DoS framework consists of six components as follows.

Service Classes. A DoS service class is qualified by a set of parameters that characterize the quality and impairments of the optical signal carried over a lightpath. These parameters are either specified in quantitative terms, such as delay, average BER, jitter, and bandwidth, or based on functional capabilities such as monitoring protection and security.

Routing and Wavelength Assignment Algorithm. In order to establish a lightpath, a dedicated wavelength has to be reserved throughout the lighpath route. An algorithm used for selecting routes and wavelengths to establish lightpaths is known as a routing and wavelength assignment (RWA) algorithm. In order to provide QoS and WR networks, it is mandatory to use an RWA algorithm that considers the QoS characteristics of different wavelength channels. An example of such a RWA algorithm is presented in reference [142]. The underlying idea behind this RWA algorithm is to employ adaptive weight functions that characterize the properties of different wavelength channels (e.g., delay, capacity).

Lightpath Groups. Lightpaths in the network are classified into groups that reflect the unique qualities of the optical transmission, such that each group corresponds to a QoS service.

Traffic Classifier. Traffic flows are classified into one of the supported classes by the network. Classification is carried out at the network ingress.

Lightpath Allocation Algorithm. A number of algorithms have been proposed in literature for allocating lightpaths to different service classes.

Admission Control. Similar to the bandwidth broker entity in the `DiffServ` architecture, an entity called an optical resource allocator is required in WDM networks to handle dynamic provisioning of lightpaths. The optical resource allocator keeps track of the resources, such as the number of wavelengths, links, crossconnects, and amplifiers, available for each lightpath, and evaluates the lightpath characteristics (BER computation) and functional capabilities (protection, monitoring, and security). The optical resource allocator is also responsible for initiating end-to-end call setup along the chain of optical resource allocators representing the different domains traversed by the lightpath.

The idea underlying most proposals for OPS is to decouple the data path from the control path. This way, routing and forwarding functions are performed using electronic chops after an O/E conversion of the packet header, while the payload is switched transparently in the optical domain without any conversion. Until now, there have been very few proposals for providing service differentiation in OPS networks. This is expected considering that OPS is a fairly new switching technique and still has many problems remaining to be solved.

In any packet-switching scenario, contention may arise when more packets are to be forwarded to the same output link at the same time. In general, QoS techniques in OPS networks aim at providing service differentiation when contention occurs by using wavelengths and FDL assignment algorithms. An overview of these algorithms as general techniques for providing QoS in OPS networks is given in reference [143].

Wavelength Allocation (WA). This technique divides the available wavelengths into disjoint subsets and assigns each subset to a different priority level such that higher priority levels get a larger share of the available wavelengths. Different WA algorithms are possible, which are similar to the LA algorithms presented eralier. WA techniques use the wavelength domain only for service differentiation and do not utilize FDI buffers.

Combined Wavelength Allocation and Thresholds Dropping (WATD). In addition to WA, this technique uses threshold dropping to differentiate among different priority classes. When the FDL buffer occupancy is above a certain threshold, lower-priority packets are discarded. By using a different dropping threshold for each priority level, different classes of service can be provided. This technique exploits both wavelength domain (WA) and time domain (FDLs) to provide service differentiation; hence, it has more computational complexity than the bufferless WA technique.

Providing QoS in optical-based switching (OBS) networks requires a signaling (reservation) protocol that supports QoS. In addition, a burst-scheduling algorithm is needed in the network core burst switches. The protocol for supporting QoS in OBS networks uses offset time as a way to provide different classes of service in bufferless optical networks.

QoS for Voice Over IP Technology

Voice over IP (VoIP), the integration of conventional telephone services with the growing number of other IP-based applications, is seen as one of the most important technologies for telecommunications providers. In addition to the cost reduction achieved by the sharing of network resources, VoIP is expected to accelerate the development of rich multimedia services.

Since quality is not generally guaranteed in an IP network, it is important that the networks and/or terminals be properly designed before providing service, the quality of the service be constantly monitored, and action be taken as necessary to maintain the level of service. In achieving these goals, methodologies for evaluating the perceptual quality of service of VoIP services are indispensable.

Figure 6.27 shows the various aspects of the perceptual QoS of a VoIP system and how these are determined. In conventional telephony until the 1980s, where the signal bandwidth was a fixed 0.3–3.4 kHz, the impairment factors were transmission loss, frequency distortion, stationary circuit noise, and, in digital systems, signal-correlated quantization noise associated with pulse code modulation (PCM) and so on. These properties are usually described in terms of a signal-to-noise ratio (SNR). Since VoIP systems are based on new coding technologies and a new transmission technology, the primary determinants of the perceptual QoS of VoIP service are distortions caused by speech coding and packet loss, loudness, delay (network and terminal delays), and echo [144].

The prime criterion for the quality of audio and video communications services is subjective quality, the user's perceptions of service quality. This can be measured through *subjective quality assessment*. The most widely used metric is the mean

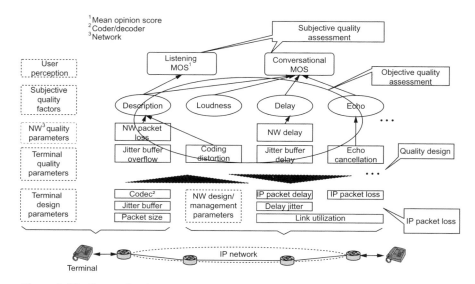

Figure 6.27 Factors for determination of the quality of a VoIP service [144]. (©2004 IEEE.)

opinion score (MOS). However, while subjective quality assessment is the most reliable method, it is also time-consuming and expensive. Methods for estimating subjective quality from physical quality parameters are thus desirable. This process is called *objective quality assessment.*

The subjective quality factors are mapped to network and terminal quality parameters. Since service providers use quality assessment technologies in order to design and manage QoS in a way that takes users' perceptions into account, they need to further map these quality parameters to parameters that are designed and/or managed.

Since subjective quality assessment is time-consuming and expensive, we need a method for estimating subjective quality by measuring the physical characteristics of the terminals and networks. In a wide sense, all such methods are forms of objective quality assessment.

Objective quality assessment methodologies can be categorized into several groups from the viewpoints of aim, measurement procedure, input information, and MOS for estimation. Objective quality assessment methodologies that exploit network and terminal quality parameters and produce estimates of conversational MOS are called *opinion models.* On the other hand, those that require speech signals as inputs and produce estimates of listening MOS are called *speech-layer objective models,* and those that exploit IP packet characteristics and produce estimates of listening MOS are called *packet-layer objective models.* Although the speech- and packet-layer objective models estimate the same thing (i.e., the listening quality), they are used in different scenarios. For example, if it is impossible or difficult to obtain actual speech samples via in-service quality monitoring, we should use packet-layer objective models. Conversely, if it is difficult to capture necessary packet information or we need to obtain quality estimates that are as accurate as possible, we should use speech-layer objective models.

QoS in Next Generation Networks

As a result of the increasing importance and widespread use of the Internet, next-generation IP networks (NGNs) are generally expected to carry not just Internet traffic but also today's telephone traffic plus other demanding services. However, for an operator to actually capitalize on the reduced operational cost of a single network versus that of multiple dedicated networks, it must ensure that the NGN supports a comparable quality of service (QoS). As a first step, its core network must be designed to meet these requirements.

Unfortunately, QoS is a quite vaguely defined term. While it is common to require low delay and especially low-delay variation as two main criteria of QoS, both requirements are likely to be met by a high-speed core network, anyway. In high-speed networks, queuing delay is negligible as long as overload is avoided. Furthermore, it turns out that the most demanding QoS requirements are those resulting from person-to-person communication, where people are accustomed to very high QoS, whereas normal data traffic or even streaming video can compensate for a lack of QoS by their own means (e.g., repetition or buffering).

Thus, the following two requirements are key to provide acceptable QoS in a high-speed NGN:

- any interruption due to link failures must be short enough to be disregarded (or not even felt) by the customer
- deterioration of quality due to overload on the network must be avoided.

The first requirement is often severally violated by the current Internet. In a typical routed network, the detection of a link failure, which together with port failures is the most widespread type of failure, and the resulting rerouting will take tens of seconds. In comparison, in a person-to-person communication, a call will typically be dropped after one or at best some very few seconds, while any interruption lasting more than roughly 100 ms will be very disturbing to most people, causing them to reiterate parts of their conversation. Thus, faster reaction to failures will be required for an NGN.

The second requirement can only be met using some kind of admission control. The customer is interested in avoiding overload effects such as packet loss while at the same time demanding a low call rejection rate. On the other hand, the operator must ensure efficient utilization of its network to achieve sufficient revenue. These somewhat contradicting requirements can be best met by ensuring an even load distribution throughout the network, thus avoiding any hot spots that would otherwise limit admissible traffic or degrade QoS. Thus, good distribution of traffic is another key ingredient for the successful operation of an NGN.

A fundamental advantage of the current Internet is that the connectionless paradigm of routed IP avoids the complexity and overhead of state management in network nodes. This is particularly important in failure situations when thousands of states have to be transferred from a failed path to a new active path. It is therefore our goal to stick with that basic concept for an NGN providing QoS services as well as best-effort service. To achieve this, the network essentially implements the following functions:

- Destination-based IP routing performed by routers handling the forwarding of both QoS and best effort traffic. QoS traffic is treated according to a differentiated strict priority regime using multiple traffic classes. Routers are expected to use novel routing strategies for better support of QoS.
- Network admission control that handles the admission of QoS traffic to the network using class-based admission budgets. These budgets include reserve capacity, so, for an arbitrary single link failure in the network, QoS traffic will still be supported without QoS impairment.
- A network management add-on dynamically determining the admission budgets according to the current traffic matrix, and performing additional functions like routine network checks.

Routing strategies supporting QoS may be found in reference [145]. Generally, QoS is a loosely defined term, and different services may require different kinds of QoS. One aspect of QoS for most services is packet loss. Although a certain amount of packet loss may be acceptable, especially for services such as voice or video, packet loss may be very disturbing for services using TCP, because beyond a certain amount of loss, TCP rapidly reduces throughput.

High availability and even distribution of traffic over the network are a prerequisite for the economical provision of QoS services.

QoS in 3G Multimedia Mobile Applications

Third-generation multimedia mobile applications will place a high emphasis on interactivity and on-demand usage, which are assumed to be the central capabilities of point-to-point telecommunication networks. However, considering the technical and marketing issues of delivering multimedia services over mobile networks, and looking at the evolution and future perspectives for storage technologies, it appears that the introduction of a broadcast/multicast mode in 3G networks can provide great opportunities for the multimedia services business. By its nature, content is aimed at the largest possible audience. With a sufficiently large storage capacity in the user equipment, broadcasting might now be able to provide on-demand and interactive applications. Push and store mechanisms make the one-to-many distribution mode that can prove to be the most efficient and cost-effective way to provide large audiences with appealing multimedia content.

The Universal Mobile Telecommunications System (UMTS) realizes 99 onward specified mechanisms for QoS support, mainly through the definition of an end-to-end QoS architecture. Another significant contribution of this release is the specification of the UMTS QoS or traffic classes. The traffic classes are intended for specific applications producing traffic that exhibits a well-known behavior (e.g., voice over IP). Having in mind the picture of the all IP future network (wired and wireless), this article focuses on appropriate networking between UMTS and the next-generation wired Internet. The need for interoperability with external IP networks has also been foreseen with the 3G standardization forum, the Third Generation Partnership Project (3GPP). The next-generation Internet will most probably provide a set of traffic classes, in the same sense as UMTS, to differentiate among the supported levels of quality.

3GPP standards propose a layered architecture for the support of end-to-end QoS for the packet domain through interaction of bearer services (BS) established among UMTS modules at different layers. Each bearer service mainly specifies, among others, the control signaling, user plane transport, and QoS management functionality. The end-to-end service may be conveyed over several networks (not only UMTS). The external BS deals with the interoperability and interworking aspects with external IP bearers, and provides the appropriate functionality to support it. It is logically situated in the GPRS Gateway Support Node (GGSN), which is the gateway of UMTS to external networks [146].

The specifications define the QoS management functions in the UMTS BS for both the control and data planes. In the control plane, in brief, the various BS managers in the different modules coordinate the overall management procedures. In particular, in the GGSN, the external BS manager controls interworking with external networks.

When the external network is an IP one, the so-called IP BS manager uses standard IP mechanisms to control IP bearer services, such as a `DiffServ` edge function or a Resource Reservation Protocol (RSVP) function. Moreover, at the edges of the UMTS BS, the translation function converts service primitives of the UMTS BS to the corresponding primitives of external networks and/or traffic engineering (TE).

In the data plane, the UMTS QoS management functions of the UMTS BS consist of the classification, mapping, and traffic conditioning functions, as well as the resource manager. The latter performs scheduling, queuing management, bandwidth management, and power control for the radio bearer, to distribute the available resources among the established services appropriately.

Establishment of QoS within a UMTS public land mobile network (PLMN) is achieved through the Packet Data Protocol (PDP) context activation procedure [147]. The UMTS specifications define four QoS classes: conversational, streaming, interactive, and background. The main distinguishing factor among these classes is delay sensitivity. The conversational class is the most sensitive, while background is the least sensitive. Conversational and streaming classes are intended for real-time traffic. They both preserve time relation (variations) among information elements of the stream, but conversation has stricter and lower delay requirements. Example applications are IP telephony for the former and streaming video for the latter. For the interactive and background classes, transfer delay is not the major factor. Instead, they both preserve the payload content. Interactive class follows a request–response pattern and defines the priorities to differentiate among bearer qualities, while it does not provide explicit quality guarantees. Background's main characteristic is that the destination does not expect the data within a certain time. Example applications are Web traffic for interactive and download of e-mails for background.

While mobile broadcast technologies still face a difficult marketing situation in Europe, hybrid satellite systems in the United States, like Sirius and XM radio, are now leading the way. Hybrid satellite systems provide a very cost-effective answer to coverage issues because they are designed to efficiently combine satellite reception in rural and suburban areas and terrestrial retransmission in urban areas where satellite signals are frequently blocked.

Satellite-based architecture undertook a feasibility analysis to evaluate the overall impact of directly reusing wideband code division multiple access (WCDMA) for the physical layer of satellite hybrid broadcast network to maximize synergy with user equipment [148].

To conclude, in parallel to the growth of the Internet, wireless mobile network technologies have also been extensively advanced to the point that mobile and wireless data services are deployed very quickly worldwide. In this context, third-generation (3G) wireless infrastructure has already adopted IP as the core network protocol in its data subsystems, and it also promises guaranteed quality for IP multimedia service, in both the access and core networks.

In the global evolution of telecommunication services to provide the consumer environment, broadcasting is now increasingly perceived as being able to optimize bandwidth usage. Two approaches to the convergence of broadcast and unicast modes within the 3G architecture have been identified: one relies on the mobile operator's radio resources, while the other takes the traffic capacity of several mobile operators. The two approaches are complementary. Satellite delivery to the edge of the 3G networks is an easy step to implement broadcast and multicast mechanisms, while Selective Directed Broadcast Mode (S-DBM), in the medium term, opens up much larger market opportunities for these concepts. In the context of the deployment of 3G networks and services, satellites can offer efficiency and cost-effective means to relieve unicast networks of the greatest and least profitable traffic, considerably increase content delivery capacity, and ensure service continuity over a wider rural coverage area.

Trends in QoS for 4G Wireless Systems

The 3G wireless systems will provide high data rates up to 2 Mbps and support a broad range of multimedia services including voice, data, and video to mobile users. In 4G systems, data rates are expected to reach as high as 20 Mbps. Because wireless systems have very scarce bandwidth of available frequency spectrum, the limited resources have to be used efficiently to provide satisfactory services to the users.

Multimedia information can exhibit highly bursty traffic rate. Packetized transmission over wireless links makes it possible to achieve a high statistical multiplexing gain. Packet flows generated by mobile users can be classified into several traffic classes. Each of these classes has its unique quality of service (QoS) requirements and traffic rate characteristics. Owing to the heterogeneous nature of multimedia traffic flows, the traditional voice-based medium access control (MAC) protocols do not perform well in a multimedia environment. A flexible MAC protocol that can efficiently accommodate multimedia traffic is required. One important MAC issue is the packet transmissions. Most packet scheduling strategies, such as first-in first-out (FIFO), round-robin, and generalized processor sharing (GPS), were originally proposed for wireless networks [149]. Random access protocols have been widely used in the past for wireless communications. A MAC protocol with bit error rate (BER) scheduling called WISPER is proposed in reference [150] for code-division multiple access (CDMA) communications, where packets with the same or similar BER requirements are transmitted in the same time slot with the same received power level for all the packets. In the 3G systems proposals, mobile terminals use random access for sending the transmission requests, to which short data bursts can be appended. For other data transmissions, the base station assigns dedicated channels to the users when the resources are sufficient. The order of channel assignments depends on the time moments when the requests are received and the priorities associated with the traffic classes. The users keep the channels as long as they have packets to transmit. As there is no specific packet scheduling, efficient statistical multiplexing cannot be achieved at the packet level. For high resources utilization, MAC should take the current packet flow loads and the users' QoS requirements into account. A MAC protocol with packet scheduling needs to be further developed to achieve efficient packet-level statistical multiplexing under QoS constraints.

A MAC protocol with fair packet loss sharing (FPLS) scheduling for 4G wireless multimedia communication is proposed in reference [151]. The MAC protocol exploits both time-division and code-division multiplexing for efficient resource utilization. FPLS is a QoS requirement based packet scheduling algorithm. The objectives of the scheduling are to provide QoS guarantees in terms of transmission delay and accuracy and to maximize the system resource utilization. In a wireless environment, a packet is expected to be delivered to the destination within a required time frame and with certain accuracy. Any violation of these two requirements will cause the packet to be useless and therefore discarded. Since QoS satisfaction and high resource utilization are in general conflicting goals, high utilization of the limited wireless bandwidth often means that the system resources cannot accommodate the traffic loads to be dropped occasionally. To support as many satisfied users as possible, fair sharing of the dropped packets among all the users is essential. The main features of the FPLS scheduler are:

- letting each have a fair share in packet loss
- assigning a minimum required received power level
- having a maximum multiplexing in the code domain for each time slot, based on the transmission rate statistics and real-time traffic load information.

For future work, we propose to develop new algorithms that integrate QoS technologies in mobile IP environment.

6.6 GENERIC NETWORKS

The emergence of digital storage and transmission of video has been driven by the availability of fast hardware at affordable prices, largely thanks to the economies obtained through standardization of compression algorithms. Digital video is being used in many applications, ranging from videoconferencing (where two or more parties can carry out an interactive communication) to video on demand (where several users can access the video information stored at a central location). Each application has different requirements in terms of bit rate, end-to-end delay, delay jitter, and so on. ATM technology is targeted to be used with BISDN and it allows flexible and efficient delivery of multimedia data, accommodating many different delays and bit rate requirements [120].

Video applications involve real-time display of the decoded sequence. Transmission over a constant bit rate (CBR) channel requires that there is a constant end-to-end delay between the time the encoder processes a frame and the time at which that same frame is available to the decoder [152]. Since the channel rate is constant, it will be necessary to buffer the variable rate information generated by the video encoder. The size of the buffer memory will depend on the acceptable end-to-end delay. It is necessary to provide a rate control mechanism that will ensure that the buffer does not overflow and thus all the information arrives at the decoder. The basic idea is to lower the video quality for scenes of higher complexity so as to avoid overflow. The simplest approach to rate control relies on deterministic mapping of each buffer occupancy level to a fixed coder mode of operation [153]. Some models of coder have been proposed to set up coding rate predictions that are used to drive the buffer control [154]. In other cases, ideas from control theory are used to devise the buffer controller [155]. In general, the buffer control is designed for a particular encoding scheme and thus scheme-dependent heuristics tend to be introduced [156, 157]. A more detailed review of the problem in connection with a survey of the algorithms proposed for rate control can be found in reference [158]. The goal of all these algorithms is to maximize the received quality given the available resources.

Whereas ISDN offers both circuit-switched and packet-switched channels, video transmission uses a circuit-switched channel. This is the case for most videoconferencing products. In this scenario, the transmission capacity available to the end user is constant throughout the duration of the call. The main advantage of this approach is its reliability since the channel capacity is guaranteed. On the other hand, in computer networks, video is manipulated just as any other type of data. Video data is packetized and routed through the network, sharing the transmission resources with other available services such as remote login, file transfer, and so on, which are also built on the top of the same transport protocols. Such systems are being implemented over LANs [159] and WANs, with both point-to-point and multipoint connections. The systems are often referred to as *best effort* because they provide no guarantees on the end-to-end transmission delay and other parameters. In a *best-effort* environment, the received video quality may change significantly over time [160, 161].

ATM networks seek to provide *the best of both worlds* by allowing the flexibility and efficiency of computer networks while providing sufficient guarantees so as to permit reliable transmission of real-time services. Using ATM techniques allows flexible use of capacity, permitting dynamic routing and reutilization of bandwidth. Because video compression algorithms produce a variable number of bits, the periods of low activity of one source can be reused by other sources. For example if N sources each require CBR channels at a rate R bits/s, it might be possible to transmit them together over a single channel with a rate less than RN. This reduction in capacity is the so-called statistical multiplexing gain (SMG). If all services are using their maximum capacity simultaneously, packets might be lost and thus transmission can also be guaranteed *most of the time*. Contrary to the best-effort characteristic of most computer networks, ATM networks are designed so as to allow QoS parameters to be met, at least statistically. ATM networks aim to accommodate very heterogeneous services, thus allowing a customized set of QoS parameters to be selected by each application. The ATM design should be able to support both the best-effort network protocols and circuit-switched connections. However, while the QoS parameters are meaningful for nonreal-time data, they are not the only factors to take into account for real-time video transmission. Video differs from other types of data in that acceptable transmission quality can be achieved even if some of the data is lost. The effect of packet losses, which in other applications results in retransmission of data, can be reduced in the video case with appropriate encoding strategies combined with error concealment techniques.

We will now examine some of the network design issues, emphasizing those aspects that directly affect video transmission. Admission control is the task of deciding whether a new connection with a given set of requested QoS parameters can be allowed into the network. The connection should be admitted if it can be guaranteed to have the required QoS without degrading the QoS of other ongoing connections. Since video encoders produce variable rate, a key factor in the admission control problem is to find statistical models for the expected bit rate of video sources. A model characterizes the bit rate of a video connection at various timescales and will attempt to capture the short- and long-term dependencies in the bit rate as well. Typical models are correlated with the number of bits for the previous frame [162], or Markov chains [163]. For a given model, the performance of the network model based on different routing and queuing strategies can be examined. The decision on whether to admit a call can be made based on the expected performance of the network. Admission control is much simpler for circuit-switched networks, since the transmission resources are constant throughout the duration of the call, and the only issue is to find out whether currently unused resources are sufficient to carry the additional call. If they are sufficient, the call can be completed, otherwise it will be rejected. In the ATM environment, the resources needed by each of the services change over time. Thus, the main problem is to estimate the likelihood that resources will be insufficient to guarantee QoS. Best-effort networks do not perform explicit admission control, although insufficient QoS during high load conditions will drive users out or make them delay their connection [164]. Admission control is part of the negotiation process between user and network to set up a connection. The result of the negotiation is a contract, which will specify the traffic parameters of the connection. Typical traffic parameters are peak cell rate and sustainable cell rate. These are operational measures of the offered bit rate and are implemented with counters.

The function called usage parameters control (UPC) or policing mechanism has the goal of preventing sources from maliciously or unwillingly exceeding the traffic parameters negotiated at call setup. Typically, the network will look at policing methods that are directly linked to the negotiated traffic parameters. For instance, if a certain peak cell rate has been agreed upon, then the policing mechanism may consist of a counter that tracks the peak rate and verifies that it does not exceed the negotiated value. One of the most popular policing mechanisms, due to its simplicity, is the so-called leaky bucket [165]. A leaky bucket is simply a counter incremented with each cell arrival and decremented at fixed intervals such that the decrement is equivalent to an average cell rate of R. The other parameter of the leaky bucket is the size of the bucket, that is, the maximum allowable value for the counter. The network can detect violations by monitoring whether the maximum value of the counter is reached. If same cells are found to be violating the policing functions, the network can decide to delete them or to just mark them for possible deletion in case of congestion. The choice of policing function is important because it may determine the type of rate that the sources will transmit through the network.

Video encoding algorithms for ATM transmission need to be robust to packet losses. This can be achieved in part by using a multiresolution encoding scheme along with different levels of priorities for the cells corresponding to each resolution. Additionally, error concealment techniques can be used to mark to some extent the perceptual effects in the decoded sequence of the loss information.

Multiresolution encoding schemes separate the information into two or more layers or resolutions. The coarse resolution contains a rough approximation of the full resolution image or sequence. The enhancement or detail resolution provides the additional information needed to reconstruct at the decoder the full resolution sequence at the targeted quality. The coarse resolution sequence is obtained by reducing the spatial or temporal resolution of the sequence or by simply having images of lower quality. A survey of multiresolution encoding techniques can be found in reference [166]. To take advantage of the multiresolution encoding, the information is packetized into two classes of packets according to whether the priority bit provided by the ATM format is set or not. The coarse resolution will be transmitted using high-priority packets while the detail resolution will be sent with the low-priority ones. Using the properties so that the packets with lower priority are discarded first in case of congestion is beneficial in terms of the end-to-end quality [167].

A further advantage of using multiresolution coding schemes is that they enable efficient error concealment techniques [168–170]. The idea is to use the available information, that is, packets that were not cost, to interpolate the missing information. When multiresolution coding is used, the information decoded from only the lower resolution layer may be sufficiently good. Other approaches that have been proposed involve interleaving the information so that a cell loss causes minor perceptual degradation in several image blocks rather than severe degradation in just a few.

ATM transmission provides the possibility of transporting a variable number of bits per frame and thus can seem to make unnecessary the use of rate control. This view is not realistic, because each connection will be specified by a series of traffic parameters, which will be monitored by the network. Transmission over the limits set by the traffic contract may result in lost packets and thus rate control is still necessary. We can make the distinction between rate control and rate shaping. Rate control entails changing the rate produced by the encoder, while rate shaping only affects the times at which cells are resent to the network, but not the total amount of information transmitted.

6.6.1 Layered Media Streams

An often cited approach for coping with receiver heterogeneity in real-time multimedia transmission is the use of layer media streams. In this model, the source distributes multiple levels of quality simultaneously across multiple network channels. In turn, each receiver individually tunes its reception rate by adjusting the number of layers that it receives. The net effect is that the signal is delivered to a heterogeneous set of receivers at different levels of quality using a heterogeneous set of rates. To fully realize this architecture, we must solve two subproblems: the layered compression problem and the layered transmission problem. That is, we must develop a compression scheme that allows us to generate multiple levels of quality using multiple layers simultaneously with a network delivery mode that allows us to

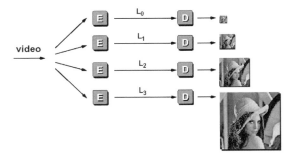

Figure 6.28 Simulcast coder.

selectively deliver a subset of layers to individual receivers. We first address the layered compression problem.

One approach for delivering multiple levels of quality across multiple network connections is to encode the video signal with a set of independent encoders, each producing a different output rate. This approach, often called simulcast, has the advantage that we can use existing codecs and/or compression algorithms as system components. Figure 6.28 illustrates the simplicity of a simulcast coder. It produces a multirate set of signals that are independent of one another. Each layer provides improved quality but does not depend on subordinate layers. Here, we show an image at multiple resolutions, but the refinement can occur across dimensions of frame rate or signal-to-noise ratio. A video signal is duplicated across the inputs to a bulk of independent encoders. These encoders compress the signal to a different rate and different quality. Finally, the decoder receives the signal independent of the other layers. In simulcast coding, each layer of video representing a resolution or quality is coded independently. Thus, a single-layer (nonscalable) decoder can decode any layer. In simulcast coding, total available bandwidth is simply portioned depending on the quality desired for each independent layer that

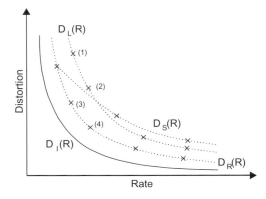

Figure 6.29 Rate distortion characteristics.

needs to be coded. It is assumed that independent decoders would be used to decode each layer [171].

In contrast, a layered coder exploits correlation across subflows to achieve better overall compression. The input signal is compressed into a number of discrete layers, arranged in a hierarchy that provides progressive refinement. For example, if only the first layer is received, the decoder will produce the lowest quality version of the signal. On the other hand, if the decoder receives two layers, it will combine the second layer information with the first layer to produce improved quality. Overall, the quality progressively improves with the number of layers that are received and decoded.

Figure 6.29 gives a rough sketch of the trade-off between the simulcast and layered approaches from the perspective of rate distortion theory. Each curve traces out the distortion incurred for imperfectly coding an information source at the given rate. Distortion rate functions for an ideal coder $D_I(R)$, a real coder $D_R(R)$, a layered coder $D_L(R)$, and a simulcast coder $D_S(R)$ are presented. The distortion measures the quality degradation between the reconstructed and original signals. The ideal curve $D_I(R)$ represents the theoretical lower bound on distortion achievable as a function of rate. A real coder $D_R(R)$ can perform close to the ideal curve, but never better. The advantage of layer representation is that both the encoder and decoder can travel along the distortion rate curve. That is, to move from point (1) to (2) on $D_L(R)$, the encoder carries out incremental computation and produces a new output that can be appended to the previous output. Conversely, to move from point (3) to (4) on $D_R(R)$ the encoder must start from scratch and compute a completely new output string. Finally, a simulcast coder $D_S(R)$ incurs the most overhead because each operating point redundantly contains all of the operating points of lesser rate.

The structure of a layered video coder is given in Figure 6.30. The input video is compressed by a layered coder that produces a set of logically distinct output strings. The decoder module D is capable of decoding any cumulative set of bit strings. Each additional string produces an improvement in reconstruction quality.

By combining the approach of layered compression with a layered transmission system, we can solve the multicast heterogeneity problem. In this architecture the simulcast source produces a layered stream, where each layer is transmitted on a

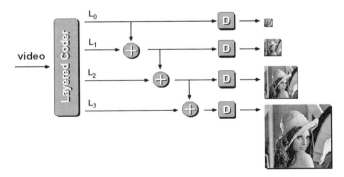

Figure 6.30 Conceptual structure of a layered video.

different network channel. The network forwards only the number of layers that each physical link can support. Each user receives the best quality signal that the network can deliver. The network must be able to selectively drop layers at each bottleneck link. The concept of layered video coding was first introduced in the context of ATM networks [3]. The video information is divided into several layers, with lower layers containing low-resolution information and higher layers containing the fine information. Such a model enables integration of video telephony and broadcast video services. In the former case, where bandwidth is at a premium, lower layers can provide the desired quality. In broadcast applications, a variable number of higher layers can be integrated with the lower ones to provide the quality and the bit rate that is compatible with the receiver.

6.6.2 Error Resilience Approach

Error resilience techniques for real-time video transport over unreliable networks include protocol and network environments and their characteristics, encoder error resilience tools, decoder error concealment techniques, as well as techniques that require cooperation among encoder, decoder, and the network. A typical video transmission system involves five steps, as shown in Figure 6.31. The video is first compressed by a video encoder to reduce the data rate. The compressed bit stream is then segmented into fixed or variable length packets and multiplexed with other data types such as audio. The packets may be sent directly over the network, if the network guarantees bit-error-free transmission. Otherwise, they usually undergo a channel encoding stage, typically using forward error correction (FEC), to protect them from transmission errors. At the receiver end, the received packets are FEC decoded and unpacked. The resulting bit stream is then input to the video decoder to reconstruct the original video. In practice, many applications embed packetization and channel encoding in the source coder as an adaptation layer to the network.

To make the compressed bit stream resilient to transmission errors, one must add redundancy into the stream, so that it is possible to detect and correct errors. Such redundancy can be added in either the source or channel coder. The classical Shannon information theory states that one can separately design the source and channel coders, to achieve error-free delivery of a compressed bit stream, as long as the source is represented by a rate below the channel capacity. Therefore, the source

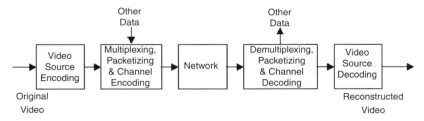

Figure 6.31 Video communication system.

coder should compress a source as much as possible for a specified distortion. Then the channel coder can add redundancy through FEC to the compressed stream to enable the correlation of transmission errors. All the error-resilient encoding methods make the source coder less efficient than it can be, so that erroneous or missing bits in a compressed stream will not have a disastrous effect in the reconstructed video quality. This is usually accomplished by carefully designing both the predictive coding loop and variable length coder, to limit the extent of error propagation.

Mechanisms devised for combating transmission errors can be categorized into three groups:

- those introduced at the source and channel encoder to make the bit stream more resilient to potential errors
- those invoked at the decoder upon detection of errors to conceal the effect of errors
- those that require interactions between the source encoder and decoder so that the encoder can adapt its operations based on the loss conditions detected at the decoder.

We will refer to all of them as error resilience (ER) techniques.

Error-Resilient Encoding

In this approach, the encoder operates in such a way that transmission errors on the coded bit stream will not adversely affect the decoder operation and lead to unacceptable distortions in the reconstructed video quality. Compared to coders that are optimized for coding efficiency, ER coders typically are less efficient in that they use more bits to obtain the same video quality in the absence of any transmission errors. The extra bits are called redundancy bits, and they are introduced to enhance the video quality when the bit stream is corrupted by transmission errors. The design goal in ER coders is to achieve a maximum gain in error resilience with the smallest amount of redundancy.

There are many ways to introduce redundancy in the bit stream. Some of the techniques help to prevent error propagation, while others enable the decoder to perform better error concealment upon detection of errors. Yet another group of techniques is aimed at guaranteeing a basic level of quality and providing a graceful degradation upon the occurrence of transmission errors [172].

One main cause for the sensitivity of a compressed video stream to transmission errors is that a video coder uses VLC to represent various symbols. Any bit errors or lost bits in the middle of a code word can not only make this code word undecodable, but also make the following code words undecodable, even if they are received correctly.

One simple and effective approach for enhancing encoder error resilience is by inserting resynchronization markers periodically. Usually, some header information is attached immediately after the resynchronization information. Obviously, insertion of resynchronization markers will reduce the coding efficiency [173]. In practical video coding systems, relatively long synchronization codewords are used.

With reversible variable length coding (RVLC), the decoder can not only decode bits after a resynchronization code word, but also decode the bits before the next resynchronization code word, from the backward direction. Thus, with RVLC, fewer correctly received bits will be discarded, and the area affected by a transmission error will be reduced. RVLC can help the decoder to detect errors that are not detectable when nonreversible VLC is used, or provide more information on the position of the errors and thus decrease the amount of data unnecessarily discarded. RVLC has been adapted in both MPEG-4 and H.263 in conjunction with insertion of synchronization markers.

Because of the syntax constraint present in compressed video bit streams, it is possible to recover data from a corrupted bit stream by making the corrected stream conform to the right syntax. Such techniques are very much dependent on the particular coding scheme. The use of synchronization codes, RVLC, and other sophisticated entropy coding means such as error resilient entropy coding can all make such repair more feasible and more effective.

Another major cause for the sensitivity of a compressed video to transmission errors is the use of temporal prediction. Once an error occurs so that a reconstructed frame at the decoder differs from that assumed at the encoder, the reference frames used in the decoder from there onwards will differ from those used at the encoder, and consequently all subsequent reconstructed frames will be in error. The use of spatial prediction for the DC coefficients and motion vectors will also cause error propagation, although it is confined within the same frame. In most video coding standards, such spatial prediction, and therefore error propagation, is further limited to a subregion in a frame.

One way to stop temporal error propagation is by periodically inserting intracoded pictures. For real-time applications the use of intraframes is typically not possible due to delay constraints. However, the use of a sufficiently high number of intracoded pictures has turned out to be an efficient and highly scalable tool for error resilience. When applying intracoded pictures for error resilience purposes, both the number of such intracoded pictures and their spatial placement have to be determined. The number of necessary intraframes is obviously dependent on the quality of the connection. The currently best known way for determining both the correct number and placement of intraframes for error resilience purposes is the use of a loss-aware rate distortion optimization scheme [174].

Another approach to limit the extent of error propagation is to split the data domain into several segments and perform temporal/spatial prediction only without the same segment. In this way, the error in one segment will not affect another segment. One such approach is to include even-indexed frames in one segment and odd-indexed frames in another segment. Even frames are only predicted from even frames. This approach is called video redundancy coding [175]. It can also be considered as an approach for accomplishing multiple description coding. Another approach is to divide a frame into regions and a region can only be predicted from the same region of the previous frame. This is known as independent segment decoding (ISD) in H.263.

By itself, layer coding is a way to enable users with different bandwidth capacities or decoding powers to access the same video at different quality levels. To serve as an error resilience (ER) tool, layer coding (LC) must be paired with unequal error protection (UEP) in the transport system, so that the base layer is protected more strongly, for example, by assigning a more reliable subchannel, using stronger FEC codes, or allowing more retransmissions [176]. There are many ways to divide a video signal into two or more layers in the standard block-based hybrid video coder. For example, a video can be temporally downsampled, and the base layer can include the bit stream for the low frame rate video, whereas the enhancement layers can include the error between the original video and the upsampled one from the low frame rate coded video. The same approach can be applied to the spatial resolution, so that the base layer contains a small frame size video. The base layer can also encode the DCT coefficients of each block with a coarser quantizer, leaving the fine details to be specified in the enhancement layers. Finally, the base layer may include the header and motion information, leaving the remaining information for the enhancement layer. In the MPEG and H.263 terminologies, the first three options are known as temporal, spatial, and SNR scalabilities, respectively, and the last one as data partitioning.

As with LC, multiple description coding (MDC) also codes a service into several substreams, known as descriptions, but the decomposition is such that the resulting descriptions are correlated and have similar importance. For each description to provide a certain degree of quality, all the descriptions must share some fundamental information about the source, and thus must be correlated. On the other hand, this correlation is also the rate of redundancy in MDC. An advantage of MDC over LC is that it does not require special provisions in the network to provide a reliable channel. To accomplish their respective goals, LC uses a hierarchical, decorrelating decomposition, whereas MDC uses a nonhierarchical, correlating decomposition [177].

The objective of error-resilient encoding is to enhance robustness of compressed video to packet loss. The standardized error-resilient encoding schemes include resynchronization marking, data partitioning, and data recovery. For video transmission over the Internet, the boundary of a packet already provides a synchronization point in the variable-length coded bit stream at the receiver side. With MDC we have robustness to loss of enhanced quality. If a receiver gets only one description (other descriptions being lost), it can still reconstruct video with acceptable quality. If a receiver gets multiple descriptions, it can combine them to produce a better reconstruction than that produced from any one of them. To make each description provide acceptable usual quality, each description must carry sufficient information about the original video. This will reduce the compression efficiency compared to conventional single description coding (SDC). In addition, although more combined descriptions provide a better visual quality, a certain degree of correlation among the multiple descriptions has to be embedded in each description, resulting in further reduction of the compressed efficiency.

Decoder Error Concealment

Decoder error concealment refers to the recovery or estimation of lost information due to transmission errors. Given the block-based hybrid coding paradigm, there are three types of information that may need to be estimated in a damaged macroblock (MB): the texture information, including the pixel and DCT coefficients values for either an original image block or a predicted error block; the motion estimation, consisting of motion vectors (MV) for a macroblock coded in either P- or B-mode; and finally the coding mode of MB. A simple and yet very effective approach to recover a damaged MB in the decoder is by copying the corresponding MB in the previously decoded frame, based on the MV for this MB. The recovery performance by this approach is critically dependent on the availability of the MV. To reduce the impact of the error in the estimated motion vectors, temporal prediction may be combined with spatial interpolation. Another simple approach is to interpolate pixels in a damaged block from pixels in adjacent correctly received blocks as all blocks or macroblocks in the same row are put into the same packet. The only available neighboring blocks are those in the current row and the row above. Because most pixels in these blocks are too far away from the missing samples, usually only the boundary pixels in neighboring blocks are used for interpolation [178]. Instead of interpolating individual pixels, a simple approach is to estimate the DC coefficient (i.e., the mean value) of a damaged block and replace the damaged block by a constant equal to the estimated DC value. The DC value can be estimated by averaging the DC values of surrounding blocks [179]. One way to facilitate such spatial interpolation is by an interleaved packetization mechanism so that the loss of one packet will damage only every other block.

A problem with the spatial interpolation approach is how to determine an appropriate interpolation filter. Another shortcoming is that it ignores received DCT coefficients, if any. These problems are resolved by requiring the recovered pixels in a damaged block to be smoothly connected with its neighboring pixels both spatially in the same frame and temporally in the previous/following frames.

Another way of accomplishing spatial interpolation is by using the spatial interpolation using projection onto convex set (POCS) method [180]. The general idea behind POCS-based estimation methods is to formulate each constraint about the unknowns as a convex set. The optimal solution is the intersection of all the convex sets, which can be obtained by recursively projecting a previous solution onto individual convex sets. When applying POCS for recovering an image block, the spatial smoothness criterion is formulated in the frequency domain, by requiring the discrete Fourier transform (DFT) of the recovered block to have energy only in several low-frequency coefficients. If the damaged block is believed to contain an edge in a particular direction, then one can require the DFT coefficients to be distributed along a narrow strip orthogonal to edge direction, that is, low-pass along the edge direction, and all-pass in the orthogonal direction. Because the solution can only be obtained through an iterative procedure, this approach may not be suitable for real-time applications.

Error-Resilient Entropy Code

Video coders encode the video data using variable length codes (VLC). Thus, in an error-prone environment, any error would propagate throughout the bit stream unless we provide a means of resynchronization. The traditional way of providing resynchronization is to insert special synchronization code words into the bit stream. These code words should have a length that exceeds the maximum VLC code length and also be robust to errors. Thus, a synchronization code should be recognized even in the presence of errors. The error-resilient entropy code (EREC) is an alternative way of providing synchronization. It works by rearranging variable-length blocks into fixed-length slots of data prior to transmission. The EREC is applicable to variable-length codes. For example, these blocks can be macroblocks in H.263. Thus, the output of the coding scheme is variable-length blocks of data. Each variable-length block must be a prefix code. This means that in the presence of errors, the block can be decoded without reference to previous or future blocks. The decoder should also be able to know when it has finished decoding a block. The EREC frame structure consists of N slots of length s_i bits. This, the total length of the frame is $T = \sum_{i=1}^{N} s_i$ [bits]. It is assumed that the values of T, N, and s_i are known to both the encoder and the decoder. Thus, the N slots of data can be transmitted sequentially without risk of loss of synchronization. EREC reorganizes the bits of each block into the EREC slots. The decoding can be performed by relying on the ability to determine the end of each variable-length block. In the absence of errors, the decoder starts decoding each slot. If it finds the block end before the slot end, it knows that the rest of the bits in that slot belong to other blocks. If the slot ends before the end of the block is found, the decoder has to look for the rest of the bits in another slot. In case one slot is corrupted, the location of the beginning of the rest of the slots is still known and the decoding of them can be attempted.

6.7 ACCESS BROADBAND NETWORKS

To most people, access broadband networks is the only part of the network they ever see or care about, because it represents the broadband solution and is the gateway that connects the home and office to multiple services. The access network, throughout its evolution, was tuned to the specific requirements of the applications for which it was built. This timing resulted in an improved service, but made the network inefficient at delivering an integrated digital service without a significant investment to either undo the tuning, deploy solutions that compensate for the limitations caused by the timing, or build entirely new networks.

Obviously, service providers have a vested interest in leveraging their existing investment, thereby rendering the third option the least attractive. It is a lot more tenable to develop solutions that would compensate for the limitations of the network or require the least retrofitting possible. We consider ISDN to be the first in the line of technologies and services called DSL (Digital Subscriber Line). ISDN held the promise of the Integrated Services Digital Network and was developed to address telephony services and lower-speed data applications. It was the vision of the ITU that ISDN would be a catalyst for the total end-to-end digitization of the Public Switched Telephone Network (PSTN) – from end user to end user and end device to end device.

6.7.1 DSL Access Networks

The development and deployment of ISDN, plus the advances in digital signal processing (DSP), lead to development of a new type of technology and service, Digital Subscriber Line (DSL). Many of the improvements and advanced features of DSL are a direct result of the experience gained with ISDN. There are many different types of DSL technologies, each having its own distinct features. Two collective terms are used to describe this generic family of technologies – DSP and DSL. We have chosen to use the term DSL. As we discuss the different DSL variants, we will sometimes make reference to its chronology. In this context, the contemporary definition of DSL will be used and basic-rate ISDN will not be factored into the chronology of the service, even though we consider it to be part of the family.

DSL service is being launched in most of the major countries of the worlds and is also beginning to appear in many less-developed ones. Worldwide, the residential market represents the larger installed base, but the business market generates the greater portion of the revenue. The growth of DSL is believed to be a direct result of an increased reliance on the Internet as a tool that facilitates communications, business, and entertainment.

DSL technology falls into two different camps, those that provide symmetric data rates (the same upstream as downstream) and those that use different speeds downstream and upstream, called asymmetric. Upstream communication is always from the perspective of the user to the service provider, and downstream is from the service provider to the user.

At first glance, asymmetric data rates may appear to be problem or inferior, but in actuality they closely match the way data typically flows between a user and a server. In most applications, the bulk of the data flows from the server to the user, with communications from the user to the server consisting of short commands. An asymmetric solution is not necessarily bad from a user perspective unless an application like FTP (File Transfer Protocol) is being used. In a file transfer, the user could be uploading a file (as opposed to downloading from the server), which would be a problem with an asymmetric data rate. When uploading, most of the data flows upward, with short commands flowing back to the user. In this type of application, asymmetric data rates are a problem because the bandwidth allocation is opposite to what the user requires. Other than applications such as FTP, a basic principle in selecting a DSL service is: if you are a user, an asymmetric technology will be fine most of the time; if, however, you are hosting a server, a symmetric solution should be employed.

All the different types of DSL use *framed transport*. Framed transport means the link sends a series of frames with no pause in between. If there is no data to be sent, the frames are still sent but are populated with a bit pattern that is understood to mean no data present. The concept is similar to a conveyor belt with attached boxes. If there is something to be sent, it is placed in a box. If there is nothing to be sent, the conveyor belt still transports the empty boxes. This is different from the techniques used in packet-based transport, where data is sent *only* when data is available to be sent.

High Bit Rate DSL (HDSL)

High Bit Rate DSL, also called High Data Rate DSL, was the first DSL technology to be put into operation [187]. Developed by Bellcore in the late 1980s, it was intended to be an economical solution to the growing corporate demand for T1 carriers. The first version of the technology was placed into service in March 1992, and since then most major telephone companies around the world have adopted the technology. HDSL operates at symmetric speeds of 1.544 Mbps and 2.048 Mbps.

HDSL is not ideal for residential broadband service for two reasons. First, it uses two copper pairs, which are not cost-effective for the service provider, especially in light of the fact that other DSL solutions available require only a single pair. Secondly, it cannot coexist with voice services on the same copper pairs.

Some common uses of HDSL include:

- Internet access from the server. The symmetrical data rates allow for the same speed in both directions, so the user accessing a server benefits from the same data rate regardless of whether or not he is doing an upload or a download of a file.
- Connecting PBX and packet-based data networks to a public network.
- Campus solutions. Many University campuses, for example, have an extensive copper network throughout the campus. HDSL is used to get T1 speeds between buildings.
- Connecting wireless-based stations into landline networks.
- Any corporate solution that requires the use of a T1 circuit.

Symmetric DSL (SDSL)

Symmetric DSL, or single-line DSL as it is sometimes called, is distinct from HDSL in that it uses a single pair. The technology, however, also has a lot common with HDSL in that it started out as a technology based on the same chip sets as HDSL. Using multiple copper pairs to provide residential digital services is not an ideal solution to a service provider. If one is to digitize the loop to the home, it makes more sense to use the existing copper than to waste an additional pair.

SDSL was intended to solve this problem. Instead of using two transceivers and two copper pairs, SDSL uses a single transceiver and a single copper pair to provide fractional T1 services. In many cases, customers require higher data speeds but not a full T1 service, and this is where SDSL fits in. Advances in technology now make it possible to gain data rates higher than the 784 kbps of the original HDSL transceivers. Different vendors now have solutions that extend the range of the SDSL modem to 1.5 Mbps and 2 Mbps.

ISDN DSL (IDSL)

Between 1982 and 1988, ANSI (the American National Standards Institute) developed standards that defined ISDN DSL (IDSL). The technology functions in much the same way as basic rate interface ISDN in that it uses the same 2B + D, which provides for a data capacity of 144 kbps. In ISDN, the two B-channels are circuit switched, each capable of carrying voice or data in both directions. The D-channel carriers control signals and customer call data in a packet-switched mode and operate at 16 kbps. Remaining throughput is absorbed by operational, administrative, maintenance, and provisioning channels operating at 16 kbps.

IDSL runs on a single pair of wires at a maximum distance of 18 kft (about 3.4 miles or 5.4 km). In traditional ISDN applications, the ISDN link requires a connection to a voice switch in the central office. IDSL eliminates this requirement and the entire connection is provided by IDSL equipment; for this reason, the technology is sometimes called BRI *without the switch* or *switchless* BRI. Some versions of IDSL allow for the full use of the 144 kbps bandwidth or full 2B+D operation, while others allow only 2B operation or 128 kbps.

A variation on IDSL exists that is based on the primary rate interface ISDN model. This version of the technology achieves higher speeds through the bonding of the B channels.

Asymmetric DSL (ADSL)

The distinguishing feature of ADSL is its ability to transport plain old telephone service along with broadband services. This was achieved by using a guard band to separate the voice-band and broadband frequencies. The broadband frequencies are used for digital services including voice, video, and data.

One of the motivating factors for the development of ADSL was the desire of telephone carriers to compete with cable service providers in the delivery of video on demand services. This was a major challenge because of loop conditions already discussed, and the optimization of the network for voice made the infrastructure more challenging to provide high data rate services. In the early 1990s, Bell Atlantic conducted the first VoD trials in New Jersey. The lessons learned from the early trials indicated that a downstream rate of about 1.5 Mbps was adequate for the delivery of MPEG-1 video streams. And upstream rates up to 64 kbps were more than adequate to allow users to issue commands like start, stop, pause, rewind, and fast forward to the video server. Speed requirements of this and other common applications are summarized in Table 6.3.

The equipment and operational cost for the delivery of VoD service priced the service out of reach for many consumers, and so the ramp-up of the service never got the traction that was hoped for. This left the carriers and cable companies in search of new markets for the technology. The Internet phenomenon, which was taking off around that time, proved to be one such market. The symmetric data streaming profile of the ADSL technology matched the data flow profile of Web browsing. The downstream data rate was significantly higher

Table 6.3 Common applications and their speed requirements for ADSL

Application	Downstream	Upstream
Video on Demand (VoD)	1.5–3 Mbps	64 kbps
Near VoD	1.5–3 Mbps	64 kbps
Computer gaming	1.5 Mbps	64 kbps
Video games	64 kbps to 2.8 Mbps	64 kbps
Video conferencing	384 kbps to 1.5 Mbps	384 kbps to 1.5 Mbps
Broadcast TV	6–8 Mbps	64 kbps
Internet access	64 kbps–1.5 Mbps	>10% of downstream
Remote LAN access	64 kbps–1.5 Mbps	>10% of downstream
Distance learning	64 kbps–1.5 Mbps	64 kbps to 384 kbps
POTS	4 kHz	4 kHz
ISDN	160 kbps	160 kbps

than the upstream rate. To address the new market, the following changes were made:

- The ratio of the downstream rate to the upstream rate was optimized for TCP/IP traffic. Using a 10:1 ratio, the downstream rate was increased to between 6 and 8 Mbps, and the upstream to 640 kbps.
- Rate adaptation was included to allow the two modems on an ADSL link to adjust their rates according to line conditions.
- ADSL was marketed as an *always on* solution like cable.

In 1994, a consortium of companies decided to form the ADSL Forum (now called the DSL Forum), which by charter decided to avoid being embroiled in line coding debates and focus on the architecture of the technology. A lot is owned to this group for the advancement of open specifications, documentation, and service availability.

The DSL Forum is a consortium of over 400 companies from various networking, service provider, and equipment industries. Established in 1994, it provides input to international standards bodies and seeks to develop technical guidelines for architecture, interfaces, and protocols for networks that incorporate DSL transceivers.

Rate-Adaptive DSL (RADSL)

The architecture of RADSL, the maximum speed, and distances supported are all essentially the same as ADSL. The differences between the two lie mainly in the earlier versions of ADSL that needed to be balanced to the conditions of the line. Technicians on both ends of the connection had to fix the speed of the link to match the conditions that existed on the line at the time of the installation. Any variation in line conditions after installation was not addressed.

RADSL addressed this shortcoming, but so does modern ADSL equipment. RADSL and current ADSL technology automatically adapt to changing line conditions each time the link becomes active. RADSL theoretically has an additional feature that is not inherent in modern ADSL. It has the capability to adapt to changing line conditions on the fly for both the upstream and downstream channels (which may be a problem to some types of applications).

In reality, the RADSL products that are deployed do not make allowance for this feature. ATM uses two variables to enforce QoS: the peak cell rate (maximum number of cells transmitted per unit time) and the average cell rate, both of which are dependent on the data rate of the link. How then is ATM to know the data rate, much less enforce QoS, if the data rate is not known until after the link becomes active? Work is ongoing to find an effective way for RADSL to communicate to ATM the data rate of the link at the time of start-up and any changes during operation − a key issue if ATM is to enforce any quality of service across the link.

Very High Data Rate DSL (VDSL)

Fiber all the way to the home (FTTH) is still prohibitively expensive, as it entails rewiring the entire neighbourhood with fiber. An alternative is to use a combination of fiber cables feeding neighbourhood optical network units (ONUs) and leverage the existing copper loops to the home or business. VDSL is a technology that gets closer to that dream.

VDSL depends on very short runs over copper of up to 6000 feet in order to maximize the available frequency range of the wire, with the remaining loop to the local exchange being served by fiber. The ONU serves as a central distribution point where the fiber from the local exchange terminates and the many VDSL copper loops aggregate. The DSL Forum refers to this arrangement as fiber to the neighbourhood (FTTN) and extends the concept to include fiber to the basement (FTTB) for high-rise buildings with vertical drops and fiber to the curb (FTTC) for short drops. The use of FTTB and FTTC gives a good indication of the target market that the technology hopes to address: places with a high concentration of people.

Different documentation uses different terminology to describe the way VDSL is delivered. The common component of all descriptions is the fiber link back to the local exchange and short copper runs. In addition to those previously mentioned, fiber to the cabinet (FTTCab) is also used to describe the scenario where the VDSL loop terminates in a cabinet close to the homes served, with a fiber feedback to the backbone network. Fiber to the exchange (FTTEx) has been used to describe those situations where the local exchange serves subscribers in the immediate vicinity.

VDSL is being touted as a full-service access network that addresses the full range of service from POTS and ISDN to linking high-speed LANs. The growing demand for high-bandwidth multimedia solutions and the proliferation of fiber has created a fertile environment for the ideas of the technology to seed and grow. The fundamental basis of its application competes, head to head, with FTTH solutions. Solutions for multidwelling and multitenant units are somewhat limited, but VDSL is positioned to be a viable solution for this space, as it leverages the use for existing telephone wires, thereby creating hope for buildings.

VDSL is asymmetric, with downstream speeds that range from 13 Mbps to 52 Mbps across copper loops ranging from 1000 feet to 4000 feet. The upstream rates range from 1.5 to 6 Mbps. This is probably an appropriate place to make reference to the different organizations that are working on the specifications, as different modes of operation, speeds, and reaches have been proposed and are currently under review.

VDSL defines five end-to-end transport modes that are combinations of STM, ATM, and packet modes of operations. STM (Synchronous Transfer Mode) operation is essentially a time-division multiplexing scheme, where each channel is given a fixed bandwidth. ATM (Asynchronous Transfer Mode) takes a more statistical approach to where the channels are multiplexed and offers some amount of discrimination in terms of bandwidth allocation and service quality. The packet mode of operation assembles the data into variable-length packets and sends them on a single channel of the maximum bandwidth.

The combinations for the different transport modes are:

- STM mode, which extends from the user terminal equipment all the way across the copper and fiber links.
- ATM mode from the user terminal equipment across both the copper and fiber.
- Packet mode from the user terminal equipment across both the copper and the fiber.

6.7.2 Cable Access Networks

Cable networks have an advantage over telephone companies and other service providers in delivering broadband services. There were, however, some obstacles that had to be overcome. CATV networks were built to deliver broadcast television in one direction only, so the network had to be upgraded to facilitate two-way communication before it could be used for broadband services. A frequency band had long been set aside for a reverse path; the hurdle, therefore, was to find a way to use it. The equipment used in cable plant was designed for one-way communication. Amplifiers were used to amplify the forward signals only, and anything flowing in the opposite direction was considered noise.

The topology and access methodology also presented challenges. The fact that cable networks used a shared medium presented security concerns, so the data from multiple subscribers had to have some level of security applied. Finally, the application of CATV networks made it more prevalent in residential than business use. So even though CATV networks had an edge over telco solutions for the delivery of broadband services, they were not necessarily the solutions of choice for corporations. This is an undesirable position to be in, because as the data for DSL usage show, worldwide residential installations outnumber business installs, but it is business that generates the greater revenue for DSL service.

A basic cable system is composed of four major components: (1) the headend; (2) the trunk link; (3) feeder and drop cables; and (4) a terminal equipment of network terminating unit that represents the subscriber. It is as simple as that. An operator of a cable system is called a *cable operator*. Companies that operate multiple cable systems are called multiple systems operators (MSOs), and the systems owned may extend into multiple states. Obviously, there are better economies of scale if the systems owned are contiguous, so MSO tend to expand by acquiring or merging with cable operators that are adjacent to systems that they own.

The distribution of the cables in a cable system has a tree and branch structure. The headend forms the root; the trunk, feeder, and drop cables, the branches; and the subscriber, the leaves. The trunk connects directly to the headend with feeder cables branching from it through the use of splitters. The feeder cable extends the system for miles from the headend and deep into local neighbourhoods. Drop cables connect subscribers by tapping into a feeder cable or the trunk.

The center of the cable system is the headend. It is the brains behind the operation, where broadast and satellite-delivered signals are received and readied for distribution. The headend can receive and process programming in various formats, analog (AM and FM) or digital. It also hosts the equipment used to descramble incoming feeds from satellites and broadcast networks. These feeds are assigned channels that match the channel on which they were originally transmitted before being transmitted over the cable lines. Finally, the headend also has the capability to encrypt signals for security purposes and play local content such as advertising and locally originated programming.

Trunk lines are high-capacity lines that carry signals from the headend to feeder cables that serve the local communities. Trunk lines are usually untapped, which means that drop cables that service subscribers are usually not directly tapped into a trunk.

There is no way to dictate to subscribers the perfect placement for their homes, which sometimes are situated in very inconvenient spots, like the very end of a cable segment served by an amplifier. To accommodate such users, the cable provider may choose to turn up the power of the serving amplifier. The customer at the end of the segment may be perfectly served by this, but the boosted power levels might adversely affect another subscriber closer to the amplifier. To address this problem, a device called a *pad* is used to induce attenuation. The pad is placed on the drop cable of the subscriber affected by the increased power levels and serves to reduce power to that subscriber.

Hybrid Fiber-Coax

By the 1990s, cable operators began to see direct broadcast satellite (DBS) service as a serious threat and, in a defensive mode, began the process of improving the cable system. In addition to needing to remain competitive with DBS services, which had double the channel capacity, the cable plant in general was ready for a major overhaul. The problems associated with wire lines and amplifiers meant high maintenance costs.

The problem of wire attenuation over long distances and the need to amplify the signal every 3000 feet or so were major headaches. The failure of a single amplifier could degrade or disrupt service to hundreds of subscribers. Take, for instance, the failure of a single amplifier close to the headends – subscribers downstream for as far as the cable extends (which could be 30 or 40 miles away) could be affected.

The quality of the final signal was also a problem. To overcome the problems of the cable plant, a hybrid version of the network called a hybrid fiber coax (HFC) is developed. In this scheme, fiber optic cables were used to replace the coaxial trunks and were run from the headend to a fiber distribution node. Coaxial feeder cables were attached to the fiber node distribution node, which converted the analog optical signals into electronics for downstream transmission along the coax feeder cables.

The HFC design transformed a single cable system into a series of smaller cable systems with individual serving areas of as few as 200 to 500 subscribers, with the larger areas serving about 500 to 25,000 subscribers. The fiber connection back to the headend meant a significant increase in available bandwidth, signal reliability, and quality of service. The fiber also meant a reduction in emission and so less interference for the existing coax cables. The segmenting of the system into smaller groups also meant that a problem in one neighborhood system would not affect another.

In an HFC system, amplifiers are needed only on the coaxial branches, so the number of amplifiers was significantly reduced. The reduction of amplifiers also simplified the task of equalizing the power levels and reduced maintenance, resulting in lower operational costs.

The use of fiber accomplished several things: it significantly reduced a number of transmission issues, it simplified the design, it created smaller and more manageable coax segments, it increased channel capacity by providing a broader range of frequencies, and it set the stage for upstream communication. Cable systems had been predominantly one-way broadcast systems even though the bandwidth was available for upstream communication. If a new two-way service like Internet access or VoD were to be rolled out, the network had to be able to facilitate communication along the return path.

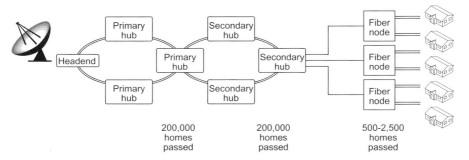

Figure 6.32 An example of hybrid fiber-coax (HFC).

The architectures of cable systems and HFC networks are evolving to allow for the integration of multiple two-way services. The architecture shown in Figure 6.32 incorporates the features of a well-designed data network, including redundancy, fiber backbones, and interconnectivity of hubs for reliability. The design of the network looks like any other robust network, and this is necessary if the HFC network is to be used as a reliable transport for mission-critical data applications. If the industry wants to attract corporate accounts in addition to its large residential customer base, a reliable and robust infrastructure is needed along with strong operations and support organization.

Many MSOs are deploying regional hubs using fiber rings to extend the reach of the HFC network and to enable sharing of a headend or multiple headends. The architecture allows multiple operators to share the cost and benefits of a fiber ring. The added benefit of hubs is that they allow sharing the cost and benefits of a fiber ring. The added benefit of hubs is that they allow sharing headend equipment among companies. As the digital era unfolds and more applications and services are launched, the capital required to build a headend and outfit it with necessary equipment becomes increasingly high. Being able to share the servers, compression and insertion equipment or maybe even a telecommunication switch becomes increasingly attractive as the equipment list and the processing requirements grow [181].

The upgrade of the cable system to HFC was in part a defensive move to stave off competition from direct broadcast satellite and new telco initiatives. A second motivation was to be able to offer new, enhanced services. With the upgrade complete and the capacity in place, new incremental applications could be supported with minimal additional costs. Several applications were targeted, high-speed data services, basic telephone services, and digital video services being a few.

Normally, basic telephone service would not be categorized as a broadband service. It is, however, an important basic service on which we all rely. For this reason, it must be considered in any discussion on integrated digital services. For this reason, we have covered it enough to provide an appreciation of developments in this area.

Cable Telephony. The cable industry has invested billions in infrastructure upgrades since 1996. These upgrades, which allow cable companies the ability to offer high-speed Internet access and digital services, also position them to provide regular telephone services. The initial venture into the residential telephone market has been through standard circuit-switched digital technology.

To offer circuit-switched telephone service over a cable system, cable operators must install a telecommunications switch within the headend and negotiate interconnection terms with the local phone carriers so that cable telephone users can call other users that are on the regular PSTN network. The switch functions in the same way as a PSTN switch – a path is temporarily set up between the caller and the called party for the duration of the call. If the called party is also a cable telephone subscriber on the same cable system, the call never leaves the switch at the headend. If, however, the called party is on the PSTN or another cable provider's cable telephone service, the call is routed through the switch at the headend to the local phone service switch on the PSTN.

Once the investment has been made in the infrastructure to support two-way data and digital services, it is a lot easier to layer incremental applications. If, however, it is determined that there is a need and a market for circuit-switched telephony, the decision to invest in the equipment to enable the service could be made at that time, on the basis of the supporting market data and projected return on investment. This was the reasoning that led many to take a more cautious approach in providing circuit-switched telephony across their HFC networks.

Deploying a separate voice telephony architecture from that of data is expensive. It requires different equipments and separate downstream and upstream channels, creating spectral inefficiencies. Cost of operations also increases because separate operational and management platforms are required with the appropriate staffing.

IP Telephony. IP telephony is needed as a much more cost-effective way for cable operators to transport voice. It leverages the existing hardware and the service becomes just data application. It will, of course, require some thought and investment to assure a level of QoS exceeding that of other types of data applications. Other factors such as call management and billing must also be addressed, but that is viewed as a small price to pay. The investment made to provide QoS, a reliable service, and the other features required for call management and billing can be leveraged for other applications. IP telephony also has the potential to integrate with other data services, which makes it even more attractive for voice-mail and e-mail without additional expensive equipment.

To provide IP-based telephone services, technical challenges need to be overcome and issues such as call management, signaling, billing, security, QoS, and provisioning must be addressed. The technical issues of IP telephony are well understood as a result of the work done on VoIP. Circuit-switched telephony uses a connection-oriented approach that creates a point-to-point circuit between the caller and the called party. Creating a circuit between the two parties is important to provide the consistent level of latency that voice requires. One problem with this approach is the inefficient use for the bandwidth of the circuit. During a telephone call, the port that terminates the loop connecting the subscriber's phone to the switch cannot be used for anything other than the call that is in progress or to alert the subscriber that there is another call waiting. Three-way calling is possible, but the point is that no other application can utilize that port or the circuit. During periods of silence, the circuit must remain dedicated to the call in progress, a waste of bandwidth.

A packet-based approach is a lot more efficient in that it allows the link to be shared by multiple applications. The trade-off with the approach is that you sacrifice some control over latency. It is a lot more difficult to control when the link will be available for use, and so a critical packet may be delayed while it waits for the link to become free. This is one of the challenges of VoIP. Using packet-based technology, voice calls are subject to delays, which adversely affects the quality of the voice call. IP telephony requires strict QoS and prioritization to minimize delays.

Digital Audiovisual Service Over Cable. The U.S. Congress has set December 31, 2006, as a target date for broadcasters to switch from analog TV systems to digital delivery, a move that is supposed to offer customers more programming choices and higher-quality sound and pictures. This move is being touted as the biggest technological advance in television since the days when color ousted the grainy black-and-white pictures. Between now and then, television will be aired in both analog and digital formats in some markets. After this date − or sometime thereafter, as the date will most likely slip − the spectrum used for analog television will be returned to the government. Cable operators will continue to distribute both analog and digital TV programs until analog television is no longer broadcast over the air.

Before upgrade, the frequency band used for broadcast television was about 400 MHz. After the upgrade to a hybrid fiber-coax system, the available frequency range about doubled. The extra bandwidth above the original 400 MHz is being used for digital services. The math, however, does not appear to add up. If the original frequency band was about 400 MHz, which yielded about 55 channels, it follows that if the band is doubled, the number of channels also doubled, to about 110 channels.

6.7.3 Wireless Access Networks

Broadband wireless is not the same as mobile wireless. Mobile wireless, like cell phones, allows users the service mobility while using the service. Broadband wireless is communications delivered from a ground antenna or satellite to buildings or fixed sites without a wired connection at high data rates.

The appeal of wireless networks lies in their reach without a dependence on a wired infrastructure. They enable the provider to deliver high-speed broadband solutions to places that lack a wired infrastructure, as well as those that have one. In our review of DSL and cable networks, we stated that a feature of these services that is heavily marketed is the fact that they are always-on. In reality, wireless is the best-positioned technology to stake a claim to that description. Unlike wired solutions, which require a subscriber to be physically connected to the network in order to be always-on, wireless networks are always on and the subscriber is always connected.

The tight economic climate of the beginning years of the 2000 decade has not been favorable for new ventures. Coming off the highs of the late 1990s, when any project that appeared to be Internet-related almost guaranteed a flock of private investments, the economic downturn in the new decade turned investors more cautious. New ventures had to have a solid business plan with a solid projection for profitability before they would be considered for funding by burnt venture capital investors.

In spite of this, the fixed wireless broadband access market now represents a substantial portion of the broadband access (wireline and wireless) market. The fixed wireless market, at the time of this writing, is segmented into three distinctly different markets based on the spectrum allocations of the service. These markets are the Local Multipoint Distribution System (LMDS), the Multichannel Multipoint Distribution System (MMDS), and license-free wireless services that comprise the

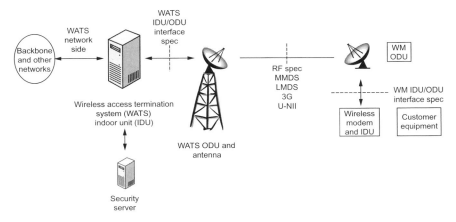

Figure 6.33 Fixed wireless architecture.

Unlicensed National Information Infrastructure (U-NII) and the industrial, scientific and medical (SM) bands.

Figure 6.33 is an architectural reference model for fixed wireless systems. It depicts the components that are common to these systems. In addition to the physical equipment such as antennas, base station, indoor and outdoor units, the figure depicts the different interfaces required for fixed wireless systems. As shown, the RF specifications for MMDS, LMDS, 3G, and U-NII define the interface between the base station and the receiving antenna.

Satellite services are also contenders for providing wireless broadband access.

LMDS (Local Multipoint Distribution System)

LMDS is a point-to-multipoint fixed broadband wireless service that can be used for two-way communication at data rates beyond 155 Mbps. Typical systems have data rates of 45 Mbps downstream and 10 Mbps upstream. The fact that a single transceiver can reach thousands of subscribers within a cell makes this service an attractive solution with an attractive price point for delivering broadband service. The incremental expense of the solution is realized at the time a subscriber subscribes and is limited to the cost of the subscriber equipment.

This is the main attraction of wireless from a service provider's perspective. A capital commitment is necessary only once the sale has been made. This is in contrast to wired solutions, where the infrastructure must already be in place, thereby requiring upfront capital investment before any revenue can be generated.

An LMDS system uses point-to-multipoint distribution of signals in cells that are roughly 2–3 miles in diameter and is a potential choice as a last-mile solution in areas where copper, cable, or fiber may not be convenient or economical. Some benefits of the technology as a fixed broadband solution are:

- lower deployment costs
- ease and speed to deployment
- return on investment realized a lot sooner
- capital outlay at the time the service is sold
- deployment feasible to hard-to-reach places and dispersed consumer base
- absence of regulation at local and state levels
- cost 80 percent electronic, not labor and structural materials (no trenches to dig, no amplifiers to fix, etc.).

MMDS (Multichannel Multipoint Distribution System)

The MMDS is a point-to-multipoint fixed wireless technology that operates in the 2.5 GHz range of the spectrum in the United States and Canada and in the 3.5 GHz range in many international markets. Sometimes called Wireless DSL, the technology is viewed as a viable solution for providing broadband services to a widely dispersed area. MMDS systems are able to achieve speeds of up to 10 Mbps.

Unlike LMDS, which is limited in reach, MMDS is able to serve a 35 mile area, with hubs that are typically located on mountaintops and other high places. A single tower can provide coverage to a huge and heavily populated area at a much lower cost than LMDS. A single MMDS hub can cover an area that would require 50 to well over 100 LMDS hubs.

The relatively longer wavelength also makes MMDS less susceptible to interference from weather and vegetation. Its wide coverage area also makes it a cost-effective solution for reaching dispersed populations in rural areas – a market that would be much more expensive to cover using LMDS because of the impact of terrain on its millimeter waves.

U-NII (Unlicensed National Information Infrastructure)

In January 1997, the Federal Communications Commission (FCC) set aside 300 MHz of spectrum in the 5 GHz band for an unlicensed service called U-NII. Three bands were defined: two were defined in the 5.15–5.25 GHz and 5.25–5.35 GHz ranges; the third was defined in the 5.725–5.825 range.

The unlicensed bands were set aside to speed up the deployment of wireless broadband access for business, schools, and hospitals. In recognition of the potential for interference, the FCC defined rules that limit the power levels of antennas. These power levels are strong enough to transmit data up to 25 Mbps or beyond, but they limit the range of the transmitted signals to about three miles.

Using this band, service providers can deploy new services at a faster rate and at lower startup costs. For holders of wireless broadband licenses, spectrum can be used to supplement the limitations of the existing licensed service, or may be used as an interim solution before the main service is launched. In essence, the major benefits of U-NII are flexibility, lower costs, and speed of service deployment.

The flip side of the coin is the fact that the foregoing benefits have the potential to attract many operators in this space. While this is good from the perspective of competitive service pricing generated by competition among many competitors, the added traffic creates the power limits imposed by the FCC.

Third Generation (3G)
To the average person, the mention of 3G wireless technology evokes mobile cellular communication. Less publicized is its potential as a fixed wireless solution – the technology is capable of speeds up to 2 Mbps in fixed mode.

First-generation wireless refers to the analog cellular transmissions that predated the digital types of service that are common today. Second-generation (2G) wireless is the current digital cellular and personal communications that we use today for voice communication and short text messages.

Third-generation wireless technology is capable of supporting circuit and packet data at high bit rates: 144 kbps for high mobility (vehicular) traffic; 384 kbps for pedestrian traffic; and 2 Mbps for fixed indoor traffic. It is a worldwide standard that is also known by the ITU designation of IMT-2000, the International Mobile Telecommunications 2000 initiative.

Of the different wireless broadband access technologies, 3G is better positioned than most to make the wireless broadband promise a reality. It has the benefit of global awareness, it is backed by standards, and it shares a commonality of design with existing wireless services that paves an easier path for making the transition into new services. In contrast, most other options, including MMDS and LMDS, are virtually unknown standards. With 3G, there is a point of reference in the mind of the customer, who understands wireless communications in terms of mobile cellular communications and therefore finds it easier to appreciate 3G as a logical next step for a technology he or she uses daily – the cognitive association is already there [185].

The business hurdles, too, are a lot less challenging with 3G. The technology requires a lot less marketing if it is positioned as the next generation of wireless service. Other wireless access options – with maybe the exception of DBS – require a more extensive marketing plan because of the need to educate on new terminology and positioning of the product. For existing wireless operators, it is easier to upgrade than to build a new infrastructure.

Direct Broadcast Satellite (DBS)
The DSB service is in the best position to grab market share in the broadband wireless access market. The development of DBS service from the perspective of new service launch has been in lockstep with cable operators. As the largest competitors to cable, both industry groups maintain a competitive posture to gain market dominance. It should, therefore, not be surprising that DBS companies have been forging ahead in the rollout of two-way high-speed services. Initially, as with cable, the focus was on Internet access with a telephone return path. While the target market – Internet access – has not changed, a satellite return path is now available in addition to telephone returns.

DBS companies have been wary of the consolidation that has been happening in the cable industry. DBS operators have gained market share in cable markets at a rate that has surprised many who thought the service would succeed only in areas that did not have cable. Any change within the cable industry that has the potential to increase the cable operators' dominance is viewed with caution.

DBS uses geostationary satellites operating in the Ku band with a 12 GHz downlink and 14 GHz uplink. The architecture of the service for broadband wireless is very straightforward. It uses a *bent-pipe* approach, a term used to describe the signal path when satellites are used: signals go up and are reflected to the target earth station.

The basic architecture of the service is very similar to cable: a headend receives feeds from the different services and networks and encodes them into MPEG for digital transmission. The composite signal is transmitted to a geostationary satellite, which in turn broadcasts it to the subscriber antennas. The subscriber antenna is connected to a receiver that connects to the television set.

For Internet access, requests for a Web page are sent from the subscriber to the provider's headend (DBS companies sometimes use the term network operations center (NOC) when the service is being used in the context of data). The NOC is connected to the Internet via a wireline facility. The response from the Internet is sent to the NOC, which combines the data along with other television feeds for transmission to the satellite and back to the subscriber.

6.8 CORE BROADBAND NETWORKS

The technologies inside the metro area and long-haul (core) networks are a lot different from those found in the access network. The technologies in these networks are also very different from those used in the enterprise and homes. These networks being predominantly a fiber optic domain, the technologies are predominantly SONET/SDH and ATM, and other emerging technologies such as MPLS and optical solutions such as DWDM are also becoming more prominent.

Figure 6.34 illustrates a network hierarchy that forms a point of reference for the networks we have been discussing. As is evident from the figure, the hierarchy is logically segmented into an access portion, a metropolitan area portion, and a long-haul core portion.

There is a natural tendency to regard the metro area network as a scaled-down version of the long-haul core network. In some perspective this is true, as both are now predominantly fiber-based networks, with the metro area having shorter runs of fiber. However, an important differentiation is that the long-haul networks tend to be a lot more stable in topology, whereas at the metro area level, the topology keeps changing. Another difference is at the metro area level, where the design and management are geared more toward the different types of traffic streams and services. In the long-haul portion of the network, the emphasis is on big pipes and fast bit pumps that move data as quickly and efficiently as possible.

6.8.1 SONET/SDH

SONET (Synchronous Optical Network) and SDH (Synchronous Digital Hierarchy) are two standards that are very closely related. In the United States, the standards for

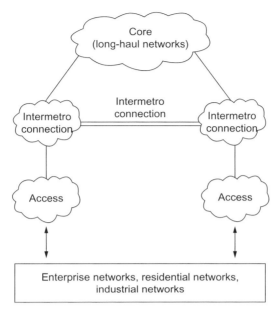

Figure 6.34 Network hierarchy reference model.

SONET are used; other places use SDH. Both standards specify interface parameters, data rate framing formats, multiplexing methods, and management methods for synchronous time-division multiplexing (TDM) over fiber.

Based on principles of TDM, multiple digital bit streams are input into a SONET/SDH system and are multiplexed into a composite signal over an optical fiber infrastructure. The input stream usually runs at a speed of 2.5 Gbps, with the resulting output signal being 2.5 Gbps times the number of input streams. So if four signals are multiplexed, as shown in Figure 6.35, the resulting output onto the fiber is 10 Gbps.

The arrival of SONET/SDH brought order to a somewhat chaotic approach to optical communication. Fiber optics being a new technology, equipment vendors recognizing its potential sought to capitalize on it by introducing proprietary equipment in the absence of standards. With the introduction of any new technology, the market for the technology emerges long before any standards. As a forerunner to standards, the market becomes the basis on which a decision can be made as to whether or not a standard is needed.

So prior to the development of the SONET/SDH standard, many optical solutions existed that were mutually incompatible. In fact, there were no guarantees that a piece of equipment would be able to benefit from future technology improvements such as higher speeds, for example. With the introduction of SONET/SDH, telecommunications companies readily embraced the standard because it provides the means for interoperability between different vendor systems and it provided an upgrade path for any new technological enhancements.

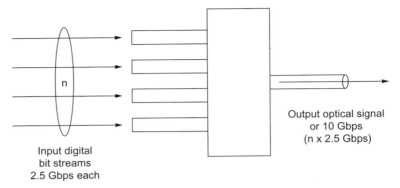

Figure 6.35 An example of SONET/SDH multiplexing.

At the turn of the 1990s, when the Internet became more prevalent and the requirement for more data grew, SONET was seen as a major enabler of the build-out of the infrastructure. The features of SONET that are now recognized as limitations were not a problem then. Its built-in reliance on a single wavelength of 1310 nm and its top speed of approximately 10 Gbps appeared adequate at the time.

Standardizing on a single wavelength had its benefits, in that it limited the choice of components, in turn eliminating the complexity of different variables. Having a smaller number of choices made sense, as it simplified the life of the telecommunications operator. The only types of laser that were required were those that operated at 1310 nm. The number of fiber types was also limited, which created a more predictable operational environment with a single-wavelength system of a single bit stream.

The upper limits of the technology also appeared to be more than adequate for applications that were available then. However, the speed limitation and the single-wavelength design of the technology bring it to major crossroads today.

Within a SONET/SDH network, there are many devices that perform different functions. An introduction to some of the more common elements of the network is appropriate, so we will list the components and provide a brief description of the functions of each.

- *Terminal multiplexer.* This device is an end-point device that is meant to reside at the customer premises but instead is mainly found in the central office. It functions as an entry-level path-terminating terminal multiplexer, acting as a concentrator.
- *Regenerator.* This is used to redefine and amplify a fiber signal that has degraded because of attenuation. The optical signal is converted to the electrical domain for processing and then sent back to optical for forwarding. A regenerator clocks itself off the received signal and replaces the section overhead bytes before retransmitting the signal.

- *Add/drop multiplexer (ADM)*. This provides interfaces between the different network signals and SONET signals and allows signals to be added or dropped off at any point in the network. At an add/drop site, only those signals that need to be accessed are dropped or inserted. The remaining traffic continues through the network element without requiring special pass-through units or other signal processing.
- *Drop and repeat (D + R)*. Also known as drop and continue, this is a key capability in both telephone and cable TV applications. With drop and repeat, a signal terminates at one node, is duplicated (repeated), and is then sent to the next and subsequent nodes.
- *Digital loop carrier (DLC)*. Used in what is known as a carrier serving area (CSA), this connects a high number of customers using ordinary copper wire.
- *Matched nodes (MN)*. Used to interconnect SONET rings, these devices provide an alternate path for the SONET signals in case of equipment failure. This feature is also known as signal protection.

6.8.2 Future of Asynchronous Transfer Mode (ATM)

Asynchronous transfer mode (ATM) was developed to bridge the gap between the requirements of applications that benefit from a circuit-switched approach and those that handle any kind of information – voice, data, image, text, and video – in an integrated manner. Many of the broadband access technologies discussed throughout the chapters of this book demonstrate that it has found a place in many different platforms, technologies, and protocols. A connection-oriented, packet-switching technique, ATM uses a field-length 53-byte cell with 48 bytes of payload and 5 bytes of overhead. The following are the requirements that ATM was designed to address:

- support all existing devices as well as emerging services
- utilize network resources efficiently
- minimize the switching complexity
- minimize the processing and buffer requirements at intermediate nodes
- support very high transmission speeds
- guarantee performance requirements of existing and emerging applications.

As a high-speed packet-switching technology, ATM was designed to be the final network technology that we would ever need. It was built to function equally in a LAN, MAN, or WAN environment regardless of the application or traffic type. In reality, the deployment of the technology has not been as pervasive as it was once billed. Today, it is a networking solution and it is now being positioned as a technology that complements competing technologies.

The constant transmission delay and guaranteed capacity of circuit switching are maintained in ATM. These features are combined with the flexibility and efficiency of packet-switching networks, which are more suited to intermittent and bursty traffic. The ATM cell consists of 5 bytes of header and 48 bytes of payload for a total size of 53 bytes. Using the information contained in the header, ATM switches are able to determine the correct output port that connects it to the next switch in the path to the cell's ultimate destination.

Based on asynchronous time-division multiplexing, ATM uses time slots that are allocated and are available, on demand, to user data. This is an important distinction of ATM. TDM is less efficient than statistical multiplexing because it results in wasted bandwidth. If there is nothing to send, the time slots in TDM are transmitted empty and cannot be used by another application. In this case, the application has to transmit at the pace at which its time slots become available, even though the remaining time slots may be empty because of no activity from the other applications that are sharing the link.

With ATM, time slots are used but made available on demand. This means that if that same application has a lot to send, it will not be limited to the time slots to which it is assigned but will be allowed to use any available time slots.

The decision to use a fixed-length cell, 53 bytes long, was based on a compromise between the requirements of voice traffic and those of data. In fact, the requirements of voice and data in themselves have nothing to do with the packet sizes, which instead are based on the effects the characteristics of one traffic type have on the performance of the other. Voice samples are short and of a fixed length, while data has variable lengths that range from short to relatively very large when compared to voice samples. Under normal circumstances, this is not really a problem, but when one looks at the latency requirements of voice, it becomes a lot clearer why mixing two on an unchannelized link can be a problem.

As a technology that once promised to be the final networking solution that would transcend the local and wide area divide, ATM has lost some of the fervor that it once stirred. The future of the technology is not as clear now as it was once perceived to be. The advent of DWDM and the higher speeds that it enables make many of the issues that ATM was built to address moot. Class of service, for instance, is extremely important when multiple data sources and traffic types are vying for a limited resource – bandwidth. If the limitations are removed and an abundant supply of bandwidth is made available, which is what DWDM enables, the concept of class of service becomes moot because each traffic type can be provisioned with the bandwidth that it requires.

ATM is now being positioned as a complementary technology rather than one that replaces. On the ATM Forum's Web site, it is presented as a technology that interworks with IP, Frame Relay, Gigabit Ehternet, DSL, wireless and SONET/SDH. The next-generation network is described as depending on ATM and complementary technologies to handle the simultaneous traffic of voice, data, and images. We believe this to be true. ATM will most likely not deliver on its promise to be the single networking technology for the local, access, and core networks that it was once hoped to be in conjuction with B-ISDN. However, elements and features of the technology will still be found in many of the access technologies that are emerging and those that are already deployed. The stronger features of the technology will also be adopted in new and emerging metro area technologies.

6.8.3 Trends in Wireless Broadband Networking

During the last couple of years, the combination of Internet technologies and mobile communications has been successfully considered the major vehicle toward the next phase of telecommunications networks. Especially in the case of cellular network integration, this is typically referred as all-IP networking. Traditionally, data-communications-driven Internet and voice-dominated mobile communications were separate disciplines, both enjoying exponential growth. If we define this as the first step of the convergence, we can argue that most of the basic technology building blocks already exist for it. The basic framework for wireless Internet is already reasonably well understood in both mobile radio (cellular) systems and more open wireless LANs (WLANs). However, the practical integration task between traditional (cellular) radio systems and TCP/IP-based architectures remains formidable with all its details and performance tuning. A lot of work is needed before transparent all-IP networking becomes reality [182].

Although the near-term research and development seems quite straightforward, economic and business environment uncertainties make it hard to predict any precise timing. It is clear that cellular networks will continue to improve in terms of capacity, services, and coverage. The most urgent development is enabling heterogeneous networking, including support for vertical handovers, seamless roaming, and micromobility with IP. Some issues will be solved at the application layer, and we will have to accept inherent limits of unreliable wireless channels.

Software-defined radio (SDR) technology will become available, and reconfiguration time will not be an issue. There is a long research road ahead before this is a reality. Not only will wireless terminals and base stations become SDR-based, but the core backbone network will also gain more adaptability and reconfigurability. The proportion of software in telecommunications systems will increase everywhere. Terminals, phones, routers, and other devices will be able to process and hold more and more programs.

WLANs have started to become more popular, not only within companies and homes, but also as a public hotspot technology. The reason for the popularity of IEEE 802.11-based networks are simple: the technology is cheap; for upper layers and applications it performs just as wireless Ethernet; and as unlicensed radio technology it is easy to deploy. It is also the only widely available wireless technology to quickly build broadband wireless networks. The development of WLAN technology from early 1 and 2 Mbps models to present-day direct sequence spread spectrum (DSSS) and orthogonal frequency-division multiplexing (OFDM) based systems up to 54 Mbps has been quite rapid, and work done toward future extensions on upper layers, QoS, and innovative new physical layer technologies is still going on.

There have occasionally been suggestions that third-generation (3G) cellular technology, which has been promising first 384 kbps and later up to 2 Mbps connections to a user, is going to break through commercially because of the emergence of

WLANs. However, it is early and drastic to predict that. It is clear that 3G has problems. It has not reached the quick and strong global success most of its ardent supporters claimed in would, and it is hard to predict its commercial progress; but 3G systems will emerge. There was perhaps unavoidable hype surrounding 3G, and we should not make the same mistake with WLAN hotspots. WLAN will be an inevitable and important part of the future wireless and mobile broadband infrastructure. However, we also need large-scale cellular networks to support high-velocity mobility and provide geographical coverage. One also should not overemphasize only raw radio bit rates, as it is well known that the actual application bit rates with WLANs are lower (e.g., with TCP the maximum data rate would be about 6.1 Mbps in 11 Mbps mode in ideal conditions); and of course, the capacity is shared between users. The overall aggregate capacity of cellular technologies is also high. The simplified battle between the *best technologies* does not make sense, since different technologies have their optimal usage in different places.

As for fourth generation (4G), a number of different interpretations exist. Some people see it as a new radio interface in the traditional movement from 3G to a new *full* standard, which presumably would provide at least higher data rates and better adaptivity. Another often used interpretation refers to heterogeneous (integrated, IP-enabled) wireless networks. Probably the truth is a combination of both, as the future hierarchical and heterogeneous overlay network infrastructure will also *definitely require* new air interfaces – and enhanced adaptivity – in order to provide better scalability, QoS, and wireless capabilities.

A safe bet is to see 4G as a system of systems that brings in standardized capabilities and technologies to make composable and autoconfigurable networks. There is a lot of interesting research being carried out in this direction; one new large initiative is the Ambient Networks integrated project funded in part by the European Union project 6th Framework Research Program. However, although combination of networks is the key, the actual deployed 4G networks will also include new radio technologies.

The core cellular technology will be based on 3G, including Universal Mobile Telecommunications System (UMTS) and Code-Division Multiple Access 2000 (CDMA2000); Global Systems for Communications (GSM) evolutions such as General Packet Service (GPRS) and Enhanced Data for GSM Evolution (EDGE) will play an important role. Even before full SDR capability, early versions of software radios can be used to provide multimode terminal technology for users; of course, multimode terminals are already available. The key issue is to use all different aspects and capabilities of networks to provide a good user experience.

In fact, cellular networks such as existing 2.5G and 3G networks have excellent global reach and good customer management (most notably billing and authentication) functionalities. Many key players in the field, among them Ericsson and Nokia, have lately demonstrated cooperation and vertical handover capabilities between WLAN, 3G, and 2G networks. The interworking architecture between WLAN and 3G/2G systems can provide fast deployment for global roaming and billing. Hence, the controversy between 3G against WLAN is partially already being tackled, and the 3G Partnership Project (3GPP) is working on interoperability [183].

There is still a need to provide better harmonization of WLAN and cellular network interfaces with operating systems and other higher-layer software. The present very heterogeneous situation is slowing down some interesting work from application developers who have no interest in developing different versions of the same software base for all different wireless access technologies. Various existing emerging wireless technologies characteristics are shown in Table 6.4.

Mesh networking combined with end user mobility is probably an important key ingredient to provide extensive and ubiquitous wireless broadband access for everyone. Especially in city areas, the wireless mesh network overlay structure can be used to feed information toward local ad hoc and peer-to-peer networks.

6.8.4 Toward the Fourth Generation (4G) System

In the area of information society technologies, the last decade has shown how technologies such as the Global System for Mobile Communications (GSM) in the wireless area, and the Internet in the network and service areas, can change society's profile regarding access to information, but also how critical the cost of deployment of such technologies, and their complexity or unavailability can be to the overall industry.

The grand challenge for the next decade toward the fourth generation (4G) is for the European research community to continue the design and creation of new technologies while ensuring that their deployment will be realized with reduced capital and operational expenditures in order to maintain sustainable growth of the whole industry and society.

Table 6.4 Various wireless technologies characteristics [182]

Network	Standard	Radio Basic Rate	Frequency Band	Mobility
WLAN	IEEE802.11b	1, 2, 5.5, 11 Mbps	ISM 2.4 GHz	Low
	IEEE802.11a	Up to 54 Mbps	ISM/UNI 5 GHz	Low
	IEEE802.11g	Up to 54 Mbps	ISM 2.4 GHz	Low
Bluetooth	IEEE802.15.1	1 Mbps	ISM 2.4 GHz	Low
WMAN	IEEE802.16	134 Mbps	10–66 GHz	N/A
	IEEE802.16a	70 Mbps	2–11 GHz	N/A
	GSM	9.6/57.6 kbps		High
2G	GPRS	115 kbps	900/1800/1900 MHz	High
	EDGE	384 kbps		High
3G	UMTS/WCDMA	Up to 2 Mbps	1900–2025 MHz	High
WLANs (next generation)		Up to 1 Gbps (indoor)		
		150–250 Mbps (outdoor)		Low

©2004 IEEE.

Cooperation of heterogeneous access networks (cellular and broadcast in particular) is an area that has been investigated for some time through a number of EU projects, with the aim of setting up technical foundations, developing specific services and architectures, and addressing network management aspects [184].

Most of the technical barriers have been identified, but need now to adapt to the current regulatory and business context characterized by openness of systems, system diversification, and the search for productive investment. Network cooperation is probably one of the main clues for addressing the 4G technological landscape, but it needs to be driven by a number of requirements that are meaningful from the technological viewpoints as well as from the regulatory and business ones.

In recent years, an excess of available technologies' convincing *killer applications* have created profitability issues for many companies, hence leading to rethinking requirements not only on the technical side, but also from the business and end-to-end perspectives.

In parallel, there has been increasing interest in the *push* paradigm, in particular to groups of users, supported by the spectrum efficiency of broadcast bearers, and the attractiveness of broadcast TV interactive services. The push towards an Information Society has motivated the development of new wireless access technologies, services, and applications. This happens in a continuous process, by

- identifying future service needs
- identifying potentially available spectrum matching the capacity needs
- agreeing on the technical specifications to enable access to new services.

This has led to the emergence of wide range of wireless digital transmission technologies and service platforms to comply with new user needs requiring more capacity, support of multimedia traffic, extended support for mobility, and so on. This is why, for instance., GSM/GPRS/EDGE are available for wide-area mobile communications. WLAN technologies were developed for local area wireless connectivity (e.g., in the enterprise and home), and DAB/DVB-T were created for audio and digital broadcast services. To capture the end-to-end dimension of the overall system, the delivery dimension is typically looked at, but not so much at application provisioning and system operations as part of the system.

In essence, the requirements must lead to efficient use of available resources (technologies, capital expenditures, services, etc.) in order to enable fulfilling the right need at the right time, and at the right cost. These are key domains to consider with specific effort.

On the path towards 4G, evolutionary rather than revolutionary progress may be preferred, as a means of ensuring maximum reuse of service platforms and access infrastructures while evolving them. 4G has to be seen as the next-generation communications systems, which may include new wireless access technologies, but in any case will be able to provide a unified framework to both ends of the communications system. This leads to defining a service infrastructure able to provide an interworking framework among various actors in the value chain.

Concretely, 2G/3G and DVB infrastructures, as well as service platforms, are good starting points because of their established popularity and complementaries in terms of coverage, operations, services, and content. The objective is to provide a unified framework such that service providers can address both cellular and broadcast terminals with applications that have similar application logic. The advent of integrated broadcast/cellular terminals would then benefit from such an existing infrastructure.

Service Framework Conception

The main objective of this service framework is to be adapted to the peculiarities of 2G/3G and broadcast platforms (i.e., complying with legacy systems), while offering opportunities for enhanced and/or new types of services, or new access and delivery methods to services. In essence, it reuses known concepts, but adapted to the coexistence of heterogeneous access networks with specific features. The service framework aims to combine the interests of the value chain members. Such a value chain is articulated along the assets with which each provider markets its added value, as illustrated in Figure 6.36. At the very end of the chain, the end user is the driver for such a system.

Services are combinations of audio, video, and data, applications running on servers, and applications to be downloaded into user terminals. Creators (i.e., content providers, application designers) of these components have to face the increasing diversification of targets (multiplicity of networks, terminals, and contexts) as well as the risk of seeing their creation misused.

Service providers combine these assets into attractive packages to be purchased by end users. This requires settings up authentication and billing mechanisms, and ensuring that the service is made available as expected by the creators in order to meet user expectations. A service provider's interest is to have the most appropriate service instances delivered to the largest number of terminals while optimizing the required bandwidth.

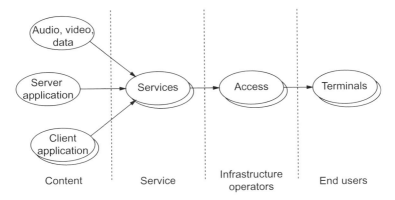

Figure 6.36 Value chain articulation [184]. [©2004 IEEE.]

700 NETWORK LAYER

Finally, infrastructure operators have the role of ensuring proper service access and asset delivery to end users, more or less irrespective of the service type. Service access and delivery requirements are agreed and captured into a service-level agreement (SLA) including quality of service (QoS), security and mobility support aspects, at an aggregated level and for specified geographical areas and time.

The client/server paradigm has proven to be very efficient in offering innovative services, based on the fact that the server application is upgradeable, and/or that client applications can be downloaded over the air. These are the features that are reused in this service framework and adapted to the case of hybrid cellular/broadcast networks.

The applications running on the server and the related client applications are the assets that will enable further reuse of audio/video/data assets, and adaptation to multiple types of terminals, for users on multiple platforms to access proposed services. This is achieved by means of a number of interactions among the various assets, as depicted in Figure 6.37.

Client applications can be generic into terminals (e.g., browser) or specific to a service. The nature of interactions among client and server applications deals with:

- *network profile* – to determine terminal dynamic connectivity characteristics such as available networks and QoS classes
- *terminal profile* – to determine terminal application execution capabilities
- *user profile* – to facilitate the personalization of the user experience, and discovery of new services
- *authentication/billing* – to support the controlled use of services, when required.

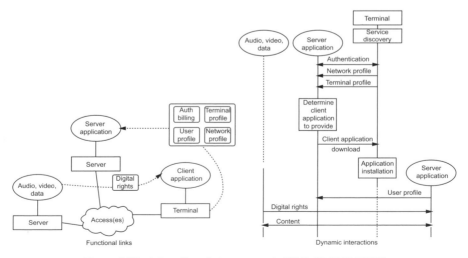

Figure 6.37 Interactions between assets [184]. (©2004 IEEE.)

Profiles can be resident in either the network (e.g., 3G concepts for offline personalization of services in support of the virtual home environment) or the terminal with motivation of preserving privacy or when a profile is likely to change fast (case of network profile). Both cases will coexist and the server application deals with both.

Note that digital rights aspects can be involved in the interactions as a means to ensure proper use or reuse of the audio/video/data.

From the application provider point of view, the process is in the business-to-business domain in the sense that it must be aware of the application in the form of what the service provider can advertise to end users. Associated with the service, there can be an application to be downloaded to the terminal, in order to enable improved user experience. Figure 6.38 summarizes the process.

The service descriptions and the client applications can both be pushed to the terminal in a typical client/server interaction. In the case of a broadcast bearer, the objects are indexed with sufficient information for the terminals to filter out unwanted ones, and played out, for instance, cyclically for permanent availability over the air. Object organization into cycles is optimized, for instance, to minimize the access time to a given object.

Hybrid QoS Management

Delivery over hybrid networks requires a framework for QoS such that guarantees can be given to service providers to ensure that content is made available to end users as expected at application design. Concretely, the service provider has to know in which form service must be *packaged* (in terms of QoS attributes) to adapt to network limitations. This typically deals with the setup of different instances of the same end user service in order to comply with the network types or QoS classes and likely context in which users are connected to a given network.

In Figure 6.39 the process is built around the idea of making terminals discover services they can actually run or access through the network to which they are

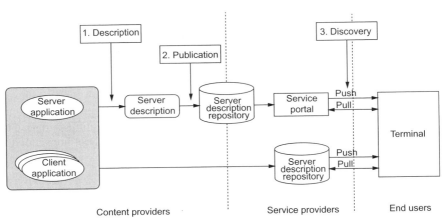

Figure 6.38 Application deployment [184]. (©2004 IEEE.)

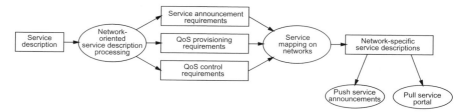

Figure 6.39 Service description based QoS management [184]. (©2004 IEEE.)

connected. This implies some preprocessing consisting of generating the network-specific service descriptions under the constraints of the QoS the targeted networks can actually offer.

At service registration, a service description is available for determining how the service should be made known to end users, and what the QoS requirements are to properly run the service. Note that the service description must include if the service can be available under different forms (e.g., video, still picture formats, and bit rates selectable for various instances). Next, the network-specific service descriptions are generated and made accessible to end users over the networks on which they are applicable.

When a service is mapped on a broadcast network, the service descriptions are pushed using an announcement protocol. Terminals filter out descriptions that do not match capabilities or use preferences.

Hybrid QoS management framework relies on a semidistributed approach and builds on intrinsic transport and QoS control mechanisms. The added value of this framework resides in the signaling links defined between the entities that take part in the end-to-end content delivery, with help of the mediation entity. The pragmatic assumption is that the QoS bottlenecks are on the wireless access side. Since there are a limited number of such access networks, there is no significant scalability issue.

Support for this framework is in the form of signaling gateways available at each administrative domain (DVB networks, 2G/3F networks) and interconnected to exchange the necessary information. The signaling gateways also host measurement probes for reporting the information at the appropriate level.

QoS provisioning consists of provisionally reserving bandwidth for a given service, while QoS control aims to allocate resources so that the QoS requirements are met as long as the service session for a given user is active. SLA optimization can be admitted in a given domain. Support of a given QoS class has a direct impact on the necessary bandwidth at a given location. Consequently, in some cases it is desirable to index the service to be deployed with some geographical information (cell identification, area, coordinates, etc.). The mediation entity can keep track of this information and link it to the service announcement process, where the announced services will take into account the available resources in different areas.

6.9 CONTENT DELIVERY NETWORKS

Content delivery network (CDN) is a comprehensive, end-to-end solution for optimizing global networks for Web content delivery. Users requesting information from a popular Web site may well have those requests served from a location closer to them than the original server on which it is generated. By serving content from points a lot closer to the user, a CDN reduces the likelihood of hot spots by dispersing the different points of convergence and by distributing the workload among multiple servers.

Delivering content *from the edge of the network* instead of the original server also has the added benefit of additional reliability. The probability of lost packets is decreased, pages load faster, and the performance of streaming audio and video clips is improved. The user experience is improved overall.

CDNs employ a network Web and caching servers that are less expensive than the hosting server but are efficient at delivering Web content. The typical arrangement is for an owner of a Web site to outsource the content of his/her site to an external content delivery network. The CDN provider takes the content and distributes it across the network to all edge servers, a process that can take from minutes to two to three hours, depending on the size of the CDN. Once loaded, the CDN intercepts all IP requests and serves the content from the available cache that is physically closest to the user.

Factors that need to be considered in the design and deployment of a CDN include the following:

- The content must be delivered to the edge servers if the service is to be of any use to the users. This is a fundamental requirement of a CDN.
- A basic service requirement is the ability to identify the location of the requesting user and respond from the edge server that is closest to the user.
- The performance of the edge server must be monitored for performance degradation, and the CDN must be able to load-balance in the event a server becoming overloaded.
- The Internet is a network of networks, and so the CDN provider must deploy enough servers to address users from different service provider networks. Issues of content management across different service provider networks must also be addressed to ensure the reliable delivery of content from the original server to the edge servers in different service provider networks.
- Consideration must also be given to geographic placement. Pockets of population by city, region, and country must be considered in the deployment of edge servers. Because the objective is to improve the user experience by placing the content closer to the user, edge servers must be deployed on a global basis.
- The content also must be kept fresh and synchronized with the serving server. This can be a very challenging task, and the more frequently the source data changes, the more complex the task becomes.

- Fault tolerance is a fundamental requirement for the delivery of a reliable service. The solution must factor failure and automatically adjust to prevent service interruption.
- A well-designed CDN must track real-time conditions of the network and avoid hot spots. If congestion is detected, it should have the capability to quickly reroute content in response to the congestion and outages.
- Reporting and billing are both requirements. Real-time logging and billing and detailed reporting on traffic patterns, user location, Internet conditions, and other reporting elements must be tracked.

From this list, it is evident that the efficient operation of a CDN and its ultimate usefulness to a distributed user base are dependent on multiple factors. It is virtually impossible for a single CDN to meet all these requirements without a significant investment, and even then, the likelihood of that CDN having enough edge servers to satisfy a global user base is very low.

From the perspective of the user and that of user's ISP, the question arises whether or not the user will be able to gain the benefit of fast access to an edge server if the only CDN provider in the region is affiliated with a competing ISP. Obviously, these questions raise issues that cannot be addressed entirely by technology. At the core of these questions are contractual issues that can be addressed only through a business arrangement among the different ISPs, CDN providers, and content owners. These questions, technical and business, are being addressed through the information of the Content Alliance. In August 2000, CiscoSystems announced the formation for the Content Alliance to speed the adoption of compatible CDN technology formed to help develop standards and protocols to advance content networking. The alliance is not a standards organizations; it generates proposals for standards and depends on traditional standards bodies such as the IETF to gain broad industry acceptance. The initial focus of the group was on *content peering*, a term that describes the process that enables the CDNs of multiple independent service providers to work in cooperation. In addition to the development of technologies, the alliance is also focused on defining specifications to address issues of authorization of the use of content among networks and the sharing of logging or billing information for charge settlement.

However, no individual CDN has the reach and coverage to span all the geographies and networks across the Internet. Likewise, a given CDN cannot take advantage of all the value-added functionality – speed, security, QoS, on-premises delivery, and other features – that an individual service provider may implement. Given the collection of independent networks that form the Internet, content peering allows CDNs to interoperate, ensuring fast performance by delivering Web content from devices located close to the viewing audience.

With content peering, a Web site owner can work with a preferred hosting service provider, but gain the reach of the combined peering networks. By leveraging the reach, power, and features of different CDNs, content peering creates the ability to deliver the benefits of content delivery networks to a global user base regardless of where the server is hosted and by whom. Content peering requires the CDNs to share information in three areas:

- *content distribution* – the process of moving files to the remote delivery devices
- *content request-routing* – the process whereby a viewer's page request is redirected to the appropriate delivery devices
- *accounting* – the process for collecting usage and billing data.

6.9.1 Content Delivery Evolution

In the earlier years of the World Wide Web, the Web was seen as a means by which an organization could publish information for public consumption, or to a closed user group. Users accessing the content would connect directly with the server that served the content. As the Web sites grew and more functions were added, the problems of content management and the addition of dynamic features became more and more challenging.

Application servers emerged and were used for content management, as shown in Figure 6.40. These servers were placed between the user and the back-end business systems applications, databases, and modern and legacy servers. With this approach, the back-end systems could continue to operate in the way they were designed, and continue to address the functions for which they were intended, without having to understand HTTP requests and HTML functions. The application server was used as a translator; on one side, it understood the Web-based structure of the user requests, and on the other hand the native structure of the serving database or server.

In recent years, this intermediate layer – occupied by the application server – has grown in function and components. This layer – the components of which are often referred to as middleware – has seen significant development in an attempt to minimize the complexity of client programs and improve performance. Functions of security were also added to this layer to ensure the security of data and user traffic.

Contemporary data centers are lot more complex than their original counterparts. The growth of e-business and other e-enabled services has placed demands on content hosting and delivery that have resulted in a more complex infrastructure, built to deliver personalized and dynamic content. Within a typical content hosting data

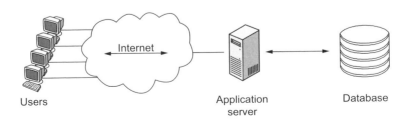

Figure 6.40 The application server position.

center, it is now common to find components that include routers, switches, firewalls, reverse proxy caches, devices that do load balancing, Web servers, application servers, database servers, and storage appliances. Each component may be duplicated to ensure redundancy, provide adequate capacity, and improve reliability and availability. The entire site is usually homed to the Internet through different access points and sometimes through different service providers, where possible.

The architecture of a Web hosting service is shown in Figure 6.41. Requests coming in from the Internet are first directed to the policy server, which validates the request against a set of rules and authorized user lists. Valid requests are then sent to the load balancer for distribution to a Web server. These Web servers interact with an application server, which in turn interacts with the appropriate content source: a database server, a transaction server, or some other server type.

This architecture provides a level of redundancy and, when viewed as a whole, is just an extension of the client/server approach, where the server function is now being delivered through the combination of many components acting as a single resource to the client. Taking this view of this system, one can easily see that it is still a centralized approach to the delivery of content, and a centralized approach is subject to many weaknesses, a few of which are listed here:

- A centralized approach is not scalable. Issues of performance and capacity are addressed by adding more components, further complicating operation and management.
- A centralized approach does not address latency introduced with the network. A user request and the response must still traverse the full path across the network and be subject to whatever problems the network may be experiencing.

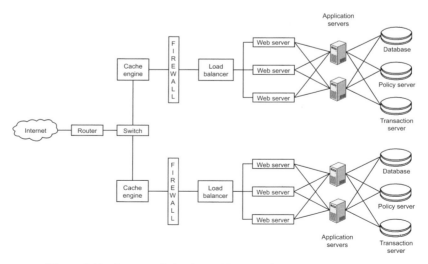

Figure 6.41 An example for the architecture of a Web hosting service.

- A centralized approach limits global reach. Users in other parts of the world may have adequate infrastructure for accessing local content, but the links back to the country of the hosting service may be limited.
- A centralized approach does not solve the problems created by flash crowds or denial of service attacks without having additional capacity that goes unused during normal operation.

6.9.2 Content Delivery Network (CDN) Functions

A CDN addresses the shortcomings of both the centralized approach and the model that employs mirroring by using a distributed model that separates the content generation tier from the tier that is responsible for assembly and delivery, as shown in Figure 6.42. This model logically divides the functions of the centralized model into three tiers, a content generation tier, a integration tier, and an assembly and delivery tier.

Content Generation Tier

This is usually centrally located in a full redundant data center. The primary function of this tier is the generation of the requested information from legacy systems, and transaction servers. Within this tier, a policy-based server may also be used to administer rules and security policies, for users and user requests. The policy server works in conjunction with the firewall to administer these security policies. The following paragraphs look at some of the key features, components, and functions of this tier in more detail.

The application server understands the different data structures of the different servers that are used to serve content. A user request received by an application server is translated into a set of requests that must be sent to the appropriate content servers, in a structure that these servers understand. The response must then be converted from the native structure of the serving platform to one that can be processed by the Web-site functions.

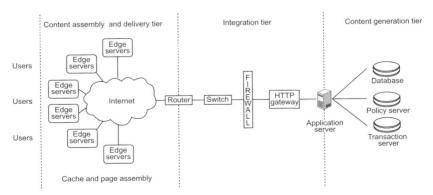

Figure 6.42 Content delivery network (CDN) distributed model.

In the centralized model, the application server may also be involved in the page assembly and user handling process. In the distributed model of a CDN, these functions are offloaded to the content delivery and page assembly tiers, greatly enhancing the performance of the network. Earlier versions of CDNs were capable of handling only static pages. The current versions of CDNs are now capable of serving personalized and dynamic pages through the use of a specification called edge side included (ESI). This is an open language for creating a uniform programming model that facilitates interoperation of ESI-compliant systems from different vendors.

ESI enabled a Web site developer to break down Web pages into fragments of different cacheable profiles. Maintaining these fragments as separate elements on the content delivery network enables dynamically generated content to be cached, then reassembled and delivered from the edge to the network. The freshness of the information is managed through a process of sending invalidation messages from the application server to the edge servers, informing them to overwrite outdated objects residing on them. In this way, changing content can be controlled much as it was when only static pages were served.

In a CDN, the initial request is always sent to the edge server, which checks its internal cache to see if this contains the page as a result of a previous request. What is cached, how long it is stored in cache, and other general rules for caching content are defined by the content provider via a metadata configuration file. A component of this rule is the time to live (TTL). This value determines how long a page fragment stays in cache and is based on how frequently the information on the fragment changes. A particular piece of information may be updated every 6 hours; another may be updated every 24 hours; each would have a different TTL value.

The ESI markup language includes the following features.

- *Inclusion*. ESI provides the ability to fetch and include files that are used to make up a Web page. Each file is subject to its own properties, such as TTL, configuration and control, and revalidation instructions, among others. Each file may include other files up to three levels of recursion.
- *Conditional inclusion*. ESI supports conditional processing based on Boolean comparisons or environmental variables.
- *Environmental variables*. ESI supports the use of a subset of standard common gateway interface (CGI) environment variables such as cookie information. These variables can be used inside ESI statements or outside of the ESI block.
- *Exception and error handling*. ESI enables developers to specify alternative pages and default behavior, such as serving a default HTML page in the event that an original site or document is not available. Further, it provides an explicit exception-handling statement set. If a severe error is encountered while processing a document with ESI markup, the content returned to the end user can be specified in a *failure action* configuration option associated with the ESI document.

Detailed information on ESI and its capabilities can be found at www.esi.org.

Directory and Policy Servers. Security solutions vary from basic user ID and password pairing to sophisticated solutions that employ elements of authorization, authentication, and accounting. Policy servers work in conjunction with directory servers and firewalls to validate a user request against predefined rules.

Database and Transaction Servers and Legacy Systems. These form the repository of data and the handlers of requests. These systems are usually not Web-aware and function in ways that are native to their own internal architecture. These back-end systems work in conjunction with other back-end systems to provide a consistent data repository and transaction management.

Storage Management. This is extremely critical to any operation. Once a centrally managed function, storage systems are taking on a new structure and architecture of their own. The introduction of terabit systems and storage appliances has enabled storage to be deployed at any point in the network. We have depicted it here in the content generation tier because this is where the host critical storage function resides. However, storage is also a function that resides within the content assembly and delivery tier.

Integration Tier

This lies between the components that provide content generation and those that do page assembly and delivery. The two main components of this tier are the HTTP gateway and the firewall. The HTTP gateway is just a fancier version of the Web server of the centralized model. The gateway's basic functions are to forward requests from the edge servers – for fragments and fields – to the applications and then send back the results. The gateway supports a multitude of different Web-based languages, such as the Common Gateway Interface (CGI), Java Server Pages (JSP), and Microsoft's Active Server Pages (ASP).

Content Assembly and Delivery Tier

This embodies the major benefit of a CDN: moving the content closer to the user, resulting in a much improved user experience. This tier forms a distributed and fully managed edge network; the bigger the network, the better the user community is served. The edge network may be further divided into a logical hierarchy consisting of edge servers and regional edge servers forming a core. The edge servers generate and serve content. If they do not have a particular file in cache, they may request copies from the core servers, which in turn may request the information from other regional servers. Using this method, the traffic to the original site is further reduced and issues of flash crowds become less a factor.

6.10 CONCLUDING REMARKS

The approach adopted by standardization projects concerning the network layer has been to resolve the open networking problem. They have designed and introduced layers of functionality that make the lower layers transparent, providing well-defined services at each layer. This has enabled selecting OSI-based products from a variety of vendors rather than a single manufacturer. There are now many companies that are developing network management products.

This chapter presents a management platform, the adopted layered approach, and its implementation issues, addressing the demanding needs of QoS-enabled IP networks for continuous changes in their behavior. Our driving forces have been the requirements of interoperability with existing management solutions, independence of this realization with respect to the implementation language and the operating system, distribution of the management functionality, abstraction of the management information from the underlying infrastructure, and finally, portability and readability of this information by humans as well as software components.

Many reliable multicast protocols have been designed that respond to a wide range of user requirements and network environments. These protocols can be differentiated by examining the set of requirements they meet, the architecture they use for interaction among the participants, the mechanisms used to effect this interaction, and the policies that determine how and when to use the mechanisms. Using several representative reliable multicast protocols as examples, the mapping between requirements and the other dimensions of the taxonomy has been explored. This permits identification of the dependencies, and is expected to lead to the development of additional building blocks that will aid in the design of future reliable multicast protocols.

By examining both the routing faults that occur in the operational Internet and the mechanisms to protect the routing infrastructure, we make the following observations.

- In a system as large as today's Internet, faults are the norm rather than the exception.
- Cryptographic protection mechanisms can be effective in guarding against specific faults, but they cannot detect or prevent all types of faults, especially those due to implementation bugs, configuration errors, or compromised routers. Furthermore, cryptographic mechanisms themselves are also subject to faults.
- A number of detection mechanisms have been developed recently to detect faults in the Internet routing system. Although each has limited detection power, collectively they can provide a stronger overall protection against faults.

As the Internet continues to grow, it faces an increasingly hostile environment. The collection of imperfect components operated by different administrative entities will increase not only the frequency of physical failures, but also the number of operational errors and unexpected faults. Furthermore, the importance of the Internet in society will attract more intentional attacks. In such a complex and hostile environment, no single protection or detection mechanisms can be adequate. Instead, we must build a multifence defense system to ensure a resilient Internet routing infrastructure.

Multimedia communication networks are being driven by the needs of users, the availability of new technologies, and the development of products and services from many different providers and suppliers. All communication networks require security analysis and consideration, and multimedia networks are no different. In fact, these networks reflect an increasing level of interactivity and need for interconnection, an aspect which may demand that even stronger security policies and mechanisms be implemented. This is especially true in light of the use of public networks (e.g., the Internet) for transport of proprietary corporate data, electronic commerce transactions, and private personal data, as well as the delivery of critical signaling/control and network management data. In many ways multimedia has users focused on content, regardless of the location of the material or the provider of the service. As such, security issues should be seriously considered by corporate and private users alike. If an analysis is not performed to determine what level of protection resources are required and what levels of user privileges are required, then it becomes quite difficult to properly secure and protect critical aspects of multimedia networks and content. Adequate security policies need to be described in order to best implement them.

The ultimate goal of Internet infrastructure security is to protect the Internet protocol suites against both known and unknown security attacks. This ambitious goal cannot be achieved in a single stroke since there are several intricacies associated with each attack, and the vulnerability caused by an attack is protocol-dependent. A pragmatic approach to solve this problem is to develop secure versions of the protocols in an evolutionary manner as given below.

Repeat

- Identify specific vulnerabilities and threats in the current implementation of the protocols.
- Develop realistic threat models based on the threats analyzed.
- Develop countermeasures based on the threat models developed. Countermeasures should aim at combating both known and unknown threats.
- Carry out quantitative and qualitative evaluation of the countermeasures.

until robust solution is achieved.

This approach will not only enable developing robust protocols, but will also provide significant insight into the nature of the security attacks leading to sustained development of better protocols.

In modeling various real-world phenomena, the fact that observations made on different scales carry essential information must be utilized. Network traffic is complex, as it exhibits strong dependencies and self-similarity, and therefore classical models of time series such as Poisson and Markov processes (relying heavily on independence or weak independence assumptions) are not appropriate for its modeling. Instead, multiresolution techniques provide a natural framework for modeling and analyzing the network traffic. For example, fractal and multifractal models have been utilized to model the long-range dependencies, while the multifractal spectrum has been utilized to model the burstiness of the data (local scaling analysis capturing overall behavior, but also rare events). As previously mentioned, traffic generated by the different services will not only increase traffic loads on the networks, but will also require different quality of service (QoS) requirements (e.g., cell loss rate, delay, and jitter) for different streams (e.g., video, voice, and data). Delivering multiple QoS to different types of traffic while maintaining high utilization of the bandwidth is the objective if efficient traffic management, which encompasses techniques like call admission control, policing, scheduling, buffer management, and congestion control, is to be maintained.

In ATM-based networks, call admission control makes the important decision of whether or not to accept a new call based upon the availability of *enough* bandwidth to support it without degrading the QoS of the existing calls. Based on simple traffic parameters (e.g., peak and average bit rate) that can hopefully describe the traffic behavior and the required QoS parameters, the controller will compute the bandwidth that should be allocated to the call. If the bandwidth is available, then the call is accepted; otherwise it is rejected. The network must subsequently monitor the traffic to ensure that the agreed-upon parameters are not violated. Examples of bandwidth allocation mechanisms in ATM networks for a cell, a burst, a virtual connection, and a virtual path are source-cell rate control in available bit rate, bandwidth reservation in ATM block transfer, and virtual path provisioning. Statistical allocation is based on a reference model given *a priori*. This model is usually simple and can be used to calculate the amount of bandwidth to be allocated offline. On the other hand, dynamic allocation monitors traffic or network conditions, and reallocates or modifies the amount of bandwidth according to the monitored results. As a result, dynamic allocation is adaptive to changes in traffic. Therefore, generally speaking, dynamic allocation is effective when the reference model is invalid.

We have covered in this chapter many of the current research topics in the field of traffic management and control of multimedia networks: from source traffic modeling to actual traffic management issues such as policing, scheduling, flow control, connection admission control, and routing. In spite of a significant amount of research in this area, the provision of optimal networks from the point of view of reliability and efficiency is still a challenge. The reader will appreciate that many of the traffic management and control decisions are performed under uncertainty due to the burstiness and unpredictability of the traffic. The question of whether to really model the traffic descriptors specified by the user, or also to use real-time traffic measurements, is still debatable among vendors and network providers. In telephony, we could rely for many years on Erlang theory to provide support for congestion prediction and efficient network design and traffic management. The challenge of finding an equivalent theory for the far more complex multimedia networks is still with us.

The diversity of network assumptions indicates that QoS remains an increasingly important issue in modern networking. In the business climate, it will be disadvantageous for a service provider to ignore the demands on quality specified by customers. A methodology that could simplify the identification and management of tasks associated with QoS is favorable. The service provider who offers credible quality at a lower price wins. However, service providers still need to learn how to achieve a viable balance between cost and quality.

Topics for future work are to incorporate a more detailed network with dynamic usage parameter control, admission control, and the ATM access switch, and to estimate the actual statistical multiplexing gain. Through the implementation of VBR coding, we expect to better understand the trade-offs among video quality, network components on the service quality, and other possible service types. On the system structure, communication with the media server, dispatching packets, and storage management can be included. To significantly improve system performance, new disk technology has to be used in the future. The future evolution phase involves the integration of advanced intelligent networks into ATM.

We provide a high-level overview of the main solutions available in the literature for constraint-based path selection. Naturally, these solutions provide different trade-offs between computational complexity and accuracy. An important property of multidimensional routing is that a nonlinear function is required to obtain exact results. QoS routing algorithms that use a linear definition for the path length will only prove useful when the link weights are positively correlated. In all other cases a nonlinear function is necessary, which significantly complicates the problem, since no simple shortest path algorithm is available to minimize such a nonlinear function. As a consequence, multiple paths must be evaluated, requiring the use of the k-shortest path algorithm. The other important techniques are nondominance, look-ahead, search-space reducing, rounding and scaling the weights, and the constraint values themselves. Depending on the availability of resources, these techniques allow for devising efficient tailormade QoS algorithms.

Today the Internet is not perceived as reliable enough for critical missions. But this is not because of a lack of advanced mechanisms such as per-flow shaping/policing/shaping. Instead, the challenge lies in how to maintain a clean network and make the right trade-off between simplicity and more control. Good network design, simplicity, high availability, and protection are the keys for providing QoS in the Internet backbone. Good network design plus a certain degree of overprovisioning not only makes a network more robust against failure, but also prevents many QoS-related problems from happening and eliminates the need for complex mechanisms designed to solve these problems. This keeps the network simple and increases its availability. Three traffic classes (premium, assured, and best-effort) are sufficient to meet foreseeable customer needs. Different traffic classes will be handled differently, especially under adverse network conditions. MPLS fast reroute or other protection schemes can be used to protect premium traffic during router or link failure. When failure happens in one part of the network, traffic engineering should be used to move traffic to other parts of the network. `DiffServ`-aware traffic engineering can be used to prevent concentration of high-priority traffic at any link so that high-priority traffic will have low delay and jitter, and can be treated preferably at the expense of other classes of traffic if necessary. In the backbone, traffic management schemes such as policing and shaping should be treated as micro control and be used when traffic engineering is insufficient. Traffic management schemes are more appropriate for the access network before the last mile circuits and the congested uplinks are upgraded.

A user charging policy based on the congestion state of the network would also motivate users to regulate traffic to improve network QoS. Through the proper deployment of technology and policy, the best-effort network has a lot to offer, and more complicated solutions may not be necessary in many situations. Higher QoS, on the other hand, will inevitably lead to higher user demands of service, which, for video applications, translates to bigger picture sizes and higher reproduced picture quality. For the television industry, HDTV is one of the most important developments since the introduction of color TV. Once the public gets used to that resolution and quality, the demand for good quality compressed video will be much higher than it is today. We, therefore, believe that compressed video in general, and error-resilient compressed video in general, will continue to be important research topics.

ADSL is the most documented and widely available version of the DSL family. Many service providers' DSL offerings begin and end with ADSL. The required splitter for the support of POTS and the high data rates do make it less cost-effective for the casual user who may be just looking for a speed connection to browse the Net. For this class of user, ADSL light is more appropriate. Its high-end speeds are more than suitable for this type of use, and the elimination of the splitter makes it a lot easier to work with. Very high data rate DSL is the newest of the DSL families and the one that hopes to facilitate the complete convergence of narrowband and broadband services. It embodies the promise of a full-service access network – a single access technology that delivers service from POTS to e-commerce to campus LAN interconnections. Its target market, MDUs and MTUs, should see substantial growth over the next few years. These trends will benefit the technology and, as it matures, will hopefully realize its promise of being a full-service access network.

Cable companies are using their upgraded broadband networks to offer a wide array of new digital services in order to meet the competition posed by DBS, wireless cable, broadcasters, and telephone companies. As of August 2001, an estimated 12.2 million homes have subscribed to digital cable, a service that offers extended channel offerings, CD-quality, and commercial-free music. The number of digital cable customers is expected to increase to 48.6 million homes by 2006. The industry is also expanding its competitive offerings to include business and residential telephone services delivered over its fiber optic infrastructure. At the close of 2001, at least nine of the nation's largest multiple system operators (MSOs) offered residential and/or commercial phone service in more than 45 markets, serving more than one million customers. Once upgraded to a two-way HFC network, cable operators are able to offer converged services that include telephone service, digital audio and video, high-speed Internet, and other advanced services to consumers. With the new fiber-based network, the dream of a single connection delivering multiple services is a lot closer to reality than it was in 1996 when the telecommunications act was passed. The new infrastructure has also enabled cable operators to explore new markets. History has shown that the cable industry has always been a tenacious group of operators. They have thrived in the face of competition and at times, regulatory challenges. In the broadband market, they lead DSL in the number of subscribers and are projected to maintain that lead in the foreseeable future.

Wireless communication offers the greatest potential for broadband access because it offers the greatest freedom. However, it also poses the greatest challenge because the medium that it uses is a finite resource, is publicly owned, and must be shared by many groups for many purposes. It is used by technologies that save our lives, protect, entertain, communicate, and support many other applications. Who then should, and by what measure do we decide which application is more important, which should go, and which should stay? The answer is, we cannot. We just have to find an equitable way of sharing. Other access media do not have these challenges – they exist within their own sealed domains. For these reasons, new technologies must be efficient in the way they utilize the spectra in which they operate. Technologies tend to be more expensive when they need to get more from a lot less, and because there is less, there is a constant need to find new ways to be more efficient, and the quest for efficiency leads to further change. The problem that this introduces is knowing when the right set of features and functions has been reached and when reasons contribute to a more deliberate approach to defining and publishing new standards. And standards are what wireless broadband access would benefit from most. Many wireless technologies are emerging for broadband access, but they are not as far along as wireline solutions. We will, however, eventually see many of the technologies discussed become more commonplace because as market projection shows, a market exists for these technologies, and where there is a market, entrepreneurs have always found a way. Wireless networks and Internet working applications will overall become a part of our society, and some aspects of wireless technology will be present almost everywhere. This will lead to great challenges and the need to reevaluate many technical assumptions and designs we are using today. The social effects of technology will become more important. Although the roadmap towards the near future seems reasonably clear, one should not underestimate the integration challenges – and inevitable surprises and technological breakthroughs. When we visualize further towards real ubiquitous computing and intelligent software agents with broadband wireless communications and terabit-per-second optical networking, the research challenges will be demanding. The challenges in networking and wireless communications have not ended, and the field is not in decline, although some observers have been claiming so. On the contrary, the long-term future seems interesting. The main challenge will be the complexity of our technology and science overall.

Content delivering networks (CDNs) are highly scalable and provide improved performance and reliability that directly benefit the user. The distributed model of their architecture protects against issues of flash crowds and network hot spots. The formation of the Content Alliance has helped lay the groundwork for a cooperative model between independent CDN operators, which has made the content owner's life a lot simpler. Through a single contract with a single CDN provider, the content owners gain the benefits of the reach and coverage provided by their CDN provider and the reach and coverage of any other CDN with which there is an agreement.

One of the more interesting questions raised is whether or not data will follow the path of voice. Since 1996, voice has experienced a rapid decline in its revenue-generating power. This has been attributed to increased competition and aggressive pricing by voice carriers, to lure subscribers form one carrier to another. Another contributing factor is the emergence of packetized voice over the Internet, specifically VoIP. As the Internet benefited from an influx of investment, a period of innovation ensued that resulted in the development of many new and innovative applications and services, VoIP being just one. Web sites began offering free long-distance voice calls to attract users to their site, with the hope of using the increased traffic to sell ads to advertisers. With the potential for free long distance across the Internet, the entrenched voice providers began to see erosion in the number of minutes used. Granted, the performance and quality of voice across the Internet was substantially less than that for a circuit-switched telephone call, but a degradation in quality is easily tolerated when the service is free. Today, the quality of voice over the Internet has improved, and now there are service providers that specialize in providing long-distance services across the Internet at a deep discount to circuit-switched solutions.

There is a population that believes the introduction of DWDM and future optical solutions will result in enough bandwidth becoming available that data will eventually see a price decline similar to that of voice. It is further argued that, with the abundance of bandwidth, issues of quality of service will become moot. After all, QoS is an issue only in those instances in which different traffics compete for limited resources. By being abundantly available, the resource is no longer limited, and so there is no need for QoS. This argument has some compelling points. It is true that QoS is an issue only when there is competition for resources. It is also true that DWDM holds a potential that is, as yet, untapped. After all, there have already been demonstrations of optical fiber operating at seven trillion bits per second, and the technology is still in the early stages of development, especially when compared with the long history of wire-based transmission. When one considers that the fiber in the network core is currently less than five percent utilized, one appreciates the amount of bandwidth that is yet to be tapped.

The years from 1990 till now have seen many changes. The rate of technological achievements, innovation, and expansion is unprecedented. During this time, the way we traditionally viewed the world and its telecommunication, data, and broadcasting infrastructure has significantly changed. During this time, there has been a distinct blurring of lines between the telephone, data, and audio/video broadcast network industries. The providers of service, who once operated within the confines of their respective industries, now compete across industries to deliver all three services.

As a result of this change, there is evidence of convergence. It has brought us a lot closer to the promise of an integrated network than we have ever been, and to all the benefits to be gained from this arrangement. For the consumer, it means that a single provider will be able to provide an integrated service. It also provides the potential for lower prices for a combined service, as a result of competition. To the provider, it means additional revenues and a single infrastructure from which multiple services can be delivered.

As we continue along this journey of convergence, new technologies and services will continue to evolve, which makes the future unpredictable, but instills a sense of anticipation. The access and metro area networks are the two networks that will see the most changes in the coming years, because these are the networks that are furthest from their full potential. The metro area networks will eventually be upgraded, and issues of bandwidth will be addressed as more fiber is deployed. Today's dependence on ATM and SONET/SDH will eventually yield to newer and improved schemes. Today it is MPLS; tomorrow it may be new and improved versions of packet/cell switching that are a lot more application aware. Copper will eventually be replaced by fiber, and issues created by limited bandwidth will become less of a concern.

Today, there are many competing access technologies, and the reality is that we may never get to a single solution. The only thing that appears certain is the inevitability of wireless playing a more significant role and the assurance that fiber will be deployed a lot closer to the home.

But even so, we can only speculate, and your guess is as good as ours. So our best advice – if you would care to attempt to predict the future of broadband – is to follow the money, because the flow of investment capital is our best indicator of our technological future.

REFERENCES

1. M. Tatipamula and B. Khasnabish (Eds.), *Multimedia Communication Networks Technologies and Services*, Artech House, Boston, 1998.
2. ISO8348 *OSI Data Communication – Network Service Definition*, 1997.
3. K. R. Rao and Z. S. Bojkovic, *Packet Video Communications Over ATM Networks*, Prentice Hall PTR, Upper Saddle River, NJ, 2000.
4. ITU MEDIACOM2004, *Project Description – Version 3.0*, March 2002.
5. ISO7498/1-4 OSI, *Information Processing Systems – Basic Reference Model of OSI*, 1998.
6. N. Modiri, The ISO reference model entities, *IEEE Network Magazine*, 5, 24–33 (1991).
7. ISO8072, *Information Processing Systems – Open Systems Interconnection – Oriented Transport Service Definition*, 1987.
8. A. R. Modarressi and S. Mohan, Control and management in next-generation networks: challenges and opportunities, *IEEE Comm. Magazine*, 38, 94–102 (2000).
9. M. Mampaey, TINA for services and advanced signaling and control in next-generation networks, *IEEE Comm. Magazine*, 38, 104–110 (2000).

10. G. Karlson, *Asynchronous Transfer of Video*, SICS Research report R95:14, Sweden, 1997.
11. M. R. Pickering and J. F. Arnold, A perceptually efficient VBR rate control algorithm, *IEEE Trans. Image Processing*, 3, 527–531 (1994).
12. A. Ortega et al., Rate constraints for video transmission over ATM networks based on joint source/network criteria, *Annales des Telecommunications*, 50, 603–616 (1995).
13. Y. Shoham and A. Gersho, Efficient bit allocation for an arbitrary set of quantizers, *IEEE Trans ASSP*, 36, 1445–1453 (1988).
14. K. Ramchandran, A. Ortega, and M. Vetterli, Bit allocation for dependent quantization with applications to multiresolution and MPEG video coders, *IEEE Trans Image Processing*, 3, 533–545 (1994).
15. J. Choi and D. Park, A stable feedback control of the buffer state using the controlled Lagrange multiplier method, *IEEE Trans. Image Processing*, 3, 546–558 (1994).
16. Y. L. Lin and A. Ortega, Bit rate control using piecewise approximated rate-distortion characteristics, *IEEE Trans CSVT*, 8, 446–459 (1998).
17. W. Ding, Rate control of MPEG-video coding and recording by rate quantization modeling, *IEEE Trans CSVT*, 6, 12–20 (1966).
18. B. Tao, H. A. Peterson, and B. W. Dickinson, A rate-quantization model for MPEG encoders, *Proc. IEEE ICIP*, 1, 338–341 1997.
19. K. H. Yang, A. Jacquin, and N. S. Jayant, A normalized rate distortion model for H.263-compatible codecs and its application to quantizer selection, *Proc. IEEE ICIP*, 1, 41–44 (1997).
20. A. Velio, H. F. Sun, and Y. Wang, MPEG-4 rate control for multiple video objects, *IEEE Trans. CSVT*, 9, 186–199 (1999).
21. Y. Ribos-Corbero and S. M. Lei, JTC1/SC29/WG11 MPEG96/M1820, *Contribution to Rate Control Q2 Experiment: A Quantization Control Tool for Achieving Target Bitrate Accurately*, Sevilla, Spain (1997).
22. H. J. Lee, T. Chiang, and Y. Q. Zhang, Scalable rate control for MPEG-2 video, *IEEE Trans CSVT*, 10, 878–894 (2000).
23. H. J. Lee, T. Chiang, and Y. Q. Zhang, Scalable rate control for very low bit rate video, *Proc. IEEE ICIP*, 2, 768–771 (1997).
24. I. E. G. Richardson, *H.264 and MPEG-4 Video Compression – Video Coding for Next-Generation Multimedia*, Wiley, Chichester, 2003.
25. Z. Li et al., ISO/IEC JTC1/SC29/WG11 and ITU-T SG16 Q.6 Document JVT-H014, *Adaptive Rate Control with HRD Consideration*, May 2003.
26. ISO/IEC 14496-2, *Coding of Audio-Visual Objects, Part 2: Visual, Annex L*, 2001.
27. Y. S. Saw, *Rate-Quality Optimized Video Coding*, Kluwer, Norwood, MA, November 1998.
28. H. Lin and P. Mouchatias, Voice over IP signaling: H.323 and beyond, *IEEE Comm. Magazine*, 38, 142–148 (2000).
29. H. Schulzrine et al., *RTP: A Transport Protocol for Real-Time Applications*, IETF RFC1889, IETF, January 1996.
30. M. Handley et al., *SIP Session Initiation Protocol*, IETF RFC2543, IETF, March 1999.
31. M. Mampaey, TINA for services and signaling and control in next-generation networks, *IEEE Comm. Magazine*, 38, 104–110 (2000).

32. L. Dimopolou et al., QM tool: An XML-based management platform for QoS-aware IP networks, *IEEE Network*, 17, 8–14 (2003).
33. G. Pavlov et al., On management technologies and the potential of Web services, *IEEE Comm. Magazine*, 42, 58–66 (2004).
34. J. Case et al., *A Simple Network Management Protocol (SNMP)*, IETF RFC1157, IETF, May 1990.
35. *Common Information Model (CMIP) Version 2.2*, Distributed Management Task Force, June 1999.
36. X. Xiao and L. Ni, Internet QoS: a big picture, *IEEE Network*, 13, 8–18 (1999).
37. *Extensible Markup Language (XML) 1.0*, W#C Recomm. REC-XML-2001006, October 2000.
38. G. Pavlov, From protocol-based to distributed object-based management architectures, in *Proc. 8th IFIP/IEEE Int. Workshop Distributed Systems and Management*, 25–40, Sydney, Australia, October 1997.
39. ITU-T Recomm. X.701, *Information Technology – Open System Interconnection, System Management Overview*, 1992.
40. J. Y. Kim et al., Towards TMN-based integrated network management using CORBA and Java technologies, *IEICE Trans Commun.*, E82-B, 1729–1741 (1999).
41. ITU-T Recomm. M.3010, *Principles for a Telecommunications Management Network*, ITU, Geneva, Switzerland, May 1996.
42. ITU-T Recomm. X.900, *Information Technologies – Open Distributed Processing, Basic Reference Model of Open Distributed Processing*, ITU, 1995.
43. W3D, *Web Services Activity Docs.*, available at http://www.w3c.org/2002/ws.
44. K. B. Song et al., Dynamic spectrum management for next-generation DSL systems, *IEEE Comm. Magazine*, 40, 101–109 (2002).
45. T. Starr et al., *DSL Advances*, Prentice Hall, Upper Saddle River, NJ, 2003.
46. K. Kerpez et al., Advanced DSL management, *IEEE Commun. Magazine*, 41, 116–123 (2003).
47. 3GPP TS29.198, *Open Service Access (OSA): Application Programming Interface (API)*, Part 1–2.
48. Parlay Group, *Parlay API Spec. 3.0*, December 2001, available at http://www.parlay.org/specs/index.asp.
49. S. Panagiotis, A. Alonistioti, and L. Merkas, An advanced location information management scheme for supporting flexible service provisioning in reconfigurable mobile networks, *IEEE Commun. Magazine*, 41, 88–98 (2003).
50. 3GPP TS23.271, *Functional, Stage 2 Description of LCS*.
51. 3GPP TS23.240, *3GPP Generic User Profile – Architecture, Stage 2*.
52. 3GPP TS22.071, *Location Services (LCS): Service Description, Stage 1*.
53. B. Shafiq et al., Wireless network resource management for web-based multimedia documents services, *IEEE Commun. Magazine*, 41, 138–145 (2003).
54. S. Baqai, M. Woo, and A. Ghafoor, Network resource management for enterprise-wide multimedia services, *IEEE Commun. Magazine*, 34, 78–85 (1996).
55. T. D. C. Little and A. Ghafoor, Multimedia synchronization protocols for integrated services, *IEEE J. Selected Areas in Comm.*, 9, 1368–1382 (1991).

56. ITU-T Recomm. G.709, *Synchronous Multiplexing Structure*, ITU, March 1993.
57. K. Sato, S. Okomoto, and H. Hadama, Network performance and integrity enhancement with optical path layer technologies, *IEEE J. Selected Areas in Commun.*, 12, 159–170 (1994).
58. ISO/IEC 13818 MPEG-2, *Information Technology: Generic Coding of Moving Pictures and Associate Audio Information*, 1995.
59. IETF RFC1889, *RTP: A Transport Protocol for Real-Time Applications*, IETF, January 1996.
60. IETF RFC793, *Transmission Control Protocol*, IETF, September 1981.
61. IETF RFC2357, *IETF Criteria for Evaluating Reliable Multicast Transport and Application Protocols*, IETF, June 1998.
62. J. W. Atwood, A classification of reliable multicast protocols, *IEEE Network*, 18, 24–34 (2004).
63. IETF RFC2887, *The Reliable Multicast Design Space for Bulk Data Transfer*, IETF, August 2000.
64. T. W. Strayer, B. J. Dempsey, and A. C. Weaver, *XTP – The Xpress Transfer Protocol*, Addison-Wesley, 1992.
65. W. T. Strayer, *Xpress Transport Protocol (XTP) Specification Version 4.0b*, Technical Report, XTP Forum, June 1998.
66. S. Floyd et al., A reliable multicast framework for lightweight sessions and application level framing, *IEEE/ACM Trans. Network*, 5, 784–803 (1997).
67. IETF RFC3208, *PGM Reliable Transport Protocol Specification*, IETF, December 2001.
68. M. Hofmann, Enabling group communications in global networks, in *Proc. Global Networking*, II, 321–330 (1997).
69. S. Paul et al., Reliable multicast transport protocol (RMTP), *IEEE J. Selected Areas in Comm.*, 15, 407–421 (1997).
70. K. R. Rao, Z. S. Bojkovic, and D. A. Milovanic, *Multimedia Communication Systems*, Prentice Hall, Upper Saddle River, NJ, 2002.
71. P. H. Ho and H. T. Mouftah, A novel distributed control protocol in dynamic wavelength-routed optical networks, *IEEE Commun. Magazine*, 40, 38–45 (2002).
72. I. Chlamac, A. Ferego, and T. Zhang, Light path (wavelength) routing in large WDM networks, *IEEE J. Selected Areas in Comm.*, 14, 909–913 (1996).
73. P. H. Ho and H. T. Mouftah, A framework of service guaranteed shared protection for optical networks, *IEEE Commun. Magazine*, 40, 97–103 (2002).
74. P. H. Ho and H. T. Mouftah, Capacity-balanced alternate routing for MPLS traffic engineering, in *Proc. IEEE Int. Symp. Comp. Commun.*, 927–932, Taormina, Italy, July 2002.
75. L. C. Wolf, C. Griwodz, and R. Steinmetz, Multimedia communication, *Proc. IEEE*, 85, 1915–1933 (1997).
76. D. Pei and L. Zhang, A framework for resilient internet routing protocols, *IEEE Network*, 18, 5–12 (2004).
77. M. Tatipamula and B. Khasnabish (Eds.), *Multimedia Communications Networks Technologies and Services*, Artech House, Boston, 1998.
78. ITU-T Q29/11, Recomm. Q.NSEC, ITU, July 1995.
79. A. Chakrabarti and G. Mammaran, Internet infrastructure security: a taxonomy, *IEEE Network*, 16, 13–21 (2002).

80. C. P. Pfleeger, *Security in Computing*, Prentice Hall, Upper Saddle River, NJ, 1996.
81. GSM 03.60, *GPRS Service Description, Stage 2*, 1998.
82. 3GPP TS 33.120, *3G Security: Security Principles and Objectives*, May 1999.
83. Ch. Xenakis and L. Merakos, On demand network-wide VPN deployment in GPRS, *IEEE Network*, 16, 28–37 (2002).
84. IETF RFC3261, *SIP Session Initiation Protocol*, IETF, June 2002.
85. S. Salseno, L. Veltri, and D. Papalilo, SIP security issues: the SIP authentication procedure and its processing load, *IEEE Network*, 16, 38–45 (2002).
86. P. Abry et al., Multiscale nature of network traffic, *IEEE Signal Processing Magazine*, 19, 28–46 (2002).
87. H. W. Braun and K. C. Clafly, Web traffic characterizations: an assessment of the impact of caching documents from NCSA022 Web server, *Computer Networks*, 28, 37–52 (1999).
88. H. Saito, Dynamic resource allocation in ATM networks, *IEEE Commun. Magazine*, 35, 146–153 (1997).
89. L. Kleiwrock, *Queuing Systems, II: Computer Applications*, Wiley, New York, NY, 1976.
90. D. E. McDysand and D. L. Spolin, *ATM Theory and Applications*, McGraw-Hill, New York, NY, 1995.
91. G. Gallassi, G. Rigolio, and L. Verri, Resource management and dimensioning in ATM networks, *IEEE Network*, 4, 8–15 (1990).
92. J. K. Shapiro, D. Towsley, and J. Kurose, Optimization-based congestion control for multicast communications, *IEEE Commun. Magazine*, 40, 90–95 (2002).
93. R. G. Addie and M. Zukerman, Performance evaluation of a single server autoregressive queue, *Aust. Telecommun. Res.*, 28, 25–32 (1994).
94. R. G. Addie and M. Zukerman, An approximation of performance evaluation of stationary server queues, *IEEE Trans. Commun.*, 42, 3150–3160 (1994).
95. B. Maglaus et al., Performance models of statistical multiplexing in packet video communications, *IEEE Trans. Comm.*, 36, 834–844 (1988).
96. D. M. Lucantoni et al., Methods or performance evaluating VBR video traffic models, *IEEE/ACM Trans. Networking*, 2, 176–180 (1994).
97. B. Jabbari et al., Statistical characterization and block-based modeling of motion-adaptive coded video, *IEEE Trans CSVT*, 3, 199–207 (1993).
98. D. Hutchison et al., Quality of service management in distributed systems, Chapt. 11 in M. Sloman, Ed., *Network and Distributed Systems Management*, Addison-Wesley, Reading, MA, 1994, Chapter 11.
99. C. Nikolaou, An architecture for real-time multimedia communication systems, *IEEE J. Selected Areas in Comm.*, 8, 387–396 (1990).
100. G. Coulson, A. Campbell, and P. Robin, Design of a QoS controlled ATM based communication system in chorus, *IEEE J. Selected Areas in Comm.*, 13, 686–699 (1995).
101. W. Doeringer et al., A survey of light-weight transport protocol for high-speed networks, *IEEE Trans Commun.*, 38, 2025–2039 (1990).
102. D. Ferrari and D. C. Verma, A scheme for real-time channel establishment in wide-area networks, *IEEE J. Selected Areas in Comm.*, 8, 368–377 (1990).
103. J. A. Stukovic et al., Implications of classical scheduling results for real-time systems, *IEEE Computer*, 28, 16–25 (1995).

104. T. D. C. Little and A. Ghafoor, Synchronization properties and storage models for multimedia objects, *IEEE J. Selected Areas in Commun.*, 8, 229–238 (1990).
105. J. F. Kurose, Open issues and challenges in providing quality of service guarantees in high speed networks, *ACM Computer Commun. Rev.*, 23, 6–15 (1993).
106. ISO/IEC JTC1/SC21/WG1 N9680, *Quality of Service Framework*, UK, 1995.
107. A. T. Campbell et al., Integrated quality of service for multimedia communications, in *Proc. IEEE INFOCOM*, 732–739, San Francisco, CA, April 1993.
108. A. A. Lazar, A real-time control management and information transport architecture for broadband networks, *Proc. Int. Zurich Sem. Digital Communications*, 281–295, May 1992.
109. D. Tenkenhouse, Layered multiplexing considered harmful, in *Protocols for High-Speed Network*, Elsevier Science Publishers, New York, NY, 1990.
110. M. Zilterbart, B. Stiller, and A. Tantewy, A model for flexible high performance communication subsystems, *IEEE J. Selected Areas in Commun.*, 11, 507–518 (1992).
111. D. D. Clark, S. Sheuder, and L. Zhang, Supporting real-time applications in an integrated services packet network: architecture and mechanisms, *Proc. ACM SIGCOMM*, 17–26, Baltimore, MD, August 1992.
112. R. Gonndan and D. P. Anderson, Scheduling and IPC mechanisms for continuous media, *Proc. ACM Symp. Operating Systems Principles*, 25, 68–80, Pacific Grove, CA, 1991.
113. A. T. Campbell et al., A continuous media transport and orchestration service, in *Proc. ACM SIGCOMM*, 99–110, Baltimore, MD, 1992.
114. V. O. K. Li and W. Liao, Distributed multimedia systems, *Proc. IEEE*, 85, 1063–1108 (1997).
115. ISO/IEC JTC1/SC21/WG1 N9680, *Quality of Service Framework*, UK, 1995.
116. D. J. Wright, Assessment of alternative transport options for video distribution and retrieval over ATM on residential broadband, *IEEE Comm. Magazine*, 35, 78–87 (1997).
117. N. Ohta, *Packet Video: Modeling and Signal Processing*, Artech House, Norwood, MA, 1994.
118. M. Hamidi, J. W. Roberts, and P. Rolin, Rate control for VBR video coders in broadband networks, *IEEE J. Selected Areas in Comm.*, 15, 1040–1051 (1997).
119. R. Steinmetz and K. Nahrstedt, *Multimedia Computing, Communications and Applications*, Prentice-Hall, Englewood Cliffs, NJ, 1995.
120. ATM Forum, *ATM User-Network Interface Specification – Version 3*, Mountain View, CA, 1996.
121. N. B. Seitz et al., User-oriented measures of telecommunication quality, *IEEE Comm. Magazine*, 32, 56–66 (1994).
122. J. Croworft et al., The global internet, *IEEE J. Selected Areas in Comm.*, 13, 1366–1370 (1995).
123. IETF RFC1633, *Integrated Services in the Internet Architecture: An Overview*, ITU, June 1994.
124. ITU-T Recomm. I.356, *B-ISDN ATM Layer Cell Transfer Performance*, ITU, Geneva, Switzerland, November 1993.
125. F. Guillemin, C. Levert, and C. Rosenberg, Cell conformance testing with respect to the peak cell rate in ATM networks, *Computer Networks and ISDN Systems*, 27, 703–725 (1995).

126. C. J. Gallego and R. Grunenfelder, Testing and measurement problems in ATM networks, *Proc. IEEE ICC*, 653–657, Dallas, TX, June 1996.
127. J. Tarnet, New directions in communications (or which way to the information age?), *IEEE Commun. Magazine*, 24, 8–15 (1986).
128. F. Kupers et al., An overview of constraint-based path selection algorithms for QoS routing, *IEEE Commun. Magazine*, 40, 50–55 (2002).
129. D. Medhi, QoS routing computations with path coding: a framework and network performance, *IEEE Commun. Magazine*, 40, 106–113 (2002).
130. J. Moy, *OS-PF-Anatomy of Internet Routing Protocol*, Addison-Wesley, Reading, MA, 1998.
131. IETF RFC2386, *A Framework for QoS-Based Routing in the Internet*, IETF, August 1998.
132. X. Xiao et al., A practical approach for providing QoS in the Internet backbone, *IEEE Commun. Magazine*, 40, 56–62 (2002).
133. IETF RFC3031, *Multiprotocol Label Switching Architecture*, IETF, January 2001.
134. IETF RFC3272, *Overview and Principles of Internet Traffic Engineering*, IETF, May 2002.
135. IETF RFC2475, *An Architecture for Differentiated Services*, IETF, December 1998.
136. E. Mykoniati et al., Admission control for providing QoS in DiffServ IP networks: the TEQUILA approach, *IEEE Commun. Magazine*, 41, 38–44 (2003).
137. E. Knightly and N. Shroff, Admission control for statistical QoS: theory and practice, *IEEE Network*, 13, 20–29 (1999).
138. IETF RFC3168, *The Addition of Explicit Congestion Notification (ECN) to IP*, IETF, September 2001.
139. K. Rege et al., QoS management in trunk-and-branch switched Ethernet networks, *IEEE Commun. Magazine*, 40, 30–37 (2002).
140. A. Kaheel et al., Quality of service mechanisms in IP over WDM networks, *IEEE Commun. Magazine*, 40, 38–43 (2002).
141. N. Golume, T. D. Ndonsse, and D. H. Su, A differentiated optical services model for WDM networks, *IEEE Commun. Magazine*, 38, 68–73 (2000).
142. A. Jukan, *QoS-Based Wavelength Routing in Multiservice WDM Networks*, Springer, 2001.
143. F. Callegati, G. Gorrazza, and C. Raffaelli, Exploitation of DWDM for optical packet switching with quality of service guarantees, *IEEE J. Selected Areas in Comm.*, 20, 190–201 (2002).
144. A. Takabashi, H. Yoshino, and N. Kitanaki, Perceptual QoS assessment technologies for VoIP, *IEEE Commun. Magazine*, 42, 28–34 (2004).
145. G. Schollmeier and Ch. Winkler, Providing sustainable QoS in next-generation networks, *IEEE Commun. Magazine*, 42, 102–107 (2004).
146. 3GPP TS23.107, *End-to-End QoS Concept for Architecture – Realize 5*, January 2002.
147. Z. Perisic and Z. Bojkovic, Quality of multimedia services in 3G mobile networks, *Proc. TELSIKS*, 104–107, Serbia and Montenegro, October 2003.
148. C. Nussli and A. Bertout, Satellite-based architecture for multimedia services, *Alcatel Telecommunications*, 2, 91–97 (2002).

149. A. K. Parekh and R. G. Gallager, A generalized processor sharing approach to flow control in integrated services networks: the single-node case, *IEEE/ACM Trans. Network*, 1, 344–357 (1993).
150. A. E. Brand and A. H. Aghvami, Multidimensional PRMA with prioritized Bayesian broadcast – A MAC strategy for multiservice traffic over UMTS, *IEEE Trans Vehic. Tech.*, VT-47, 1148–1161 (1998).
151. V. Huang and W. Zhuang, QoS-oriented access control for 4G mobile multimedia CDMA communications, *IEEE Commun. Magazine*, 40, 118–125 (2002).
152. A. R. Reibman and B. G. Haskell, Constraints on variable rate video for ATM networks, *IEEE Trans CSVT*, 2, 361–372 (1992).
153. B. G. Haskell, Buffer and channel sharing by several inter-frame picturephone coders, *Bell Systems Tech. J.*, 51, 261–289 (1972).
154. Y. Zdepski, D. Raychaudhuri, and K. Joseph, Statistically based buffer control policies for constant rate transmission of compressed digital video, *IEEE Trans Comm.*, 39, 947–957 (1999).
155. J. P. Leduc and S. D'Agostino, Universal VBR video codecs for ATM networks in the Belgian broadband experiment, *Signal Processing: Image Comm.*, 3, 157–165 (1991).
156. C. T. Chen and A. Wong, A self-governing rate buffer control strategy for pseudo-constant bit rate video coding, *IEEE Trans Image Proc.*, 2, 50–59 (1993).
157. K. H. Tzou, An intrafield DCT-based HDTV coding for ATM networks, *IEEE Trans CSVT*, 1, 184–196 (1991).
158. A. Ortega, K. Ramchandran, and M. Vetterli, Optimal trellis-based buffered compression and fast approximations, *IEEE Trans. Image Proc.*, 3, 26–40 (1994).
159. A. Eleftheriadis, S. Petajan, and D. Anastassiou, Algorithms and performance evaluation of the Xphone multimedia communication system, *Proc. of the ACM Multimedia Conf.*, Anaheim, CA, pp. 311–320, August 1993.
160. Y. C. Bolot and T. Turletti, A rate control mechanism for packet video in the Internet, *Proc. of Infocom*, pp. 1.216–1.223, Toronto, Canada, June 1994.
161. M. Macedonia and D. Brutzman, MBONE provides audio and video across the Internet, *Computer*, 27, 30–36 (1994).
162. B. Maglaris et al., Performance models of statistical multiplexing in packet video communications, *IEEE Trans Commun.*, 30, 834–843 (1988).
163. D. P. Heyman, A. Tabatabai, and T. Lakshman, Statistical analysis and simulation study of video teleconferencing traffic in ATM networks, *IEEE Trans. CSVT*, 2, 49–59 (1992).
164. A. Ortega, Video transmission over ATM networks, in Shen et al., Eds., *Microsystems Technology for Multimedia Applications*, IEEE Press, Piscataway, NJ, May 1995.
165. E. P. Rathgeb, Modeling and performance comparison of policing mechanisms for ATM networks, *IEEE J. Selected Areas in Comm.*, 3, 325–334 (1991).
166. M. Vetterli and K. M. Uz, Multiresolution coding techniques for digital television: a review, *Special issue on Multidimensional Processing of Video Signals, Multidimensional Systems and Signal Processing*, pp. 161–187, March 1992.
167. W. M. Garret and M. Vetterli, Joint source/channel coding of statistically multiplexed real time services on packet networks, *IEEE/ACM Trans. Networking*, 1, 71–80 (1993).

168. K. Ramchandran et al., Multiresolution broadcast for digital HDTV using joint source-channel coding, *IEEE J. Selected Areas in Comm.*, 11, 6–23, Jan. 1993.
169. Q.F. Zhu, Y. Wang, and L. Shaw, Coding and cell loss recovery in DCT-based packet video, *IEEE Trans CSVT*, 3, 248–258 (1993).
170. M. Ghanbari and V. Seferidis, Cell-loss concealment in ATM video codecs, *IEEE Trans CSVT*, 3, 238–247 (1993).
171. T. Chiang and D. Anastassiou, Hierarchical coding of digital television, *IEEE Comm. Magazine*, 32, 38–45 (1994).
172. Y. Wang and Q. Zhu, Error control and concealment for video communication: an overview, *Proc. IEEE*, 86, 974–997 (1998).
173. Y. Wang et al., Error resilient video coding techniques, *IEEE Signal Proc. Magazine*, 17, 61–82 (2000).
174. G. Cote, S. Shrirami, and F. Kossentini, Optimal mode selection and synchronization for robust video communications over error phone networks, *IEEE J. Selected Areas in Comm.*, 18, 952–965 (2000).
175. S. Wenger et al., Error resilience support in H.263+, *IEEE Trans CSVT*, 8, 867–877 (1998).
176. J. Kondi, F. Ishtig, and A. K. Katsaggelos, Joint source/channel coding for scalable video, *Proc. SPIE Conf. Visual Communications and Image Processing*, pp. 324–335, San Jose, CA, January 2000.
177. Q. Zhu and Y. Wang, Error concealment in visual communication, in A. R. Reibman and M. T. Sun, Eds., *Compressed Video Networks*, Marcel Dekker, New York, NY, 2000.
178. S. Agni, Error concealment for MPEG-2 video, in A. K. Katsaggelos and N. P. Galatsanos, Eds., *Signal Recovery Techniques for Image and Video Compression and Transmission*, Kluwer, Norwell, MA, 1998, pp. 235–268.
179. M. C. Hong et al., Video error concealment techniques, *Signal Processing: Image Comm.*, 14, 437–492 (1999).
180. D. W. Redmill and N. G. Kingsbury, The EREC: an error resilient technique for coding variable length block of data, *IEEE Trans Image Processing*, 5, 565–574 (1996).
181. G. Carty, *Broadband Networking*, McGraw-Hill, New York, NY, 2000.
182. P. Mahonene et al., Hop-by-hop toward future mobile broadband IP, *IEEE Commun. Magazine*, 42, 138–146 (2004).
183. K. Ahmavaara, H. Haverinen, and R. Pichua, Interworking architecture between 3GPP and WLAN systems, *IEEE Commun. Magazine*, 41, 74–81 (2003).
184. O. Benali et al., A framework for an evolutionary path toward 4G by means of cooperation of networks, *IEEE Commun. Magazine*, 42, 82–88 (2004).
185. P. Demestichas et al., Wireless beyond 3G: managing services and network resources, *IEEE Commun. Magazine*, 40, 80–82 (2002).
186. H. Saito, *Teletraffic Technologies in ATM Networks*, Artech House, Boston, MA, 1994.
187. E. Gelenbe, X. Mang, and R. Onnural, Bandwidth allocation and call admission control in high-speed networks, *IEEE Commun. Magazine*, 35, 122–129 (1997).
188. B. Wydrowski and M. Zukerman, QoS in best-effort networks, *IEEE Commun. Magazine*, 40, 44–49 (2002).

INDEX

Applications
  A projects, 366
  data broadcast, 312
  interactive broadcasting, 166, 184
  media distribution and consumption, 205
  media streaming, 184
  mobile, 330, 340
  multimedia conferencing, 184
  networked, 19
  sensor networks, 518
  video streaming, 439
Advanced Television System Committee (ATSC), 143, 260
  audio subsystem, 262
  multiplex, 264
  services, 267
  transport system, 263
  video subsystem, 261
Asynchronous transfer mode (ATM), 29, 693
  Forum, 121
  networks, 665
  QoS classes, 635
  QoS parameters, 633
  security, 596
  statistical multiplexing gain (SMG), 665
  switching, 6
  traffic model, 619
ATSC, *see* Advanced Television System Committee

Broadband access
  cellular radio networks, 467
  digital subscriber line (DSL), 466
    asymmetrical DSL, 678
    high bit rate DSL, 677
    integrated DSL, 678
    rate-adaptive DSL, 679
    symmetric DSL, 677
    very high rate DSL, 680
  hybrid fiber coaxial (HFC) cable, 681
  passive optical networks, 464
  wireless access
    LMDS, WLAN, 465
Business models
  3G services, 329
  content creation, 339
  content packing, 339
  datacast, 315, 318
  market making, 339
  m-commerce, 336
  media commerce, 515

Common Object Request Broker Architecture (CORBA)
  IETF Object Management Group, 78
  network management, 557
  TINA, 127
Communication model
  client-server, 20
  cognitive, 9
  emotional, 9
  horizontal, 23
  human, 9
  knowledge, 15
  layers, 10
  middleware, 518

Introduction to Multimedia Communications, By K. R. Rao, Zoran S. Bojkovic, and Dragorad A. Milovanovic
Copyright © 2006 John Wiley & Sons, Inc.

728 INDEX

Communication model (*Continued*)
  mission, 9
  mobile networks, 574
  open horizontal interfaces, 23
  peer-to-peer, 20
  three-level, 19
  vertical, 22
Communication services, 277
  audiovisual, 49
  best effort type, 6
  connectionless, 6
  connection-oriented, 6
  guaranteed type, 6
  multicast, 6
  personal, 566
Conditional access (CA) systems
  architecture, 477
  scrambling algorithm, 250
  simulcrypt/multicrypt, 479
Connection admission control, 638
  peak rate allocation, 610
  rate envelope multiplexing, 610
  rate shaping, 611
Copy protection
  DVD, 504
  encryption/watermarking-based, 481
  Secure digital music initiative (SDMI), 506
Copyright protection, 476
  fingerprinting, 509
  MPEG-2, 175
  unauthorized duplication and distribution, 509
  watermarking, 497
Core broadband networks
  ATM, 693
  SONET/SDH, 690
  WMAN, 2G, 3G, 695
Cryptography
  access control, 593
  authentication, 593
  digital signature, 493, 495
  hash function, 592
  integrity, 593
  public key, 508, 592
  secret key, 592

Datacasting, 306, 317
  audio/video (A/V) streaming, 309
  business model, 318
  cached content, 313
  data streaming, 312, 320
  Internet Protocol Datacast Forum, 314
  services, 319, 320
Digital radio broadcasting, 216
  audio coding, 220

Digital Radio Mondiale (DRM), 217
  EUREKA-147, 217
  in-band on-channel (IBOC), 230
  satellite, 234
  satellite audio/radio services (SDARS), 218, 236
  speech coding, 219
  terrestrial, 228
  XM/Sirius, 237
Digital rights management (DRM)
  applications, 508
  e-commerce, 516
  protocols, 482
  systems, 205, 506
Digital video broadcasting (DVB)
  architectural framework, 246
  base band processing, 253
  datacasting, 306, 314
  DVB-C, 250
  DVB-RCS, 301
  DVB-RCT, 304
  DVB-S, 251
  DVB-T, 250
  family of standards, 139
  interactive services, 290, 296
  interoperability, 240
  project, 241
  program specific information (PSI), 257
  service information (SI), 249
  transport stream multiplex, 256
DVB – internet protocol infrastructure (IPI)
  home reference model, 272
  layer model, 269
  protocol layers, 271
  services, 274
DVB – multimedia home platform (MHP)
  applications, 285
  DVB-J API, 281
  layers, 281
  reference model, 279
  virtual machine (VM), 282

Encryption, 592
  content, 478
  copy protection, 481
  data encryption standard (DES), 478
  encryption controlled message (ECM), 175
  encryption management message (EMM), 175
Entropy coding
  context-based adaptive binary arithmetic coding (CABAC), 399, 402
  context-based adaptive variable length coding (CAVLC), 399

INDEX **729**

universal variable length coding (UVLC), 399
Error resilience, 381
    data partitioning, 673
    data recovery, 673
    decoder concealment, 446, 674
    encoding, 446, 671
    entropy coding, 675
    forward error correction, 670
    graceful degradation, 671
    resynchronization marking, 671
    unequal error protection, 673
European Information Infrastructure (EII)
    standardization projects
    applications, 100
    middleware, 371
    network, 540
Extensible Markup Language (XML), 332
    MPEG-21 DDL, 204
    MPEG-7 DSs, 483
    Web services, 556

Global Information Infrastructure (GII)
    application
        aspects, 148
        functions, 83
    functional model, 83
    implementation model, 85
    middleware
        aspects, 365
        functions, 83, 369
    network
        aspects, 366
        functions, 84
    scenario, 86
    standards categorization, 78
    technological building blocks, 89, 537

Industrial fora and consortia, 121
    3GPP/3GPP2, 136
    ATM Forum, 121
    ATSC, 143
    DAVIC, 141
    DVB, 137
    EBU, 143
    IMTC, 133
    ISMA, 135
    M4IF, 134
    TINA, 124
Information technology (IT), 71
Intellectual property management and protection (IPMP), 111
    MPEG-21 Part 4, 207
    MPEG-4 Systems, 184
    universal multimedia access (UMA), 350

Interactive television, 158
    compression capabilities, 292
    content, 292
    storage hierarchy, 292
    subscriber management, 295
    transmission, 293
    watermarking, 505
International Telecommunications Union (ITU)
    ITU-R SG6 WP 6M
        interactivity in broadcasting, 166
    ITU-T H.323, 52, 163, 554
    ITU-T JVT H.264/AVC, 55
        advanced video coding (AVC), 382
        applications, 382
        deblocking filter, 397
        fidelity range extensions (FRExt), 389
        intra/interprediction
            B picture, 395
            I slice, 393
            P picture, 395
            SI/SP picture, 396
        MPEG-4 Part 10, 388
        network abstraction layer (NAL), 389
        parameter set, 382
        profiles and levels, 400
        reference frames, 398
        service categories, 413
        video coding layer (VCL), 389
    ITU-T MEDIACOM2004 standards
        network design, 153
    ITU-T SG16 multimedia services and systems, 371
        questions, 160
Internet Engineering Task Force (IETF), 119
    applications areas, 149
    Internet Architecture Board (IAB), 119
    Internet Engineering Steering Group (IESG), 119
    Internet Society (ISOC), 119
    standards, 98
Internet protocol (IP) multicast, 36, 308
    applications-level, 448
    IP Simulcast, 320
    M Bone, 36
    protocols, 578

JPEG2000, 62, 431

Media coding, 105, 153, 540
    content representation, 374
    core compression technologies, 380
    quality measurement, 541
    transcoding, 418

730 INDEX

Media commerce
  business models, 515
  media for commerce, 513
  media in commerce, 513
  media lifecycle, 514
Media streaming, 431
  content replication, 449
  media synchronization, 450
  protocols, 453
  servers, 450
Media synchronization
  intermedia, 455
  interobject, 451
  interstream, 451
  intrastream, 451
  mechanisms, 437
Metadata, 186, 245, 482
  MPEG-21, 206
  MPEG-7, 59
  specification, 248
  stream, 349
  XML, 348
Middleware, 24, 363
  component-based systems, 517
  peer-to-peer, 521
Mobile communications technologies, 468, 697
  4G, 697
  Bluetooth, 332
  EDGE, 137
  GPRS, 213, 601
  GSM, 213, 331
  UMTS, 136, 213, 331
Mobile services
  applications, 333
  applications adoption, 340
  business models, 334
  categories, 332
  m-commerce, 336
  media provider value chain, 335
  sources of revenues, 337
MPEG-1, 58, 408
  audio, 174
  constraints, 174
  systems, 172
  video, 173
MPEG-2, 58, 408
  audio, 180
  DSM-CC, 181
  packetized elementary stream (PES), 255
  profiling, 183
  systems, 176
  transport stream (TS), 256
  video, 178

MPEG-4, 59, 408
  audio, 198
    natural audio, 198
    synthetic audio, 199
  delivery multimedia integration framework (DMIF), 432
    delivery application services (DAI), 434
    delivery integration network interface (DNI), 434
  fine granularity scalability (FGS), 407, 430, 438
  multiplexing
    FlexMux, 190
    TransMux, 190
  profiling, 200
  scene description, 189
  systems, 186
  visual, 191
    natural video, 193
    synthetic video, 196
  visual texture coding (VTC), 61
MPEG-7, 59, 489
  coding schemes, 483
  description definition language, 204
  description schemes, 203
  descriptors, 202
  system, 205
  tools, 202
MPEG-21, 60, 106
  digital items, 109
  file format, 114
  IPMP, 111
  MEDIACOM, 210
  parts, 206
  reference software, 114
  rights data dictionary (RDD), 112
  rights expression language (REL), 111
  universal multimedia access (UMA), 208
  use case scenario, 114
  user model, 108
Multimedia, 1, 538, 539
  conferencing, 34
  interactive, 171
  networking, 5, 32
Multimedia applications
  DVB services, 275
  e-commerce, 116
  financial data distribution, 116
  interactive television, 291
  mailing systems, 324
  news distribution, 115, 116
  personal computing, 172
  picture archive, 117
  video-on-demand, 305

INDEX **731**

Multimedia authentication
  authenticator transfer, 492
  hard, 492
  information leakage, 492
  soft
    content-based, 495
    quality-based, 494
  speech, 496
  system, 490
  undected modification, 491
Multimedia communication, 17
  convergence, 26
  networking technology, 368
  technology framework, 26
Multimedia content
  aggregator, 335
  consumer, 117
  delivery, 352
  description, identification, protection, 40
  distribution, 482
    authentication, 489
    description, 484
    management, 489
    watermarking, 497
  intellectual property, 350
  on-demand, 134
  producer, 335
  providers, 108, 269, 320
  publisher, 335
  representation
    file/stream formats, 347
  technologies, 38
  value chain, 39
    devices, 339
    infrastructure, 339
    services, 339
Multimedia implementation
  embedded systems, 430
  microprocessor architecture, 429
  programmable platforms, 429
  strategies, 429
Multimedia middleware, 364
  conditional access, 477
  content distribution, 482
  content management, 489
  EII projects, 371
  GII aspects, 366
  M projects, 366
  media coding, 374
  media streaming, 431
  transcoding, 418
Multimedia security
  conditional access, 477
  copy protection, 480

  copyright, 476, 508
  digital rights management, 506
  item identification, 511
  streaming content, 481
Multimedia systems, 162, 374, 536
  authoring, 290
  security, 475
Multipoint video conferencing, 426
  multipoint control unit (MCU), 420

Network management
  application, 558
  evolution, 557
  Forum, 129
  Internet framework, 556
  ISO framework, 546, 559
  location information, 566
  manager-agent model, 558
  mobility, 574
  policies, 606
  quality of presentation (QoP), 570
  simple network management protocol (SNMP), 558
  technologies, 560
  Telecommunications Management Network (TMN) model, 538
Networking
  access networks, 675
  content delivery networks, 703
  functions, 542
  generic networks, 664
  ISO reference model, 543
  MEDIACOM design, 540
  N projects, 366
  protocols, 578
  quality of service (QoS), 622
  routing
    adaptive, 585
    fixed, 586
    fixed alternate, 586
  security, 589
  signaling, 554
  technologies, 538
  traffic, 604
  transport mechanisms, 577
Networking security
  criteria, 595
  cryptography, 592
  data confidentiality, 591
  denial, loss of service, and interruption, 591
  GPRS security, 601
  internet infrastructure attacks
    denial of service, 601
    domain name system hacking, 600

Networking security (*Continued*)
   packet mistreatment, 600
   routing table poisoning, 600
  mechanisms, 589
  policy, 594
  repudiation, 591
  technologies, 590
  threats, 590
  unauthorized modification/usage, 591

Perceptual audio coding, 180, 198
  embedded/multistream, 226
  joint coding, 224
  masking effects, 220
  variable bit-rate, 222
Protocols, 10, 89
  H-stack, 50
  internet routing
    distance vector, 588
    link state, 588
    path vector, 588
  media streaming stacks, 453
  multicast, 578
  network-layer, 453
  session control
    RTSP, SIP, 456
  TCP/IP suite, 556
  TransMux stack, 432
  transport
    UDP, TCP, RTP, RTCP, 454
Providers
  application service providers (ASPs), 449
  content, 108, 205, 269
  content delivery service providers (CSP), 449
  internet service providers (ISPs), 36
  service, 151, 270

Quality of service (QoS), 18, 659
  3GPP, 661
  architecture, 471, 623, 628
  best-effort, 650
  cellular networks, 473
  classes, 635
  control, 437, 441, 626, 630
  DiffServ, 642, 648
  end-to-end, 472, 623, 655
  filtering, 628
  framework, 469
  guaranteed, 548, 648
  IntServ, 648
  maintenance, 627
  management, 624, 627, 701
  mechanisms, 625
  monitoring, 627, 638
  negotiation, 82
  parameters, 633
  provision, 625
  QoS-based routing, 587, 638
  resource management, 471, 570, 625, 643
  viewpoints, 631

Rate distortion (RD)
  operational, 549
  optimized packet scheduling, 417
  performance, 417
  PSNR, 386

Standardization framework, 41, 71
  academia, 45
  competition, 47
  convergence, 45
  EII, 78
  GII, 70, 78
  industry, 48
  licensing, 46
  market, 46
  NII, 78
  regulation, 44
  research, 43
  technology cycle, 77
Standardization process
  IETF, 98, 119
  ISO/IEC approach, 95
  ITU-T strategies, 92
Standardization projects
  EII, 370
  ETSI, 99
  GII, 366
  IP Cablecom, 164
  JTC1, 96
  MEDIACOM2004, 102
Standards bodies, 42
  industry, 148
  international, 148
  national, 148
  prestandardization, 148
  regional, 148

Telecommunications traffic, 608
  bandwidth allocation, 613
  categories, 605
  management, 608
  resource allocation, 612
  traffic engineering (TE), 606
Traffic models, 606
  aggregate, 622
  autoregressive processes, 619
  Markov modulated, 619

on/off, 618
self-similar, 620
Transcoding, 347
  architectures, 421
  bit-rate reduction, 418
  codec concatenation, 419
  multipoint video bridging, 426
  temporal resolution reduction, 418

Universal multimedia access (UMA), 114
  content adaptation, 349
  intellectual property and protection, 350
  limitations of terminals, 345
  user environment description, 348
Universal multimedia experience (UME), 346

Video coding
  4x4 integer transform, 404
  common video formats, 390
  complexity-scalable, 430
  discrete wavelet transform (DWT), 61, 197, 431
  fine granularity scalability (FGS), 439
  inverse transform mismatch, 406
  layered, 670
  multicast, 443
  multiplier-free, 405
  rate control, 551
  scalable, 179, 195, 409, 438
Video coding structure
  block, 390
  group of pictures (GOP), 390
  macroblock (MB), 390
  picture, 390
  sequence, 390
  slice, 390
  subblocks, 390
Video over Internet
  MPEG-4 arhitecture, 459
  MPEG-4 transport stack, 460
  QoS framework, 469

RTP/UDP/IP layers, 459
Video streaming
  congestion control, 441
    network filtering, 447
  delay-constraint retransmission, 445
  error control, 444
  QoS control, 441
  rate adaptive encoding, 457
  rate control
    receiver-based, 441
    source-based
      probe/model-based, 442
  rate shaping, 379, 444, 457
  single-channel multicast, 443
  taxonomy, 457
Video-on-demand (VoD)
  broadcast environment, 305
  true VoD, 37
Virtual reality modeling language (VRML)
  binary format for scenes (BIFS), 189
Voice over IP (VoIP)
  bridge, 35
  H.323, 554
  IP telephony, 685
  QoS, 658

Watermarking, 481, 499
  audio, 504
  complements cryptography, 497
  data hiding, 506
  fragile, 492
  image, 502
  key-dependent check sums, 493
  lossless, 493
  requirements, 498
  robust, 495
  side information, 498
  text, 501
  video, 503